DIGITAL SIGNAL PROCESSING

A Computer-Based Approach

DIGITAL SIGNAL PROCESSING
A Computer-Based Approach

Sanjit K. Mitra

Department of Electrical and Computer Engineering
University of California, Santa Barbara

The McGraw-Hill Companies, Inc.

New York   St. Louis   San Francisco   Auckland   Bogotá   Caracas
Lisbon   London   Madrid   Mexico City   Milan   Montreal
New Delhi   San Juan   Singapore   Sydney   Tokyo   Toronto

McGraw-Hill

A Division of The McGraw·Hill Companies

This book was set in Times Roman, by Aksen Associates, who also handled all production services.
The sponsoring editor was Lynn B. Cox.
The production supervisor was Louis Swaim.
R.R. Donnelley & Sons, Crawfordsville, was the printer and binder.

Library of Congress Cataloging-in-Publication Data

Mitra, Sanjit Kumar.
　　　Digital signal processing: a computer-based approach / Sanjit K. Mitra
　　　　　　p.　　cm. – (McGraw-Hill series in electrical and computer
　　engineering
　　Includes bibliographical references and index.
　　ISBN 0-07-042953-7
　　1. Signal processing–Digital techniques–Data processing
2. MATLAB. I. Series.
TK5102 . 9 . M57 1998
621.382 ' 2–dc21

96-49125
CIP

http://www.mhcollege.com

Communications and Signal Processing

Senior Consulting Editor

Stephen W. Director, University of Michigan, Ann Arbor

Antoniou: *Digital Filters: Analyses and Design*
Auñón/Chandrasekar: *Introduction to Probability and Random Processes*
Candy: *Signal Processing: The Model-Based Approach*
Candy: *Signal Processing: The Modern Approach*
Carlson: *Communications Systems: An Introduction to Signal and Noise in Electrical Communication*
Cherin: *An Introduction to Optical Fibers*
Collin: *Antennas and Radiowave Propagation*
Collin: *Foundations for Microwave Engineering*
Cooper and McGillem: *Modern Communications and Spread Spectrum*
Davenport: *Probability and Random Processes: An Introduction for Applied Scientists and Engineers*
Drake: *Fundamentals of Applied Probability Theory*
Huelsman and Allen: *Introduction to the Theory and Design of Active Filters*
Jong: *Method of Discrete Signal and System Analysis*
Keiser: *Local Area Networks*
Keiser: *Optical Fiber Communications*
Kershenbaum: *Telecommunications Network Design Algorithms*
Kraus: *Antennas*
Kuc: *Introduction to Digital Signal Processing*
Mitra: *Digital Signal Processing: A Computer-Based Approach*
Papoulis: *Probability, Random Variables, and Stochastic Principles*
Papoulis: *Signal Analysis*
Papoulis: *The Fourier Integral and Its Applications*
Proakis: *Digital Communications*
Schwartz: *Information Transmission, Modulation, and Noise*
Schwartz and Shaw: *Signal Processing*
Siebert: *Circuits, Signals, and Systems*
Smith: *Modern Communication Circuits*
Taub and Schilling: *Principles of Signals and Systems*
Taylor: *Principles of Signals and Systems*

About the Author

Sanjit K. Mitra received his M.S. and Ph.D. in electrical engineering from the University of California, Berkeley. After holding the position of assistant professor at Cornell University until 1965 and working at AT&T Bell Laboratories, Holmdel, New Jersey, until 1967, he joined the faculty of the University of California at Davis. Dr. Mitra then transferred to the Santa Barbara campus in 1977, where he served as department chairman from 1979 to 1982 and is now a Professor of Electrical and Computer Engineering. Dr. Mitra has published more than 180 journal papers and 9 books, and he holds 4 patents. He served as President of the IEEE Circuits and Systems Society in 1986 and is currently a member of the editorial boards for five journals: *Circuits, Systems and Signal Processing; Multidimensional Systems and Signal Processing; Signal Processing: Journal of the Franklin Institute;* and *Automatika.* Dr. Mitra has received many distinguished industry and academic awards, including the 1973 F. E. Terman Award, the 1985 AT&T Foundation Award of the American Society of Engineering Education, the 1989 Education Award of the IEEE Circuits and Systems Society, the 1989 Distinguished Senior U.S. Scientist Award from the Alexander von Humboldt Foundation of Germany, and the 1996 Technical Achievement Award of the IEEE Signal Processing Society. In 1987 Dr. Mitra was awarded an Honorary Doctorate of Technology from Tampere University of Technology in Finland. Dr. Mitra is a Fellow of the IEEE, AAAS, and SPIE and is a member of EURASIP and the ASEE.

To my students

Contents

Preface

The field of digital signal processing (DSP) has seen explosive growth during the past three decades, as phenomenal advances both in research and application have been made. Fueling this growth have been the advances in digital computer technology and software development. Almost every electrical and computer engineering department in this country and abroad now offers one or more courses in digital signal processing, with the first course usually being offered at the senior level. This book is intended for a two-semester course on digital signal processing for seniors or first-year graduate students. It is also written at a level suitable for self-study by the practicing engineer or scientist.

A key feature of this book is the extensive use of MATLAB®-based[1] examples that illustrate the program's powerful capability to solve signal processing problems. The book uses a three-stage pedagogical structure designed to take full advantage of MATLAB and to avoid the pitfalls of a "cookbook" approach to problem solving. First, each chapter begins by developing the essential theory and algorithms. Second, the material is illustrated with examples solved by hand calculation. And third, solutions are derived using MATLAB. From the beginning, MATLAB codes are provided with enough details to permit the students to repeat the examples on their computers. In addition to conventional theoretical problems requiring analytical solutions, each chapter also includes a large number of problems requiring solution via MATLAB. This book requires a minimal knowledge of MATLAB. I believe students learn the intricacies of problem solving with MATLAB faster by using tested, complete programs, and then writing simple programs to solve specific problems that are included at the end of Chapters 2 to 11.

Altogether there are 76 MATLAB programs in the text that have been tested under version 4.2d of MATLAB and version 3 of the Signal Processing Toolbox. Some of the programs listed in this book are not necessarily the fastest with regard to their execution

[1]MATLAB is a registered trademark of The MathWorks, Inc., 24 Prime Park Way, Natick, MA 01760-1500, Phone: 508-647-7000, **http://www.mathworks.com**

speeds, nor are they the shortest. They have been written for maximum clarity without detailed explanations. All MATLAB programs included in this book are available from the web site **http://www.mhcollege/engin/ee/ electrical.html** or via anonymous FTP from the Internet site **ece.ucsb.edu** in the directory **/pub/mitra**.

A second attractive feature of this book is the inclusion of 224 simple but practical examples that expose the reader to real-life signal processing problems which has been made possible by the use of computers in solving practical design problems. This book also covers many topics of current interest not normally found in an upper division text. Additional topics are also introduced to the reader through problems at the end of each chapter. Finally, the book concludes with a chapter that focuses on several important, practical applications of digital signal processing. These applications are easy to follow and do not require knowledge of other advanced level courses.

The prerequisite for this book is a junior level course in linear continuous-time and discrete-time systems, which is usually required in most universities. A minimal review of linear systems and transforms is provided in the text, and basic materials from linear system theory are included, with important materials summarized in tables. This approach permits the inclusion of more advanced materials without significantly increasing the length of the book.

The book is divided into 11 chapters. Chapter 1 presents an introduction to the field of signal processing and provides an overview of signals and signal processing methods. Chapter 2 discusses the time-domain representations of discrete-time signals and discrete-time systems as sequences of numbers and describes classes of such signals and systems commonly encountered. Several basic discrete-time signals that play important roles in the time-domain characterization of arbitrary discrete-time signals and discrete-time systems are then introduced. Next, a number of basic operations to generate other sequences from one or more sequences are described. A combination of these operations is also used in developing a discrete-time system. The problem of representing a continuous-time signal by a discrete-time sequence is examined for a simple case. Finally, the time-domain characterization of discrete-time random signals is discussed.

Chapter 3 is devoted to the transform-domain representations of a discrete-time sequence. Specifically discussed are the discrete-time Fourier transform, the discrete Fourier transform (DFT), and the z transform. Properties of each of these transforms are reviewed and a few simple applications outlined. The chapter ends with a discussion of the transform-domain representation of a random signal.

This book concentrates almost exclusively on the linear time-invariant discrete-time systems, and Chapter 4 discusses their transform-domain representations. Specific properties of such transform-domain representations are investigated, and several simple applications are considered.

Chapter 5 is concerned primarily with the discrete-time processing of continuous-time signals. The conditions for discrete-time representation of a bandlimited continuous-time signal under ideal sampling and its exact recovery from the sampled version are first derived. Several interface circuits are used for the discrete-time processing of continuous-time signals. Two of these circuits are the anti-aliasing filter and the reconstruction filter, which are analog lowpass filters. As a result, a brief review of the basic theory behind some commonly used analog filter design methods is included, and their use is illustrated with MATLAB. Other interface circuits discussed in this chapter are the sample-and-hold circuit, the analog-to-digital converter, and the digital-to-analog converter.

A structural representation using interconnected basic building blocks is the first step in the hardware or software implementation of an LTI digital filter. The structural representation provides the relations between some pertinent internal variables with the input and the output, which in turn provides the keys to the implementation. There are various forms of the structural representation of a digital filter, and two such representations are reviewed in Chapter 6, followed by a discussion of some popular schemes for the realization of real causal IIR and FIR digital filters. In addition, it describes a method for the realization of IIR digital filter structures that can be used for the generation of a pair of orthogonal sinusoidal sequences.

Chapter 7 considers the digital filter design problem. First, it discusses the issues associated with the filter design problem. Then it describes two popular approaches to IIR filter design, based on the conversion of a prototype analog transfer function to a digital transfer function. The transformation of one type of IIR transfer function into another type is discussed. Then two different approaches to FIR filter design are described. Finally, the chapter reviews computer-aided design of both IIR and FIR digital filters. The use of MATLAB in digital filter design is illustrated.

Chapter 8 is concerned with the implementation aspects of DSP algorithms. Two major issues concerning implementation are discussed first. The software implementations of digital filtering and DFT algorithms on a computer using MATLAB are reviewed to illustrate the main points. This is followed by a discussion of various schemes for the representation of number and signal variables on a digital machines, which is basic to the development of methods for the analysis of finite wordlength effects considered in Chapter 9. Algorithms used to implement addition and multiplication, the two key arithmetic operations in digital signal processing, are reviewed next, along with operations developed to handle overflow. Finally, the chapter outlines two general methods for the design and implementation of tunable digital filters, followed by a discussion of algorithms for the approximation of certain special functions.

Chapter 9 is devoted to analysis of the effects of the various sources of quantization errors; it describes structures that are less sensitive to these effects. Included here are discussions on the effect of coefficient quantization.

Chapter 10 discusses multirate discrete-time systems with unequal sampling rates at various parts. The chapter includes a review of the basic concepts and properties of sampling rate alteration, design of decimation and interpolation digital filters, and multirate filter bank design.

The final chapter, Chapter 11, reviews a few simple practical applications of digital signal processing to provide a glimpse of its potential.

The materials in this book have been used in a two-quarter course sequence on digital signal processing at the University of California, Santa Barbara, and have been extensively tested in the classroom for about 9 years. Basically, Chapters 2 through 6 form the basis of an upper division course, while Chapters 7 through 10 form the basis of a graduate level course.

Many topics included in this text can be omitted from class discussion, depending on the coverage of other courses in the curriculum. For example, Section 6.2 on signal flow graphs can be omitted from discussion because this topic is usually covered in the junior level linear systems course, which is a prerequisite for this text. Because a senior level course on random signals and systems is required of all electrical and computer engineering majors in most universities, materials in Sections 2.7, 3.10, and 4.9 can be excluded from an upper division course on digital signal processing. However, these topics are important in the analysis of wordlength effects discussed in Chapter 9, and readers not familiar with this subject are encouraged to review these sections before reading Chapter 9. Likewise, materials on state-space representation of discrete-time systems (Sections 2.6 and 6.6) and its applications (Sections 9.8.2 and 9.9.3) can be excluded from the course if time does not permit their discussion in class. Section 8.4 on number representation and Section 5.5 on arithmetic operations can similarly be omitted from discussion since most students taking a digital signal processing course usually take a course on digital hardware design.

This text contains more than 500 problems and 150 MATLAB exercises. A solutions manual prepared by Rajeev Gandhi that contains solutions to all problems and MATLAB exercises is available to instructors from the publisher.

The preliminary versions of the complete manuscript were reviewed by Dr. Hrvojc Babic of the University of Zagreb, Croatia; Dr. James F. Kaiser of Duke University; Dr. Wolfgang F.G. Mecklenbräuker of the Technical University of Vienna, Austria; and Dr. P. P. Vaidyanathan of the California Institute of Technology. A later version has been reviewed by Dr. Roberto H. Bambmerger of Washington State University; Dr. Charles Boumann of Purdue University; Dr. Kevin Buckley of the University of Minnesota; Drs. John A. Flemming and Jerry D. Gibson of Texas A&M University; Dr. John Gowdy of Clemson University; Drs. James Harris and Mahmood Nahvi of the California Polytechnic University, San Louis Obispo; Dr. Yih-Chyun Jenq of Portland State University; Dr. Troung Q. Ngyuen of the University of Wisconsin, Madison; and Dr. Andreas Spanias of Arizona State University. Various parts of the manuscript were reviewed by Dr. C. Sydney Burrus of Rice University; Drs. Richard V. Cox and Nikil S. Jayant of Lucent Technologies, Inc., Murray Hill, New Jersey; Dr. Richard Galton of the University of California, San Diego; Dr. Tor Ramstad of the Norwegian University of Science and Technology, Trondheim, Norway; Dr. B. Ananth Shenoi of Wright State University; Dr. Hans W. Schüssler of the University of Erlangen-Nuremberg, Germany; and Drs. Richard Schreier and Gabor C. Temes of Oregon State University. I thank all of them for their valuable comments, which have improved the book tremendously. I also thank Dr. P. P. Vaidyanathan for arousing my interest in multirate signal processing and assisting me in the development of Chapter 10.

Many of my former and present research students reviewed various portions of the manuscript and tested a number of the MATLAB programs. In particular, I would like to thank Drs. Charles D. Creusere, Michael Lightstone, Ing-Song Lin, and Stefan Thurhofer, and Messrs. Rajeev Gandhi, Eric Leipnik, Michael Moore, Debargha Mukherjee, and Norbert Strobel. I am also indebted to all former students in my ECE 158 and ECE 258A classes at the University of California, Santa Barbara, for their feedback over the years, which helped refine the book. Two of these students, Kenan Ezal and Stephanie Lockwood, read selected chapters and provided helpful comments. I would like to thank my son Goutam for his contribution to the development of the cover design of my book.

Every attempt has been made to ensure the accuracy of all materials in this book, including the MATLAB programs. I would, however, appreciate readers bringing to my attention any errors that may appear in the printed version for reasons beyond my control and that of the publisher. These errors and any other comments can be communicated to me by e-mail addressed to: **mitra@ece.ucsb.edu**.

Finally, I have been particularly fortunate to have had the opportunity to work with the outstanding students who were in my research group during all my teaching career, which spans over 30 years. I have benefited immensely, and continue to do so, both professionally and personally, from my friendship and association with them, and to them I dedicate this book.

Sanjit K. Mitra

CHAPTER 1

Signals and Signal Processing

Signals play an important role in our daily life. Examples of signals that we encounter frequently are speech, music, picture, and video signals. A signal is a function of independent variables such as time, distance, position, temperature, pressure. For example, speech and music signals represent air pressure as a function of time at a point in space. A black-and-white picture is a representation of light intensity as a function of two spatial coordinates. The video signal in television consists of a sequence of images, called frames, and is a function of three variables: two spatial coordinates and time.

Most signals we encounter are generated by natural means. However, a signal can also be generated synthetically or by computer simulation. A signal carries information, and the objective of signal processing is to extract the information carried by the signal. The method of information extraction depends on the type of signal and the nature of information being carried by the signal. Thus, roughly speaking, signal processing is concerned with the mathematical representation of the signal and the algorithmic operation carried out on it to extract the information present. The representation of the signal can be in terms of basis functions in the domain of the original independent variable(s) or it can be in terms of basis functions in a transformed domain. Likewise, the information extraction process may be carried out in the original domain of the signal or in a transformed domain. This book is concerned with discrete-time representation of signals and their discrete-time processing.

This chapter provides an overview of signals and signal processing methods. The mathematical characterization of the signal is first discussed along with a classification of signals. Next, some typical signals are discussed in detail and the type of information carried by them is described. Then a review of some commonly used signal processing operations is provided and illustrated through examples. Advantages and disadvantages of digital processing of signals are then discussed. Finally, a brief review of some typical signal processing applications is included.

1

1.1   Characterization and Classification of Signals

Depending on the nature of the independent variables and the value of the function defining the signal, various types of signals can be defined. For example, independent variables can be continuous or discrete. Likewise, the signal can either be a continuous or a discrete function of the independent variables. Moreover, the signal can be either a real-valued function or a complex-valued function.

A signal can be generated by a single source or by multiple sources. In the former case, it is a scalar signal and in the latter case it is a vector signal, often called a multichannel signal.

A one-dimensional (1-D) signal is a function of a single independent variable. A two-dimensional (2-D) signal is a function of two independent variables. A multidimensional (M-D) signal is a function of more than one variable. The speech signal is an example of a 1-D signal where the independent variable is time. An image signal, such as a photograph, is an example of a 2-D signal where the two independent variables are two spatial variables. Each frame of a black-and-white video signal is a 2-D image signal that is a function of two discrete spatial variables, with each frame occurring sequentially at discrete instants of time. Hence, the black-and-white video signal can be considered an example of a three-dimensional (3-D) signal where the three independent variables are the two spatial variables and time. A color video signal is a three-channel signal composed of three 3-D signals representing the three primary colors: red, green, and blue (RGB). For transmission purpose, the RGB television signal is transformed into another type of a three-channel signal composed of a luminance component and two chrominance components.

The value of the signal at specific values of the independent variable(s) is called its *amplitude*. The variation of the amplitude as a function of the independent variable(s) is called its *waveform*.

For a 1-D signal, the independent variable is usually labeled as *time*. If the independent variable is continuous, the signal is called a *continuous-time signal*. If the independent variable is discrete, the signal is called a *discrete-time signal*. A continuous-time signal is defined at every instant of time. On the other hand, a discrete-time signal is defined at discrete instants of time, and hence, it is a sequence of numbers.

A continuous-time signal with a continuous amplitude is usually called an *analog signal*. A speech signal is an example of an analog signal. Analog signals are commonly encountered in our daily life and are usually generated by natural means. A discrete-time signal with discrete-valued amplitudes represented by a finite number of digits is referred to as a *digital signal*. An example of a digital signal is the digitized music signal stored in a CD-ROM disk. A discrete-time signal with continuous-valued amplitudes is called a *sampled-data signal*. This last type of signal occurs in switched-capacitor (SC) circuits. A digital signal is thus a quantized sampled-data signal. Finally, a continuous-time signal with discrete-valued amplitudes has been referred to as a *quantized boxcar signal* [Ste93]. Figure 1.1 illustrates the four types of signals.

The functional dependence of a signal in its mathematical representation is often explicitly shown. For a continuous-time 1-D signal, the continuous independent variable is usually denoted by t, whereas for a discrete-time 1-D signal, the discrete independent variable is usually denoted by n. For example, $u(t)$ represents a continuous-time 1-D signal and $\{v[n]\}$ represents a discrete-time 1-D signal. Each member, $v[n]$, of a discrete-time

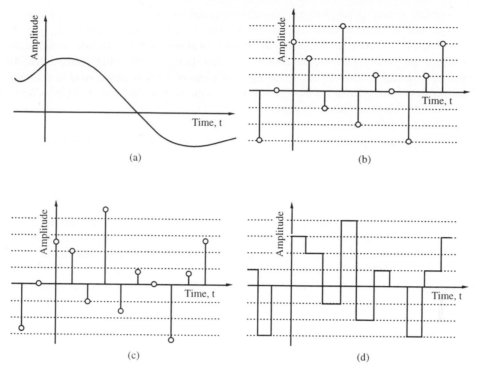

Figure 1.1   (a) A continuous-time signal, (b) a digital signal, (c) a sampled-data signal, and (d) a quantized boxcar signal.

signal is called a *sample*. In many applications, a discrete-time signal is generated from a parent continuous-time signal by sampling the latter at uniform intervals of time. If the discrete instants of time at which a discrete-time signal is defined are uniformly spaced, the independent discrete variable n can be normalized to assume integer values.

In the case of a continuous-time 2-D signal, the two independent variables are the spatial coordinates, which are usually denoted by x and y. For example, the intensity of a black-and-white image can be expressed as $u(x, y)$. On the other hand, a digitized image is a 2-D discrete-time signal, and its two independent variables are discretized spatial variables often denoted by m and n. Hence, a digitized image can be represented as $v[m, n]$. Likewise, a black-and-white video sequence is a 3-D signal and can be represented as $u(x, y, t)$ where x and y denote the two spatial variables and t denotes the temporal variable time. A color video signal is a vector signal composed of three signals representing the three primary colors: red, green, and blue:

$$\mathbf{u}(x, y, t) = \begin{bmatrix} r(x, y, t) \\ g(x, y, t) \\ b(x, y, t) \end{bmatrix}.$$

There is another classification of signals that depends on the certainty by which the signal can be uniquely described. A signal that can be uniquely determined by a well-defined process such as a mathematical expression or rule, or table look-up, is called a *deterministic signal*. A signal that is generated in a random fashion and cannot be predicted ahead of time is called a *random signal*. In this text we are primarily concerned with the processing of discrete-time deterministic signals. However, since practical discrete-time systems employ finite word-lengths for the storing of signals and the implementation of the signal processing algorithms, it is necessary to develop tools for the analysis of finite word-length effects on the performance of discrete-time systems. To this end, it has been found convenient to represent certain pertinent signals as random signals and employ statistical techniques for their analysis.

Some typical signal processing operations are reviewed in the following section.

1.2   Typical Signal Processing Operations

Various types of signal processing operations are employed in practice. In the case of analog signals, most signal processing operations are usually carried out in the time-domain, whereas, in the case of discrete-time signals, both time-domain and frequency-domain operations are employed. In either case, the desired operations are implemented by a combination of some elementary operations. These operations are also usually implemented in real-time or near real-time, even though, in certain applications, they may be implemented off-line.

1.2.1   Elementary Time-Domain Operations

The three most basic time-domain signal operations are scaling, delay, and addition. *Scaling* is simply the multiplication of the signal by a positive or a negative constant. In the case of analog signals, this operation, is usually called *amplification* if the magnitude of the multiplying constant, called *gain*, is greater than one. If the magnitude of the multiplying constant is less than one, the operation is called *attenuation*. Thus, if $x(t)$ is an analog signal, the scaling operation generates a signal $y(t) = \alpha x(t)$ where α is the multiplying constant. Two other elementary operations are integration and differentiation. The *integration* of an analog signal $x(t)$ generates a signal $y(t) = \int_{-\infty}^{t} x(\tau)\, d\tau$, while its *differentiation* results in a signal $w(t) = dx(t)/dt$.

The *delay* operation generates a signal that is a delayed replica of the original signal. For an analog signal $x(t)$, $y(t) = x(t - t_0)$ is the signal obtained by delaying $x(t)$ by the amount t_0 which is assumed to be a positive number. If t_0 is negative, then it is an *advance* operation.

Many applications require operations involving two or more signals to generate a new signal. For example, $y(t) = x_1(t) + x_2(t) - x_3(t)$ is the signal generated by the *addition* of the three analog signals $x_1(t)$, $x_2(t)$, and $x_3(t)$. Another elementary operation is the *product* of two signals. Thus, the product of two signals $x_1(t)$ and $x_2(t)$ generates a signal $y(t) = x_1(t)x_2(t)$.

The elementary operations mentioned above are also carried out on discrete-time signals and are discussed in later parts of this text. Next we review some commonly used complex signal processing operations that are implemented by a combining two or more of the elementary operations.

1.2.2   Filtering

One of the most widely used complex signal processing operations is *filtering*. Filtering is used to pass certain frequency components in a signal through the system without any distortion and to block other frequency components. The system implementing this operation is called a *filter.* The range of frequencies that is allowed to pass through the filter is called the *passband*, and the range of frequencies that is blocked by the filter is called the *stopband*. Various types of filters can be defined, depending on the nature of the filtering operation. In most cases, the filtering operation for analog signals is linear and is described by the convolution integral

$$y(t) = \int_{-\infty}^{\infty} h(t-\tau)x(\tau)\,d\tau, \qquad (1.1)$$

where $x(t)$ is the input signal and $y(t)$ is the output of the filter characterized by an impulse response $h(t)$.

A *lowpass filter* passes all low-frequency components below a certain specified frequency f_c, called the *cutoff frequency*, and blocks all high-frequency components above f_c. A *highpass filter* passes all high-frequency components above a certain cutoff frequency f_c and blocks all low-frequency components below f_c. A *bandpass filter* passes all frequency components between two cutoff frequencies f_{c1} and f_{c2}, and blocks all frequency components below the frequency f_{c1} and above the frequency f_{c2}. A *bandstop filter* blocks all frequency components between two cutoff frequencies f_{c1} and f_{c2}, and passes all frequency components below the frequency f_{c1} and above the frequency f_{c2}. Figure 1.2(a) shows a signal composed of three sinusoidal components of frequencies 50 Hz, 110 Hz, and 210 Hz, respectively. Figures 1.2(b) to (e) show the results of the above four types of filtering operations with appropriately chosen cutoff frequencies.

A bandstop filter designed to block a single frequency component is called a *notch filter*. A *multiband filter* has more than one passband and more than one stopband. A *comb filter* is designed to block frequencies that are integral multiples of a low frequency.

A signal may get corrupted unintentionally by an interfering signal called *interference* or *noise*. In many applications the desired signal occupies a low-frequency band from dc to some frequency f_L Hz, and it is corrupted by a high-frequency noise with frequency components above f_H Hz with $f_H > f_L$. In such cases, the desired signal can be recovered from the noise-corrupted signal by passing the latter through a lowpass filter with a cutoff frequency f_c where $f_L < f_c < f_H$. A common source of noise is power lines radiating electric and magnetic fields. The noise generated by power lines appears as a 60-Hz sinusoidal signal corrupting the desired signal and can be removed by passing the corrupted signal through a notch filter with a notch frequency at 60 Hz.[1]

1.2.3   Generation of Complex Signals

As indicated earlier, a signal can be real-valued or complex-valued. For convenience, the former is usually called a *real signal* while the latter is called a *complex signal*. All naturally generated signals are real-valued. In some applications, it is desirable to develop a complex signal from a real signal having more desirable properties. A complex signal can

---

[1] In many countries, power lines generate 50-Hz noise.

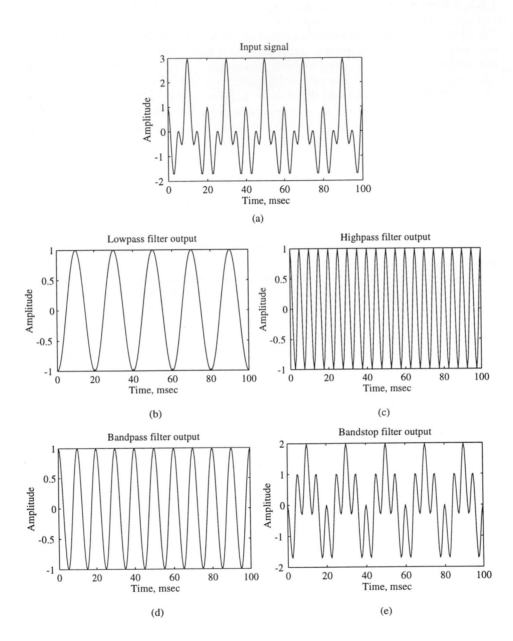

Figure 1.2   (a) Input signal, (b) output of a lowpass filter with a cutoff at 80 Hz, (c) output of a highpass filter with a cutoff at 150 Hz, (d) output of a bandpass filter with cutoffs at 80 Hz and 150 Hz, and (e) output of a bandstop filter with cutoffs at 80 Hz and 150 Hz.

be generated from a real signal by employing a *Hilbert transformer* that is characterized by an impulse response $h_{HT}(t)$ given by [Fre94], [Opp83]

$$h_{HT}(t) = \frac{1}{\pi t}. \tag{1.2}$$

To illustrate the method, consider a real analog signal $x(t)$ with a continuous-time Fourier transform $X(j\Omega)$ given by

$$X(j\Omega) = \int_{-\infty}^{\infty} x(t)e^{-j\Omega t}\, dt. \tag{1.3}$$

$X(j\Omega)$ is called the spectrum of $x(t)$. The magnitude spectrum of a real signal exhibits even symmetry while the phase spectrum exhibits odd symmetry. Thus, the spectrum $X(j\Omega)$ of a real signal $x(t)$ contains both positive and negative frequencies and can therefore be expressed as

$$X(j\Omega) = X_p(j\Omega) + X_n(j\Omega), \tag{1.4}$$

where $X_p(j\Omega)$ is the portion of $X(j\Omega)$ occupying the positive frequency range and $X_n(j\Omega)$ is the portion of $X(j\Omega)$ occupying the negative frequency range. If $x(t)$ is passed through a Hilbert transformer, its output $\hat{x}(t)$ is given by the linear convolution of $x(t)$ with $h_{HT}(t)$:

$$\hat{x}(t) = \int_{-\infty}^{\infty} h_{HT}(t-\tau)x(\tau)\, d\tau. \tag{1.5}$$

The spectrum $\hat{X}(j\Omega)$ of $\hat{x}(t)$ is given by the product of the Fourier transform of $x(t)$ and $h_{HT}(t)$. Now the continuous-time Fourier transform $H_{HT}(j\Omega)$ of $h_{HT}(t)$ of Eq. (1.2) is given by

$$H_{HT}(j\Omega) = \begin{cases} -j, & \Omega > 0, \\ j, & \Omega < 0. \end{cases} \tag{1.6}$$

Hence,

$$\hat{X}(j\Omega) = H_{HT}(j\Omega)X(j\Omega) = -jX_p(j\Omega) + jX_n(j\Omega). \tag{1.7}$$

As the magnitude and the phase of $\hat{X}(j\Omega)$ are an even and odd function, respectively, it follows from Eq. (1.7) that $\hat{x}(t)$ is also a real signal. Consider the complex signal $y(t)$ formed by the sum of $x(t)$ and $\hat{x}(t)$:

$$y(t) = x(t) + j\hat{x}(t). \tag{1.8}$$

The signals $x(t)$ and $\hat{x}(t)$ are called, respectively, the *in-phase* and *quadrature components* of $y(t)$. By making use of Eqs. (1.4) and (1.7) in the continuous-time Fourier transform of $y(t)$, we obtain

$$Y(j\Omega) = X(j\Omega) + j\hat{X}(j\Omega) = 2X_p(j\Omega). \tag{1.9}$$

In other words, the complex signal $y(t)$, called an *analytic signal*, has only positive frequency components.

A block diagram representation of the scheme for the analytic signal generation from a real signal is sketched in Figure 1.3.

Figure 1.3   Generation of an analytic signal using a Hilbert transformer.

1.2.4   Modulation and Demodulation

For transmission of signals over long distances, a transmission media such as cable, optical fiber, or the atmosphere is employed. Each such medium has a bandwidth that is more suitable for the efficient transmission of signals in the high-frequency range. As a result, for the transmission of a low-frequency signal over a channel, it is necessary to transform the signal to a high-frequency signal by means of a modulation operation. At the receiving end, the modulated high-frequency signal is demodulated, and the desired low-frequency signal is then extracted by further processing. There are four major types of modulation of analog signals: amplitude modulation, frequency modulation, phase modulation, and pulse amplitude modulation. Of these schemes, amplitude modulation is conceptually simple and is discussed here [Fre94], [Opp83].

In the *amplitude modulation* scheme, the amplitude of a high-frequency sinusoidal signal $A \cos(\Omega_0 t)$, called the *carrier signal*, is varied by the low-frequency bandlimited signal $x(t)$, called the *modulating signal*, generating a high-frequency signal, called the *modulated signal*, $y(t)$ according to

$$y(t) = Ax(t)\cos(\Omega_0 t). \qquad (1.10)$$

Thus, amplitude modulation can be implemented by forming the product of the modulating signal with the carrier signal. To demonstrate the frequency-translation property of the amplitude modulation process, let $x(t) = \cos(\Omega_1 t)$ where Ω_1 is much smaller than the carrier frequency Ω_0, i.e., $\Omega_1 << \Omega_0$. From Eq. (1.10) we therefore obtain

$$
\begin{aligned}
y(t) &= A\cos(\Omega_1 t) \cdot \cos(\Omega_0 t) \\
&= \tfrac{A}{2}\cos\left((\Omega_0 + \Omega_1)t\right) + \tfrac{A}{2}\cos\left((\Omega_0 - \Omega_1)t\right). \qquad (1.11)
\end{aligned}
$$

Thus, the modulated signal $y(t)$ is composed of two sinusoidal signals of frequencies $\Omega_0 + \Omega_1$ and $\Omega_0 - \Omega_1$ which are close to Ω_0 as Ω_1 has been assumed to be much smaller than the carrier frequency Ω_0.

It is instructive to examine the spectrum of $y(t)$. From the properties of the continuous-time Fourier transform it follows that the spectrum $Y(j\Omega)$ of $y(t)$ is given by

$$Y(j\Omega) = \frac{A}{2}X\left(j(\Omega - \Omega_0)\right) + \frac{A}{2}X\left(j(\Omega - \Omega_0)\right), \qquad (1.12)$$

where $X(j\Omega)$ is the spectrum of the modulating signal $x(t)$. Figure 1.4 shows the spectra of the modulating signal and that of the modulated signal under the assumption that the carrier frequency Ω_0 is greater than Ω_m, the highest frequency contained in $x(t)$. As seen

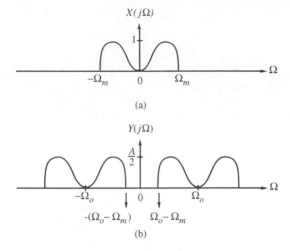

Figure 1.4 (a) Spectrum of the modulating signal $x(t)$, and (b) spectrum of the modulated signal $y(t)$. For convenience, both spectra are shown as real functions.

from this figure, $y(t)$ is now a bandlimited high-frequency signal with a bandwidth $2\Omega_m$ centered at Ω_0.

The portion of the amplitude-modulated signal between Ω_0 and $\Omega_0 + \Omega_m$ is called the *upper sideband,* whereas the portion between Ω_0 and $\Omega_0 - \Omega_m$ is called the *lower sideband.* Because of the generation of two sidebands and the absence of a carrier component in the modulated signal, the process is called *double sideband suppressed carrier* (DSB-SC) *modulation.*

The *demodulation* of $y(t)$, assuming $\Omega_0 > \Omega_m$, is carried out in two stages. First, the product of $y(t)$ with a sinusoidal signal of the same frequency as the carrier is formed. This results in

$$r(t) = y(t) \cos \Omega_0 t = Ax(t) \cos^2 \Omega_0 t \qquad (1.13)$$

which can be rewritten as

$$r(t) = y(t) \cos \Omega_0 t = \frac{A}{2}x(t) + \frac{A}{2}x(t) \cos(2\Omega_0 t). \qquad (1.14)$$

This result indicates that the product signal is composed of the original modulating signal scaled by a factor 1/2 and an amplitude-modulated signal with a carrier frequency $2\Omega_0$. The spectrum $R(j\Omega)$ of $r(t)$ is as indicated in Figure 1.5. The original modulating signal can now be recovered from $r(t)$ by passing it through a lowpass filter with a cutoff frequency Ω_c satisfying the relation $\Omega_m < \Omega_c < 2\Omega_0 - \Omega_m$. The output of the filter is then a scaled replica of the modulating signal.

Figure 1.6 shows the block diagram representations of the amplitude modulation and demodulation schemes. The underlying assumption in the demodulation process outlined above is that a sinusoidal signal identical to the carrier signal can be generated at the receiving end. In general, it is difficult to ensure that the demodulating sinsusoidal signal

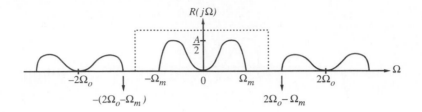

Figure 1.5   Spectrum of the product of the modulated signal and the carrier.

Figure 1.6   Schematic representations of the amplitude modulation and demodulation schemes: (a) modulator, and (b) demodulator.

has a frequency identical to that of the carrier all the time. To get around this problem, in the transmission of amplitude-modulated radio signals, the modulation process is modified so that the transmitted signal includes the carrier signal. This is achieved by redefining the amplitude-modulation operation as follows:

$$y(t) = A[1 + mx(t)]\cos(\Omega_0 t), \tag{1.15}$$

where m is a number chosen to ensure that $[1+mx(t)]$ is positive for all t. Figure 1.7 shows the waveforms of a modulating sinusoidal signal of frequency 2 kHz and the amplitude-modulated carrier obtained according to Eq. (1.15) for a carrier frequency of 40 kHz and $m = 0.5$. Note that the envelope of the modulated carrier is essentially the waveform of the modulating signal. As here the carrier is also present in the modulated signal, the process is called simply *double-sideband* (DSB) *modulation*. At the receiving end, the carrier signal is separated first and then used for demodulation.

1.2.5   Multiplexing and Demultiplexing

For an efficient utilization of a wideband transmission channel, many narrow-bandwidth low-frequency signals are combined to form a composite wideband signal that is transmitted as a single signal. The process of combining these signals is called *multiplexing* which is implemented to ensure that a replica of the original narrow-bandwidth low-frequency signals can be recovered at the receiving end. The recovery process is called *demultiplexing*.

One widely used method of combining different voice signals in a telephone communication system is the *frequency-division multiplexing* (FDM) scheme [Cou83], [Opp83]. Here, each voice signal, typically bandlimited to a low-frequency band of width $2\Omega_m$, is

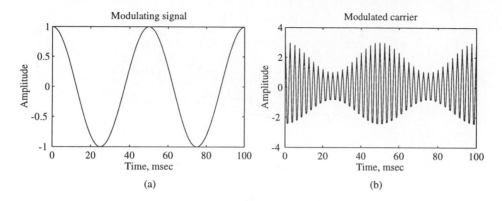

(a)   (b)

Figure 1.7   (a) A sinusoidal modulating signal of frequency 2 kHz, and (b) modulated carrier with a carrier frequency of 40 kHz based on the DSB modulation.

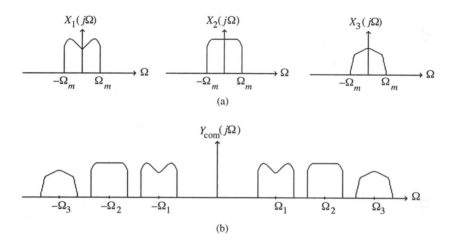

Figure 1.8   Illustration of the frequency-domain multiplexing operation.

frequency-translated into a higher frequency band using the amplitude-modulation method of Eq. (1.10). The carrier frequency of adjacent amplitude-modulated signals is separated by Ω_0, with $\Omega_0 > 2\Omega_m$ to ensure that there is no overlap in the spectra of the individual modulated signals after they are added to form a baseband composite signal. This signal is then modulated onto the main carrier developing the FDM signal and transmitted. Figure 1.8 illustrates the frequency-division multiplexing scheme.

At the receiving end, the composite baseband signal is first derived from the FDM signal by demodulation. Then each individual frequency-translated signal is first demultiplexed by passing the composite signal through a bandpass filter with a center frequency of identical value as that of the corresponding carrier frequency and a bandwidth slightly greater than $2\Omega_m$. The output of the bandpass filter is then demodulated using the method of Figure 1.6(b) to recover a scaled replica of the original voice signal.

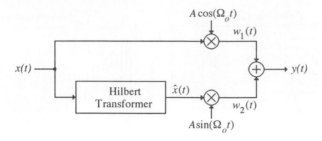

Figure 1.9   Single sideband modulation scheme employing a Hilbert transformer.

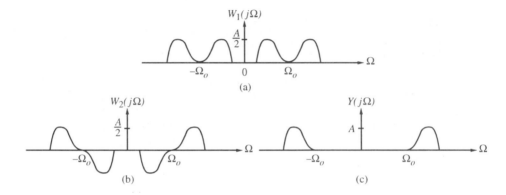

Figure 1.10   Spectra of pertinent signals in Figure 1.9.

In the case of the conventional amplitude modulation, as can be seen from Figure 1.4, the modulated signal has a bandwidth of $2\Omega_m$, whereas the bandwidth of the modulating signal is Ω_m. To increase the capacity of the transmission medium, a modified form of the amplitude modulation is often employed in which either the upper sideband or the lower sideband of the modulated signal is transmitted. The corresponding procedure is called *single-sideband modulation* (SSB) to distinguish it from the double-sideband modulation scheme of Figure 1.6(a).

One way to implement single-sideband amplitude modulation is indicated in Figure 1.9, where the Hilbert transformer is defined by Eq. (1.6). The spectra of pertinent signals in Figure 1.9 are shown in Figure 1.10.

1.2.6   Quadrature Amplitude Modulation

We observed earlier that double-sideband (DSB) amplitude modulation is half as efficient as single-sideband (SSB) amplitude modulation with regard to utilization of the spectrum. The *quadrature amplitude modulation* (QAM) method uses DSB modulation to modulate two different signals so that they both occupy the same bandwidth; thus QAM only takes up as much bandwidth as the SSB modulation method. To understand the basic idea behind the QAM approach, let $x_1(t)$ and $x_2(t)$ be two bandlimited low-frequency signals

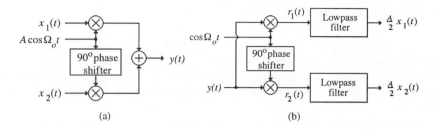

Figure 1.11   Schematic representations of the quadrature amplitude modulation and demodulation schemes: (a) modulator, and (b) demodulator.

with a bandwidth of Ω_m as indicated in Figure 1.4(a). The two modulating signals are individually modulated by the two carrier signals $A\cos(\Omega_0 t)$ and $A\sin(\Omega_0 t)$, respectively, and are summed, resulting in a composite signal $y(t)$ given by

$$y(t) = Ax_1(t)\cos(\Omega_0 t) + Ax_2(t)\sin(\Omega_0 t). \tag{1.16}$$

Note that the two carrier signals have the same carrier frequency Ω_0 but have a phase difference of 90°. In general, the carrier $A\cos(\Omega_0 t)$ is called the *in-phase component* and the carrier $A\sin(\Omega_0 t)$ is called the *quadrature component*. The spectrum $Y(j\Omega)$ of the composite signal $y(t)$ is now given by

$$Y(j\Omega) = \tfrac{A}{2}\{X_1(j(\Omega - \Omega_0)) + X_1(j(\Omega + \Omega_0))\}$$
$$+ \tfrac{A}{2j}\{X_2(j(\Omega - \Omega_0)) - X_2(j(\Omega + \Omega_0))\}, \tag{1.17}$$

and is seen to occupy the same bandwidth as the modulated signal obtained by a DSB modulation.

To recover the original modulating signals, the composite signal is multiplied by both the in-phase and the quadrature components of the carrier separately, resulting in two signals:

$$r_1(t) = y(t)\cos(\Omega_0 t),$$
$$r_2(t) = y(t)\sin(\Omega_0 t). \tag{1.18}$$

Substituting $y(t)$ from Eq. (1.16) in Eq. (1.18), we obtain after some algebra

$$r_1(t) = \tfrac{A}{2}x_1(t) + \tfrac{A}{2}x_1(t)\cos(2\Omega_0 t) + \tfrac{A}{2}x_2(t)\sin(2\Omega_0 t),$$
$$r_2(t) = \tfrac{A}{2}x_2(t) + \tfrac{A}{2}x_1(t)\sin(2\Omega_0 t) - \tfrac{A}{2}x_2(t)\cos(2\Omega_0 t), \tag{1.19}$$

Lowpass filtering of $r_1(t)$ and $r_2(t)$ by filters with a cutoff at Ω_m yield the two modulating signals. Figure 1.11 shows the block diagram representations of the quadrature amplitude-modulation and demodulation schemes.

As in the case of the DSB suppressed carrier modulation method, the QAM method also requires at the receiver an exact replica of the carrier signal employed in the transmitting end for accurate demodulation. It is therefore not employed in the direct transmission of analog signals, but finds applications in the transmission of discrete-time data sequences and in the transmission of analog signals converted into discrete-time sequences by sampling and A/D conversion.

1.2.7   Signal Generation

An equally important part of signal processing is synthetic signal generation. One of
the simplest such signal generators is a device generating a sinusoidal signal, called an
oscillator. Such a device is an integral part of the amplitude-modulation and demodulation
system described in the previous two sections. It also has various other signal processing
applications.

There are applications that require the generation of other types of periodic signals such
as square waves, triangular waves, etc. Certain types of random signals with a spectrum
of constant amplitude for all frequencies, called *white noise*, often find applications in
practice. One such application, for example, is in the generation of discrete-time synthetic
speech signals.

1.3   Examples of Typical Signals[2]

To better understand the breadth of the signal processing task, we now examine a number of
examples of some typical signals and their subsequent processing in typical applications.

Electrocardiography (ECG) signal.   The electrical activity of the heart is represented
by the ECG signal [Sha81]. A typical ECG signal trace is shown in Figure 1.12(a). The
ECG trace is essentially a periodic waveform. One such period of the ECG waveform as
depicted in Figure 1.12(b) represents one cycle of the blood transfer process from the heart
to the arteries. This part of the waveform is generated by an electrical impulse originating
at the sinoatrial node in the atrium of the heart. The impulse causes contraction of the atria,
forcing the blood in each atrium to squeeze into its corresponding ventricle. The resulting
signal is called the P-wave. The atrioventricular node delays the excitation impulse until
the blood transfer from the atria to the ventricle is completed, resulting in the P-R interval
of the ECG waveform. The excitation impulse then causes contraction of the ventricle,
squeezing the blood into the arteries and generating the QRS part of the ECG waveform.
During this phase the atria are relaxed and filled with blood. The T-wave of the waveform
represents the relaxation of the ventricles. The complete process is repeated periodically,
generating the ECG trace.

Each portion of the ECG waveform carries various types of information for the physician analyzing a patient's heart condition [Sha81]. For example, the amplitude and timing
of the P and QRS portions indicate the condition of the cardiac muscle mass. Loss of amplitude indicates muscle damage, whereas increased amplitude indicates abnormal heart
rates. Too long a delay in the atrioventricular node is indicated by a very long P-R interval.
Likewise, blockage of some or all of the contraction impulses is reflected by intermittent
synchronization between P- and QRS-waves. Most of these abnormalities can be treated
with various drugs and the effectiveness of the drugs can again be monitored by observing
the new ECG waveforms taken after the drug treatment.

In practice, there are various types of externally produced artifacts that appear in the
ECG signal [Tom81]. Unless these interferences are removed, it is difficult for a physician
to make a correct diagnosis. A common source of noise is the 60-Hz power lines whose

[2]This section has been adapted from *Handbook for Digital Signal Processing*, Sanjit K. Mitra and James F.
Kaiser, Eds., ©1993, John Wiley & Sons. Adapted by permission of John Wiley & Sons.

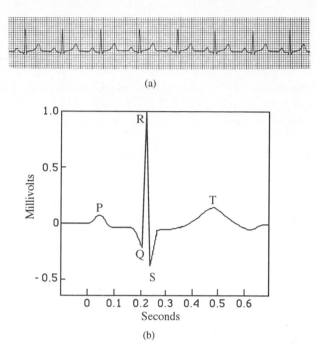

Figure 1.12   (a) A typical ECG trace, and (b) one cycle of a ECG waveform.

radiated electric and magnetic fields are coupled to the ECG instrument through capacitive coupling and/or magnetic induction. Other sources of interference are the electromyographic signals that are the potentials developed by contracting muscles. These and other interferences can be removed with careful shielding and signal processing techniques.

Electroencephalogram (EEG) Signal.   The summation of the electrical activity caused by the random firing of billions of individual neurons in the brain is represented by the EEG signal [Coh86], [Tom81]. In multiple EEG recordings, electrodes are placed at various positions on the scalp with two common electrodes placed on the earlobes, and potential differences between the various electrodes are recorded. A typical bandwidth of this type of EEG ranges from 0.5 to about 100 Hz, with the amplitudes ranging from 2 to 100 mV. An example of a multiple EEG trace is shown in Figure 1.13.

Both frequency-domain and time-domain analyses of the EEG signal have been used for the diagnosis of epilepsy, sleep disorders, psychiatric malfunctions, etc. To this end, the EEG spectrum is subdivided into the following five bands: (1) the *delta* range, occupying the band from 0.5 to 4 Hz; (2) the *theta* range, occupying the band from 4 to 8 Hz; (3) the *alpha* range, occupying the band from 8 to 13 Hz; (4) the *beta* range, occupying the band from 13 to 22 Hz; and (5) the *gamma* range, occupying the band from 22 to 30 Hz.

The delta wave is normal in the EEG signals of children and sleeping adults. Since it is not common in alert adults, its presence indicates certain brain diseases. The theta wave is usually found in children even though it has been observed in alert adults. The alpha wave is common in all normal humans and is more pronounced in a relaxed and awake

Figure 1.13   Multiple EEG signal traces.

subject with closed eyes. Likewise, the beta activity is common in normal adults. The EEG exhibits rapid, low-voltage waves, called *rapid-eye-movement* (REM) waves, in a subject dreaming during sleep. Otherwise, in a sleeping subject, the EEG contains bursts of alpha-like waves, called *sleep spindles.* The EEG of an epileptic patient exhibits various types of abnormalities, depending on the type of epilepsy that is caused by uncontrolled neural discharges.

Seismic Signals.   These types of signals are caused by the movement of rocks resulting from an earthquake, a volcanic eruption, or an underground explosion [Bol93]. The ground movement generates elastic waves that propagate through the body of the earth in all directions from the source of movement. Three basic types of elastic waves are generated by the earth movement. Two of these waves propagate through the body of the earth, one moving faster with respect to the other. The faster moving wave is called the *primary* or *P-wave*, while the slower moving one is called the *secondary* or *S-wave*. The third type of wave is known as the *surface wave*, which moves along the ground surface. These seismic waves are converted into electrical signals by a seismograph and are recorded on a strip chart recorder or a magnetic tape.

Because of the three-dimensional nature of ground movement, a seismograph usually consists of three separate recording instruments that provide information about the movements in the two horizontal directions and one vertical direction and develops three records as indicated in Figure 1.14. Each such record is a one-dimensional signal. From the recorded signals it is possible to determine the magnitude of the earthquake or nuclear explosion and the location of the source of the original earth movement.

Seismic signals also play an important role in the geophysical exploration for oil and gas [Rob80]. In this type of applications linear arrays of seismic sources, such as high-energy explosives, are placed at regular intervals on the ground surface. The explosions cause seismic waves to propagate through the subsurface geological structures and reflect back to the surface from interfaces between geological strata. The reflected waves are

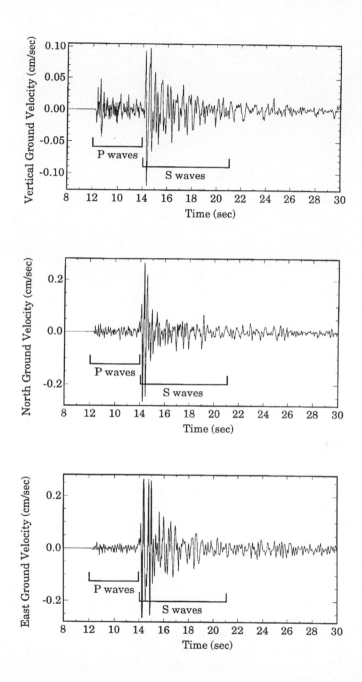

Figure 1.14 Seismograph record of the Northridge aftershock, January 29, 1994. Recorded at Stone Canyon Reservoir, Los Angeles, CA [Courtesy of Institute for Crustal Research, University of California, Santa Barbara, CA].

Figure 1.15   A typical seismic signal trace gather. [Courtesy of Institute for Crustal Research, University of California, Santa Barbara, CA].

converted into electrical signals by a composite array of geophones laid out in certain patterns and displayed as a two-dimensional signal that is a function of time and space, called a *trace gather*, as indicated in Figure 1.15. Before these signals are analyzed, some preliminary time and amplitude corrections are made on the data to compensate for different physical phenomena. From the corrected data, the time differences between reflected seismic signals are used to map structural deformations, whereas the amplitude changes usually indicate the presence of hydrocarbons.

Diesel engine signal.   Signal processing is playing an important role in the precision adjustment of diesel engines during production [Jur81]. Efficient operation of the engine requires the accurate determination of the topmost point of piston travel (called the *top dead center*) inside the cylinder of the engine. Figure 1.16 shows the signals generated by a dual probe inserted into the combustion chamber of a diesel engine in place of the glow plug. The probe consists of a microwave antenna and a photodiode detector. The microwave probe captures signals reflected from the cylinder cavity caused by the up and down motion of the piston while the engine is running. Interestingly, the waveforms of these signals exhibit a symmetry around the top dead center independent of the engine speed, temperature, cylinder pressure, or air-fuel ratio. The point of symmetry is determined automatically by a microcomputer, and the fuel-injection pump position is then adjusted by the computer accurately to within ±0.5 degree using the luminosity signal sensed by the photodiode detector.

Figure 1.16 Diesel engine signals. (Reproduced with permission from R. K. Jurgen, "Detroit bets on electronics to stymie Japan," IEEE Spectrum, Vol. 18, July 1981, pp. 29—32 ©1981 IEEE).

Speech signal.   The linear acoustic theory of speech production has led to mathematical models for the representation of speech signals. A speech signal is formed by exciting the vocal tract and is composed of two types of sounds: *voiced* and *unvoiced* [Rab78], [Slu84]. The voiced sound, which includes the vowels and a number of consonants such as B, D, L, M, N, and R, is excited by the pulsatile airflow resulting from the vibration of the vocal folds. On the other hand, the unvoiced sound is produced downstream in the forward part of the oral cavity (mouth) with the vocal cords at rest and includes sounds like F, S, and SH.

Figure 1.17(a) depicts the speech waveform of a male utterance "every salt breeze comes from the sea" [Fla79]. The total duration of the speech waveform is 2.5 seconds. Magnified versions of the "A" and "S" segments in the word "salt" are sketched in Figure 1.17(b) and (c), respectively. The slowly varying low-frequency voiced waveform of "A" and the high-frequency unvoiced fricative waveform of "S" are evident from the magnified waveforms. The voiced waveform in Figure 1.17(b) is seen to be quasi-periodic and can be modeled by a sum of a finite number of sinusoids. The lowest frequency of oscillation in this representation is called the *fundamental frequency* or *pitch frequency.* The unvoiced waveform in Figure 1.17(c) has no regular fine structure and is more noise-like.

One of the major applications of digital signal processing techniques is in the general area of speech processing. Problems in this area are usually divided into three groups: (1) speech analysis, (2) speech synthesis, and (3) speech analysis and synthesis [Opp78].

Figure 1.17   Speech waveform example. (a) sentence-length segment, (b) magnified version of the voiced segment (the letter A), and (c) magnified version of the unvoiced segment (the letter S). (Reproduced with permission from J. L. Flanagan et al, "Speech coding," IEEE Trans. on Communications, vol. COM-27, April 1979, pp. 710—737 ©1979 IEEE.)

Digital speech analysis methods are used in automatic speech recognition, speaker verification, and speaker identification. Applications of digital speech synthesis techniques include reading machines for the automatic conversion of written text into speech, and retrieval of data from computers in speech form by remote access through terminals or telephones. One example belonging to the third group is voice scrambling for secure transmission. Speech data compression for an efficient use of the transmission medium is another example of the use of speech analysis followed by synthesis. A typical speech signal after conversion into a digital form contains about 64,000 bits per second (bps). Depending on the desired quality of the synthesized speech, the original data can be compressed considerably, e.g., down to about 1000 bps.

Musical sounds.   The electronic synthesizer is an example of the use of modern signal processing techniques [Moo77], [Ler83]. The natural sound generated by most musical instruments is generally produced by mechanical vibrations caused by activating some form of mechanical oscillator that then causes other parts of the instrument to vibrate. All these vibrations together in a single instrument generate the musical sound. For example, in a violin, the primary oscillator is a stretched piece of string (cat gut). Its movement is caused by drawing a bow across it; this sets the wooden body of the violin vibrating, which in turn sets up vibrations of the air inside as well as outside the instrument. Likewise, in a piano, the primary oscillator is a stretched steel wire that is set into vibratory motion by the hitting of a hammer, which in turn causes vibrations in the wooden body (sounding board) of the piano. In wind or brass instruments, the vibration occurs in a column of air, and a mechanical change in the length of the air column by means of valves or keys regulates the rate of vibration.

The sound of orchestral instruments can be classified into two groups: *quasi-periodic* and *aperiodic*. Quasi-periodic sounds can be described by a sum of a finite number of sinusoids with independently varying amplitudes and frequencies. The sound waveforms of two different instruments, the cello and the bass drum, are indicated in Figure 1.18(a) and (b), respectively. In each figure, the top waveform is the plot of an entire isolated note, whereas the bottom plot shows an expanded version of a portion of the note: 10 ms for the cello and 80 ms for the bass drum. The waveform of the note from a cello is seen to be quasi-periodic. On the other hand, the bass drum waveform is clearly aperiodic. The tone of an orchestral instrument is commonly divided into three segments called the *attack* part, the *steady-state* part, and the *decay* part. Figure 1.18 illustrates this division for the two tones. Note that the bass drum tone of Figure 1.18(b) shows no steady-state part. A reasonable approximation of many tones is obtained by splicing together these parts. However, high-fidelity reproduction requires a more complex model.

Time series.   The signals described thus far are continuous functions with time as the independent variable. In many cases the signals of interest are naturally discrete functions of the independent variables. Often such signals are of finite duration. Examples of such signals are the yearly average number of sunspots, daily stock prices, the value of total monthly exports of a country, the yearly population of animal species in a certain geographic area, the annual yields per acre of crops in a country, the monthly totals of international airline passengers, etc., over certain periods. This type of finite extent signal, usually called a *time series*, occurs in business, economics, physical sciences, social sciences, engineering, medicine, and many other fields. Plots of some typical time series are shown in Figures 1.19 to 1.21.

There are many reasons for analyzing a particular time series [Box70]. In some applications, there may be a need to develop a model to determine the nature of the dependence of the data on the independent variable and use it to forecast the future behavior of the series. As an example, in business planning, reasonably accurate sales forecasts are necessary. Some types of series possess seasonal or periodic components, and it is important to extract these components. The study of sunspot numbers is important for predicting climate variations. Invariably, the time series data are noisy, and their representations require models based on their statistical properties.

Images.   As indicated earlier, an image is a two-dimensional signal whose intensity at any point is a function of two spatial variables. Common examples are photographs, still video images, radar and sonar images, and chest and dental x-rays. An image sequence, such as that seen in a television, is essentially a three-dimensional signal for which the

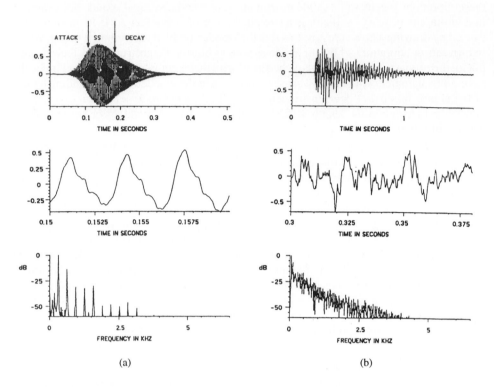

(a)   (b)

Figure 1.18   Waveforms of (a) the cello and (b) the bass drum. (Reproduced with permission from J. A. Moorer, "Signal processing aspects of computer music: A survey," Proceedings of the IEEE, Vol. 65, August 1977, pp. 1108—1137 ©1977 IEEE).

Figure 1.19   Seasonally adjusted quarterly Gross National Product of United States in 1982 dollars from 1976 to 1986 (Adapted from H. Lütkepohl, *Introduction to Multiple Time Series Analysis*, Springer-Verlag, New York, NY, 1991).

Figure 1.20   Monthly mean St. Louis, Missouri temperature in degrees Celsius for the years 1975 to 1978 (Adapted from S. Lawrence Marple, Jr., *Digital Spectral Analysis with Applications,* Prentice-Hall, Englewood Cliffs, NJ, 1987).

Figure 1.21   Monthly gasoline demand in Ontario, Canada (in millions of gallons) from January 1971 to December 1975 (Adapted from B. Abraham and J. Ledolter, *Statistical Methods for Forcasting,* Wiley, New York, NY, 1983).

image intensity at any point is a function of three variables: two spatial variables and time. Figure 1.22(a) shows the photograph of a digital image.

The basic problems in image processing are image signal representation and modeling, enhancement, restoration, reconstruction from projections, analysis, and coding [Jai89].

Each picture element in a specific image represents a certain physical quantity; a characterization of the element is called the *image representation.* For example, a photograph represents the luminances of various objects as seen by the camera. An infrared image taken by a satellite or an airplane represents the temperature profile of a geographical area. Depending on the type of image and its applications, various types of image models are usually defined. Such models are also based on perception, and on local or global characteristics. The nature and performance of the image processing algorithms depend on the image model being used.

Image enhancement algorithms are used to emphasize specific image features to improve the quality of the image for visual perception or to aid in the analysis of the image for feature extraction. These include methods for contrast enhancement, edge detection, sharpening, linear and nonlinear filtering, zooming, noise removal. Figure 1.22(b) shows the contrast-enhanced version of the image of Figure 1.22(a) developed using a nonlinear filter [Str94].

The algorithms used for elimination or reduction of degradations in an image, such as blurring, geometric distortion, etc., caused by the imaging system and/or its surroundings are known as *image restoration. Image reconstruction* from projections involves the development of a two-dimensional image slice of a three-dimensional object from a number of planar projections obtained from various angles. By creating a number of contiguous slices, a three-dimensional image giving an inside view of the object is developed.

Image analysis methods are employed to develop a quantitative description and classification of one or more desired objects in an image.

For digital processing, an image needs to be sampled and quantized using an A/D converter. A reasonable size digital image in its original form takes a considerable amount of memory space for storage. For example, a image of size 512×512 samples with 8-bit resolution per sample contains over 2 million bits. *Image coding* methods are used to reduce the total number of bits in an image without any degradation in visual perception quality as in speech coding, e.g., down to about 1 bit per sample on the average.

1.4   Typical Signal Processing Applications[3]

There are numerous applications of signal processing that we often encounter in our daily life without being aware of them. Due to space limitations, it is not possible to discuss all of these applications. However, an overview of selected applications is presented.

1.4.1   Sound Recording Applications

The recording of most musical programs nowadays is usually made in an acoustically inert studio. The sound from each instrument is picked up by its own microphone closely placed to the instrument and is recorded on a single track in a multitrack tape recorder containing as many as 48 tracks. The signals from individual tracks in the master recording are then edited and combined by the sound engineer in a *mix-down* system to develop a two-track stereo recording. There are a number of reasons for following this approach. First, the closeness of each individual microphone to its assigned instrument provides a high degree of separation between the instruments and minimizes the background noise in the recording. Second, the sound part of one instrument can be rerecorded later if necessary. Third, during the mix-down process the sound engineer can manipulate individual signals by using a variety of signal processing devices to alter the musical balances between the sounds generated by the instruments, can change the timbre, and can add natural room acoustics effects and other special effects [Ble78], [Ear76].

Various types of signal processing techniques are utilized in the mix-down phase. Some of these techniques are used to modify the spectral characteristics of the sound signal and to add special effects, whereas others are used to improve the quality of the transmission medium. The signal processing circuits most commonly used are: (1) compressors and limiters, (2) expanders and noise gates, (3) equalizers and filters, (4) noise reduction systems, (5) delay and reverberation systems, and (6) circuits for special effects [Ble78], [Ear76], [Hub89], [Wor89]. These operations are usually performed on the original analog audio signals and are implemented using analog circuit components. However, there is

---

[3]This section has been adapted from *Handbook for Digital Signal Processing*, Sanjit K. Mitra and James F. Kaiser, Eds., ©1993, John Wiley & Sons. Adapted by permission of John Wiley & Sons.

Figure 1.22   (a) A digital image, and (b) its contrast enhanced version.

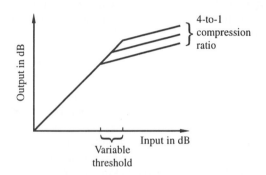

Figure 1.23   Transfer characteristic of a typical compressor.

a growing trend toward all digital implementation and its use in the processing of the digitized versions of the analog audio signals [Ble78].

Compressors and limiters.   These devices are used for the compression of the dynamic range of an audio signal. The compressor can be considered as an amplifier with two gain levels: the gain is unity for input signal levels below a certain threshold and less than unity for signals with levels above the threshold. The threshold level is adjustable over a wide range of the input signal. Figure 1.23 shows the transfer characteristic of a typical compressor.

The parameters characterizing a compressor are its compression ratio, threshold level, attack time, and release time, which are illustrated in Figure 1.24.

When the input signal level suddenly rises above a prescribed threshold, the time taken by the compressor to adjust its normal unity gain to the lower value is called the *attack time*. Because of this effect, the output signal exhibits a slight degree of overshoot before the desired output level is reached. A zero attack time is desirable to protect the system from

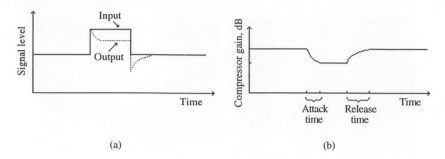

Figure 1.24   Parameters characterizing a typical compressor.

sudden high-level transients. However, in this case, the impact of sharp musical attacks is eliminated, resulting in a dull "lifeless" sound [Wor89]. A longer attack time causes the output to sound more percussive than normal.

Similarly, the time taken by the compressor to reach its normal unity gain value when the input level suddenly drops below the threshold is called the *release time* or *recovery time.* If the input signal fluctuates rapidly around the threshold in a small region, the compressor gain also fluctuates up and down. In such a situation, the rise and fall of background noise results in an audible effect called *breathing* or *pumping,* which can be minimized with a longer release time for the compressor gain.

There are various applications of the compressor unit in musical recording [Ear76]. For example, it can be used to eliminate variations in the peaks of an electric bass output signal by clamping them to a constant level, thus providing an even and solid bass line. To maintain the original character of the instrument, it is necessary to use a compressor with a long recovery time compared to the natural decay rate of the electric bass. The device is also useful to compensate for the wide variations in the signal level produced by a singer who moves frequently, changing the distance from the microphone.

A compressor with a compression ratio of 10-to-1 or greater is called a *limiter* since its output levels are essentially clamped to the threshold level. The limiter is used to prevent overloading of amplifiers and other devices caused by signal peaks exceeding certain levels.

Expanders and noise gates.   The expander's function is opposite that of the compressor. It is also an amplifier with two gain levels: the gain is unity for input signal levels above a certain threshold and less than unity for signals with levels below the threshold. The threshold level is again adjustable over a wide range of the input signal. Figure 1.25 shows the transfer characteristic of a typical expander. The *expander* is used to expand the dynamic range of an audio signal by boosting the high-level signals and attenuating the low-level signals. The device can also be used to reduce noise below a threshold level.

The expander is characterized by its expansion ratio, threshold level, attack time, and release time. Here, the time taken by the device to reach the normal unity gain for a sudden change in the input signal to a level below the threshold is defined as the *attack time.* Likewise, the time required by the device to lower the gain from its normal value of one for a sudden increase in the input signal level is called the *release time.*

The noise gate is a special type of expander that heavily attenuates signals with levels below the threshold. It is used, for example, to totally cutoff a microphone during a musical pause so as not to pass the noise being picked up by the microphone.

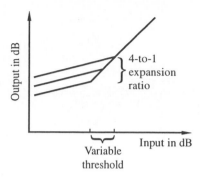

Figure 1.25   Transfer characteristic of a typical expander.

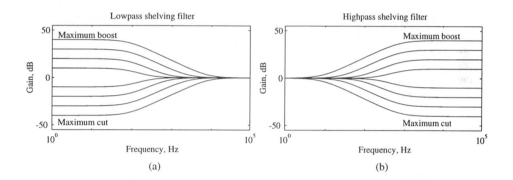

Figure 1.26   Frequency responses of (a) low-frequency shelving filter and (b) high-frequency shelving filter.

Equalizers and filters.   Various types of filters are used to modify the frequency response of a recording or the monitoring channel. One such filter, called the *shelving filter*, provides boost (rise) or cut (drop) in the frequency response at either the low or at the high end of the audio frequency range while not affecting the frequency response in the remaining range of the audio spectrum, as shown in Figure 1.26. *Peaking filters* are used for midband equalization and are designed to have either a bandpass response to provide a boost or a bandstop response to provide a cut, as indicated in Figure 1.27.

The parameters characterizing a low-frequency shelving filter are the two frequencies f_{1L} and f_{2L}, where the magnitude response begins tapering up or down from a constant level and the low-frequency gain levels in dB. Likewise, the parameters characterizing a high-frequency shelving filter are the two frequencies f_{1H} and f_{2H}, where the magnitude response begins tapering up or down from a constant level and the high-frequency gain levels in dB. In the case of a peaking filter, the parameters of interest are the center frequency f_0, the 3-dB bandwidth Δf of the bell-shaped curve, and the gain level at the center frequency. Most often, the quality factor $Q = f_0/\Delta f$ is used to characterize the shape of the frequency response instead of the bandwidth Δf.

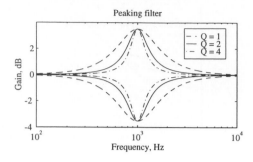

Figure 1.27   Peaking filter frequency response.

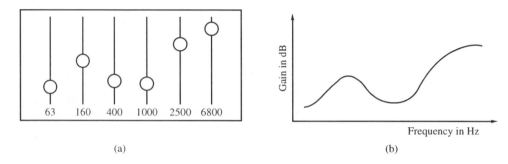

Figure 1.28   Graphic equalizer: (a) control panel settings, and (b) corresponding frequency response (Adapted from [Ear76]).

A typical equalizer consists of a cascade of a low-frequency shelving filter, a high-frequency shelving filter, and three or more peaking filters with adjustable parameters to provide adjustment of the overall equalizer frequency response over a broad range of frequencies in the audio spectrum. In a *parametric equalizer*, each individual parameter of its constituent filter blocks can be varied independently without affecting the parameters of the other filters in the equalizer.

The *graphic equalizer* consists of a cascade of peaking filters with fixed center frequencies but adjustable gain levels that are controlled by vertical slides in the front panel. The physical position of the slides reasonably approximates the overall equalizer magnitude response, as shown schematically in Figure 1.28.

Other types of filters that also find applications in the musical recording and transfer processes are the *lowpass, highpass,* and *notch filters*. Their corresponding frequency responses are indicated in Figure 1.29. The notch filter is designed to attenuate a particular frequency component and has a narrow notch width so as not to affect the rest of the musical program.

Two major applications of equalizers and filters in recording are to correct certain types of problems that may have occurred during the recording or the transfer process and to alter the harmonic or timbral contents of a recorded sound purely for musical or creative purposes [Ear76]. For example, a direct transfer of a musical recording from old 78 rpm

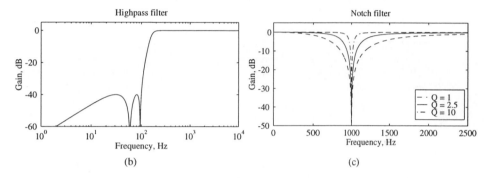

Figure 1.29 Frequency responses of other types of filters: (a) lowpass filter, (b) highpass filter, and (c) notch filter.

disks to a wideband playback system will be highly noisy due to the limited bandwidth of the old disks. To reduce this noise, a bandpass filter with a passband matching the bandwidth of the old records is utilized. Often, older recordings are made more pleasing by adding a broad high-frequency peak in the 5- to 10-kHz range and by shelving out some of the lower frequencies. The notch filter is particularly useful in removing 60-Hz power supply hum.

In creating a program by mixing down a multichannel recording, the recording engineer usually employs equalization of individual tracks for creative reasons [Ear76]. For example, a "fullness" effect can be added to weak instruments such as the acoustical guitar, by boosting frequency components in the range 100 to 300 Hz. Similarly, by boosting the 2- to 4-kHz range, the transients caused by the fingers against the string of an acoustical guitar can be made more pronounced. A high-frequency shelving boost above the 1- to 2-kHz range increases the "crispness" in percussion instruments such as the bongo or snare drums.

Noise reduction system.   The overall dynamic range of human hearing is over 120 dB. However, most recording and transmission mediums have a much smaller dynamic range. The music to be recorded must be above the sound background or noise. If the background noise is around 30 dB, the dynamic range available for the music is only 90 dB, requiring dynamic range compression for noise reduction.

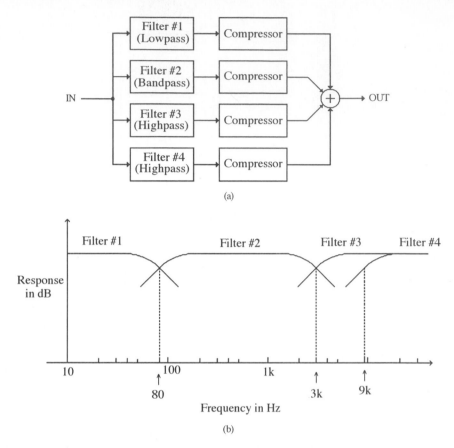

(a)

(b)

Figure 1.30 The Dolby A-type noise reduction scheme for the recording mode. (a) Block diagram, and (b) Frequency responses of the four filters with cutoff frequencies as shown.

A noise reduction system consists of two parts. The first part provides the compression during the recording mode while the second part provides the complementary expansion during the playback mode. To this end, the most popular methods in musical recording are the Dolby noise reduction schemes, of which there are several types [Ear76], [Hub89], [Wor89].

In the Dolby A-type method used in professional recording, for the recording mode, the audio signal is split into four frequency bands by a bank of four filters; separate compression is provided in each band and the outputs of the compressors are combined, as indicated in Figure 1.30(a). Moreover, the compression in each band is restricted to a 20-dB input range from −40 to −20 dB. Below the lower threshold (−40 dB), very-low-level signals are boosted by 10 dB, and above the upper threshold (−20 dB), the system has unity gain, passing the high-level signals unaffected. The transfer characteristic for the record mode is thus as shown in Figure 1.31.

In the playback mode, the scheme is essentially the same as that in the recording mode, except here the compressors are replaced by expanders with complementary transfer characteristics, as indicated in Figure 1.31. Here, the expansion is limited to a 10-dB input range from −30 to −20 dB. Above the upper threshold (−20 dB), very-high-level signals

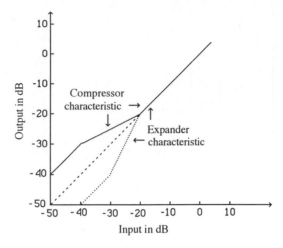

Figure 1.31 Compressor and expander transfer characteristic for the Dolby A-type noise reduction scheme.

are cut by 10 dB, while below the lower threshold (-30 dB), the system has unity gain passing the low-level signals unaffected.

Note that for each band, a 2-to-1 compression is followed by a 1-to-2 complementary expansion such that the dynamic range of the signal at the input of the compressor is exactly equal to that at the expander output. This type of overall signal processing operation is often called *companding.* Moreover, the companding operation in one band has no effect on a signal in another band and may often be masked by other bands with no companding.

Delay and reverberation systems. Music generated in an inert studio does not sound natural compared to the music performed inside a room, such as a concert hall. In the latter case, the sound waves propagate in all directions and reach the listener from various directions and at various times, depending on the distance traveled by the sound waves from the source to the listener. The sound wave coming directly to the listener, called the *direct sound,* reaches first and determines the listener's perception of the location, size, and nature of the sound source. This is followed by a few closely spaced echoes, called *early reflections,* generated by reflections of sound waves from all sides of the room and reaching the listener at irregular times. These echoes provide the listener subconscious cues as to the size of the room. After these early reflections, more and more densely packed echoes reach the listener due to multiple reflections. The latter group of echoes is referred to as the *reverberation.* The amplitude of the echoes decay exponentially with time as a result of attenuation at each reflection. Figure 1.32 illustrates this concept. The period of time in which the reverberation falls by 60 dB is called the *reverberation time.* Since the absorption characteristics of different materials are not the same at different frequencies, the reverberation time varies from frequency to frequency.

Delay systems with adjustable delay factors are employed to artificially create the early reflections. Electronically generated reverberation combined with artificial echo reflections are usually added to the recordings made in a studio. The block diagram

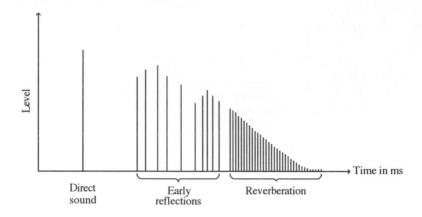

Figure 1.32   Various types of echoes generated by a single sound source in a room.

Figure 1.33   Block diagram of a complete delay-reverberation system in a monophonic system.

representation of a typical delay-reverberation system in a monophonic system is depicted in Figure 1.33.

There are various other applications of electronic delay systems, some of which are described next [Ear76].

Special effects.   By feeding in the same sound signal through adjustable delay and gain control, as indicated in Figure 1.34, it is possible to vary the localization of the sound source from the left speaker to the right for a listener located on the plane of symmetry. For example, in Figure 1.34, a 0-dB loss in the left channel and a few milliseconds delay in the right channel gives the impression of a localization of the sound source at the left. However, lowering of the left-channel signal level by a few-dB loss results in a phantom image of the sound source moving toward the center. This scheme can be further extended to provide a degree of *image broadening* by phase shifting one channel with respect to the other through allpass networks[4] as shown in Figure 1.35.

Another application of the delay-reverberation system is in the processing of a single track into a pseudo-stereo format while simulating a natural acoustical environment, as illustrated in Figure 1.36.

---

[4] An allpass network is characterized by a magnitude spectrum which is equal to one for all frequencies.

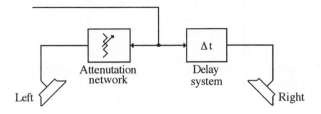

Figure 1.34   Localization of sound source using delay systems and attenuation network.

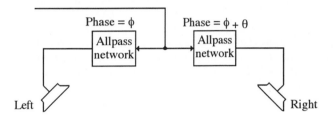

Figure 1.35   Image broadening using allpass networks.

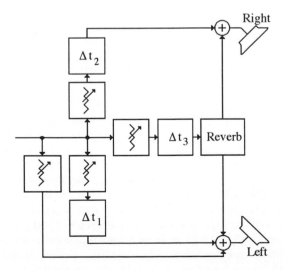

Figure 1.36   A possible application of delay systems and reverberation in a stereophonic system.

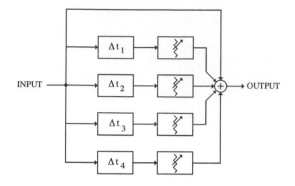

Figure 1.37   A scheme for implementing chorus effect.

The delay system can also be used to generate a chorus effect from the sound of a soloist. The basic scheme used is illustrated in Figure 1.37. Each of the delay units has a variable delay controlled by a low-frequency pseudo-random noise source to provide a random pitch variation [Ble78].

It should be pointed out here that additional signal processing is employed to make the stereo submaster developed by the sound engineer more suitable for the record-cutting lathe or the cassette tape duplicator.

1.4.2   Telephone Dialing Applications

Signal processing plays a key role in the detection and generation of signaling tones for push-button telephone dialing [Dar76]. In telephones equipped with TOUCH-TONE® dialing, the pressing of each button generates a unique set of two-tone signals that are processed at the telephone central office to identify the number pressed by determining the two associated tone frequencies. Seven frequencies are used to code the 10 decimal digits and the two special buttons marked "* " and "# ". The low-band frequencies are 697 Hz, 770 Hz, 852 Hz, and 941 Hz. The remaining three frequencies belonging to the highband are 1209 Hz, 1336 Hz, and 1477 Hz. The fourth high-band frequency of 1633 Hz is not presently in use and has been assigned for future applications to permit the use of four additional push-buttons for special services. The frequency assignments used in the TOUCH-TONE® dialing scheme are shown in Figure 1.38 [ITU84].

The scheme used to identify the two frequencies associated with the button that has been pressed is shown in Figure 1.39. Here, the two tones are first separated by a lowpass and a highpass filter. The passband cutoff frequency of the lowpass filter is slightly above 1000 Hz, whereas that of the highpass filter is slightly below 1200 Hz. The output of each filter is next converted into a square wave by a limiter and then processed by a bank of bandpass filters with narrow passbands. The four bandpass filters in the low-frequency channel have center frequencies at 697 Hz, 770 Hz, 852 Hz, and 941 Hz. The four bandpass filters in the high-frequency channel have center frequencies at 1209 Hz, 1336 Hz, 1477 Hz, and 1633 Hz. The detector following each bandpass filter develops the necessary dc switching signal if its input voltage is above a certain threshold.

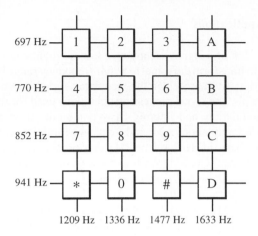

Figure 1.38   The tone frequency assignments for TOUCH-TONE® dialing.

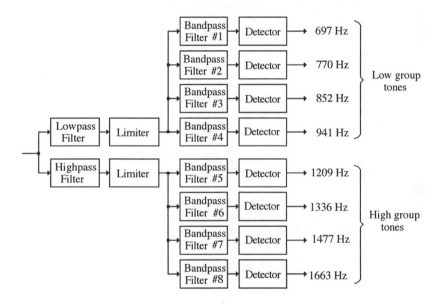

Figure 1.39   The tone detection scheme for TOUCH-TONE® dialing.

All the signal processing functions described above are usually implemented in practice in the analog domain. However, increasingly, these functions are being implemented using digital techniques.[5]

---

[5]See Section 11.1.

1.4.3   FM Stereo Applications

For wireless transmission of a signal occupying a low-frequency range, such as an audio signal, it is necessary to transform the signal to a high-frequency range by modulating it onto a high-frequency carrier. At the receiver, the modulated signal is demodulated to recover the low frequency signal. The signal processing operations used for wireless transmission are modulation, demodulation, and filtering. Two commonly used modulation schemes for radio are amplitude modulation (AM) and frequency modulation (FM).

We next review the basic idea behind the FM stereo broadcasting and reception scheme as used in the United States [Cou83]. An important feature of this scheme is that at the receiving end, the signal can be heard over a standard monoaural FM radio with a single speaker or over a stereo FM radio with two speakers. The system is based on the frequency-division multiplexing (FDM) method described earlier in Section 1.2.5.

The block diagram representations of the FM stereo transmitter and the receiver are shown in Figure 1.40(a) and (b), respectively. At the transmitting end, the sum and the difference of the left and right channel audio signals, $s_L(t)$ and $s_R(t)$, are first formed. Note that the summed signal $s_L(t) + s_R(t)$ is used in monoaural FM radio. The difference signal $s_L(t) - s_R(t)$ is modulated using the double sideband suppressed carrier (DSB-SC) scheme[6] using a subcarrier frequency f_{sc} of 38 kHz. The summed signal, the modulated difference signal, and a 19-kHz pilot tone signal are then added, developing the composite baseband signal $s_B(t)$. The spectrum of the composite signal is shown in Figure 1.40(c). The baseband signal is next modulated onto the main carrier frequency f_c using the frequency modulation (FM) method. At the receiving end, the FM signal is demodulated to derive the baseband signal $s_B(t)$ which is then separated into the low- frequency summed signal and the modulated difference signal using a lowpass filter and a bandpass filter. The cutoff frequency of the lowpass filter is around 15 kHz, whereas the center frequency of the bandpass filter is at 38 kHz. The 19-kHz pilot tone is used in the receiver to develop the 38 kHz reference signal for a coherent subcarrier demodulation for recovering the audio difference signal. The sum and difference of the two audio signals create the desired left audio and right audio signals.

1.4.4   Electronic Music Synthesis

The generation of the sound of a musical instrument using electronic circuits is another example of the application of signal processing methods [All80], [Moo77]. The basis of such music synthesis is the following representation of the sound signal $s(t)$:

$$s(t) = \sum_{k=1}^{N} A_k(t) \sin(2\pi f_k(t)t) \tag{1.20}$$

where $A_k(t)$ and $f_k(t)$ are the time-varying amplitude and frequency of the kth component of the signal. The frequency function $f_k(t)$ varies slowly with time. For an instrument playing an isolated tone, $f_k(t) = k f_o$, i.e.

$$s(t) = A_k(t) \sin(2\pi k f_0 t) \tag{1.21}$$

---

[6]See Section 1.2.4.

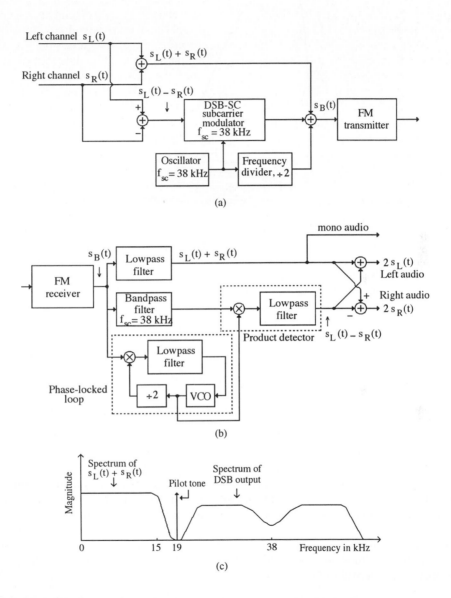

Figure 1.40 The FM stereo system: (a) transmitter, (b) receiver, and (c) spectrum of the composite baseband signal $s_B(t)$.

Figure 1.41   Perspective plot of the amplitude functions $A_k(t)$ for an actual note from a clarinet. (Reproduced with permission from J. A. Moorer, "Signal processing aspects of computer music: A survey," Proceedings of the IEEE, Vol. 65, August 1977, pp. 1108—1137 ©1977 IEEE).

where f_0 is called the *fundamental frequency.* In a musical sound with many tones, all other frequencies are usually integer multiples of the fundamental and are called *partial frequencies,* also called *harmonics.* Figure 1.41 shows, for example, the perspective plot of the amplitude functions as a function of time of 17 partial frequency components of an actual note from a clarinet. The aim of the synthesis is to produce electronically the $A_k(t)$ and $f_k(t)$ functions. To this end, the two most popular approaches followed are described next.

Subtractive synthesis.   This approach, which nearly duplicates the sound generation mechanism of a musical instrument, is based on the generation of a periodic signal containing all required harmonics and the use of filters to selectively attenuate (i.e., *subtract*) unwanted partial frequency components. The frequency-dependent gain of the filters can also be used to boost certain frequencies. The desired variations in the amplitude functions are generated by an analog multiplier or a voltage-controlled amplifier. Additional variations in the amplitude functions can be provided by dynamically adjusting the frequency response characteristics of the filters.

Additive synthesis.   Here, partial frequency components are generated independently by oscillators with time-varying oscillation frequencies. The amplitudes of the required signals are then individually modified, approximating the actual variations obtained by analysis and combined (i.e., *added*) to produce the desired sound signal. For example, a piecewise linear approximation of the clarinet note of Figure 1.41 is sketched in

Figure 1.42 A piecewise-linear approximation to the amplitude functions of Figure 1.41. (Reproduced with permission from J. A. Moorer, "Signal processing aspects of computer music: A survey," Proceedings of the IEEE, Vol. 65, August 1977, pp. 1108—1137 ©1977 IEEE).

Figure 1.42 and can be used to generate a reasonable replica of the note. Usually, some alterations to the amplitude and frequency functions may be needed before the music that is generated sounds as close as possible to that of the original instrument.

1.4.5 Echo Cancellation in Telephone Networks

In a telephone network the central offices perform the necessary switching to connect two subscribers [Dut80], [Fre78], [Mes82]. For economic reasons, a "two-wire" circuit is used to connect a subscriber to his/her central office, whereas the central offices are connected using "four-wire" circuits. A two-wire circuit is bidirectional and carries signals in both directions. A four-wire circuit uses two separate unidirectional paths for signal transmission in both directions. The latter is preferred for long-distance trunk connections since signals at intermediate points in the trunk can be equalized and amplified using repeaters and, if necessary, multiplexed easily. A hybrid coil in the central office provides the interface between a two-wire circuit and a four-wire circuit, as shown in Figure 1.43. The hybrid circuit ideally should provide a perfect impedance match to the two-wire circuit by impedance balancing so that the incoming four-wire receive signal is passed directly to the two-wire circuit connected to the hybrid with no portion appearing in the four-wire transmit path. However, to save cost, a hybrid coil is shared among several subscribers. Thus, it is not possible to provide a perfect impedance match in every case since the length of the subscriber lines vary. The resulting imbalance causes a large portion of the incoming receive signal from the distance talker to appear in the transmit path, and it is returned to the talker as an echo. Figure 1.44 illustrates the normal transmission between a talker and a listener as well as two possible major echo paths.

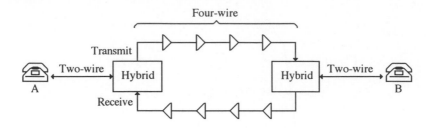

Figure 1.43   Basic 2/4-wire interconnection scheme.

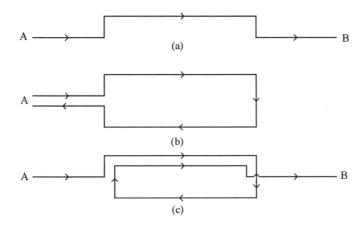

Figure 1.44   Various signal paths in a telephone network. (a) Transmission path from talker A to listener B, (b) echo path for talker A, and (c) echo path for listener B.

The effect of the echo can be annoying to the talker, depending on the amplitude and delay of the echo, i.e., on the length of the trunk circuit. The effect of the echo is worst for telephone networks involving geostationary satellite circuits, where the echo delay is about 540 ms.

Several methods are followed to reduce the effect of the echo. In trunk circuits up to 3000 km in length, adequate reduction of the echo is achieved by introducing additional signal loss in both directions of the four-wire circuit. In this scheme, an improvement in the signal-to-echo ratio is realized since the echo undergoes loss in both directions while the signals are attenuated only once.

For distances greater than 3000 km, echoes are controlled by means of an echo suppressor inserted in the trunk circuit, as indicated in Figure 1.45. The device is essentially a voice-activated switch implementing two functions. It first detects the direction of the conversation and then blocks the opposite path in the four-wire circuit. Even though it introduces distortion when both subscribers are talking by clipping parts of the speech signal, the echo suppressor has provided a reasonably acceptable solution for terrestrial transmission.

Figure 1.45 Echo suppression scheme.

Figure 1.46 Echo cancellation scheme.

For telephone conversation involving satellite circuits, an elegant solution is based on the use of an echo canceler. The circuit generates a replica of the echo using the signal in the receive path and subtracts it from the signal in the transmit path, as indicated in Figure 1.46. Basically, it is an adaptive filter structure whose parameters are adjusted using certain adaptation algorithms until the residual signal is satisfactorily minimized.[7] Typically, an echo reduction of about 40 dB is considered satisfactory in practice. To eliminate the problem generated when both subscribers are talking, the adaptation algorithm is disabled when the signal in the transmit path contains both the echo and the signal generated by the speaker closer to the hybrid coil.

1.5 Why Digital Signal Processing?[8]

In some sense, the origin of digital signal processing techniques can be traced back to the seventeenth century when finite difference methods, numerical integration methods, and numerical interpolation methods were developed to solve physical problems involving continuous variables and functions. The more recent interest in digital signal processing arose in the 1950s with the availability of large digital computers. Initial applications were primarily concerned with the simulation of analog signal processing methods. Around the beginning of the 1960s, researchers began to consider digital signal processing as a separate field by itself. Since then, there have been significant and myriad developments and breakthroughs in both theory and applications of digital signal processing.

Digital processing of an analog signal consists basically of three steps: conversion of the analog signal into a digital form, processing of the digital version, and finally, conversion of the processed digital signal back into an analog form. Figure 1.47 shows the overall scheme in a block diagram form.

---

[7]For a review of adaptive filtering methods see [Cio93].

[8]This section has been adapted from *Handbook for Digital Signal Processing*, Sanjit K. Mitra and James F. Kaiser, Eds., ©1993, John Wiley & Sons. Adapted by permission of John Wiley & Sons.

Figure 1.47   Scheme for the digital processing of an analog signal.

Since the amplitude of the analog input signal varies with time, a *sample-and-hold* (S/H) circuit is used first to sample the analog input at periodic intervals and hold the sampled value constant at the input of the *analog-to-digital* (A/D) *converter* to permit accurate digital conversion. The input to the A/D converter is a staircase-type analog signal if the S/H circuit holds the sampled value until the next sampling instant. The output of the A/D converter is a binary data stream that is next processed by the digital processor implementing the desired signal processing algorithm. The output of the digital processor, another binary data stream, is then converted into a staircase type analog signal by the *digital-to-analog* (D/A) *converter.* The lowpass filter at the output of the D/A converter then removes all undesired high-frequency components and delivers at its output the desired processed analog signal. Figure 1.48 illustrates the waveforms of the pertinent signals at various stages in the above process, where for clarity the two levels of the binary signals are shown as a positive and a negative pulse, respectively.

In contrast to the above, a direct analog processing of an analog signal is conceptually much simpler since it involves only a single processor, as illustrated in Figure 1.49. It is therefore natural to ask what the advantages are of digital processing of an analog signal.

There are of course many advantages in choosing digital signal processing. The most important ones are discussed next [Bel84], [Pro92].

Unlike analog circuits, the operation of digital circuits do not depend on precise values of the digital signals. As a result, a digital circuit is less sensitive to tolerances of component values and is fairly independent of temperature, aging, and most other external parameters. A digital circuit can be reproduced easily in volume quantities and does not require any adjustments either during construction or later while in use. Moreover, it is amenable to full integration, and with the recent advances in *very large scale integrated* (VLSI) circuits, it has been possible to integrate highly sophisticated and complex digital signal processing systems on a single chip.

In a digital processor, the signals and the coefficients describing the processing operation are represented as binary words. Thus, any desirable accuracy can be achieved by simply increasing the wordlength, subject to cost limitation. Moreover, the dynamic ranges for signals and coefficients can be increased still further by using floating-point arithmetic if necessary .

Digital processing allows the sharing of a given processor among a number of signals by timesharing, thus reducing the cost of processing per signal. Figure 1.50 illustrates the concept of timesharing where two digital signals are combined into one by time-division multiplexing. The multiplexed signal can then be fed into a single processor. By switching the processor coefficients prior to the arrival of each signal at the input of the processor, the processor can be made to look like two different systems. Finally, by demultiplexing the output of processor, the processed signals can be separated.

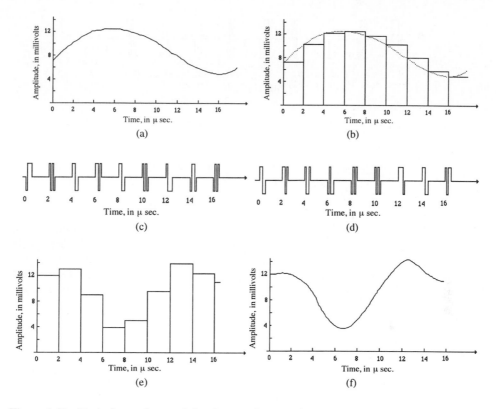

Figure 1.48 Typical waveforms of signals appearing at various stages in Figure 1.45. (a) Analog input signal, (b) Output of the S/H circuit, (c) A/D converter output, (d) Output of the digital processor, (e) D/A converter output, and (f) Analog output signal. In (c) and (d), the digital HIGH and LOW levels are shown as positive and negative pulses for clarity.

Figure 1.49 Analog processing of analog signals.

Digital implementation permits easy adjustment of processor characteristics during processing, such as that needed in implementing adaptive filters. Such adjustments can be simply carried out by periodically changing the coefficients of the algorithm representing the processor characteristics. Another application of the changing of coefficients is in the realization of systems with programmable characteristics, such as frequency selective filters with adjustable cutoff frequencies. Filter banks with guaranteed complementary frequency response characteristics are easily implemented in digital form.

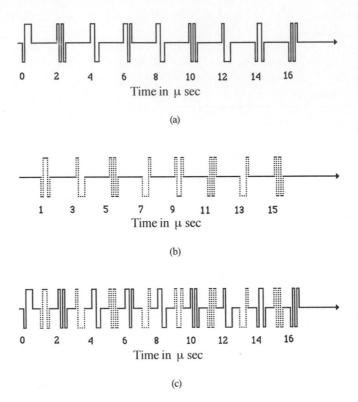

Figure 1.50   Illustration of the time-sharing concept. The signal shown in (c) has been obtained by time-multiplexing the signals shown in (a) and (b).

Digital implementation allows the realization of certain characteristics not possible with analog implementation, such as exact linear phase and multirate processing. Digital circuits can be cascaded without any loading problems unlike analog circuits. Digital signals can be stored almost indefinitely without any loss of information on various storage media such as magnetic tapes and disks, optical disks. Such stored signals can later be processed off-line, such as in the compact disk player, the digital video disk player, the digital audio tape player, or simply by using a general purpose computer as in seismic data processing. On the other hand, stored analog signals deteriorate rapidly as time progresses and cannot be recovered in their original forms.

Another advantage is the applicability of digital processing to very low-frequency signals, such as those occurring in seismic applications, where inductors and capacitors needed for analog processing would be physically very large in size.

Digital signal processing is also associated with some disadvantages. One obvious disadvantage is the increased system complexity in the digital processing of analog signals because of the need for additional pre- and postprocessing devices such as the A/D and D/A converters and their associated filters and complex digital circuitry.

A second disadvantage associated with digital signal processing is the limited range of frequencies available for processing. This property limits its application particularly in the digital processing of analog signals. As shown later, in general, an analog continuous-time signal must be sampled at a frequency that is at least twice the highest frequency component present in the signal. If this condition is not satisfied, then signal components with frequencies above half the sampling frequency appear as signal components below this particular frequency, totally distorting the input analog signal waveform. The available frequency range of operation of a digital signal processor is primarily determined by the sample-and-hold (S/H) circuit and the A/D converter, and as a result is limited by the state-of-the-art of the technology. The highest sampling frequency reported in the literature presently is around 1 GHz [Pou87]. Such high sampling frequencies are not usually used in practice since the achievable resolution of the A/D converter, given by the wordlength of the digital equivalent of the analog sample, decreases with an increase in the speed of the converter. For example, the reported resolution of an A/D converter operating at 1 GHz is 6 bits [Pou87]. On the other hand, in most applications, the required resolution of an A/D converter is from 12 bits to around 16 bits. Consequently, a sampling frequency of at most 10 MHz is presently a practical upper limit. This upper limit, however, is getting larger and larger with advances in technology.

The third disadvantage stems from the fact that digital systems are constructed using active devices that consume electrical power. For example, the WE DSP32C Digital Signal Processor chip contains over 405,000 transistors and dissipates around 1 watt. On the other hand, a variety of analog processing algorithms can be implemented using passive circuits employing inductors, capacitors, and resistors that do not need any power. Moreover, active devices are less reliable than passive components.

However, the advantages far outweigh disadvantages in various applications, and with the continuing decrease in the cost of digital processor hardware, applications of digital signal processing are increasing rapidly.

CHAPTER 2

Time-Domain Representations of Discrete-Time Signals and Systems

The signals arising in digital signal processing are basically discrete-time signals, and discrete-time systems are used to process these signals. As indicated in Figure 1.1(c), a discrete-time signal in its most basic form is defined at equally spaced discrete values of time, the independent variable, with the signal amplitude at these discrete times being continuous. Consequently, a discrete-time signal can be represented as a sequence of numbers, with the independent time variable represented as an integer in the range from $-\infty$ to $+\infty$. Discrete-time signal processing then involves the processing of a discrete-time signal by a discrete-time system to develop another discrete-time signal with more desirable properties or to extract certain information about the original discrete-time signal.

In many applications, it is increasingly becoming more attractive to process a continuous-time signal by discrete-time signal processing methods. To this end, the continuous-time signal is first converted into an "equivalent" discrete-time signal by periodic sampling, the discrete-time signal is then processed by a discrete-time system to generate another discrete-time signal, and the latter is converted into an equivalent continuous-time signal, if necessary. As we shall show later in this book, under certain (ideal) conditions, the conversion of a continuous-time signal can be carried out such that the discrete-time equivalent has all the information contained in the original continuous-time signal, and if necessary, can be converted back into the original continuous-time signal without any distortion.

Thus, to understand the theory of digital signal processing and the design of discrete-time systems, we need to know the characterization of discrete-time signals and systems in the time-domain, a subject we discuss in this chapter. It turns out that it is often convenient to characterize the discrete-time signals and systems in a transformed domain. This alternative representation is considered in the following two chapters.

In this chapter, we first discuss the time-domain representation of a discrete-time signal as a sequence of numbers and its various classifications. We then describe several basic discrete-time signals or sequences that play important roles in the time-domain

Figure 2.1   Graphical representation of a discrete-time sequence $\{x[n]\}$.

characterization of arbitrary discrete-time signals and discrete-time systems. A number of basic operations that generate other sequences from one or more sequences are described next. As we show later, a discrete-time system is composed of a combination of these basic operations. The problem of representing a continuous-time signal by a discrete-time sequence is examined for a simple case. A more thorough mathematical treatment for the general case is deferred until Chapter 5 since it is based on a transform-domain representation of the discrete-time signal discussed in Chapter 3.

In the latter half of this chapter, we introduce the general concept of the processing of a discrete-time signal by a discrete-time system and the classification of such systems. Of these systems, the class of linear, time-invariant type is of exclusive interest in this book, and we describe its time-domain characterization in several different forms.

Most of the book deals with the processing of signals that are deterministic in nature. However, in some instances, the signals encountered could be random, and a discussion of the time-domain representation of discrete-time random signals is also included in this chapter.

Throughout this chapter and successive chapters, we make extensive use of MATLAB to illustrate through computer simulations the various concepts introduced.

2.1   Discrete-Time Signals

2.1.1   Time-Domain Representation

As indicated earlier, in digital signal processing, signals are represented as sequences of numbers, called *samples*. A sample value of a typical discrete-time signal or sequence is denoted as $x[n]$ with the argument n being an integer in the range $-\infty$ and ∞. It should be noted that $x[n]$ is defined only for integer values of n and is undefined for noninteger values of n. The discrete-time signal is represented by $\{x[n]\}$. The graphical representation of a sequence $\{x[n]\}$ with real-valued samples is illustrated in Figure 2.1.

In some applications a discrete-time sequence $\{x[n]\}$ is generated by periodically sampling a continuous-time signal $x_a(t)$ at uniform time intervals:

$$x[n] = x_a(t)|_{t=nT} = x_a(nT), \qquad n = \dots, -2, -1, 0, 1, 2, \dots \qquad (2.1)$$

as illustrated in Figure 2.2. The spacing T between two consecutive samples in Eq. (2.1) is called the *sampling interval* or *sampling period*. The reciprocal of the sampling interval

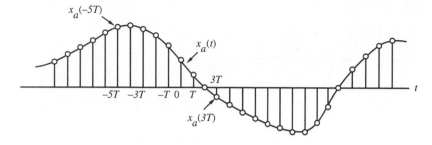

Figure 2.2 Sequence generated by sampling a continuous-time signal $x_a(t)$.

T, denoted as F_T, is called the *sampling frequency*:

$$F_T = \frac{1}{T}. \tag{2.2}$$

The unit of sampling frequency is cycles per second, or hertz (Hz), if the sampling period is in seconds (sec).

It should be noted that, whether or not a sequence $\{x[n]\}$ has been obtained by sampling, the quantity $x[n]$ is called the *n-th sample* of the sequence. For a sequence $\{x[n]\}$, the n-th sample value $x[n]$ can, in general, take any real or complex value. If $x[n]$ is real for all values of n, then $\{x[n]\}$ is a *real sequence*. On the other hand, if the n-th sample value is complex for one or more values of n, then it is a *complex sequence*. By separating the real and imaginary parts of $x[n]$, we can write a complex sequence $\{x[n]\}$ as

$$\{x[n]\} = \{x_{re}[n]\} + j\,\{x_{im}[n]\}, \tag{2.3}$$

where $x_{re}[n]$ and $x_{im}[n]$ are the real part and the imaginary part of $x[n]$, respectively, and are thus real sequences. The complex conjugate sequence of $\{x[n]\}$ is usually denoted by $\{x^*[n]\}$.[1] Often the braces are ignored to denote a sequence if there is no ambiguity.

As defined in the previous chapter, there are basically two types of discrete-time signals: sampled-data signals in which the samples are continuous-valued and digital signals in which the samples are discrete-valued. The pertinent signals in a practical digital signal processing system are digital signals. Figure 2.3 shows a typical digital signal. For digital processing of a continuous-time signal, it is first converted into an equivalent digital signal by means of a sample-and-hold circuit followed by an analog-to-digital converter. The processed digital signal is then converted back into an equivalent continuous-time signal by a digital-to-analog converter followed by an analog reconstruction filter. Chapter 5 is concerned with the digital processing of continuous-time signals. It develops the mathematical foundation of the sampling process and describes the operations of various interface circuits between the continuous-time domain and the digital domain. Chapter 9 considers the effect of discretization of the amplitudes.

---

[1]The complex conjugation operation is denoted by the symbol *.

Figure 2.3   A digital signal with amplitudes taking discrete integer values in the range from -3 to 3.

2.1.2   Types of Sequences

A discrete-time signal can be classified in various ways. One classification is in terms of the number of samples defining the sequence. Another classification is with respect to the symmetry exhibited by the samples with respect to the time index $n = 0$. It can also be classified in terms of its other properties such as periodicity, summability, energy, and power.

Classification Based on Length.   The discrete-time signal may be a *finite-length* or an *infinite-length sequence*. A finite-length (also called *finite-duration* or *finite-extent*) sequence is defined only for a finite time interval:

$$N_1 \leq n \leq N_2, \tag{2.4}$$

where $-\infty < N_1$ and $N_2 < \infty$ with $N_2 \geq N_1$. The *length or duration* N of the above finite-length sequence is

$$N = N_2 - N_1 + 1. \tag{2.5}$$

A length-N discrete-time sequence consists of N samples and is often referred to as an N-point sequence. A finite-length sequence can also be considered as an infinite-length sequence by assigning zero values to samples whose arguments are outside the above range.

Example 2.1 An example of a finite-length sequence of length 7 defined for $-2 \leq n \leq 4$ is

$$\{h[n]\} = \{3.75, -9.816, 0.0056, 345.8, 123.106, 0, 667\},$$
$$\uparrow$$

where the arrow $\uparrow$ under the sample value 0.0056 indicates that this sample value is associated with the time index $n = 0$, and the sample values to its left are for negative values of n , whereas the sample values to its right are for positive values of n.[2] For this sequence the seven specific sample values are:

$$h[-2] = 3.75, \quad h[-1] = -9.816, \quad h[0] = 0.0056, \quad h[1] = 345.8,$$
$$h[2] = 123.106, \quad h[3] = 0, \quad h[4] = 667.$$

---

[2]The arrow will be left out from the explicit description of a sequence, if the location of the sample associated with the time index $n = 0$ is evident.

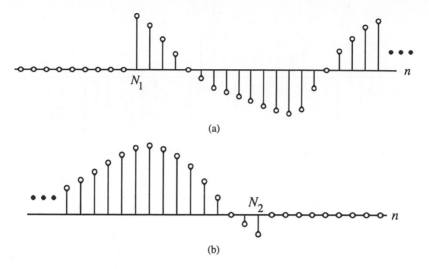

(a)

(b)

Figure 2.4   (a) A right-sided sequence, and (b) a left-sided sequence.

Note that we can treat the above sequence as an infinite-length sequence by defining in addition

$$h[n] = 0 \qquad \text{for } n < -2 \text{ and } n > 4.$$

The process of lengthening a sequence by adding zero-valued samples is called *appending* or *padding* with zeros.

There are three types of infinite-length sequences. A *right-sided sequence* $x[n]$ has zero-valued samples for $n < N_1$, i.e.

$$x[n] = 0 \qquad \text{for } n < N_1, \tag{2.6}$$

where N_1 is a finite integer that can be positive or negative. If $N_1 \geq 0$, a right-sided sequence is usually called a *causal sequence*.[3] Likewise, a *left-sided sequence* $x[n]$ has zero-valued samples for $n > N_2$, i.e.

$$x[n] = 0 \qquad \text{for } n > N_2, \tag{2.7}$$

where N_2 is a finite integer which can be positive or negative. If $N_2 \leq 0$, a left-sided sequence is usually called an *anti-causal sequence*. A general *two-sided sequence* is defined for all values of n in the range $-\infty \leq n \leq \infty$. Figure 2.4 illustrates the above two types of one-sided sequences.

Classification Based on Symmetry.   A sequence $x[n]$ is *conjugate-symmetric* if $x[n] = x^*[-n]$. A real conjugate-symmetric sequence is called an *even sequence*. A sequence $x[n]$ is *conjugate-antisymmetric* if $x[n] = -x^*[-n]$. A real conjugate-antisymmetric sequence is called an *odd sequence*. Examples of even and odd sequences are shown in Figure 2.5.

---

[3]The term *causal sequence* has its origin in the causality condition of a discrete-time system as discussed in Section 2.5.4.

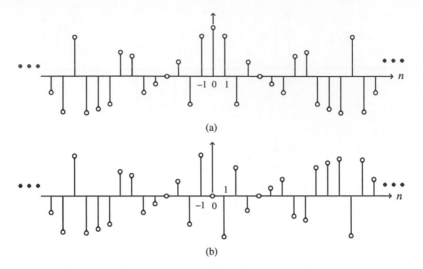

(a)

(b)

Figure 2.5   (a) An even sequence, and (b) an odd sequence.

Any complex sequence $x[n]$ can be expressed as a sum of its conjugate-symmetric part $x_{cs}[n]$ and its conjugate-antisymmetric part $x_{ca}[n]$:

$$x[n] = x_{cs}[n] + x_{ca}[n], \tag{2.8}$$

where

$$x_{cs}[n] = \frac{1}{2}\left(x[n] + x^*[-n]\right), \tag{2.9a}$$

$$x_{ca}[n] = \frac{1}{2}\left(x[n] - x^*[-n]\right). \tag{2.9b}$$

Likewise, any real sequence $x[n]$ can be expressed as a sum of its even part $x_{ev}[n]$ and its odd part $x_{od}[n]$:

$$x[n] = x_{ev}[n] + x_{od}[n], \tag{2.10}$$

where

$$x_{ev}[n] = \frac{1}{2}\left(x[n] + x[-n]\right), \tag{2.11a}$$

$$x_{od}[n] = \frac{1}{2}\left(x[n] - x[-n]\right). \tag{2.11b}$$

For a finite-length sequence, the above definitions of symmetry are not useful. For example, for a real sequence $x[n]$ of length N defined for $0 \le n \le N - 1$, its even part $x_{ev}[n]$ based on the definition of Eq. (2.11a) is of length $2N - 1$ and is defined for $-N + 1 \le n \le N - 1$. Alternative, more useful definitions of symmetry in the case of finite-length sequences are thus given using a modulo operation with all such symmetric and antisymmetric parts of a sequence of length N being also of length N and defined for the same range of values of the time index n. Thus, a length-N sequence $x[n]$ can be expressed as

$$x[n] = x_{pcs}[n] + x_{pca}[n], \quad 0 \le n \le N - 1, \tag{2.12}$$

Figure 2.6   An example of a periodic sequence.

where $x_{pcs}[n]$ and $x_{pca}[n]$ denote, respectively, the *periodic conjugate-symmetric part* and the *periodic conjugate-antisymmetric part*, defined by [4]

$$x_{pcs}[n] = \frac{1}{2}\left(x[n] + x^*[\langle -n \rangle_N]\right), \quad 0 \le n \le N-1, \tag{2.13a}$$

$$x_{pca}[n] = \frac{1}{2}\left(x[n] - x^*[\langle -n \rangle_N]\right), \quad 0 \le n \le N-1. \tag{2.13b}$$

For a real sequence $x[n]$, the periodic conjugate-symmetric and the periodic conjugate-antisymmetric parts are real.

A length-N sequence $x[n]$, defined for $0 \le n \le N-1$ is said to be *periodic conjugate-symmetric* if $x[n] = x^*[\langle -n \rangle_N]$, and is said to be *periodic conjugate-antisymmetric* if $x[n] = -x^*[\langle -n \rangle_N]$.

The symmetric properties of sequences often simplify their respective frequency-domain representations and can be exploited in signal analysis. Implications of the symmetry conditions are considered in Chapter 3.

Classification Based on Periodicity.   A sequence $\tilde{x}[n]$ satisfying

$$\tilde{x}[n] = \tilde{x}[n+kN] \qquad \text{for all } n \tag{2.14}$$

is called a *periodic* sequence with a period N where N is a positive integer and k is any integer. An example of a periodic sequence that has a period $N = 7$ samples is shown in Figure 2.6. A sequence is called an *aperiodic sequence* if it is not periodic. To distinguish a periodic sequence from an aperiodic sequence, we shall denote the former with a "~" on top. The *fundamental period* N_f of a periodic signal is the smallest value of N for which Eq. (2.14) holds.

Other Types of Classification.   A sequence $x[n]$ is said to be *bounded* if each of its samples is of magnitude less than or equal to a finite positive number B_x, i.e.,

$$|x[n]| \le B_x < \infty. \tag{2.15}$$

The periodic sequence of Figure 2.6 is a bounded sequence with a bound $B_x = 2$.

A sequence $x[n]$ is said to be *absolutely summable* if

$$\sum_{n=-\infty}^{\infty} |x[n]| < \infty. \tag{2.16}$$

---

[4] $\langle k \rangle_N = k$ modulo N.

A sequence is said to be *square-summable* if

$$\sum_{n=-\infty}^{\infty} |x[n]|^2 < \infty \tag{2.17}$$

The *energy* of a sequence $x[n]$ is defined by

$$\mathcal{E} = \sum_{n=-\infty}^{\infty} |x[n]|^2. \tag{2.18}$$

A square-summable sequence therefore has finite energy. Note that a finite-length sequence with finite sample values always has a finite energy, whereas an infinite-length sequence with finite sample values may or may not have finite energy, as illustrated in the following example.

Example 2.2 The infinite-length sequence $x_1[n]$ defined by

$$x_1[n] = \begin{cases} \frac{1}{n}, & n \geq 1, \\ 0, & n \leq 0, \end{cases} \tag{2.19}$$

has an energy equal to

$$\mathcal{E} = \sum_{n=1}^{\infty} \left(\frac{1}{n}\right)^2,$$

which converges to $\pi^2/6$, indicating that $x_1[n]$ has a finite energy. However, the infinite-length sequence $x_2[n]$ defined by

$$x_2[n] = \begin{cases} \frac{1}{\sqrt{n}}, & n \geq 1, \\ 0, & n \leq 0, \end{cases} \tag{2.20}$$

has an energy equal to

$$\mathcal{E} = \sum_{n=1}^{\infty} \frac{1}{n},$$

which does not converge. As a result, the sequence $x_2[n]$ has an infinite energy.

The *average power* of an aperiodic sequence $x[n]$ is defined by

$$\mathcal{P}_{av} = \lim_{K \to \infty} \frac{1}{2K+1} \sum_{n=-K}^{K} |x[n]|^2. \tag{2.21}$$

The average power of a sequence can be related to its energy by defining its energy over a finite interval $-K \leq n \leq K$ as

$$\mathcal{E}_K = \sum_{n=-K}^{K} |x[n]|^2. \tag{2.22}$$

Then

$$\mathcal{P}_{av} = \lim_{K \to \infty} \frac{1}{2K+1} \mathcal{E}_K. \tag{2.23}$$

The average power of a periodic sequence $\tilde{x}[n]$ with a period N is given by

$$\mathcal{P}_{av} = \frac{1}{N} \sum_{n=0}^{N-1} |\tilde{x}[n]|^2. \tag{2.24}$$

Note from the above that the average power of a finite-length sequence is always finite. However, the average power of an infinite-length sequence may be finite or infinite.

Example 2.3 Consider the causal sequence defined by

$$x[n] = \begin{cases} 2(-1)^n, & n \geq 0, \\ 0, & n < 0. \end{cases}$$

It follows from Eq. (2.18) that $x[n]$ has infinite energy. On the other hand, from Eq. (2.21) its average power is given by [5]

$$\mathcal{P}_{av} = \lim_{K \to \infty} \frac{1}{K+1} \left(4 \sum_{n=0}^{K} 1 \right) = \lim_{K \to \infty} \frac{4(K+1)}{K+1} = 4,$$

which is finite.

2.1.3   Some Basic Sequences

We now consider several special types of sequences that play an important role in the analysis and design of discrete-time systems. For example, an arbitrary sequence can be expressed in terms of some of these basic sequences. Another fundamental application that is the key behind discrete-time signal processing is the representation of a class of discrete-time systems in terms of the response of the system to certain basic sequences. This representation permits the computation of the response of the discrete-time system to arbitrary discrete-time signals if the latter can be expressed in terms of these basic sequences.

Unit Sample Sequence.   The simplest and one of the most useful sequences is the *unit sample sequence*, often called the *discrete-time impulse* or the *unit impulse*, as shown in Figure 2.7. It is denoted by $\delta[n]$ and defined by

$$\delta[n] = \begin{cases} 1, & n = 0, \\ 0, & n \neq 0. \end{cases} \tag{2.25}$$

We shall show later, in Section 2.5.1, that a certain class of discrete-time systems is completely characterized in the time-domain by its output response to a unit impulse input. Furthermore, knowing this particular response of the system, we can compute its response to any arbitrary input sequence.

---

[5]Note the change in the definition of average power for a causal sequence.

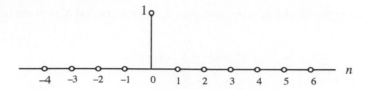

Figure 2.7   The unit sample sequence $\{\delta[n]\}$.

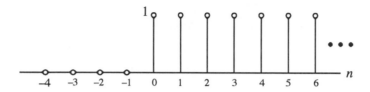

Figure 2.8   The unit step sequence $\{\mu[n]\}$.

Unit Step Sequence.   A second basic sequence is the *unit step sequence* shown in Figure 2.8. It is denoted by $\mu[n]$ and is defined by

$$\mu[n] = \begin{cases} 1, & n \geq 0, \\ 0, & n < 0. \end{cases} \tag{2.26}$$

The unit sample and the unit step sequences are related as follows (Problem 2.11):

$$\mu[n] = \sum_{k=-\infty}^{n} \delta[k], \qquad \delta[n] = \mu[n] - \mu[n-1]. \tag{2.27}$$

Exponential and Sinusoidal Sequences.   Another set of basic sequences is formed by taking the n-th sample value to be the n-th power of a real or complex constant. Such sequences are termed *exponential sequences* and their most general form is given by

$$x[n] = A\alpha^{n}, \tag{2.28}$$

where A and α are real or complex numbers. By expressing

$$\alpha = e^{(\sigma_0 + j\omega_0)}, \qquad A = |A|\, e^{j\phi},$$

we can rewrite Eq. (2.28) as

$$x[n] = Ae^{(\sigma_0 + j\omega_0)n} = |A|\, e^{\sigma_0 n} e^{j(\omega_0 n + \phi)} \tag{2.29a}$$
$$= |A|\, e^{\sigma_0 n} \cos(\omega_0 n + \phi) + j\,|A|\, e^{\sigma_0 n} \sin(\omega_0 n + \phi), \tag{2.29b}$$

to arrive at an alternative general form of a *complex exponential sequence* where σ_0, ϕ and ω_0 are now real numbers. If we write $x[n] = x_{re}[n] + jx_{im}[n]$, then from Eq. (2.29b):

$$x_{re}[n] = |A|\, e^{\sigma_0 n} \cos(\omega_0 n + \phi),$$

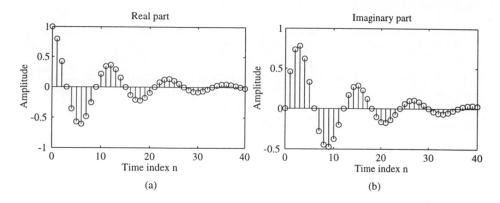

Figure 2.9   A complex exponential sequence $x[n] = e^{(-1/12+j\pi/6)n}$.   (a) Real part and (b) Imaginary part.

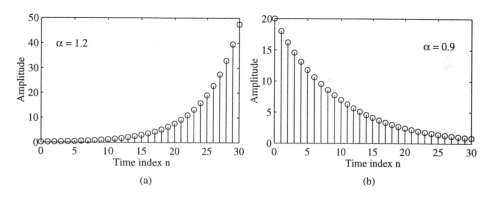

Figure 2.10   Examples of real exponential sequences: (a) $x[n] = 0.2(1.2)^n$, (b) $x[n] = 20(0.9)^n$.

$$x_{im}[n] = |A|\, e^{\sigma_0 n}\, \sin(\omega_0 n + \phi).$$

Thus the real and imaginary parts of a complex exponential sequence are real sinusoidal sequences with constant ($\sigma_0 = 0$), growing ($\sigma_0 > 0$), or decaying ($\sigma_0 < 0$) amplitudes for $n > 0$. Figure 2.9 depicts a complex exponential sequence with a decaying amplitude. Note that in the display of a complex exponential sequence, its real and imaginary parts are shown separately.

With both A and α real, the sequence of Eq. (2.28) reduces to a *real exponential sequence*. For $n \geq 0$ such a sequence with $|\alpha| < 1$ decays exponentially as n increases and with $|\alpha| > 1$ grows exponentially as n increases. Examples of real exponential sequences obtained for two values of α are shown in Figure 2.10.

Another commonly encountered sequence is the *real sinusoidal sequence* with constant amplitude of the form

$$x[n] = A\cos(\omega_0 n + \phi), \tag{2.30}$$

where A, ω_0, and ϕ are real numbers. The parameters A, ω_0, and ϕ are called, respectively, the amplitude, the angular frequency, and the initial phase of the sinusoidal sequence $x[n]$.

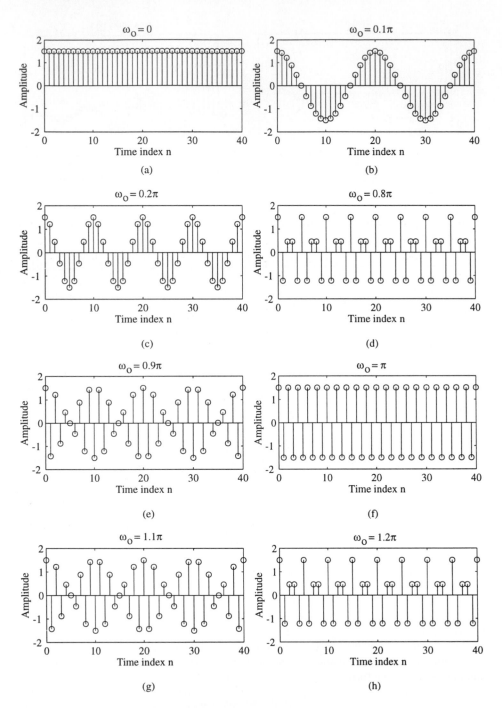

Figure 2.11   A family of sinusoidal sequences given by $x[n] = 1.5 \cos \omega_0 n$: (a) $\omega_0 = 0$, (b) $\omega_0 = 0.1\pi$, (c) $\omega_0 = 0.2\pi$, (d) $\omega_0 = 0.8\pi$, (e) $\omega_0 = 0.9\pi$, (f) $\omega_0 = \pi$, (g) $\omega_0 = 1.1\pi$, and (h) $\omega_0 = 1.2\pi$.

Figure 2.11 shows different types of sinusoidal sequences. The real sinusoidal sequence of Eq. (2.30) can be written alternatively as

$$x[n] = x_i[n] + x_q[n], \tag{2.31}$$

where $x_i[n]$ and $x_q[n]$ are, respectively, the *in-phase* and the *quadrature components* of $x[n]$, and are given by

$$x_i[n] = A\cos\phi\cos(\omega_0 n), \qquad x_q[n] = -A\sin\phi\sin(\omega_0 n). \tag{2.32}$$

We shall show later in Section 3.1 that a large class of sequences can be expressed in terms of complex sequences of the form $e^{j\omega n}$.

Note that the complex exponential sequence of Eq. (2.29a) with $\sigma_0 = 0$ and the sinusoidal sequence of Eq. (2.30) are periodic sequences of period N as long as $\omega_0 N$ is an integer multiple of 2π, i.e. $\omega_0 N = 2\pi r$ where N is a positive integer. The smallest possible N satisfying this condition is the *fundamental period* of the sequence. To verify this, consider the two sinusoidal sequences $x_1[n] = \cos(\omega_0 n + \phi)$ and $x_2[n] = \cos(\omega_0(n + N) + \phi)$. Now

$$x_2[n] = \cos((\omega_0(n + N) + \phi) = \cos(\omega_0 n + \phi)\cos\omega_0 N - \sin(\omega_0 n + \phi)\sin\omega_0 N$$

which will be equal to $\cos(\omega_0 n + \phi) = x_1[n]$ only if $\omega_0 N$ is an integer multiple of 2π, i.e.,

$$\omega_0 N = 2\pi r \tag{2.33a}$$

or

$$\frac{2\pi}{\omega_0} = \frac{N}{r}, \tag{2.33b}$$

where r is an integer. If $2\pi/\omega_0$ is a noninteger rational number, then the period will be a multiple of $2\pi/\omega_0$. If $2\pi/\omega_0$ is not a rational number, then the sequence is aperiodic even though it has a sinusoidal envelope. For example, $x[n] = \cos(\sqrt{3}n + \phi)$ is an aperiodic sequence.

Example 2.4 Let us determine the periods of the sinusoidal sequences of Figure 2.11. In Figure 2.11(a), $\omega_0 = 0$, and Eq. (2.33a) is satisfied with $r = 0$ and any value of N. Here, the smallest value of N is equal to 1. Likewise, in Figure 2.11(b), $\omega_0 = 0.1\pi$, and hence, applying Eq. (2.33a) we find that $N = 20$. Following similar lines we arrive at $N = 10$ for Figure 2.11(c), $N = 5$ for Figure 2.11(d), $N = 20$ for Figure 2.11(e), $N = 2$ for Figure 2.11(f), $N = 20$ for Figure 2.11(g), and $N = 5$ for Figure 2.11(h). These numbers are also evident from the figures.

The number ω_0 in the two sequences of Eqs. (2.29a) and (2.30) is called the *angular frequency*. Since the time instant n is dimensionless, the unit of angular frequency ω_0 and *phase* ϕ is simply radians. If the unit of n is designated as samples, then the unit of ω_0 and ϕ is radians per sample. Often in practice, the angular frequency ω is expressed as

$$\omega = 2\pi f, \tag{2.34}$$

where f is *frequency* in cycles per sample.

Two interesting properties of these sequences are discussed next. Consider two complex exponential sequences $x_1[n] = e^{j\omega_1 n}$ and $x_2[n] = e^{j\omega_2 n}$ with $0 \leq \omega_1 < 2\pi$ and $2\pi k \leq \omega_2 < 2\pi(k+1)$, where k is any positive or negative integer. If

$$\omega_2 = \omega_1 + 2\pi k, \tag{2.35}$$

then

$$x_2[n] = e^{j\omega_2 n} = e^{j(\omega_1 + 2\pi k)n} = e^{j\omega_1 n} = x_1[n].$$

Thus these two sequences are indistinguishable. Likewise, two sinusoidal sequences $x_1[n] = \cos(\omega_1 n + \phi)$ and $x_2[n] = \cos(\omega_2 n + \phi)$ with $0 \leq \omega_1 < 2\pi$ and $2\pi k \leq \omega_2 < 2\pi(k+1)$, where k is any positive or negative integer, are indistinguishable from one another if $\omega_2 = \omega_1 + 2\pi k$.

The second interesting feature of discrete-time sinusoidal signals can be seen from Figure 2.11. The frequency of oscillation of the discrete-time sinusoidal sequence $x[n] = A\cos(\omega_0 n)$ increases as ω_0 increases from 0 to π, and then the frequency of oscillation decreases as ω_0 increases from π to 2π. As a result of the first property, a frequency ω_0 in the neighborhood of $\omega = 2\pi k$ is indistinguishable from a frequency $\omega_0 - 2\pi k$ in the neighborhood of $\omega = 0$, and a frequency ω_0 in the neighborhood of $\omega = \pi(2k+1)$ is indistinguishable from a frequency $\omega_0 - \pi(2k+1)$ in the neighborhood of $\omega = \pi$ for any integer value of k. Therefore, frequencies in the neighborhood of $\omega_0 = 2\pi k$ are usually called *low frequencies*, and frequencies in the neighborhood of $\omega_0 = \pi(2k+1)$ are called *high frequencies*. For example, $v_1[n] = \cos(0.1\pi n) = \cos(1.9\pi n)$ shown in Figure 2.11(b) is a low-frequency signal, whereas, $v_2[n] = \cos(0.8\pi n) = \cos(1.2\pi n)$ shown in Figure 2.11(d) and (h) is a high-frequency signal.

2.1.4 Sequence Generation Using MATLAB

MATLAB includes a number of functions that can be used for signal generation. Some of these functions of interest are

exp,   sin,   cos,   square,   sawtooth

For example, Program 2_1 given below can be employed to generate a complex exponential sequence of the form shown in Figure 2.9.

```
% Program 2_1
% Generation of complex exponential sequence
%
a = input('Type in real exponent = ');
b = input('Type in imaginary exponent = ');
c = a + b*i;
K = input('Type in the gain constant = ');
N = input ('Type in length of sequence = ');
n = 1:N;
x = K*exp(c*n);
```

```
stem(n-1,real(x));
xlabel('Time index n');ylabel('Amplitude');
title('Real part');
disp('PRESS RETURN for imaginary part');
pause
stem(n-1,imag(x));
xlabel('Time index n');ylabel('Amplitude');
title('Imaginary part');
```

Likewise, Program 2_2 listed below can be employed to generate a real exponential sequence of the form shown in Figure 2.10.

```
% Program 2_2
% Generation of real exponential sequence
%
a = input('Type in exponent = ');
K = input('Type in the gain constant = ');
N = input ('Type in length of sequence = ');
n = 1:N;
x = K*a.^n;
stem(n-1,x);
xlabel('Time index n');ylabel('Amplitude');
title(['alpha =  ',num2str(a)]);
```

Another type of sequence generation using MATLAB is given later in Example 2.14.

2.2 Operations on Sequences

A single-input, single-output discrete-time system operates on a sequence, called the *input sequence*, according to some prescribed rules and develops another sequence, called the *output sequence*, usually with more desirable properties. For example, the input may be a signal corrupted by an additive noise, and the discrete-time system is designed to remove the noise component from the input. An M-input, N-output discrete-time system operates on M input signals generating N output signals. The FM stereo transmission system is a two-input, single-output system since here the left and right channel audio signals are combined into a high-frequency composite baseband signal. In most cases, the operation defining a particular discrete-time system is composed of some basic operations that we describe next.

2.2.1 Basic Operations

Let $x[n]$ and $y[n]$ be two known sequences. By forming the *product* of the sample values of these two sequences at each instant, we form a new sequence $w_1[n]$:

$$w_1[n] = x[n] \cdot y[n]. \tag{2.36}$$

In some applications, the product operation is also known as *modulation*. An application of the modulation operation is to transform a sequence with low-frequency sinusoidal components to a sequence with high-frequency components by modulating the former with a sinusoidal sequence of very high frequency, as illustrated in the following example.

Example 2.5 Let $x_1[n] = \cos(\omega_1 n)$ and $x_2[n] = 2\cos(\omega_2 n)$ with $\pi > \omega_2 \gg \omega_1 > 0$. The product sequence $y[n] = x_1[n]x_2[n]$ is then given by

$$y[n] = 2\cos(\omega_1 n) \cdot \cos(\omega_2 n).$$

Using a trigonometric identity we arrive at

$$y[n] = \cos((\omega_1 + \omega_2)n) + \cos((\omega_2 - \omega_1)n).$$

The new sequence $y[n]$ is thus composed of two sinusoidal sequences of frequencies $\omega_1 + \omega_2$ and $\omega_2 - \omega_1$. It should be noted that because of the property given by Eq. (2.35) if $\omega_1 + \omega_2 > \pi$, the high-frequency sinusoidal sequence $\cos((\omega_1 + \omega_2)n)$ appears as a low-frequency sinusoidal sequence $-\cos((\omega_1 + \omega_2 - \pi)n)$ with an angular frequency of value smaller than π.[6]

Another application of the product operation is in forming a finite-length sequence from an infinite-length sequence by multiplying the latter with a finite-length sequence called a *window sequence*. This process of forming the finite-length sequence is usually called *windowing*, which plays an important role in the design of certain types of digital filters (Section 7.7).

The second basic operation is the *addition* by which a new sequence $w_2[n]$ is obtained by adding the sample values of two sequences $x[n]$ and $y[n]$:

$$w_2[n] = x[n] + y[n]. \qquad (2.37)$$

Applications of this operation are given in Eqs. (2.8) to (2.11b).

The third basic operation is the scalar *multiplication*, whereby a new sequence is generated by multiplying each sample of a sequence $x[n]$ by a scalar A:

$$w_3[n] = Ax[n]. \qquad (2.38)$$

Equations (2.9a), (2.9b), (2.11a), (2.11b), (2.28), (2.29a), (2.29b), and (2.30) illustrate the use of the scalar multiplication operation.

The *time-reversal* operation is another useful scheme to develop a new sequence. An example is:

$$w_4[n] = x[-n], \qquad (2.39)$$

which is the time-reversed version of the sequence $x[n]$. The time-reversal operation has been employed in Eqs. (2.9a), (2.9b), (2.11a) and (2.11b).

The *delay* operation illustrated below in Eq. (2.40) shows the relation between $x[n]$ and its delayed version $w_5[n]$:

$$w_5[n] = x[n - N], \qquad (2.40)$$

where N is a positive integer. Note that if N is a negative integer, then the above operation becomes an *advance* of a sequence. Schematic representations of some of the above operations are shown in Figure 2.12(a) to (e).

---

[6]The appearance of a high-frequency signal $\cos((\omega_1 + \omega_2)n)$ as a low-frequency signal $-\cos((\omega_1 + \omega_2 - \pi)n)$ is called aliasing (See Section 2.3).

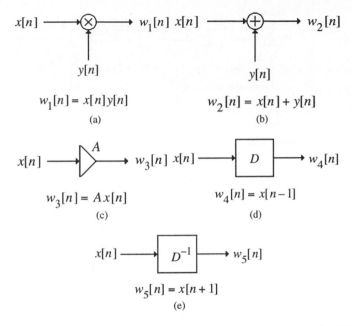

Figure 2.12   Schematic representations of basic operations on sequences: (a) Modulation, (2) addition, (c) multiplication, (d) unit delay and (e) unit advance.

Example 2.6   Consider the following two sequences of length 5 defined for $0 \le n \le 4$:

$$c[n] = \{3.2, \quad 41, \quad 36, \quad -9.5, \quad 0\},$$
$$\uparrow$$
$$d[n] = \{1.7, \quad -0.5, \quad 0, \quad 0.8, \quad 1\}.$$
$$\uparrow$$

Several new sequences of length 5 generated from above are given by

$$w_1[n] = c[n] \cdot d[n] = \{5.44, \quad -20.5, \quad 0, \quad -7.6, \quad 0\},$$
$$\uparrow$$
$$w_2[n] = c[n] + d[n] = \{4.9, \quad 40.5, \quad 36, \quad -8.7, \quad 1\},$$
$$\uparrow$$
$$w_3[n] = \frac{7}{2}c[n] = \{11.2, \quad 143.5, \quad 126, \quad -33.25, \quad 0\}.$$
$$\uparrow$$

For simplicity, for finite-length sequences defined for positive values of n beginning at $n = 0$, the first sample in the sequence will always be assumed to be the one associated with $n = 0$ without the arrow being explicitly shown under it.

As indicated by the above example, operations on two or more sequences to generate a new sequence tacitly assume all sequences to be of the same length. However, in some situations, this problem can be circumvented by appending zero-valued samples to the sequence(s) of smaller lengths to make all sequences of equal length. This process is illustrated in the following example.

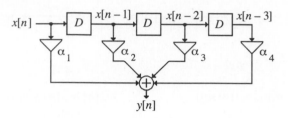

Figure 2.13   Discrete-time system of Example 2.8.

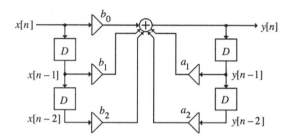

Figure 2.14   Discrete-time system of Example 2.9.

Example 2.7   Consider a sequence $\{g[n]\}$ of length 3 defined for $0 \leq n \leq 2$ given by

$$\{g[n]\} = \{-21, \quad 1.5, \quad 3\}.$$

It is clear that we cannot develop another sequence by operating on this sequence and any one of the length-5 sequences of Example 2.6. However, it is possible to treat $\{g[n]\}$ as a sequence of length 5 by appending it with two zero-valued samples:

$$\{g_e[n]\} = \{-21, \quad 1.5, \quad 3, \quad 0, \quad 0\}.$$

Examples of new sequences generated from $\{g_e[n]\}$ and $c[n]$ of the previous example are indicated below:

$$\{w_4[n]\} = \{c[n] \cdot g_e[n]\} = \{-67.2, \quad 61.5, \quad 108, \quad 0, \quad 0\},$$
$$\{w_5[n]\} = \{c[n] + g_e[n]\} = \{-17.8, \quad 42.5, \quad 39, \quad -9.5, \quad 0\}.$$

In most applications, combinations of the above basic operations are used. Illustrations of such combinations are given in the following two examples.

Example 2.8   Consider the block diagram of Figure 2.13. We now analyze this figure to determine how the sequence $y[n]$ is generated from the sequence $x[n]$. From the definition of the unit delay block given in Figure 2.12(d), it follows that the leftmost delay block generates a sequence $x[n-1]$, the middle delay block develops $x[n-2]$ and finally, the rightmost delay block generates $x[n-3]$. The multipliers labeled α_1, α_2, α_3 and α_4 generate sequences $\alpha_1 x[n]$, $\alpha_2 x[n-1]$, $\alpha_3 x[n-2]$, and $\alpha_4 x[n-3]$, respectively. The sequence $y[n]$ is then obtained by adding $\alpha_1 x[n]$, $\alpha_2 x[n-1]$, $\alpha_3 x[n-2]$, and $\alpha_4 x[n-3]$:

$$y[n] = \alpha_1 x[n] + \alpha_2 x[n-1] + \alpha_3 x[n-2] + \alpha_4 x[n-3]. \tag{2.41}$$

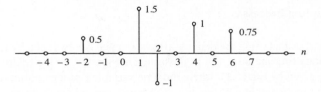

Figure 2.15   An arbitrary sequence $x[n]$.

Example 2.9 We now analyze the block diagram of Figure 2.14. Observe first that the two leftmost delay blocks generate the sequences $x[n-1]$ and $x[n-2]$, whereas the two rightmost delay blocks develop the sequences $y[n-1]$ and $y[n-2]$. These delayed sequences along with $x[n]$ are then applied to the five multipliers, labeled b_0, b_1, b_2, a_1, and a_2 developing the sequences $b_0x[n]$, $b_1x[n-1]$, $b_2x[n-2]$, $a_1y[n-1]$ and $a_2y[n-2]$ which are then added to yield $y[n]$:

$$y[n] = b_0x[n] + b_1x[n-1] + b_2x[n-2] + a_1y[n-1] + a_2y[n-2]. \tag{2.42}$$

Another quite useful operation is the sampling rate alteration that is employed to generate a new sequence with a sampling rate higher or lower than that of a given sequence. Thus, if $x[n]$ is a sequence with a sampling rate of F_T Hz and it is used to generate another sequence $y[n]$ with a desired sampling rate of F_T' Hz, then the *sampling rate alteration ratio* is given by

$$\frac{F_T'}{F_T} = R. \tag{2.43}$$

If $R > 1$, the process is called *interpolation* and results in a sequence with a higher sampling rate. On the other hand, if $R < 1$, the sampling rate is decreased by a process called *decimation*.

The basic operations employed in the sampling rate alteration process are called *up-sampling* and *down-sampling*. These operations play important roles in multirate discrete-time systems and are discussed in Section 10.1.

2.2.2   Representation of an Arbitrary Sequence

An arbitrary sequence can be represented in the time-domain as a weighted sum of some basic sequence and its delayed versions. A commonly used basic sequence in the representation is the unit sample sequence. For example, the sequence $x[n]$ of Figure 2.15 can be expressed as

$$x[n] = 0.5\delta[n+2] + 1.5\delta[n-1] - \delta[n-2] + \delta[n-4] + 0.75\delta[n-6]. \tag{2.44}$$

An implication of this type of representation is considered later in Section 2.5.1, where we develop the general expression for calculating the output sequence of certain types of discrete-time systems for an arbitrary input sequence.

Since the unit step sequence and the unit sample sequence are simply related through Eq. (2.27), it is also possible to represent an arbitrary sequence as a weighted combination of delayed unit step sequences (Problem 2.17).

2.3   The Sampling Process

We indicated earlier that often the discrete-time sequence is developed by uniformly sampling a continuous-time signal $x_a(t)$, as illustrated in Figure 2.2. The relation between the two signals are given by Eq. (2.1), where the time variable t of the continuous-time signal is related to the time variable n of the discrete-time signal only at discrete-time instants t_n given by

$$t_n = nT = \frac{n}{F_T} = \frac{2\pi n}{\Omega_T}, \tag{2.45}$$

with $F_T = 1/T$ denoting the sampling frequency and $\Omega_T = 2\pi F_T$ denoting the sampling angular frequency. For example, if the continuous-time signal is

$$x_a(t) = A\cos(2\pi f_0 t + \phi) = A\cos(\Omega_0 t + \phi), \tag{2.46}$$

the corresponding discrete-time signal is given by

$$\begin{aligned} x[n] &= A\cos(\Omega_0 nT + \phi) \\ &= A\cos(\frac{2\pi \Omega_0}{\Omega_T} n + \phi) = A\cos(\omega_0 n + \phi), \end{aligned} \tag{2.47}$$

where

$$\omega_0 = \frac{2\pi \Omega_0}{\Omega_T} = \Omega_0 T \tag{2.48}$$

is the normalized digital angular frequency of the discrete-time signal $x[n]$. The units of the normalized digital angular frequency ω_0 is radians per sample, while the units of the normalized analog angular frequency Ω_0 is radians per second and the units of the analog frequency f_0 is hertz if the units of the sampling period T is in seconds.

Example 2.10 Consider the three sequences generated by uniformly sampling the three cosine functions of frequencies 3 Hz, 7 Hz and 13 Hz, respectively: $g_1(t) = \cos(6\pi t)$, $g_2(t) = \cos(14\pi t)$ and $g_3(t) = \cos(26\pi t)$ with a sampling rate of 10 Hz, i.e., with $T = 0.1$ sec. The derived sequences are therefore:

$$g_1[n] = \cos(0.6\pi n), \qquad g_2[n] = \cos(1.4\pi n), \qquad g_3[n] = \cos(2.6\pi n).$$

A plot of these sequences (shown with circles) and their parent time functions are given in Figure 2.16. Note from these plots that each sequence has exactly the same sample value for any given n. This also can be verified by observing that

$$g_2[n] = \cos(1.4\pi n) = \cos((2\pi - 0.6\pi)n) = \cos(0.6\pi n),$$

and

$$g_3[n] = \cos(2.6\pi n) = \cos((2\pi + 0.6\pi)n) = \cos(0.6\pi n).$$

As a result, all three sequences above are identical and it is difficult to associate a unique continuous-time function with this sequence. In fact, all cosine waveforms of frequencies given by $(10k \pm 3)$ Hz, with k being any nonnegative integer, lead to the sequence $g_1[n] = \cos(0.6\pi n)$ when sampled at a 10 Hz rate.

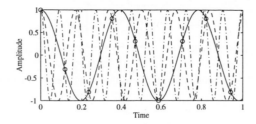

Figure 2.16   Ambiguity in the discrete-time representation of continuous-time functions. $g_1(t)$ is shown with the solid line, $g_2(t)$ is shown with the dashed line, $g_3(t)$ is shown with the dashed-dot line, and the sequence obtained by sampling is shown with circles.

In the general case, the family of continuous-time sinusoids

$$x_{a,k}(t) = A\cos((\Omega_0 + k\Omega_T)t + \phi), \qquad k = 0, \pm1, \pm2, \dots \qquad (2.49)$$

lead to identical sampled signals:

$$x_{a,k}(nT) = A\cos\left((\Omega_0 + k\Omega_T)nT + \phi\right) = A\cos\left(\frac{2\pi(\Omega_0 + k\Omega_T)n}{\Omega_T} + \phi\right)$$

$$= A\cos\left(\frac{2\pi\Omega_0 n}{\Omega_T} + \phi\right) = A\cos(\omega_0 n + \phi) = x[n]. \qquad (2.50)$$

The above phenomenon of a continuous-time sinusoidal signal of higher frequency acquiring the identity of a sinusoidal sequence of lower frequency after sampling is called *aliasing*. Since there are an infinite number of continuous-time functions that can lead to a given sequence when sampled periodically, additional conditions need to be imposed so that the sequence $\{x[n]\}$ can uniquely represent the parent continuous-time function $x_a(t)$, or in other words, $x_a(t)$ can be fully recovered from a knowledge of $\{x[n]\}$.

It follows from Eq. (2.48) that if $\Omega_T > 2\Omega_0$, then the corresponding normalized digital frequency ω_0 of the discrete-time signal $x[n]$ obtained by sampling the parent continuous-time signal $x_a(t)$ will be in the range $-\pi < \omega < \pi$, implying no aliasing. On the other hand, if $\Omega_T < 2\Omega_0$, the normalized digital frequency will fold into a lower digital frequency in the range $-\pi < \omega < \pi$ because of aliasing. Hence, to prevent aliasing, the sampling frequency Ω_T should be greater than 2 times the frequency Ω_0 of the sinusoidal signal being sampled. Generalizing the above result, we observe that if we have an arbitrary continuous-time signal $x_a(t)$ that can be represented as a weighted sum of a number of sinusoidal signals, then $x_a(t)$ can also be represented uniquely by its sampled version $\{x[n]\}$ if the sampling frequency Ω_T is chosen to be greater than 2 times the highest frequency contained in $x_a(t)$. The condition to be satisfied by the sampling frequency to prevent aliasing is called the *sampling theorem*, which is formally derived later in Section 5.2.

Example 2.11   Determine the discrete-time signal $v[n]$ obtained by uniformly sampling at a sampling rate of 200 Hz a continuous-time signal $v_a(t)$ composed of a weighted sum of five sinusoidal signals of frequencies 30 Hz, 150 Hz, 170 Hz, 250 Hz, and 330 Hz, as given below:

$$v_a(t) = 6\cos(60\pi t) + 3\sin(300\pi t) + 2\cos(340\pi t)$$
$$+ 4\cos(500\pi t) + 10\sin(660\pi t).$$

Figure 2.17   Schematic representation of a discrete-time system.

The sampling period $T = 1/200 = 0.005$ sec. Hence, the generated discrete-time signal $v[n]$ is given by

$$v[n] = 6\cos(0.3\pi n) + 3\sin(1.5\pi n) + 2\cos(1.7\pi n) + 4\cos(2.5\pi n)$$
$$+10\sin(3.3\pi n)$$
$$= 6\cos(0.3\pi n) + 3\sin\left((2\pi - 0.5\pi)n\right) + 2\cos\left((2\pi - 0.3\pi)n\right)$$
$$+4\cos\left((2\pi + 0.5\pi)n\right) + 10\sin\left((4\pi - 0.7\pi)n\right)$$
$$= 6\cos(0.3\pi n) - 3\sin(0.5\pi n) + 2\cos(0.3\pi n) + 4\cos(0.5\pi n)$$
$$-10\sin(0.7\pi n)$$
$$= 8\cos(0.3\pi n) + 5\cos(0.5\pi n + 0.6435) - 10\sin(0.7\pi n).$$

The discrete-time signal $v[n]$ is composed of a weighted sum of three discrete-time sinusoidal signals of normalized angular frequencies: 0.3π, 0.5π, and 0.7π.

It should be noted that an identical discrete-time signal is generated by uniformly sampling at a 200-Hz sampling rate the following continuous-time signal composed of a weighted sum of three sinusoidal continuous-time signals of frequencies 30 Hz, 50 Hz, and 70 Hz:

$$w_a(t) = 8\cos(60\pi t) + 5\cos(100\pi t + 0.6435) - 10\sin(140\pi t).$$

The discrete-time signal obtained by sampling a continuous-time signal $x_a(t)$ may be represented as a sequence $\{x_a[nT]\}$. However, we shall use the more common notation $\{x[n]\}$ for simplicity (with T assumed to be normalized to 1 sec). It should be noted that when dealing with sampled version of continuous-time functions, it is essential to know the numerical value of the sampling period T.

2.4   Discrete-Time Systems

The function of a discrete-time system is to process a given *input sequence* to generate an *output sequence*. In most applications, the discrete-time system used is a single-input, single-output system, as shown schematically in Figure 2.17. The output sequence is generated sequentially, beginning with a certain value of the time index n, and thereafter progressively increasing values of n. If the beginning time index is n_0, the output $y[n_0]$ is first computed, then $y[n_0 + 1]$ is computed, and so on. We restrict our attention in this text to this class of discrete-time systems with certain specific properties as described later in this section.

Two examples of discrete-time systems are given in Examples 2.8 and 2.9. Some additional examples of discrete-time systems are given below.

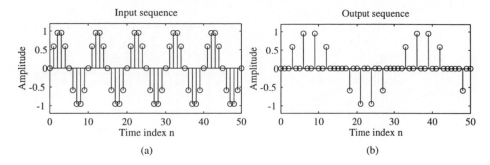

Figure 2.18   Illustration of the up-sampling process.

Example 2.12 A very simple example of a discrete-time system is the *accumulator* defined by the input-output relation

$$y[n] = \sum_{\ell=-\infty}^{n} x[\ell] \qquad (2.51a)$$

$$= \sum_{\ell=-\infty}^{-1} x[\ell] + \sum_{\ell=0}^{n} x[\ell] = y[-1] + \sum_{\ell=0}^{n} x[\ell]. \qquad (2.51b)$$

The output $y[n]$ at time instant n is the sum of the input sample value $x[n]$ at time instant n and the previous output $y[n-1]$ at time instant $n-1$, which is the sum of all previous input sample values from $-\infty$ to the time instant $n-1$. The system therefore cumulatively adds, i.e., it accumulates all input sample values from $-\infty$ to n. The second form of the accumulator is used for a causal input sequence, in which case $y[-1]$ is called the *initial condition*.

Example 2.13 The *up-sampler* is a discrete-time system defined by the input-output relation

$$y[n] = \begin{cases} x[n/L], & n = 0, \pm L, \pm 2L, \ldots, \\ 0, & \text{otherwise}, \end{cases} \qquad (2.52)$$

where L, called the *up-sampling factor*, is a positive integer greater than 1. Basically it operates by inserting $L-1$ zero-valued equispaced samples between each consecutive pair of input samples. This device finds applications in the sampling rate alteration process, to be discussed in detail in Chapter 10. Figure 2.18 illustrates the up-sampling operation on a sinusoidal sequence of frequency 0.12 rad/sec for an up-sampling factor of $L = 3$.

We next illustrate the operation of a discrete-time system by two examples.

Example 2.14 Another simple example of a discrete-time system is the M-point *moving average system* defined by

$$y[n] = \frac{1}{M} \sum_{k=0}^{M-1} x[n-k]. \qquad (2.53)$$

Such a system is often used in smoothing random variations in data. Consider for example a signal $s[n]$ corrupted by a noise $d[n]$ for $n \geq 0$, resulting in a measured data given by

$$x[n] = s[n] + d[n].$$

Figure 2.19   (a) The original uncorrupted sequence $s[n]$, and (b) the noise sequence $d[n]$.

We would like to reduce the effect of the noise $d[n]$ and get a better estimate of $s[n]$ from $x[n]$. To this end, the moving average discrete-time system of Eq. (2.53) often gives reasonably good results [Sch75]. We illustrate this approach using MATLAB and assume for simplicity the original uncorrupted signal is given by

$$s[n] = 2[n(0.9)^n]. \tag{2.54}$$

The MATLAB program to generate $s[n]$ and $d[n]$ is given below where the noise $d[n]$ is computed using the function rand. Both sequences are shown in Figure 2.19.

```
% Program 2_3
%   Illustration of Signal Generation
%
% Random noise generation
R = 50;
d = rand(R,1)-0.5;
% Generate the uncorrupted signal
m = 0:1:R-1;
s = 2*m.*(0.9^m);
subplot(2,1,1)
stem(m,s)
xlabel('Time index n'); ylabel('Amplitude')
title('Original uncorrupted sequence')
subplot(2,1,2)
stem(m,d)
xlabel('Time index n'); ylabel('Amplitude')
title('Noise')
```

Program 2_4 has been used to generate the smoothed output $y[n]$ from the noise corrupted signal $x[n]$ using the moving average system of Eq. (2.53). Figure 2.20 shows the plots of pertinent signals generated by this program for $M = 3$. During execution, the program requests the input data, which is the desired number M of input samples to be added. It should be noted that to illustrate the effect of noise smoothing more clearly, the discrete-time signals have been plotted as continuous curves using the function plot.

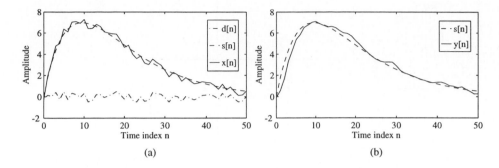

Figure 2.20   Pertinent signals of Example of 2.14: $s[n]$ is the original uncorrupted sequence, $d[n]$ is the noise sequence, $x[n] = s[n] + d[n]$, and $y[n]$ is the output of the moving average filter.

```
% Program 2_4
% Signal Smoothing by a Moving Average Filter
%
R = 50;
d = rand(R,1)-0.5;
m = 0:1:R-1;
s = 2*m.*(0.9^m);
x  = s + d ;
plot(m,d,'r-',m,s,'b--',m,x,'g:')
xlabel('Time index n'); ylabel('Amplitude')
legend('r-','d[n]','b--','s[n]','g:','x[n]');
pause
M = input('Number of input samples = ');
b = ones(M,1)/M;
y = filter(b,1,x);
plot(m,s,'r-',m,y,'b--')
legend('r-','s[n]','b--','y[n]');
xlabel ('Time index n');ylabel('Amplitude')
```

Note that in Figure 2.20(b), the output $y[n]$ of the 3-point moving average system is nearly equal to the desired uncorrupted input $s[n]$, except that it is delayed by one sample. We shall show later that a delay of $(M-1)/2$ samples is inherent in an M-point moving average filter.

Example 2.15   Another example of a discrete-time system is the linear interpolation method often employed to estimate sample values between pairs of adjacent sample values of a discrete-time sequence. The linear interpolation is implemented by first passing the input sequence $x[n]$ to be interpolated through an up-sampler whose output $x_u[n]$ is then passed through a second discrete-time system that "fills in" the zero-valued samples inserted by the up-sampler with values obtained by a linear interpolation of the pair of input samples surrounding the zero-valued samples. The interpolated samples thus lie on a straight line joining the pair of input samples, as illustrated in Figure 2.21 for a factor-of-4 interpolation.

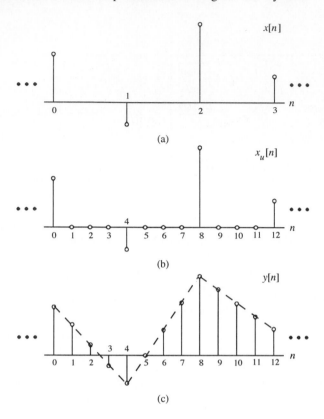

Figure 2.21   Illustration of the linear interpolation method.

If we denote the output of the factor-of-2 interpolator as $y[n]$, then it is computed using the relation

$$y[n] = x_u[n] + \frac{1}{2} \left(x_u[n-1] + x_u[n+1] \right).$$ (2.55)

Likewise, for a factor-of-3 interpolator, the output is obtained using the relation

$$y[n] = x_u[n] + \frac{1}{3} \left(x_u[n-1] + x_u[n+2] \right) + \frac{2}{3} \left(x_u[n-2] + x_u[n+1] \right).$$ (2.56)

2.4.1   Classification

There are various types of classification of discrete-time systems that are described next. These classifications are based on the input-output relation of the system.

Linearity.   The most widely used discrete-time system, and the one that we shall be exclusively concerned with in this text, is a *linear system* for which the superposition principle always holds. More precisely, for a linear discrete-time system, if $y_1[n]$ and $y_2[n]$ are the responses to the input sequences $x_1[n]$ and $x_2[n]$, respectively, then for an input

$$x[n] = \alpha x_1[n] + \beta x_2[n],$$

the response is given by

$$y[n] = \alpha y_1[n] + \beta y_2[n].$$

The superposition property must hold for any arbitrary constants, α and β, and for all possible inputs, $x_1[n]$ and $x_2[n]$. The above property makes it very easy to compute the response to a complicated sequence that can be decomposed as a weighted combination of some simple sequences, such as the unit sample sequences or the complex exponential sequences. In this case, the desired output is given by a similarly weighted combination of the outputs to the simple sequences.

Example 2.16 Consider the discrete-time accumulator of Example 2.12. From its input-output relation given by Eq. (2.51a) we observe that the outputs $y_1[n]$ and $y_2[n]$ for inputs $x_1[n]$ and $x_2[n]$ are given by

$$y_1[n] = \sum_{\ell=-\infty}^{n} x_1[\ell],$$

$$y_2[n] = \sum_{\ell=-\infty}^{n} x_2[\ell].$$

The output $y[n]$ due to an input $\alpha x_1[n] + \beta x_2[n]$ is then given by

$$y[n] = \sum_{\ell=-\infty}^{n} (\alpha x_1[\ell] + \beta x_2[\ell]) = \alpha \sum_{\ell=-\infty}^{n} x_1[\ell] + \beta \sum_{\ell=-\infty}^{n} x_2[\ell] = \alpha y_1[n] + \beta y_2[n].$$

Hence, the discrete-time system of Eq. (2.51a) is a linear system.

Consider the modified form of the accumulator defined by Eq. (2.51b) with a causal input applied at $n = 0$. Now, the outputs $y_1[n]$ and $y_2[n]$ for inputs $x_1[n]$ and $x_2[n]$ are given by

$$y_1[n] = y_1[-1] + \sum_{\ell=0}^{n} x_1[\ell],$$

$$y_2[n] = y_2[-1] + \sum_{\ell=0}^{n} x_2[\ell].$$

Hence, the output $y[n]$ for an input $\alpha x_1[n] + \beta x_2[n]$ is of the form

$$y[n] = y[-1] + \sum_{\ell=0}^{n} (\alpha x_1[\ell] + \beta x_2[\ell]) = y[-1] + \alpha \sum_{\ell=0}^{n} x_1[\ell] + \beta \sum_{\ell=0}^{n} x_2[\ell]. \qquad (2.57)$$

On the other hand,

$$\alpha y_1[n] + \beta y_2[n] = \alpha y_1[-1] + \alpha \sum_{\ell=0}^{n} x_1[\ell] + \beta y_2[-1] + \beta \sum_{\ell=0}^{n} x_2[\ell], \qquad (2.58)$$

which is equal to $y[n]$ of Eq. (2.57) if

$$y[-1] = \alpha y_1[-1] + \beta y_2[-1].$$

For the system of Eq. (2.51b) to be linear, the above condition must hold for all initial conditions $y[-1]$, $y_1[-1]$, and $y_2[-1]$ and all constants α and β. This condition cannot be satisfied unless the system of Eq. (2.51b) is initially at rest with a zero initial condition. For nonzero initial conditions the discrete-time system of Eq. (2.51b) is nonlinear.

It can be easily verified that the discrete-time systems of Eqs. (2.41), (2.42), (2.52), (2.53), and (2.55) are linear systems (Problem 2.19).

Example 2.17 Consider the discrete-time system given by [Kai80]

$$y[n] = x^2[n] - x[n-1]x[n+1].$$

The outputs $y_1[n]$ and $y_2[n]$ for inputs $x_1[n]$ and $x_2[n]$ are given by

$$y_1[n] = x_1^2[n] - x_1[n-1]x_1[n+1],$$

$$y_2[n] = x_2^2[n] - x_2[n-1]x_2[n+1].$$

The output $y[n]$ due to an input $\alpha x_1[n] + \beta x_2[n]$ is given by

$$
\begin{aligned}
y[n] &= \{\alpha x_1[n] + \beta x_2[n]\}^2 \\
&\quad - \{\alpha x_1[n-1] + \beta x_2[n-1]\}\{\alpha x_1[n+1] + \beta x_2[n+1]\} \\
&= \alpha^2 \left\{ x_1^2[n] - x_1[n-1]x_1[n+1] \right\} + \beta^2 \left\{ x_2^2[n] - x_2[n-1]x_2[n+1] \right\} \\
&\quad + \alpha\beta \left\{ 2x_1[n]x_2[n] - x_1[n-1]x_2[n+1] - x_1[n+1]x_2[n-1] \right\}.
\end{aligned}
$$

On the other hand,

$$
\begin{aligned}
\alpha y_1[n] + \beta y_2[n] &= \alpha \left\{ x_1^2[n] - x_1[n-1]x_1[n+1] \right\} \\
&\quad + \beta \left\{ x_2^2[n] - x_2[n-1]x_2[n+1] \right\} \neq y[n].
\end{aligned}
$$

Therefore, the system is *nonlinear*.

Shift-Invariance.   The *shift-invariance* property is the second condition imposed on most digital filters in practice. For a shift-invariant discrete-time system, if $y_1[n]$ is the response to an input $x_1[n]$, then the response to an input

$$x[n] = x_1[n - n_0]$$

is simply

$$y[n] = y_1[n - n_0],$$

where n_0 is any positive or negative integer. This relation between the input and output must hold for any arbitrary input sequence and its corresponding output. In the case of sequences and systems with indices n related to discrete instants of time, the above restriction is more commonly called the *time-invariance* property. The time-invariance property ensures that for a specified input, the output of the system is independent of the time the input is being applied.

A *linear time-invariant* (LTI) discrete-time system satisfies both the linearity and the time-invariance properties. Such systems are mathematically easy to analyze, and characterize, and as a consequence, easy to design. In addition, highly useful signal processing algorithms have been developed utilizing this class of systems during the last 30 years. In this text, we consider almost entirely this type of discrete-time systems.

Example 2.18 The up-sampler of Eq. (2.52) is a time-varying system. To show this, we observe from Eq. (2.52) that its output $y_1[n]$ for an input $x_1[n] = x[n - n_0]$ is given by

$$y_1[n] = \begin{cases} x_1[n/L], & n = 0, \pm L, \pm 2L, \ldots, \\ 0, & \text{otherwise,} \end{cases}$$

$$= \begin{cases} x[(n - Ln_0)/L], & n = 0, \pm L, \pm 2L, \ldots, \\ 0, & \text{otherwise.} \end{cases}$$

But from Eq. (2.52)

$$y[n - n_0] = \begin{cases} x[(n - n_0)/L], & n = n_0, n_0 \pm L, n_0 \pm 2L, \\ 0, & \text{otherwise,} \end{cases}$$

$$\neq y_1[n].$$

Causality.   In addition to the above two properties, we impose, for practicality, additional restrictions of causality and stability on the class of discrete-time systems we deal with in this text. In a *causal* discrete-time system, the n_0-th output sample $y[n_0]$ depends only on input samples $x[n]$ for $n \leq n_0$ and does not depend on input samples for $n > n_0$. Thus, if $y_1[n]$ and $y_2[n]$ are the responses of a causal discrete-time system to the inputs $u_1[n]$ and $u_2[n]$, respectively, then

$$u_1[n] = u_2[n] \quad \text{for } n < N$$

implies also that

$$y_1[n] = y_2[n] \quad \text{for } n < N.$$

Simply speaking, for a causal system, changes in output samples do not precede changes in the input samples. It should be pointed out here that the definition of causality given above can be applied only to discrete-time systems with the same sampling rate for the input and the output.[7]

It can be easily shown that the discrete-time systems of Eqs. (2.41), (2.42), (2.51a), (2.51b), and (2.53) are causal systems. However, the discrete-time systems defined by Eqs. (2.55) and (2.56) are noncausal systems. It should be noted that these two noncausal systems can be implemented as causal systems by simply delaying the output by one and two samples, respectively.

Stability.   There are various definitions of stability. We define a discrete-time system to be *stable* if and only if, for every bounded input, the output is also bounded. This implies that, if the response to $x[n]$ is the sequence $y[n]$ and if

$$|x[n]| < B_x$$

for all values of n, then

$$|y[n]| < B_y$$

for all values of n where B_x and B_y are finite constants. This type of stability is usually referred to as *bounded-input, bounded-output (BIBO) stability*.

---

[7]If the input and output sampling rates are not the same, the definition of causality has to be modified.

Example 2.19 Consider the M-point moving average system of Example 2.14. It follows from Eq. (2.53) that for a bounded input $|x[n]| < B_x$, the magnitude of n-th sample of output is given by

$$|y[n]| = \left| \frac{1}{M} \sum_{k=0}^{M-1} x[n-k] \right| \leq \frac{1}{M} \sum_{k=0}^{M-1} |x[n-k]| \leq \frac{1}{M}(M)B_x \leq B_x,$$

indicating that the system is BIBO stable.

Passivity. A discrete-time system is said to be *passive* if, for every finite energy input sequence $x[n]$, the output sequence $y[n]$ has, at most, the same energy, i.e.,

$$\sum_{n=-\infty}^{\infty} |y[n]|^2 \leq \sum_{n=-\infty}^{\infty} |x[n]|^2 < \infty. \tag{2.59}$$

If the above inequality is satisfied with an equal sign for every input sequence, the discrete-time system is said to be *lossless*. For example, the discrete-time system defined by $y[n] = \alpha x[n-N]$ with N a positive integer, is a passive system if $|\alpha| < 1$ and is a lossless system if $|\alpha| = 1$.

As we shall see later, in Section 9.7, the passivity and the losslessness properties are crucial to the design of discrete-time systems with very low sensitivity to changes in the filter coefficients.

2.4.2   Impulse and Step Responses

The response of a digital filter to a unit sample sequence $\{\delta[n]\}$ is called the *unit sample response*, or simply, the *impulse response*, and is denoted as $\{h[n]\}$. Correspondingly, the response of a discrete-time system to a unit step sequence $\{\mu[n]\}$, denoted as $\{s[n]\}$, is its *unit step response* or simply, the *step response*. As we show next, a linear time-invariant digital filter is completely characterized in the time-domain by its impulse response or its step response.

Example 2.20 The impulse response $\{h[n]\}$ of the discrete-time accumulator of Eq. (2.51a) is obtained by setting $x[n] = \delta[n]$ resulting in

$$h[n] = \sum_{\ell=-\infty}^{n} \delta[\ell],$$

which from Eq. (2.27) is precisely the unit step sequence $\mu[n]$.

Example 2.21 The impulse response $\{h[n]\}$ of the factor-of-2 interpolator of Eq. (2.55) is obtained by setting $x_u[n] = \delta[n]$ and is given by

$$h[n] = \delta[n] + \frac{1}{2}(\delta[n-1] + \delta[n+1]).$$

The impulse response is seen to be a finite length sequence of length 3 and can be written alternately as

$$h[n] = \{0.5,\ 1,\ 0.5\}.$$
$$\uparrow$$

2.5   Time-Domain Characterization of LTI Discrete-Time Systems

In most cases an LTI discrete-time system is designed as an interconnection of simple subsystems. Each subsystem in turn is implemented with the aid of the basic building blocks discussed earlier in Section 2.2.1. In order to be able to analyze such systems in the time-domain, we need to develop the pertinent relationships between the input and the output of an LTI discrete-time system, and the characterization of the interconnected system.

2.5.1   Input-Output Relationship

A consequence of the linear, time-invariance property is that an LTI discrete-time system is completely specified by its impulse response, i.e., knowing the impulse response, we can compute the output of the system to any arbitrary input. We develop this relationship now.

Let $h[n]$ denote the impulse response of the LTI discrete-time system of interest, i.e., the response to an input $\delta[n]$. We first compute the response of this filter to the input $x[n]$ of Eq. (2.44). Since the discrete-time system is time-invariant, its response to $\delta[n-1]$ will be $h[n-1]$. Likewise, the responses to $\delta[n+2]$, $\delta[n-4]$ and $\delta[n-6]$ will be, respectively, $h[n+2]$, $h[n-4]$, and $h[n-6]$. Because of linearity, the response of the LTI discrete-time system to the input

$$x[n] = 0.5\delta[n+2] + 1.5\delta[n-1] - \delta[n-2] + \delta[n-4] + 0.75\delta[n-6]$$

will be simply

$$y[n] = 0.5h[n+2] + 1.5h[n-1] - h[n-2] + h[n-4] + 0.75h[n-6].$$

We now generalize the above result for an arbitrary input sequence $x[n]$ that can be expressed as

$$x[n] = \sum_{k=-\infty}^{\infty} x[k]\delta[n-k]. \tag{2.60}$$

In the above expression, $x[k]$ denotes specifically the k-th sample value. The response of the LTI discrete-time system to the sequence $x[k]\delta[n-k]$ will be $x[k]h[n-k]$. As a result, the response $y[n]$ of the discrete-time system to $x[n]$ will be given by

$$y[n] = \sum_{k=-\infty}^{\infty} x[k]h[n-k], \tag{2.61a}$$

which can be alternately written as

$$y[n] = \sum_{k=-\infty}^{\infty} x[n-k]h[k] \tag{2.61b}$$

by a simple change of variables. The above sum in Eqs. (2.61a) and (2.61b) is called the *convolution sum* of the sequences $x[n]$ and $h[n]$, and represented compactly as:

$$y[n] = x[n] \circledast h[n] \tag{2.62}$$

where the notation $\circledast$ denotes the convolution sum.[8]

The convolution sum operation satisfies several useful properties. First, the operation is *commutative*, i.e.,

$$x_1[n] \circledast x_2[n] = x_2[n] \circledast x_1[n]. \tag{2.63}$$

Second, the convolution operation, for stable and single-sided sequences, is *associative*, i.e.,

$$(x_1[n] \circledast x_2[n]) \circledast x_3[n] = x_1[n] \circledast (x_2[n] \circledast x_3[n]). \tag{2.64}$$

and last, the operation is *distributive*, i.e.,

$$x_1[n] \circledast (x_2[n] + x_3[n]) = x_1[n] \circledast x_2[n] + x_1[n] \circledast x_3[n]. \tag{2.65}$$

Proof of these properties is left as an exercise (Problems 2.27 to 2.29).

It is clear from the above discussion that the impulse response $\{h[n]\}$ completely characterizes an LTI discrete-time system in the time-domain because, knowing the impulse response, we can compute, in principle, the output sequence $y[n]$ for any given input sequence $x[n]$ using the convolution sum of Eq. (2.61a) or (2.61b). The computation of an output sample is simply a sum of products involving fairly simple arithmetic operations such as additions, multiplications, and delays. However, in practice, the convolution sum can be employed to compute the output sample at any instant only if either the impulse response sequence and/or the input sequence is of finite length, resulting in a finite sum of products. Note that if both the input and the impulse response sequences are of finite length, the output sequence is also of finite length. In the case of a discrete-time system with an infinite-length impulse response, it is obviously not possible to compute the output using the convolution sum if the input is also of infinite length. We shall therefore consider alternative time-domain descriptions of such systems that involve only finite sums of products.

To understand the process involved in generating the new sequence $y[n]$, let us first compute the sample $y[0]$ that is given by

$$y[0] = \sum_{k=-\infty}^{\infty} x[k]h[-k].$$

Thus, to compute the output sample $y[0]$ we first time-reverse the sequence $h[k]$ to arrive at $h[-k]$. Then we form the product of the k-th sample of $h[-k]$ with that of $x[k]$ and sum it over all values of k. Next, to compute $y[1]$ we note

$$y[1] = \sum_{k=-\infty}^{\infty} x[k]h[1-k],$$

which indicates that first the time-reversed sequence $h[-k]$ is shifted to the right by one sampling period to arrive at the sequence $h[1-k]$. Then the k-th sample of this sequence is multiplied with the k-th input sample $x[k]$ and summed over all values of k to arrive at $y[1]$. This process is continued for increasing values of n.

---

[8]In the literature, the symbol for the convolution sum is $*$ without the circle. However, as $*$ is always used for denoting the complex conjugation operation, in this text we have adopted the symbol $\circledast$ to denote the convolution sum operation.

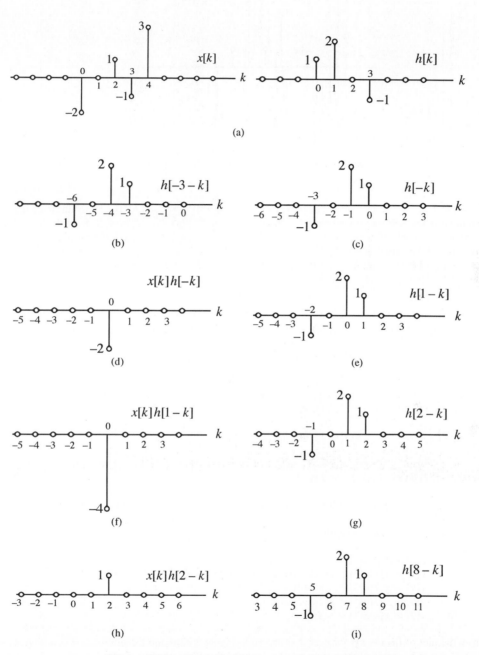

Figure 2.22 Illustration of the convolution process.

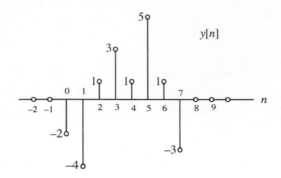

Figure 2.23   Sequence generated by the convolution.

Example 2.22 We systematically develop the sequence $y[n]$ generated by the convolution of the two finite-length sequences $x[n]$ and $h[n]$ shown in Figure 2.22(a). As can be seen from the plot of the shifted time-reversed version $\{h[n-k]\}$ for $n < 0$ sketched in Figure 2.22(b) for $n = -3$, for any value of the sample index k, the kth sample of either $\{x[k]\}$ or $\{h[n-k]\}$ is zero. As a result the product of the k-th samples is always zero for any k, and consequently, the convolution sum of Eq. (2.61a) leads to

$$y[n] = 0 \qquad \text{for } n < 0.$$

Consider now the calculation of $y[0]$. We form $\{h[-k]\}$ as shown in Figure 2.22(c). The product sequence $\{x[k]h[-k]\}$ is plotted in Figure 2.22(d), which has a single nonzero sample for $k = 0$, $x[0]h[0]$. Thus,

$$y[0] = x[0]h[0] = -2.$$

For the computation of $y[1]$, we shift $\{h[-k]\}$ to the right by one sample period to form $\{h[1-k]\}$ as sketched in Figure 2.22(e). The product sequence $\{x[k]h[1-k]\}$ shown in Figure 2.22(f) has one nonzero sample for $k = 0$, $x[0]h[1]$. As a result,

$$y[1] = x[0]h[1] + x[1]h[0] = -4 + 0 = -4.$$

To calculate $y[2]$, we form $\{h[2-k]\}$, as indicated in Figure 2.22(g). The resulting product sequence $\{x[k]h[2-k]\}$ is plotted in Figure 2.22(h) ,which leads to

$$y[2] = x[0]h[2] + x[1]h[1] + x[2]h[0] = 0 + 0 + 1 = 1.$$

This process is continued, resulting in

$$y[3] = x[0]h[3] + x[1]h[2] + x[2]h[1] + x[3]h[0] = 2 + 0 + 0 + 1 = 3,$$
$$y[4] = x[1]h[3] + x[2]h[2] + x[3]h[1] + x[4]h[0] = 0 + 0 - 2 + 3 = 1,$$
$$y[5] = x[2]h[3] + x[3]h[2] + x[4]h[1] = -1 + 0 + 6 = 5,$$
$$y[6] = x[3]h[3] + x[4]h[2] = 1 + 0 = 1,$$
$$y[7] = x[4]h[3] = -3.$$

From the plot of $\{h[n-k]\}$ for $n > 7$ as shown in Figure 2.22(i) and the plot of $\{x[k]\}$ in Figure 2.22(a), it can be seen that there is no overlap between these two sequences. As a result,

$$y[n] = 0 \qquad \text{for } n > 7.$$

The sequence $\{y[n]\}$ as obtained above is sketched in Figure 2.23.

As seen from Example 2.22, the convolution of two finite-length sequences results in a finite-length sequence. In this example, the convolution of a sequence $\{x[n]\}$ of length 5 with a sequence $\{h[n]\}$ of length 4 resulted in a sequence $\{y[n]\}$ of length 8. In general, if the lengths of the two sequences being convolved are M and N, then the resulting sequence after convolution is of length $M + N - 1$ (Problem 2.30).

In MATLAB, the function $c = \text{conv(a,b)}$ implements the convolution of two finite-length sequences a and b, generating the finite-length sequence c. The process is illustrated in the following example.

Example 2.23 We repeat the problem of the previous example using Program 2_5 given below.

```
% Program 2_5
% Illustration of Convolution
%
a = input('Type in the first sequence = ');
b = input('Type in the second sequence = ');
c = conv(a, b);
M = length(c)-1;
n = 0:1:M;
disp('output sequence =');disp(c)
stem(n,c)
xlabel('Time index n'); ylabel('Amplitude');
```

During execution, the input data requested are the two sequences to be convolved that are entered in a vector format inside square brackets as indicated below:

```
a = [-2    0    1    -1    3]
b = [ 1    2    0    -1]
```

The program then computes the convolution and displays the resulting sequence shown below:

```
output sequence =
 -2    -4    1    3    1    5    1    -3
```

and provides a plot, as indicated in Figure 2.24. As expected, the result is identical to that derived in the previous example.

2.5.2 Simple Interconnection Schemes

Two widely used schemes for developing complex LTI discrete-time systems from simple LTI discrete-time system sections are described next.

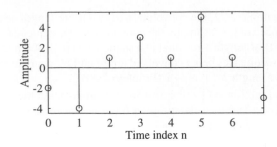

Figure 2.24   Sequence generated by convolution using MATLAB.

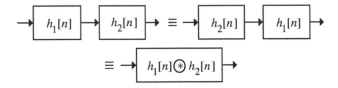

Figure 2.25   The cascade connection.

Cascade Connection.   In Figure 2.25, the output of one filter is connected to the input of a second filter and the two filters are said to be connected in *cascade*. The overall impulse response $h[n]$ of the cascade of two filters of impulse responses $h_1[n]$ and $h_2[n]$ is given by

$$h[n] = h_1[n] \circledast h_2[n]. \tag{2.66}$$

Note that, in general, the ordering of the filters in the cascade has no effect on the overall impulse response because of the commutative property of convolution.

It can be shown that the cascade connection of two stable systems is stable. Likewise, the cascade connection of two passive (lossless) systems is passive (lossless).

The cascade connection scheme is employed in the development of an *inverse system*. If the two LTI systems in the cascade connection of Figure 2.25 are such that

$$h_1[n] \circledast h_2[n] = \delta[n], \tag{2.67}$$

then the LTI system $h_2[n]$ is said to be the inverse of the LTI system $h_1[n]$, and vice-versa. As a result of the above relation, if the input to the cascaded system is $x[n]$, its output is also $x[n]$. An application of this concept is in the recovery of a signal from its distorted version appearing at the output of a transmission channel. This is accomplished by designing an inverse system if the impulse response of the channel is known.

The following example illustrates the development of an inverse system.

Example 2.24 From Example 2.20 the impulse response of the discrete-time accumulator is the unit step response $\mu[n]$. Therefore, from Eq. (2.67) the inverse system must satisfy the condition

$$\mu[n] \circledast h_2[n] = \delta[n]. \tag{2.68}$$

It follows from Eq. (2.68) that $h_2[n] = 0$ for $n < 0$ and

$$h_2[0] = 1,$$

Figure 2.26   The parallel connection.

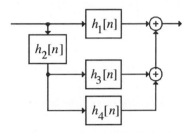

Figure 2.27   The discrete-time system of Example 2.25.

$$\sum_{\ell=0}^{n} h_2[\ell] = 0, \qquad n \geq 1.$$

As a result,

$$h_2[1] = -1$$

and

$$h_2[n] = 0, \qquad \text{for } n \geq 2.$$

Thus the impulse response of the inverse system is given by

$$h_2[n] = \delta[n] - \delta[n-1],$$

which is called a *backward difference system*.

Parallel Connection.   The connection scheme of Figure 2.26 is called the *parallel* connection, and here the outputs of the two filters are added to form the new output while the same input is fed to both filters. The impulse response of the overall filter is given here by

$$h[n] = h_1[n] + h_2[n]. \tag{2.69}$$

It is a simple exercise to show that the parallel connection of two stable systems is stable. However, the parallel connection of two passive (lossless) systems may or may not be passive (lossless).

Example 2.25 Consider the discrete-time system of Figure 2.27 composed of an interconnection of four simple discrete-time systems with impulse responses given by

$$h_1[n] = \delta[n] + \tfrac{1}{2}\delta[n-1], \qquad h_2[n] = \tfrac{1}{2}\delta[n] - \tfrac{1}{4}\delta[n-1],$$
$$h_3[n] = 2\delta[n], \qquad\qquad\quad h_4[n] = -2\left(\tfrac{1}{2}\right)^{n}\mu[n].$$

The overall impulse response $h[n]$ is given by

$$h[n] = h_1[n] + h_2[n]\circledast(h_3[n] + h_4[n]) = h_1[n] + h_2[n]\circledast h_3[n] + h_2[n]\circledast h_4[n].$$

Now

$$h_2[n]\circledast h_3[n] = (\tfrac{1}{2}\delta[n] - \tfrac{1}{4}\delta[n-1])\circledast 2\delta[n] = \delta[n] - \tfrac{1}{2}\delta[n-1]$$

and

$$h_2[n]\circledast h_4[n] = \left(\tfrac{1}{2}\delta[n] - \tfrac{1}{4}\delta[n-1]\right)\circledast\left(-2\left(\tfrac{1}{2}\right)^n\mu[n]\right)$$

$$= -\left(\tfrac{1}{2}\right)^n\mu[n] + \tfrac{1}{2}\left(\tfrac{1}{2}\right)^{n-1}\mu[n-1] = -\left(\tfrac{1}{2}\right)^n\mu[n] + \left(\tfrac{1}{2}\right)^n\mu[n-1]$$

$$= -\left(\tfrac{1}{2}\right)^n\delta[n] = -\delta[n].$$

Therefore

$$h[n] = \delta[n] + \tfrac{1}{2}\delta[n-1] + \delta[n] - \tfrac{1}{2}\delta[n-1] - \delta[n] = \delta[n].$$

2.5.3   Stability Condition for LTI Discrete-Time Systems

Recall from Section 2.4.1 that a discrete-time system is defined to be stable, or precisely, bounded-input, bounded-output (BIBO) stable, if the output sequence $y[n]$ of the system remains bounded for all bounded input sequences $x[n]$. We now develop the stability condition for an LTI discrete-time system. We show that an LTI digital filter is BIBO stable if and only if its impulse response sequence $\{h[n]\}$ is absolutely summable, i.e.,

$$S = \sum_{n=-\infty}^{\infty} |h[n]| < \infty. \tag{2.70}$$

We prove the above statement for a real impulse response $h[n]$. The extension of the proof for a complex impulse response sequence is left as an exercise (Problem 2.43). Now, if the input sequence $x[n]$ is bounded, i.e., $|x[n]| \le B_x < \infty$, then the output amplitude, from Eq. (2.61b), is

$$|y[n]| = \left|\sum_{k=-\infty}^{\infty} h[k]x[n-k]\right| \le \sum_{k=-\infty}^{\infty} |h[k]|\,|x[n-k]|$$

$$\le B_x \sum_{k=-\infty}^{\infty} |h[k]| = B_x S < \infty. \tag{2.71}$$

Thus, $S < \infty$ implies $|y[n]| \le B_y < \infty$, indicating that $y[n]$ is also bounded. To prove the converse, assume $y[n]$ is bounded, i.e., $|y[n]| \le B_y$. Now, consider the input given by

$$x[n] = \begin{cases} \mathrm{sgn}\,(h[-n]), & \text{if } h[-n] \neq 0, \\ K, & \text{if } h[-n] = 0, \end{cases} \tag{2.72}$$

where $\mathrm{sgn}(c) = +1$ if $c > 0$ and $\mathrm{sgn}(c) = -1$ if $c < 0$, and $|K| \le 1$. Note that since $|x[n]| \le 1$, $\{x[n]\}$ is obviously bounded. For this input, $y[n]$ at $n = 0$ is

$$y[0] = \sum_{k=-\infty}^{\infty} \mathrm{sgn}\,(h[k])h[k] = S \le B_y < \infty. \tag{2.73}$$

Therefore $|y[n]| \le B_y$ implies $S < \infty$.

Example 2.26 Consider a causal LTI discrete-time system with an impulse response given by

$$h[n] = (\alpha)^n \mu[n].$$

For this system

$$S = \sum_{n=-\infty}^{\infty} |\alpha^n| \, \mu[n] = \sum_{n=0}^{\infty} |\alpha^n| = \frac{1}{1 - |\alpha|} \qquad \text{if } |\alpha| < 1.$$

Therefore $S < \infty$ if $|\alpha| < 1$ for which the above system is BIBO stable. On the other hand, if $|\alpha| = 1$ the above system is not BIBO stable.

Example 2.27 Now an LTI discrete-time system with an impulse response given by

$$h[n] = \begin{cases} \alpha^n, & N_1 \le n \le N_2, \\ 0, & \text{otherwise}, \end{cases} \tag{2.74}$$

is always BIBO stable independent of the value of α as long as it is not infinite since the impulse response sequence is absolutely summable for finite values of N_1 and N_2.

2.5.4 Causality Condition

We now develop the condition for an LTI discrete-time system to be causal. Let $x_1[n]$ and $x_2[n]$ be two input sequences with

$$x_1[n] = x_2[n] \qquad \text{for } n \le n_0. \tag{2.75}$$

From Eq. (2.61b) the corresponding output samples at $n = n_0$ of an LTI discrete-time system with an impulse response $\{h[n]\}$ are then given by

$$y_1[n_0] = \sum_{k=-\infty}^{\infty} h[k]x_1[n_0 - k]$$

$$= \sum_{k=0}^{\infty} h[k]x_1[n_0 - k] + \sum_{k=-\infty}^{-1} h[k]x_1[n_0 - k], \tag{2.76a}$$

$$y_2[n_0] = \sum_{k=-\infty}^{\infty} h[k]x_2[n_0 - k]$$

$$= \sum_{k=0}^{\infty} h[k]x_2[n_0 - k] + \sum_{k=-\infty}^{-1} h[k]x_2[n_0 - k]. \tag{2.76b}$$

If the LTI discrete-time system is also causal, then $y_1[n_0]$ must be equal to $y_2[n_0]$. Now, because of Eq. (2.75), the first sum on the right-hand side of Eq. (2.76a) is equal to the first sum on the right-hand side of Eq. (2.76b). This implies that the second sums on the right-hand side of the above two equations must be equal. As $x_1[n]$ may not be equal to

$x_2[n]$ for $n > n_0$, the only way these two sums will be equal is if they are each equal to zero, which is satisfied if

$$h[k] = 0 \qquad \text{for } k < 0. \tag{2.77}$$

As a result, an LTI discrete-time system is *causal* if and only if its impulse response sequence $\{h[n]\}$ satisfies the condition of Eq. (2.77).

It follows from Example 2.20 that the discrete-time accumulator of Eq. (2.51a) is a causal system since its impulse response satisfies the causality condition of Eq. (2.77). On the other hand, from Example 2.21 we observe that the factor-of-2 linear interpolator defined by Eq. (2.55) is a noncausal system because its impulse response does not satisfy the causality condition of Eq. (2.77). However, a noncausal discrete-time system with a finite-length impulse response can often be realized as a causal system by delaying the output sequence by an appropriate amount. For example, a causal version of the discrete-time factor-of-2 linear interpolator is obtained by delaying the output by one sample period with an input-output relation given by

$$y[n] = x_u[n-1] + \frac{1}{2}\left(x_u[n-2] + x_u[n]\right).$$

2.5.5   Finite-Dimensional LTI Discrete-Time Systems

An important subclass of LTI discrete-time systems is characterized by a linear constant coefficient difference equation of the form

$$\sum_{k=0}^{N} d_k y[n-k] = \sum_{k=0}^{M} p_k x[n-k], \tag{2.78}$$

where $x[n]$ and $y[n]$ are, respectively, the input and the output of the system, and $\{d_k\}$ and $\{p_k\}$ are constants. The *order* of the discrete-time system is said to be N (assuming $N \geq M$), which is the order of the difference equation characterizing the system. It is possible to implement an LTI system characterized by Eq. (2.78) since the computation here involves two finite sum of products. In general, such a system may not be causal, as illustrated by the following example.

Example 2.28   The filter described by

$$y[n] - a y[n-1] = x[n]$$

has an impulse response $h[n]$ (obtained by setting $x[n] = \delta[n]$):

$$h[n] = a^n \mu[n] + c a^n,$$

where c is any arbitrary constant. Consider the two solutions:

$$h_1[n] = a^n \mu[n], \qquad h_2[n] = -a^n \mu[-n-1].$$

The first solution $h_1[n]$, obtained for $c = 0$, represents a causal system and is stable for $|a| < 1$, whereas the second solution $h_2[n]$, obtained for $c = -1$, represents a noncausal system that is stable for $|a| > 1$.

Example 2.29 There exist infinite-impulse response LTI discrete-time systems that cannot be characterized by the difference equation form of Eq. (2.78). The system defined by the impulse response

$$h[n] = \frac{1}{n^2}\mu[n-1]$$

is such an example. Note that the above system is causal and also BIBO stable.

If we assume the system to be causal, then we can rewrite Eq. (2.78) to express $y[n]$ explicitly as a function of $x[n]$:

$$y[n] = -\sum_{k=1}^{N} \frac{d_k}{d_0} y[n-k] + \sum_{k=0}^{M} \frac{p_k}{d_0} x[n-k], \tag{2.79}$$

The output $y[n]$ can be computed for all $n \geq n_0$ knowing $x[n]$ and the initial conditions $y[n_0 - 1]$, $y[n_0 - 2]$, . . . , $y[n_0 - N]$. Since the impulse response $h[n]$ of a causal discrete-time system is a causal sequence, Eq. (2.79) can be used to calculate recursively the impulse response for $n \geq 0$ by setting initial conditions to zero values, i.e., by setting $y[-1] = y[-2] = \ldots = y[-N] = 0$, and using a unit sample sequence $\delta[n]$ as the input $x[n]$. The step response of a causal LTI system can similarly be computed recursively by setting zero initial conditions and applying a unit step sequence as the input. It should be noted that the causal discrete-time system of Eq. (2.79) is linear only for zero initial conditions (Problem 2.36). A system with all zero initial conditions is often called a *relaxed* system.

The causal LTI system of the form of Eq. (2.79) can be simulated in MATLAB using the function `filter` already made use of in Program 2_4. In one of its forms, the function

```
y = filter (p,d,x)
```

processes the input data vector x using the system characterized by the coefficient vectors p and d to generate the output vector y assuming zero initial conditions. The length of y is the same as the length of x. Since the function implements Eq. (2.79), the coefficient d_0 must be nonzero.

The following example illustrates the computation of the impulse and step responses of an LTI system described by Eq. (2.79).

Example 2.30 Program 2_6 given below can be employed to compute the impulse response of a causal finite-dimensional LTI discrete-system of the form of Eq. (2.79). The program calls for the following input data: desired length of the impulse response, and the filter coefficient vectors p and d that must be entered inside square brackets. The program then plots the impulse response sequence. To illustrate its application we determine the first 41 samples of the impulse response of the causal LTI system defined by

$$y[n] + 0.7y[n-1] - 0.45y[n-2] - 0.6y[n-3]$$
$$= 0.8x[n] - 0.44x[n-1] + 0.36x[n-2] + 0.02x[n-3]. \tag{2.80}$$

Figure 2.28   Impulse response of the system of Eq. (2.80).

```
% Program 2_6
% Illustration of Impulse Response Computation
%
N = input('Desired impulse response length = ');
p = input('Type in the vector p =');
d = input('Type in the vector d =');
x = [1 zeros(1,N-1)];
y = filter(p,d,x);
k = 0:1:N-1;
stem(k,y)
xlabel('Time index n'); ylabel('Amplitude')
```

The input data entered are:

```
N = 41
p = [0.8  -0.44  0.36  0.02]
d = [1   0.7  -0.45 -0.6]
```

The impulse response sequence generated by the program is then plotted as indicated in Figure 2.28. Note that the impulse response of a discrete-time finite-dimensional system can also be computed in MATLAB using the function impz (Problem M2.9).

To determine the step response we replace in the above program the statement x = [1 zeros(1,N-1)] with the statement x = [ones(1,N)]. The computed first 41 samples of the step response are indicated in Figure 2.29.

It should be noted that the impulse response samples of a stable LTI system decays to zero values as the time index n becomes very large. Likewise, the step response samples of a stable LTI system approach a constant value as n becomes very large. From the plots of Figures 2.28 and 2.29 we can conclude that most likely the LTI system of Eq. (2.80) is BIBO stable. However, it is impossible to check the stability of a system just by examining only a finite segment of its impulse or step response as in these figures.

Figure 2.29   Step response of the system of Eq. (2.80).

2.5.6   Classification of LTI Discrete-Time Systems

Linear time-invariant (LTI) discrete-time systems are usually classified either according to the length of their impulse response sequences or according to the method of calculation employed to determine the output samples.

Classification Based on Impulse Response Length.   If $h[n]$ is of finite length, i.e.,

$$h[n] = 0 \qquad \text{for } n < N_1 \qquad \text{and } n > N_2 \qquad \text{with } N_1 < N_2, \qquad (2.81)$$

then it is known as a *finite impulse response* (FIR) discrete-time system, for which the convolution sum reduces to

$$y[n] = \sum_{k=N_1}^{N_2} h[k]x[n-k]. \qquad (2.82)$$

Note that the above convolution sum, being a finite sum, can be used to calculate $y[n]$ directly. The basic operations involved are simply multiplication and addition. Note that when $N_1 = 0$, the calculation of the present value of the output sequence involves the present value and $N_2 - N_1$ previous values of the input sequence and involves the $N_2 - N_1 + 1$ impulse response samples describing the FIR discrete-time system.

Examples of FIR discrete-time systems are the moving average system of Eq. (2.53) and the linear interpolators of Eqs. (2.55) and (2.56).

If $h[n]$ is of infinite length, then it is known as an *infinite impulse response* (IIR) discrete-time system. In this case the convolution sum cannot be used to numerically calculate the output sequence samples due to the infinite nature of the sum unless the input sequence is of finite length. The class of IIR filters we are concerned with in this text is the causal system characterized by the linear constant coefficient difference equation of Eq. (2.79). Note that here also the basic operations needed in the output calculations are multiplication and addition. An example of an IIR system is the accumulator of Eqs. (2.51a) and (2.51b). Another example is described below.

Example 2.31 Examples of an IIR filter are the familiar numerical integration formulas that are used to numerically solve integrals of the form

$$y(t) = \int_0^t x(\tau)\,d\tau.$$

If we divide the interval of integration into n equal parts of length T, then the above integral can be rewritten as

$$y(nT) = y((n-1)T) + \int_{(n-1)T}^{nT} x(\tau)\,d\tau, \qquad (2.83)$$

where we have set $t = nT$ and used the notation

$$y(nT) = \int_0^{nT} x(\tau)\,d\tau.$$

Using the trapezoidal rule we integrate the integral on the right-hand side of Eq. (2.83) and arrive at (with $y(0) = 0$)

$$y(nT) = y((n-1)T) + \frac{T}{2}\left\{x((n-1)T) + x(nT)\right\}.$$

Denoting $y(nT) = y[n]$ and $x(nT) = x[n]$ we rewrite the above as

$$y[n] = y[n-1] + \frac{T}{2}\left\{x[n] + x[n-1]\right\}, \qquad (2.84)$$

which is recognized as the difference equation representation of a first-order IIR discrete-time system.

Classification Based on the Output Calculation Process.   If the output sample can be calculated sequentially, knowing only the present and past input samples, the filter is said to be a *nonrecursive* discrete-time system. If, on the other hand, the computation of the output involves past output samples in addition to the present and past input samples, it is known as a *recursive* discrete-time system. An example of a nonrecursive system is the FIR discrete-time system implemented using Eq. (2.82). The IIR discrete-time system implemented using the difference equation of Eq. (2.79) is an example of the recursive system. This equation permits the recursive computation of the output response beginning at some instant $n = n_0$ and progressively for higher values of n provided the initial conditions $y[n_0 - 1]$ through $y[n_0 - N]$ are known. However, it is possible to implement an FIR system using a recursive computational scheme and an IIR system using a nonrecursive computational scheme. The former case is illustrated in the example below.

Example 2.32 Consider the FIR discrete-time system given by the impulse response

$$\{h[n]\} = \{1, 1, 1, 1, 1, 1\}.$$
$$\uparrow$$

Substituting the above in Eq. (2.82) we get the input-output relation

$$y[n] = x[n] + x[n-1] + x[n-2] + x[n-3] + x[n-4] + x[n-5]. \qquad (2.85)$$

To develop the recursive version of this filter, we note from the above

$$y[n-1] = x[n-1] + x[n-2] + x[n-3] + x[n-4] + x[n-5] + x[n-6],$$

which can be rewritten as

$$y[n-1] - x[n-6] = x[n-1] + x[n-2] + x[n-3] + x[n-4] + x[n-5].$$

Substituting the above in Eq. (2.85) we arrive at

$$y[n] = x[n] - x[n-6] + y[n-1], \qquad (2.86)$$

which is a difference equation of the form of Eq. (2.79).

(a) (b)

Figure 2.30   Representations of an LTI IIR discrete-time system.

Classification Based on the Coefficients.   A third classification scheme is based on the real or complex nature of the impulse response sequence. Thus, a discrete-time system with a real impulse response is defined as a *real discrete-time system*. Likewise, for a *complex discrete-time system*, the impulse response is a complex sequence.

2.6   State-Space Representation of LTI Discrete-Time Systems

An alternative time-domain representation of a causal LTI discrete-time system is by means of the state-space equations. They can be obtained by reducing the N-th order difference equation to a system of N first-order equations.

2.6.1   Development of the State Equations

Without any loss of generality, consider a fourth-order system characterized by the difference equation

$$y[n] = p_0 x[n] + p_1 x[n-1] + p_2 x[n-2] + p_3 x[n-3] + p_4 x[n-4]$$
$$-d_1 y[n-1] - d_2 y[n-2] - d_3 y[n-3] - d_4 y[n-4]. \tag{2.87}$$

We can rewrite the above as a set of two equations:

$$v[n] = p_0 x[n] + p_1 x[n-1] + p_2 x[n-2] + p_3 x[n-3] + p_4 x[n-4], \tag{2.88a}$$
$$y[n] = v[n] - d_1 y[n-1] - d_2 y[n-2] - d_3 y[n-3] - d_4 y[n-4], \tag{2.88b}$$

which can be interpreted as the representation of the overall fourth-order discrete-time system as a cascade of two subsystems $\mathcal{N}_1$ and $\mathcal{N}_2$ as indicated in Figure 2.30(a). Without changing the input-output relationship, we can reverse the ordering of the two systems in the cascade representation, as shown in Figure 2.30(b), because of the commutative property of the convolution operation (See Eq. (2.63)).

Now the output $g[n]$ of the subsystem $\mathcal{N}_2$ for an input $x[n]$ can be calculated using Eq. (2.88b) and is given by

$$g[n] = x[n] - d_1 g[n-1] - d_2 g[n-2] - d_3 g[n-3] - d_4 g[n-4]. \tag{2.89}$$

We define a set of new variables:

$$\begin{aligned} s_1[n] &= g[n-4], \\ s_2[n] &= g[n-3], \\ s_3[n] &= g[n-2], \\ s_4[n] &= g[n-1]. \end{aligned} \tag{2.90}$$

It follows from the above that

$$
\begin{aligned}
s_1[n+1] &= g[n-3] = s_2[n], \\
s_2[n+1] &= g[n-2] = s_3[n], \\
s_3[n+1] &= g[n-1] = s_4[n], \\
s_4[n+1] &= g[n] = x[n] - d_4 s_1[n] - d_3 s_2[n] - d_2 s_3[n] - d_1 s_4[n].
\end{aligned}
\tag{2.91}
$$

We can write the above in matrix form as

$$
\begin{bmatrix} s_1[n+1] \\ s_2[n+1] \\ s_3[n+1] \\ s_4[n+1] \end{bmatrix}
=
\begin{bmatrix} 0 & 1 & 0 & 0 \\ 0 & 0 & 1 & 0 \\ 0 & 0 & 0 & 1 \\ -d_4 & -d_3 & -d_2 & -d_1 \end{bmatrix}
\begin{bmatrix} s_1[n] \\ s_2[n] \\ s_3[n] \\ s_4[n] \end{bmatrix}
+
\begin{bmatrix} 0 \\ 0 \\ 0 \\ 1 \end{bmatrix}
x[n].
\tag{2.92}
$$

If $g[n]$ is now fed as an input to the subfilter $\mathcal{N}_1$, its output $y[n]$, from Eq. (2.88a), is given by

$$
\begin{aligned}
y[n] &= p_0 g[n] + p_1 g[n-1] + p_2 g[n-2] + p_3 g[n-3] + p_4 g[n-4] \\
&= p_0 \{ x[n] - d_4 s_1[n] - d_3 s_2[n] - d_2 s_3[n] - d_1 s_4[n] \} \\
&\quad + p_1 s_4[n] + p_2 s_3[n] + p_3 s_2[n] + p_4 s_1[n] \\
&= (p_4 - p_0 d_4) s_1[n] + (p_3 - p_0 d_3) s_2[n] + (p_2 - p_0 d_2) s_3[n] \\
&\quad + (p_1 - p_0 d_1) s_4[n] + p_0 x[n],
\end{aligned}
\tag{2.93}
$$

which can be written in matrix form as:

$$
y[n] = \begin{bmatrix} p_4 - p_0 d_4 & p_3 - p_0 d_3 & p_2 - p_0 d_2 & p_1 - p_0 d_1 \end{bmatrix}
\begin{bmatrix} s_1[n] \\ s_2[n] \\ s_3[n] \\ s_4[n] \end{bmatrix}
+ p_0 x[n].
\tag{2.94}
$$

Denoting[9]

$$
\mathbf{s}[n] = \begin{bmatrix} s_1[n] & s_2[n] & s_3[n] & s_4[n] \end{bmatrix}^T,
\tag{2.95}
$$

$$
\mathbf{A} = \begin{bmatrix} 0 & 1 & 0 & 0 \\ 0 & 0 & 1 & 0 \\ 0 & 0 & 0 & 1 \\ -d_4 & -d_3 & -d_2 & -d_1 \end{bmatrix}, \qquad
\mathbf{B} = \begin{bmatrix} 0 \\ 0 \\ 0 \\ 1 \end{bmatrix},
$$

$$
\mathbf{C} = \begin{bmatrix} p_4 - p_0 d_4 & p_3 - p_0 d_3 & p_2 - p_0 d_2 & p_1 - p_0 d_1 \end{bmatrix}, \qquad D = p_0,
\tag{2.96}
$$

we can write Eqs. (2.92) and (2.94) in a compact form as

$$
\mathbf{s}[n+1] = \mathbf{A}\,\mathbf{s}[n] + \mathbf{B}x[n],
\tag{2.97}
$$

$$
y[n] = \mathbf{C}\,\mathbf{s}[n] + Dx[n].
\tag{2.98}
$$

The above equations are the state-space representation of the fourth-order LTI discrete-time system given by Eq. (2.87), where $\mathbf{s}[n]$, given by Eq. (2.95), is called the *state vector*

---

[9] In this text, row and column vectors, in general, are indicated by boldface lowercase letters, while, the matrices are represented by boldface uppercase letters.

with its elements $s_i[n]$ known as the *state variables*. The matrix $\mathbf{A}$ given in Eq. (2.96) is called the *state-transition matrix*.

The above development can be generalized easily, leading to a representation of a causal N-th order LTI discrete-time system by a system of equations (known as the *state equations*) of the form of Eqs. (2.97) and (2.98), where $\mathbf{s}[n]$ is now a column vector of order N, $\mathbf{A}$ is an $N \times N$ matrix, $\mathbf{B}$ is a column vector of order N, $\mathbf{C}$ is a row vector of order N and D is a scalar.

The *Signal Processing Toolbox* of MATLAB has a number of M-files suitable for state-space-based analysis of discrete-time systems. Two of the functions relevant to this chapter are tf2ss, and ss2tf. The function tf2ss generates the state-space parameter matrices $\mathbf{A}, \mathbf{B}, \mathbf{C}$, and D from the difference equation description of the discrete-time system. The reverse operation is accomplished using the function ss2tf. The following two examples illustrate the application of these two functions.

Example 2.33 In this example we consider the development of the state-space description from the difference equation representation of Eq. (2.80). To this end, Program 2_7 given below can be employed.

```
% Program 2_7
% State-Space Parameter Matrices from
% Difference Equation Representation
%
p = input('Type in the vector p =');
d = input('Type in the vector d =');
[A,B,C,D] = tf2ss(p,d);
disp('A = ');disp(A); disp('B = ');disp(B);
disp('C = ');disp(C); disp('D = ');disp(D);
```

As it is run, the program requests the input data consisting of the p and d vectors, which are entered inside square brackets with both being the same size, as indicated below:

```
p = [0.8   -0.44   0.36   0.02]
d = [1    0.7   -0.45   -0.6]
```

If the number of p coefficients is less than that of the d coefficients, the former must be padded with zeros to make it the same length as the latter. The program then computes the state-space parameter matrices and displays them as indicated below.

```
State-space Parameter Matrices are

A =
    -0.7000   0.4500   0.6000
     1.0000        0        0
          0   1.0000        0

B =
      1
```

```
        0
        0

  C =
      -1.0000   0.7200   0.5000

  D =
       0.8000
```

Example 2.34 Program 2_8 performs the reverse operation, i.e., converts a state-space description to a difference equation representation. The discrete-time system used as an example here is given by

$$\mathbf{A} = \begin{bmatrix} 0.5 & 0.3 \\ 1.75 & -0.16 \end{bmatrix}, \quad \mathbf{B} = \begin{bmatrix} 1 \\ -0.5 \end{bmatrix}, \quad \mathbf{C} = [0 \quad -1], \quad D = 2. \tag{2.99}$$

```
% Program 2_8
% Difference Equation Description from
% State-Space Representation
%
A = input('Type in the matrix A = ');
B = input('Type in the vector B = ');
C = input('Type in the vector C = ');
D = input('Type in the scalar D = ');
[p, d] = ss2tf(A,B,C,D,1);
disp('Difference equation coefficients are');
disp('numerator = ');disp(p);
disp('denominator = ');disp(d);
```

As before, the input data must be entered in the appropriate format inside square brackets. The program then determines the coefficients of the difference equation, as indicated below for the system of Eq. (2.99):

```
Difference equation coefficients are
numerator =
    2.0   -0.18   -3.21

denominator =
    1.0   -0.34   -0.605
```

As a result, the difference equation representation of the system described by Eq. (2.99) is given by

$$y[n] = 0.34y[n-1] + 0.605y[n-2] + 2x[n] - 0.18x[n-1] - 3.21x[n-2].$$

2.6.2 Output Calculation

The first equation in the set of state equations describes the value of the state vector at the next time-instant from the knowledge of the present value of the state vector and the input sample. As a result, knowing the initial value of the state vector at a given instant, say, $n = 0$, and the the input sequence $x[n]$ for $n \geq 0$, we can determine recursively the state values at $n = 1, 2, 3, \ldots$:

$$\mathbf{s}[1] = \mathbf{A}\,\mathbf{s}[0] + \mathbf{B}x[0],$$
$$\mathbf{s}[2] = \mathbf{A}\,\mathbf{s}[1] + \mathbf{B}x[1] = \mathbf{A}^2 s[0] + \mathbf{AB}x[0] + \mathbf{B}x[1],$$
$$\mathbf{s}[3] = \mathbf{A}\,\mathbf{s}[2] + \mathbf{B}x[2] = \mathbf{A}^3 s[0] + \mathbf{A}^2\mathbf{B}x[0] + \mathbf{AB}x[1] + \mathbf{B}x[2],$$
$$\vdots$$

In general,

$$\mathbf{s}[n] = \mathbf{A}^n\mathbf{s}[0] + \sum_{k=0}^{n-1} \mathbf{A}^{n-1-k}\mathbf{B}x[k], \qquad n > 0, \tag{2.100}$$

where $\mathbf{A}^0$ is an $N \times N$ identity matrix. Substituting the above in Eq. (2.98), we arrive at the general expression for the output, given by

$$y[n] = \begin{cases} \mathbf{C}\mathbf{s}[0] + Dx[0], & n = 0, \\ \mathbf{C}\mathbf{A}^n\mathbf{s}[0] + \mathbf{C}\sum_{k=0}^{n-1}\mathbf{A}^{n-1-k}\mathbf{B}x[k] + Dx[n], & n > 0. \end{cases} \tag{2.101}$$

The above system is linear only if $\mathbf{s}[0] = \mathbf{0}$. The system is said to be *initially relaxed* if $\mathbf{s}[0] = \mathbf{0}$.

For an initially relaxed system with a causal input sequence $x[n]$, the output sample $y[n]$ is given by

$$y[n] = \begin{cases} Dx[0], & n = 0, \\ \mathbf{C}\sum_{k=0}^{n-1}\mathbf{A}^{n-1-k}\mathbf{B}x[k] + Dx[n], & n > 0. \end{cases} \tag{2.102}$$

2.6.3 Impulse Response

The impulse response of an initially relaxed LTI discrete-time system can be expressed in terms of the state equation parameters, $\mathbf{A}$, $\mathbf{B}$, $\mathbf{C}$, and D, by letting $x[n] = \delta[n]$ in Eq. (2.102).

For $n = 0$,
$$y[0] = h[0] = D\delta[0] = D = p_0,$$
For $n > 0$,
$$y[n] = h[n] = \mathbf{C}\mathbf{A}^{n-1}\mathbf{B}\delta[0] + \mathbf{C}\mathbf{A}^{n-2}\mathbf{B}\delta[1] + \ldots + D\delta[n]$$
$$= \mathbf{C}\mathbf{A}^{n-1}\mathbf{B}.$$

As a result,

$$h[n] = \begin{cases} D, & \text{for } n = 0, \\ \mathbf{C}\mathbf{A}^{n-1}\mathbf{B}, & \text{for } n > 0. \end{cases} \tag{2.103}$$

2.6.4   Nonuniqueness of State Representation

The state-space representation of a causal LTI discrete-time system is not unique. To show the nonuniqueness of state representation, consider the state-space representation of an N-th order system given by Eqs. (2.97) and (2.98). Define a set of new variables $\hat{s}[n]$ according to

$$\hat{s}[n] = \mathbf{Q}s[n], \tag{2.104}$$

where $\mathbf{Q}$ is a nonsingular $N \times N$ real matrix. Now, using Eq. (2.104) in Eq. (2.97), we obtain

$$\hat{s}[n+1] = \mathbf{Q}s[n+1] = \mathbf{Q}\mathbf{A}s[n] + \mathbf{Q}\mathbf{B}x[n] = \mathbf{Q}\mathbf{A}\mathbf{Q}^{-1}\hat{s}[n] + \mathbf{Q}\mathbf{B}x[n], \tag{2.105}$$

and using Eq. (2.104) in Eq. (2.98), we obtain

$$y[n] = \mathbf{C}s[n] + Dx[n] = \mathbf{C}\mathbf{Q}^{-1}\hat{s}[n] + Dx[n]. \tag{2.106}$$

Denoting

$$\hat{\mathbf{A}} = \mathbf{Q}\mathbf{A}\mathbf{Q}^{-1}, \quad \hat{\mathbf{B}} = \mathbf{Q}\mathbf{B}, \quad \hat{\mathbf{C}} = \mathbf{C}\mathbf{Q}^{-1}, \tag{2.107}$$

we arrive at a new state-space representation given by

$$\hat{s}[n+1] = \hat{\mathbf{A}}\hat{s}[n] + \hat{\mathbf{B}}x[n], \tag{2.108}$$
$$y[n] = \hat{\mathbf{C}}\hat{s}[n] + Dx[n]. \tag{2.109}$$

It can be easily shown that the input-output relation of the LTI discrete-time system characterized by Eqs. (2.108) and (2.109) is identical to that of the system given by Eqs. (2.97) and (2.98). To verify this property, we observe from Eq. (2.100) that the output $y[n]$ of the system defined by the new state equations of Eqs. (2.108) and (2.109) is of the form

$$\begin{aligned}
y[n] &= \hat{\mathbf{C}}\hat{\mathbf{A}}^n\hat{s}[0] + \hat{\mathbf{C}}\sum_{k=0}^{n-1}\hat{\mathbf{A}}^{n-1-k}\hat{\mathbf{B}}x[k] + Dx[n] \\
&= \left(\mathbf{C}\mathbf{Q}^{-1}\right)\left(\mathbf{Q}\mathbf{A}\mathbf{Q}^{-1}\right)^n \mathbf{Q}s[0] \\
&\quad + \left(\mathbf{C}\mathbf{Q}^{-1}\right)\sum_{k=0}^{n-1}\left(\mathbf{Q}\mathbf{A}\mathbf{Q}^{-1}\right)^{n-1-k}(\mathbf{Q}\mathbf{B})x[k] + Dx[n] \\
&= \mathbf{C}\mathbf{A}^n s[0] + \mathbf{C}\sum_{k=0}^{n-1}\mathbf{A}^{n-1-k}\mathbf{B}x[k] + Dx[n], \tag{2.110}
\end{aligned}$$

which is identical to that given in Eq. (2.101) based on the original state-space description. In deriving the last expression in the above equation, we have used the result $\left(\mathbf{Q}\mathbf{A}\mathbf{Q}^{-1}\right)^\ell = \mathbf{Q}\mathbf{A}^\ell\mathbf{Q}^{-1}$.

2.7 Random Signals

The underlying assumption on the discrete-time signals we have considered so far is that they can be uniquely determined by well-defined processes such as a mathematical expression or a rule or a look-up table. Such a signal is usually called a *deterministic signal* since all sample values of the sequence are well defined for all values of the time index. For example, the exponential sequence of Eq. (2.28) and the sinusoidal sequence of Eq. (2.30) are deterministic sequences.

Signals for which each sample value is generated in a random fashion and cannot be predicted ahead of time comprise another class of signals. Such a signal, called a *random signal* or a *stochastic signal*, cannot be reproduced at will, not even using the process generating the signal; and therefore needs to be modeled using statistical information about the signal. Some common examples of a random signal are speech, music, seismic signals, etc. The error signal generated by forming the difference between the ideal sampled version of a continuous-time signal and its quantized version generated by a practical analog-to-digital converter is usually modeled as a random signal for analysis purposes.[10] The noise sequence $d[n]$ of Figure 2.19(b) generated using the rand function of MATLAB is also an example of a random signal.

The discrete-time *random signal* or process consists of a, typically infinite, collection or ensemble of discrete-time sequences $\{X[n]\}$.[11] One particular sequence in this collection $\{x[n]\}$ is called a *realization* of the random process. At a given time index n, the observed sample value $x[n]$ is the value taken by the random variable $X[n]$. Thus, a random process is a family of random variables $\{X[n]\}$. In general, the range of sample values is a continuum. We review in this section the important statistical properties of the random variable and the random process.

2.7.1 Statistical Properties of a Random Variable

The statistical properties of a random variable depend on its probability distribution function or, equivalently, on its probability density function, which are defined next. The probability that the random variable X takes a value in a specified range from $-\infty$ to α is given by its *probability distribution function*

$$P_X(\alpha) = \text{probability}[X \leq \alpha]. \tag{2.111}$$

The *probability density function* of X is defined by

$$p_X(\alpha) = \frac{\partial P_X(\alpha)}{\partial \alpha}, \tag{2.112}$$

if X can assume a continuous range of values. From Eq. (2.112) the probability distribution function is therefore given by

$$P_X(\alpha) = \int_{-\infty}^{\alpha} p_X(u)\,du. \tag{2.113}$$

---

[10] See Section 9.4.1.

[11] Unless otherwise stated, we shall assume the random signal to be a complex discrete-time sequence.

The probability density function satisfies the following two properties:

$$p_X(\alpha) \geq 0, \tag{2.114a}$$

$$\int_{-\infty}^{\infty} p_X(\alpha)\, d\alpha = 1. \tag{2.114b}$$

Likewise, the probability distribution function satisfies the following properties, which follow from Eqs. (2.113) and (2.114a), (2.114b):

$$0 \leq P_X(\alpha) \leq 1, \tag{2.115a}$$

$$P_X(\alpha_1) \leq P_X(\alpha_2) \qquad \text{for all } \alpha_2 \geq \alpha_1, \tag{2.115b}$$

$$P_X(-\infty) = 0, \qquad P_X(+\infty) = 1, \tag{2.115c}$$

$$\text{Probability}[\alpha_1 < X \leq \alpha_2] = P_X(\alpha_2) - P_X(\alpha_1), \tag{2.115d}$$

A random variable is characterized by a number of statistical properties. For example, the r-th *moments* are defined by

$$\mu_r = E(X^r) = \int_{-\infty}^{\infty} \alpha^r p_X(\alpha)\, d\alpha, \tag{2.116}$$

where r is any nonnegative integer and $E(\cdot)$ denotes the expectation operator. A random variable is completely characterized by all its moments. In most cases, all such moments are not known a priori or are difficult to evaluate. Three more commonly used statistical properties characterizing a random variable are the *mean* or *expected value* m_X, the *mean-square value* $E(X^2)$, and the *variance* σ_X^2 as defined below:

$$m_X = E(X) = \int_{-\infty}^{\infty} \alpha p_X(\alpha)\, d\alpha, \tag{2.117a}$$

$$E(X^2) = \int_{-\infty}^{\infty} \alpha^2 p_X(\alpha)\, d\alpha, \tag{2.117b}$$

$$\sigma_X^2 = E\left([X - m_X]^2\right) = \int_{-\infty}^{\infty} (\alpha - m_X)^2 p_X(\alpha) d\alpha. \tag{2.117c}$$

These three properties provide adequate information about a random variable in most practical cases. It can be easily shown that

$$\sigma_X^2 = E(X^2) - (m_X)^2. \tag{2.118}$$

The square root of the variance, σ_X, is called the *standard deviation* of the random variable X. It follows from Eq. (2.118) that the variance and the mean-square value are equal for a random variable with zero mean.

It can be shown that the mean value m_X is the best constant representing a random variable X in a minimum mean-squared error sense, i.e. $E([X - \kappa]^2)$ is a minimum for $\kappa = m_X$ and the minimum mean-square error is given by its variance σ_X^2 (Problem 2.64). This implies that if the variance is small then the value assumed by X likely to be close to m_X and if the variance is large the value assumed by X is likely to be far from m_X.

We illustrate the concepts introduced so far by means of an example.

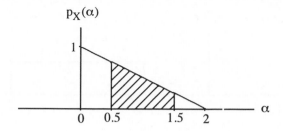

Figure 2.31 Probability density function of Eq. (2.119).

Example 2.35 Let a random variable X be characterized by a probabality density function

$$p_X(\alpha) = \begin{cases} 1 - \frac{\alpha}{2}, & \text{for } 0 \leq \alpha \leq 2, \\ 0, & \text{otherwise,} \end{cases} \qquad (2.119)$$

as shown in Figure 2.31. Its probability distribution function is thus given by

$$P_X(\alpha) = \int_{-\infty}^{\alpha} (1 - \frac{u}{2})\, du = \int_{0}^{\alpha} (1 - \frac{u}{2})\, du = \alpha - \frac{\alpha^2}{4}, \qquad \text{for } 0 < \alpha \leq 2. \qquad (2.120)$$

From Eqs. (2.115d) and (2.120) we can compute the probability that X is in a specified range. For example, the probability that X is in the range $0.5 < X \leq 1.5$ is given by

$$\text{Probability } [0.5 < X \leq 1.5] = P_X(1.5) - P_X(0.5) = \frac{15}{16} - \frac{7}{16} = \frac{1}{2},$$

which is the area of the shaded portion in Figure 2.31.

From Eq. (2.117a) the mean value is obtained as

$$m_X = \int_{0}^{2} \alpha(1 - \frac{\alpha}{2})\, d\alpha = \frac{2}{3},$$

and from Eq. (2.117b) the mean-square value is computed as

$$E(X^2) = \int_{0}^{2} \alpha^2(1 - \frac{\alpha}{2})\, d\alpha = \frac{2}{3}.$$

Substituting the above results in Eq. (2.118) we arrive at the variance given by

$$\sigma_X^2 = E(X^2) - (m_X)^2 = \frac{2}{9}.$$

Two probability density functions, commonly encountered in digital signal processing applications, are the *uniform density function* defined by

$$p_X(\alpha) = \begin{cases} \frac{1}{b-a}, & \text{for } a \leq \alpha \leq b, \\ 0, & \text{otherwise,} \end{cases} \qquad (2.121)$$

and the *Gaussian density function*, also called the *normal density function*, defined by

$$p_X(\alpha) = \frac{1}{\sigma_X \sqrt{2\pi}} e^{-(\alpha - m_X)^2 / 2\sigma_X^2}, \qquad (2.122)$$

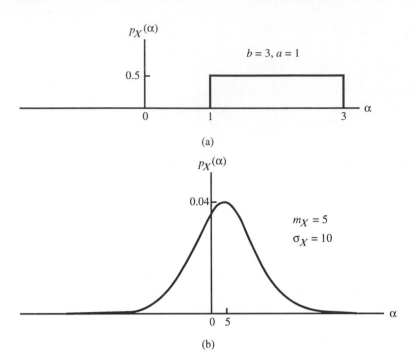

Figure 2.32   (a) Uniform and (b) Gaussian probability density functions.

where the parameters m_X and σ_X are, respectively, the mean value and the standard deviation of X and lie in the range $-\infty < m_X < \infty$ and $\sigma_X > 0$. These density functions are plotted in Figure 2.32. Various other density functions are defined in the literature (Problem 2.59).

Example 2.36 Determine the mean and the variance of an uniformly distributed random variable X defined by Eq. (2.121).
 From Eqs. (2.117a) and (2.117b), we arrive at

$$m_X = \frac{1}{b-a}\int_a^b \alpha \, d\alpha = \frac{b+a}{2}, \tag{2.123}$$

and

$$E\left(X^2\right) = \frac{1}{b-a}\int_a^b \alpha^2 \, d\alpha = \frac{b^2+a^2+ab}{3}.$$

Substituting the above values in Eq. (2.118), we obtain

$$\sigma_X^2 = \frac{(b-a)^2}{12}. \tag{2.124}$$

 In the case of two random variables X and Y their joint statistical properties as well as their individual statistical properties are of practical interest. The probability that X takes a value in a specified range from $-\infty$ to α and that Y takes a value in a specified range from $-\infty$ to β is given by their *joint probability distribution function*

$$P_{XY}(\alpha, \beta) = \text{Probability } [X \le \alpha, Y \le \beta] \tag{2.125}$$

or, equivalently, by their *joint probability density function*

$$p_{XY}(\alpha, \beta) = \frac{\partial^2 P_{XY}(\alpha, \beta)}{\partial \alpha\, \partial \beta}. \tag{2.126}$$

The joint probability distribution function is thus given by

$$P_{XY}(\alpha, \beta) = \int_{-\infty}^{\alpha} \int_{-\infty}^{\beta} p_{XY}(u, v)\, du\, dv. \tag{2.127}$$

The joint probability density function satisfies the following two properties:

$$p_{XY}(\alpha, \beta) \geq 0, \tag{2.128a}$$

$$\int_{-\infty}^{\infty} \int_{-\infty}^{\infty} p_{XY}(\alpha, \beta)\, d\alpha\, d\beta = 1. \tag{2.128b}$$

The joint probability distribution function satisfies the following properties, which are a direct consequence of Eqs. (2.127) and (2.128a), (2.128b):

$$0 \leq P_{XY}(\alpha, \beta) \leq 1, \tag{2.129a}$$

$$P_{XY}(\alpha_1, \beta_1) \leq P_{XY}(\alpha_2, \beta_2) \quad \text{for } \alpha_2 \geq \alpha_1 \text{ and } \beta_2 \geq \beta_1, \tag{2.129b}$$

$$P_{XY}(-\infty, -\infty) = 0, \quad P_{XY}(+\infty, +\infty) = 1, \tag{2.129c}$$

The joint statistical properties of two random variables X and Y are described by their cross-correlation and cross-covariance, as defined by

$$\phi_{XY} = E(XY) = \int_{-\infty}^{\infty} \int_{-\infty}^{\infty} \alpha\beta p_{X,Y}(\alpha, \beta)\, d\alpha\, d\beta, \tag{2.130}$$

$$\gamma_{XY} = E([X - m_X][Y - m_Y])$$

$$= \int_{-\infty}^{\infty} \int_{-\infty}^{\infty} (\alpha - m_X)(\beta - m_Y) p_{X,Y}(\alpha, \beta)\, d\alpha d\beta$$

$$= \phi_{XY} - m_X m_Y, \tag{2.131}$$

where m_X and m_Y are, respectively, the mean of the random variables X and Y. The two random variables X and Y are said to be *linearly independent or uncorrelated* if

$$E(XY) = E(X)E(Y), \tag{2.132a}$$

and *statistically independent* if

$$P_{X,Y}(\alpha, \beta) = P_X(\alpha) P_Y(\beta). \tag{2.132b}$$

It can be shown that if the random variables X and Y are statistically independent then they are also linearly independent (Problem 2.60). However, if X and Y are linearly independent they may not be statistically independent.

The statistical independence property makes it easier to compute the statistical properties of a random variable that is a function of several independent random variables. For example, if X and Y are statistically independent random variables with means m_X and m_Y, respectively, then it can be shown that the mean of the random variable $V = aX + bY$, where a and b are constants, is given by $m_V = am_X + bm_Y$. Likewise, if the variances of X and Y are σ_X^2 and σ_Y^2, respectively, the variance of V is given by $\sigma_V^2 = a^2\sigma_X^2 + b^2\sigma_Y^2$ (Problem 2.65).

Figure 2.33   Range of the random variables with the joint probability density function of Eq. (2.133).

Example 2.37 Consider the two random variables X and Y described by a uniformly distributed joint probability density function given by

$$p_{X,Y}(\alpha, \beta) = \begin{cases} A, & -1 \le \alpha \le 1, 0 \le \beta \le 1, \\ 0 & \text{otherwise.} \end{cases} \qquad (2.133)$$

Determine the value of the constant A, and then compute the probability that X and Y lie in the range $(0.5 \le \alpha \le 1, 0 \le \beta \le 1)$ shown by the shaded region in Figure 2.33.

Invoking the property of Eq. (2.128b), we obtain

$$A \int_{-1}^{1} \int_{0}^{1} d\alpha \, d\beta = A \left[\int_{-1}^{1} d\alpha \right] \left[\int_{0}^{1} d\beta \right] = 2A = 1,$$

and hence, $A = 1/2$. Next, applying Eq. (2.127), we get

$$\text{Probability} \, [0.5 \le X \le 1, 0 \le Y \le 1] = \frac{1}{2} \int_{0}^{1} \int_{0.5}^{1} d\alpha \, d\beta = \frac{1}{4}.$$

2.7.2   Statistical Properties of a Random Signal

As indicated earlier, the random discrete-time signal is a sequence of random variables and consists of a typically infinite, collection or ensemble of discrete-time sequences. Figure 2.34 shows four possible realizations of a random sinusoidal signal

$$\{X[n]\} = \{A \cos(\omega_0 n + \Phi)\}, \qquad (2.134)$$

where the amplitude A and the phase Φ are statistically independent random variables with uniform probability distribution in the range $0 \le \alpha \le 4$ for the amplitude and in the range $0 \le \phi \le 2\pi$ for the phase.

The statistical properties of the random signal $\{X[n]\}$ at time index n are given by the statistical properties of the random variable $X[n]$. Thus, the *mean* or *expected value* of $\{X[n]\}$ at time index n is given by:

$$m_{X[n]} = E(X[n]) = \int_{-\infty}^{\infty} \alpha p_{X[n]}(\alpha; n) \, d\alpha. \qquad (2.135)$$

The *mean-square value* of $\{X[n]\}$ at time index n is given by

$$E\left(X[n]^2\right) = \int_{-\infty}^{\infty} \alpha^2 p_{X[n]}(\alpha; n) \, d\alpha. \qquad (2.136)$$

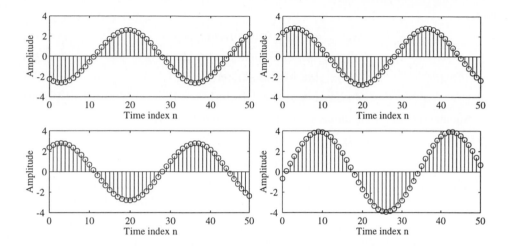

Figure 2.34 Sample realizations of the random sinusoidal signal of Eq. (2.134) for $\omega_0 = 0.03$.

The *variance* $\sigma^2_{X[n]}$ of $\{X[n]\}$ at time index n is defined by

$$\sigma^2_{X[n]} = E\left(\{X[n] - m_{X[n]}\}^2\right) = E\left(X[n]^2\right) - \left(m_{X[n]}\right)^2 \qquad (2.137)$$

In general, the mean, mean-square value, and variance of a random discrete-time signal are functions of the time index n, and can be considered as sequences.

So far we have assumed the random variables and the random signals to be real-valued. It is straightforward to generalize the treatment to complex-valued random variables and random signals. For example, the n-th sample of a complex-valued random signal $\{X[n]\}$ is of the form

$$X[n] = X_{re}[n] + jX_{im}[n], \qquad (2.138)$$

where $\{X_{re}[n]\}$ and $\{X_{im}[n]\}$ are real-valued sequences called the real and imaginary parts of $\{X[n]\}$, respectively. The mean-value of a complex sequence at time index n is thus given by

$$m_{X[n]} = E(X[n]) = E(X_{re}[n]) + jE(X_{im}[n]) = m_{X_{re}[n]} + jm_{X_{im}[n]}. \qquad (2.139)$$

Likewise, the variance $\sigma^2_{X[n]}$ of $\{X[n]\}$ at time index n is given by

$$\sigma^2_{X[n]} = E\left(|X[n] - m_{X[n]}|^2\right) = E\left(|X[n]|^2\right) - \left(|m_{X[n]}|\right)^2 \qquad (2.140)$$

Often, the statistical relation of the samples of a random discrete-time signal at two different time indices m and n is of interest. One such relation is the *autocorrelation*, which for a complex random discrete-time signal $\{X[n]\}$ is defined by

$$\phi_{XX}[m, n] = E\left(X[m]X^*[n]\right), \qquad (2.141)$$

where * denotes complex conjugation. Substituting Eq. (2.138) in Eq. (2.141) we obtain the expression for autocorrelation of $X[n]$:

$$\phi_{XX}[m, n] = \phi_{X_{re}X_{re}}[m, n] + \phi_{X_{im}X_{im}}[m, n]$$
$$-j\phi_{X_{re}X_{im}}[m, n] + j\phi_{X_{im}X_{re}}[m, n], \qquad (2.142)$$

where

$$\phi_{X_{re}X_{re}}[m, n] = E\left(X_{re}[m]X_{re}[n]\right) \tag{2.143a}$$
$$\phi_{X_{im}X_{im}}[m, n] = E\left(X_{im}[m]X_{im}[n]\right), \tag{2.143b}$$
$$\phi_{X_{re}X_{im}}[m, n] = E\left(X_{re}[m]X_{im}[n]\right), \tag{2.143c}$$
$$\phi_{X_{im}X_{re}}[m, n] = E\left(X_{im}[m]X_{re}[n]\right). \tag{2.143d}$$

Another relation is the *autocovariance* of $\{X[n]\}$, defined by

$$\gamma_{XX}[m, n] = E\left((X[m] - m_{X[m]})(X[n] - m_{X[n]})^*\right)$$
$$= \phi_{XX}[m, n] - m_{X[m]}(m_{X[n]})^*. \tag{2.144}$$

As can be seen from the above, both the autocorrelation and the autocovariance are functions of two time indices m and n and can be considered as two-dimensional sequences.

Example 2.38 We compute in this example the statistical properties of the random signal of Eq. (2.134). Now the probability density functions of the amplitude A and the phase Φ are given by

$$p_A(\alpha) = \begin{cases} \frac{1}{4}, & 0 \leq \alpha \leq 4, \\ 0, & \text{otherwise} \end{cases} \tag{2.145}$$

and

$$p_\Phi(\phi) = \begin{cases} \frac{1}{2\pi}, & 0 \leq \phi \leq 2\pi, \\ 0, & \text{otherwise}, \end{cases} \tag{2.146}$$

respectively. Since the two random variables are statistically independent, their joint probability density function is

$$p_{A\Phi}(\alpha, \phi) = \begin{cases} \frac{1}{8\pi}, & 0 \leq \alpha \leq 4, 0 \leq \phi \leq 2\pi, \\ 0, & \text{otherwise}. \end{cases} \tag{2.147}$$

The mean-value of the random process $\{X[n]\}$ is thus

$$m_{X[n]} = \frac{1}{8\pi} \int_0^4 \int_0^{2\pi} \alpha \cos(\omega_0 n + \phi) \, d\alpha \, d\phi$$
$$= \frac{1}{8\pi} \left(\int_0^4 \alpha \, d\alpha\right)\left(\int_0^{2\pi} \cos(\omega_0 n + \phi) \, d\phi\right)$$
$$= \frac{1}{\pi}\left(\sin(\omega_0 n + 2\pi) - \sin(\omega_0 n)\right) = 0. \tag{2.148}$$

The mean-square value is given by

$$E\left(X^2[n]\right) = \frac{1}{8\pi} \int_0^4 \int_0^{2\pi} \alpha^2 \cos^2(\omega_0 n + \phi) \, d\alpha \, d\phi$$
$$= \frac{1}{8\pi}\left(\int_0^4 \alpha^2 \, d\alpha\right)\left(\int_0^{2\pi} \cos^2(\omega_0 n + \phi) \, d\phi\right) = \frac{8}{3}, \tag{2.149}$$

which is also the variance since the random process is of zero mean.

The autocorrelation function is given by

$$\phi_{XX}[m, n] = E\left(X[m]X[n]\right)$$
$$= \frac{1}{8\pi} \int_0^4 \alpha^2 \, d\alpha \int_0^{2\pi} \cos(\omega_0 m + \phi) \cos(\omega_0 n + \phi) \, d\phi$$
$$= \frac{8}{3} \cos\left(\omega_0(m - n)\right). \tag{2.150}$$

The correlation between two different random discrete-time signals $\{X[n]\}$ and $\{Y[n]\}$ is described by the *cross-correlation function*

$$\phi_{XY}[m, n] = E\left(X[m]Y^*[n]\right) = \int_{-\infty}^{\infty} \int_{-\infty}^{\infty} \alpha \beta^* p_{X[m], Y[n]}(\alpha, m, \beta, n) \, d\alpha \, d\beta, \quad (2.151)$$

and the *cross-covariance function*

$$\gamma_{XY}[m, n] = E\left((X[m] - m_{X[m]})(Y[n] - m_{Y[n]})^*\right)$$
$$= \phi_{XY}[m, n] - m_{X[m]}(m_{Y[n]})^*, \quad (2.152)$$

where $p_{X[m], Y[n]}(m, \alpha, n, \beta)$ is the joint probability density of $X[n]$ and $Y[n]$. Both the cross-correlation and the cross-covariance functions can also be considered as two-dimensional sequences. The two random discrete-time signals $\{X[n]\}$ and $\{Y[n]\}$ are uncorrelated if $\gamma_{XY}[m, n] = 0$ for all values of the time indices m and n.

2.7.3 Wide-Sense Stationary Random Signal

In general, the statistical properties of a random discrete-time signal $\{X[n]\}$, such as the mean and variance of the random variable $X[n]$, and the autocorrelation and the autocovariance functions, are time-varying functions. The class of random signals often encountered in digital signal processing applications are the so-called *wide-sense stationary* (WSS) random processes for which some of the key statistical properties are either independent of time or of the time origin. More specifically, for a wide-sense stationary random process $\{X[n]\}$, the mean $E(X[n])$ has the same constant value m_X for all values of the time index n, and the autocorrelation and the autocovariance functions depend only on the difference of the time indices m and n, i.e.,

$$m_X = E(X[n]), \text{ for all } n, \quad (2.153)$$
$$\Phi_{XX}[\ell] = \Phi_{XX}[n + \ell, n] = E(X[n + \ell]X^*[n]), \text{ for all } n \text{ and } \ell, \quad (2.154)$$
$$\gamma_{XX}[\ell] = \gamma_{XX}[n + \ell, n] = E\left((X[n + \ell] - m_X)(X[n] - m_X)^*\right)$$
$$= \phi_{XX}[\ell] - |m_X|^2 \text{ for all } n \text{ and } \ell. \quad (2.155)$$

Note that in the case of a WSS random process, the autocorrelation and the autocovariance functions are one-dimensional sequences.

The mean-square value of a WSS random process $\{X[n]\}$ is given by

$$E\left(|X[n]|^2\right) = \phi_{XX}[0], \quad (2.156)$$

and the variance is given by

$$\sigma_X^2 = \gamma_{XX}[0] = \phi_{XX}[0] - |m_X|^2. \quad (2.157)$$

It follows from Eqs. (2.148) and (2.150) that the random process of Eq. (2.134) is a wide-sense stationary signal.

The cross-correlation and cross-covariance functions between two WSS random processes $\{X[n]\}$ and $\{Y[n]\}$ are given by

$$\phi_{XY}[\ell] = E\left(X[n+\ell]Y^*[n]\right), \tag{2.158}$$

$$\gamma_{XY}[\ell] = E\left((X[n+\ell] - m_X)(Y[n] - m_Y)^*\right)$$
$$= \phi_{XY}[\ell] - m_X(m_Y)^*. \tag{2.159}$$

The symmetry properties satisfied by the autocorrelation, autocovariance, cross-correlation, and the cross-covariance functions are:

$$\phi_{XX}[-\ell] = \phi_{XX}^*[\ell], \tag{2.160a}$$

$$\gamma_{XX}[-\ell] = \gamma_{XX}^*[\ell], \tag{2.160b}$$

$$\phi_{XY}[-\ell] = \phi_{YX}^*[\ell], \tag{2.160c}$$

$$\gamma_{XY}[-\ell] = \gamma_{YX}^*[\ell]. \tag{2.160d}$$

From the above symmetry properties it can be seen that sequences $\phi_{XX}[\ell]$, $\gamma_{XX}[\ell]$, $\phi_{XY}[\ell]$, and $\gamma_{XY}[\ell]$ are always two-sided sequences.

Some additional useful properties concerning these functions are:

$$\phi_{XX}[0]\phi_{YY}[0] \geq |\phi_{XY}[\ell]|^2, \tag{2.161a}$$

$$\gamma_{XX}[0]\gamma_{YY}[0] \geq |\gamma_{XY}[\ell]|^2, \tag{2.161b}$$

$$\phi_{XX}[0] \geq |\phi_{XX}[\ell]|, \tag{2.161c}$$

$$\gamma_{XX}[0] \geq |\gamma_{XX}[\ell]|. \tag{2.161d}$$

A consequence of the above properties is that the autocorrelation and autocovariance functions of a WSS random process assume their maximum values at $\ell = 0$. In addition, it can be shown that, for a WSS signal with nonzero mean, i.e., $m_{X[n]} \neq 0$, and with no periodic components,

$$\lim_{|\ell| \to \infty} \phi_{XX}[\ell] = \left|m_{X[n]}\right|^2. \tag{2.162}$$

If $X[n]$ has a periodic component, then $\phi_{XX}[\ell]$ will contain the same periodic component as illustrated in Example 2.38.

Example 2.39 Determine the mean and variance of a WSS real signal with an autocorrelation function given by

$$\phi_{XX}[\ell] = \frac{10 + 21\ell^2}{1 + 3\ell^2}. \tag{2.163}$$

Applying Eq. (2.162), we obtain $\left|m_{X[n]}\right|^2 = 7$, and hence, the mean-value is $m_{X[n]} = \pm\sqrt{7}$. Next, from Eqs. (2.156) and (2.163) we arrive at the mean-square value as 10 and as a result, the variance is equal to $\sigma_{X[n]}^2 = 10 - 7 = 3$.

2.7.4 Concept of Power in a Random Signal

The average power of a deterministic sequence $x[n]$ was defined earlier and is given by Eq. (2.21). To compute the power associated with a random signal $\{X[n]\}$ we use instead the following definition:

$$\mathcal{P}_X = E\left(\lim_{N\to\infty} \frac{1}{2N+1} \sum_{n=-N}^{N} |X[n]|^2\right). \tag{2.164}$$

In most practical cases, the expectation and summation operators in Eq. (2.164) can be interchanged, resulting in a more easier-to-use expression given by

$$\mathcal{P}_X = \lim_{N\to\infty} \frac{1}{2N+1} \sum_{n=-N}^{N} E\left(|X[n]|^2\right). \tag{2.165}$$

In addition, if the random signal has a constant mean-square value for all values of n, as in the case of a WSS signal, then Eq. (2.165) reduces to

$$\mathcal{P}_X = E\left(|X[n]|^2\right) \tag{2.166}$$

From Eqs. (2.156) and (2.157) it follows that for a WSS signal, the average power is given by

$$\mathcal{P}_X = \phi_{XX}[0] = \sigma_X^2 + |m_X|^2. \tag{2.167}$$

2.7.5   Ergodic Signal

In many practical situations, the random signal of interest cannot be described in terms of a simple analytical expression, as in Eq. (2.134), to permit computations of its statistical properties, which invariably involves the evaluation of definite integrals or summations. Often a finite portion of a single realization of the random signal is available, from which some estimation of the statistical properties of the ensemble must be made. Such an approach can lead to meaningful results if the ergodicity condition is satisfied. More precisely, a stationary random signal is defined to be an *ergodic signal* if all its statistical properties can be estimated from a single realization of sufficiently large finite length.

For an ergodic signal, time averages equal ensemble average derived via the expectation operator in the limit as the length of the realization goes to infinity. For example, for a real ergodic signal we can compute the mean-value, variance, and autocovariance as:

$$m_X = \lim_{M\to\infty} \frac{1}{2M+1} \sum_{n=-M}^{M} x[n], \tag{2.168a}$$

$$\sigma_X^2 = \lim_{M\to\infty} \frac{1}{2M+1} \sum_{n=-M}^{M} (x[n] - m_X)^2, \tag{2.168b}$$

$$\gamma_{XX}[\ell] = \lim_{M\to\infty} \frac{1}{2M+1} \sum_{n=-M}^{M} (x[n] - m_X)(x[n+\ell] - m_X). \tag{2.168c}$$

The limiting operation required to compute the ensemble averages by means of time averages are still not practical in most situations and are therefore replaced with a finite sum to provide an estimate of the desired statistical properties. For example, approximations to Eq. (2.168a)–(2.168c) that are often used are:

$$\hat{m}_X = \frac{1}{M+1} \sum_{n=0}^{M} x[n], \tag{2.169a}$$

$$\hat{\sigma}_X^2 = \frac{1}{M+1} \sum_{n=0}^{M} (x[n] - m_X)^2, \tag{2.169b}$$

$$\hat{\gamma}_{XX}[\ell] = \frac{1}{M+1} \sum_{n=0}^{M} (x[n] - m_X)(x[n+\ell] - m_X). \tag{2.169c}$$

2.8 Summary

In this chapter we introduced some important and fundamental concepts regarding the characterization of discrete-time signals and systems in the time-domain. Certain basic discrete-time signals that play important roles in discrete-time signal processing have been defined, along with basic mathematical operations used for generating more complex signals and systems. The relation between a continuous-time signal and the discrete-time signal generated by sampling the former at uniform time intervals has been examined.

This text deals almost exclusively with linear, time-invariant (LTI) discrete-time systems that find numerous applications in practice. These systems are defined and their convolution sum representation in the time-domain is derived. The concepts of causality and stability of LTI systems are introduced. Also discussed is an important class of LTI systems described by an input-output relation composed of a linear constant coefficient difference equation with zero initial conditions. The LTI discrete-time system is usually classified in terms of the length of its impulse response. The state-space representation of an LTI discrete-time system is considered. Finally, the chapter concludes with a review of the time-domain characterization of a discrete-time random signal in terms of some of its statistical properties.

For further details on discrete-time signals and systems, we refer the reader to the texts by Cadzow [Cad73], Gabel and Roberts [Gab87], Jackson [Jac91], Lathi [Lat87], Oppenheim and Willsky [Opp83], Sturm and Kirk [Stu88], and Ziemer et al [Zie83]. Additional materials on probability theory and statistical properties of random discrete-time signals can be found in Cadzow [Cad87], Papoulis [Pap65], Peebles [Pee87], Stark and Woods [Sta94], and Therrien [The92].

Further insights can often be obtained by considering the frequency-domain representations of discrete-time signals and LTI discrete-time systems. These are discussed in the following two chapters.

2.9   Problems

2.1 Determine the even and odd parts of the following real sequences:

(a) $x[n] = \{3 \quad -4 \quad 2 \quad 0 \quad 6 \quad 3 \quad 9 \quad 5\}$,
$\qquad\qquad\quad \uparrow$

(b) $u[n] = \{-2 \quad 1 \quad 2 \quad 5 \quad 0 \quad 0.1 \quad 6 \quad 3\}$,
$\qquad\qquad\qquad\qquad \uparrow$

(c) $v[n] = A\cos(\omega_0 n) + B\sin(\omega_0 n)$,

(d) $g[n] = n^2$,

(e) $h[n] = n^3$.

2.2 Determine the periodic conjugate symmetric and periodic conjugate antisymmetric parts of the following sequences:

(a) $\{x[n]\} = \{A\alpha^n\}, 0 \le n \le N - 1$, where A and α are complex numbers,

(b) $\{h[n]\} = \{3 + j2 \quad -4 - j5 \quad 2 + j3 \quad 5 + j \quad -6 + j4\}$.
$\qquad\qquad\qquad\qquad\qquad\qquad\qquad \uparrow$

2.3 Which ones of the following sequences are bounded sequences?

(a) $\{x[n]\} = \{A\alpha^n\}$, where A and α are complex numbers, and $|\alpha| < 1$,

(b) $\{y[n]\} = A\alpha^n \mu[n]$, where A and α are complex numbers, and $|\alpha| < 1$,

(c) $\{h[n]\} = C\beta^n \mu[n]$, where C and β are complex numbers, and $|\beta| > 1$,

(d) $\{g[n]\} = 4\sin^2(\omega_a n)$,

(e) $\{v[n]\} = 3\cos(\omega_b n^2)$.

2.4 Show that a causal real sequence $x[n]$ can be fully recovered from its even part $x_{ev}[n]$ for all n, whereas it can be recovered from its odd part $x_{od}[n]$ for all $n > 0$.

2.5 Is it possible to fully recover a causal complex sequence $y[n]$ from its conjugate antisymmetric part $y_{ca}[n]$? Can $y[n]$ be fully recovered from its conjugate symmetric part $y_{cs}[n]$? Justify your answers.

2.6 Show that the even and odd parts of a real sequence are, respectively, even and odd sequences.

2.7 Show that the periodic conjugate symmetric part $x_{pcs}[n]$ and the periodic conjugate antisymmetric part $x_{pca}[n]$ of a length-N sequence $x[n]$, $0 \le n \le N - 1$, as defined in Eqs. (2.13a) and (2.13b) can be alternately expressed as

$$x_{pcs}[n] = x_{cs}[n] + x_{cs}[n - N], \quad 0 \le n \le N - 1 \qquad (2.170a)$$
$$x_{pca}[n] = x_{ca}[n] + x_{ca}[n - N], \quad 0 \le n \le N - 1 \qquad (2.170b)$$

where $x_{cs}[n]$ and $x_{ca}[n]$ are, respectively, the conjugate symmetric and conjugate antisymmetric parts of $x[n]$.

2.8 Show that the periodic conjugate symmetric part $x_{pcs}[n]$ and the periodic conjugate antisymmetric part $x_{pca}[n]$ of a length-N sequence $x[n]$, $1 \leq n \leq N - 1$ as defined in Eqs. (2.13a) and (2.13b) can also be expressed as

$$x_{pcs}[n] = \tfrac{1}{2}(x[n] + x^*[N - n]), \qquad 1 \leq n \leq N - 1, \qquad (2.171a)$$
$$x_{pcs}[0] = \text{Re}(x[0]), \qquad (2.171b)$$
$$x_{pca}[n] = \tfrac{1}{2}(x[n] - x^*[N - n]), \qquad 1 \leq n \leq N - 1 \qquad (2.171c)$$
$$x_{pca}[0] = \text{Im}(x[0]). \qquad (2.171d)$$

2.9 Compute the energy of the length-N sequence

$$x[n] = \cos\left(\frac{2\pi k n}{N}\right), \qquad 0 \leq n \leq N - 1.$$

2.10 Determine the power of the following sequences:

(a) $\mu[n]$,

(b) $A \cos((2\pi n/M) + \phi)$.

2.11 Verify the relation between the unit sample sequence $\delta[n]$ and the unit step sequence $\mu[n]$ given in Eq. (2.27).

2.12 The following sequences represent one period of sinusoidal sequence of the form $x[n] = A \sin(\omega_0 n + \phi)$:

(a) $\{0 \quad 1.5 \quad 0 \quad -1.5\}$,

(b) $\{\sqrt{2} \quad \sqrt{2} \quad -\sqrt{2} \quad -\sqrt{2}\}$,

(c) $\{3 \quad -3\}$,

(d) $\{2 \quad \sqrt{2} \quad 0 \quad -\sqrt{2} \quad -2 \quad -\sqrt{2} \quad 0 \quad \sqrt{2}\}$.

Determine the values of the parameters A, ω_0, and ϕ for each case.

2.13 Determine the fundamental period of the following periodic sequences:

(a) $\tilde{x}_1[n] = e^{j0.25\pi n}$,

(b) $\tilde{x}_2[n] = \cos(0.2\pi n)$,

(c) $\tilde{x}_3[n] = 2\cos(0.1\pi n) + 2\sin(0.2\pi n)$,

(d) $\tilde{x}_4[n] = 3\sin(0.8\pi n) - 4\cos(0.1\pi n)$,

(e) $\tilde{x}_5[n] = 5\sin(0.1\pi n) + 4\sin(0.9\pi n) - \cos(0.8\pi n)$,

(f) $\tilde{x}_6[n] = n$ modulo 5.

2.14 Determine the fundamental period of the sinusoidal sequence $x[n] = A\cos(\omega_0 n)$ for the following values of the angular frequency ω_0:

(a) 0.15π, (b) 0.18π, (c) 0.225π, (d) 0.3π, (e) 0.45π.

2.15 Let $g[n]$ and $h[n]$ be even and odd real sequences, respectively. Which one of the following sequences is even and which one is odd?

(a) $x[n] = g[n]g[n]$,

(b) $u[n] = g[n]h[n]$,

(c) $v[n] = h[n]h[n].$

2.16 Analyze the block diagrams of Figure P2.1 and develop the relation between $y[n]$ and $x[n]$.

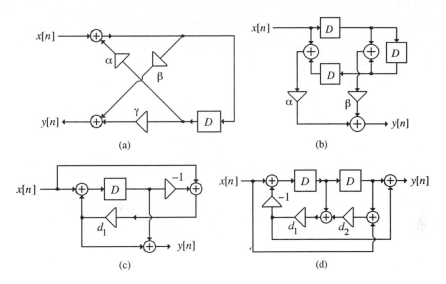

(a)   (b)

(c)   (d)

Figure P2.1

2.17 Express the sequence $x[n]$ of Figure 2.15 as a linear combination of delayed unit step sequences.

2.18 Consider the sequence $w[n]$ shown in Figure P2.2.

  (a) Express $w[n]$ as a weighted sum of delayed unit sample sequences.

  (b) Express $w[n]$ as a weighted sum of delayed unit step sequences.

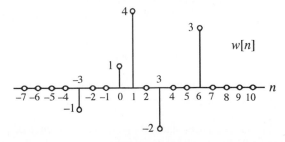

Figure P2.2

2.19 Show that the discrete-time systems described by Eqs. (2.41), (2.42), (2.52), (2.53), (2.55), and (2.56) are linear systems.

2.20 For each the following discrete-time systems, where $y[n]$ and $x[n]$ are, respectively, the output and the input sequences, determine whether or not the system is (1) linear, (2) causal, (3) stable, and (4) shift-invariant:

(a) $y[n] = nx[n]$,

(b) $y[n] = x^3[n]$,

(c) $y[n] = \beta + \sum_{\ell=0}^{5} x[n - \ell]$, β is a nonzero constant,

(d) $y[n] = \beta + \sum_{\ell=-5}^{5} x[n - \ell]$, β is a nonzero constant,

(e) $y[n] = \alpha x[-n]$, α is a nonzero constant,

(f) $y[n] = x[n + 6]$.

2.21 The second derivative $y[n]$ of a sequence $x[n]$ at time instant n is usually approximated by

$$y[n] = x[n + 1] - 2x[n] + x[n - 1].$$

If $y[n]$ and $x[n]$ denote the output and input of a discrete-time system, is the system linear? Is it time-invariant? Is it causal?

2.22 The *median filter* is often used for smoothing of signals corrupted by impulse noise [Reg93]. It is implemented by sliding a window of odd length over the input sequence $x[n]$ one sample at a time. At the nth instant, the input samples inside the window are rank ordered from the largest to the smallest in values, and the sample at the middle is the median value. The output $y[n]$ of the median filter is then given

$$y[n] = \text{med}\{x[n - K], \ldots, x[n - 1], x[n], x[n1], \ldots, x[n + K]\}.$$

For example, med$\{2, -3, 10, 5, -1\} = 2$. Is the median filter a linear or nonlinear discrete-time system? Is it time-invariant? Justify your answer.

2.23 Consider the discrete-time system characterized by the input-output relation [Cad87]

$$y[n] = \frac{1}{2}\left(y[n - 1] + \frac{x[n]}{y[n - 1]}\right), \tag{2.172}$$

where $x[n]$ and $y[n]$ are, respectively, the input and output sequences. Show that the output $y[n]$ of the above system for an input $x[n] = \alpha\mu[n]$ with $y[-1] = 1$ converges to $\sqrt{\alpha}$ as $n \to \infty$ when α is a positive number. Is the above system linear or nonlinear? Is it time-invariant? Justify your answer.

2.24 Develop a general expression for the output $y[n]$ of an LTI discrete-time system in terms of its input $x[n]$ and the unit step response $s[n]$ of the system.

2.25 Determine the linear convolution of the following two sequences: $h[n] = 0.5\delta[n - 2] + \delta[n - 1] + 0.5\delta[n]$, $x[n] = \delta[n - 3] + \delta[n - 2] + \delta[n - 1] + \delta[n]$.

2.26 Determine the following sequences obtained by a linear convolution of a pair of finite-length sequences shown in Figure P2.3:

(a) $y_1[n] = x_1[n] \circledast h_1[n]$,

(b) $y_2[n] = x_2[n] \circledast h_2[n]$,

(c) $y_3[n] = x_1[n] \circledast h_2[n]$,

(d) $y_4[n] = x_2[n] \circledast h_1[n]$.

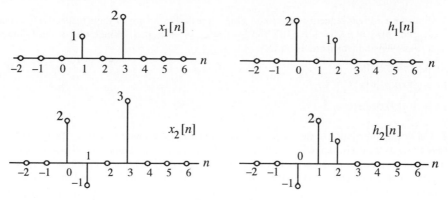

Figure P2.3

2.27 Prove that the convolution sum operation is commutative and distributive.

2.28 Consider the following three sequences:

$$x_1[n] = A \text{ (a constant)}, \qquad x_2[n] = \mu[n], \qquad x_3[n] = \begin{cases} 1, & \text{for } n = 0 \\ -1, & \text{for } n = 1, \\ 0, & \text{otherwise.} \end{cases}$$

Show that $x_3[n] \circledast x_2[n] \circledast x_1[n] \neq x_2[n] \circledast x_3[n] \circledast x_1[n]$.

2.29 Prove that the convolution operation is associative only for stable and single-sided sequences.

2.30 Show that the convolution of a length-M sequence with a length-N sequence leads to a sequence of length $(M + N - 1)$.

2.31 Let $x[n]$ be a length-N sequence given by

$$x[n] = \begin{cases} 1, & 0 \le n \le N - 1, \\ 0, & \text{otherwise.} \end{cases}$$

Determine $y[n] = x[n] \circledast x[n]$ and show that it is a triangular sequence with a maximum sample value of N. Determine the locations of the samples with the following values: $N/4$, $N/2$, and N.

2.32 Let $x[n]$ be a length-N sequence given by

$$x[n] = \begin{cases} 1, & 0 \le n \le \frac{N}{2} - 1, \\ -1, & \frac{N}{2} \le n \le N - 1, \\ 0, & \text{otherwise.} \end{cases}$$

Determine the locations and the values of the largest positive and largest negative samples of the sequence $y[n] = x[n] \circledast x[n]$.

2.33 Consider two real sequences $h[n]$ and $g[n]$ expressed as a sum of their respective even and odd parts, i.e., $h[n] = h_{\text{ev}}[n] + h_{\text{od}}[n]$, and $g[n] = g_{\text{ev}}[n] + g_{\text{od}}[n]$. Which one of the following sequences is even and which one is odd?

(a) $h_{\text{ev}}[n] \circledast g_{\text{ev}}[n]$,

(b) $h_{\text{od}}[n] \circledast g_{\text{ev}}[n]$,

(c) $h_{\text{od}}[n] \circledast g_{\text{od}}[n]$.

2.34 The linear convolution of two sequences $x[n]$ and $h[n]$ is defined by $y[n] = h[n] \circledast x[n]$. The process of recovering $x[n]$ knowing $y[n]$ and $h[n]$ is called *deconvolution*. A recursive algorithm for deconvolution can be developed from a sequence obtained by a convolution of two finite-length sequences. Let the lengths of $x[n]$ and $h[n]$ be N and M, respectively. Show that $x[n]$ can be computed recursively for $0 \le n \le N - 1$ from $h[n]$ and the first N samples of $y[n]$ by a linear constant coefficient difference equation.

2.35 Let $y[n]$ be the sequence obtained by a linear convolution of two causal finite-length sequences $h[n]$ and $x[n]$. For each pair of $y[n]$ and $h[n]$ listed below, determine $x[n]$. The first sample in each sequence is its value at $n = 0$.

(a) $\{y[n]\} = \{2,\ 8,\ 20,\ 40,\ 60,\ 68,\ 62,\ 40\}$, $\{h[n]\} = \{2,\ 4,\ 6,\ 8\}$,

(b) $\{y[n]\} = \{-3,\ 8,\ -20,\ 10,\ -12,\ 12,\ 16,\ -58,\ 35\}$, $\{h[n]\} = \{-1,\ 2,\ -4,\ -4,\ 5\}$,

(c) $\{y[n]\} = \{j5,\ -4 - j6,\ 17 + j24,\ -15 - j,\ 8 - j4\}$, $\{h[n]\} = \{2 + j,\ -3 + j4,\ 2 - j2\}$.

2.36 Consider a causal discrete-time system characterized by a first-order linear, constant coefficient difference equation given by

$$y[n] = ay[n - 1] + bx[n], \quad n \ge 0,$$

where $y[n]$ and $x[n]$ are, respectively, the output and input sequences. Compute the expression for the output sample $y[n]$ in terms of the initial condition $y[-1]$ and the input samples.

(a) Is the system time-invariant if $y[-1] = 1$? Is the system linear if $y[-1] = 1$?

(b) Repeat part (a) if $y[-1] = 0$.

(c) Generalize the results of parts (a) and (b) to the case of an Nth order causal discrete-time system given by Eq. (2.79).

2.37 Develop a difference equation representation of a bank account that permits only a single withdrawal and/or a single deposit on the first day of the month. The deposit can be of any amount, but the withdrawal can only be less than the balance. Denote the account balance at any month as $y[n]$ and the initial deposit as $y[-1]$. The two input data are the amount of withdrawal $x_1[n]$ and the amount of deposit $x_2[n]$.

2.38 The sequence of Fibonacci numbers $f[n]$ is a causal sequence defined by

$$f[n] = f[n - 1] + f[n - 2], \quad n > 2$$

with $f[0] = 0$ and $f[1] = 1$.

(a) Develop an exact formula to calculate $f[n]$ directly for any n.

(b) Show that $f[n]$ is the impulse response of a causal LTI system described by the difference equation [Joh89]

$$y[n] = y[n - 1] + y[n - 2] + x[n - 1].$$

2.39 Let $y[n]$ denote the outstanding principal left after the nth payment $x[n]$ of a loan charging an interest rate compunded at the rate of R dollars per year. Describe the loan payment process by means of a difference equation [Cad73]. If the initial loan is $y[0] = D$ dollars and the yearly payment is made in equal amount of P dollars, determine the numbers of years N needed to repay the loan fully.

2.40 Consider a first-order complex digital filter characterized by a difference equation

$$y[n] = \alpha y[n-1] + x[n],$$

where $x[n]$ is the real input sequence, $y[n] = y_{re}[n] + j y_{im}[n]$ is the complex output sequence with $y_{re}[n]$ and $y_{im}[n]$ denoting its real and imaginary parts, and $\alpha = a + jb$ is a complex constant. Develop an equivalent two-output, single-input real difference equation representation of the above complex digital filter. Show that the single-input, single-output digital filter relating $y_{re}[n]$ to $x[n]$ is described by a second-order difference equation.

2.41 Consider a cascade of two causal stable LTI systems characterized by impulse responses $\alpha^n \mu[n]$ and $\beta^n \mu[n]$, where $0 \le \alpha < 1$ and $0 \le \beta < 1$. Determine the expression for the impulse response $h[n]$ of the cascade.

2.42 A digital filter is designed by cascading two identical FIR digital filters with an impulse response

$$h[n] = \begin{cases} n+1, & 0 \le n \le 3, \\ 0 & \text{otherwise.} \end{cases}$$

(a) Determine the impulse response $g[n]$ of the composite filter.

(b) The composite filter is to be implemented using the convolution sum form, where the filter coefficients are precisely the impulse response samples $g[n]$. However, due to hardware constraints, the filter coefficients $g[n]$ need to be scaled to new values $\hat{g}[n]$ such that $|\hat{g}[n]| \le 1$ without altering the shape of the magnitude response. Determine $\hat{g}[n]$.

2.43 Prove that the BIBO stability condition of Eq. (2.70) also holds for an LTI digital filter with a complex impulse response.

2.44 Prove that the cascade of two stable LTI system is also stable.

2.45 Prove that the cascade of two passive (lossless) LTI system is also passive (lossless).

2.46 Determine the impulse response $g[n]$ of the inverse system of the LTI discrete-time system of Example 2.26.

2.47 Determine the impulse response $g[n]$ characterizing the inverse system of the LTI discrete-time system of Problem 2.36.

2.48 Consider the causal LTI system described by the difference equation

$$y[n] = p_0 x[n] + p_1 x[n-1] - d_1 y[n-1],$$

where $x[n]$ and $y[n]$ denote, respectively, its input and output. Determine the difference equation representation of its inverse system.

2.49 Let $h[0]$, $h[1]$, and $h[2]$ denote the first three impulse response samples of the first-order causal LTI system of Problem 2.48. Show that the coefficients of the difference equation characterizing this system can be uniquely determined from these impulse response samples.

2.50 Let a causal IIR digital filter be described by the difference equation

$$\sum_{k=0}^{N} d_k y[n-k] = \sum_{k=0}^{M} p_k x[n-k], \qquad (2.173)$$

where $y[n]$ and $x[n]$ denote the output and the input sequences, respectively. If $h[n]$ denotes its impulse response, show that

$$p_k = \sum_{n=0}^{k} h[n] d_{k-n}, \qquad k = 0, 1, \ldots, M.$$

From the above result, show that $p_n = h[n] \circledast d_n$.

2.51 Show that the causal IIR filter of Problem 2.48 can be realized as a cascade of an FIR filter with an impulse response $\{p_0, p_1, \ldots, p_M\}$ and the inverse of a causal FIR filter with an impulse response $\{d_0, d_1, \ldots, d_N\}$.

2.52 Consider a causal FIR filter of length $L+1$ with an impulse response given by $\{g[n]\}$, $n = 0, 1, \ldots, L$. Develop the difference equation representation of the form of Eq. (2.173) where $M + N = L$ of a causal finite-dimensional IIR digital filter with an impulse response $\{h[n]\}$ such that $h[n] = g[n]$ for $n = 0, 1, \ldots, L$.

2.53 Determine the expression for the impulse response of each of the LTI systems shown in Figure P2.4.

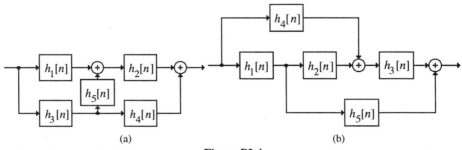

(a)   (b)

Figure P2.4

2.54 Determine the overall impulse response of the system of Figure P2.5, where

$$h_1[n] = \delta[n-1] + 3\delta[n], \qquad h_2[n] = \delta[n-2] + 2\delta[n],$$

$$h_3[n] = 6\delta[n-6] + 7\delta[n-4] - 3\delta[n-1] + \delta[n].$$

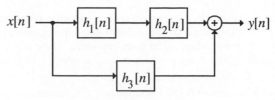

Figure P2.5

2.55 Develop a state space description of a causal LTI system described by the difference equation

$$y[n] + 0.8y[n-1] + 0.6y[n-2] + 0.4y[n-3]$$
$$= 0.4x[n] + 0.6x[n-1] + x[n-2] + 0.8x[n-3].$$

2.56 Develop a difference equation representation of a causal LTI system characterized by the state-space description

$$\mathbf{A} = \begin{bmatrix} 0.5 & 0.4 \\ -0.2 & 0.3 \end{bmatrix}, \ \mathbf{B} = \begin{bmatrix} 1 \\ 3 \end{bmatrix}, \ \mathbf{C} = [2 \quad 3], \ D = 0.4.$$

2.57 Prove Eq. (2.118).

2.58 Let X and Y be two random variables. Show that $E(X+Y) = E(X) + E(Y)$ and $E(cX) = cE(X)$, where c is a constant.

2.59 Compute the mean value and the variance of the random variables with the probability density functions listed below [Par60].

(a) *Cauchy distribution*: $p_X(x) = \frac{\alpha/\pi}{x^2+\alpha^2}$,

(b) *Laplacian distribution*: $p_X(x) = \frac{\alpha}{2}e^{-\alpha|x|}$,

(c) *Binomial distribution*: $p_X(x) = \sum_{\ell=0}^{n} \binom{n}{\ell} p^{\ell}(1-p)^{n-\ell}\delta(x-\ell)$,

(d) *Poisson distribution*: $p_X(x) = \sum_{\ell=0}^{\infty} \frac{e^{-\alpha}\alpha^{\ell}}{\ell!}\delta(x-\ell)$.

(e) *Rayleigh distribution*: $p_X(x) = \frac{x}{\alpha^2}e^{-x^2/2\alpha^2}\mu(x)$.

In above, $\delta(x)$ is the Dirac delta function and $\mu(x)$ is the unit step function.

2.60 Show that if the two random variables X and Y are statistically independent, then they are also linearly independent.

2.61 Prove the symmetry properties of Eqs. (2.160a) through (2.160d).

2.62 Verify the inequalities of Eqs. (2.161a) through (2.161d).

2.63 If there are no periodicities in $\{X[n]\}$, then prove that

$$\lim_{\ell \to \infty} \phi_{XX}[\ell] = |m_X|^2.$$

2.64 Determine the value of the constant κ that minimizes the mean-square error $E([X-\kappa]^2)$, and then find the minimum value of the mean-squared error.

2.65 Let X and Y be two statistically independent random variables with means m_X and m_Y and variances σ_X^2 and σ_Y^2, respectively. Consider the random variable V obtained by a linear combination of X and Y, i.e. $V = aX + bY$, where a and b are constants. Show that the mean m_V and the variance σ_V^2 of V are given by $m_V = am_X + bm_Y$ and $\sigma_V^2 = a^2\sigma_X^2 + b\sigma_Y^2$, respectively.

2.66 The *autocorrelation sequence* $r[n]$ of a deterministic discrete-time sequence $x[n]$ is defined by

$$r[n] = \sum_{\ell=-\infty}^{\infty} x[\ell]x[\ell + n].$$

If $x[n]$ is a periodic or a random sequence, then its autocorrelation sequence is defined by

$$r[n] = \lim_{M\to\infty} \frac{1}{2M + 1} \sum_{\ell=-M}^{M} x[\ell]x[\ell + n]. \tag{2.174}$$

(a) For a periodic sequence $x[n]$ with a period N, show that its autocorrelation function $r[n]$ is also a periodic sequence with a period N.

(b) Show that $r[n]$ is an even function of n, i.e., $r[n] = r[-n]$.

(c) Show that $|r[n]| \le r[0]$ for all values of n.

(d) Show that $r[0]$ is equal to the energy of an aperiodic sequence $x[n]$, i.e., $r[0] = \sum_{\ell=-\infty}^{\infty} x^2[\ell]$ and is equal to the average power of a periodic or a random sequence $x[n]$.

(e) Develop a method for determining the period N of periodic sequence from its autocorrelation sequence.

2.67 Determine and plot the autocorrelation sequence of each of the following sequences. From the computed autocorrelation sequences, find the sequences that are periodic and their periods.

(a) $x_1[n] = \sin(2\pi n/M)$, where M is a positive integer,

(b) $x_2[n] = n$ modulo 7,

(c) $x_3[n] = (-1)^n$.

2.10   MATLAB Exercises

M 2.1 Write a MATLAB program to generate the following sequences and plot them using the function stem: (a) unit sample sequence $\delta[n]$, (b) unit step sequence $\mu[n]$, and (c) ramp sequence $n\mu[n]$. The input parameters specified by the user are the desired length L of the sequence and the sampling frequency F_T in Hz. Using this program generate the first 100 samples of each of the above sequences with a sampling rate of 20 kHz.

M 2.2 The square wave and the sawtooth wave are two periodic sequences as sketched in Figure P2.6. Using the functions sawtooth and square write a MATLAB program to generate the above two sequences and plot them using the function stem. The input data specified by the user are: desired length L of the sequence, peak value A, and the period N. For the square wave sequence an additional user-specified parameter is the duty cycle, which is the percent of the period for which the signal is positive. Using this program generate the first 100 samples of each of the above sequences with a sampling rate of 15 kHz, a peak value of 8, a period of 11, and a duty cycle of 40% for the square wave.

Figure P2.6

M 2.3   (a) Using Program 2_1 generate the sequences shown in Figure 2.9.

(b) Generate and plot the complex exponential sequence $1.5e^{(-0.3+j\pi/3)n}$ for $0 \leq n \leq 50$ using Program 2_1.

M 2.4   (a) Write a MATLAB program to generate a sinusoidal sequence $x[n] = A\cos(\omega_0 n + \phi)$ and plot the sequence using the stem function. The input data specified by the user are the desired length L, amplitude A, the angular frequency ω_0 and the phase ϕ where $0 < \omega_0 < \pi$ and $0 \leq \phi \leq 2\pi$. Using this program generate the sinusoidal sequence shown in Figure 2.11.

(b) Generate sinusoidal sequences with angular frequencies given in Problem 2.14. Determine the period of each sequence from the plot and verify the result theoretically.

M 2.5 Write a MATLAB program to plot a continuous-time sinusoidal signal and its sampled version and verify Figure 2.16. You need to use the hold function to keep both plots.

M 2.6 Using the program developed in the previous problem, verify experimentally that the family of continuous-time sinusoids given by Eq. (2.50) lead to identical sampled signals.

M 2.7 Using Program 2_4 investigate the effect of signal smoothing by a moving average filter of length 4 and higher. Comment on your results.

M 2.8 Write a MATLAB program implementing the discrete-time system of Eq. (2.172) in Problem 2.23 and show that the output $y[n]$ of this system for an input $x[n] = \alpha\mu[n]$ with $y[-1] = 1$ converges to $\sqrt{\alpha}$ as $n \to \infty$.

M 2.9 Using the function impz write a MATLAB program to compute and plot the impulse response of a causal finite-dimensional discrete-time system characterized by a difference equation of the form of Eq. (2.78). The input data to the program are the desired length of the impulse response, and the constants $\{p_k\}$ and $\{d_k\}$ of the difference equation. Generate and plot the first 41 samples of the impulse response of the system of Eq. (2.80).

M 2.10 The deconvolution operation defined in Problem 2.34 can be easily performed using MAT-LAB. To this end the function employed is `deconv`. Solve the deconvolution problems described in Problem 2.35 using MATLAB.

CHAPTER 3

Transform-Domain Representations of Discrete-Time Signals

In Section 2.2.2 we pointed out that any arbitrary sequence can be represented in the time-domain as a weighted linear combination of delayed unit sample sequencess $\{\delta[n-k]\}$. An important consequence of this representation, derived in Section 2.5.1, is the input-output characterization of an LTI digital filter in the time-domain by means of the convolution sum describing the output sequence in terms of a weighted linear combination of its delayed impulse responses. We consider in this chapter an alternate description of a sequence in terms of complex exponential sequences of the form $\{e^{-j\omega n}\}$ and $\{z^{-n}\}$ where z is a complex variable. This leads to three particularly useful representations of discrete-time sequences and LTI discrete-time systems in a transform domain.[1] These transform-domain representations are reviewed here along with the conditions for their existence and their properties. MATLAB has been used extensively to illustrate various concepts and implement a number of useful algorithms. Application of these concepts are discussed in the following chapters.

The first transform-domain representation of a discrete-time sequence we discuss is the discrete-time Fourier transform by which a time-domain sequence is mapped into a continuous function of a frequency variable. Because of the periodicity of the discrete-time Fourier transform, the parent discrete-time sequence can be simply obtained by computing its Fourier series representation. We then show that for a length-N sequence, N equally-spaced samples of its discrete-time Fourier transform are sufficient to describe the frequency-domain representation of the sequence and from these N frequency samples, the original N samples of the discrete-time sequence can be obtained by a simple inverse operation. These N frequency samples constitute the discrete Fourier transform of a length-N sequence, a second transform-domain representation. We next consider a generalization of the discrete-time Fourier transform, called the z-transform, the third type

---

[1] Periodic sequences can be represented in the frequency-domain by means of a discrete-time Fourier series (see Problem 3.20)

of transform-domain representation of a sequence. Finally, the transform-domain representation of a random signal is discussed. Each of these representations is an important tool in signal processing and is used often in practice. A thorough understanding of these three transforms is therefore very important to make best use of the signal processing algorithms discussed in this book.

3.1   The Discrete-Time Fourier Transform

The *discrete-time Fourier transform* (DTFT) or, simply, the Fourier transform of a discrete-time sequence $x[n]$ is a representation of the sequence in terms of the complex exponential sequence $\{e^{-j\omega n}\}$ where ω is the real frequency variable. The DTFT representation of a sequence, if it exists, is unique and the original sequence can be computed from its DTFT by an inverse transform operation. We define the forward transform and derive its inverse transform. We then describe the condition for its existence and summarize its important properties.

3.1.1   Definition

The discrete-time Fourier transform $X(e^{j\omega})$ of a sequence $x[n]$ is defined by

$$X(e^{j\omega}) = \sum_{n=-\infty}^{\infty} x[n]e^{-j\omega n}. \tag{3.1}$$

In general $X(e^{j\omega})$ is a complex function of the real variable ω and can be written as

$$X(e^{j\omega}) = X_{\text{re}}(e^{j\omega}) + jX_{\text{im}}(e^{j\omega}), \tag{3.2}$$

where $X_{\text{re}}(e^{j\omega})$ and $X_{\text{im}}(e^{j\omega})$ are, respectively, the real and imaginary parts of $X(e^{j\omega})$, and are real functions of ω. $X(e^{j\omega})$ can alternately be expressed in the form

$$X(e^{j\omega}) = |X(e^{j\omega})|e^{j\theta(\omega)}, \tag{3.3}$$

where

$$\theta(\omega) = \arg\{X(e^{j\omega})\}. \tag{3.4}$$

The quantity $|X(e^{j\omega})|$ is called the *magnitude function* and the quantity $\theta(\omega)$ is called the *phase function* with both functions again being real functions of ω. In many applications, the Fourier transform is called the *Fourier spectrum* and likewise, $|X(e^{j\omega})|$ and $\theta(\omega)$ are referred to as the *magnitude spectrum* and *phase spectrum*, respectively. The complex conjugate of $X(e^{j\omega})$ is denoted as $X^*(e^{j\omega})$.

Note from Eq. (3.3) that if we replace $\theta(\omega)$ with $\theta(\omega) + 2\pi k$, where k is any integer, $X(e^{j\omega})$ remains unchanged implying that the phase function cannot be uniquely specified for any Fourier transform. Unless otherwise stated, we will assume that the phase function $\theta(\omega)$ is restricted to the following range of values:

$$-\pi \leq \theta(\omega) < \pi,$$

called the *principal value*. As illustrated in Example 3.3, the Fourier transforms of some sequences exhibit discontinuities of 2π in their phase responses. In such cases, it is often

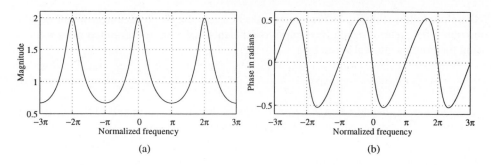

Figure 3.1 Magnitude and phase of $X(e^{j\omega}) = 1/(1 - 0.5e^{-j\omega})$.

useful to consider an alternate type of phase function that is a continuous function of ω derived from the original phase function by removing the discontinuities of 2π. The process of removing the discontinuities is called "*unwrapping the phase*" and the new phase function will be denoted as $\theta_c(\omega)$ with the subscript "c" indicating that it is a continuous function of ω.[2]

Example 3.1 Consider a causal sequence $x[n]$ where

$$x[n] = (0.5)^n \mu[n].$$

Its Fourier transform $X(e^{j\omega})$ obtained using Eq. (3.1) is given by

$$X(e^{j\omega}) = \sum_{n=-\infty}^{\infty} (0.5)^n \mu[n] e^{-j\omega n} = \sum_{n=0}^{\infty} (0.5)^n e^{-j\omega n}$$

$$= \sum_{n=0}^{\infty} (0.5e^{-j\omega})^n = \frac{1}{1 - 0.5e^{-j\omega}}.$$

The magnitude and phase of the above Fourier transform have been plotted in Figure 3.1.

It should be noted here that for most practical discrete-time sequences, their DTFTs can be expressed in terms of a sum of a convergent geometric series which can be summed in a simple closed form as illustrated by the above example. We take up the issue of the convergence of a general DTFT later in this section.

As can be seen from the definition and also from Figure 3.1, the Fourier transform $X(e^{j\omega})$ of a sequence $x[n]$ is a continuous function of ω. It is also a periodic function in ω with a period 2π. To verify this latter property observe that

$$X(e^{j(\omega_1 + 2\pi k)}) = \sum_{n=-\infty}^{\infty} x[n] e^{-j(\omega_1 + 2\pi k)n} = \sum_{n=-\infty}^{\infty} x[n] e^{-j\omega_1 n} e^{-j2\pi kn}$$

$$= \sum_{n=-\infty}^{\infty} x[n] e^{-j\omega_1 n} = X(e^{j\omega_1}).$$

---

[2]In some cases, discontinuities of π may still be present after phase unwrapping (see Table 3.1 for an example).

It therefore follows that Eq. (3.1) represents the Fourier series representation of the periodic function $X(e^{j\omega})$. As a result, the Fourier coefficients $x[n]$ can be computed from $X(e^{j\omega})$ using the Fourier integral given by

$$x[n] = \frac{1}{2\pi} \int_{-\pi}^{\pi} X(e^{j\omega}) e^{j\omega n} \, d\omega, \qquad (3.5)$$

called the *inverse discrete-time Fourier transform*. Equations (3.1) and (3.5) constitute a discrete-time Fourier transform (DTFT) pair for the sequence $x[n]$.

3.1.2   Convergence Condition

Now, an infinite series of the form of Eq. (3.1) may or may not converge. The Fourier transform $X(e^{j\omega})$ of $x[n]$ is said to exist if the series in Eq. (3.1) converges in some sense. It can be shown that if $x[n]$ is an *absolutely summable* sequence, i.e. if

$$\sum_{n=-\infty}^{\infty} |x[n]| < \infty, \qquad (3.6)$$

the series of Eq. (3.1) converges uniformly to a continuous function of ω, and is said to be *absolutely convergent*. Equation (3.6) guarantees that

$$|X(e^{j\omega})| < \infty \qquad \text{for all values of } \omega, \qquad (3.7)$$

and is therefore a sufficient condition for the existence of the Fourier transform (Problem 3.1). Note the sequence $x[n] = (0.5)^n \mu[n]$ of Example 3.1 is absolutely summable as

$$\sum_{n=-\infty}^{\infty} |(0.5)^n| \mu[n] = \sum_{n=0}^{\infty} |(0.5)^n| = \frac{1}{1 - 0.5} = 2,$$

and its Fourier transform $X(e^{j\omega})$ therefore converges to $1/(1 - 0.5e^{-j\omega})$ uniformly.

Since

$$\sum_{n=-\infty}^{\infty} |x[n]|^2 \leq \left(\sum_{n=-\infty}^{\infty} |x[n]| \right)^2,$$

an absolutely summable sequence has always a finite energy. However, a finite energy sequence is not necessarily absolutely summable. The sequence $x_1[n]$ of Example 2.2 is such a sequence. To represent such sequences by a Fourier transform, it is convenient to consider the convergence of the infinite series of Eq. (3.1) in a *mean-square sense*. If we denote

$$X_K(e^{j\omega}) = \sum_{n=-K}^{K} x[n] e^{-j\omega n},$$

then for mean-square convergence of $X(e^{j\omega})$, the total energy of the error $(X(e^{j\omega}) - X_K(e^{j\omega}))$ must approach zero at each value of ω as K goes to ∞, i.e.

$$\lim_{K \to \infty} \int_{-\pi}^{\pi} \left| X(e^{j\omega}) - X_K(e^{j\omega}) \right|^2 d\omega = 0.$$

In such a case, the DTFT is no longer bounded, i.e. Eq. (3.7) does not hold.

Figure 3.2   Unit area pulse function.

Note that the unit step sequence of Eq. (2.26), the complex exponential sequence of Eq. (2.28), and the sinusoidal sequence of Eq. (2.30) are neither absolutely summable nor square-summable. For this type of sequences a Fourier transform representation is possible by using Dirac delta functions. A Dirac delta function $\delta(\omega)$ is a function of ω with infinite height, zero width and unit area. It is the limiting form of an unit area pulse function $p_\Delta(\omega)$ shown in Figure 3.2 as Δ goes to 0, i.e.

$$\delta(\omega) = \lim_{\Delta \to 0} p_\Delta(\omega). \tag{3.8}$$

The Fourier transforms resulting from the use of Dirac delta functions are not continuous and infinitely differentiable functions of ω. Table 3.1 lists the Fourier transforms of some commonly encountered sequences.

A second example illustrating the DTFT computation is considered next.

Example 3.2   Consider the factor-of-L up-sampler of Eq. (2.52) repeated below for convenience:

$$y[n] = \begin{cases} x[n/L], & n = 0, \pm L, \pm 2L, \ldots \\ 0, & \text{otherwise.} \end{cases} \tag{3.9}$$

We would like to develop its input-output relation in the frequency-domain. Applying Eq. (3.1) to Eq. (3.9), we obtain

$$Y(e^{j\omega}) = \sum_{n=\text{mult.of } L}^{\infty} x[n/L]e^{-j\omega n}. \tag{3.10}$$

Substituting $m = n/L$ on the right-hand side of the above equation we arrive at

$$Y(e^{j\omega}) = \sum_{m=-\infty}^{\infty} x[m]e^{-j\omega m L} = X(e^{jL\omega}). \tag{3.11}$$

3.1.3   Properties

There are a number of important properties of the discrete-time Fourier transform which are useful in digital signal processing applications. These are listed here without proof. However, their proofs are quite straight-forward and have been left as exercises. We list the general properties in Table 3.2, and the symmetry properties in Tables 3.3 and 3.4.

Table 3.1   Commonly used Fourier transform pairs.

Sequence	Fourier transform
$\delta[n]$	1
$\mu[n]$	$\dfrac{1}{1 - e^{-j\omega}} + \displaystyle\sum_{k=-\infty}^{\infty} \pi\delta(\omega + 2\pi k)$
$e^{j\omega_0 n}$	$\displaystyle\sum_{k=-\infty}^{\infty} 2\pi\delta(\omega - \omega_0 + 2\pi k)$
$\alpha^n \mu[n], \quad (|\alpha| < 1)$	$\dfrac{1}{1 - \alpha e^{-j\omega}}$

3.1.4   DTFT Computation Using MATLAB

The *Signal Processing Toolbox* in MATLAB includes a number of *M*-file functions to aid in the DTFT based analysis of discrete-time signals. Specifically, the functions that can be used are freqz, abs, angle, and unwrap. In addition, the built-in MATLAB functions real and imag are also useful in some applications.

The function freqz can be used to compute the values of the DTFT of a sequence, described as a rational function in $e^{j\omega}$ in the form

$$X(e^{j\omega}) = \frac{P(e^{j\omega})}{D(e^{j\omega})} = \frac{p_0 + p_1 e^{-j\omega} + \cdots + p_M e^{-j\omega M}}{d_0 + d_1 e^{-j\omega} + \cdots + d_N e^{-j\omega N}}, \qquad (3.12)$$

at a prescribed set of discrete frequency points $\omega = \omega_\ell$. For a reasonably accurate plot, a fairly large number of frequency points should be selected. There are various forms of this function:

```
      H   = freqz(num,den,w),        H   = freqz(num,den,f,FT),
    [H,w] = freqz(num,den,k),   [H,f] = freqz(num,den,k,FT),
    [H,w] = freqz(num,den,k,'whole'),
    [H,f] = freqz(num,den,k,'whole',FT),  freqz(num,den) .
```

The function freqz returns the frequency response values as a vector H of a DTFT defined in terms of the vectors num and den containing the coefficients $\{p_i\}$ and $\{d_i\}$, respectively, at a prescribed set of frequency points. In H = freqz(num,den,w), the prescribed set of frequencies between 0 and 2π are given by the vector w. In H = freqz(num,den,f,FT) the vector f is used to provide the prescribed frequency points whose values must be in the range 0 to FT/2 with FT being the sampling frequency. The total number of frequency points can be specified by k in the argument of freqz. In this case, the DTFT values H are computed at k equally spaced points between 0 and π, and

Table 3.2   General properties of the Fourier transform

Type of property	Length-N Sequence	N-point DFT
	$g[n]$	$G(e^{j\omega})$
	$h[n]$	$H(e^{j\omega})$
Linearity	$\alpha g[n] + \beta h[n]$	$\alpha G(e^{j\omega}) + \beta H(e^{j\omega})$
Time-shifting	$g[n - n_0]$	$e^{-j\omega n_0} G(j\omega)$
Frequency-shifting	$e^{j\omega_0 n} g[n]$	$G\left(e^{j(\omega - \omega_0)}\right)$
Differentiation in frequency	$ng[n]$	$j\dfrac{dG(e^{j\omega})}{d\omega}$
Convolution	$g[n] \circledast h[n]$	$G(e^{j\omega})H(e^{j\omega})$
Modulation	$g[n]h[n]$	$\frac{1}{2\pi}\int_{-\pi}^{\pi} G(e^{j\theta})H(e^{j(\omega-\theta)})\, d\theta$
Parseval's relation	$\displaystyle\sum_{n=-\infty}^{\infty} g[n]h^*[n] = \frac{1}{2\pi}\int_{-\pi}^{\pi} G(e^{j\omega})H^*(e^{j\omega})\, d\omega$	

Table 3.3   Symmetry relations of the Fourier transform of a complex sequence.

Sequence	Fourier Transform
$x[n]$	$X(e^{j\omega})$
$x[-n]$	$X(e^{-j\omega})$
$x^*[-n]$	$X^*(e^{j\omega})$
$\mathrm{Re}\{x[n]\}$	$X_{cs}(e^{j\omega}) = \frac{1}{2}\{X(e^{j\omega}) + X^*(e^{-j\omega})\}$
$j\mathrm{Im}\{x[n]\}$	$X_{ca}(e^{j\omega}) = \frac{1}{2}\{X(e^{j\omega}) - X^*(e^{-j\omega})\}$
$x_{cs}[n]$	$X_{re}(e^{j\omega})$
$x_{ca}[n]$	$jX_{im}(e^{j\omega})$

Note: $X_{cs}(e^{j\omega})$ and $X_{ca}(e^{j\omega})$ are the conjugate-symmetric and conjugate antisymmetric parts of $X(e^{j\omega})$, respectively. Likewise, $x_{cs}[n]$ and $x_{ca}[n]$ are the conjugate symmetric and conjugate antisymmetric parts of $x[n]$, respectively.

Table 3.4   Symmetry relations of the Fourier transform of a real sequence.

Sequence	Fourier transform
$x[n]$	$X(e^{j\omega}) = X_{\text{re}}(e^{j\omega}) + jX_{\text{im}}(e^{j\omega})$
$x_{\text{ev}}[n]$	$X_{\text{re}}(e^{j\omega})$
$x_{\text{od}}[n]$	$jX_{\text{im}}(e^{j\omega})$
Symmetry Relations	$X(e^{j\omega}) = X^*(e^{-j\omega})$ $X_{\text{re}}(e^{j\omega}) = X_{\text{re}}(e^{-j\omega})$ $X_{\text{im}}(e^{j\omega}) = -X_{\text{im}}(e^{-j\omega})$ $|X(e^{j\omega})| = |X(e^{-j\omega})|$ $\arg\{X(e^{j\omega})\} = -\arg\{X(e^{-j\omega})\}$

Note: $x_{\text{ev}}[n]$ and $x_{\text{od}}[n]$ denote the even and odd parts of $x[n]$, respectively.

returned as the output data vector w or computed at k equally spaced points between 0 and FT/2, and returned as the output data vector f. For faster computation, it is recommended that the number k be chosen as a power of 2, such as 256 or 512. By including 'whole' in the argument of freqz, the range of frequencies becomes 0 to 2π or 0 to FT, as the case may be. After the DTFT values have been determined, they can be plotted either showing their real and imaginary parts using the functions real and imag or in terms of their magnitude and phase components using the functions abs and angle. The function angle computes the phase angle in radians. If desired, the phase can be unwrapped using the function unwrap. freqz(num,den) with no output arguments, computes and plots the magnitude and phase response values as a function of frequency in the current figure window.

We illustrate the DTFT computation in the following example.

Example 3.3   Program 3_1 can be employed to determine the values of the DTFT of a real sequence described as a rational function in $e^{-j\omega}$. The input data requested by the program are the number of frequency points k at which the DTFT is to be computed, the vectors num and den containing the coefficients of the numerator, and the denominator of the DTFT given in ascending powers of $e^{-j\omega}$. These vectors must be entered inside square brackets. The program computes the DTFT values at the prescribed frequency points and plots the real and imaginary parts, and the magnitude and phase spectrums. It should be noted that because of the symmetry relations of the DTFT of a real sequence as indicated in see Table 3.3, the DTFT is evaluated only at k equally spaced values of ω between 0 and π. As an example we consider the plotting of the following DTFT:

$$X(e^{j\omega}) = \frac{0.008 - 0.033e^{-j\omega} + 0.05e^{-j2\omega} - 0.033e^{-j3\omega} + 0.008e^{-j4\omega}}{1 + 2.37e^{-j\omega} + 2.7e^{-j2\omega} + 1.6e^{-j3\omega} + 0.41e^{-j4\omega}}.$$

```
% Program 3_1
% Discrete-Time Fourier Transform Computation
%
% Read in the desired length of DFT
k = input('Type in the number of frequency points = ');
% Read in the numerator and denominator coefficients
num = input('Type in the numerator coefficients = ');
den = input('Type in the denominator coefficients = ');
% Compute the frequency response
w = 0:pi/k:pi;
h = freqz(num, den, w);
% Plot the frequency response
subplot(2,2,1)
plot(w/pi,real(h));grid
title('Real part')
xlabel('Normalized angular frequency')
ylabel('Amplitude')
subplot(2,2,2)
plot(w/pi,imag(h));grid
title('Imaginary part')
xlabel('Normalized angular frequency')
ylabel('Amplitude')
subplot(2,2,3)
plot(w/pi,abs(h));grid
title('Magnitude Spectrum')
xlabel('Normalized angular frequency')
ylabel('Magnitude')
subplot(2,2,4)
plot(w/pi,angle(h));grid
title('Phase Spectrum')
xlabel('Normalized angular frequency')
ylabel('Phase, radians')
```

The input data used for the DTFT given above are as follows:

```
k = 256
num = [0.008    -0.033    0.05    -0.033    0.008]
den = [1    2.37    2.7    1.6    0.41].
```

The program first computes the DTFT at the specified 256 discrete frequency points equally spaced between 0 and π. It then computes the real and imaginary parts, and the magnitude and phase of the DTFT at these frequencies, which are plotted as shown in Figure 3.3. As can be seen from this figure, the phase spectrum displays a discontinuity of 2π at around $\omega = 0.72$. This discontinuity can be removed using the function unwrap and the unwrapped phase plotted as shown in Figure 3.4.

Figure 3.3   Plots of the real and imaginary parts, and the magnitude and phase spectrums of the DTFT of Example 3.3.

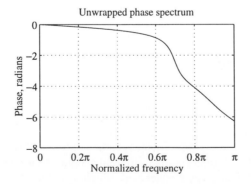

Figure 3.4   Unwrapped phase spectrum of the DTFT of Example 3.3.

3.1.5   Linear Convolution Using DTFT

An important property of the DTFT is given by the convolution theorem in Table 3.2, which states that the DTFT $Y(e^{j\omega})$ of a sequence $y[n]$ generated by the linear convolution of two sequences, $g[n]$ and $h[n]$, is simply given by the product of their respective DTFTs, $G(e^{j\omega})$ and $H(e^{j\omega})$. This implies that the linear convolution $y[n]$ of two sequences, $g[n]$ and $h[n]$, can be implemented by computing first their DTFTs, $G(e^{j\omega})$ and $H(e^{j\omega})$, forming the product $Y(e^{j\omega}) = G(e^{j\omega})H(e^{j\omega})$, and then computing the inverse DTFT of the product. In some applications, particularly in the case of infinite length sequences, this DTFT based approach may be more convenient to carry out than the direct convolution.

3.2   Discrete Fourier Transform

In the case of a finite-length sequence $x[n]$, $0 \le n \le N - 1$, there is a simpler relation between the sequence and its discrete-time Fourier transform $X(e^{j\omega})$. In fact, for a length-N sequence, only N values of $X(e^{j\omega})$, called the *frequency samples,* at N distinct frequency points, $\omega = \omega_k$, $k = 0, 1, \ldots, N - 1$, are sufficient to determine $x[n]$, and hence, $X(e^{j\omega})$, uniquely [Mit92a]. This leads to the concept of discrete Fourier transform, a second transform-domain representation that is applicable only to a finite-length sequence. In this section we define the discrete Fourier transform, usually known as DFT, and develop the inverse transformation, often abbreviated as IDFT. We then summarize its major properties and study especially two of its unique properties. Several important applications of the DFT, such as the numerical computation of the DTFT and implementation of linear convolution, are also discussed here.

3.2.1   Definition

The simplest relation between a finite-length sequence $x[n]$, defined for $0 \le n \le N - 1$, and its DTFT $X(e^{j\omega})$ is obtained by uniformly sampling $X(e^{j\omega})$ on the ω-axis between $0 \le \omega \le 2\pi$ at $\omega_k = 2\pi k/N$, $k = 0, 1, \ldots, N - 1$. From Eq. (3.1),

$$X[k] = X(e^{j\omega})\big|_{\omega=2\pi k/N} = \sum_{n=0}^{N-1} x[n]e^{-2\pi kn/N} \qquad k = 0, 1, \ldots, N - 1. \qquad (3.13)$$

Note that $X[k]$ is also a finite-length sequence in the frequency domain and is of length N. The sequence $X[k]$ is called the *discrete Fourier transform* (DFT) of the sequence $x[n]$.[3] Using the commonly used notation

$$W_N = e^{-j2\pi/N}, \qquad (3.14)$$

we can rewrite Eq. (3.13) as

$$X[k] = \sum_{n=0}^{N-1} x[n]W_N^{kn}, \qquad k = 0, 1, \ldots, N - 1. \qquad (3.15)$$

---

[3] A generalization of the DFT concept is the *nonuniform discrete Fourier transform* (NDFT) obtained by sampling the DTFT at nonuniformly spaced frequency points [Mit92a]. The NDFT is investigated in Problem 3.72.

The *inverse discrete Fourier transform* (IDFT) is given by

$$x[n] = \frac{1}{N} \sum_{k=0}^{N-1} X[k] W_N^{-kn}, \qquad n = 0, 1, \dots, N - 1. \tag{3.16}$$

To verify the above relation we multiply both sides of Eq. (3.16) by $W_N^{\ell n}$ and sum the result from $n = 0$ to $n = N - 1$, resulting in

$$\sum_{n=0}^{N-1} x[n] W_N^{\ell n} = \sum_{n=0}^{N-1} \left(\frac{1}{N} \sum_{k=0}^{N-1} X[k] W_N^{-kn} \right) W_N^{\ell n}$$

$$= \frac{1}{N} \sum_{n=0}^{N-1} \sum_{k=0}^{N-1} X[k] W_N^{-(k-\ell)n}. \tag{3.17}$$

An interchange of the order of summation on the right-hand side of Eq. (3.17) yields

$$\sum_{n=0}^{N-1} x[n] W_N^{\ell n} = \frac{1}{N} \sum_{k=0}^{N-1} \sum_{n=0}^{N-1} X[k] W_N^{-(k-\ell)n}$$

$$= \frac{1}{N} \sum_{k=0}^{N-1} X[k] \left[\sum_{n=0}^{N-1} W_N^{-(k-\ell)n} \right].$$

The right-hand side of above reduces to $X[\ell]$ by virtue of the following identity (Problem 3.16):

$$\sum_{n=0}^{N-1} W_N^{-(k-\ell)n} = \begin{cases} N, & \text{for } k - \ell = rN, r \text{ an integer,} \\ 0, & \text{otherwise.} \end{cases} \tag{3.18}$$

thus verifying Eq. (3.16) is indeed the IDFT of $X[k]$.

Example 3.4 Consider the length-N sequence defined for $n = 0, 1, \dots, N - 1$:

$$x[n] = \begin{cases} 1, & n = 0, \\ 0, & \text{otherwise.} \end{cases} \tag{3.19}$$

Its N-point DFT is obtained by applying Eq. (3.15) resulting in

$$X[k] = 1. \tag{3.20}$$

Now consider the length-N sequence defined for $n = 0, 1, \dots, N - 1$

$$y[n] = \begin{cases} 1, & n = m, \\ 0, & \text{otherwise.} \end{cases} \tag{3.21}$$

Its DFT is given by

$$Y[k] = W_N^{km}. \tag{3.22}$$

Example 3.5 Compute the N-point DFT of the length-N sequence

$$x[n] = \cos(2\pi r n/N), \qquad 0 \le n \le N-1, \ 0 \le r \le N-1. \tag{3.23}$$

Note that using a trigonometric identity and the notation of Eq. (3.14), we can rewrite $x[n]$,

$$x[n] = \tfrac{1}{2}\left(e^{j2\pi rn/N} + e^{-j2\pi rn/N}\right) = \tfrac{1}{2}\left(W_N^{-rn} + W_N^{rn}\right). \tag{3.24}$$

Substituting the above in Eq. (3.15), we arrive at its DFT,

$$X[k] = \frac{1}{2}\left[\sum_{n=0}^{N-1} W_N^{-(r-k)n} + \sum_{n=0}^{N-1} W_N^{(r+k)n}\right]. \tag{3.25}$$

Making use of the identity of Eq. (3.18) in Eq. (3.25) we obtain the DFT of the length-N sequence $x[n]$ of Eq. (3.23):

$$X[k] = \begin{cases} N/2, & \text{for } k = r, \\ N/2, & \text{for } k = N-r, \\ 0, & \text{otherwise.} \end{cases} \tag{3.26}$$

As can be seen from Eqs. (3.15) and (3.16), the computation of the DFT and the IDFT require, respectively, approximately N^2 complex multiplications and $N(N-1)$ complex additions. However, elegant methods have been developed to reduce the computational complexity to about $N(\log_2 N)$ operations. These techniques are usually called fast Fourier transform (FFT) algorithms and are discussed in Section 8.3.2. As a result of the availability of these fast algorithms, the DFT and the IDFT, and their variations, are often used in digital signal processing applications for various purposes.

3.2.2 DTFT from DFT by Interpolation

As indicated by Eq. (3.13), the N-point DFT $X[k]$ of a length-N sequence $x[n]$ is simply the frequency samples of its DTFT $X(e^{j\omega})$ evaluated at N uniformly spaced frequency points, $\omega = \omega_k = 2\pi k/N$ radians, $k = 0, 1, \ldots, N-1$. Given the N-point DFT $X[k]$ of a length-N sequence, it is also possible to determine its DTFT $X(e^{j\omega})$ uniquely. To this end, we first determine $x[n]$ using the IDFT relation of Eq. (3.16) and then compute its DTFT using Eq. (3.1), resulting in

$$X(e^{j\omega}) = \sum_{n=0}^{N-1} x[n]e^{-j\omega n} = \frac{1}{N}\sum_{n=0}^{N-1}\left[\sum_{k=0}^{N-1} X[k]W_N^{-kn}\right]e^{-j\omega n}$$

$$= \frac{1}{N}\sum_{k=0}^{N-1} X[k]\sum_{n=0}^{N-1} e^{j2\pi kn/N}e^{-j\omega n}, \tag{3.27}$$

where we have used Eq. (3.14). Now the right-hand summation in the above expression can be rewritten as

$$\sum_{n=0}^{N-1} e^{-j(\omega-2\pi k/N)n} = \frac{1 - e^{-j(\omega N - 2\pi k)}}{1 - e^{-j[\omega-(2\pi k/N)]}}$$

$$= \frac{e^{-j[(\omega N - 2\pi k)/2]}}{e^{-j[(\omega N - 2\pi k)/2N]}} \cdot \frac{\sin\left(\frac{\omega N - 2\pi k}{2}\right)}{\sin\left(\frac{\omega N - 2\pi k}{2N}\right)}$$

$$= \frac{\sin\left(\frac{\omega N - 2\pi k}{2}\right)}{\sin\left(\frac{\omega N - 2\pi k}{2N}\right)} \cdot e^{-j[\omega - (2\pi k/N)][(N-1)/2]} \qquad (3.28)$$

Substituting Eq. (3.28) in Eq. (3.27) we obtain the desired relation expressing $X(e^{j\omega})$ in terms of $X[k]$:

$$X(e^{j\omega}) = \frac{1}{N} \sum_{k=0}^{N-1} X[k] \frac{\sin\left(\frac{\omega N - 2\pi k}{2}\right)}{\sin\left(\frac{\omega N - 2\pi k}{2N}\right)} \cdot e^{-j[\omega - (2\pi k/N)][(N-1)/2]} \qquad (3.29)$$

It should be noted that the DTFT $X(e^{j\omega})$ of a length-N sequence $x[n]$ can be fully recovered from its M frequency samples $X[k]$ evaluated at uniformly spaced frequency points, $\omega = \omega_k = 2\pi k/M$, $k = 0, 1, \ldots, M - 1$, only if $M \geq N$, but cannot be recovered if $M < N$ (Problem 3.51). As a result, if $x[n]$ is an infinite-length sequence, its DTFT $X(e^{j\omega})$ cannot be obtained from a finite number of its frequency samples.

3.2.3 Matrix Relations

The DFT samples defined in Eq. (3.15) can be expressed in matrix form as

$$\mathbf{X} = \mathbf{D}_N \mathbf{x}, \qquad (3.30)$$

where $\mathbf{X}$ is the vector composed of the N DFT samples,

$$\mathbf{X} = [\, X[0] \quad X[1] \quad \cdots \quad X[N-1] \,]^T, \qquad (3.31)$$

$\mathbf{x}$ is the vector of N input samples,

$$\mathbf{x} = [\, x[0] \quad x[1] \quad \cdots \quad x[N-1] \,]^T, \qquad (3.32)$$

and $\mathbf{D}_N$ is the $N \times N$ *DFT matrix* given by

$$\mathbf{D}_N = \begin{bmatrix} 1 & 1 & 1 & \cdots & 1 \\ 1 & W_N^1 & W_N^2 & \cdots & W_N^{N-1} \\ 1 & W_N^2 & W_N^4 & \cdots & W_N^{2(N-1)} \\ \vdots & \vdots & \vdots & \ddots & \vdots \\ 1 & W_N^{N-1} & W_N^{2(N-1)} & \cdots & W_N^{(N-1)\times(N-1)} \end{bmatrix}. \qquad (3.33)$$

Likewise, the IDFT relations can be expressed in matrix form as

$$\begin{bmatrix} x[0] \\ x[1] \\ \vdots \\ x[N-1] \end{bmatrix} = \mathbf{D}_N^{-1} \begin{bmatrix} X[0] \\ X[1] \\ \vdots \\ X[N-1] \end{bmatrix}, \qquad (3.34)$$

where $\mathbf{D}_N^{-1}$ is the $N \times N$ *IDFT matrix* given by

$$\mathbf{D}_N^{-1} = \frac{1}{N} \begin{bmatrix} 1 & 1 & 1 & \cdots & 1 \\ 1 & W_N^{-1} & W_N^{-2} & \cdots & W_N^{-(N-1)} \\ 1 & W_N^{-2} & W_N^{-4} & \cdots & W_N^{-2(N-1)} \\ \vdots & \vdots & \vdots & \ddots & \vdots \\ 1 & W_N^{-(N-1)} & W_N^{-2(N-1)} & \cdots & W_N^{-(N-1) \times (N-1)} \end{bmatrix}. \tag{3.35}$$

It follows from Eqs. (3.33) and (3.35) that

$$\mathbf{D}_N^{-1} = \frac{1}{N} \mathbf{D}_N^*. \tag{3.36}$$

3.2.4 Numerical Computation of the DTFT Using the DFT

The DFT provides a practical approach to the numerical computation of the DTFT of a finite-length sequence, particularly if fast algorithms are available for the computation of the DFT. Let $X(e^{j\omega})$ be the DTFT of a length-N sequence $x[n]$. We wish to evaluate $X(e^{j\omega})$ at a dense grid of frequencies $\omega_k = 2\pi k/M$, $k = 0, 1, \ldots, M-1$, where $M >> N$:

$$X(e^{j\omega_k}) = \sum_{n=0}^{N-1} x[n] e^{-j\omega_k n} \doteq \sum_{n=0}^{N-1} x[n] e^{-j2\pi kn/M}, \tag{3.37}$$

Define a new sequence $x_e[n]$ obtained from $x[n]$ by augmenting with $M - N$ zero-valued samples:

$$x_e[n] = \begin{cases} x[n], & 0 \leq n \leq N-1, \\ 0, & N \leq n \leq M-1. \end{cases} \tag{3.38}$$

Making use of $x_e[n]$ in Eq. (3.37) we arrive at

$$X(e^{j\omega_k}) = \sum_{n=0}^{M-1} x_e[n] e^{-j2\pi kn/M}, \tag{3.39}$$

which is seen to be an M-point DFT $X_e[k]$ of the length-M sequence $x_e[n]$. The DFT $X_e[k]$ can be computed very efficiently using the FFT algorithm if M is an integer power of 2.

The MATLAB function `freqz` employs the above approach to evaluate the frequency response of a rational DTFT expressed as a rational function in $e^{-j\omega}$ at a prescribed set of discrete frequencies. It computes the DFTs of the numerator and the denominator separately by considering each as finite-length sequences, and then expresses the ratio of the DFT samples at each frequency point to evaluate the DTFT.

3.2.5 DFT Computation Using MATLAB

There are four built-in functions in MATLAB for the computation of the DFT and the IDFT:

```
fft(x),      fft(x,N),      ifft(X),      ifft(X,N)
```

The function fft(x) computes the R-point DFT of a vector x, with R being the length of x. For computing the DFT of a specific length N, the function fft(x,N) is used. Here, if $R > N$, it is truncated to the first N samples, whereas, if $R < N$, the vector x is zero-padded at the end to make it into a length-N sequence. Likewise, the function ifft(X) computes the R-point IDFT of a vector X, where R is the length of X, while ifft(X,N) computes the IDFT of X, with the size N of the IDFT being specified by the user. As before, if $R > N$, it is automatically truncated to the first N samples, whereas, if $R < N$, the DFT vector X is zero-padded at the end by the program to make it into a length-N DFT sequence.

In addition, the function dftmtx(N) in the Signal Processing Toolbox of MATLAB can be used to compute the $N \times N$ DFT matrix $\mathbf{D}_N$ defined in Eq. (3.33). To compute the inverse of the $N \times N$ DFT matrix, one can use the function CONJ(DFTMTX(N))/N.

We illustrate the application of the above M-file functions in the following three examples.

Example 3.6 Using MATLAB we determine the M-point DFT $U[k]$ of the following N-point sequence

$$u[n] = \begin{cases} 1, & 0 \le n \le N-1, \\ 0, & \text{otherwise.} \end{cases} \tag{3.40}$$

To this end we can use Program 3_2 given below. During execution the program requests the input data N and M. To ensure correct DFT values M must be greater than or equal to N. After they are entered, it computes the M-point DFT and plots the original N-point sequence, and the magnitude and phase of the DFT sequence as indicated in Figure 3.5 for $N = 8$ and $M = 16$.

```
% Program 3_2
% Illustration of DFT Computation
%
% Read in the length N of sequence and the desired
% length M of the DFT
N = input('Type in the length of the sequence = ');
M = input('Type in the length of the DFT = ');
%   Generate the length-N time-domain sequence
u = [ones(1,N)];
% Compute its M-point DFT
U = fft(u,M);
% Plot the time-domain sequence and its DFT
t = 0:1:N-1;
stem(t,u)
title('Original time-domain sequence')
xlabel('Time index n'); ylabel('Amplitude')
pause
subplot(2,1,1)
k = 0:1:M-1;
stem(k,abs(U))
title('Magnitude of the DFT samples')
xlabel('Frequency index k'); ylabel('Magnitude')
subplot(2,1,2)
stem(k,angle(U))
title('Phase of the DFT samples')
xlabel('Frequency index k'); ylabel('Phase')
```

Figure 3.5   (a) Original length-N sequence of Eq. (3.40) and (b) its M-point DFT for $N = 8$ and $M = 16$.

Example 3.7 We next illustrate the computation of the IDFT using MATLAB. The K-point DFT sequence is given by

$$V[k] = \begin{cases} k/K, & 0 \le k \le K - 1, \\ 0, & \text{otherwise.} \end{cases} \tag{3.41}$$

We make use of Program 3_3 to determine its IDFT sequence $v[n]$.

```
% Program 3_3
% Illustration of IDFT Computation
%
% Read in the length K of the DFT and the desired
% length N of the IDFT
K = input('Type in the length of the DFT = ');
N = input('Type in the length of the IDFT = ');
%  Generate the length-K DFT sequence
k = 1:K;
U = (k-1)/K;
```

```
%  Compute its N-point IDFT
u = ifft(U,N);
% Plot the DFT and its IDFT
k=1:K;
stem(k-1,U)
xlabel('Frequency index k'); ylabel('Amplitude')
title('Original DFT samples')
pause
subplot(2,1,1)
n = 0:1:N-1;
stem(n,real(u))
title('Real part of the time-domain samples')
xlabel('Time index n'); ylabel('Amplitude')
subplot(2,1,2)
stem(n,imag(u))
title('Imaginary part of the time-domain samples')
xlabel('Time index n'); ylabel('Amplitude')
```

As the program is run, it calls for the input data consisting of the length of the DFT and the length of the IDFT. It then computes the IDFT of the ramp DFT sequence of Eq. (3.41) and plots the original DFT sequence and its IDFT as indicated in Figure 3.6. Note that even though the DFT sequence is real, its IDFT is a complex time-domain sequence as expected.

Example 3.8 Let $r = 3$ and $N = 16$ for the finite-length sequence $x[n]$ of Eq. (3.23). From Eq. (3.26) its 16-point DFT is therefore given by

$$X[k] = \begin{cases} 8, & \text{for } k = 3, \\ 8, & \text{for } k = 13, \\ 0, & \text{otherwise.} \end{cases}$$

We determine the DTFT $X(e^{j\omega})$ of the length-16 $x[n]$ by computing a 512-point DFT using the MATLAB program indicated below.

```
% Program 3_4
% Numerical Computation of DTFT Using DFT
%
% Generate the length-16 sinusoidal sequence
k = 0:15;
x = cos(2*pi*k*3/16);
% Compute its 512-point DFT
X = fft(x);
XE = fft(x,512);
% Plot the frequency response
L = 0:511;
plot(L/512,abs(XE))
hold
plot(k/16,abs(X),'o')
xlabel('Normalized angular frequency')
ylabel('Magnitude')
```

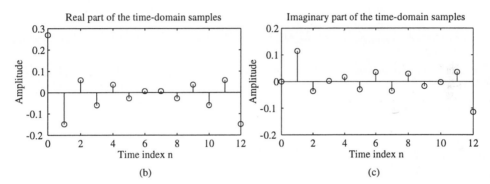

Figure 3.6   (a) Original DFT sequence of length $K = 8$, and (b) its 13-point IDFT.

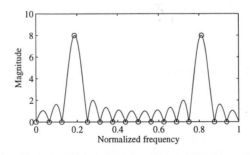

Figure 3.7   The magnitudes of the DTFT $X(e^{j\omega})$ and the DFT $X[k]$ of the sequence $x[n]$ of Eq. (3.26) with $r = 3$ and $N = 16$. The DTFT is plotted as a solid line and the DFT samples are shown by circles.

Figure 3.7 shows the plot of DTFT $X(e^{j\omega})$ along with the DFT samples $X[k]$. As indicated in this figure the DFT values $X[k]$, indicated by circles, are precisely the frequency samples of the DTFT $X(e^{j\omega})$ at $\omega = \pi k/8$, $k = 0, 1, \ldots, 15$.

Table 3.5 General properties of the DFT.

Type of Property	Length-N Sequence	N-point DFT
	$g[n]$	$G[K]$
	$h[n]$	$H[k]$
Linearity	$\alpha g[n] + \beta h[n]$	$\alpha G[k] + \beta H[k]$
Circular Time-shifting	$g[\langle n - n_0\rangle_N]$	$W_N^{kn_0} G[k]$
Circular Frequency-shifting	$W_N^{-k_0 n} g[n]$	$G[\langle k - k_0\rangle_N]$
Duality	$G[n]$	$N[g\langle -k\rangle_N]$
N-point Circular Convolution	$\sum_{m=0}^{N-1} g[m]h[\langle n - m\rangle_N]$	$G[k]H[k]$
Modulation	$g[n]h[n]$	$\frac{1}{N}\sum_{m=0}^{N-1} G[m]H[\langle k - m\rangle_N]$

3.3   Discrete Fourier Transform Properties

Like the DTFT, the DFT also satisfies a number of properties that are useful in signal processing applications. Some of these properties are essentially identical to those of the DTFT, while some others are somewhat different. A summary of the DFT properties are included in Tables 3.5, 3.6, and 3.7. Their proofs are again quite straightforward and have been left as exercises. Most of these properties can also be verified using MATLAB. We discuss next those properties that are different from their counterparts for the DTFT.

3.3.1   Circular Shift of a Sequence

This property is analogous to the time-shifting property of the DTFT as given in Table 3.2, but with a subtle difference. Let us consider length-N sequences defined for $0 \leq n \leq N-1$. Such sequences have sample values equal to zero for $n < 0$ and $n \geq N$. If $x[n]$ is such a sequence then, for any arbitrary integer n_0, the shifted sequence $x_1[n] = x[n - n_0]$ is no longer defined for the range $0 \leq n \leq N - 1$. We therefore need to define another type of a shift that will always keep the shifted sequence in the range $0 \leq n \leq N - 1$. This is achieved by defining a new type of shift, called the *circular shift*, according to[4]

$$x_c[n] = x\left[\langle n - n_0\rangle_N\right]. \tag{3.42}$$

---

[4] $\langle m\rangle_N = m$ modulo N.

Table 3.6   Symmetry properties of the DFT of a complex sequence.

Length-N Sequence	N-point DFT
$x[n]$	$X[K]$
$x^*[n]$	$X^*[\langle -k \rangle_N]$
$x^*[\langle -n \rangle_N]$	$X^*[k]$
$\mathrm{Re}\{x[n]\}$	$X_{\mathrm{pcs}}[k] = \frac{1}{2}\{X[\langle k \rangle_N] + X^*[\langle -k \rangle_N]\}$
$j\,\mathrm{Im}\{x[n]\}$	$X_{\mathrm{pca}}[k] = \frac{1}{2}\{X[\langle k \rangle_N] - X^*[\langle -k \rangle_N]\}$
$x_{\mathrm{pcs}}[n]$	$\mathrm{Re}\{X[k]\}$
$x_{\mathrm{pca}}[n]$	$j\,\mathrm{Im}\{X[k]\}$

Note: $x_{\mathrm{pcs}}[n]$ and $x_{\mathrm{pca}}[n]$ are the periodic conjugate-symmetric and periodic conjugate-antisymmetric parts of $x[n]$, respectively. Likewise, $X_{\mathrm{pcs}}[k]$ and $X_{\mathrm{pca}}[k]$ are the periodic conjugate-symmetric and periodic conjugate-antisymmetric parts of $X[k]$, respectively.

Table 3.7   Symmetry properties of the DFT of a real sequence.

Length-N Sequence	N-point DFT
$x[n]$	$X[k] = \mathrm{Re}\{X[k]\} + j\,\mathrm{Im}\{X[k]\}$
$x_{\mathrm{pe}}[n]$	$\mathrm{Re}\{X[k]\}$
$x_{\mathrm{po}}[n]$	$j\,\mathrm{Im}\{X[k]\}$
Symmetry Relations	$X[k] = X^*[\langle -k \rangle_N]$
	$\mathrm{Re}\,X[k] = \mathrm{Re}\,X[\langle -k \rangle_N]$
	$\mathrm{Im}\,X[k] = -\,\mathrm{Im}\,X[\langle -k \rangle_N]$
	$|X[k]| = |X[\langle -k \rangle_N]|$
	$\arg X[k] = \arg X[\langle -k \rangle_N]$

Note: $x_{\mathrm{pe}}[n]$ and $x_{\mathrm{po}}[n]$ are the periodic even and periodic odd parts of $x[n]$, respectively.

(a)   (b)   (c)

Figure 3.8   Illustration of circular shift of a finite-length sequence (a) $x[n]$, (b) $x[\langle n-1 \rangle_6] = x[\langle n+5 \rangle_6]$, and (c) $x[\langle n-4 \rangle_6] = x[\langle n+2 \rangle_6]$.

For $n_0 > 0$ (right circular shift), above equation implies

$$x_c[n] = \begin{cases} x[n - n_0], & \text{for } n_0 \le n \le N-1, \\ x[N - n_0 + n], & \text{for } 0 \le n < n_0. \end{cases} \tag{3.43}$$

The concept of a circular shift of a finite-length sequence is illustrated in Figure 3.8. Figure 3.8(a) shows a length-6 sequence $x[n]$. Figure 3.8(b) shows its circularly shifted version shifted to the right by 1 sample period, whereas Figure 3.8(c) depicts its circularly shifted version shifted to the right by 4 sample periods.

As can be seen from Figure 3.8(b) and (c), a right circular shift by n_0 is equivalent to a left circular shift by $N - n_0$ sample periods. Moreover, a circular shift by an integer number n_0 greater than N is equivalent to a circular shift by $\langle n_0 \rangle_N$.

If we view the length-N sequence displayed on the circumference of a cylinder at N equally spaced points, then the circular shift operation can be considered as a clockwise or anticlockwise rotation of the sequence by n_0-sample spacings on the cylinder.

3.3.2   Circular Convolution

This property is analogous to the linear convolution of Eq. (2.61), but with a subtle difference. Consider two length-N sequences, $g[n]$ and $h[n]$, respectively. Their linear convolution results in a length-$(2N-1)$ sequence $y_L[n]$ given by

$$y_L[n] = \sum_{m=0}^{N-1} g[m]h[n-m], \qquad n = 0, 1, \ldots 2N - 2, \tag{3.44}$$

where we have assumed that both N-length sequences have been zero-padded to extend their lengths to $2N - 1$.[5] The longer length of $y_L[n]$ results from the time-reversal of the sequence $h[n]$ and its linear shifting to the right. The first nonzero value of $y_L[n]$ is $y_L[0] = g[0]h[0]$, and the last nonzero value of $y_L[n]$ is $y_L[2N-2] = g[N-1]h[N-1]$.

---

[5]Note that the sum of the indices of each sample product inside the summation is equal to the index of the sample being generated by the linear convolution operation.

Figure 3.9   Two length-4 sequences.

To develop a convolution-like operation resulting in a length-N sequence $y_C[n]$, we need to define a circular time-reversal and then apply a circular time-shift. The resulting operation, called a *circular convolution,* is defined below:[6]

$$y_C[n] = \sum_{m=0}^{N-1} g[m]h\left[\langle n - m\rangle_N\right]. \tag{3.45}$$

Since the above operation involves two length-N sequences, it is often referred to as an *N-point circular convolution,* denoted as

$$y_C[n] = g[n]\text{\textcircled{N}}h[n]. \tag{3.46}$$

Like the linear convolution, the circular convolution is commutative (Problem 3.39), i.e.,

$$g[n]\text{\textcircled{N}}h[n] = h[n]\text{\textcircled{N}}g[n]. \tag{3.47}$$

We illustrate the concept of circular convolution through several examples.

Example 3.9   Determine the 4-point circular convolution of the two length-4 sequences $g[n]$ and $h[n]$ given by

$$g[n] \;=\; \{1 \quad 2 \quad 0 \quad 1\}, \quad h[n] \;=\; \{2 \quad 2 \quad 1 \quad 1\}, \tag{3.48}$$

and sketched in Figure 3.9. The result is a length-4 sequence $y_C[n]$ given by

$$y_C[n] = g[n]\text{\textcircled{4}}h[n] = \sum_{m=0}^{3} g[m]h[\langle n - m\rangle_4], \qquad 0 \le n \le 3. \tag{3.49}$$

From above, $y_C[0]$ is given by

$$y_C[0] = \sum_{m=0}^{3} g[m]h[\langle -m\rangle_4]. \tag{3.50}$$

The circularly time-reversed sequence $h[\langle -m\rangle_4]$ is shown in Figure 3.10(a). By forming the products of $g[m]$ with that of $h[\langle -m\rangle_4]$ for each value of m in the range $0 \le m < 3$ and summing the products, we arrive at

$$
\begin{aligned}
y_C[0] &= g[0]h[0] + g[1]h[3] + g[2]h[2] + g[3]h[1] \\
&= (1 \times 2) + (2 \times 1) + (0 \times 1) + (1 \times 2) = 6.
\end{aligned} \tag{3.51}
$$

---

[6]Note that here the sum of the indices of each sample product inside the summation modulo N is equal to the index of the sample being generated by the circular convolution operation.

$$h[<-m>_4]\qquad h[<1-m>_4]\qquad h[<2-m>_4]\qquad h[<3-m>_4]$$

Figure 3.10 (a) $h[\langle -m\rangle_4]$, (b)–(d) circularly shifted versions of $h[\langle -m\rangle_4]$.

Next, from Eq. (3.49) we compute $y_C[1]$ as

$$y_C[1] = \sum_{m=0}^{3} g[m]h[\langle 1 - m\rangle_4]. \tag{3.52}$$

The sequence $h[\langle 1 - m\rangle_4]$ obtained by circularly time-shifting $h[\langle -m\rangle_4]$ to the right by one sample period is shown in Figure 3.10(b). Summing the products $g[m]h[\langle 1 - m\rangle_4]$ for each value of m, we arrive at

$$
\begin{aligned}
y_C[1] &= g[0]h[1] + g[1]h[0] + g[2]h[3] + g[3]h[2]\\
&= (1 \times 2) + (2 \times 2) + (0 \times 1) + (1 \times 1) = 7.
\end{aligned} \tag{3.53}
$$

Continuing this process, we determine the remaining two samples of $y_C[n]$ as:

$$
\begin{aligned}
y_C[2] &= g[0]h[2] + g[1]h[1] + g[2]h[0] + g[3]h[3]\\
&= (1 \times 1) + (2 \times 2) + (0 \times 2) + (1 \times 1) = 6.
\end{aligned} \tag{3.54}
$$

$$
\begin{aligned}
y_C[3] &= g[0]h[3] + g[1]h[2] + g[2]h[1] + g[3]h[0]\\
&= (1 \times 1) + (2 \times 1) + (0 \times 2) + (1 \times 2) = 5.
\end{aligned} \tag{3.55}
$$

The length-4 sequence $y_C[n]$ obtained by a 4-point circular convolution of the two length-4 sequences $g[n]$ and $h[n]$ is sketched in Figure 3.11(a).

The circular convolution of the two length-N sequences can also be computed by forming the product of their N-point DFTs and then applying an N-point IDFT, as indicated in Table 3.5. The process is illustrated in the following example.

Example 3.10 We compute the circular convolution of the two length-4 sequences of Example 3.9 via a DFT-based approach.

Now the 4-point DFT $G[k]$ of the length-4 sequence $g[n]$ of Eq. (3.48) is given by

$$
\begin{aligned}
G[k] &= g[0] + g[1]e^{-j2\pi k/4} + g[2]e^{-j4\pi k/4} + g[3]e^{-j6\pi k/4}\\
&= 1 + 2e^{-j\pi k/2} + e^{-j3\pi k/2}, \qquad k = 0, 1, 2, 3.
\end{aligned} \tag{3.56}
$$

Therefore,

$$
\begin{aligned}
G[0] &= 1 + 2 + 1 = 4,\\
G[1] &= 1 - j2 + j = 1 - j,\\
G[2] &= 1 - 2 - 1 = -2,\\
G[3] &= 1 + j2 - j = 1 + j.
\end{aligned} \tag{3.57}
$$

The above DFT samples can also be computed using the matrix relation of Eq. (3.30):

$$
\begin{bmatrix} G[0] \\ G[1] \\ G[2] \\ G[3] \end{bmatrix} = \mathbf{D}_4 \begin{bmatrix} g[0] \\ g[1] \\ g[2] \\ g[3] \end{bmatrix} = \begin{bmatrix} 1 & 1 & 1 & 1 \\ 1 & -j & -1 & j \\ 1 & -1 & 1 & -1 \\ 1 & j & -1 & -j \end{bmatrix} \begin{bmatrix} 1 \\ 2 \\ 0 \\ 1 \end{bmatrix} = \begin{bmatrix} 4 \\ 1-j \\ -2 \\ 1+j \end{bmatrix}, \tag{3.58}
$$

where $\mathbf{D}_4$ is the 4×4 DFT matrix.

Likewise, the 4-point DFT of the length-4 sequence $h[n]$ of Eq. (3.48) is given by

$$
\begin{bmatrix} H[0] \\ H[1] \\ H[2] \\ H[3] \end{bmatrix} = \mathbf{D}_4 \begin{bmatrix} h[0] \\ h[1] \\ h[2] \\ h[3] \end{bmatrix} = \begin{bmatrix} 1 & 1 & 1 & 1 \\ 1 & -j & -1 & j \\ 1 & -1 & 1 & -1 \\ 1 & j & -1 & -j \end{bmatrix} \begin{bmatrix} 2 \\ 2 \\ 1 \\ 1 \end{bmatrix} = \begin{bmatrix} 6 \\ 1-j \\ 0 \\ 1+j \end{bmatrix}. \tag{3.59}
$$

Applying the 4-point IDFT to the product $Y_C[k] = G[k]H[k]$,

$$
\begin{bmatrix} Y_C[0] \\ Y_C[1] \\ Y_C[2] \\ Y_C[3] \end{bmatrix} = \begin{bmatrix} G[0]H[0] \\ G[1]H[1] \\ G[2]H[2] \\ G[3]H[3] \end{bmatrix} = \begin{bmatrix} 24 \\ -j2 \\ 0 \\ j2 \end{bmatrix}, \tag{3.60}
$$

we arrive at the desired circular convolution result:

$$
\begin{bmatrix} y_C[0] \\ y_C[1] \\ y_C[2] \\ y_C[3] \end{bmatrix} = \frac{1}{4}\mathbf{D}_4^* \begin{bmatrix} 24 \\ -j2 \\ 0 \\ j2 \end{bmatrix} = \frac{1}{4} \begin{bmatrix} 1 & 1 & 1 & 1 \\ 1 & j & -1 & -j \\ 1 & -1 & 1 & -1 \\ 1 & -j & -1 & j \end{bmatrix} \begin{bmatrix} 24 \\ -j2 \\ 0 \\ j2 \end{bmatrix} = \begin{bmatrix} 6 \\ 7 \\ 6 \\ 5 \end{bmatrix}, \tag{3.61}
$$

which is identical to the results given in Eqs. (3.51) to (3.55).

Example 3.11 Now let us extend the two length-4 sequences of Figure 3.8 to length 7 by appending each with three zero-valued samples, i.e., by forming

$$
g_e[n] = \begin{cases} g[n], & 0 \le n \le 3, \\ 0, & 4 \le n \le 6, \end{cases} \tag{3.62}
$$

$$
h_e[n] = \begin{cases} h[n], & 0 \le n \le 3, \\ 0, & 4 \le n \le 6. \end{cases} \tag{3.63}
$$

We next determine the 7-point circular convolution of $g_e[n]$ and $h_e[n]$:

$$
y[n] = \sum_{m=0}^{6} g_e[m]h_e[\langle n-m \rangle_7]. \tag{3.64}
$$

From the above,

$$
y[0] = g_e[0]h_e[0] + g_e[1]h_e[6] + g_e[2]h_e[5] + g_e[3]h_e[4] \\ + g_e[4]h_e[3] + g_e[5]h_e[2] + g_e[6]h_e[1]. \tag{3.65}
$$

Substituting the values from Eqs. (3.62) and (3.63) in Eq. (3.65), we arrive at

$$
y[0] = g[0]h[0] = (1 \times 2) = 2.
$$

Next we compute $y[1]$. From Eq. (3.64),

$$
y[1] = g_e[0]h_e[1] + g_e[1]h_e[0] + g_e[2]h_e[6] + g_e[3]h_e[5] \\ + g_e[4]h_e[4] + g_e[5]h_e[3] + g_e[6]h_e[2], \tag{3.66}
$$

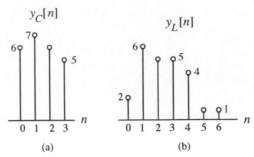

Figure 3.11   Results of convolution of the two sequences of Figure 3.9. (a) Circular convolution, and (b) linear convolution.

which leads to

$$y[1] = g[0]h[1] + g[1]h[0] = (1 \times 2) + (2 \times 2) = 6$$

using Eqs. (3.62) and (3.63). Continuing the above procedure, we arrive at the remaining samples of $y[n]$:

$$y[2] = g[0]h[2] + g[1]h[1] + g[2]h[0] = (1 \times 1) + (2 \times 2) + (0 \times 2) = 5,$$
$$y[3] = g[0]h[3] + g[1]h[2] + g[2]h[1] + g[3]h[0]$$
$$= (1 \times 1) + (2 \times 1) + (0 \times 2) + (1 \times 2) = 5,$$
$$y[4] = g[1]h[3] + g[2]h[2] + g[3]h[0] = (2 \times 1) + (0 \times 1) + (1 \times 2) = 4,$$
$$y[5] = g[2]h[3] + g[3]h[2] = (0 \times 1) + (1 \times 1) = 1,$$
$$y[6] = g[3]h[3] = (1 \times 1) = 1.$$

It is evident from the above that $y[n]$ is precisely the result $y_L[n]$, shown in Figure 3.11(b), obtained by a linear convolution of $g[n]$ and $h[n]$.

The N-point circular convolution operation of Eq. (3.46) can be written in matrix form as

$$
\begin{bmatrix} y_C[0] \\ y_C[1] \\ y_C[2] \\ \vdots \\ y_C[N-1] \end{bmatrix} =
\begin{bmatrix}
h[0] & h[N-1] & h[N-2] & \cdots & h[1] \\
h[1] & h[0] & h[N-1] & \cdots & h[2] \\
h[2] & h[1] & h[0] & \cdots & h[3] \\
\vdots & \vdots & \vdots & \ddots & \vdots \\
h[N-1] & h[N-2] & h[N-3] & \cdots & h[0]
\end{bmatrix}
\begin{bmatrix} g[0] \\ g[1] \\ g[2] \\ \vdots \\ g[N-1] \end{bmatrix}.
$$

$$(3.67)$$

The $N \times N$ matrix of the type shown in Eq. (3.67) is called a *circulant matrix*.

3.4   Computation of the DFT of Real Sequences

In most practical applications, sequences of interest are real. In such cases, the symmetry properties of the DFT given in Table 3.7 can be exploited to make the DFT computations more efficient.

3.4.1   N-Point DFTs of Two Real Sequences Using a Single N-Point DFT

Let $g[n]$ and $h[n]$ be two real sequences of length N each with $G[k]$ and $H[k]$ denoting their respective N-point DFTs. These two N-point DFTs can be computed efficiently using a single N-point DFT $X[k]$ of a complex length-N sequence $x[n]$ defined by

$$x[n] = g[n] + jh[n]. \tag{3.68}$$

From the above, $g[n] = \mathrm{Re}\{x[n]\}$ and $h[n] = \mathrm{Im}\{x[n]\}$.
    From Table 3.6 we arrive at

$$G[k] = \tfrac{1}{2}\left\{X[k] + X^*[\langle -k\rangle_N]\right\}, \tag{3.69}$$

$$H[k] = \tfrac{1}{2j}\left\{X[k] - X^*[\langle -k\rangle_N]\right\}. \tag{3.70}$$

Note that $X^*[\langle -k\rangle_N] = X^*[\langle N - k\rangle_N]$.

Example 3.12   We illustrate the computation of the 4-point DFTs of the two real sequences of Example 3.9 using a single 4-point DFT.
    From Eq. (3.48) the complex sequence $x[n] = g[n] + jh[n]$ is given by

$$x[n] = \{1 + j2 \quad\quad 2 + j2 \quad\quad j \quad\quad 1 + j\}.$$
$$\uparrow$$

Its DFT is then

$$
\begin{bmatrix} X[0] \\ X[1] \\ X[2] \\ X[3] \end{bmatrix} =
\begin{bmatrix} 1 & 1 & 1 & 1 \\ 1 & j & -1 & -j \\ 1 & -1 & 1 & -1 \\ 1 & -j & -1 & j \end{bmatrix}
\begin{bmatrix} 1 + j2 \\ 2 + j2 \\ j \\ 1 + j \end{bmatrix} =
\begin{bmatrix} 4 + j6 \\ 2 \\ -2 \\ j2 \end{bmatrix} \tag{3.71}
$$

From the above,
$$X^*[k] = \{4 - j6 \quad\quad 2 \quad\quad -2 \quad\quad -j2\}.$$

Therefore
$$X^*[\langle 4 - k\rangle_4] = \{4 - j6 \quad\quad -j2 \quad\quad -2 \quad\quad 2\}. \tag{3.72}$$

Substituting Eqs. (3.71) and (3.72) in Eqs. (3.69) and (3.70), we get

$$G[k] = \{4 \quad\quad 1 - j \quad\quad -2 \quad\quad 1 + j\},$$
$$H[k] = \{6 \quad\quad 1 - j \quad\quad 0 \quad\quad 1 + j\},$$

verifying the results obtained in Example 3.9.

3.4.2   2N-Point DFT of a Real Sequence Using a Single N-Point DFT

Let $v[n]$ be a real sequence of length $2N$ with $V[k]$ denoting its $2N$-point DFT. Define two real sequences $g[n]$ and $h[n]$ of length N each as

$$g[n] = v[2n], \quad\quad h[n] = v[2n + 1], \quad\quad 0 \le n < N, \tag{3.73}$$

with $G[k]$ and $H[k]$ denoting their N-point DFTs. Now define a complex length-N sequence $x[n]$ according to Eq. (3.68). The DFTs $G[k]$ and $H[k]$ can be computed from the N-point DFT $X[k]$ of the sequence $x[n]$ by means of Eqs. (3.69) and (3.70).

Now,

$$
\begin{aligned}
V[k] &= \sum_{n=0}^{2N-1} v[n]W_{2N}^{nk} = \sum_{n=0}^{N-1} v[2n]W_{2N}^{2nk} + \sum_{n=0}^{N-1} v[2n+1]W_{2N}^{(2n+1)k} \\
&= \sum_{n=0}^{N-1} g[n]W_N^{nk} + \sum_{n=0}^{N-1} h[n]W_N^{nk}W_{2N}^{k} \\
&= \sum_{n=0}^{N-1} g[n]W_N^{nk} + W_{2N}^{k} \sum_{n=0}^{N-1} h[n]W_N^{nk}.
\end{aligned}
$$

Note that the first sum on the last expression is simply an N-point DFT $G[k]$ of the length-N sequence $g[n]$, whereas, the second sum is an N-point DFT $H[k]$ of the length-N sequence $h[n]$. Therefore, we can express the $2N$-point DFT $V[k]$ as

$$
V[k] = G[\langle k \rangle_N] + W_{2N}^{k} H[\langle k \rangle_N], \qquad 0 \le k \le 2N - 1, \tag{3.74}
$$

where we have used the identity $W_{2N}^{2k} = W_N^{k}$ and used the modulo sign for the argument of the two N-point DFTs on the right-side since k here is in the range $0 \le k \le 2N - 1$.

3.5   Linear Convolution Using the DFT

Linear convolution is a key operation in most signal processing applications. Since an N-point DFT can be implemented very efficiently using approximately $N(\log_2 N)$ arithmetic operations, it is of interest to investigate methods for the implementation of the linear convolution using the DFT. Earlier in Example 3.10, we have already illustrated for a very specific case how to implement a linear convolution using a circular convolution. We first generalize this example for the linear convolution of two finite-length sequences of unequal lengths. Later we consider the implementation of the linear convolution of a finite-length sequence with an infinite-length sequence.

3.5.1   Linear Convolution of Two Finite-Length Sequences

Let $g[n]$ and $h[n]$ be finite-length sequences of lengths N and M, respectively. Denote $L = M + N - 1$. Define two length-L sequences,

$$
g_e[n] = \begin{cases} g[n], & 0 \le n \le N-1, \\ 0, & N \le n \le L, \end{cases} \tag{3.75}
$$

$$
h_e[n] = \begin{cases} h[n], & 0 \le n \le M-1, \\ 0, & M \le n \le L, \end{cases} \tag{3.76}
$$

obtained by appending $g[n]$ and $h[n]$ with zero-valued samples. Then

$$
y_L[n] = g[n] \circledast h[n] = y_C[n] = g_e[n] \textcircled{L} h_e[n]. \tag{3.77}
$$

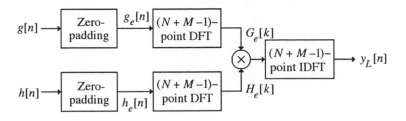

Figure 3.12   DFT based implementation of the linear convolution of two finite-length sequences

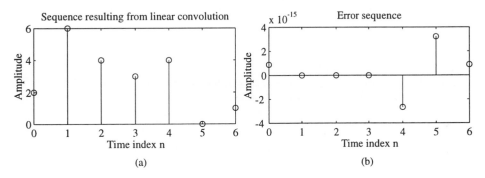

(a)   (b)

Figure 3.13   Plots of the output and the error sequences.

To implement Eq. (3.77) using the DFT, we first zero-pad $g[n]$ with $(M-1)$ zeros to obtain $g_e[n]$, and zero-pad $h[n]$ with $(N-1)$ zeros to obtain $h_e[n]$. Then we compute the $(N+M-1)$-point DFTs of $g_e[n]$ and $h_e[n]$, respectively, resulting in $G_e[k]$ and $H_e[k]$. An $(N+M-1)$-point IDFT of the product $G_e[k]H_e[k]$ results in $y_L[n]$. The process involved is sketched in Figure 3.12.

The following example uses MATLAB to illustrate the above approach.

Example 3.13 Program 3_5 can be used to determine the linear convolution of two finite-length sequences via the DFT-based approach and to compare the result using a direct linear convolution. The input data to be entered in vector format inside square brackets are the two sequences to be convolved. The program plots the result of the linear convolution obtained using the DFT based approach, and the difference between this result and the sequence obtained using a direct linear convolution using the M-file function conv. Using this program we verify the result of Example 3.11 as demonstrated in Figure 3.13.

```
% Program 3_5
% Linear Convolution Via the DFT
%
% Read in the two sequences
x = input('Type in the first sequence = ');
h = input('Type in the second sequence = ');
% Determine the length of the result of convolution
L = length(x)+length(h)-1;
% Compute the DFTs by zero-padding
```

```
XE = fft(x,L);
HE = fft(h,L);
% Determine the IDFT of the product
y1 = ifft(XE.*HE);
% Plot the sequence generated by circular
% convolution and the error from direct linear
% convolution
k = 0:1:L-1;
subplot(2,1,1)
stem(k,y1)
xlabel('Time index n');ylabel('Amplitude');
title('Sequence resulting from linear convolution')
y2 = conv(x,h);
error = y1-y2;
subplot(2,1,2)
stem(k,abs(error))
xlabel('Time index n');ylabel('Amplitude')
title('Magnitude of error sequence')
```

3.5.2 Linear Convolution of a Finite-Length Sequence with an Infinite-Length Sequence

We consider now the DFT-based implementation of

$$y[n] = \sum_{\ell=0}^{M-1} h[\ell]x[n-\ell] = h[n] \circledast x[n], \tag{3.78}$$

where $h[n]$ is a finite-length sequence of length M and $x[n]$ is of infinite length (or a finite-length sequence of length much greater than M). There are two different approaches to solving this problem, as described below [Sto66].

Overlap-add method

In this method, we first segment $x[n]$, assumed to be a causal sequence here without any loss of generality, into a set of contiguous finite-length subsequences $x_m[n]$ of length N each:

$$x[n] = \sum_{m=0}^{\infty} x_m[n-mN], \tag{3.79}$$

where

$$x_m[n] = \begin{cases} x[n+mN], & 0 \le n \le N-1, \\ 0, & \text{otherwise.} \end{cases} \tag{3.80}$$

Substituting Eq. (3.79) in Eq. (3.78) we get

$$y[n] = \sum_{m=0}^{\infty} y_m[n-mN], \tag{3.81}$$

where $y_m[n] = h[n] \circledast x_m[n]$. Since $h[n]$ is of length M and $x_m[n]$ is of length N, the linear convolution $h[n] \circledast x_m[n]$ is of length $(N + M - 1)$. As a result, the desired linear convolution of Eq. (3.74) has been broken up into a sum of infinite number short-length linear convolutions of length $(N + M - 1)$ each. Each of these short convolutions can be implemented using the method outlined in Figure 3.12, where now the DFTs (and the IDFT) are computed on the basis of $(N + M - 1)$ points. There is one more subtlety to take care of before we can implement Eq. (3.81) using the DFT-based method.

Now the first short convolution in Eq. (3.81) given by $h[n] \circledast x_0[n]$, which is of length $(N + M - 1)$, is defined for $0 \le n \le N + M - 2$. The second short convolution in Eq. (3.81), given by $h[n] \circledast x_1[n]$, is also of length $(N + M - 1)$ but is defined for $N \le n \le 2N + M - 2$. This implies that there is an overlap of $M - 1$ samples between these two short linear convolutions in the range $N \le n \le N + M - 2$. Likewise, the third convolution in Eq. (3.81), given by $h[n] \circledast x_2[n]$, is defined for $2N \le n \le 3N + M - 2$, causing an overlap between the samples of $h[n] \circledast x_1[n]$ and $h[n] \circledast x_2[n]$ for $2N \le n \le 2N + M - 2$. In general, there will be an overlap of $M - 1$ samples between the samples of the short convolutions $h[n] \circledast x_{r-1}[n]$ and $h[n] \circledast x_r[n]$ for $rN \le n \le rN + M - 2$.

This process is illustrated in Figure 3.14. Figure 3.14(b) shows the first three length-7 $(N = 7)$ segments $x_m[n]$ of the sequence $x[n]$ of Figure 3.14(a). Each of these segments is convolved with a length-5 $(M = 5)$ sequence $h[n]$, resulting in length-11 $(N + M - 1 = 11)$ short linear convolutions $y_m[n]$ shown in Figure 3.14(c). As can be seen from Figure 3.14(c), the last $M - 1 = 4$ samples of $y_0[n]$ overlap with the first 4 samples of $y_1[n]$. Likewise, the last $M - 1 = 4$ samples of $y_1[n]$ overlap with the first 4 samples of $y_2[n]$, and so on. Therefore, the desired sequence $y[n]$ obtained by a linear convolution of $x[n]$ and $h[n]$ is given by

$$
\begin{aligned}
y[n] &= y_0[n], & 0 &\le n \le 6, \\
y[n] &= y_0[n] + y_1[n - 7], & 7 &\le n \le 10, \\
y[n] &= y_1[n - 7], & 11 &\le n \le 13, \\
y[n] &= y_1[n - 7] + y_2[n - 14], & 14 &\le n \le 17, \\
y[n] &= y_2[n - 14], & 18 &\le n \le 20, \\
&\;\;\vdots
\end{aligned}
$$

The above procedure is called the *overlap-add method* since the results of the short linear convolutions overlap and the overlapped portions are added to get the correct final result.

The M-file function fftfilt can be used to implement the above method. There are two different forms of this function:

$$ \texttt{y = fftfilt(h,x)} \quad \text{and} \quad \texttt{y = fftfilt(h,x,r)}, $$

where h is the impulse response vector of the FIR filter, x is the input vector segmented into successive sections, and y is the filtered output. In the first form the input data x is segmented into successive sections of length 512 each, whereas in the second form the input is segmented into sections of length r specified by the user. We illustrate its use in the following example.

Example 3.14 We consider the filtering of the noise-corrupted signal of Example 2.14 using a length-3 moving average filter. To this end, we use Program 3_6 given below. The computer-generated results are indicated in Figure 3.15.

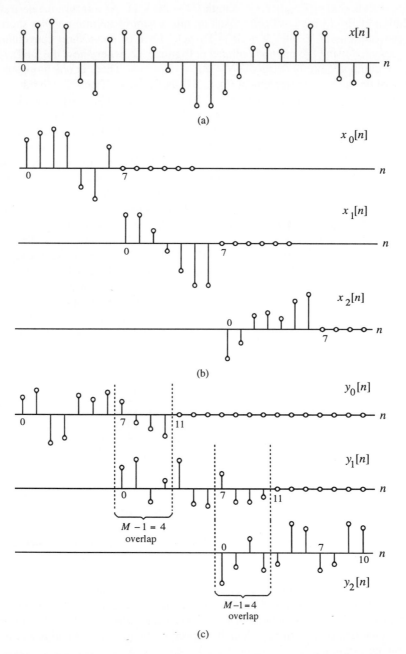

Figure 3.14   (a) Original $x[n]$, (b) segments $x_m[n]$ of $x[n]$, and (c) linear convolution of $x_m[n]$ with $h[n]$.

Figure 3.15   Uncorrupted input signal $s[n]$ (shown with solid line) and the filtered noisy signal $y[n]$ (shown with dashed line).

```
% Program 3_6
% Illustration of Overlap-Add Method
%
% Generate the noise sequence
R = 64;
d = rand(R,1)-0.5;
% Generate the uncorrupted sequence and add noise
for m = 1:1:R;
      s(m)  = 2*(m-1)*((0.9)^(m-1));
      x(m)  = s(m) + d(m);
end
k = 0:1:R-1;
% Read in the length of the moving average filter
M = input('Length of moving average filter = ');
% Generate the moving average filter coefficients
h = ones(1,M)/M;
% Perform the overlap-add filtering operation
y = fftfilt(h,x,4);
% Plot the results
plot(k,s,'r-',k,y,'b--')
xlabel('Time index n');ylabel('Amplitude')
legend( 'r-','s[n]','b--','y[n]')
```

Overlap-save method

In implementing the previous method using the DFT, we need to compute $(N+M-1)$-point DFTs and IDFT since the overall linear convolution of Eq. (3.78) was expressed as a sum of short length linear convolutions of length $(N + M - 1)$ each. It is possible to implement the linear convolution of Eq. (3.78) by performing instead circular convolutions of length shorter than $(N + M - 1)$. To this end, it is necessary to segment $x[n]$ into overlapping blocks $x_m[n]$, keep the terms of the circular convolution of $h[n]$ with $x_m[n]$ that corresponds to the terms obtained by a linear convolution of $h[n]$ and $x_m[n]$, and throw away the other parts of the circular convolution.

To understand the correspondence between the linear and circular convolutions, consider a length-4 sequence $x[n]$ and a length-3 sequence $h[n]$. Let $y_L[n]$ denote the result of a linear convolution of $x[n]$ with $h[n]$. The six samples of $y_L[n]$ are given by

$$
\begin{aligned}
y_L[0] &= h[0]x[0], \\
y_L[1] &= h[0]x[1] + h[1]x[0], \\
y_L[2] &= h[0]x[2] + h[1]x[1] + h[2]x[0], \\
y_L[3] &= h[0]x[3] + h[1]x[2] + h[2]x[1], \\
y_L[4] &= h[1]x[3] + h[2]x[2], \\
y_L[5] &= h[2]x[3].
\end{aligned}
\tag{3.82}
$$

If we append $h[n]$ with a single zero sample and convert it into a length-4 sequence $h_e[n]$, the 4-point circular convolution $y_C[n]$ of $h_e[n]$ and $x[n]$ is given by

$$
\begin{aligned}
y_C[0] &= h[0]x[0] + h[1]x[3] + h[2]x[2], \\
y_C[1] &= h[0]x[1] + h[1]x[0] + h[2]x[3], \\
y_C[2] &= h[0]x[2] + h[1]x[1] + h[2]x[0], \\
y_C[3] &= h[0]x[3] + h[1]x[2] + h[2]x[1].
\end{aligned}
\tag{3.83}
$$

Comparing Eqs. (3.82) and (3.83), we observe that the first two terms of the circular convolution do not correspond to the first two terms of the linear convolution, whereas the last two terms of the circular convolution are precisely the same as the third and fourth terms of the linear convolution, i.e.,

$$
\begin{array}{ll}
y_L[0] \neq y_C[0], & y_L[1] \neq y_C[1], \\
y_L[2] = y_C[2], & y_L[3] = y_C[3].
\end{array}
$$

In the general case of a circular convolution of a length-M sequence $h[n]$ with a length-N sequence $x[n]$ where $N > M$, the first $M - 1$ samples of the circular convolution are incorrect and are rejected, while the remaining $N - M + 1$ samples correspond to the correct samples of the linear convolution of $h[n]$ and $x[n]$.

Now consider an infinitely long or a very long sequence $x[n]$. We break it up as a collection of smaller length (length-4) sequences $x_m[n]$ as indicated below:

$$
x_m[n] = x[n + 2m], \qquad 0 \leq n \leq 3, 0 \leq m \leq \infty.
\tag{3.84}
$$

Next, we form

$$
w_m[n] = h[n]④x_m[n],
$$

or equivalently,

$$
\begin{aligned}
w_m[0] &= h[0]x_m[0] + h[1]x_m[3] + h[2]x_m[2], \\
w_m[1] &= h[0]x_m[1] + h[1]x_m[0] + h[2]x_m[3], \\
w_m[2] &= h[0]x_m[2] + h[1]x_m[1] + h[2]x_m[0], \\
w_m[3] &= h[0]x_m[3] + h[1]x_m[2] + h[2]x_m[1].
\end{aligned}
\tag{3.85}
$$

Computing the above for $m = 0, 1, 2, 3, \ldots$, and substituting the values of $x_m[n]$ from Eq. (3.84), we arrive at

$$w_0[0] = h[0]x[0] + h[1]x[3] + h[2]x[2], \qquad \leftarrow \quad Reject$$
$$w_0[1] = h[0]x[1] + h[1]x[0] + h[2]x[3], \qquad \leftarrow \quad Reject$$
$$w_0[2] = h[0]x[2] + h[1]x[1] + h[2]x[0] = y[2], \qquad \leftarrow \quad Save$$
$$w_0[3] = h[0]x[3] + h[1]x[2] + h[2]x[1] = y[3], \qquad \leftarrow \quad Save$$

$$w_1[0] = h[0]x[2] + h[1]x[5] + h[2]x[4], \qquad \leftarrow \quad Reject$$
$$w_1[1] = h[0]x[3] + h[1]x[2] + h[2]x[5], \qquad \leftarrow \quad Reject$$
$$w_1[2] = h[0]x[4] + h[1]x[3] + h[2]x[2] = y[4], \qquad \leftarrow \quad Save$$
$$w_1[3] = h[0]x[5] + h[1]x[4] + h[2]x[3] = y[5], \qquad \leftarrow \quad Save$$

$$w_2[0] = h[0]x[4] + h[1]x[5] + h[2]x[6], \qquad \leftarrow \quad Reject$$
$$w_2[1] = h[0]x[5] + h[1]x[4] + h[2]x[7], \qquad \leftarrow \quad Reject$$
$$w_2[2] = h[0]x[6] + h[1]x[5] + h[2]x[4] = y[6], \qquad \leftarrow \quad Save$$
$$w_2[3] = h[0]x[7] + h[1]x[6] + h[2]x[5] = y[7], \qquad \leftarrow \quad Save$$

It should be noted that to determine $y[0]$ and $y[1]$, we need to form $x_{-1}[n]$:

$$x_{-1}[0] = 0, \quad x_{-1}[1] = 0, \quad x_{-1}[2] = x[0], \quad x_{-1}[3] = x[1],$$

and compute $w_{-1}[n] = h[n] \textcircled{4} x_{-1}[n]$ for $0 \leq n \leq 3$, reject $w_{-1}[0]$ and $w_{-1}[1]$, and save $w_{-1}[2] = y[0]$ and $w_{-1}[3] = y[1]$.

Generalizing above, let $h[n]$ be a sequence of length M, and $x_m[n]$, the mth section of an infinitely long sequence $x[n]$ defined by

$$x_m[n] = x[n + m(N - M + 1)], \qquad 0 \leq n \leq N - 1, \qquad (3.86)$$

be of length N with $M \leq N$. If $w_m[n]$ denotes the N-point circular convolution of $h[n]$ and $x_m[n]$, i.e., $w_m[n] = h[n] \textcircled{N} x_m[n]$, then we reject the first $M - 1$ samples of $w_m[n]$ and "abut" the remaining $N - M + 1$ saved samples of $w_m[n]$ to form $y_L[n]$, the linear convolution of $h[n]$ and $x[n]$. If we denote the saved portion of $w_m[n]$ as $y_m[n]$, i.e.,

$$y_m[n] = \begin{cases} 0, & 0 \leq n \leq M - 2, \\ w_m[n], & M - 1 \leq n \leq N - 1, \end{cases} \qquad (3.87)$$

then,

$$y_L[n + m(N - M)] = y_m[n], \qquad M - 1 \leq n \leq N - 1. \qquad (3.88)$$

The above process is illustrated in Figure 3.16. The approach is called the *overlap-save method* since the input is segmented into overlapping sections and part of the results of the circular convolution are saved and abutted to determine the linear convolution result.

3.6   The z-Transform

The Fourier transform provides a frequency-domain representation of discrete-time signals and LTI systems. However, because of the convergence condition, in many cases, the Fourier transform of a sequence may not exist, and as a result, it is not possible to make

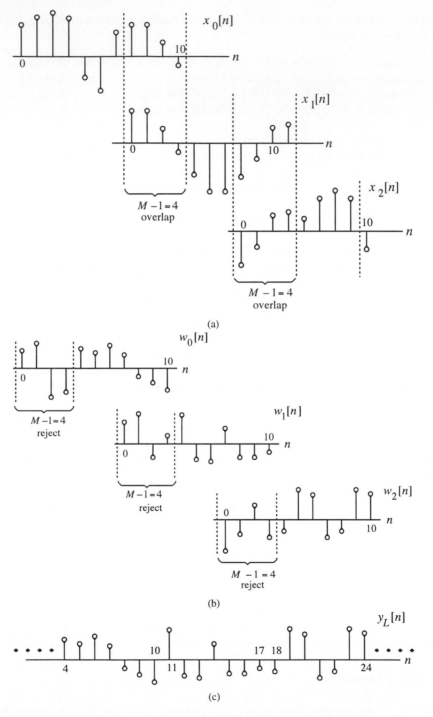

Figure 3.16   Illustration of the overlap-save method. (a) Overlapped segments of the sequence $x[n]$ of Figure 3.7(a), (b) sequences generated by an 11-point circular convolution, and (c) sequence obtained by rejecting first 4 samples of $w_i[n]$ and abutting the remaining samples.

use of such frequency-domain characterization in these cases. A generalization of the Fourier transform defined in Eq. (3.1) leads to the z-transform, which may exist for many sequences for which the Fourier transform does not exist.[7] Moreover, the use of z-transform techniques permits simple algebraic manipulations. Consequently, the z-transform has become an important tool in the analysis and design of digital filters.

We first define the z-transform of a sequence and study its properties by treating it as a generalization of its discrete-time Fourier transform. This leads to the concept of the region of the convergence of a z-transform that is investigated in detail. We then describe the inverse transform operation and point out two straightforward approaches for the computation of the inverse of a real rational z-transform. Next, the properties of the z-transform are reviewed.

For a given sequence $g[n]$, its z-transform $G(z)$ is defined as

$$G(z) = \mathcal{Z}\{g[n]\} = \sum_{n=-\infty}^{\infty} g[n]z^{-n}, \tag{3.89}$$

where z is a complex variable. If we let $z = re^{j\omega}$, then the right-hand side of the above expression reduces to

$$G(re^{j\omega}) = \sum_{n=-\infty}^{\infty} g[n]r^{-n}e^{-j\omega n}, \tag{3.90}$$

which can be interpreted as the Fourier transform of the modified sequence $\{g[n]r^{-n}\}$. For $r = 1$ (i.e., $|z| = 1$), the z-transform of $g[n]$ reduces to its Fourier transform, provided the latter exists. The contour $|z| = 1$ is a circle in the z-plane of unity radius and is called the *unit circle*.

Like the Fourier transform, there are conditions on the convergence of the infinite series of Eq. (3.89). For a given sequence, the set $\mathcal{R}$ of values of z for which its z-transform converges is called the *region of convergence* (ROC). It follows from our earlier discussion on the uniform convergence of the Fourier transform that the series of Eq. (3.90) converges if $g[n]r^{-n}$ is absolutely summable, i.e., if

$$\sum_{n=-\infty}^{\infty} \left| g[n]r^{-n} \right| < \infty. \tag{3.91}$$

In general, the region of convergence $\mathcal{R}$ of a z-transform of a sequence $g[n]$ is an annular region of the z-plane:

$$R_{g-} < |z| < R_{g+} \tag{3.92}$$

where $0 \le R_{g-} < R_{g+} \le \infty$. It should be noted that the z-transform as defined by Eq. (3.89) is a form of Laurent series and is an analytic function at every point in the ROC. This in turn implies that the z-transform and all its derivatives are continuous functions of the complex variable z in the ROC.

It follows from above that the Fourier transform $X(e^{j\omega})$ of a sequence $x[n]$ converges uniformly if and only if the ROC of the z-transform $X(z)$ of the sequence includes the unit circle.

---

[7]For some sequences, the Fourier transform may exist while the z transform may not exist.

Table 3.8   Some commonly used z-transform pairs.

Sequence	z-transform	ROC
$\delta[n]$	1	All values of z
$\mu[n]$	$\dfrac{1}{1-z^{-1}}$	$\|z\| > 1$
$\alpha^n \mu[n]$	$\dfrac{1}{1-\alpha z^{-1}}$	$\|z\| > \|\alpha\|$
$(r^n \cos\omega_0 n)\mu[n]$	$\dfrac{1-(r\cos\omega_0)z^{-1}}{1-(2r\cos\omega_0)z^{-1}+r^2 z^{-2}}$	$\|z\| > r$
$(r^n \sin\omega_0 n)\mu[n]$	$\dfrac{(r\sin\omega_0)z^{-1}}{1-(2r\cos\omega_0)z^{-1}+r^2 z^{-2}}$	$\|z\| > r$

Example 3.15 Let us determine the z-transform $\mu(z)$ of the unit step sequence $\mu[n]$ defined in Eq. (2.26) and its region of convergence.

Applying the definition of Eq. (3.89), we obtain

$$\mu(z) = \sum_{n=-\infty}^{\infty} \mu[n]z^{-n} = \sum_{n=0}^{\infty} z^{-n}.$$

The above power series converges to

$$\mu(z) = \frac{1}{1-z^{-1}}, \qquad \text{for } |z| > 1, \tag{3.93}$$

indicating that the ROC is the annular region $1 < |z| \le \infty$. Note that the unit step sequence is not absolutely summable, and as a result, its Fourier transform does not exist in the sense defined earlier.

Some commonly used z-transform pairs are listed in Table 3.8.

3.7   Region of Convergence of a Rational z-Transform

The ROC of a z-transform is an important concept for a variety of reasons. As we shall show later, without the knowledge of the ROC, there is no unique relationship between a sequence and its z-transform. Hence, the z-transform must always be specified with its ROC. Moreover, if the ROC of a z-transform of sequence includes the unit circle, the Fourier transform of the sequence is obtained simply by evaluating the z-transform on the unit circle. In the following chapter, we shall point out the relationship between the ROC of the z-transform of the impulse response of a causal LTI system and its BIBO stability. It is thus of interest to investigate the ROC more throughly.

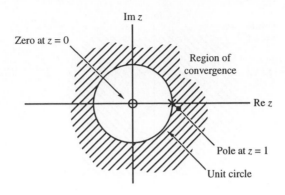

Figure 3.17   The pole-zero plot and the region of convergence of $\mathcal{Z}\{\mu[n]\}$.

In the case of LTI discrete-time systems we are concerned with in this text, all pertinent z-transforms are rational functions of z^{-1}, i.e., are ratios of two polynomials in z^{-1}:

$$G(z) = \frac{P(z)}{D(z)} = \frac{p_0 + p_1 z^{-1} + \cdots + p_{M-1} z^{-(M-1)} + p_M z^{-M}}{d_0 + d_1 z^{-1} + \cdots + d_{N-1} z^{-(N-1)} + d_N z^{-N}}, \qquad (3.94)$$

which can be alternately written in factored form as

$$G(z) = \frac{p_0}{d_0} \frac{\prod_{\ell=1}^{M} (1 - \xi_\ell z^{-1})}{\prod_{\ell=1}^{N} (1 - \lambda_\ell z^{-1})} = \frac{p_0}{d_0} z^{(N-M)} \frac{\prod_{\ell=1}^{M} (z - \xi_\ell)}{\prod_{\ell=1}^{N} (z - \lambda_\ell)}. \qquad (3.95)$$

At a root $z = \xi_\ell$ of the numerator polynomial, $G(\xi_\ell) = 0$, and as a result, these values of z are known as the *zeros* of $G(z)$. Likewise, at a root $z = \lambda_\ell$ of the denominator polynomial, $G(\lambda_\ell) \to \infty$, and these points in the z-plane are called the *poles* of $G(z)$. It also follows from the expression in Eq. (3.95) that there are additional $(N - M)$ zeros at $z = 0$ (the origin in the z-plane) if $N > M$ or additional $(N - M)$ poles at $z = 0$ if $N < M$. For example, the z-transform $\mu(z)$ of Eq. (3.93) can be rewritten as

$$\mu(z) = \frac{z}{z - 1}, \qquad \text{for } |z| > 1, \qquad (3.96)$$

which has a zero at $z = 0$ and a pole at $z = 1$.

Now, the ROC of a rational z-transform is bounded by the locations of its poles. To understand this relationship between the poles and the ROC, it is instructive to examine the plot of the poles and the zeros of a z-transform. Figure 3.17 shows the pole-zero plot of the z-transform $\mu(z)$ of Eq. (3.96), where the location of the pole is indicated by a cross "×" and the location of the zero is indicated by a circle "o". In this figure the ROC, shown as the cross-hatched area is the region of the z-plane just outside the circle centered at the origin and going through the pole at $z = 1$, and extending all the way to $z = \infty$.

Example 3.16 Determine the ROC of the z-transform $H(z)$ of the sequence $h[n] = (-0.6)^n \mu[n]$. Applying Eq. (3.89) we arrive at

$$H(z) = \sum_{n=0}^{\infty} (-0.6)^n z^{-n} = \sum_{n=0}^{\infty} (-0.6 z^{-1})^n,$$

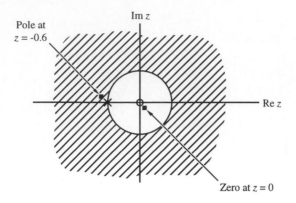

Figure 3.18   Pole-zero plot of $\mathcal{Z}\{(-0.6)^n \mu[n]\}$.

which simplifies to

$$H(z) = \frac{1}{1 + 0.6\,z^{-1}} = \frac{z}{z + 0.6}, \tag{3.97}$$

provided $|z| > 0.6$. This implies that the ROC is just outside the circle going through the point $z = -0.6$ and extending all the way to $z = \infty$, as shown in Figure 3.18. Note that $H(z)$ has a zero at $z = 0$ and a pole at $z = -0.6$.

In general, the ROC depends on the type of the sequence of interest as defined earlier in Section 2.1.2. We examine next the ROCs of the z-transforms of several different types of sequences.

Example 3.17 Consider a finite-length sequence $g[n]$ defined for $-M \le n \le N$, where M and N are nonnegative integers and $|g[n]| < \infty$. Its z-transform is given by

$$G(z) = \sum_{n=-M}^{N} g[n]z^{-n} = \frac{\sum_{n=0}^{N+M} g[n-M]z^{N+M-n}}{z^N}. \tag{3.98}$$

Note from Eq. (3.98) that $G(z)$ has M poles at $z = \infty$ and N poles at $z = 0$. As can be seen from the above, in general, the z-transform of a finite-length bounded sequence converges everywhere in the z-plane except possibly at $z = 0$ and/or at $z = \infty$.

Example 3.18 A right-sided sequence $u_1[n]$ with nonzero sample values only for $n \ge 0$ is sometimes called a causal sequence. Its z-transform is given by

$$U_1(z) = \sum_{n=0}^{\infty} u_1[n]z^{-n}. \tag{3.99}$$

It can be shown that $U_1(z)$ converges exterior to a circle $|z| = R_1$, including the point $z = \infty$. On the other hand, a right-sided sequence $u_2[n]$ with nonzero sample values only for $n \ge -M$ with M nonnegative has a z-transform $U_2(z)$ with M poles at $z = \infty$, and therefore its ROC is exterior to a circle $|z| = R_2$, excluding the point $z = \infty$.

Example 3.19 A left-sided sequence $v_1[n]$ with nonzero sample values only for $n \leq 0$, often called an anticausal sequence, has a z-transform given by

$$V_1(z) = \sum_{n=-\infty}^{0} v_1[n]z^{-n}, \tag{3.100}$$

and converges interior to a circle $|z| = R_3$, including the point $z = 0$. However, a leftsided sequence $v_2[n]$ with nonzero sample values only for $n \leq N$ with N nonnegative has a z-transform $V_2(z)$ with N poles at $z = 0$. As a result, the ROC of $V_2(z)$ is interior to a circle $|z| = R_4$ excluding the point $z = 0$.

The z-transform of a two-sided sequence $w[n]$ can be expressed as

$$W(z) = \sum_{n=-\infty}^{\infty} w[n]z^{-n} = \sum_{n=0}^{\infty} w[n]z^{-n} + \sum_{n=-\infty}^{-1} w[n]z^{-n}. \tag{3.101}$$

The first term on the right-hand side of Eq. (3.101) can be interpreted as the z-transform of a right-sided sequence, and it converges exterior to the circle $|z| = R_5$. The second term, on the other hand, is the z-transform of a left-sided sequence, and it converges interior to the circle $|z| = R_6$. As a result, if $R_5 < R_6$, there is an overlapping ROC given by $R_5 < |z| < R_6$. If, however, $R_6 < R_5$, the z-transform does not exist, as illustrated by the following example.

Example 3.20 The two-sided sequence defined by

$$u[n] = \alpha^n,$$

where α can be a real or complex number, does not have a z-transform, regardless of the absolute value $|\alpha|$. This follows by noting that the z-transform expression can be rewritten as

$$U(z) = \sum_{n=0}^{\infty} \alpha^n z^{-n} + \sum_{n=-\infty}^{-1} \alpha^n z^{-n}. \tag{3.102}$$

The first term on the right-hand side of Eq. (3.102) converges for $|z| > |\alpha|$, whereas the second term converges for $|z| < |\alpha|$, and hence, there is no overlap of the two ROCs.

For a sequence with a rational z-transform, the ROC of the z-transform cannot contain any poles and is bounded by the poles. This property of such z-transforms can be seen from the z-transform of a unit step sequence given by Eq. (3.96) and illustrated by the pole-zero plot of Figure 3.17. Another example is the sequence defined in Example 3.16 and its z-transform given by Eq. (3.97), with the corresponding pole-zero plot given in Figure 3.18.

To show that it is bounded by the poles, assume that the z-transform $X(z)$ has simple poles at α and β, with $|\alpha| < |\beta|$. If the sequence is also assumed to be a right-sided sequence, then it is of the form

$$x[n] = (r_1(\alpha)^n + r_2(\beta)^n)\mu[n - N_0], \tag{3.103}$$

where N_0 is a positive or negative integer. Now the z-transform of a right-sided sequence $(\gamma)^n \mu[n - N_o]$ exists if

$$\sum_{n=N_0}^{\infty} \left| (\gamma)^n z^{-n} \right| < \infty$$

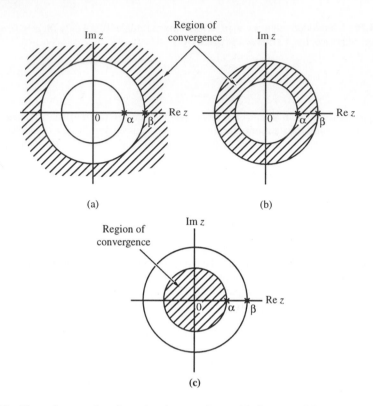

Figure 3.19   The pole-zero plot of a rational z-transform with three possible ROCs corresponding to three different sequences. (a) Right-sided sequence, (b) two-sided sequence, and (c) left-sided sequence

for some z. It can be seen that the above holds for $|z| > |\gamma|$ but not for $|z| \leq |\gamma|$. The right-sided sequence of Eq. (3.103) has thus a ROC defined by $|\beta| < |z| \leq \infty$. A similar argument shows that if $X(z)$ is the z-transform of left-sided sequence of the form of Eq. (3.103), with $\mu[n - N_0]$ replaced by $\mu[-n - N_0]$ then its ROC is defined by $0 \leq |z| < |\alpha|$. Finally, for a two-sided sequence, some of the poles contribute to terms for $n < 0$ and the others to terms for $n \geq 0$. The ROC is thus bounded on the outside by the pole with the smallest magnitude that contributes for $n < 0$ and on the inside by the pole with the largest magnitude that contributes for $n \geq 0$.

Figure 3.19 shows the three possible ROCs of a rational z-transform with poles at $z = \alpha$, and $z = \beta$ and with each ROC associated with a unique sequence. In general, if the rational z-transform has N poles with R distinct magnitudes, then it has $R + 1$ ROCs and, as a result, $R + 1$ distinct sequences having the same rational z-transform. Consequently, a rational z-transform with a specified ROC has a unique sequence as its inverse z-transform. A rational z-transform without a specified ROC is thus not meaningful.

MATLAB can be used to determine the ROCs of a rational z-transform. To this end, several functions need to be used. The statement `[z,p,k] = tf2zp(num,den)` determines the zeros, poles, and the gain constant of a rational z-transform expressed as a ratio of polynomials in descending powers of z as in Eq. (3.94). The input arguments are the row vectors `num` and `den` containing the coefficients of the numerator and the denominator polynomials in descending powers of z.[8] The output files are the column vectors z and p containing the zeros and poles of the rational z-transform, and the gain constant k. The statement [num,den] = zp2tf(z,p,k) is used to implement the reverse process.

From the zero-pole description, the factored form of the transfer function can be obtained using the function sos = zp2sos(z,p,k). The statement computes the coefficients of each second-order factor given as an $L \times 6$ matrix sos, where

$$\texttt{sos} = \begin{bmatrix} b_{01} & b_{11} & b_{21} & a_{01} & a_{11} & a_{21} \\ b_{02} & b_{12} & b_{22} & a_{02} & a_{12} & a_{22} \\ \vdots & \vdots & \vdots & \vdots & \vdots & \vdots \\ b_{0L} & b_{1L} & b_{2L} & a_{0L} & a_{1L} & a_{2L} \end{bmatrix},$$

where the kth row contains the coefficients of the numerator and the denominator of the kth second-order factor of the z-transform $G(z)$:

$$G(z) = \prod_{k=1}^{L} \frac{b_{0k} + b_{1k}z^{-1} + b_{2k}z^{-2}}{a_{0k} + a_{1k}z^{-1} + a_{2k}z^{-2}}.$$

The pole-zero plot of a rational z-transform can also be plotted by using the *M*-file function `zplane`. The z-transform can be described either in terms of its zeros and poles given as vectors `zeros` and `poles` or in terms of the numerator and the denominator polynomials entered as vectors `num` and `den` containing coefficients in descending powers of z:

$$\texttt{zplane(zeros,poles)}, \qquad \texttt{zplane(num,den)}$$

It should be noted that the argument zeros and poles must be entered as column vectors, whereas, the argument num and den need to be entered as row vectors. The function `zplane` plots the zeros and poles in the current figure window, with the zero indicated by the symbol "o" and the pole indicated by the symbol "×." The unit circle is also included in the plot for reference. The automatic scaling included in the function can be overwritten if necessary.[9]

The following two examples illustrate the application of the above functions.

Example 3.21 Express the following z-transform in factored form, plot its poles and zeros, and then determine its ROCs.

$$G(z) = \frac{2z^4 + 16z^3 + 44z^2 + 56z + 32}{3z^4 + 3z^3 - 15z^2 + 18z - 12}. \tag{3.104}$$

---

[8] Or, equivalently, in ascending powers of z^{-1}.

[9] See the manual for the *Signal Processing Toolbox* for details [Kra94].

To carry out the factorization we use the following MATLAB program:

```
% Program 3_7
% Determination of the factored form
% of a rational z-transform
%
num = input('Type in the numerator coefficients = ');
den = input('Type in the denominator coefficients = ');
[z,p,k] = tf2zp(num,den);
m = abs(p);
disp('Zeros are at'); disp(z);
disp('Poles are at'); disp(p);
disp('Gain constant'); disp(k);
disp('Radius of poles');disp(m);
sos = zp2sos(z,p,k);
disp('Second-order sections');disp(real(sos));
zplane(num,den)
```

.   During execution the program asks for the coefficients of the numerator and denominator polynomials. These are entered as row vectors inside square brackets in descending powers of z. The program then computes and prints the locations of the poles and zeros, the value of the gain constant, the radius of the poles, and the coefficients of the second-order sections as indicated below:[10]

```
Zeros are at
  -4.0000
  -2.0000
  -1.0000 + 1.0000i
  -1.0000 - 1.0000i

Poles are at
  -3.2361
   1.2361
   0.5000 + 0.8660i
   0.5000 - 0.8660i

Gain constant
    0.6667
Radius of poles
    3.2361
    1.2361
    1.0000
    1.0000
  Second-order sections
   0.0667    0.4000    0.5333    1.0000    2.0000   -4.0000
  10.0000   20.0000   20.0000    1.0000   -1.0000    1.0000
```

----

[10]Note that in MATLAB, the usual notation for $\sqrt{-1}$ is "i" instead of "j."

Figure 3.20   Pole-zero plot of the z-transform of Eq. (3.104).

Therefore the factored form of the z-transform of Eq. (3.104) is given by

$$G(z) = \frac{(0.0667 + 0.4z^{-1} + 0.5333z^{-2})(10.0 + 20.0z^{-1} + 20.0z^{-2})}{(1.0 + 2.0z^{-1} - 4.0z^{-2})(1.0 - 1.0z^{-1} + 1.0z^{-2})}$$

$$= 0.667\frac{(1 + 6z^{-1} + 8z^{-2})(1 + 2z^{-1} + 2z^{-2})}{(1 + 2z^{-1} - 4z^{-2})(1 - z^{-1} + z^{-2})} \qquad (3.105)$$

The pole-zero plot developed by the program is shown in Figure 3.20. From Eq. (3.105) the four regions of convergences are thus seen to be

$$\mathcal{R}_1 : \infty \geq |z| > 3.2361,$$
$$\mathcal{R}_2 : 3.2361 > |z| > 1.2361,$$
$$\mathcal{R}_3 : 1.2361 > |z| > 1,$$
$$\mathcal{R}_4 : 1 > |z| \geq 0.$$

Example 3.22   We now consider the determination of the rational z-transform from its zero and pole locations. The zeros are at $\xi_1 = 0.21$, $\xi_2 = 3.14$, $\xi_3 = -0.3 + j0.5$, $\xi_4 = -0.3 - j0.5$, the poles are at $\lambda_1 = -0.45$, $\lambda_2 = 0.67$, $\lambda_3 = 0.81 + j0.72$, $\lambda_4 = 0.81 - j0.72$, and the gain constant k is 2.2.

The MATLAB program that is employed to compute the corresponding rational z-transform is as follows:

```
% Program 3_8
% Determination of the rational z-transform
% from its poles and zeros
%
format long
zr = input('Type in the zeros as a row vector = ');
pr = input('Type in the poles  as a row vector = ');
% Transpose zero and pole row vectors
```

```
z = zr'; p = pr';
k = input('Type in the gain constant = ');
[num, den] = zp2tf(z, p, k);
disp('Numerator polynomial coefficients'); disp(num);
disp('Denominator polynomial coefficients'); disp(den);
```

The program during execution asks for the input data, the zero and pole locations, and the gain constant. The program then displays the coefficients of the numerator and the denominator polynomials in descending powers of z, as indicated below:

```
Numerator polynomial coefficients
  Columns 1 through 4

   2.2   -5.126   -4.57028   -3.062048

  Column 5

  -0.4932312

Denominator polynomial coefficients
  Columns 1 through 4

   1.0   -1.840   1.22940   0.23004

  Column 5
  -0.35411175
```

From the above we thus arrive at the desired expression:

$$G(z) = \frac{2.2z^4 - 5.126z^3 - 4.57028z^2 - 3.062048z - 0.4932312}{z^4 - 1.84z^3 + 1.2294z^2 + 0.23004z - 0.35411175}.$$

3.8   The Inverse z-Transform

We now derive the expression for the inverse z-transform and outline two methods for its computation.

3.8.1   General Expression

Recall that, for $z = re^{j\omega}$, the z-transform $G(z)$ given by Eq. (3.89) is merely the Fourier transform of the modified sequence $g[n]r^{-n}$. Accordingly, by the inverse Fourier transform relation of Eq. (3.5), we have

$$g[n]r^{-n} = \frac{1}{2\pi} \int_{-\pi}^{\pi} G(re^{j\omega})e^{j\omega n}\, d\omega. \qquad (3.106)$$

By making the change of variable $z = re^{j\omega}$, the above equation can be converted into a contour integral given by

$$g[n] = \frac{1}{2\pi j} \oint_{C'} G(z)z^{n-1}\, dz, \qquad (3.107)$$

where C' is a counterclockwise contour of integration defined by $|z| = r$. But the integral remains unchanged when C' is replaced with any contour C encircling the point $z = 0$ in the ROC of $G(z)$. The contour integral in Eq. (3.107) can be evaluated using the Cauchy's residue theorem resulting in

$$g[n] = \sum \left[\text{residues of } G(z)z^{n-1} \text{ at the poles inside } C \right]. \qquad (3.108)$$

Note that Eq. (3.108) needs to be evaluated at all values of n which can be quite complicated in most cases. Two simple methods for the inverse transform computation are reviewed next.

3.8.2   Inverse Transform by Partial-Fraction Expansion

The expression of Eq. (3.107) can be computed in a number of ways. A rational z-transform $G(z)$ with a causal inverse transform $g[n]$ has an ROC that is exterior to a circle. In this case it is more convenient to express $G(z)$ in a partial-fraction expansion form and then determine $g[n]$ by summing the inverse transforms of the individual simpler terms in the expansion.

A rational $G(z)$ can be expressed as

$$G(z) = \frac{P(z)}{D(z)}, \qquad (3.109)$$

where $P(z)$ and $D(z)$ are polynomials in z^{-1}. If the degree M of the numerator polynomial $P(z)$ is greater than or equal to the degree N of the denominator polynomial $D(z)$, we can divide $P(z)$ by $D(z)$ and re-express $G(z)$ as

$$G(z) = \sum_{\ell=0}^{M-N} \eta_\ell z^{-\ell} + \frac{P_1(z)}{D(z)}, \qquad (3.110)$$

where the degree of the polynomial $P_1(z)$ is less than that of $D(z)$. The rational function $P_1(z)/D(z)$ is called a proper fraction. In most practical cases, $G(z)$ is a proper fraction with simple poles. Let the poles of $G(z)$ be at $z = \lambda_k$, $k = 1, 2, \ldots, N$, where λ_k are distinct. A partial-fraction expansion of $G(z)$ then is of the form

$$G(z) = \sum_{\ell=1}^{N} \frac{\rho_\ell}{1 - \lambda_\ell z^{-1}}, \qquad (3.111)$$

where the constants ρ_ℓ in the above expression, called the *residues*, are given by

$$\rho_\ell = (1 - \lambda_\ell z^{-1}) G(z) \Big|_{z=\lambda_\ell}. \qquad (3.112)$$

Each term of the sum on the right-hand side of Eq. (3.111) has an ROC given by $z > |\lambda_\ell|$ and, thus, an inverse transform of the form $\rho_\ell(\lambda_\ell)^n \mu[n]$. Therefore, the inverse transform $g[n]$ of $G(z)$ is given by

$$g[n] = \sum_{\ell=1}^{N} \rho_\ell(\lambda_\ell)^n \mu[n]. \tag{3.113}$$

Note that the above approach with slight modifications can also be used to determine the inverse z-transform of a noncausal sequence with a rational z-transform.

Example 3.23 Let the z-transform of a causal sequence $h[n]$ be given by

$$H(z) = \frac{z(z + 2.0)}{(z - 0.2)(z + 0.6)} = \frac{1 + 2.0z^{-1}}{(1 - 0.2z^{-1})(1 + 0.6z^{-1})}. \tag{3.114}$$

A partial-fraction expansion of $H(z)$ is then of the form

$$H(z) = \frac{\rho_1}{1 - 0.2z^{-1}} + \frac{\rho_2}{1 + 0.6z^{-1}}. \tag{3.115}$$

Using Eq. (3.112), we obtain

$$\rho_1 = (1 - 0.2z^{-1})H(z)\Big|_{z=0.2} = \frac{1 + 2.0z^{-1}}{1 + 0.6z^{-1}}\Big|_{z=0.2} = 2.75,$$

and

$$\rho_2 = (1 + 0.6z^{-1})H(z)\Big|_{z=-0.6} = \frac{1 + 2.0z^{-1}}{1 - 0.2z^{-1}}\Big|_{z=-0.6} = -1.75.$$

Substituting the above in Eq. (3.115), we arrive at

$$H(z) = \frac{2.75}{1 - 0.2z^{-1}} - \frac{1.75}{1 + 0.6z^{-1}}.$$

The inverse transform of the above is given by

$$h[n] = 2.75(0.2)^n \mu[n] - 1.75(-0.6)^n \mu[n]. \tag{3.116}$$

If $G(z)$ has multiple poles, the partial-fraction expansion is slightly of different form. For example, if the pole at $z = \nu$ is of multiplicity L and the remaining $N - L$ poles are simple and at $z = \lambda_\ell$, $\ell = 1, \ldots, N - L$, then the general partial-fraction expansion of $G(z)$ takes the form

$$G(z) = \sum_{\ell=0}^{M-N} \eta_\ell z^{-\ell} + \sum_{\ell=1}^{N-L} \frac{\rho_\ell}{1 - \lambda_\ell z^{-1}} + \sum_{i=1}^{L} \frac{\gamma_i}{(1 - \nu z^{-1})^i}, \tag{3.117}$$

where the constants γ_i (no longer called the residues for $i \neq 1$) are computed using the formula

$$\gamma_i = \frac{1}{(L-i)!(-\nu)^{L-i}} \frac{d^{L-i}}{d(z^{-1})^{L-i}} \Big[(1 - \nu z^{-1})^L G(z)\Big]\Big|_{z=\nu}, \qquad i = 1, \ldots, L, \tag{3.118}$$

and the residues ρ_k are calculated using Eq. (3.112). Techniques for determining the inverse z-transform of terms like $\gamma_i/(1 - \nu z^{-1})^i$ are described later.

MATLAB includes M-file functions to develop the partial-fraction expansion of a rational z-transform and to convert a z-transform expressed in a partial-fraction form to its rational form. For the former case, the statement is `[r,p,k] = residuez(num,den)`, where the input data are the vectors num and den containing the coefficients of the numerator and the denominator polynomials, respectively, expressed in descending powers of z, and the output files are the vector r of the residues and the numerator constants, the vector p of corresponding poles, and the vector k containing the constants η_ℓ. The statement `[num,den] = residuez(r,p,k)` is employed to carry out the reverse operation. The applications of these functions are considered in the following two examples.

Example 3.24 Using MATLAB we determine the partial-fraction expansion of the z-transform $G(z)$ given by

$$G(z) = \frac{18z^3}{18z^3 + 3z^2 - 4z - 1}. \tag{3.119}$$

To this end we use the following MATLAB program:

```
Program 3_9
% Partial fraction expansion of rational z-transform
%
num = input('Type in numerator coefficients = ');
den = input('Type in denominator coefficients = ');
[r,p,k] = residuez(num,den);
disp('Residues');disp(r')
disp('Poles');disp(p')
disp('Constants');disp(k)
```

During execution the program calls for the input data that are the vectors num and den of the coefficients of the numerator and the denominator polynomials, respectively, in descending powers of z. These data are entered inside square brackets as follows:

```
num = [18]
den = [18   3   -4   -1]
```

The output data are the residues and the constants, the poles of the expansion of $G(z)$ in the form of Eq. (3.117). For our above example, these are as given below.

```
Residues
     0.2400     0.4000     0.3600

Poles
    -0.3333    -0.3333     0.5000

Constants
```

Note also that the z-transform of Eq. (3.119) has double poles at $z = -1/3 = -0.3333$. The first entry in both the residues and poles given above corresponds to the simple pole factor $(1 + 0.3333z^{-1})$, the second entry corresponds to the factor $(1 + 0.3333z^{-1})^2$ and the third entry corresponds to the simple pole factor $(1 - 0.5z^{-1})$ in the partial fraction expansion. Thus the desired expansion is given by

$$G(z) = \frac{0.24}{1 + 0.3333z^{-1}} + \frac{0.4}{(1 + 0.3333z^{-1})^2} + \frac{0.36}{1 - 0.5z^{-1}}. \tag{3.120}$$

Example 3.25 We now consider the determination of the rational form of a z-transform from its partial-fraction expansion representation as given by Eq. (3.120). The MATLAB program that can be used to find the rational form is given below.

```
% Program 3_10
% Partial fraction expansion to rational z-transform
%
r = input('Type in the residues = ');
p = input('Type in the poles = ');
k = input('Type in the constants = ');
[num, den] = residuez(r,p,k);
disp('Numerator polynomial coefficients'); disp(num)
disp('Denominator polynomial coefficients'); disp(den)
```

During execution, the program requests the input data vector r of residues, the vector p of pole locations, and the vector k of constants, with each entered using square brackets as follows:

```
r = [0.24 0.4 0.36]
p = [-0.3333  -0.3333  0.5]
k = [0]
```

It then prints the coefficients of the numerator and the denominator polynomials, respectively, in descending powers of z as indicated below. It can be seen that the coefficients are exactly same as in Eq. (3.119) if we multiply each coefficient by 18.

```
Numerator polynomial coefficients
    1.0000    -0.0000         0            0

Denominator polynomial coefficients
    1.0000     0.1667    -0.2222    -0.0556
```

3.8.3   Inverse Transform via Long Division

For causal sequences, the z-transform $G(z)$ can be expanded into a power series in z^{-1}. In the series expansion, the coefficient multiplying the term z^{-n} is then the nth sample

$g[n]$. For a rational $G(z)$, a convenient way to determine the power series is to express the numerator and the denominator as polynomials in z^{-1}, and then obtain the power series expansion by long division.

Example 3.26 Consider the causal $H(z)$ of Example 3.22 as given in Eq. (3.114). Multiplying the numerator and the denominator by z^{-2}, we get

$$H(z) = \frac{1 + 2.0\,z^{-1}}{1 + 0.4\,z^{-1} - 0.12\,z^{-2}}. \tag{3.121}$$

Long division of the numerator by the denominator yields

$$
\begin{array}{r}
1 + 1.6z^{-1} - 0.52z^{-2} + 0.4z^{-3} - 0.2224z^{-4} + \cdots \\
\hline
\end{array}
$$

$1 + 0.4z^{-1} - 0.12z^{-2}$ ⟌ $1 + 2.0z^{-1}$

$\qquad\qquad\qquad\quad\; 1 + 0.4z^{-1} - 0.12z^{-2}$

$\qquad\qquad\qquad\quad \overline{1.6z^{-1} + 0.12z^{-2}}$

$\qquad\qquad\qquad\quad\; 1.6z^{-1} + 0.64z^{-2} - 0.192z^{-3}$

$\qquad\qquad\qquad\qquad\quad \overline{-0.52z^{-2} + 0.192z^{-3}}$

$\qquad\qquad\qquad\qquad\quad -0.52z^{-2} - 0.208z^{-3} + 0.0624z^{-4}$

$\qquad\qquad\qquad\qquad\qquad\quad \overline{0.400z^{-3} - 0.0624z^{-4}}$

$\qquad\qquad\qquad\qquad\qquad\quad 0.400z^{-3} + 0.1600z^{-4} - 0.0480z^{-5}$

$\qquad\qquad\qquad\qquad\qquad\qquad\quad \overline{-0.2224z^{-4} + 0.0480z^{-5}}$

$\qquad\qquad\qquad\qquad\qquad\qquad\qquad \cdots\;\cdots$

$\qquad\qquad\qquad\qquad\qquad\qquad\qquad \cdots\;\cdots$

As a result, $H(z)$ can be expressed as

$$H(z) = 1.0 + 1.6z^{-1} - 0.52z^{-2} + 0.4z^{-3} - 0.2224z^{-4} + \cdots$$

which leads to

$$\{h[n]\} = \{\underset{\uparrow}{1.0}, \quad 1.6, \quad -0.52, \quad 0.4, \quad -0.2224, \ldots\} \qquad \text{for } n \geq 0.$$

By computing the values of $h[n]$ from Eq. (3.116) for $n = 0, 1, 2, 3,$ and 4, the correctness of the sample values obtained via long division method can be easily verified.

The inverse of a rational z-transform can also be readily calculated using MATLAB. The function `impz` can be utilized for this purpose. Three versions of this function are as follows:

```
[h,t] = impz(num,den),   [h,t] = impz(num,den,L),
[h,t] = impz(num,den,L,FT)
```

where the input data consist of the vectors `num` and `den` containing the coefficients of the numerator and the denominator polynomials of the z-transform given in a descending powers of z, the output impulse response vector `h`, and the time index vector `t`. In the first form, the length `L` of `h` is determined automatically by the computer with `t = 0:L-1`, whereas in the remaining two forms it is supplied by the user through the input data `L`. In the last form, the time interval is scaled so that the sampling interval is equal to 1/FT. The default value of FT is 1.

Another way of arriving at this result using MATLAB is by making use of the *M*-file function

```
y = filter(num,den,x)
```

where y is the output vector containing the coefficients of the power series representation of $H(z)$ in increasing power of z^{-1}. The numerator and denominator polynomials of $H(z)$ expressed in ascending powers of z^{-1} are two of the input data vectors num and den. The length of x is the same as that of y, and its elements are all zeros except for the first one being a 1.

We present next two examples to illustrate the use of both functions.

Example 3.27 We first illustrate the use of the function impz to determine the first 11 coefficients of the inverse *z*-transform of Eq. (3.114). To this end, we use the following MATLAB program:

```
% Program 3_11
% Power series expansion of a rational z-transform
%
% Read in the number of inverse z-transform coefficients
% to be computed
L = input('Type in the length of output vector = ');
% Read in the numerator and denominator coefficients of
% the z-transform
num = input('Type in the numerator coefficients = ');
den = input('Type in the denominator coefficients = ');
% Compute the desired number of inverse transform
% coefficients
[y,t] = impz(num,den,L);
disp('Coefficients of the power series expansion');
disp(y')
```

The input data called by the program are the desired number of coefficients and the vectors of the numerator and denominator coefficients of the rational *z*-transform, with these last entered inside square brackets. In our example, these are as follows:

```
L = 11
num = [1  2]
den = [1  0.4  -0.12]
```

The output data are the desired inverse *z*-transform coefficients as indicated below:

```
Coefficients of the power series expansion

   Columns 1 through 7
```

```
1.0000   1.6000   -0.5200   0.4000   -0.2224   0.1370
-0.0815

Columns 8 through 11

0.0490   -0.0294   0.0176   -0.0106
```

whose first five coefficients are seen to be identical to that derived in Example 3.25 using the long division approach.

Example 3.28 We determine the inverse z-transform of Eq. (3.114). The MATLAB program that can be used to determine the inverse z-transform of a rational z-transform is given below.

```
% Program 3_12
% Power series expansion of a rational z-transform
%
% Read in the number of inverse z-transform coefficients
% to be computed
N = input('Type in the length of output vector = ');
% Read in the numerator and denominator coefficients of
% the z-transform
num = input('Type in the numerator coefficients = ');
den = input('Type in the denominator coefficients = ');
% Compute the desired number of inverse transform
% coefficients
x = [1 zeros(1, N-1)];
y = filter(num, den, x);
disp('Coefficients of the power series expansion');
disp(y)
```

As the program is run, it asks for the desired length of the output vector y followed by the vectors containing the numerator and denominator coefficients entered inside square brackets and in descending powers of z. After computing the output vector, it prints its elements that are the coefficients of the power series expansion of the rational z-transform. The program assumes that the z-transform is a proper fraction with the degree of the denominator being greater than or equal to that of the numerator.

An application of the above program to the z-transform of Eq. (3.114) yields the following output:

```
Coefficients of the power series expansion
   1.0000   1.6000   -0.5200   0.4000   -0.2224
```

which is seen to be identical to that derived in Example 3.25 using the long division approach.

Table 3.9   Some useful properties of the z-transform.

Property	Sequence	z-transform	ROC
	$g[n]$	$G(z)$	$\mathcal{R}_g$
	$h[n]$	$H(z)$	$\mathcal{R}_h$
Conjugation	$g^*[n]$	$G^*(z^*)$	$\mathcal{R}_g$
Time-reversal	$g[-n]$	$G(1/z)$	$1/\mathcal{R}_g$
Linearity	$\alpha g[n] + \beta h[n]$	$\alpha G(z) + \beta H(z)$	Includes $\mathcal{R}_g \cap \mathcal{R}_h$
Time shifting	$g[n - n_0]$	$z^{-n_0} G(z)$	$\mathcal{R}_g$, except possibly the point $z = 0$ or ∞
Multiplication by an exponential sequence	$\alpha^n g[n]$	$G(z/\alpha)$	$\lvert\alpha\rvert\mathcal{R}_g$
Differentiation of $G(z)$	$ng[n]$	$-z\dfrac{dG(z)}{dz}$	$\mathcal{R}_g$, except possibly the point $z = 0$ or ∞
Convolution	$g[n]\circledast h[n]$	$G(z)H(z)$	Includes $\mathcal{R}_g \cap \mathcal{R}_h$
Modulation	$g[n]h[n]$	$\frac{1}{2\pi j} \oint_C G(v)H(z/v)v^{-1}\,dv$	Includes $\mathcal{R}_g\mathcal{R}_h$
Parseval's relation	$\sum\limits_{n=-\infty}^{\infty} g[n]h^*[n] = \frac{1}{2\pi j} \oint_C G(v)H^*(1/v^*)v^{-1}\,dv$		

Note: $\mathcal{R}_g$ denotes the region $R_{g-} < \lvert z\rvert < R_{g+}$, $\mathcal{R}_h$ denotes the region $R_{h-} < \lvert z\rvert < R_{h+}$, and $1/\mathcal{R}_g$ denotes the region $1/R_{g+} < \lvert z\rvert < 1/R_{g-}$.

3.9   Properties of the z-Transform

We summarize in Table 3.9 some specific properties of the z-transform. An understanding of these properties makes the application of the z-transform techniques to the analysis and design of digital filters often easier. The reader is therefore encouraged to verify these properties. We consider the applications of some of these properties in the next two examples. In the following example we illustrate the use of the conjugation and linearity properties in developing the z-transform pair given by the fourth entry in Table 3.8.

Example 3.29 Determine the z-transform $X(z)$ of the causal real sequence $x[n] = r^n(\cos\omega_o n)\mu[n]$ and its ROC.

Now we can express $x[n]$ as a sum of two causal exponential sequences:

$$x[n] = \tfrac{1}{2}r^n e^{j\omega_0 n}\mu[n] + \tfrac{1}{2}r^n e^{-j\omega_0 n}\mu[n].$$

We can write the above expression as $x[n] = v[n] + v^*[n]$, where

$$v[n] = \tfrac{1}{2}\alpha^n \mu[n],$$

with $\alpha = re^{j\omega_0}$. The z-transform $V(z)$ of $v[n]$ from Table 3.8 is given by

$$V(z) = \frac{1}{2} \cdot \frac{1}{1 - \alpha z^{-1}} = \frac{1}{2} \cdot \frac{1}{1 - r e^{j\omega_0} z^{-1}}, \qquad |z| > |\alpha| = r$$

Using Table 3.9, we obtain the z-transform of $v^*[n]$ as $V^*(z^*)$:

$$V^*(z^*) = \frac{1}{2} \cdot \frac{1}{1 - \alpha^* z^{-1}} = \frac{1}{2} \cdot \frac{1}{1 - r e^{-j\omega_0} z^{-1}}, \qquad |z| > |\alpha| = r$$

Therefore, by the linearity property of the z-transform, we obtain

$$X(z) = V(z) + V^*(z^*) = \frac{1}{2}\left(\frac{1}{1 - r e^{j\omega_0} z^{-1}} + \frac{1}{1 - r e^{-j\omega_0} z^{-1}}\right)$$

$$= \frac{1 - (r\cos\omega_0)\, z^{-1}}{1 - (2r\cos\omega_0)\, z^{-1} + r^2 z^{-2}}, \qquad |z| > r.$$

An application of the differentiation property of the z-transform is considered in the following example.

Example 3.30 Determine the inverse z-transform of

$$G(z) = \frac{z^3}{\left(z - \frac{1}{2}\right)\left(z + \frac{1}{3}\right)^2}, \qquad |z| > \frac{1}{2}. \qquad (3.122)$$

As the ROC is exterior to a circle of radius $1/2$, the inverse transform is a right-sided sequence. The partial-fraction expansion of $G(z)$ has been determined in Example 3.24 and is given by Eq. (3.120) where the ROC is given by $|z| > 1/2$. From Table 3.8, the inverse z-transform of $1/(1 - 0.5z^{-1})$ is $(0.5)^n \mu[n]$, and the inverse z-transform of $1/\left(1 + \frac{1}{3}z^{-1}\right)$ is $\left(-\frac{1}{3}\right)^n \mu[n]$. To determine the inverse z-transform of $1/\left(1 + \frac{1}{3}z^{-1}\right)^2$ we observe from Table 3.9 that the z-transform of $n\left(-\frac{1}{3}\right)^n \mu[n]$ is given by $-z\frac{d}{dz}\left(1/\left(1 + \frac{1}{3}z^{-1}\right)\right) = -\frac{1}{3}/\left(1 + \frac{1}{3}z^{-1}\right)^2$. As a result the inverse z-transform of $1/\left(1 + \frac{1}{3}z^{-1}\right)^2$ is given by $-3n\left(-\frac{1}{3}\right)^n \mu[n]$. Therefore the inverse z-transform of $G(z)$ of Eq. (3.122) is

$$g[n] = \left[0.24\left(-\frac{1}{3}\right)^n - 0.12n\left(-\frac{1}{3}\right)^n + 0.36\left(\frac{1}{2}\right)^n\right]\mu[n].$$

Parseval's relation given in Table 3.9 can be used to compute the energy of a sequence. To establish the required formula, we let $g[n] = h[n]$, where $g[n]$ is a real sequence, in the expression given in Table 3.9. This leads to

$$\sum_{n=-\infty}^{\infty} g^2[n] = \frac{1}{2\pi j}\oint_C G(z)G(z^{-1})z^{-1}\, dz. \qquad (3.123a)$$

where C is a closed contour in the ROC of $G(z)G(z^{-1})$. Note that if the ROC of $G(z)$ includes the unit circle, then that of $G(z^{-1})$ will also include the unit circle. In fact, for an absolutely summable sequence $g[n]$, the ROC of its z-transform $G(z)$ must include the unit circle. In this case, we can let $z = e^{j\omega}$ in Eq. (3.123a), which then reduces to

$$\sum_{n=-\infty}^{\infty} g^2[n] = \frac{1}{2\pi}\int_{-\pi}^{\pi} \left|G(e^{j\omega})\right|^2 d\omega. \qquad (3.123b)$$

3.10   Transform Domain Representations of Random Signals

The notion of random discrete-time signals was introduced in Section 2.7 along with their statistical characterizations in the time-domain. These infinite length signals have infinite energy and do not have a transform-domain characterizations like the deterministic signals. However, the autocorrelation and the autocovariance sequences of stationary random signals, defined by Eqs. (2.141) and (2.144), are of finite energy and, in most practical cases, their transform-domain representations do exist. We discuss these representations in this section.

3.10.1   Fourier Transform Representation

The Fourier transform of the autocorrelation sequence $\phi_{XX}[\ell]$ given by

$$\Phi_{XX}(e^{j\omega}) = \sum_{\ell=-\infty}^{\infty} \phi_{XX}[\ell]e^{-j\omega\ell}, \qquad |\omega| < \pi, \tag{3.124}$$

is often referred to as its *power spectral density* and is denoted by $P_{XX}(\omega)$. Likewise, the Fourier transform of the autocovariance sequence $\gamma_{XX}[\ell]$ is given by

$$\Gamma_{XX}(e^{j\omega}) = \sum_{\ell=-\infty}^{\infty} \gamma_{XX}[\ell]e^{-j\omega\ell}, \qquad |\omega| < \pi, \tag{3.125}$$

Applying the inverse Fourier transform to Eq. (3.124) and using the notation $P_{XX}(\omega) = \Phi_{XX}(e^{j\omega})$, we arrive at

$$\phi_{XX}[\ell] = \frac{1}{2\pi} \int_{-\pi}^{\pi} P_{XX}(\omega)e^{j\omega\ell}\,d\omega. \tag{3.126}$$

It follows from Eqs. (3.126) and (2.156) that

$$E\left(|X[n]|^2\right) = \phi_{XX}[0] = \frac{1}{2\pi} \int_{-\pi}^{\pi} P_{XX}(\omega)\,d\omega. \tag{3.127}$$

Similarly, the inverse transform of Eq. (3.125) yields

$$\gamma_{XX}[\ell] = \frac{1}{2\pi} \int_{-\pi}^{\pi} \Gamma_{XX}(e^{j\omega})e^{j\omega\ell}\,d\omega, \tag{3.128}$$

from which and Eq. (2.157) we get

$$\sigma_X^2 = \gamma_{XX}[0] = \frac{1}{2\pi} \int_{-\pi}^{\pi} \Gamma_{XX}(e^{j\omega})\,d\omega$$

$$= \frac{1}{2\pi} \int_{-\pi}^{\pi} P_{XX}(\omega)\,d\omega - |m_X|^2. \tag{3.129}$$

Applying the Fourier transform to both sides of Eq. (2.160a), we can show that the power spectral density $P_{XX}(\omega)$ of a WSS random discrete-time signal $\{X[n]\}$ is a real-valued function of ω. In addition, if $\{X[n]\}$ is a real random signal, $P_{XX}(\omega)$ is an even function of ω, i.e., $P_{XX}(\omega) = P_{XX}(-\omega)$. We shall demonstrate later in Section 4.11 that for a real-valued WSS random signal, $P_{XX}(\omega) \geq 0$.

Likewise, the Fourier transform of the cross-correlation sequence $\gamma_{XY}[\ell]$ of two jointly stationary random signals $\{X[n]\}$ and $\{Y[n]\}$ given by

$$\Phi_{XY}(e^{j\omega}) = \sum_{\ell=-\infty}^{\infty} \phi_{XY}[\ell]e^{-j\omega\ell}, \qquad |\omega| < \pi \qquad (3.130)$$

is usually referred to as the *cross-power spectral density* $P_{XY}(\omega)$ which is, in general, a complex function of ω. Applying the Fourier transform to both sides of Eq. (2.160c), we can show that $P_{XY}(\omega) = P_{YX}^*(\omega)$. Similarly, we can define the Fourier transform of the cross-covariance sequence $\gamma_{XY}[\ell]$ as

$$\Gamma_{XY}(e^{j\omega}) = \sum_{\ell=-\infty}^{\infty} \gamma_{XY}[\ell]e^{-j\omega\ell}, \qquad |\omega| < \pi. \qquad (3.131)$$

The relation between the Fourier transforms of the autocorrelation sequence and the autocovariance sequence can be derived from Eq. (2.155) and is given by

$$\Gamma_{XX}(e^{j\omega}) = P_{XX}(\omega) - 2\pi |m_X|^2 \delta(\omega), \qquad |\omega| < \pi, \qquad (3.132)$$

where we have used the notation $P_{XX}(\omega) = \Phi_{XX}(e^{j\omega})$. Likewise, the relation between the Fourier transforms of the cross-correlation sequence and the cross-covariance sequence follows from Eq. (2.159) and is given by

$$\Gamma_{XY}(e^{j\omega}) = P_{XY}(\omega) - 2\pi m_X m_Y^* \delta(\omega), \qquad |\omega| < \pi, \qquad (3.133)$$

where we have used the notation $P_{XY}(\omega) = \Phi_{XY}(e^{j\omega})$.

3.10.2 z-Transform Representation

As can be seen from Eqs. (3.132) and (3.133), the Fourier transforms of the sequences $\gamma_{XX}[\ell]$ and $\gamma_{XY}[\ell]$ contain impulse functions. As a result, their z-transforms do not exist in general. However, for zero-mean stationary random signals, the z-transform of the autocorrelation sequence, $\Phi_{XX}(z)$, and that of the cross-correlation sequence, $\Phi_{XY}(z)$, may exist under certain conditions. Since the autocorrelation and cross-correlation sequences are two-sided sequences, their region of convergence must be an annular region of the form

$$R_1 < |z| < \frac{1}{R_1}. \qquad (3.134)$$

We can generalize some of the results of the previous section if the z-transforms exist. For example, from the symmetry properties of the power spectral density $P_{XX}(\omega)$ and cross-power spectral density $P_{XY}(\omega)$, it follows that $\Phi_{XX}(z) = \Phi_{XX^*}(1/z^*)$ and $\Phi_{XY}(z) = \Phi_{YX^*}(1/z^*)$. It also follows from Eqs. (3.129) and (3.132) that

$$\sigma_X^2 = \frac{1}{2\pi j} \oint_C \Phi_{XX}(z)z^{-1}\, dz, \qquad (3.135)$$

where C is a closed counterclockwise contour in the ROC of $\Phi_{XX}(z)$.

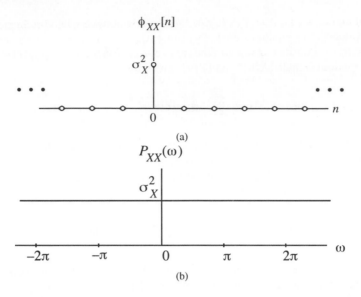

Figure 3.21   (a) Autocorrelation sequence and (b) power spectral density of a white noise.

3.10.3   White Noise

A random process $\{X[n]\}$ for which any pair of samples, $X[m]$ and $X[n]$ are uncorrelated, i.e., $E(X[m]X[n]) = E(X[m])E(X[n])$ with $m \neq n$ is called a white random process. For a WSS white random process, the autocorrelation sequence is given by

$$\phi_{XX}[\ell] = \sigma_X^2 \delta[\ell] + m_X^2, \tag{3.136}$$

and the corresponding power spectral density is given by

$$P_{XX}(\omega) = \sigma_X^2 + 2\pi m_X^2 \delta(\omega), \qquad |\omega| < \pi. \tag{3.137}$$

A zero-mean white WSS random process has an autocorrelation sequence that is an impulse sequence of area σ_x^2 and a power spectrum $P_{XX}(\omega)$ that is of constant value σ_x^2 for all values of ω, as indicated in Fig. 3.21. Such a random process is more commonly called *white noise* and plays an important role in digital signal processing.

3.11   Summary

Three different frequency-domain representations of an aperiodic discrete-time sequence have been introduced and their properties reviewed. Two of these representations, the discrete-time Fourier transform (DTFT) and the z-transform, are applicable to any arbitrary sequences, whereas the third one, the discrete Fourier transform (DFT), can be applied only to finite-length sequences. Each of these representations consists of a pair of expressions: the analysis equation and the synthesis equation. The analysis equation is used to convert from the time-domain representation to the frequency-domain representation, while the synthesis equation is used for the reverse process.

Relations between these three transforms have been established. The chapter ends with a discussion on the transform-domain representation of a random discrete-time sequence.

For future convenience we summarize below these three frequency-domain representations.

Discrete-time Fourier transform (DTFT)

Analysis equation:

$$X(e^{j\omega}) = \sum_{n=-\infty}^{\infty} x[n]e^{-j\omega n}. \tag{3.138}$$

Synthesis equation:

$$x[n] = \frac{1}{2\pi} \int_{-\pi}^{\pi} X(e^{j\omega})e^{j\omega n}\, d\omega. \tag{3.139}$$

Discrete Fourier Transform (DFT)

Analysis Equation:

$$X[k] = \sum_{n=0}^{N-1} x[n]W_N^{kn}, \qquad 0 \le k \le N-1, \tag{3.140}$$

Synthesis Equation:

$$x[n] = \frac{1}{N} \sum_{k=0}^{N-1} X[k]W_N^{-kn}, \qquad 0 \le n \le N-1, \tag{3.141}$$

where

$$W_N = e^{-j2\pi/N}. \tag{3.142}$$

z-Transform

Analysis Equation:

$$X(z) = \sum_{n=-\infty}^{\infty} x[n]z^{-n}. \tag{3.143}$$

Synthesis Equation:

$$x[n] = \frac{1}{2\pi j} \oint_C X(z)z^{n-1}\, dz, \quad C \text{ in ROC of } X(z). \tag{3.144}$$

We apply the concepts developed in this chapter to the representation, analysis, and design of the so-called linear, time-invariant (LTI) discrete-time systems in the following chapter.

3.12 Problems

3.1 Show that a sufficient condition for the discrete-time Fourier transform $X(e^{j\omega})$ of a sequence $x[n]$ to converge is that it be absolutely summable.

3.2 Derive the discrete-time Fourier transform pairs given in Table 3.1.

3.3 Prove the general properties of the discrete-time Fourier transform listed in Table 3.2.

3.4 Prove the symmetry properties of the discrete-time Fourier transform of a complex sequence listed in Table 3.3.

3.5 Prove the symmetry properties of the discrete-time Fourier transform of a real sequence listed in Table 3.4.

3.6 Let $X(e^{j\omega})$ denote the discrete-time Fourier transform of a real sequence $x[n]$.

(a) Show that $x[n]$ can be computed from $X(e^{j\omega})$ using the following expression if $x[n]$ is even:

$$x[n] = \frac{1}{\pi} \int_0^\pi X(e^{j\omega}) \cos \omega n \, d\omega.$$

(b) If $x[n]$ is odd, then show that $x[n]$ can be computed from $X(e^{j\omega})$ using the expression

$$x[n] = \frac{j}{\pi} \int_0^\pi X(e^{j\omega}) \sin \omega n \, d\omega.$$

3.7 Determine the discrete-time Fourier transform of the causal sequence

$$x[n] = A\alpha^n \cos(\omega_0 n + \phi)\mu[n],$$

where A, α, ω_0, and ϕ are real.

3.8 Determine the discrete-time Fourier transform of each of the following sequencës.

(a) $x_1[n] = \alpha^n \mu[n], \quad |\alpha| < 1,$ (b) $x_2[n] = \alpha^n \mu[-n], \quad |\alpha| > 1,$

(c) $x_3[n] = \begin{cases} \alpha^{|n|}, & |n| \le M, \\ 0, & \text{otherwise}, \end{cases}$ (d) $x_4[n] = \alpha^n \mu[n+3], \quad |\alpha| < 1,$

(e) $x_5[n] = n\alpha^n \mu[n], \quad |\alpha| < 1.$

3.9 Determine the discrete-time Fourier transform of each of the following sequences.

(a) $y_1[n] = \begin{cases} 1 & -N \le n \le N, \\ 0, & \text{otherwise}, \end{cases}$

(b) $y_2[n] = \begin{cases} 1 - \frac{|n|}{N}, & -N \le n \le N, \\ 0, & \text{otherwise}, \end{cases}$

(c) $y_3[n] = \begin{cases} \cos(\pi n/2N), & -N \le n \le N, \\ 0, & \text{otherwise}. \end{cases}$

3.10 Evaluate the inverse discrete-time Fourier transform of each of the following Fourier transforms.

(a) $X_a(e^{j\omega}) = \sum_{k=-\infty}^{\infty} \delta(\omega + 2\pi k)$,   (b) $X_b(e^{j\omega}) = \dfrac{1 - e^{j\omega(N+1)}}{1 - e^{j\omega}}$,

(c) $X_c(e^{j\omega}) = 1 + 2\sum_{\ell=0}^{N} \cos \omega\ell$,   (d) $X_d(e^{j\omega}) = \dfrac{j\alpha e^{j\omega}}{(1 - \alpha e^{j\omega})^2}$, $|\alpha| < 1$.

3.11 Let $x[n]$ be a finite-length sequences given by

$$\{x[n]\} \quad = \quad \{2 \quad 1 \quad -1 \quad 0 \quad 3 \quad 2 \quad 0 \quad -3 \quad 4\}$$
$$\uparrow$$

with a discrete-time Fourier transform $X(e^{j\omega})$. Evaluate the following functions of $X(e^{j\omega})$ without computing the transform itself:

(a) $X(e^{j0})$,   (b) $X(e^{j\pi})$,

(c) $\int_{-\pi}^{\pi} X(e^{j\omega})\, d\omega$,   (d) $\int_{-\pi}^{\pi} |X(e^{j\omega})|^2\, d\omega$,

(e) $\int_{-\pi}^{\pi} \left| \dfrac{dX(e^{j\omega})}{d\omega} \right|^2 d\omega$

3.12 Repeat Problem 3.11 for the sequence

$$\{x[n]\} \quad = \{1 \quad 5 \quad -2 \quad 1 \quad 3 \quad 4 \quad 2 \quad 0 \quad 5\}$$
$$\uparrow$$

3.13 Let $G_1(e^{j\omega})$ denote the discrete-time Fourier transform of the sequence $g_1[n]$ shown in Figure P3.1(a). Express the discrete-time Fourier transforms of the remaining sequences in Figure P3.1 in terms of $G_1(e^{j\omega})$.

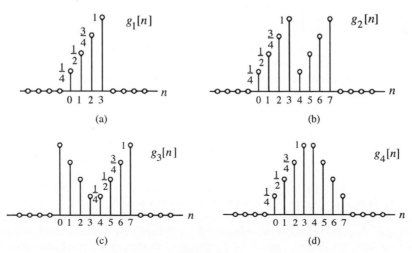

Figure P3.1

3.14 Let $y[n]$ denote the linear convolution of two sequences, $g[n]$ and $h[n]$, i.e., $y[n] = g[n] \circledast h[n]$. Show that

(a) $\displaystyle\sum_{n=-\infty}^{\infty} y[n] = \left(\sum_{n=-\infty}^{\infty} g[n]\right)\left(\sum_{n=-\infty}^{\infty} h[n]\right),$

(b) $\displaystyle\sum_{n=-\infty}^{\infty} (-1)^n y[n] = \left(\sum_{n=-\infty}^{\infty} (-1)^n g[n]\right)\left(\sum_{n=-\infty}^{\infty} (-1)^n h[n]\right).$

3.15 Let $x[n]$ be a causal and absolutely summable real sequence with a discrete-time Fourier transform $X(e^{j\omega})$. If $X_{re}(e^{j\omega})$ and $X_{im}(e^{j\omega})$ denote the real and imaginary parts of $X(e^{j\omega})$, show that they are related as

$$X_{im}(e^{j\omega}) = -\frac{1}{2\pi}\int_{-\pi}^{\pi} X_{re}(e^{j\phi})\cot\left(\frac{\omega-\phi}{2}\right)d\phi, \tag{3.145a}$$

$$X_{re}(e^{j\omega}) = \frac{1}{2\pi}\int_{-\pi}^{\pi} X_{im}(e^{j\phi})\cot\left(\frac{\omega-\phi}{2}\right)d\phi + x[0]. \tag{3.145b}$$

The above pair of equations are called the *discrete Hilbert transform relations*.

3.16 Verify the identity of Eq. (3.18).

3.17 The *periodic convolution* of two periodic sequences, $\tilde{x}[n]$ and $\tilde{h}[n]$, of period N each, is defined by

$$\tilde{y}[n] = \sum_{r=0}^{N-1} \tilde{x}[r]\tilde{h}[n-r]. \tag{3.146}$$

Show that $\tilde{y}[n]$ is also a periodic sequence of period N.

3.18 Determine the periodic sequence $\tilde{y}[n]$ obtained by a periodic convolution of the following two periodic sequences of period 4 each:

$$\tilde{x}[n] = \begin{cases} 0, & \text{for } n = 0, \\ 1, & \text{for } n = 1, \\ 0, & \text{for } n = 2, \\ 2, & \text{for } n = 3, \end{cases} \qquad \tilde{h}[n] = \begin{cases} 2, & \text{for } n = 0, \\ 0, & \text{for } n = 1, \\ 1, & \text{for } n = 2, \\ 0, & \text{for } n = 3. \end{cases}$$

3.19 Determine the periodic sequence $\tilde{y}[n]$ obtained by a periodic convolution of the following two periodic sequences of period 4 each:

$$\tilde{x}[n] = \begin{cases} 2, & \text{for } n = 0, \\ -1, & \text{for } n = 1, \\ 0, & \text{for } n = 2, \\ 3, & \text{for } n = 3, \end{cases} \qquad \tilde{h}[n] = \begin{cases} -1, & \text{for } n = 0, \\ 2, & \text{for } n = 1, \\ 1, & \text{for } n = 2, \\ 0, & \text{for } n = 3. \end{cases}$$

3.20 Let $\tilde{x}[n]$ be a periodic sequence with period N, i.e., $\tilde{x}[n] = \tilde{x}[n + \ell N]$, where ℓ is any integer. The sequence $\tilde{x}[n]$ can be represented by a Fourier series given by a weighted sum of periodic complex exponential sequences $\tilde{\Psi}_k[n] = e^{j2\pi kn/N}$. Show that, unlike the Fourier series representation of a periodic continuous-time signal, the Fourier series representation of a periodic discrete-time sequence requires only N of the periodic complex exponential sequences $\tilde{\Psi}_k[n]$, $k = 0, 1, \ldots, N - 1$, and is of the form

$$\tilde{x}[n] = \frac{1}{N}\sum_{k=0}^{N-1} \tilde{X}[k]e^{j2\pi kn/N}, \tag{3.147a}$$

where the Fourier coefficients $\tilde{X}[k]$ are given by

$$\tilde{X}[k] = \sum_{k=0}^{N-1} \tilde{x}[n] e^{-j2\pi kn/N}. \tag{3.147b}$$

Show that $\tilde{X}[k]$ is also a periodic sequence in k with a period N. The set of equations in Eq. (3.147) represents the *discrete Fourier series* pair.

3.21 Determine the discrete Fourier series coefficients of the following periodic sequences:

(a) $\tilde{x}_1[n] = \sin(\pi n/4)$,

(b) $\tilde{x}_2[n] = 2\sin(\pi n/4) + \cos(\pi n/3)$.

3.22 Show that the periodic impulse train

$$\tilde{p}[n] = \sum_{\ell=-\infty}^{\infty} \delta[n + \ell N]$$

can be expressed in the form

$$\tilde{p}[n] = \frac{1}{N} \sum_{\ell=-\infty}^{\infty} e^{j2\pi kn/N}.$$

3.23 The z-transform $X(z)$ of the length-6 sequence

$$x[n] \quad = \quad \{1 \quad 2 \quad 3 \quad 4 \quad 5 \quad 6\}$$
$$\uparrow$$

is sampled at four points $\omega_k = \pi k/2, k = 0, 1, 2, 3,$ on the unit circle yielding the frequency samples

$$\tilde{X}[k] = X(z)|_{z=e^{j\pi k/2}}, \qquad k = 0, 1, 2, 3.$$

Determine, without evaluating $\tilde{X}[k]$, the periodic sequence $\tilde{x}[n]$ whose discrete Fourier series coefficients are given by $\tilde{X}[k]$. What is the period of $\tilde{x}[n]$?

3.24 Let $\tilde{x}[n]$ and $\tilde{y}[n]$ be two periodic sequences with period N. Denote their discrete Fourier series coefficients as $\tilde{X}[k]$ and $\tilde{Y}[k]$, respectively.

(a) Let $\tilde{g}[n] = \tilde{x}[n]\tilde{y}[n]$ with $\tilde{G}[k]$ denoting its discrete Fourier series coefficients. Show that $\tilde{G}[k]$ can be expressed in terms of $\tilde{X}[k]$ and $\tilde{Y}[k]$ as

$$\tilde{G}[k] = \frac{1}{N} \sum_{\ell=0}^{N-1} \tilde{X}[\ell]\tilde{Y}[k-\ell]. \tag{3.148}$$

(b) Let $\tilde{H}[k] = \tilde{X}[k]\tilde{Y}[k]$ denote the discrete Fourier series coefficients of a periodic sequence $\tilde{h}[n]$. Show that $\tilde{h}[n]$ can be expressed in terms of $\tilde{x}[n]$ and $\tilde{y}[n]$ as

$$\tilde{h}[k] = \sum_{r=0}^{N-1} \tilde{x}[r]\tilde{y}[n-r]. \tag{3.149}$$

3.25 Prove the general properties of the DFT listed in Table 3.5.

3.26 Let $X[k]$ denote the N-point DFT of a length-N sequence $x[n]$. Prove Parseval's relation for the DFT given by

$$\sum_{n=0}^{N-1} |x[n]|^2 = \frac{1}{N} \sum_{k=0}^{N-1} |X[k]|^2 \qquad (3.150)$$

3.27 Prove the symmetry properties of the DFT of a complex sequence listed in Table 3.6.

3.28 Prove the symmetry properties of the DFT of a real sequence listed in Table 3.7.

3.29 Figure P3.2 shows four length-N sequences for $N = 8$.

(a) Which sequences have real-valued DFTs?

(b) Which sequences have imaginary-valued DFTs?

Figure P3.2

3.30 Let $x[n]$ be a length-N sequence with $X[k]$ denoting its N-point DFT.

(a) If $x[n]$ is a symmetric sequence satisfying the condition $x[n] = x[N - 1 - n]$, show that $X[N/2] = 0$ for N even.

(b) If $x[n]$ is a antisymmetric sequence satisfying the condition $x[n] = -x[N - 1 - n]$, show that $X[0] = 0$.

(c) If $x[n]$ is a sequence satisfying the condition $x[n] = -x[n + M]$ with $N = 2M$, show that $X[2\ell] = 0$ for $\ell = 0, 1, \ldots, M - 1$.

3.31 Show that for a real length-N sequence with equal valued samples, its N-point DFT is real-valued and its DTFT has a linear phase.

3.32 Let $G[k]$ and $H[k]$ denote the 6-point DFTs of two length-6 sequences, $g[n]$ and $h[n]$, respectively.

(a) If $G[k] = \{1 + j \quad -2.1 + j3.2 \quad -1.2 - j2.4 \quad 0 \quad 0.9 + j3.1 \quad -0.3 + j1.1\}$ and $h[n] = g[\langle n - 4 \rangle_6]$, determine $H[k]$ without computing the DFT.

(b) If $g[n] = \{4.1 \quad 3.5 \quad 1.2 \quad 5 \quad 2 \quad 3.3\}$ and $H[k] = G[\langle k - 3 \rangle_6]$, determine $h[n]$ without computing the DFT.

3.33 Let $Y[k]$ denote MN-point DFT of a length-sequence $x[n]$ appended with $(M - 1)N$ zeros. Show that the N-point DFT $X[k]$ can be simply obtained from $Y[k]$ as follows:

$$X[k] = Y[kM], \qquad 0 \leq k \leq N - 1.$$

3.34 Let $x[n]$ be a finite-length sequence given by

$$\{x[n]\} \;=\; \{2 \quad 1 \quad 1 \quad 0 \quad 3 \quad 2 \quad 0 \quad 3 \quad 4 \quad 6\}$$
$$\uparrow$$

with a 10-point DFT given by $X[k]$. Evaluate the following functions of $X[k]$ without computing the DFT:

(a) $X[0]$,   (b) $X[5]$,   (c) $\displaystyle\sum_{k=0}^{9} X[k]$,   (d) $\displaystyle\sum_{k=0}^{9} e^{-j(4\pi k/5)} X[k]$,   (e) $\displaystyle\sum_{k=0}^{9} |X[k]|^2$.

3.35 Let $X[k]$ be a 12-point DFT of a length-12 real sequence $x[n]$. The first 7 samples of $X[k]$ are given by

$$X[0] = 10, \qquad X[1] = -5 - j4, \qquad X[2] = 3 - j2, \qquad X[3] = 1 + j3,$$
$$X[4] = 2 + j5, \qquad X[5] = 6 - j2, \qquad X[6] = 12,$$

Evaluate the following functions of $x[n]$ without computing the IDFT of $X[k]$:

(a) $x[0]$,   (b) $x[6]$,   (c) $\displaystyle\sum_{n=0}^{11} x[n]$,   (d) $\displaystyle\sum_{n=0}^{11} e^{j(2\pi n/3)} x[n]$,   (e) $\displaystyle\sum_{n=0}^{11} |x[n]|^2$.

3.36 Let $g[n]$ and $h[n]$ be two finite-length sequences of length 6 each. If $y_L[n]$ and $y_C[n]$ denote the linear and 6-point circular convolutions of $g[n]$ and $h[n]$, respectively, express $y_C[n]$ in terms of $y_L[n]$.

3.37 Let $g[n]$ and $h[n]$ be two finite-length sequences as goven below:

$$\{g[n]\} \;=\; \{5 \quad 2 \quad 4 \quad -1\}, \qquad \{h[n]\} \;=\; \{-3 \quad 4 \quad 0 \quad 2 \quad -1 \quad 2\}$$
$$\uparrow \qquad\qquad\qquad\qquad\qquad\qquad \uparrow$$

(a) Determine $y_L[n] = g[n]\circledast h[n]$.

(b) Extend $g[n]$ to length-6 sequence $g_e[n]$ by zero-padding and compute $y_C[n] = g_e[n]\,⑥\,h[n]$.

(c) Determine $y_C[n]$ using the DFT-based approach.

3.38 Compute the 6-point DFTs of $g_e[n]$ and $h[n]$ of Problem 3.37 using a single 6-point DFT.

3.39 Show that the circular convolution is commutative.

3.40 Let $y[n]$ denote the N-point circular convolution of two length-N sequences $g[n]$ and $h[n]$ defined for $0 \le n \le N - 1$, i.e., $y[n] = g[n]\,Ⓝ\,h[n]$. Prove the following equalities:

(a) $\displaystyle\sum_{n=0}^{N-1} y[n] = \left(\sum_{n=0}^{N-1} g[n]\right)\left(\sum_{n=0}^{N-1} h[n]\right),$

(b) $\displaystyle\sum_{n=0}^{N-1} (-1)^n y[n] = \left(\sum_{n=0}^{N-1} (-1)^n g[n]\right)\left(\sum_{n=0}^{N-1} (-1)^n h[n]\right)$ for N even.

3.41 Let $X[k]$ denote the N-point DFT of a length-N sequence $x[n]$. Determine the N-point DFT $Y[k]$ of the N-point sequence $y[n] = \cos((2\pi \ell n)/N)x[n]$.

3.42 The even samples of the 9-point DFT of a length-9 real sequence are given by $X[0] = 3.1$, $X[2] = 2.5 + j4.6$, $X[4] = -1.7 + j5.2$, $X[6] = 9.3 + j6.3$, $X[8] = 5.5 - j8.0$. Determine the missing odd samples of the DFT.

3.43 The following five samples of the 9-point DFT $X[k]$ of a real length-9 sequence are given: $X[0] = 23$, $X[1] = 2.2426 - j$, $X[4] = -6.3794 + j4.1212$, $X[6] = 6.5 + j2.5981$, and $X[7] = -4.1527 + j0.2645$. Determine the remaining four samples.

3.44 The samples of the 8-point DFT $X[k]$ of a length-8 sequence $x[n]$ are all real. The first five samples of $x[n]$ are given by: $x[0] = 1.125$, $x[1] = -0.6402 - j0.0795$, $x[2] = 0.25 + j0.125$, $x[3] = -0.1098 + j1.6705$, and $x[4] = 0.875$. Find the remaining three samples of $x[n]$.

3.45 A length-7 sequence $x[n]$ is given by $x[0] = 5$, $x[1] = 3$, $x[2] = -2$, $x[3] = -4$, $x[4] = 1$, $x[5] = 2$, $x[6] = 0$. Let $X[k]$ denote the 7-point DFT of $x[n]$. Without computing the IDFT, determine the sequence $y[n]$ whose 7-point DFT is given by $Y[k] = W_7^{4k} X[k]$

3.46 Let $X[k]$ denote the 6-point DFT of the length-6 real sequence $x[n]$ shown in Figure P3.3. Without computing the IDFT, determine the length-6 sequence $g[n]$ whose 6-point DFT is given by $G[k] = W_6^{3k} X[k]$.

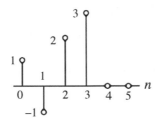

Figure P3.3

3.47 Let $x[n]$ be a length-N sequence with an N-point DFT given by $X[k]$. Assume N is divisible by 5. Define a sequence

$$y[n] = x[5n], \qquad 0 \le n \le \frac{N}{5} - 1.$$

Express the $(N/5)$-point DFT $Y[k]$ of $y[n]$ in terms of $X[k]$.

3.48 The 8-point DFT of a length-8 complex sequence $v[n] = x[n] + jy[n]$ is given by

$$V[0] = 1 - j3, \quad V[1] = -2 + j4, \quad V[2] = 3 + j7, \quad V[3] = -4 - j5,$$

$$V[4] = 2 + j5, \quad V[5] = 3 - j, \quad V[6] = 4 - j8, \quad V[7] = j6.$$

Without computing the IDFT of $V[k]$, determine the 8-point DFTs $X[k]$ and $Y[k]$ of the real sequences $x[n]$ and $y[n]$, respectively.

3.49 Determine the 4-point DFTs of the following pair of length-4 sequences by computing a single DFT:

$$\{g[n]\} = \{3 \quad 2 \quad 1 \quad 4\}, \qquad \{h[n]\} = \{2 \quad 1 \quad 1 \quad 3\}$$

3.50 Let $P[k]$ denote the M-point DFT of the numerator coefficients and $D[k]$ denote the N-point DFT of the denominator coefficients of a rational discrete-time Fourier transform $X(e^{j\omega})$ of the form of Eq. (3.12). For the following sets of 4-point DFTs, $P[k]$ and $D[k]$, determine the expressions of their corresponding DTFTs:

(a) $\{P[k]\} = \{9, -2 - j3, 3, -2 + j3\}$, $\{D[k]\} = \{0.7, 0.4 + j0.9, 2.5, 0.4 - j0.9\}$,

(b) $\{P[k]\} = \{9, -2 + j, -2, -2 - j\}$, $\{D[k]\} = \{0.5, 0.5 + j0.2, 2.5, 0.5 - j0.2\}$.

3.51 Consider a length-N sequence $x[n]$ with a DTFT $X(e^{j\omega})$. Define an M-point DFT $\hat{X}[k] = X(e^{j\omega_k})$, where $\omega_k = 2\pi k/M, k = 0, 1, \ldots, M-1$. Denote the inverse DFT of $\hat{X}[k]$ as $\hat{x}[n]$, which is a length-M sequence. Express $x[n]$ in terms of $\hat{x}[n]$ and show that $x[n]$ can be fully recovered from $\hat{x}[n]$ only if $M \geq N$.

3.52 Let $x[n]$ be a length-N sequence with $X[k]$ denoting its N-point DFT. We represent the DFT operation as $X[k] = \mathcal{F}\{x[n]\}$. Determine the sequence $y[n]$ obtained by applying the DFT operation 4 times to $x[n]$, i.e,

$$y[n] = \mathcal{F}\{\mathcal{F}\{\mathcal{F}\{\mathcal{F}\{x[n]\}\}\}\}.$$

3.53 Let $x[n]$ and $h[n]$ be two length-30 sequences defined for $0 \leq n \leq 29$. It is known that $h[n] = 0$ for $0 \leq n \leq 9$ and $20 \leq n \leq 29$. Denote the 30-point circular convolution of these two sequences as $u[n]$ and their linear convolution as $y[n]$. Determine the range of n for which $y[n] = u[n]$.

3.54 The linear convolution of a length-60 sequence with a length-1200 sequence is to be computed using 128-point DFTs and IDFTs.

(a) Determine the smallest number of DFTs and IDFTs needed to compute the above linear convolution using the overlap-add approach.

(b) Determine the smallest number of DFTs and IDFTs needed to compute the above linear convolution using the overlap-save approach.

3.55 Let $X(z)$ denote the z-transform of the length-10 sequence $x[n]$ of Problem 3.34. Let $X_0[k]$ represent the samples of $X(z)$ evaluated on the unit circle at 7 equally spaced points given by $z = e^{j(2\pi k/7)}, k = 0, 1, \ldots, 6$, i.e.,

$$X_0[k] = X(z)|_{z=e^{j(2\pi k/7)}}, \qquad k = 0, 1, \ldots, 6.$$

Determine the 7-point IDFT $x_0[n]$ of $X_0[k]$ without computing the latter function.

3.56 (a) Consider a length-N sequence $x[n]$, $0 \leq n \leq N - 1$, with an N-point DFT $X[k]$. Define a sequence $y[n]$ of length $LN, 0 \leq n \leq NL - 1$, given by

$$y[n] = \begin{cases} x[n/L], & n = 0, L, 2L, \ldots, (N-1)L, \\ 0, & \text{otherwise,} \end{cases} \qquad (3.151)$$

where L is a positive integer. Express the NL-point DFT $Y[k]$ of $y[n]$ in terms of $X[k]$.

(b) The 7-point DFT $X[k]$ of a length-7 sequence $x[n]$ is shown in Figure P3.4. Sketch the 21-point DFT $Y[k]$ of a length-21 sequence $y[n]$ generated using Eq. (3.151).

Figure P3.4

3.57 Consider two real, symmetric length-N sequences, $x[n]$ and $y[n]$, $0 \leq n \leq N - 1$ with N even. Define the length-$(N/2)$ sequences:

$$x_0[n] = x[2n + 1] + x[2n], \qquad x_1[n] = x[2n + 1] - x[2n],$$
$$y_0[n] = y[2n + 1] + y[2n], \qquad y_1[n] = y[2n + 1] - y[2n],$$

where $0 \leq n \leq \frac{N}{2} - 1$. It can be easily shown that $x_0[n]$ and $y_0[n]$ are real, symmetric sequences of length-$(N/2)$ each. Likewise, the sequences $x_1[n]$ and $y_1[n]$ are real and antisymmetric sequences. Denote the $(N/2)$-point DFTs of $x_0[n]$, $x_1[n]$, $y_0[n]$, and $y_1[n]$ by $X_0[k]$, $X_1[k]$, $Y_0[k]$, and $Y_1[k]$, respectively. Define a length-$(N/2)$ sequence $u[n]$:

$$u[n] = x_0[n] + y_1[n] + j(x_1[n] + y_0[n])$$

Determine $X_0[k]$, $X_1[k]$, $Y_0[k]$, and $Y_1[k]$ in terms of the $(N/2)$-point DFT $U[k]$ of $u[n]$.

3.58 Let $X[k]$ denote the N-point DFT of a length-N sequence $x[n]$ with N even. Define two length-$(N/2)$ sequences given by

$$g[n] = \tfrac{1}{2}(x[2n] + x[2n + 1]), \quad h[n] = \tfrac{1}{2}(x[2n] - x[2n + 1]), \quad 0 \leq n \leq \frac{N}{2} - 1.$$

If $G[k]$ and $H[k]$ denote $(N/2)$-point DFTs of $g[n]$ and $h[n]$, respectively, determine the N-point DFT $X[k]$ from these two $(N/2)$-point DFTs.

3.59 Let $X[k]$ denote the N-point DFT of a length-N sequence $x[n]$ with N even. Define two length-$(N/2)$ sequences given by

$$g[n] = a_1 x[2n] + a_2 x[2n + 1], \quad h[n] = a_3 x[2n] + a_4 x[2n + 1], \quad 0 \leq n \leq \frac{N}{2} - 1,$$

where $a_1 a_4 \neq a_2 a_3$. If $G[k]$ and $H[k]$ denote $(N/2)$-point DFTs of $g[n]$ and $h[n]$, respectively, determine the N-point DFT $X[k]$ from these two $(N/2)$-point DFTs.

3.60 Consider the z-transform

$$G(z) = \frac{(z - 0.5)(z - 0.78)(z + 2)(z - 3)}{(z^2 + 2z + 5)(z^2 - 4z + 13)}. \tag{3.152}$$

There are three nonoverlapping regions of convergence (ROCs) of this z-transform. Discuss the type of inverse z-transform (left-sided, right-sided, or two-sided sequences) associated with each of the three ROCs. It is not necessary to compute the exact inverse transform.

3.61 Consider the following sequences:

(i) $x_1[n] = (0.3)^n \mu[n]$

(ii) $x_2[n] = (-0.5)^n \mu[n]$,

(iii) $x_3[n] = (0.2)^n \mu[n - 5]$,

(iv) $x_4[n] = (-0.2)^n \mu[-n - 1]$.

(a) Determine the regions of convergence (ROC)s of the z-transform of each of the above sequences.

(b) From the ROCs determined in part (a), determine the ROCs of the following sequences:

(i) $y_1[n] = x_1[n] + x_2[n]$,

(ii) $y_2[n] = x_1[n] + x_3[n]$,

(iii) $y_3[n] = x_1[n] + x_4[n]$,

(iv) $y_4[n] = x_2[n] + x_3[n]$,

(v) $y_5[n] = x_2[n] + x_4[n]$.

3.62 Derive the z-transform pairs given in Table 3.8.

3.63 Determine the z-transforms of the sequences of Problem 3.8 and their regions of convergences (ROCs). Show that the ROC includes the unit circle for each z-transform. Evaluate the z-transform evaluated on the unit circle for each sequence and show that it is precisely the discrete-time Fourier transform of the respective sequence computed in Problem 3.8.

3.64 Determine the z-transforms of the sequences of Problem 3.9 and their regions of convergences (ROCs). Show that the ROC includes the unit circle for each z-transform. Evaluate the z-transform evaluated on the unit circle for each sequence and show that it is precisely the discrete-time Fourier transform of the respective sequence computed in Problem 3.9.

3.65 Evaluate the inverse z-transforms in closed-forms of the following functions:

(a) $Y_1(z) = \dfrac{1}{1 - 4z}$, $|z| < \dfrac{1}{4}$,

(b) $Y_2(z) = \dfrac{1}{1 - z^{-1} + 0.5z^{-2}}$, $|z| > 1$,

(c) $Y_3(z) = \dfrac{12 + 8z^{-1} - 3z^{-2}}{12 - 7z^{-1} + z^{-2}}$, $|z| > \dfrac{1}{3}$,

(d) $Y_4(z) = \dfrac{1 - z^{-1} + \frac{1}{6}z^{-2}}{(1 - \frac{1}{3}z^{-1})(1 - \frac{1}{2}z^{-1})}$, $\dfrac{1}{3} < |z| < \dfrac{1}{2}$.

3.66 Find the inverse z-transform of the following functions using the partial-fraction approach and the long division approach:

(a) $X_a(z) = z^2 + 3z + 5 + \dfrac{2}{z^2 + 5z + 4}$, $|z| > 4$.

(b) $X_b(z) = \dfrac{1 + 0.5z^{-1}}{1 - 0.75z^{-1} + 0.125z^{-2}}$, $|z| > 0.5$.

(c) $X_c(z) = \dfrac{2}{(z+0.5)(z+1)}$, $0.5 < |z| < 1$.

(d) $X_d(z) = \dfrac{z^4}{(z-0.5)^2(z-0.2)(z+0.6)}$, $|z| > 0.6$.

3.67 Show that the inverse z-transform $h[n]$ of the following rational z-transform

$$H(z) = \frac{1}{1 - 2r(\cos\theta)z^{-1} + r^2 z^{-2}}, \quad |z| > r > 0$$

is given by

$$h[n] = \frac{r^n \sin(n+1)\theta}{\sin\theta} \cdot \mu[n].$$

3.68 Prove the properties of the z-transform listed in Table 3.9.

3.69 Determinethe inverse z-transforms, $x_1[n]$ and $x_2[n]$, of the following rational z-transforms

(a) $X_1(z) = \dfrac{1}{1+z^{-3}}$, $|z| > 1$,

(b) $X_2(z) = \dfrac{1}{1+z^{-2}}$, $|z| > 1$,

by expanding each in a power series and computing the inverse z-transform of the individual terms in the power series. Compare the results with that obtained using a partial-fraction approach.

3.70 Determine the inverse z-transforms of the following z-transforms

(a) $X_1(z) = \log(1 - az^{-1})$, $|z| > |\alpha|$,

(b) $X_2(z) = \log(1 - \beta z)$, $|z| < 1/|\beta|$.

3.71 The z-transform of a right-sided sequence $h[n]$ is given by

$$H(z) = \frac{z+1}{(z-0.5)(z+0.25)}.$$

Find its inverse z-transform $h[n]$ via the partial-fraction approach. Verify the partial-fraction using MATLAB.

3.72 A generalization of the DFT concept leads to the *nonuniform discrete Fourier transform* (NDFT) $X_{\text{NDFT}}[k]$ defined by [Mit92a]

$$X_{\text{NDFT}}[k] = X(z_k) = \sum_{n=0}^{N-1} x[n]z_k^{-n}, \quad k = 0, 1, \ldots, N-1, \tag{3.153}$$

where $z_0, z_1, \ldots, z_{N-1}$ are N distinct points located arbitrarily in the z-plane. The NDFT can be expressed in a matrix form as

$$\begin{bmatrix} X_{\text{NDFT}}[0] \\ X_{\text{NDFT}}[1] \\ \vdots \\ X_{\text{NDFT}}[N-1] \end{bmatrix} = \mathbf{D}_N \begin{bmatrix} x[0] \\ x[1] \\ \vdots \\ x[N-1] \end{bmatrix}, \tag{3.154}$$

where

$$\mathbf{D}_N = \begin{bmatrix} 1 & z_0^{-1} & z_0^{-2} & \cdots & z_0^{-(N-1)} \\ 1 & z_1^{-1} & z_1^{-2} & \cdots & z_1^{-(N-1)} \\ 1 & z_2^{-1} & z_2^{-2} & \cdots & z_2^{-(N-1)} \\ \vdots & \vdots & \vdots & \ddots & \vdots \\ 1 & z_{N-1}^{-1} & z_{N-1}^{-2} & \cdots & z_{N-1}^{-(N-1)} \end{bmatrix} \qquad (3.155)$$

is the $N \times N$ NDFT matrix. The matrix $\mathbf{D}_N$ is known as the *Vandermonde matrix*. Show that it is nonsingular provided the N sampling points z_k are distinct. In which case, the inverse NDFT is given by

$$\begin{bmatrix} x[0] \\ x[1] \\ \vdots \\ x[N-1] \end{bmatrix} = \mathbf{D}_N^{-1} \begin{bmatrix} X_{\text{NDFT}}[0] \\ X_{\text{NDFT}}[1] \\ \vdots \\ X_{\text{NDFT}}[N-1] \end{bmatrix}, \qquad (3.156)$$

3.73 In general, for large N, the Vandermonde matrix is usually ill-conditioned (except for the case when the NDFT reduces to the conventional DFT), and a direct inverse computation is not advisable. A more efficient way is to directly determine the z-transform $X(z)$,

$$X(z) = \sum_{n=0}^{N-1} x[n] z^{-n}, \qquad (3.157)$$

and hence, the sequence $x[n]$, from the given N-point NDFT $X_{\text{NDFT}}[k]$ by using some type of polynomial interpolation methods. One popular method is the Lagrange interpolation formula, which expresses $X(z)$ as

$$X(z) = \sum_{k=0}^{N-1} \frac{I_k(z)}{I_k(z_k)} X_{\text{NDFT}}[k], \qquad (3.158)$$

where

$$I_k(z) = \prod_{\substack{i=0 \\ i \neq k}}^{N-1} (1 - z_i z^{-1}). \qquad (3.159)$$

Consider the z-transform $X(z) = 2 + 2z^{-1} + z^{-2} + z^{-3}$ of a length-4 sequence $x[n]$. By evaluating $X(z)$ at $z_0 = -1/2$, $z_1 = 1$, $z_2 = 1/2$, and $z_3 = 1/3$, determine the 4-point NDFT of $x[n]$ and then use the Lagrange interpolation method to show that $X(z)$ can be uniquely determined from these NDFT samples.

3.74 Another method of evaluating $X(z)$ from its N-point NDFT $X_{\text{NDFT}}[k]$ is the Newton interpolation approach in which $X(z)$ is expressed as

$$X(z) = C_0 + C_1(1 - z_0 z^{-1}) + C_2(1 - z_0 z^{-1})(1 - z_1 z^{-1}) + \cdots + C_{N-1} \prod_{k=0}^{N-2} (1 - z_k z^{-1}). \quad (3.160)$$

The coefficients $\{C_i\}$ are recursively computed as follows:

$$\begin{aligned} C_0 &= X_{\text{NDFT}}[0], \\ C_1 &= \frac{X_{\text{NDFT}}[1] - X_{\text{NDFT}}[0]}{1 - z_0 z_1^{-1}}, \end{aligned} \qquad (3.161)$$

Table 3.10   Some properties of the nonuniform discrete Fourier transform.

Property	Sequence	NDFT
	$g[n]$	$G_{\text{NDFT}}[k] = G(z_k)$
	$h[n]$	$H_{\text{NDFT}}[k]$
1. Linearity	$\alpha g[n] + \beta h[n]$	$\alpha G_{\text{NDFT}}[k] + \beta H_{\text{NDFT}}[k]$

Properties 2 to 5 hold if the sampling points $\{z_k\}$ are chosen in complex conjugate pairs.

2. Conjugation	$g^*[n]$	$G^*(z_k^*)$
3.	$\text{Re}\,\{g[n]\}$	$G_{cs}(z_k) = \frac{1}{2}\{G(z_k) + G^*(z_k^*)\}$
4.	$j\,\text{Im}\,\{g[n]\}$	$G_{ca}(z_k) = \frac{1}{2}\{G(z_k) - G^*(z_k^*)\}$

Property 5 holds if $g[n]$ is real.

5. Symmetry properties	$G(z_k) = G^*(z_k^*)$
	$\text{Re}\,\{G(z_k)\} = \text{Re}\,\{G(z_k^*)\}$
	$\text{Im}\,\{G(z_k)\} = \text{Im}\,\{G(z_k^*)\}$
	$|G(z_k)| = |G(z_k^*)|$
	$\arg\,\{G(z_k)\} = -\arg\,\{G(z_k^*)\}$

$$C_2 = \frac{X_{\text{NDFT}}[2] - X_{\text{NDFT}}[0] - C_1(1 - z_0 z_2^{-1})}{(1 - z_0 z_2^{-1})(1 - z_1 z_2^{-1})},$$

$$\vdots \tag{3.162}$$

Using the Newton interpolation method show that $X(z) = 2 + 2z^{-1} + z^{-2} + z^{-3}$ can be uniquely determined from its NDFT samples, $X(-1/2)$, $X(1)$, $X(1/2)$, and $X(1/3)$.

3.75 Prove the properties of the NDFT listed in Table 3.10.

3.76 Let $x[n]$ be a right-sided deterministic sequence with a z-transform $X(z)$ whose ROC is given by $0 < R_{x-} < |z| < R_{x+}$. Determine the ROC of $R(z)$, the z-transform of its autocorrelation sequence $r[n]$ as defined by Eq. (2.173) in Problem 2.66. Determine an expression to compute the energy contained in $r[n]$ in terms of $X(z)$ if the ROC of $X(z)$ includes the unit circle.

3.77 The *discrete cosine transform* (DCT) is used frequently in image-coding applications based on the compression of transform coefficients [Lim90]. We develop here the DCT computation algorithm via the DFT. Let $x[n]$ be a length-N sequence defined for $0 \le n \le N - 1$. First $x[n]$ is extended to a length-$2N$ by zero-padding:

$$x_e[n] = \begin{cases} x[n], & 0 \le n \le N - 1 \\ 0, & N \le n \le 2N - 1. \end{cases}$$

Then a length-$2N$ sequence $y[n]$ is formed from $x_e[n]$ according to

$$y[n] = x_e[n] + x_e[2N - 1 - n], \qquad 0 \le n \le 2N - 1,$$

and its $2N$-point DFT $Y[k]$ is computed as

$$Y[k] = \sum_{n=0}^{2N-1} y[n]W_{2N}^{nk}, \qquad 0 \le k \le 2N - 1.$$

The N-point DCT $C_x[k]$ of $x[n]$ is then defined by

$$C_x[k] = \begin{cases} W_{2N}^{k/2}Y[k] & 0 \le k \le N - 1, \\ 0 & \text{otherwise.} \end{cases}$$

Show that

$$C_x[k] = \sum_{n=0}^{N-1} 2x[n] \cos\left(\frac{(2n + 1)k\pi}{2N}\right), \qquad 0 \le k \le N - 1. \tag{3.163}$$

Note from the above definition that the DCT coefficients of a real sequence are real, whereas, in general the DFT of a real sequence is always complex. The DCT defined by Eq. (3.163) is often referred to as the even symmetrical DCT.

3.78 To form the inverse discrete cosine transform (IDCT) of an N-point DCT $C_x[k]$, first a 2N-point DFT $Y[k]$ is formed according to

$$Y[k] = \begin{cases} W_{2N}^{-k/2}C_x[k], & 0 \le k \le N - 1, \\ 0 & k = N, \\ -W_{2N}^{-k/2}C_x[2N - k], & N + 1 \le k \le 2N - 1, \end{cases}$$

and its 2N-point IDFT $y[n]$ is computed as

$$y[n] = \frac{1}{2N} \sum_{k=0}^{2N-1} Y[k]W_{2N}^{-kn}, \qquad 0 \le n \le 2N - 1.$$

The N-point IDCT of $C_x[k]$ is then given by

$$x[n] = \begin{cases} y[n], & 0 \le n \le N - 1, \\ 0, & \text{otherwise.} \end{cases}$$

(a) Show that

$$x[n] = \begin{cases} \frac{1}{N} \sum_{k=0}^{N-1} \alpha(k)C_x[k] \cos\left(\frac{(2n+1)\pi k}{2N}\right), & 0 \le n \le N - 1, \\ 0, & \text{otherwise,} \end{cases} \tag{3.164}$$

where

$$\alpha(k) = \begin{cases} 1/2, & k = 0, \\ 1, & 1 \le k \le N - 1. \end{cases} \tag{3.165}$$

(b) Prove that $x[n]$ given by Eq. (3.164) is indeed the inverse DCT of the DCT coefficients $C_x[k]$ given by Eq. (3.163).

3.79 Let $g[n]$ and $h[n]$ be two length-N sequences with N-point DCTs given by $C_g[k]$ and $C_h[k]$, respectively. Show that the N-point DCT $C_y[k]$ of the sequence $y[n] = \alpha g[n] + \beta h[n]$ is given by $C_y[k] = \alpha C_g[k] + \beta C_h[k]$, where α and β are arbitrary constants.

3.80 If the N-point DCT of a length-N sequence $x[n]$ is given by $C_x[k]$, show that the N-point DCT of $x^*[n]$ is given by $C_x^*[k]$.

3.81 If the N-point DCT of a length-N sequence $x[n]$ is given by $C_x[k]$, show that

$$\sum_{n=0}^{N-1} |x[n]|^2 = \frac{1}{2N} \sum_{k=0}^{n-1} \alpha(k) |C_x[k]|^2,$$

where $\alpha(k)$ is given by Eq. (3.165).

3.82 The N-point *discrete Hartley transform* (DHT) $X_{\text{DHT}}[k]$ of a length-N sequence $x[n]$ is defined by [Bra83]

$$X_{\text{DHT}}[k] = \sum_{n=0}^{N-1} x[n] \left(\cos\left(\frac{2\pi nk}{N} \right) + \sin\left(\frac{2\pi nk}{N} \right) \right), \quad k = 0, 1, \ldots, N-1. \qquad (3.166)$$

As can be seen from the above, the DHT of a real sequence is also a real sequence. Show that the inverse discrete Hartley transform (DHT) is given by

$$x[n] = \frac{1}{N} \sum_{k=0}^{N-1} X_{\text{DHT}}[k] \left(\cos\left(\frac{2\pi nk}{N} \right) + \sin\left(\frac{2\pi nk}{N} \right) \right), \quad n = 0, 1, \ldots, N-1. \qquad (3.167)$$

3.83 Let $X_{\text{DHT}}[k]$ denote the N-point DHT of a length-N sequence $x[n]$.

(a) Show that the DHT of $x[\langle n - n_0 \rangle_N]$ is given by

$$X_{\text{DHT}}[k] \cos\left(\frac{2\pi n_0 k}{N} \right) + X_{\text{DHT}}[-k] \sin\left(\frac{2\pi n_0 k}{N} \right),$$

(b) Determine the N-point DHT of $x[\langle -n \rangle_N]$,

(c) Parseval's relation: Show that

$$\sum_{n=0}^{N-1} x^2[n] = \frac{1}{N} \sum_{k=0}^{N-1} X_{\text{DHT}}^2[k]. \qquad (3.168)$$

3.84 Develop the relation between the N-point DHT $X_{\text{DHT}}[k]$ and the N-point DFT $X[k]$ of a length-N sequence $x[n]$.

3.85 Let the N-point DHTs of the three length-N sequences $x[n]$, $g[n]$, and $y[n]$ be denoted by $X_{\text{DHT}}[k]$, $G_{\text{DHT}}[k]$, and $Y_{\text{DHT}}[k]$, respectively. If $y[n] = x[n] \circledN g[n]$, show that

$$Y_{\text{DHT}}[k] = \tfrac{1}{2} X_{\text{DHT}}[k](G_{\text{DHT}}[k] + G_{\text{DHT}}[\langle -k \rangle_N])$$

$$+ \tfrac{1}{2} X_{\text{DHT}}[\langle -k \rangle_N](G_{\text{DHT}}[k] - G_{\text{DHT}}[\langle -k \rangle_N]). \qquad (3.169)$$

3.86 The Hadamard transform $X_{HT}[k]$ of a length-N sequence $x[n]$, $n = 0, 1, \ldots, N-1$, is given by [Gon87]

$$X_{HT}[k] = \frac{1}{N} \sum_{n=0}^{N-1} x[n](-1)^{\sum_{i=0}^{\ell-1} b_i(n)b_i(k)}, \qquad k = 0, 1, \ldots, N-1, \qquad (3.170)$$

where $b_i(r)$ is the ith bit in the binary representation of r, and $N = 2^\ell$. In matrix form, the Hadamard transform can be represented as

$$\mathbf{X}_{HT} = \mathbf{H}_N \mathbf{x},$$

where

$$\mathbf{X}_{HT} = [X_{HT}[0] \quad X_{HT}[1] \quad \cdots \quad X_{HT}[N-1]]^T,$$
$$\mathbf{x} = [x[0] \quad x[1] \quad \cdots \quad x[N-1]]^T.$$

(a) Determine the form of the Hadamard matrix $\mathbf{H}_N$ for $N = 2, 4,$ and 8.

(b) Show that

$$\mathbf{H}_4 = \begin{bmatrix} \mathbf{H}_2 & \mathbf{H}_2 \\ \mathbf{H}_2 & -\mathbf{H}_2 \end{bmatrix}, \qquad \mathbf{H}_8 = \begin{bmatrix} \mathbf{H}_4 & \mathbf{H}_4 \\ \mathbf{H}_4 & -\mathbf{H}_4 \end{bmatrix}.$$

(c) Determine the expression for the inverse Hadamard transform.

3.87 In this problem we consider the computation of a limited number of NDFT samples $X(z_\ell)$ of a length-N sequence $x[n]$ at $z_\ell = AV^{-\ell}$, $\ell = 0, 1, \ldots, L-1$ where $A = A_0 e^{j\theta_0}$ and $V = V_0 e^{j\phi_0}$ with A_0 and V_0 being positive real numbers, and in general $L < N$. The contour spirals toward the origin as ℓ increases if $V_0 > 1$ and it spirals outward if $V_0 < 1$. The *chirp-z transform* (CZT) is then defined by [Rab69]

$$X(z_\ell) = \sum_{n=0}^{N-1} x[n] A^{-n} V^{\ell n}, \qquad \ell = 0, 1, \ldots, L-1. \qquad (3.171)$$

Convert the above expression into a convolution sum using the identity

$$\ell n = \tfrac{1}{2}(\ell^2 + n^2 - (\ell - n)^2),$$

and show that the discrete-time system of Figure P3.5 can be employed to compute the CZT.

Figure P3.5

3.88 We wish to compute the L-point chirp-z transform (CZT) samples $X(z_\ell)$, $\ell = 0, 1, 2, \ldots, L-1$, of a length-$N$ sequence $x[n]$ according to Eq. (3.171), where $z_\ell = AV^{-\ell}$ with $A = A_0 e^{j\theta_0}$ and $V = V_0 e^{j\phi_0}$. What are the values of A_0, θ_0, V_0 and ϕ_0 if the CZT needs to be calculated at points $\{z_\ell\}$ on the real axis in the z-plane such that $z_\ell = \alpha^\ell$, $0, 1, 2, \ldots, L-1$, for α real and $\alpha \neq \pm 1$?

3.89 Let $y_L[n]$ denote the length-$(2N-1)$ sequence obtained by a linear convolution of two length-N sequences $h[n]$ and $x[n]$, i.e., $y_L[n] = x[n] \circledast h[n]$. From the convolution theorem of z-transform (see Table 3.9) it is known that $Y_L(z) = X(z)H(z)$, where $Y_L(z)$, $X(z)$, and $H(z)$ denote the z-transforms of the sequences $y_L[n]$, $x[n]$, and $h[n]$, respectively. This implies that the samples of the sequence $y_L[n]$ are simply given by the coefficients of the product of the two polynomials $X(z)$ and $H(z)$. Now, let $y_C[n]$ denote the N-point circular convolution of $x[n]$ and $h[n]$, i.e., $y_C[n] = x[n] \textcircled{N} h[n]$. It can be shown that the samples of the sequence $y_C[n]$ can be obtained from the coefficients of the polynomial $Y_C(z) = Y_L(z) \bmod (1 - z^{-N})$. Verify the above result for $N = 3, 4$, and 5.

Note: $(y[N+m]z^{-N-m}) \bmod (1-z^{-N}) = y[N+m]z^{-m}$ where $0 \le m < N$.

3.90 How would you compute the dc value of a finite-length sequence from its DFT?

3.91 Consider a sequence $x[n]$ with a z-transform $X(z)$. Define a new z-transform $\hat{X}(z)$ given by the complex natural logarithm of $X(z)$, i.e., $\hat{X}(z) = \log X(z)$. The inverse z-transform of $\hat{X}(z)$ to be denoted by $\hat{x}[n]$ is called the *complex cepstrum* of $x[n]$ [Tri79]. Assume that the the ROCs of both $X(z)$ and $\hat{X}(z)$ include the unit circle.

(a) Relate the DTFT $X(e^{j\omega})$ of $x[n]$ to the DTFT $\hat{X}(e^{j\omega})$ of its complex cepstrum $\hat{x}[n]$.

(b) Show that the complex cepstrum of a real sequence is a real-valued sequence.

(c) Let $\hat{x}_{ev}[n]$ and $\hat{x}_{od}[n]$ denote, respectively, the even and odd parts of a real-valued complex cepstrum $\hat{x}[n]$. Express $\hat{x}_{ev}[n]$ and $\hat{x}_{od}[n]$ in terms of $X(e^{j\omega})$, the DTFT of $x[n]$.

3.92 Determine the complex cepstrum $\hat{x}[n]$ of a sequence $x[n] = a\delta[n] + b\delta[n-1]$, where $|b/a| < 1$. Comment on your results.

3.93 Let $x[n]$ be a sequence with a rational z-transform $X(z)$ given by

$$X(z) = K \frac{\prod_{k=1}^{N_\alpha}(1 - \alpha_k z^{-1}) \prod_{k=1}^{N_\gamma}(1 - \gamma_k z)}{\prod_{k=1}^{N_\beta}(1 - \beta_k z^{-1}) \prod_{k=1}^{N_\delta}(1 - \delta_k z)}$$

where α_k and β_k are the zeros and poles of $X(z)$ that are strictly inside the unit circle, and $1/\gamma_k$ and $1/\delta_k$ are the zeros and poles of $X(z)$ that are strictly outside the unit circle [Rab78].

(a) Determine the exact expression for the complex cepstrum $\hat{x}[n]$ of $x[n]$.

(b) Show that $\hat{x}[n]$ is a decaying bounded sequence as $|n| \to \infty$.

(c) If $\alpha_k = \beta_k = 0$, show that $\hat{x}[n]$ is an anticausal sequence.

(d) If $\gamma_k = \delta_k = 0$, show that $\hat{x}[n]$ is a causal sequence.

3.94 Let $x[n]$ be a sequence with a rational z-transform $X(z)$ with poles and zeros strictly inside the unit circle. Show that the complex cepstrum $\hat{x}[n]$ of $x[n]$ can be computed using the recursion relation [Rab78]:

$$\hat{x}[n] = \begin{cases} 0, & n < 0, \\ \log(x[0]), & n = 0, \\ \frac{x[n]}{x[0]} - \sum_{k=0}^{n-1} \frac{k}{n} \cdot \hat{x}[k] \cdot \frac{x[n-k]}{x[0]}, & n > 0. \end{cases}$$

3.13 MATLAB Exercises

M 3.1 Using Program 3_1 determine and plot the real and imaginary parts and the magnitude and phase spectra of the following DTFT for various values of r and θ,

$$G(e^{j\omega}) = \frac{1}{1 - 2r(\cos\theta)e^{-j\omega} + r^2 e^{-j2\omega}}, \qquad 0 < r < 1.$$

M 3.2 Using Program 3_1 determine and plot the real and imaginary parts and the magnitude and phase spectra of the DTFTs of the sequences of Problem 3.9.

M 3.3 Using Program 3_1 determine and plot the real and imaginary parts, and the magnitude and phase spectra of the following DTFTs:

(a) $X(e^{j\omega}) = \dfrac{0.201(1 + 0.1915e^{j\omega} - 0.1761e^{-j2\omega} + 0.1915e^{-j\omega} + e^{-j4\omega})}{1 + 0.486e^{-j\omega} + 0.8358e^{-j2\omega} + 0.284e^{-j3\omega} + 0.4247e^{-j4\omega}},$

(b) $X(e^{j\omega}) = \dfrac{1 + 3e^{-j\omega} + 3e^{-j2\omega} + e^{-j3\omega}}{0.0736 - 0.0718e^{-j\omega} + 0.0631e^{-j2\omega} - 0.0214e^{-j3\omega}}.$

M 3.4 Using MATLAB verify the general properties of the DTFT listed in Table 3.2.

M 3.5 Using MATLAB verify the symmetry relations of the DTFT of a complex sequence as listed in Table 3.3.

M 3.6 Using MATLAB verify the symmetry relations of the DTFT of a real sequence as listed in Table 3.4.

M 3.7 Write a MATLAB program to compute the $N \times N$ DFT matrix $\mathbf{D}_N$ of Eq. (3.34) and then form its inverse. Using this program verify the relation given by Eq. (3.37) for $N = 3, 4, 5$ and 6.

M 3.8 Using MATLAB compute the N-point DFTs of the length-N sequences of Problem 3.9 for $N = 3, 5, 7,$ and 10. Compare your results with that obtained by evaluating the DTFTs computed in Problem 3.9 at $\omega = 2\pi k/N, k = 0, 1, \ldots, N - 1$.

M 3.9 Write a MATLAB program to compute the circular convolution of two length-N sequences via the DFT-based approach. Using this program determine the circular convolution of the following pairs of sequences:

(a) $g[n] = \{1, -3, 4, 2, 0, -2\}, \quad h[n] = \{3, 0, 1, -1, 2, 1\},$

(b) $x[n] = \{3 + j2, -2 + j, 0 + j3, 1 + j4, -3 + j3\},$
$\quad v[n] = \{1 - j3, 4 + j2, 2 - j2, -3 + j5, 2 + j\},$

(c) $x[n] = \cos(\pi n/2), \qquad y[n] = 3^n, \qquad 0 \le n \le 5.$

M 3.10 Using MATLAB verify the general properties of the DFT listed in Table 3.5.

M 3.11 Using MATLAB verify the symmetry relations of the DFT of a complex sequence as listed in Table 3.6.

M 3.12 Using MATLAB verify the symmetry relations of the DFT of a real sequence as listed in Table 3.7.

M 3.13 Verify the results of Problem 3.34 by computing the DFT $X[k]$ of the sequence $x[n]$ given using MATLAB and then evaluate the functions of $X[k]$ listed.

M 3.14 Verify the results of Problem 3.35 by computing the IDFT $x[n]$ of the DFT $X[k]$ given using MATLAB and then evaluate the functions of $x[n]$ listed.

M 3.15 Using MATLAB determine the factored form of the following z-transforms

(a) $G_1(z) = \dfrac{2z^4 + 16z^3 + 44z^2 + 56z + 32}{3z^4 + 3z^3 - 15z^2 + 18z - 12}$,

(b) $G_2(z) = \dfrac{4z^4 - 8.68z^3 - 17.98z^2 + 26.74z - 8.04}{z^4 - 2z^3 + 10z^2 + 6z + 65}$,

and show their pole-zero plots. Determine all regions of convergence (ROCs) of each of the above z-transforms, and describe the type of their inverse z-transforms (left-sided, right-sided, two-sided sequences) associated with each of the ROCs.

M 3.16 Using Program 3_9 determine the partial-fraction expansions of the z-transforms listed in Problem 3.66 and then determine their inverse z-transforms.

M 3.17   (a) Using Program 3_10 determine the z-transform as a ratio of two polynomials in z^{-1} from each of the partial-fraction expansions listed below:

(i) $X_1(z) = 3 + \dfrac{14}{2 - z^{-1}} - \dfrac{16}{4 - z^{-1}}$, $|z| > 0.5$,

(ii) $X_2(z) = 2.5 + \dfrac{3}{1 + 0.5z^{-1}} - \dfrac{4 - z^{-1}}{1 + 0.25z^{-2}}$, $|z| > 0.5$,

(iii) $X_3(z) = \dfrac{25}{(5 + 2z^{-1})^2} - \dfrac{10}{5 + 2z^{-1}} + \dfrac{4}{1 + 0.9z^{-2}}$, $|z| > 0.4$,

(iv) $X_4(z) = 4 + \dfrac{10}{5 + 2z^{-1}} + \dfrac{z^{-1}}{6 + 5z^{-1} + z^{-2}}$, $|z| > 0.4$.

(b) From the partial-fractions developed in part (a), determine the inverse z-transform of each z-transform given.

M 3.18 Using Program 3_11 determine the first 30 samples of the inverse z-transforms of the rational z-transforms determined in part (a) of Problem M3.17. Show that these samples are identical to those obtained by explicitly evaluating the exact inverse z-transforms computed in part (b) of Problem M3.17.

M 3.19 Write a MATLAB program to compute the NDFT and the inverse NDFT using the Lagrange interpolation method. Verify your program by computing the NDFT of a length-25 sequence and reconstructing the sequence from its computed NDFT.

M 3.20 Write a MATLAB program to compute the NDFT and the inverse NDFT using the Newton interpolation method. Verify your program by computing the NDFT of a length-25 sequence and reconstructing the sequence from its computed NDFT.

CHAPTER 4

Transform-Domain Representations of LTI Discrete-Time Systems

We showed in Section 2.5 that a linear, time-invariant (LTI) discrete-time system is completely characterized in the time-domain by its impulse response sequence $\{h[n]\}$. As a result, the transform-domain representation of a discrete-time signal can equally be applied to the transform-domain representation of an LTI discrete-time system. Such transform-domain representations provide additional insight into the behavior of such systems and also make it easier to design and implement them for specific applications.

In this chapter we discuss the use of the DTFT and the z-transform in transforming the various time-domain representations of an LTI discrete-time system to alternative characterizations. Specific properties of such transform-domain representations are investigated and several simple applications are considered. As in the earlier chapters, MATLAB has been used extensively to illustrate various concepts and applications.

4.1 The Frequency Response

Most discrete-time signals encountered in practice can be represented as a linear combination of a very large, may be infinite, number of sinusoidal discrete-time signals of different angular frequencies. Thus, knowing the response of the LTI system to a single sinusoidal signal, we can determine its response to more complicated signals by making use of superposition property of the system. Since a sinusoidal signal can be expressed in terms of exponential signal, the response of the LTI system to an exponential input is of practical interest. This leads to the concept of frequency response, a transform-domain representation of the LTI discrete-time system. We first define the frequency response, investigate its properties, and describe some of its applications. Computation of the time-domain representation of the LTI system from its frequency response is outlined.

4.1.1   Definition

Recall from Section 2.5.1, the input-output relationship of an LTI discrete-time system with an impulse response $h[n]$ is given by the convolution sum of Eq. (2.61) and is of the form

$$y[n] = \sum_{k=-\infty}^{\infty} h[k]x[n-k] \tag{4.1}$$

where $y[n]$ and $x[n]$ are, respectively, the output and the input sequences. Now if the input $x[n]$ is a complex exponential of the form

$$x[n] = e^{j\omega n}, \qquad -\infty < n < \infty, \tag{4.2}$$

then from Eq. (4.1) the output is given by

$$y[n] = \sum_{k=-\infty}^{\infty} h[k]e^{j\omega(n-k)} = \left(\sum_{k=-\infty}^{\infty} h[k]e^{-j\omega k} \right) e^{j\omega n}, \tag{4.3}$$

which can be rewritten as

$$y[n] = H(e^{j\omega})e^{j\omega n}, \tag{4.4}$$

where we have used the notation

$$H(e^{j\omega}) = \sum_{n=-\infty}^{\infty} h[n]e^{-j\omega n}. \tag{4.5}$$

The quantity $H(e^{j\omega})$ defined above is called the *frequency response* of the LTI discrete-time system and it provides a frequency-domain description of the system. Note from Eq. (4.5) that $H(e^{j\omega})$ is precisely the discrete-time Fourier transform (DTFT) of the impulse response $h[n]$ of the system.

Equation (4.4) implies that for a complex sinusoidal input sequence $x[n]$ of angular frequency ω as in Eq. (4.2), the output $y[n]$ is also a complex sinusoidal sequence of the same angular frequency but weighted by a complex amplitude $H(e^{j\omega})$ that is a function of the input frequency ω and the system's impulse response coefficients $h[n]$. We shall later show in Section 4.2 $H(e^{j\omega})$ completely characterizes the LTI discrete-time system in the frequency domain.

Just like any other discrete-time Fourier transform, in general, $H(e^{j\omega})$ is also a complex function of ω with a period 2π and can be expressed in terms of its real and imaginary parts or its magnitude and phase. Thus,

$$H(e^{j\omega}) = H_{\text{re}}(e^{j\omega}) + jH_{\text{im}}(e^{j\omega}) = |H(e^{j\omega})|e^{j\theta(\omega)}, \tag{4.6}$$

where $H_{\text{re}}(e^{j\omega})$ and $H_{\text{im}}(e^{j\omega})$ are, respectively, the real and imaginary parts of $H(e^{j\omega})$, and

$$\theta(\omega) = \arg\{H(e^{j\omega})\}. \tag{4.7}$$

The quantity $|H(e^{j\omega})|$ is called the *magnitude response* and the quantity $\theta(\omega)$ is called the *phase response* of the LTI discrete-time system. Design specifications for the discrete-time systems, in many applications, are given in terms of the magnitude response or the phase response or both. In some cases, the magnitude function is specified in decibels as defined below:

$$\mathcal{G}(\omega) = 20 \log_{10} |H(e^{j\omega})| \quad \text{dB} \tag{4.8}$$

where $\mathcal{G}(\omega)$ is called the *gain function*. The negative of the gain function, $a(\omega) = -\mathcal{G}(\omega)$, is called the *attenuation* or *loss function*.

It should be noted that the magnitude and phase functions are real functions of ω, whereas, the frequency response is a complex function of ω. For a discrete-time system characterized by a real impulse response $h[n]$ it follows from Table 3.4 that the magnitude function is an even function of ω, i.e., $|H(e^{j\omega})| = |H(e^{-j\omega})|$, and the phase function is an odd function of ω, i.e., $\theta(\omega) = -\theta(-\omega)$. Likewise, $H_{\mathrm{re}}(e^{j\omega})$ is even, and $H_{\mathrm{im}}(e^{j\omega})$ is odd.

The M-file function $\texttt{freqz(h,w)}$ in MATLAB can be used to determine the values of the frequency response of a prescribed impulse response vector h at a set of given frequency points w. From these frequency response values, one can then compute the real and imaginary parts using the functions $\texttt{real}$ and $\texttt{imag}$, and the magnitude and phase using the functions $\texttt{abs}$ and $\texttt{angle}$ as illustrated in the following example.

Example 4.1 Consider the moving average filter of Eq. (2.53). Comparing Eq. (2.53) with Eq. (2.82) we note that the impulse response of the moving average filter is given by

$$h[n] = \begin{cases} \frac{1}{M}, & 0 \le n \le M - 1, \\ 0, & \text{otherwise.} \end{cases} \tag{4.9}$$

From Eq. (4.5) its frequency response is thus given by

$$
\begin{aligned}
H(e^{j\omega}) &= \frac{1}{M} \sum_{n=0}^{M-1} e^{-j\omega n} = \frac{1}{M} \left(\sum_{n=0}^{\infty} e^{-j\omega n} - \sum_{n=M}^{\infty} e^{-j\omega n} \right) \\
&= \frac{1}{M} \left(\sum_{n=0}^{\infty} e^{-j\omega n} \right) \left(1 - e^{-jM\omega} \right) = \frac{1}{M} \frac{1 - e^{-jM\omega}}{1 - e^{-j\omega}} \\
&= \frac{1}{M} \frac{\sin(M\omega/2)}{\sin(\omega/2)} e^{-j(M-1)\omega/2}.
\end{aligned}
\tag{4.10}
$$

From the above the magnitude and phase responses of the moving average filter of Eq. (4.9) are obtained as

$$|H(e^{j\omega})| = \left| \frac{1}{M} \frac{\sin(M\omega/2)}{\sin(\omega/2)} \right|, \tag{4.11}$$

$$\theta(\omega) = -\frac{(M-1)\omega}{2} + \pi \sum_{k=0}^{\lfloor M/2 \rfloor} \mu\left(\omega - \frac{2\pi k}{M}\right), \tag{4.12}$$

where $\mu(\omega)$ is a step function in ω. Figure 4.1 shows the magnitude and phase responses of the moving average filters for $M = 5$ and $M = 14$. These plots have been obtained using the following MATLAB program.

```
% Program 4_1
% Generate the filter coefficients
h1 = ones(1,5)/5;  h2 = ones(1,14)/14;
```

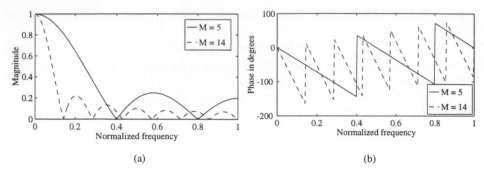

Figure 4.1   Magnitude and phase responses of the moving average filters of length 5 and 14.

```
w = 0:pi/255:pi;
%  Compute the frequency responses
H1 = freqz(h1, 1, w);
H2 = freqz(h2, 1, w);
%  Compute and plot the magnitude responses
m1 = abs(H1); m2 = abs(H2);
plot(w/pi,m1,'r-',w/pi,m2,'b--');
ylabel('Magnitude')
xlabel('Normalized angular frequency');
legend('r-','M=5','b--','M=14')
pause
%  Compute and plot the phase responses
ph1 = angle(H1)*180/pi; ph2 = angle(H2)*180/pi;
plot(w/pi,ph1,w/pi,ph2);
ylabel('Phase in degrees');
xlabel('Normalized angular frequency');
legend('r-','M=5','b--','M=14');
```

It can be seen from Figure 4.1 that in the range from $\omega = 0$ to $\omega = \pi$, the magnitude has a maximum value of unity at $\omega = 0$, and has zeros at $\omega = 2\pi k/M$ with $k = 1, 2, \ldots, \lfloor M/2 \rfloor$.[1] The phase function exhibits discontinuities of π at each zero of $H(e^{j\omega})$ and is linear elsewhere with a slope of $-(M - 1)/2$. It should be noted that both the magnitude and phase functions are periodic in ω with a period 2π.

The phase responses of discrete-time systems when determined by a computer may also exhibit jumps by an amount of 2π caused by the way the arctangent function is computed, for example, in the function angle in MATLAB. The phase response can be made a continuous function of ω by unwrapping the phase response across the jumps by adding multiples of $\pm 2\pi$. The MATLAB function unwrap can be employed to this end, provided the computed phase is in radians.[2] Figure 4.27 illustrates the application of unwrap. These jumps should not be confused with the jumps caused by the zeros of the frequency response as shown for example in Figure 4.1(b).

---

[1] $\lfloor x \rfloor$ denotes the integer part of x.

[2] Care should be taken in using unwrap as it can give wrong answers sometimes if the computed phase response is sparse with rapidly changing values.

Another parameter often used in characterizing the digital filter in the frequency domain is the *group delay function*, which provides a measure of the linearity of the phase function. If we denote the unwrapped phase function as $\theta_c(\omega)$, then the group delay is defined by

$$\tau(\omega) = -\frac{d\theta_c(\omega)}{d\omega}, \tag{4.13}$$

If the phase function is in radians, then the group delay is in seconds. For the moving average filter of Eq. (4.9) the group delay is given by

$$\tau(\omega) = \frac{M-1}{2}, \tag{4.14}$$

or in other words, the moving average filter exhibits a constant group delay.

The group delay can be determined using the function grpdelay in MATLAB. We shall illustrate its use later in this chapter.

4.1.2 Steady-State Response

Note that the frequency response function also determines the steady-state response of the LTI discrete-time system to a sinusoidal input as shown in the following example.

Example 4.2 Determine the steady-state output $y[n]$ of a real coefficient LTI discrete-time system with a frequency response $H(e^{j\omega})$ for an input

$$x[n] = A\cos(\omega_0 n + \phi), \tag{4.15}$$

where A is real.

Now using trigonometric identities, we can express the input $x[n]$ as a sum of two complex exponential sequences:

$$x[n] = g[n] + g^*[n],$$

where $g[n] = \frac{1}{2}Ae^{j\phi}e^{j\omega_0 n}$. From Eq. (4.4), the output of the LTI discrete-time system for the input $e^{j\omega_0 n}$ is simply $H(e^{j\omega_0})e^{j\omega_0 n}$. Because of linearity, the response $v[n]$ to the input $g[n]$ is then given by

$$v[n] = \frac{1}{2}Ae^{j\phi}H(e^{j\omega_0})e^{j\omega_0 n}.$$

Likewise, the output $v^*[n]$ to the input $g^*[n]$ is the complex conjugate of $v[n]$, i.e.

$$v^*[n] = \frac{1}{2}Ae^{-j\phi}H(e^{-j\omega_0})e^{-j\omega_0 n}.$$

Combining these two expressions, we arrive at the expression for the desired output $y[n]$ as

$$\begin{aligned} y[n] &= v[n] + v^*[n] \\ &= \frac{1}{2}Ae^{j\phi}H(e^{j\omega_0})e^{j\omega_0 n} + \frac{1}{2}Ae^{-j\phi}H(e^{-j\omega_0})e^{-j\omega_0 n} \\ &= \frac{1}{2}A|H(e^{j\omega_0})| \left\{ e^{j\theta(\omega_0)}e^{j\phi}e^{j\omega_0 n} + e^{-j\theta(\omega_0)}e^{-j\phi}e^{-j\omega_0 n} \right\} \\ &= A|H(e^{j\omega_0})|\cos(\omega_0 n + \theta(\omega_0) + \phi). \end{aligned} \tag{4.16}$$

Thus, the output signal $y[n]$ has the same sinusoidal waveform as the input signal $x[n]$ of Eq. (4.15) with two differences: (1) the amplitude is multiplied by $|H(e^{j\omega_0})|$, the value of the magnitude function of the discrete-time system at $\omega = \omega_0$; and (2) the output signal has a phase difference with the input by an amount $\theta(\omega_0)$, the value of the phase of the discrete-time system at $\omega = \omega_0$.

4.1.3   The Concept of Filtering

One application of an LTI discrete-time system is to pass certain frequency components in an input sequence without any distortion (if possible) and to block other frequency components. Such systems are called *digital filters* and are one of the main subjects of discussion in this text. The key to the filtering process is the inverse discrete-time Fourier transform given in Eq. (3.5) which expresses an arbitrary input sequence as a linear weighted sum of an infinite number of exponential sequences, or equivalently, as a linear weighted sum of sinusoidal sequences. As a result, by appropriately choosing the values of magnitude function of the LTI digital filter at frequencies corresponding to the frequencies of the sinusoidal components of the input, some of these sinusoidal sequences can be selectively heavily attenuated or filtered with respect to the others.

We now explain the concept of filtering and then define the most commonly desired filter characteristics. To understand the mechanism behind the design of such system, consider a real coefficient LTI discrete-time system characterized by a magnitude function

$$|H(e^{j\omega})| \cong \begin{cases} 1, & |\omega| \leq \omega_c, \\ 0, & \omega_c < |\omega| \leq \pi. \end{cases} \tag{4.17}$$

We apply an input $x[n] = A \cos \omega_1 n + B \cos \omega_2 n$ to this system, where $0 < \omega_1 < \omega_c < \omega_2 < \pi$. Because of linearity, it follows from Eq. (4.16) that the output $y[n]$ of this system is of the form

$$y[n] = A|H(e^{j\omega_1})| \cos(\omega_1 n + \theta(\omega_1)) + B|H(e^{j\omega_2})| \cos(\omega_2 n + \theta(\omega_2)). \tag{4.18}$$

Making use of Eq. (4.17) in Eq. (4.18), we get

$$y[n] \cong A|H(e^{j\omega_1})| \cos(\omega_1 n + \theta(\omega_1)),$$

indicating the LTI discrete-time system acts like a lowpass filter.

Example 4.3 In this example, we consider the design of a very simple digital filter [Ham89]. The input signal consists of a sum of two cosine sequences of angular frequencies 0.1 rad/sec and 0.4 rad/sec, respectively. We need to design a highpass filter that will pass the high-frequency component of the input but block the low-frequency part.

For simplicity we assume the filter to be an FIR filter of length 3 with an impulse response

$$h[0] = h[2] = \alpha, \quad h[1] = \beta.$$

Hence, from Eq. (4.1), the digital filtering is performed using the difference equation

$$\begin{aligned} y[n] &= h[0]x[n] + h[1]x[n-1] + h[2]x[n-2] \\ &= \alpha x[n] + \beta x[n-1] + \alpha x[n-2] \end{aligned} \tag{4.19}$$

where $y[n]$ and $x[n]$ represent, respectively, the output and the input sequences. Thus, our design objective is to choose suitable values for the filter parameters, α and β, so that the output of the filter is a cosine sequence with a frequency 0.4 rad/sec.

Now, from Eq. (4.5) the frequency response of the above FIR filter is given by

$$\begin{aligned} H(e^{j\omega}) &= h[0] + h[1]e^{-j\omega} + h[2]e^{-j2\omega} \\ &= \alpha(1 + e^{-j2\omega}) + \beta e^{-j\omega} = 2\alpha \left(\frac{e^{j\omega} + e^{-j\omega}}{2} \right) e^{-j\omega} + \beta e^{-j\omega} \\ &= (2\alpha \cos \omega + \beta)e^{-j\omega}. \end{aligned} \tag{4.20}$$

The magnitude and phase functions of this filter, assuming the amplitude function $(2\alpha \cos \omega + \beta)$ to be positive, are

$$|H(e^{j\omega})| = 2\alpha \cos \omega + \beta, \tag{4.21}$$

$$\theta(\omega) = -\omega. \tag{4.22}$$

In order to stop the low-frequency component from appearing at the output of the filter, the magnitude function at $\omega = 0.1$ should be equal to zero. Similarly, to pass the high-frequency component without any attenuation, we need to ensure that the magnitude function at $\omega = 0.4$ is equal to 1. Thus, the two conditions that must be satisfied are

$$|H(e^{j0.1})| = 2\alpha \cos(0.1) + \beta = 0,$$
$$|H(e^{j0.4})| = 2\alpha \cos(0.4) + \beta = 1.$$

Solving the above two equations, we arrive at

$$\alpha = -6.76195, \qquad \beta = 13.456335. \tag{4.23}$$

Note that the above solution guarantees the positiveness of $(2\alpha \cos \omega + \beta)$. Substituting Eq. (4.23) in Eq. (4.19) we obtain the input-output relation of the desired FIR filter as

$$y[n] = -6.76195(x[n] + x[n - 2]) + 13.456335x[n - 1], \tag{4.24}$$

with

$$x[n] = \{\cos(0.1n) + \cos(0.4n)\}\mu[n]. \tag{4.25}$$

To verify the filtering action, we implement the filter of Eq. (4.24) on MATLAB and calculate the first 100 output samples beginning from $n = 0$. Note that the input has been assumed to be a causal sequence with the first nonzero sample occurring at $n = 0$, and for calculating y[0] and y[1] we set $x[-1] = x[-2] = 0$. The MATLAB program used to calculate the output is given below.

```
% Program 4_2
% Set up the filter coefficients
b = [-6.76195 13.456335 -6.76195];
%   Set initial conditions to zero values
zi = [0 0];
% Generate the two sinusoidal sequences
n = 0:99;
x1 = cos(0.1*n);
x2 = cos(0.4*n);
%   Generate the filter output sequence
y = filter(b, 1, x1+x2, zi);
% Plot the input and the output sequences
plot(n,y,'r-',n,x2,'b--',n,x1,'g-.');grid
axis([0 100 -1.2 4]);
ylabel('Amplitude'); xlabel('Time index n')
legend('r-','y[n]','b--',x2[n]','g-.','x1[n]')
```

Figure 4.2 shows the plots of the output $y[n]$ and the two sinusoids of the input $x[n]$ where the output values have been clipped to the range $(-1.2, 4)$ to show the steady-state output more clearly.

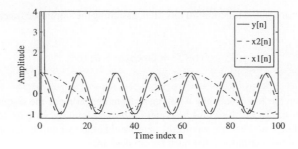

Figure 4.2   Output $y[n]$ (solid line), low-frequency input $x_1[n]$ (dash-dotted line), and high-frequency input $x_2[n]$ (dashed line) signals of the FIR filter of Eq. (4.24).

Table 4.1   Input and output sequences of the filter of Example 4.3.

n	$\cos(0.1n)$	$\cos(0.4n)$	$x[n]$	$y[n]$
0	1.0	1.0	2.0	−13.52390
1	0.9950041	0.9210609	1.9160652	13.956333
2	0.9800665	0.6967067	1.6767733	0.9210616
3	0.9553364	0.3623577	1.3176942	0.6967064
4	0.9210609	−0.0291995	0.8918614	0.3623572
5	0.8775825	−0.4161468	0.4614357	−0.0292002
6	0.8253356	−0.7373937	0.0879419	−0.4161467

The first seven samples of the output are shown in Table 4.1 along with that of the two sinusoids. From this table and also from Figure 4.2, we observe that, neglecting the least significant digit,

$$y[n] = \cos(0.4(n-1)) \quad \text{for } n = 2, 3, 4, 5 \text{ and } 6.$$

Several comments are in order here. First, computation of the present value of output requires the knowledge of the present and two previous input samples. Hence, the first two output samples are the result of the assumed zero input sample values at $n = -1$ and $n = -2$, and are therefore the transient part of the output. The steady-state is reached at $n = 2$. Second, the output is a delayed version of the high-frequency component $\cos(0.4n)$ of the input, and the delay is one sample period. This also follows from the expression for the phase function of this filter as given in Eq. (4.22). Note that this filter is a linear-phase filter with a group delay of unity.

4.1.4   The Group Delay Concept

Now consider the case of an input $x[n]$ consisting of three sinusoidal sequences of the form $x[n] = A\cos\omega_1 n + B\cos\omega_2 n + C\cos\omega_3 n$, with $0 < \omega_1 < \omega_2 < \omega_c < \omega_3 < \pi$.

Then the output $y[n]$ of the LTI digital filter of Eq. (4.17) is approximately given by

$$y[n] \cong A|H(e^{j\omega_1})|\cos(\omega_1 n + \theta(\omega_1)) + B|H(e^{j\omega_2})|\cos(\omega_2 n + \theta(\omega_2)). \qquad (4.26)$$

It follows from the above that the output waveform will be a delayed replica of the waveform of the desired portion of the input except for a change in the amplitude if

$$|H(e^{j\omega_1})| \cong |H(e^{j\omega_2}), \qquad (4.27)$$

and

$$\theta(\omega_1) = K\omega_1, \qquad \theta(\omega_2) = K\omega_2, \qquad (4.28)$$

or in other words, inside the frequency range $0 \le \omega \le \omega_c$ the magnitude function remains essentially constant and the phase function is a linear function of ω. In fact, the preservation of the waveform at the output can be achieved if the strict linear phase condition is replaced with the condition that the phase function be of the form

$$\theta(\omega) = K_1\omega + K_2, \qquad (4.29)$$

or equivalently, the LTI digital filter has a constant group delay

$$\tau(\omega) = -K_1. \qquad (4.30)$$

Figure 4.3 illustrates the meaning of constant group delay. Figure 4.3(a) shows a sinusoidal sequence $x[n] = \cos(\pi n/2)$, while Figure 4.3(b) shows another sinusoidal sequence $y[n] = \cos(\pi(n - K)/2)$. As can be seen from this figure, both sequences have the same angular frequency $\pi/2$, but the sequence $y[n]$ is not a delayed replica of the sequence $x[n]$ as long as K is not an integer. Let $x_a(t)$ and $y_a(t)$ represent the continuous-time signals whose sampled versions, sampled at $t = nT$, are precisely $x[n]$ and $y[n]$, respectively. Then as indicated in Figure 4.3(c), the delay between the waveforms of $x_a(t)$ and $y_a(t)$ is the group delay K.

It should be noted that for physical realizability, if the group delay is constant as in Eq. (4.30) it must be positive, i.e., $K_1 < 0$.

4.1.5   Frequency-Domain Characterization of the LTI Discrete-Time System

We now derive the input-output relation of an LTI discrete-time system in the frequency domain. If $Y(e^{j\omega})$ and $X(e^{j\omega})$ denote the DTFTs of the output and input sequences, $y[n]$ and $x[n]$, respectively, then from Eq. (4.1) by taking the DTFT of both sides we obtain

$$Y(e^{j\omega}) = \sum_{n=-\infty}^{\infty} y[n]e^{-j\omega n} = \sum_{n=-\infty}^{\infty} \left(\sum_{k=-\infty}^{\infty} h[k]x[n-k] \right) e^{-j\omega n}, \qquad (4.31)$$

Interchanging the summation signs on the right-hand side of Eq. (4.31) and rearranging we arrive at

$$Y(e^{j\omega}) = \sum_{k=-\infty}^{\infty} h[k] \left(\sum_{n=-\infty}^{\infty} x[n-k]e^{-j\omega n} \right)$$

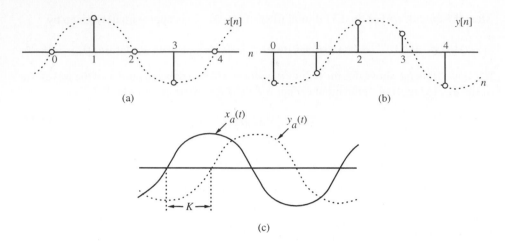

Figure 4.3   Illustration of the meaning of group delay.

$$= \sum_{k=-\infty}^{\infty} h[k] \left(\sum_{\ell=-\infty}^{\infty} x[\ell] e^{-j\omega(\ell+k)} \right)$$

$$= \sum_{k=-\infty}^{\infty} h[k] \left(\sum_{\ell=-\infty}^{\infty} x[\ell] e^{-j\omega\ell} \right) e^{-j\omega k}. \qquad (4.32)$$

The quantity inside the parentheses on the right-hand side of Eq. (4.32) is recognized as $X(e^{j\omega})$, the DTFT of the input sequence $x[n]$. Substituting this notation and rearranging, we finally obtain

$$Y(e^{j\omega}) = \left(\sum_{k=-\infty}^{\infty} h[k] e^{-j\omega k} \right) X(e^{j\omega}) = H(e^{j\omega}) X(e^{j\omega}) \qquad (4.33)$$

where $H(e^{j\omega})$ is the frequency response of the LTI system as defined in Eq. (4.5). Equation (4.33) thus relates the input and the output of an LTI system in the frequency domain.

From Eq. (4.33) we obtain

$$H(e^{j\omega}) = \frac{Y(e^{j\omega})}{X(e^{j\omega})}, \qquad (4.34)$$

Thus, the frequency response of an LTI discrete-time system is given by the ratio of the Fourier transform $Y(e^{j\omega})$ of the output sequence $y[n]$ to the Fourier transform $X(e^{j\omega})$ of the input sequence $x[n]$. For example, for an LTI system characterized by a linear constant coefficient difference equation of the form of Eq. (2.78), the expression for its frequency response $H(e^{j\omega})$ can be simply derived by taking the Fourier transform of both sides of the difference equation. From Table 3.2, the Fourier transforms of $y[n-k]$ and $x[n-k]$ are $e^{-j\omega k} Y(e^{j\omega})$ and $e^{-j\omega k} X(e^{j\omega})$, respectively. Thus, from Eq. (2.78) we obtain

$$\left(\sum_{k=0}^{N} d_k e^{-j\omega k} \right) Y(e^{j\omega}) = \left(\sum_{k=0}^{M} p_k e^{-j\omega k} \right) X(e^{j\omega}). \qquad (4.35)$$

Therefore,

$$H(e^{j\omega}) = \frac{\sum_{k=0}^{M} p_k e^{-j\omega k}}{\sum_{k=0}^{N} d_k e^{-j\omega k}}. \tag{4.36}$$

4.2   The Transfer Function

A generalization of the frequency response function, $H(e^{j\omega})$ leads to the concept of transfer function, which is defined next. As we have seen, the frequency response function does provide valuable information on the behavior of an LTI digital filter in the frequency-domain. However, being a complex function of the frequency variable ω, it is difficult to manipulate it for the realization of a digital filter. On the other hand, the z-transform of the impulse response of an LTI system, called the transfer function, is a polynomial in z^{-1}, and for a system with a real impulse response, it is a polynomial with real coefficients. Moreover, in most practical cases, the LTI digital filter of interest is characterized by a linear difference equation with constant and real coefficients. The transfer function of such a filter is a real rational function of the variable z^{-1}, i.e., a ratio of two polynomials in z^{-1} with real coefficients, and is thus more amenable for the synthesis.

We first develop the input-output relation of an LTI system in the z-domain from its various time-domain descriptions and arrive at different forms of the transfer function representation of the system. We then study its properties and in particular develop the conditions for the BIBO stability of a causal LTI system.

Consider the LTI digital discrete-time system of Figure 4.4. The input-output relation of this system is given by

$$y[n] = x[n] \circledast h[n], \tag{4.37}$$

where $x[n]$, $y[n]$ and $h[n]$ are, respectively, the input, output and impulse response sequences. From Table 3.9, using the convolution property of the z-transform, we arrive at the input-output relation of the filter in the z-domain given by

$$Y(z) = H(z)X(z). \tag{4.38}$$

From the above we arrive at the expression

$$H(z) = \frac{Y(z)}{X(z)}. \tag{4.39}$$

The quantity $H(z)$, which is the z-transform of the impulse response sequence $h[n]$ of the filter, is more commonly called the *transfer function* or the *system function*. If the ROC of $H(z)$ includes the unit circle, it is then related to the frequency response $H(e^{j\omega})$ of the LTI digital filter through

$$H(e^{j\omega}) = H(z)|_{z=e^{j\omega}}. \tag{4.40}$$

As indicated in Eq. (4.6), the frequency response $H(e^{j\omega})$ can be written in terms of its real and imaginary parts, $H_{re}(e^{j\omega})$ and $H_{im}(e^{j\omega})$, or in terms of its magnitude and phase functions, $|H(e^{j\omega})|$ and $\arg[H(e^{j\omega})]$, respectively. For a real coefficient transfer function $H(z)$, it can be shown that

$$|H(e^{j\omega})|^2 = H(e^{j\omega})H^*(e^{j\omega}) = H(e^{j\omega})H(e^{-j\omega}) = H(z)H(z^{-1})|_{z=e^{j\omega}}. \tag{4.41}$$

Figure 4.4   An LTI discrete-time system.

The inverse z-transform of the transfer function $H(z)$ yields the impulse response $h[n]$.

For a causal rational transfer function, the methods outlined in Section 3.8 can be used to compute its impulse response. For example, an analytical form of the impulse response can be determined via a partial-fraction expansion using the MATLAB Program 3_9. On the other hand, a fixed number of impulse response samples starting at $n = 0$ can be computed using the MATLAB Programs 3_11 or 3_12.

4.2.1   Derivation of the Transfer Function Expression

In the case of an FIR filter, the input-output relation in the time-domain is given by Eq. (2.82), and by taking the z-transform of both sides of this equation, we arrive at

$$Y(z) = \left(\sum_{n=N_1}^{N_2} h[n]z^{-n} \right) X(z),$$

from which we obtain

$$H(z) = \sum_{n=N_1}^{N_2} h[n]z^{-n}. \tag{4.42}$$

For a causal FIR filter, $N_1 = 0$ and $N_2 > 0$. Note that all poles of $H(z)$ of a causal FIR filter are at the origin in the z-plane, and as a result, the ROC of $H(z)$ is the entire z-plane excluding the point $z = 0$. On the other hand, in the case of an IIR digital filter, the transfer function expression in general is an infinite series. However, for the finite-dimensional IIR filter characterized by the difference equation of Eq. (2.78), the transfer function expression is a rational function of z. It is obtained by taking the z-transform of Eq. (2.78) which leads to

$$\left\{ \sum_{k=0}^{N} d_k z^{-k} \right\} Y(z) = \left\{ \sum_{k=0}^{M} p_k z^{-k} \right\} X(z), \tag{4.43}$$

from which we arrive at

$$H(z) = \frac{Y(z)}{X(z)} = \frac{p_0 + p_1 z^{-1} + p_2 z^{-2} + \cdots + p_M z^{-M}}{d_0 + d_1 z^{-1} + d_2 z^{-2} + \cdots + d_N z^{-N}}. \tag{4.44}$$

This is seen to be a rational function in z^{-1}, i.e., it is a ratio of two polynomials in z^{-1}. By multiplying the numerator and the denominator of the right-hand side by z^M and z^N, respectively, the transfer function can be expressed as a rational function in z:

$$H(z) = z^{(N-M)} \frac{p_0 z^M + p_1 z^{M-1} + p_2 z^{M-2} + \cdots + p_M}{d_0 z^N + d_1 z^{N-1} + d_2 z^{N-2} + \cdots + d_N}. \tag{4.45}$$

An alternate way to express the transfer function of Eq. (4.44) is to factor out the numerator and denominator polynomials leading to

$$H(z) = \frac{p_0}{d_0} \frac{\prod_{k=1}^{M}(1 - \xi_k z^{-1})}{\prod_{k=1}^{N}(1 - \lambda_k z^{-1})}, \tag{4.46a}$$

or to represent the transfer function of Eq. (4.45) in factored form, as

$$H(z) = \frac{p_0}{d_0} z^{N-M} \frac{\prod_{k=1}^{M}(z - \xi_k)}{\prod_{k=1}^{N}(z - \lambda_k)}, \tag{4.46b}$$

where $\xi_1, \xi_2, \ldots, \xi_N$ are the finite zeros and $\lambda_1, \lambda_2, \ldots, \lambda_N$ are the finite poles of $H(z)$. If $N > M$, there are additional $(N - M)$ zeros at $z = 0$, and if $N < M$, there are additional $(M - N)$ poles at $z = 0$.

Example 4.4 Consider the moving average filter of Example 4.1 with an impulse response $h[n]$ given by Eq. (4.9). The transfer function of the moving average FIR filter is then given by

$$H(z) = \frac{1}{M} \sum_{n=0}^{M-1} z^{-n} \tag{4.47}$$

$$= \frac{1 - z^{-M}}{M(1 - z^{-1})} = \frac{z^M - 1}{M[z^{M-1}(z - 1)]}. \tag{4.48}$$

From Eq. (4.48) it is seen that the transfer function has M zeros on the unit circle at $z = e^{j2\pi kM}$, $k = 0, 1, 2, \ldots, M - 1$. There is an $(M - 1)$th-order pole at the origin $(z = 0)$ and a single pole at $z = 1$. But the pole at $z = 1$ exactly cancels a zero at the same place resulting in a transfer function with all poles at the origin. This is always the characteristics of a causal FIR filter for which the ROC is the entire z-plane except the origin.

Example 4.5 A causal LTI IIR digital filter is characterized by a constant coefficient difference equation given by

$$y[n] = x[n - 1] - 1.2x[n - 2] + x[n - 3]$$
$$+1.3y[n - 1] - 1.04y[n - 2] + 0.222y[n - 3]. \tag{4.49}$$

From Eqs. (4.49) and (4.44) we arrive directly at the expression for the transfer function $H(z)$ as a rational function in z^{-1}:

$$H(z) = \frac{Y(z)}{X(z)} = \frac{z^{-1} - 1.2z^{-2} + z^{-3}}{1 - 1.3z^{-1} + 1.04z^{-2} - 0.222z^{-3}}. \tag{4.50}$$

Multiplying the numerator and the denominator of the right-hand-side term in the above equation by z^3 we obtain the expression for the transfer function as a rational function in z:

$$H(z) = \frac{z^2 - 1.2z + 1}{z^3 - 1.3z^2 + 1.04z - 0.222}. \tag{4.51}$$

The above transfer function can also be expressed in a factored form. To this end we make use of the MATLAB Program 3_7 of Example 3.20 whose output are the zeros, poles and gain constant as indicated below:

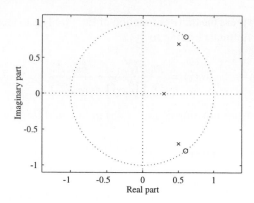

Figure 4.5   Pole-zero plot of the IIR transfer function of Eq. (4.50).

```
Zeros are at
    0.6000 + 0.8000i
    0.6000 - 0.8000i

Poles are at
    0.5000 + 0.7000i
    0.5000 - 0.7000i
    0.3000

Gain constant
    1
```

The above transfer function has three zeros: one at $z = 0.6 - j0.8$, the second at $z = 0.6 + j0.8$, and the third at $z = \infty$. Likewise, it has three poles: one at $z = 0.5 - j0.7$, the second at $z = 0.5 + j0.7$, and the third at $z = 0.3$. Thus the factored form of the transfer function is given by

$$H(z) = \frac{(z - 0.6 + j0.8)(z - 0.6 - j0.8)}{(z - 0.3)(z - 0.5 + j0.7)(z - 0.5 - j0.7)}. \tag{4.52}$$

A pole-zero plot of the above transfer function obtained using the function zplane in MATLAB is sketched in Figure 4.5.

The transfer function of a causal IIR digital filter can also be expressed in terms of the state-space parameters. From Eqs. (2.97) and (2.98), the state-space representation of a linear N-th order LTI discrete-time system is given by

$$\mathbf{s}[n + 1] = \mathbf{A}\mathbf{s}[n] + \mathbf{B}x[n], \tag{4.53a}$$

$$y[n] = \mathbf{C}\mathbf{s}[n] + Dx[n], \tag{4.53b}$$

where

$$\mathbf{s}[n] = [s_1[n] \quad s_2[n] \quad \cdots \quad s_N[n]]^T. \tag{4.54}$$

A z-transform of the state-vector $\mathbf{s}[n]$ is therefore given by

$$\mathbf{S}(z) = [S_1(z) \quad S_2(z) \quad \cdots \quad S_N(z)]^T.$$

From Eq. (4.54) the vector $\mathbf{s}[n+1]$ is given by

$$\mathbf{s}[n+1] = [s_1[n+1] \quad s_2[n+1] \cdots s_N[n+1]]^T,$$

whose z-transform is thus given by

$$\begin{aligned}
\mathcal{Z}\{\mathbf{s}[n+1]\} &= [zS_1(z) \quad zS_2(z) \cdots zS_N(z)]^T \\
&= z\,[S_1(z) \quad S_2(z) \cdots S_N(z)]^T = z\mathbf{S}(z).
\end{aligned}$$

As a result, the z-domain representation of the state equation of Eq. (4.53a) leads to

$$z\mathbf{S}(z) = \mathbf{A}\mathbf{S}(z) + \mathbf{B}X(z).$$

The above equation can be solved for $\mathbf{S}(z)$, yielding

$$\mathbf{S}(z) = [z\mathbf{I} - \mathbf{A}]^{-1}\mathbf{B}X(z), \tag{4.55}$$

where $\mathbf{I}$ denotes an identity matrix of order N. Substituting the above in the z-transform version of Eq. (4.53b) we get

$$Y(z) = \mathbf{C}\mathbf{S}(z) + DX(z) = \mathbf{C}[z\mathbf{I} - \mathbf{A}]^{-1}\mathbf{B}X(z) + DX(z),$$

which leads to the desired expression

$$H(z) = \mathbf{C}[z\mathbf{I} - \mathbf{A}]^{-1}\mathbf{B} + D. \tag{4.56}$$

Note from the above that the denominator polynomial of $H(z)$ is given by the determinant of the matrix $[z\mathbf{I} - \mathbf{A}]$. The zeros of this determinant, which are the poles of the transfer function, are also the eigenvalues of the state transition matrix $\mathbf{A}$.

Example 4.6 Let us determine the transfer function $H(z)$ of the causal digital filter described by the state-space matrices:

$$\mathbf{A} = \begin{bmatrix} -4 & 3 \\ 1 & -2 \end{bmatrix}, \quad \mathbf{B} = \begin{bmatrix} 0 \\ 1 \end{bmatrix}, \quad \mathbf{C} = [-1 \; -2], \quad D = 1. \tag{4.57}$$

We first form

$$[z\mathbf{I} - \mathbf{A}] = \begin{bmatrix} z+4 & -3 \\ -1 & z+2 \end{bmatrix}.$$

Its inverse is given by

$$[z\mathbf{I} - \mathbf{A}]^{-1} = \begin{bmatrix} z+4 & -3 \\ -1 & z+2 \end{bmatrix}^{-1} = \frac{1}{\Delta(z)}\begin{bmatrix} z+2 & 3 \\ 1 & z+4 \end{bmatrix},$$

where $\Delta(z) = z^2 + 6z + 5$ is the determinant of $[z\mathbf{I} - \mathbf{A}]$. Substituting the above and the expressions for $\mathbf{B}$, $\mathbf{C}$ and D of Eq. (4.57) in Eq. (4.56) we arrive at

$$\begin{aligned}
H(z) &= 1 + \frac{1}{\Delta(z)}[-1 \; -2]\begin{bmatrix} z+2 & 3 \\ 1 & z+4 \end{bmatrix}\begin{bmatrix} 0 \\ 1 \end{bmatrix} \\
&= 1 - \frac{2z+11}{z^2+6z+5} = \frac{z^2+4z-6}{z^2+6z+5}. \tag{4.58}
\end{aligned}$$

In factored form, $H(z)$ can be expressed as

$$H(z) = \frac{(z - 1.1623)(z + 5.1623)}{(z + 1)(z + 5)}.$$

Thus, the poles of the discrete-time system of Eq. (4.57) are at $z = -1$ and at $z = -5$.

Using the function $\texttt{eig(A)}$ in MATLAB we compute the eigen values of the state-transition matrix $\mathbf{A}$. They are found to be -5 and -1, as expected.

There are a number of M-file functions in MATLAB suitable for state-space based analysis of discrete-time systems. The pertinent ones are

```
[num,den] = ss2tf(A,B,C,D)
[A,B,C,D] = tf2ss(num,den)
[z,p,k]   = ss2zp(A,B,C,D)
[A,B,C,D] = zp2ss(z,p,k)
```

The function $\texttt{ss2tf(A,B,C,D)}$ is used to develop a rational z-transform expressed as a ratio of polynomials in positive powers of z as in Eq. (4.45) from its state-space form given by Eqs. (2.97) and (2.98). The input arguments are the state-parameter matrices A, B, and C, and the scalar D. The output files are the column vectors num and den containing the coefficients of the numerator and the denominator polynomials of the transfer function in descending positive powers of z. The function $\texttt{tf2ss(num,den)}$ is used to implement the reverse process.

The function $\texttt{ss2zp(A,B,C,D)}$ is employed to obtain the zeros, poles, and gain constant of the transfer function from its state-space representation, while the function $\texttt{zp2ss(z,p,k)}$ implements the reverse process.[3]

We illustrate next the application of the first two M-files through examples.

Example 4.7 We determine the transfer function of the LTI discrete-time system of Example 4.6 from its state-space representation given in Eq. (4.57) using MATLAB. To this end we can use Program 4_3 given below.

```
% Program 4_3
% Transfer function from state-space form
%
% Read in the state-space parameters
A = input('Type in matrix A =');
B = input('Type in column vector B =');
C = input('Type in row vector C =');
D = input('Type in scalar D =');
% Develop the transfer function
[num,den] = ss2tf(A,B,C,D);
disp('Numerator polynomial coefficients');
disp(num);
disp('Denominator polynomial coefficients');
disp(den);
```

[3]The functions $\texttt{ss2tf}$ and $\texttt{tf2ss}$ have been used earlier in Section 2.6.1.

As the program is run, it calls for the input data consisting of the state-space matrices. The matrix A and the vectors B and C must be entered inside square brackets following standard MATLAB representation; D being a scalar is entered without the square brackets. After completion of the computation, the program displays the coefficients of the numerator and the denominator polynomials in descending positive powers of z. For our example, the input data entered are as follows:

```
A = [- 4      3;   1    - 2]
B = [0;    1]
C = [- 1    - 2]
D = 1
```

The output data generated by the program are given by

```
Numerator    polynomial    coefficients
  1.0000         4.0000       -6.0000

Denominator  polynomial    coefficients
  1              6            5
```

The above coefficients are identical to those given in Eq. (4.58).

Example 4.8 We can develop the state-space representation of a LTI discrete-time system from its transfer function using the MATLAB Program 4_4 given below. We illustrate its application for the causal LTI discrete-time system of Example 4.5 with a transfer function given in Eq. (4.51).

```
% Program 4_4
% State-space representation from transfer function
%
% Read in the transfer function coefficients
num = input('Type in numerator coefficients =');
den = input('Type in denominator coefficients =');
%  Generate the state-space parameters
[A,B,C,D] = tf2ss(num,den);
disp('A =');disp(A)
disp('B =');disp(B)
disp('C =');disp(C)
disp('D =');disp(D)
```

During execution the program asks for the numerator and the denominator coefficients, which must be entered inside square brackets in descending positive powers of z:

```
num = [1     -1.2    1]
den = [1     -1.3    1.04    -0.222]
```

The program then displays the state-space parameters A, B, C, and D as indicated below:

```
A =
     1.3000    -1.0400    0.2220
     1.0000         0          0
         0     1.0000          0
```

```
B =
      1
      0
      0

C =
      1.0000    -1.2000     1.0000

D =
      0
```

4.2.2   Stability Condition

We describe in this text several methods for determining the transfer function of an FIR or IIR filter meeting the prescribed frequency domain specifications. However, before the digital filter is implemented we need to ensure that the transfer function derived will lead to a stable structure. We now establish the condition to be satisfied by a causal LTI digital filter described by a rational transfer function $H(z)$ in order that it be BIBO stable.

Recall from our earlier discussion in Section 2.5.3 that an LTI digital filter is BIBO stable if and only if its impulse response sequence $h[n]$ is absolutely summable, i.e.,

$$S = \sum_{n=-\infty}^{\infty} |h[n]| < \infty. \tag{4.59}$$

Now, the ROC of the z-transform of $h[n]$, $H(z)$, is defined by values of $|z| = r$ for which $h[n]r^{-n}$ is absolutely summable. Thus if the ROC includes the unit circle, then the digital filter is stable, and vice-versa. In addition, for a stable and causal digital filter, for which $h[n]$ is a right-sided sequence, the ROC will include the unit circle and the entire z-plane outside the unit circle including the point $z = \infty$.

As indicated earlier, an FIR digital filter with bounded impulse response coefficients is always stable. On the other hand, an IIR filter may be unstable if not designed properly. In addition, an originally stable IIR filter characterized by infinite precision coefficients may also become unstable after implemention due to the unavoidable quantization of all coefficients, as illustrated by the following example.

Example 4.9   Consider the causal second-order IIR transfer function given by

$$H(z) = \frac{1}{1 - 1.845z^{-1} + 0.850586z^{-2}}. \tag{4.60}$$

The plot of the impulse response coefficients of the above transfer function using a slightly modified Program 3_11 is shown in Figure 4.6(a). As seen from this figure, the impulse response coefficient $h[n]$ decays rapidly to zero value as n increases indicating that the absolute summability condition of Eq. (4.59) is satisfied. Hence, $H(z)$ is a stable transfer function.

Now consider the case when the transfer function coefficients are rounded up to values with two digits after the decimal point. Then the transfer function of Eq. (4.60) takes the form

$$\hat{H}(z) = \frac{1}{1 - 1.85z^{-1} + 0.85z^{-2}} \tag{4.61}$$

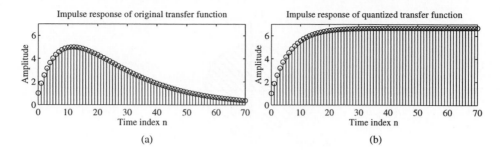

Figure 4.6 (a) Impulse response of original transfer function of Eq. (4.60), and (b) impulse response of transfer function of Eq. (4.61).

A plot of its impulse response coefficients is shown in Figure 4.6(b). In this case, the impulse response coefficient $h[n]$ increase rapidly to a constant value as n increases. As a result the absolute summability condition of Eq. (4.59) is violated, implying that $\hat{H}(z)$ is an unstable transfer function.

The testing of an IIR transfer function for stability is therefore an important problem. However, it is difficult to compute the sum $\mathcal{S}$ of Eq. (4.59) analytically in most cases. For a causal IIR transfer function, it can be computed approximately on a computer by replacing the sum with the following

$$\mathcal{S}_K = \sum_{n=0}^{K-1} |h[n]|, \tag{4.62}$$

where the number of terms K in the sum is chosen such that the absolute value of the difference $\mathcal{S}_K$ and $\mathcal{S}_{K-1}$ is smaller than some arbitrarily chosen small number, which is typically 10^{-6}. For a transfer function of very high order this approach may not be satisfactory. We now develop an alternative stability condition based on the location of the poles of the transfer function. For practical reasons, we restrict our attention to causal transfer functions.

Consider the IIR digital filter described by the rational transfer function $H(z)$ of Eq. (4.44). If the digital filter is assumed to be causal, i.e., the impulse response sequence $\{h[n]\}$ is a right-sided sequence, the ROC of $H(z)$ is to the exterior of the circle going through the pole that is farthest from the origin. But stability requires that $\{h[n]\}$ be absolutely summable, which in turn implies that the discrete-time Fourier transform of $\{h[n]\}$ exists. As a result, the unit circle must be inside the ROC. This implies that all poles of a stable causal transfer function $H(z)$ must be strictly inside the unit circle, as indicated in Figure 4.7.

Example 4.10 The factored form of the transfer function $H(z)$ of the causal LTI digital filter of Eq. (4.60) is given by

$$H(z) = \frac{1}{(1 - 0.902z^{-1})(1 - 0.943z^{-1})},$$

which has a real pole at $z = 0.902$ and a real pole at $z = 0.943$. Since the magnitude of each pole of $H(z)$ is less than 1, all poles are inside the unit circle in the z-plane. Hence, the transfer function is stable.

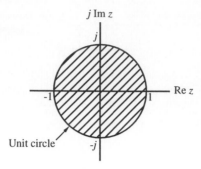

Figure 4.7   Stability region (shown cross-hatched) in the z-plane for pole locations of a stable casual transfer function.

On the other hand, the factored form of the transfer function $\hat{H}(z)$ of Eq. (4.61) is given by

$$\hat{H}(z) = \frac{1}{(1 - z^{-1})(1 - 0.85z^{-1})},$$

which now has one pole on the unit circle at $z = 1$ and the other pole inside the unit circle at $z = 0.85$, indicating that it is unstable.

4.3   Types of Transfer Functions

The time-domain classification of a digital transfer function based on the length of its impulse response sequence leads to the *finite impulse response* (FIR) and the *infinite impulse response* (IIR) transfer functions. We describe here several other types of classification. In the case of digital transfer functions with frequency-selective frequency responses, one classification is based on the shape of the magnitude function $|H(e^{j\omega})|$ or the form of the phase function $\theta(\omega)$. Based on this, four types of ideal filters are usually defined. These ideal filters have doubly infinite impulse response and are unrealizable. We describe very simple realizable FIR and IIR digital filter approximations. In a number of applications, these simple filters are quite adequate and provide satisfactory performances. The importance of transfer functions with linear phase is then pointed out and the possible realizations of these transfer functions with FIR filters discussed.

4.3.1   Ideal Filters

As pointed out in Example 4.3, a digital filter designed to pass signal components of certain frequencies without any distortion should have a frequency response of value equal to one at these frequencies, and should have a frequency response of value equal to zero at all other frequencies to totally block signal components with these frequencies. The range of frequencies where the frequency response takes the value of one is called the *passband*, and the range of frequencies where the frequency response is equal to zero is called the *stopband* of the filter.

The frequency responses of the four popular types of ideal digital filters with real impulse response coefficients are shown in Figure 4.8. For the *lowpass filter* of Figure

Figure 4.8   Four types of ideal filters: (a) ideal lowpass filter, (b) ideal highpass filter, (c) ideal bandpass filter, and (d) ideal bandstop filter.

4.8(a), the passband and the stopband are given by $0 \leq \omega \leq \omega_c$ and $\omega_c < \omega \leq \pi$, respectively. For the *highpass filter* of Figure 4.8(b), the stopband is given by $0 \leq \omega < \omega_c$, while the passband is given by $\omega_c \leq \omega \leq \pi$. The passband region of the *bandpass filter* of Figure 4.8(c) is $\omega_{c1} \leq \omega \leq \omega_{c2}$ and the stopband regions are given by $0 \leq \omega < \omega_{c1}$ and $\omega_{c2} < \omega < \pi$. Finally, for the *bandstop filter* of Figure 4.8(d), the passband regions are $0 \leq \omega \leq \omega_{c1}$ and $\omega_{c2} \leq \omega \leq \pi$, while the stopband is from $\omega_{c1} < \omega < \omega_{c2}$. The frequencies ω_c, ω_{c1}, and ω_{c2} are called the *cutoff frequencies* of their respective filters. Note from this figure that an ideal filter thus has a magnitude response equal to unity in the passband and zero in the stop band and has a zero phase everywhere.

Example 4.11   Determine the impulse response $h_{LP}[n]$ of the ideal lowpass filter of Figure 4.8(a). From the expression for the inverse DTFT of Eq. (3.5) and the ideal specification shown in Figure 4.8(a), the impulse response is obtained as

$$H_{LP}[n] = \frac{1}{2\pi} \int_{-\pi}^{\pi} H_{LP}(e^{j\omega}) e^{j\omega n}\, d\omega = \frac{1}{2\pi} \int_{-\omega_c}^{\omega_c} e^{j\omega n}\, d\omega$$

$$= \frac{1}{2\pi} \left(\frac{e^{j\omega_c n}}{jn} - \frac{e^{-j\omega_c n}}{jn} \right) = \frac{\sin \omega_c n}{\pi n}, \qquad -\infty < n < \infty. \qquad (4.63)$$

In practice, it is impossible to realize a finite dimensional LTI filter with the ideal "brick wall" characteristics of Figure 4.8 as the corresponding impulse response is not causal and is of doubly infinite length. Moreover, the impulse response is not absolutely summable, and hence, the corresponding transfer function is not BIBO stable.

In order to develop stable and realizable transfer functions, the ideal frequency response specifications of Figure 4.8 are relaxed by including a transition band between the passband and the stopband to permit the magnitude response to decay more gradually from its maximum value in the passband to the zero value in the stopband. Moreover, the magnitude response is allowed to vary by a small amount both in the passband and the stopband. Typical magnitude specifications used for the design of a lowpass filter are shown in Figure 7.1. Chapter 7 is devoted to a discussion of various filter design methods that lead to stable and realizable transfer functions meeting such relaxed specifications. In the following two sections we describe several very simple low-order FIR and IIR digital filters that exhibit selective frequency response characteristics providing a first-order approximation to the ideal characteristics of Figure 4.8. Frequency responses with sharper characteristics can often be obtained by cascading one or more of these simple filters, which in many applications are quite satisfactory.

4.3.2   Simple FIR Digital Filters

FIR digital filters considered here have integer-valued impulse response coefficients. These filters are employed in a number of practical applications, primarily because of their simplicity, which makes them amenable for inexpensive hardware implementation.

Lowpass FIR digital filters

The simplest lowpass FIR filter is the moving average filter of Eq. (4.9) with $M = 2$ which has a transfer function

$$H_0(z) = \tfrac{1}{2}(1 + z^{-1}), \tag{4.64}$$

with a frequency response given by

$$H_0(e^{j\omega}) = e^{-j\omega/2} \cos\left(\tfrac{\omega}{2}\right), \tag{4.65}$$

whose magnitude response is shown in Figure 4.9(a). The magnitude response $|H_0(e^{j\omega})| = \cos(\omega/2)$ is a monotonically decreasing function of ω from $\omega = 0$ to $\omega = \pi$. The maximum value of the magnitude function is unity at $\omega = 0$, and the minimum value is zero at $\omega = \pi$, i.e.,

$$|H_0(e^{j0})| = 1, \qquad |H_0(e^{j\pi})| = 0.$$

The frequency $\omega = \omega_c$ at which $|H_0(e^{j\omega_c})| = \frac{1}{\sqrt{2}}|H_0(e^{j0})|$ is of practical interest since here the gain $\mathcal{G}(\omega_c)$ in dB is given by

$$\mathcal{G}(\omega_c) = 20\log_{10}|H_0(e^{j\omega_c})| = 20\log_{10}|H_0(e^{j0})| - 20\log_{10}\sqrt{2}$$
$$= 0 - 3.0103 \cong -3.0\,\text{dB},$$

since the dc gain $\mathcal{G}(0) = 20\log_{10}|H_1(e^{j0})| = 0$. Thus, the gain $\mathcal{G}(\omega)$ at $\omega = \omega_c$ is approximately 3 dB less than that at zero frequency. As a result, ω_c is called the *3-dB cutoff frequency*. To determine the expression for ω_c we set $|H_0(e^{j\omega_c})|^2 = \cos^2(\omega_c/2) = 1/2$, which yields $\omega_c = \pi/2$. This result checks with that given in the plot of Figure 4.9(a). The 3-dB cutoff frequency ω_c can be considered as the passband edge frequency, and as a result, for this filter the passband width is approximately $\pi/2$.

A cascade of the simple FIR filters of Eq. (4.64) results in an improved lowpass frequency response, as illustrated in Figure 4.9(b) for a cascade of three sections. The

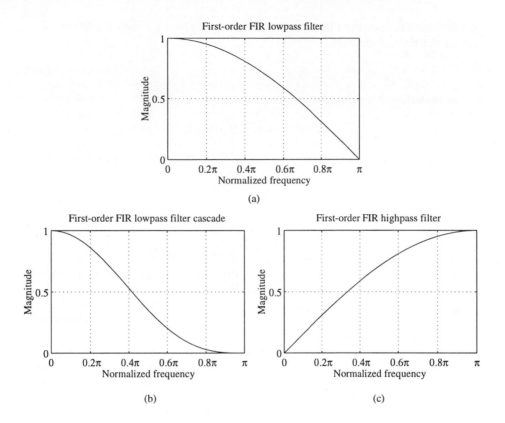

Figure 4.9 Magnitude responses of FIR filters: (a) first-order lowpass FIR filter $H_0(z)$ of Eq. (4.64), (b) cascade of 3 sections of $H_0(z)$, and (c) first-order highpass filter $H_1(z)$ of Eq. (4.67).

3-dB cutoff frequency of a cascade of M sections of the lowpass filter of Eq. (4.64) can be shown to be given by (Problem 4.47)

$$\omega_c = 2\cos^{-1}(2^{-1/2M}). \tag{4.66}$$

For $M = 3$, the above expression yields $\omega_c = 0.3002\pi$, which also checks with the plot in Figure 4.9(b). Thus, the cascade of first-order sections yields a sharper magnitude response but at the expense of a decrease in the width of the passband.

A better approximation to the ideal lowpass filter is given by a higher order moving-average filter of Eq. (4.9). Signals with rapid fluctuations in sample values are generally associated with high-frequency components that are essentially eliminated by a moving-average filter of the type of Eq. (4.9), resulting in a much smoother output waveform as illustrated earlier in Example 2.14. Example 4.16 discusses a simple approach to improve the magnitude response of a moving-average lowpass filter.

The moving-average filter is often employed as the basic building block in the design of lowpass filters used in sampling rate alteration and is considered in Section 11.12.

The simplest highpass FIR filter is obtained by replacing z with $-z$ in Eq. (4.64), resulting in a transfer function

$$H_1(z) = \tfrac{1}{2}(1 - z^{-1}). \tag{4.67}$$

The corresponding frequency response is given by

$$H_1(e^{j\omega}) = e^{-j\omega/2}\sin(\tfrac{\omega}{2}), \tag{4.68}$$

whose magnitude response is shown in Figure 4.9(c). Improved highpass frequency response can be obtained by cascading several sections of the simple highpass filter of Eq. (4.67). Alternatively, a higher order highpass filter of the form

$$H_1(z) = \frac{1}{M}\sum_{n=0}^{M-1}(-1)^n z^{-n}, \tag{4.69}$$

obtained by replacing z with $-z$ in the expression for the transfer function of a moving-average lowpass filter yields a sharper magnitude response.

An application of the FIR highpass filters is in moving-target-indicator (MTI) radars. In these radars, interfering signals, called *clutters*, are generated from fixed objects in the path of the radar beam [Sko62]. The clutter, generated mainly from ground echoes and weather returns, has frequency components primarily near the zero frequency (dc) and can be removed by filtering the radar return signal through a *two-pulse canceler*, which is the first-order highpass filter of Eq. (4.67). Often, the frequency components of the clutter occupy a small band near dc, and for a more effective removal it is necessary to use a highpass filter with a sharper magnitude response and a slightly broader stopband. To this end, a cascade of two two-pulse cancelers, called a *three-pulse canceler*, provides an improved performance. Problems 4.53 and M4.5 describe two simple IIR highpass filters proposed for clutter rejection.

4.3.3   Simple IIR Digital Filters

We now describe several simple IIR digital filters with first-order and second-order transfer functions and sketch their corresponding frequency responses. In many applications, use of such filters provides satisfactory results. Often, more complex frequency responses can be achieved by cascading these simple transfer functions.

First-order Lowpass IIR Digital Filter

A first-order lowpass IIR digital filter has a transfer function given by

$$H_{\text{LP}}(z) = \frac{1-\alpha}{2}\,\frac{1+z^{-1}}{1-\alpha z^{-1}}, \tag{4.70}$$

where $|\alpha| < 1$ for stability. From the above, the squared magnitude function can be easily derived:

$$|H_{\text{LP}}(e^{j\omega})|^2 = \frac{(1-\alpha)^2(1+\cos\omega)}{2(1+\alpha^2 - 2\alpha\cos\omega)} \tag{4.71}$$

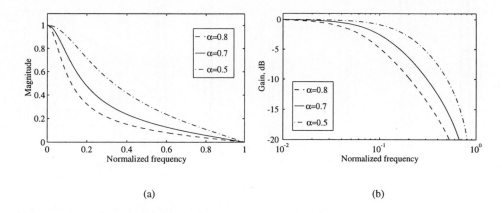

Figure 4.10 Magnitude and gain responses of the first-order lowpass filter of Eq. (4.70) for three values of α.

The derivative of $|H_{\text{LP}}(e^{j\omega})|^2$ with respect to ω is given by

$$\frac{d|H_{\text{LP}}(e^{j\omega})|^2}{d\omega} = \frac{-(1-\alpha)^2(1+2\alpha+\alpha^2)\sin\omega}{2(1+\alpha^2-2\alpha\cos\omega)^2}$$

which is always nonpositive in the range $0 \leq \omega \leq \pi$. Therefore, $|H_{\text{LP}}(e^{j\omega})|$ is a monotonically decreasing function of ω from $\omega = 0$ to $\omega = \pi$ (see Figure 4.10). The maximum value of the magnitude function is unity at $\omega = 0$, and the minimum value is zero at $\omega = \pi$, i.e.,

$$|H_{\text{LP}}(e^{j0})| = 1, \qquad |H_{\text{LP}}(e^{j\pi})| = 0.$$

To determine 3-dB cutoff frequency ω_c we set $|H_{\text{LP}}(e^{j\omega_c})|^2 = 1/2$ in Eq. (4.71) and arrive at the equation

$$\frac{(1-\alpha)^2(1+\cos\omega_c)}{2(1+\alpha^2-2\alpha\cos\omega_c)} = \frac{1}{2}$$

or

$$(1-\alpha)^2(1+\cos\omega_c) = 1+\alpha^2-2\alpha\cos\omega_c,$$

which when solved yields

$$\cos\omega_c = \frac{2\alpha}{1+\alpha^2}. \qquad (4.72\text{a})$$

The above quadratic equation can be solved for α yielding two solutions. The solution resulting in a stable transfer function $H_{\text{LP}}(z)$ is given by

$$\alpha = \frac{1-\sin\omega_c}{\cos\omega_c}. \qquad (4.72\text{b})$$

A plot of the magnitude and the gain responses of the above lowpass transfer function are sketched in Figure 4.10 for several values of α.

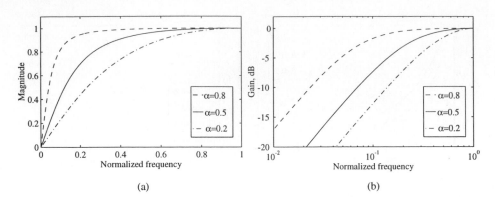

Figure 4.11   Magnitude and gain responses of the first-order highpass filter of Eq. (4.73) for three values of α.

First-order highpass digital filter

A first-order highpass transfer function $H_{HP}(z)$ is given by

$$H_{HP}(z) = \frac{1+\alpha}{2}\frac{1-z^{-1}}{1-\alpha z^{-1}}, \tag{4.73}$$

where $|\alpha| < 1$ for stability. Its 3-dB cutoff frequency ω_c is also given by Eq. (4.72a) and (4.72b). A plot of the magnitude and the gain responses of the above highpass transfer function for several values of α are shown in Figure 4.11.

Example 4.12   Design a first-order highpass digital filter with a 3-dB cutoff frequency of 0.8π.
  Now, $\sin(\omega_c) = \sin(0.8\pi) = 0.58778525$ and $\cos(\omega_c) = \cos(0.8\pi) = -0.80902$, which when substituted in Eq. (4.72b) yields $\alpha = -0.50952449$. Therefore, from Eq. (4.73), the transfer function of the desired first-order highpass filter is given by

$$H_{HP}(z) = 0.245237755\left(\frac{1-z^{-1}}{1+0.50952449z^{-1}}\right).$$

Second-order bandpass digital filter

A second-order bandpass digital filter is described by the transfer function

$$H_{BP}(z) = \frac{1-\alpha}{2}\frac{1-z^{-2}}{1-\beta(1+\alpha)z^{-1}+\alpha z^{-2}}. \tag{4.74}$$

Its squared magnitude function is given by

$$|H_{BP}(e^{j\omega})|^2 = \frac{(1-\alpha)^2(1-\cos 2\omega)}{2[1+\beta^2(1+\alpha)^2+\alpha^2-2\beta(1+\alpha)^2\cos\omega+2\alpha\cos 2\omega]}, \tag{4.75}$$

which goes to zero at $\omega = 0$ and at $\omega = \pi$. It assumes a maximum value of unity at $\omega = \omega_0$, called the *center frequency* of the bandpass filter, where

$$\omega_0 = \cos^{-1}(\beta). \tag{4.76}$$

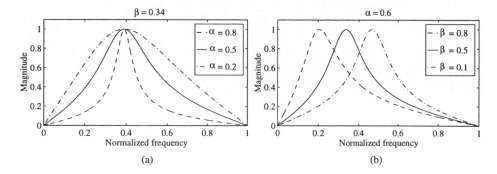

Figure 4.12 Magnitude response of the second-order bandpass filter of Eq. (4.74): (a) three specific values of α with $\beta = 0.34$, and (b) three specific values of β with $\alpha = 0.6$.

The frequencies ω_{c1} and ω_{c2} where the squared magnitude response goes to 1/2 are called the *3-dB cutoff frequencies*, and their difference, $\Delta\omega_{3\text{dB}}$, assuming $\omega_{c2} > \omega_{c1}$, called the *3-dB bandwidth*, is given by

$$\Delta\omega_{3\text{dB}} = \omega_{c2} - \omega_{c1} = \cos^{-1}\left(\frac{2\alpha}{1+\alpha^2}\right). \tag{4.77}$$

Plots of the magnitude responses of the bandpass filter of Eq. (4.74) are given in Figure 4.12 for several values of α and β.

Example 4.13 Design a second-order bandpass digital filter with center frequency at 0.4π and a 3-dB bandwidth of 0.1π.
    From Eq. (4.76),
$$\beta = \cos(\omega_0) = \cos(0.4\pi) = 0.309016994,$$
and from Eq. (4.77),

$$\frac{2\alpha}{1+\alpha^2} = \cos(\Delta\omega_{3\text{dB}}) = \cos(0.1\pi) = 0.951056516.$$

    The solution of the above quadratic equation yields two values for α: 1.37638192 and 0.726542528. The corresponding second-order bandpass transfer functions are given by

$$H'_{\text{BP}}(z) = -0.18819096\frac{1-z^{-1}}{1-0.7343423986z^{-1}+1.37638192z^{-2}}, \tag{4.78a}$$

and

$$H''_{\text{BP}}(z) = 0.136728736\frac{1-z^{-2}}{1-0.53353098z^{-1}+0.72542528z^{-2}}. \tag{4.78b}$$

The poles of the transfer function $H'_{\text{BP}}(z)$ of Eq. (4.78a) are at $z = 0.367171199 \pm j1.114256358$ and have a magnitude of 1.173193, indicating that they are outside the unit circle. Therefore this transfer function is unstable. On the other hand, the poles of the transfer function $H''_{\text{BP}}(z)$ of Eq. (4.78b) are at $z = 0.2667655 \pm j0.8095546$ and have a magnitude of 0.85237464. As a result, $H''_{\text{BP}}(z)$ is BIBO stable. In Example 4.19 we outline a simpler approach to the testing of the stability of a second-order IIR transfer function.
    Figure 4.13 shows the plots of the magnitude function and the group delay of $H''_{\text{BP}}(z)$. The group delay has been computed using the M-function `grpdelay` in MATLAB.

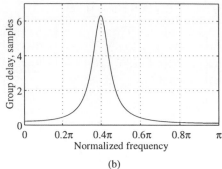

Figure 4.13   Magnitude response and the group delay of the bandpass transfer function of Eq. (4.78b).

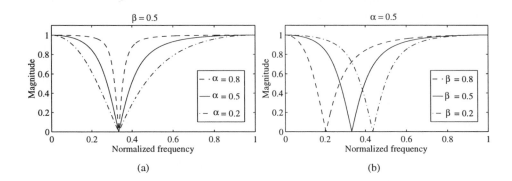

Figure 4.14   Frequency response of the second-order bandstop filter of Eq. (4.79). (a) Three specific values with α with $\beta = 0.5$, and (b) three specific values of β with $\alpha = 0.5$.

Second-order bandstop digital filter

Finally, a second-order bandstop digital filter has a transfer function of the form

$$H_{\text{BS}}(z) = \frac{1+\alpha}{2} \frac{1 - 2\beta z^{-1} + z^{-2}}{1 - \beta(1+\alpha)z^{-1} + \alpha z^{-2}}. \tag{4.79}$$

Its magnitude response is plotted in Figure 4.14 for various values of α and β. Here, the magnitude response takes the maximum value of unity at $\omega = 0$ and at $\omega = \pi$ and goes to zero at $\omega = \omega_0$, where ω_0 is given by Eq. (4.76). Since the magnitude response goes to zero at ω_0, ω_0 is called the *notch frequency*, and the digital filter of Eq. (4.79) is more commonly called a *notch filter*. The *3-dB notch bandwidth* $\Delta\omega_{\text{3dB}}$ is given by Eq. (4.77).

Higher-Order filters

By cascading the simple digital filters described above, we can implement digital filters with sharper magnitude responses. For example, for a cascade of K first-order lowpass sections characterized by the transfer function of Eq. (4.70), the overall structure has a transfer function $G_{\mathrm{LP}}(z)$ given by

$$G_{\mathrm{LP}}(z) = \left(\frac{1 - \alpha}{2} \frac{1 + z^{-1}}{1 - \alpha z^{-1}} \right)^K. \tag{4.80}$$

Using Eq. (4.71) we obtain the corresponding squared-magnitude function as

$$|G_{\mathrm{LP}}(e^{j\omega})|^2 = \left[\frac{(1 - \alpha)^2 (1 + \cos \omega)}{2(1 + \alpha^2 - 2\alpha \cos \omega)} \right]^K. \tag{4.81}$$

To determine the relation between its 3-dB cutoff frequency ω_c and the paramater α, we set

$$\left[\frac{(1 - \alpha)^2 (1 + \cos \omega_c)}{2(1 + \alpha^2 - 2\alpha \cos \omega_c)} \right]^K = \frac{1}{2}, \tag{4.82}$$

which when solved for α yields, for a stable $G_{\mathrm{LP}}(z)$,

$$\alpha = \frac{1 + (1 - B) \cos \omega_c - \sin \omega_c \sqrt{2B - B^2}}{1 - B + \cos \omega_c} \tag{4.83}$$

where

$$B = 2^{(K-1)/K}. \tag{4.84}$$

It should be noted that Eq. (4.83) reduces to Eq. (4.72b) for $K = 1$.

Example 4.14 Design a lowpass filter with a 3-dB cutoff frequency at $\omega_c = 0.4\pi$ using a single-stage realization and as a cascade of four first-order lowpass sections, and compare their gain responses.

From Eq. (4.72b) or, equivalently from Eq. (4.83) with $K = 1$, i.e., $B = 1$, and $\omega_c = 0.4\pi$, we arrive at $\alpha = 0.1584$ for a single-stage design. Likewise, from Eq. (4.84), $B = 1.6818$ for $K = 4$, which when substituted in Eq. (4.83) along with $\omega_c = 0.4\pi$, yields $\alpha = -0.251$. Figure 4.15 shows the gain responses of a single first-order lowpass filter (plot marked $K = 1$) and a cascade of four identical first-order lowpass filters (plot marked $K = 4$). As can be seen from this figure, cascading has resulted in a sharper roll-off in the gain response, as expected.

Likewise, a cascade of first-order highpass sections results in a highpass filter with a sharper roll-off in the gain response.

Figure 4.16 shows the magnitude responses of a single second-order bandpass filter (plot marked 1), a cascade of two identical second-order bandpass filters (plot marked 2), and a cascade of three identical second-order bandpass filters (plot marked 3). All bandpass sections are characterized by $\alpha = 0.2$ and $\beta = 0.34$. Since the parameter β of all second-order sections is identical, the center frequency of the cascade structure is the same as that of the single section. However, the 3-dB bandwidth decreases with an increase in the number of sections. Similarly, a cascade of identical second-order bandstop filters results in a higher order bandstop filter with an identical center frequency but with a narrower 3-dB bandwidth.

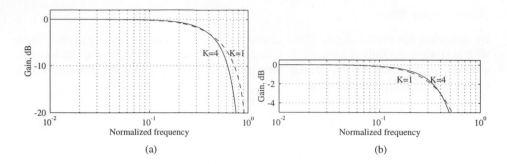

Figure 4.15   (a) Gain responses of a single first-order lowpass filter ($K = 1$) and a cascade of four identical first-order lowpass filters ($K = 4$) with a 3-dB cutoff frequency of $\omega_c = 0.4\pi$. (b) Passband details.

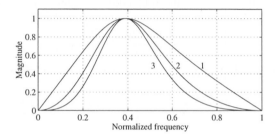

Figure 4.16   Gain responses of a single second-order bandpass filter, a cascade of two identical second-order bandpass filters, and a cascade of three identical second-order bandpass filters. All sections characterized by $\alpha = 0.2$ and $\beta = 0.34$.

4.3.4   Comb Filters

The simple filters of the previous two sections are characterized either by a single pass-band and/or a single stopband. There are a number of applications where filters with multiple passbands and stopbands are required. The *comb filter* is an example of this latter type of filter. In its most general form, a comb filter has a frequency response that is a periodic function of ω with a period $2\pi/M$, where M is a positive integer. If $H(z)$ is a filter with a single passband and/or a single stopband, a comb filter can easily be generated from it by replacing each delay in its realization with M delays, resulting in a structure with a transfer function given by $G(z) = H(z^M)$. It should be noted that a comb filter can be generated from either an FIR or an IIR prototype filter.

To illustrate the generation of the comb filter consider the prototype lowpass FIR filter of Eq. (4.64), which has a lowpass magnitude response as indicated in Figure 4.9(a). The comb filter generated from $H_0(z)$ has a transfer function

$$G_0(z) = H_0(z^M) = \tfrac{1}{2}(1 + z^{-M}), \qquad (4.85)$$

with a corresponding magnitude resonse as sketched in Figure 4.17(a). The new filter is essentially a notch filter with M notch frequencies in the range $0 \leq \omega < 2\pi$ located

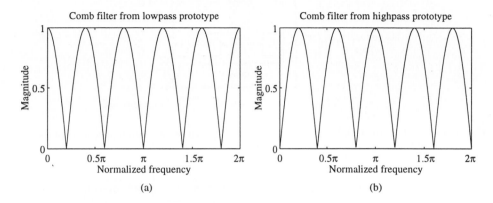

Figure 4.17 Magnitude responses of FIR comb filters: (a) generated from a prototype lowpass filter of Eq. (4.64) with $M = 5$, and (b) (a) generated from a prototype highpass filter of Eq. (4.67) with $M = 5$.

at $\omega = \omega_k = (2k + 1)\pi/M$ and has M peaks in its magnitude response located at $\omega = \omega_k = 2\pi k/M, k = 0, 1, \ldots, M - 1$.

The comb filter $G_1(z)$ generated from the prototype highpass filter of Eq. (4.67) has a transfer function given by

$$G_1(z) = H_1(z^M) = \tfrac{1}{2}(1 - z^{-M}), \tag{4.86}$$

with a magnitude response as indicated in Figure 4.17(b). This comb filter again has M notch frequencies in the range $0 \leq \omega < 2\pi$ located at $\omega = \omega_k = 2\pi k/M, k = 0, 1, \ldots, M - 1$, which are exactly at the locations of the peaks of the comb filter $G_0(z)$ of Eq. (4.85). Likewise its magnitude response has M peaks at $\omega_k = (2k + 1)\pi/M$, $k = 0, 1, \ldots, M - 1$, which are precisely the locations of the notch frequencies of $G_0(z)$.

Depending on the application, comb filters with other types of periodic magnitude responses can be easily generated by appropriately choosing the prototype filter. One of the applications of a comb filter considered in Section 11.4.1 is in creating special audio effects in musical sound processing, where both FIR and IIR prototypes are employed. Comb filters with multiple notch frequencies find applications in the cancellation of periodic inteferences. The comb filter is also employed in LORAN navigation systems for the suppression of cross-rate interferences [Jac96].

An interesting application of the comb filters of $G_0(z)$ and $G_1(z)$ of Eqs. (4.85) and (4.86), respectively, is in digital color television receivers for separating the luminance component containing the intensity information and the chrominance components containing the color information from the composite video signal [Orf96]. The basic structure for this purpose is as shown in Figure 4.18, where the delay chain is chosen to provide a line delay, i.e. the time to scan one horizontal line. However, a complete separation of the two components is not possible with this structure. Moreover, the filter $G_0(z)$ acts as a lowpass filter by averaging two successive horizontal lines of the video signal and blurs the luminance component. On the other hand, improved separation of the two components can be achieved by the structure of Figure 4.18 when the delay chain is chosen to provide

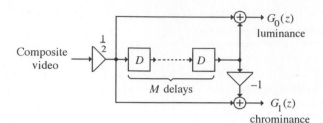

Figure 4.18   Filter structure for the separation of the luminance and chrominance components of a composite video signal.

$$u[n] = v[\pm n], \quad y[n] = w[\pm n]$$

Figure 4.19   Implementation of a zero-phase filtering scheme.

a frame delay. Unfortunately, the separation fails if there is some motion from frame to frame.

4.3.5   Zero-Phase and Linear-Phase Transfer Functions

A second classification of a transfer function is with respect to its phase characteristics.

In many applications, it is necessary to ensure that the digital filter designed does not distort the phase of the input signal components with frequencies in the passband. One way to avoid any phase distortions is to make the frequency response of the filter real and nonnegative, i.e., to design the filter with a *zero-phase* characteristic. However, it is impossible to design a causal digital filter with a zero-phase. For non-real time processing of real valued input signals of finite length, zero-phase filtering can be very simply implemented if the causality requirement is relaxed. To this end, one of two feasible schemes can be followed. In one scheme, the finite-length input data is processed through a causal real coefficient filter $H(z)$ whose output is then time-reversed and processed by the same filter, once again as indicated in Figure 4.19.

To verify the above scheme, let $v[-n] = u[n]$. Also, let $X(e^{j\omega})$, $V(e^{j\omega})$, $U(e^{j\omega})$, $W(e^{j\omega})$, and $Y(e^{j\omega})$ denote the discrete-time Fourier transforms of $x[n]$, $v[n]$, $u[n]$, $w[n]$ and $y[n]$, respectively. Now, from Figure 4.19 and making use of the symmetry relations given in Tables 3.3 and 3.4, we arrive at the relations between the various Fourier transforms as

$$V(e^{j\omega}) = H(e^{j\omega})X(e^{j\omega}), \qquad W(e^{j\omega}) = H(e^{j\omega})U(e^{j\omega})$$
$$U(e^{j\omega}) = V^*(e^{j\omega}), \qquad Y(e^{j\omega}) = W^*(e^{j\omega})$$

Combining the above equations we obtain

$$Y(e^{j\omega}) = W^*(e^{j\omega}) = H^*(e^{j\omega})U^*(e^{j\omega}) = H^*(e^{j\omega})V(e^{j\omega})$$
$$= H^*(e^{j\omega})H(e^{j\omega})X(e^{j\omega}) = |H(e^{j\omega})|^2 X(e^{j\omega}).$$

Therefore the overall arrangement of Figure 4.19 implements a zero-phase filter with a frequency response $|H(e^{j\omega})|^2$.

The function filtfilt in MATLAB implements the above scheme.

A second scheme to achieve zero-phase filtering is outlined in Problem 4.62.

In the case of a causal transfer function with a nonzero phase response, the phase distortion can be avoided by ensuring that the transfer function has a unity magnitude and a *linear-phase* characteristic in the frequency band of interest. The most general type of such a filter has a frequency response given by

$$H(e^{j\omega}) = e^{-j\omega D}, \tag{4.87}$$

which has a linear-phase response from $\omega = 0$ to $\omega = 2\pi$. Note that the above filter has a unity magnitude response and a linear phase with a group delay of amount D at all frequencies, i.e.,

$$|H(e^{j\omega})| = 1, \qquad \tau(\omega) = D. \tag{4.88}$$

The output of this filter to an input $x[n] = Ae^{j\omega n}$ is then given by

$$y[n] = Ae^{-j\omega D}e^{j\omega n} = Ae^{j\omega(n-D)}.$$

If $x_a(t)$ and $y_a(t)$ represent the continuous-time signals whose sampled versions, sampled at $t = nT$, are $x[n]$ and $y[n]$ given above, then the delay between $x_a(t)$ and $y_a(t)$ is precisely the group delay of amount D. Note that if D is an integer, then the output sequence $y[n]$ is equal to $x[n - D]$, a delayed version of the input sequence, otherwise, in general, $y[n]$ is not a delayed version of $x[n]$.

If we desire to pass input signal components in a certain frequency range undistorted in both magnitude and phase, then the transfer function should exhibit a unity magnitude response and linear-phase response in the band of interest. Figure 4.20 shows the frequency response of a lowpass transfer function with a linear-phase characteristic in the passband. Since the signal components in the stopband are blocked, the phase response in the stopband can be of any shape.

Example 4.15 Determine the impulse response of an ideal lowpass filter with a linear-phase response. The desired frequency response is now given by

$$H_{\text{LP}}(e^{j\omega}) = \begin{cases} e^{-j\omega n_0}, & 0 < |\omega| < \omega_c, \\ 0 & \omega_c \le |\omega| \le \pi. \end{cases} \tag{4.89}$$

Applying the frequency shifting property of the Fourier transform given in Table 3.2 to Eq. (4.63) we arrive at the impulse response of the linear phase lowpass filter of Eq. (4.89) as

$$h_{\text{LP}}[n] = \frac{\sin \omega_c(n - n_0)}{\pi(n - n_0)}, \qquad -\infty < n < \infty. \tag{4.90}$$

As before, the above filter is noncausal and is of doubly infinite length, and hence, unrealizable. However, by truncating the impulse response to a finite number of terms, we can develop a realizable FIR filter approximation to the ideal linear-phase lowpass filter. It should be noted that the truncated approximation may or may not exhibit linear phase, depending on the value of n_0 chosen. If we choose $n_0 = N/2$ with N a positive integer, the truncated and shifted approximation

$$\hat{h}_{\text{LP}}[n] = \frac{\sin \omega_c(n - N/2)}{\pi(n - N/2)}, \qquad 0 \le n \le N, \tag{4.91}$$

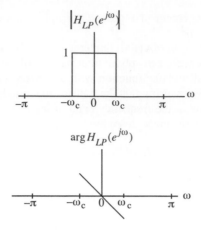

Figure 4.20   Frequency response of an ideal lowpass filter with a linear phase response in the passband.

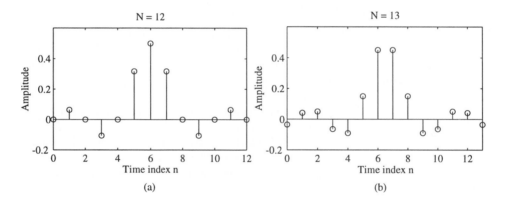

Figure 4.21   FIR approximation to the ideal linear-phase lowpass filter.

will be a length $N+1$ causal linear phase FIR filter. Figure 4.21 shows the filter coefficients obtained using the function sinc in MATLAB for two different values of N.

Because of the symmetry of the impulse response coefficients as indicated in this figure, the frequency response of the truncated approximation can be expressed as

$$\hat{H}_{\mathrm{LP}}(e^{j\omega}) = \sum_{n=0}^{N} h[n]e^{-j\omega n} = e^{-j\omega N/2}\breve{H}_{\mathrm{LP}}(\omega), \tag{4.92}$$

where $\breve{H}_{\mathrm{LP}}(\omega)$, called the *zero-phase response* or *amplitude response*, is a real function of ω. A plot of $\breve{H}_{\mathrm{LP}}(\omega)$ for several values of N is shown in Figure 4.22.

Observe from this figure that independent of the value of N, the amplitude response of the lowpass FIR filter obtained by truncating the ideal infinite-length filter of Eq. (4.90) exhibits *ripples* (i.e., oscillatory behavior) around the cutoff frequency ω_c. Moreover, an

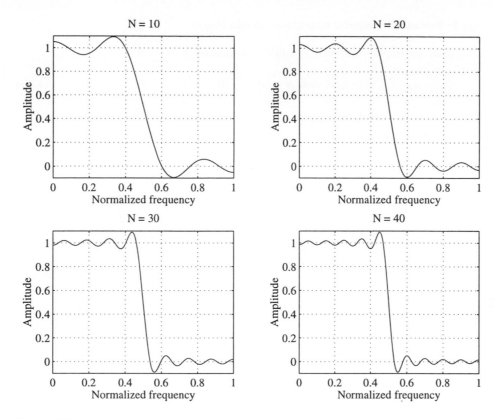

Figure 4.22 Amplitude response of the truncated ideal lowpass filter of Eq. (4.91) for four different values of N.

increase in N is associated with a sharper cutoff but with an increase in the number of ripples while maintaining the height of the largest ripple both in the passband and in the stopband. This type of behavior in the magnitude response is called the *Gibbs phenomenon*.

4.3.6   Types of Linear-Phase FIR Transfer Functions

In the previous section we pointed out why it is important to have a transfer function with a linear-phase property. It turns out it is always possible to design an FIR transfer function with an exact linear-phase response, while it is nearly impossible to design a linear phase IIR transfer function. Recall that the length-3 FIR transfer function of Example 4.3 has a linear-phase response, as indicated in Eq. (4.22). This filter was characterized by a symmetric impulse response of the form $h[0] = h[2]$. Note also the symmetry of the impulse responses of the truncated approximation to the ideal linear-phase lowpass filter shown in Figure 4.21. We show now that the symmetry in the impulse response is a necessary condition for linear-phase characteristic. We define four types of symmetry for an FIR filter with real impulse response and of length $N + 1$, and examine each of these four cases next [Rab75].

Type 1: Symmetric impulse response with odd length

In this case, the impulse response satisfies the condition

$$h[n] = h[N - n], \qquad 0 \le n \le N. \tag{4.93}$$

First, assume that $N = 8$ for simplicity. The transfer function of the corresponding filter is given by

$$H(z) = h[0] + h[1]z^{-1} + h[2]z^{-2} + h[3]z^{-3} + h[4]z^{-4}$$
$$+ h[5]z^{-5} + h[6]z^{-6} + h[7]z^{-7} + h[8]z^{-8}. \tag{4.94}$$

But from Eq. (4.93) for $N = 8$, $h[0] = h[8]$, $h[1] = h[7]$, $h[2] = h[6]$ and $h[3] = h[5]$. Then, Eq. (4.94) reduces to

$$
\begin{aligned}
H(z) &= h[0](1 + z^{-8}) + h[1](z^{-1} + z^{-7}) + h[2](z^{-2} + z^{-6}) \\
&\quad + h[3](z^{-3} + z^{-5}) + h[4]z^{-4} \\
&= h[0]z^{-4}(z^4 + z^{-4}) + h[1]z^{-4}(z^3 + z^{-3}) + h[2]z^{-4}(z^2 + z^{-2}) \\
&\quad + h[3]z^{-4}(z + z^{-1}) + h[4]z^{-4} \\
&= z^{-4}\{h[0](z^4 + z^{-4}) + h[1](z^3 + z^{-3}) + h[2](z^2 + z^{-2}) \\
&\quad + h[3](z + z^{-1}) + h[4]\}.
\end{aligned}
\tag{4.95}
$$

As a result, the corresponding frequency response is given by

$$
\begin{aligned}
H(e^{j\omega}) &= e^{-j4\omega}\{2h[0]\cos(4\omega) + 2h[1]\cos(3\omega) + 2h[2]\cos(2\omega) \\
&\quad + 2h[3]\cos(\omega) + h[4]\}.
\end{aligned}
\tag{4.96}
$$

Note that the quantity inside the braces in the above expression is a real function of ω and can assume positive or negative values in the range $0 \le |\omega| \le \pi$. Here the phase is a linear function of ω in the generalized sense. As a result, the group delay is given by

$$\tau(\omega) = 4, \tag{4.97}$$

indicating a generalized linear-phase property with a constant group delay of 4 samples. It should be noted that since in general $|H(e^{j\omega})|$ is not a constant, the output waveform is not a replica of the input waveform.

In the general case, the frequency response is of the form

$$H(e^{j\omega}) = e^{-jN\omega/2}\left\{ \sum_{n=0}^{N/2} a[n]\cos(\omega n) \right\}, \tag{4.98}$$

where

$$a[0] = h\left[\tfrac{N}{2}\right], \quad a[n] = 2h\left[\tfrac{N}{2} - n\right], \quad 1 \le n \le \tfrac{N}{2}. \tag{4.99}$$

Type 2: Symmetric impulse response with even length

In the case of N odd, say, $N = 7$, by making use of the symmetry of the impulse response coefficients given by Eq. (4.93), the transfer function of the FIR filter can be written as:

$$
\begin{aligned}
H(z) &= h[0](1 + z^{-7}) + h[1](z^{-1} + z^{-6}) + h[2](z^{-2} + z^{-5}) \\
&\quad + h[3](z^{-3} + z^{-4}) \\
&= z^{-7/2}\{h[0](z^{7/2} + z^{-7/2}) + h[1](z^{5/2} + z^{-5/2}) \\
&\quad + h[2](z^{3/2} + z^{-3/2}) + h[3](z^{1/2} + z^{-1/2})\}.
\end{aligned}
$$

The frequency response function is thus given by

$$
H(e^{j\omega}) = e^{-j7\omega/2} \left\{ 2h[0]\cos(\tfrac{7\omega}{2}) + 2h[1]\cos(\tfrac{5\omega}{2}) \right.
$$
$$
\left. + 2h[2]\cos(\tfrac{3\omega}{2}) + 2h[3]\cos(\tfrac{\omega}{2}) \right\}, \tag{4.100}
$$

As before, here, the quantity inside the braces in the above expression is a real function of ω, and as in the previous case, it can assume positive or negative values in the range $0 \le |\omega| \le \pi$. Here again the phase is a linear function of ω in the generalized sense. The corresponding group delay is constant:

$$
\tau(\omega) = -\frac{d\theta(\omega)}{d\omega} = \frac{7}{2}, \tag{4.101}
$$

indicating a group delay of $\frac{7}{2}$ sample periods.

The expression for the frequency response in the general case is of the form

$$
H(e^{j\omega}) = e^{-jN\omega/2} \left\{ \sum_{n=1}^{(N+1)/2} b[n]\cos\left(\omega\left(n - \tfrac{1}{2}\right)\right) \right\}, \tag{4.102}
$$

where

$$
b[n] = 2h\left[\frac{N+1}{2} - n\right], \qquad 1 \le n \le \frac{N+1}{2}. \tag{4.103}
$$

Example 4.16 Consider the transfer function

$$
H_0(z) = \frac{1}{6}\left[\frac{1}{2} + z^{-1} + z^{-2} + z^{-3} + z^{-4} + z^{-5} + \frac{1}{2}z^{-6}\right]. \tag{4.104}
$$

which is seen to be a slightly modified version of a length-7 moving-average FIR filter. The above transfer function has a symmetric impulse response and therefore a linear-phase characteristic. A plot of its magnitude response is shown in Figure 4.23 along with that of a length-7 conventional moving-average filter. Note the improved magnitude response obtained by simply changing the first and last impulse response coefficients of a moving-average FIR filter [Ham89].

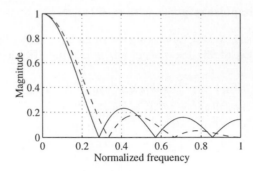

Figure 4.23   Magnitude responses of the length-7 moving average lowpass filter (solid line) and the modified lowpass filter of Eq. (4.104) (dashed line).

Type 3: Antisymmetric impulse response with odd length

Another way of achieving a constant group delay but with an "almost linear phase" property is by ensuring an antisymmetric condition on the impulse response coefficients:

$$h[n] = -h[N - n], \qquad 0 \le n \le N. \qquad (4.105)$$

We first consider the case of N even. It follows from Eq. (4.105) that for N even, $h[N/2] = 0$. Let $N = 8$. Then applying the above symmetry condition on the expression for the transfer function, we arrive at

$$H(z) = z^{-4}\{h[0](z^4 - z^{-4}) + h[1](z^3 - z^{-3}))$$
$$+h[2](z^2 - z^{-2}) + h[3](z - z^{-1})\}, \qquad (4.106)$$

where we have used the fact that $h[4] = 0$. As a result, the frequency response is given by

$$H(e^{j\omega}) = e^{-j4\omega}e^{j\pi/2}\{2h[0]\sin 4\omega + 2h[1]\sin 3\omega + 2h[2]\sin 2\omega$$
$$+2h[3]\sin \omega\}. \qquad (4.107)$$

It also exhibits a generalized linear phase with a group delay given by

$$\tau(\omega) = 4, \qquad (4.108)$$

implying a constant group delay of 4 samples.

The expression for the frequency response in the general case is of the form

$$H(e^{j\omega}) = e^{-jN\omega/2}e^{j\pi/2}\left\{\sum_{n=1}^{N/2} c[n]\sin(\omega n)\right\}, \qquad (4.109)$$

where

$$c[n] = 2h\left[\tfrac{N}{2} - n\right], \qquad 1 \le n \le \tfrac{N}{2}. \qquad (4.110)$$

Type 4: Antisymmetric impulse response with even length

In this case N is odd. For $N = 7$, the transfer function can be expressed as

$$H(z) = z^{-7/2}\{h[0](z^{7/2} - z^{-7/2}) + h[1](z^{5/2} - z^{-5/2})$$
$$+ h[2](z^{3/2} - z^{-3/2}) + h[3](z^{1/2} - z^{-1/2})\}. \tag{4.111}$$

The corresponding frequency response is thus given by

$$H(e^{j\omega}) = e^{-j7\omega/2}e^{j\pi/2}\{2h[0]\sin(7\omega/2) + 2h[1]\sin(5\omega/2)$$
$$+ 2h[2]\sin(3\omega/2) + 2h[3]\sin(\omega/2)\}. \tag{4.112}$$

It has a generalized phase response with a constant group delay given by

$$\tau(\omega) = \tfrac{7}{2}. \tag{4.113}$$

Here, the frequency response in the general case is given by

$$H(e^{j\omega}) = e^{-jN\omega/2}e^{j\pi/2}\left\{\sum_{n=1}^{(N+1)/2} d[n]\sin\left(\omega\left(n - \tfrac{1}{2}\right)\right)\right\}, \tag{4.114}$$

where

$$d[n] = 2h\left[\tfrac{N+1}{2} - n\right], \qquad 1 \le n \le \tfrac{N+1}{2}, \tag{4.115}$$

Figure 4.24 illustrates the four types of symmetry required for ensuring the generalized linear-phase property in an FIR transfer function. In each of the four types of linear-phase FIR filters, the frequency response $H(e^{j\omega})$ is of the form

$$H(e^{j\omega}) = e^{-jN\omega/2}e^{j\beta}\breve{H}(\omega), \qquad \beta = 0 \text{ or } \tfrac{\pi}{2}, \tag{4.116}$$

where the *amplitude response* $\breve{H}(\omega)$, also called the *zero-phase response*, is a real function of ω. It should be noted that $\breve{H}(\omega)$ can become negative over certain frequency ranges, typically in the stopband, as indicated in Figure 4.22. An FIR filter with a frequency response that is a real function of ω is often called a *zero-phase filter*. Such a filter must have a noncausal impulse response.

4.3.7 Zero Locations of Linear-Phase FIR Transfer Functions

Let us now study the zero locations of a linear-phase FIR transfer function. Consider first an FIR filter with a symmetric impulse response. Its transfer function $H(z)$ can be written as

$$H(z) = \sum_{n=0}^{N} h[n]z^{-n} = \sum_{n=0}^{N} h[N - n]z^{-n}, \tag{4.117}$$

using the symmetry condition of Eq. (4.93). By making a change of variable $m = N - n$, we can rewrite the rightmost expression in Eq. (4.117) as

$$H(z) = \sum_{m=0}^{N} h[m]z^{-N+m} = z^{-N}\sum_{n=0}^{N} h[m]z^{m} = z^{-N}H(z^{-1}), \tag{4.118}$$

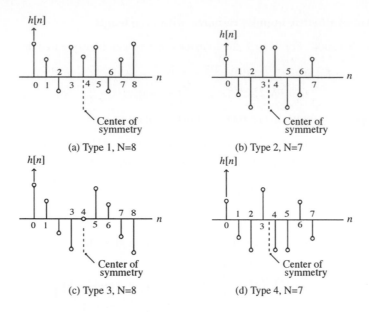

Figure 4.24   Illustration of the four types of impulse response symmetry.

Similarly, the transfer function $H(z)$ of an FIR filter with an antisymmetric response satisfying the condition of Eq. (4.105) can be expressed as

$$H(z) = \sum_{n=0}^{N} h[n]z^{-n} = -\sum_{n=0}^{N} h[N-n]z^{-n} = -z^{-N}H(z^{-1}), \qquad (4.119)$$

In either case, it follows from Eqs. (4.118) and (4.119) that if $z = \xi_0$ is a zero of $H(z)$, so is $z = 1/\xi_0$. Moreover, for an FIR filter with real impulse response, the zeros occur in complex conjugate pairs. Hence, a zero at $z = \xi_0$ is associated with a zero at $z = \xi_0^*$. Therefore, a complex zero that is not on the unit circle is associated with a set of four zeros given by

$$z = re^{\pm j\phi}, \qquad z = \frac{1}{r}e^{\pm j\phi}.$$

For a zero on the unit circle, its reciprocal is also its complex conjugate. Hence, in this case the zeros appear as a pair:

$$z = e^{\pm j\phi}.$$

A real zero is paired with its reciprocal zero appearing at

$$z = \rho, \qquad z = \frac{1}{\rho}.$$

Note that a zero at $z = \pm 1$ is its own reciprocal, implying it can appear only singly. However, a Type 2 FIR filter must have a zero at $z = -1$ since from Eq. (4.118) we note that

$$H(-1) = (-1)^N H(-1) = -H(-1),$$

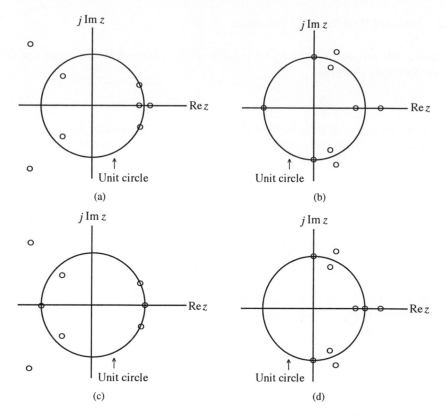

Figure 4.25 Examples of zero locations of linear phase FIR transfer functions: (a) type 1, (b) type 2, (c) type 3, and (d) type 4 cases.

implying $H(-1) = 0$. In the case of a Type 3 or 4 FIR filter, Eq. (4.119) implies that $H(1) = -H(1)$, indicating that the filter must have a zero at $z = 1$. On the other hand, only a Type 3 FIR filter is restricted to have a zero at $z = -1$ since here

$$H(-1) = -(-1)^N H(-1) = -H(-1),$$

forcing $H(-1) = 0$. Figure 4.25 shows some examples of zero locations of all four types of FIR filters.

The presence of zeros at $z = \pm 1$ leads to the following limitations on the use of these linear-phase FIR filters for designing filters. For example, since the Type 2 FIR filter always has a zero at $z = -1$, it cannot be used to design a highpass filter. Likewise, the Type 3 FIR filter has zeros at both $z = 1$ and $z = -1$ and, as a result, cannot be used to design either a lowpass or a highpass or a bandstop filter. Similarly, the Type 4 FIR filter is not appropriate to design a lowpass filter due to the presence of a zero at $z = 1$. Finally, the Type 1 FIR filter has no such restrictions and can be used to design almost any type of filter.

4.3.8   Bounded Real Transfer Functions

A causal stable real coefficient transfer function $H(z)$ is defined as a *bounded real (BR) transfer function* [Vai84] if

$$|H(e^{j\omega})| \leq 1, \qquad \text{for all values of } \omega . \tag{4.120}$$

If the input and output of a digital filter characterized by a BR transfer function $H(z)$ are given by $x[n]$ and $y[n]$, respectively, with $X(e^{j\omega})$ and $Y(e^{j\omega})$ denoting their respective Fourier transforms, then Eq. (4.120) implies that

$$|Y(e^{j\omega})|^2 \leq |X(e^{j\omega})|^2. \tag{4.121}$$

Integrating above from $-\pi$ to π, and applying Parseval's relation (see Table 3.2), we arrive at

$$\sum_{n=-\infty}^{\infty} y^2[n] \leq \sum_{n=-\infty}^{\infty} x^2[n]. \tag{4.122}$$

Or in other words, for all finite energy inputs, the output energy is less than or equal to the input energy implying that a digital filter characterized by a BR transfer function is a *passive* structure.

  If Eq. (4.120) is satisfied with an equal sign, then from Eq. (4.122), the output energy is equal to the input energy, and such a digital filter is therefore a *lossless* system. A causal stable real coefficient transfer function $H(z)$ with a frequency response $H(e^{j\omega})$ of unity magnitude is thus called a *lossless bounded real (LBR) transfer function* [Vai84].

  It follows from Eq. (4.71) that the first-order lowpass transfer function $H_{\text{LP}}(z)$ of Eq. (4.70) is a BR function if $|\alpha| < 1$. It can be shown that the first-order highpass transfer function of Eq. (4.73) is BR if $|\alpha| < 1$. Similarly, the second-order transfer functions of Eqs. (4.74) and (4.79) are BR if $|\alpha| < 1$ and $|\beta| < 1$.

  The BR and LBR transfer functions are the keys to the realization of digital filters with low coefficient sensitivity (See Section 9.7).

4.4   Allpass Transfer Function

We now turn our attention to a very special type of IIR transfer function that is characterized by unity magnitude for all frequencies. Such a transfer function, called an *allpass transfer function*, has many useful applications in digital signal processing [Reg88]. We define the allpass transfer function, examine some of its key properties, and outline one of its common applications. Later in this chapter and elsewhere in the text we discuss various other applications. One important application considered in this chapter is the development of an algebraic test for the BIBO stability of a causal IIR transfer function.

4.4.1   Definition

An IIR transfer function $A(z)$ with unity magnitude response for all frequencies, i.e.

$$|A(e^{j\omega})|^2 = 1, \qquad \text{for all } \omega \tag{4.123}$$

Figure 4.26   Pole-zero plot of the real coefficient allpass transfer function of Eq. (4.127).

is called an *allpass transfer function.* Now an Mth order causal real coefficient allpass transfer function is of the form

$$A_M(z) = \pm \frac{d_M + d_{M-1}z^{-1} + \cdots + d_1 z^{-M+1} + z^{-M}}{1 + d_1 z^{-1} + \cdots + d_{M-1}z^{-M+1} + d_M z^{-M}}. \qquad (4.124)$$

If we denote the denominator polynomial of the allpass function $A_M(z)$ as $D_M(z)$,

$$D_M(z) = 1 + d_1 z^{-1} + \cdots + d_{M-1}z^{-M+1} + d_M z^{-M}, \qquad (4.125)$$

then it follows that $A_M(z)$ can be written as

$$A_M(z) = \pm \frac{z^{-M} D_M(z^{-1})}{D_M(z)}. \qquad (4.126)$$

Note from above that if $z = re^{j\phi}$ is a pole of a real coefficient allpass transfer function, then it has a zero at $z = (1/r)e^{-j\phi}$. The numerator of an allpass transfer function is said to be the *mirror-image polynomial* of the denominator, and vice-versa. We shall use the notation $\tilde{D}_M(z)$ to denote the mirror-image polynomial of a degree-M polynomial, i.e., $\tilde{D}_M(z) = z^{-M} D_M(z^{-1})$. Equation (4.126) implies that the poles and the zeros of a real coefficient allpass function exhibit *mirror-image symmetry* in the z-plane, as shown in Figure 4.26 for the following third-order allpass function:

$$A_3(z) = \frac{-0.2 + 0.18z^{-1} + 0.4z^{-2} + z^{-3}}{1 + 0.4z^{-1} + 0.18z^{-2} - 0.2z^{-3}}. \qquad (4.127)$$

To show that the magnitude of $A_M(e^{j\omega})$ is indeed equal to one for all ω, it follows from Eq. (4.126) that

$$A_M(z^{-1}) = \pm \frac{z^M D_M(z)}{D_M(z^{-1})}.$$

Therefore,

$$A_M(z)A_M(z^{-1}) = \frac{z^{-M} D_M(z^{-1})}{D_M(z)} \frac{z^M D_M(z)}{D_M(z^{-1})} = 1.$$

Figure 4.27   (a) The principal value of the phase function, and (b) the unwrapped phase function of the allpass transfer function of Eq. (4.127).

Hence,

$$|A_M(e^{j\omega})|^2 = A_M(z)A_M(z^{-1})|_{z=e^{j\omega}} = 1 \qquad (4.128)$$

As shown earlier in Section 4.2.2, the poles of a causal stable transfer function must lie inside the unit circle. As a result, all zeros of a causal stable allpass transfer function lie outside the unit circle in a mirror-image symmetry with its poles situated inside the unit circle.

It is interesting to examine the behavior of the phase function of an allpass transfer function. Figure 4.27(a) shows the principal value of the phase of the stable third-order allpass transfer function of Eq. (4.127). Note the discontinuity by the amount of 2π in the phase $\theta(\omega)$. If we unwrap the phase by removing the discontinuity, we arrive at the unwrapped phase function $\theta_c(\omega)$ indicated in Figure 4.27(b). As can be seen from this figure, the unwrapped phase function is a nonpositive continuous function of ω in the range $0 < \omega < \pi$. This property of the unwrapped phase function holds for any arbitrary causal stable allpass function.

4.4.2   Properties

We now state three very useful and important properties of a causal stable allpass function without proof [Reg88].

Property 1.   It follows from Section 4.3.8 that a causal stable real coefficient allpass transfer function is a lossless bounded real (LBR) transfer function or, equivalently, a causal stable allpass filter is a lossless structure.

Property 2.   The second property is concerned with the magnitude of a stable allpass function $A(z)$. It can be shown very simply that (Problem 4.70)

$$|A(z)| \begin{cases} < 1 & \text{for } |z| > 1, \\ = 1 & \text{for } |z| = 1, \\ > 1 & \text{for } |z| < 1. \end{cases} \qquad (4.129)$$

Figure 4.28   Use of an allpass filter as a delay equalizer.

Property 3.   The last property of interest is with regard to the change in phase for a real stable allpass function over the frequency range $\omega = 0$ to $\omega = \pi$. Let $\tau(\omega)$ denote the group delay function of the allpass filter $A(z)$, i.e.

$$\tau(\omega) = -\tfrac{d}{d\omega}[\theta_c(e^{j\omega})],$$

where $\theta_c(\omega)$ is the unwrapped form of the phase function $\theta(\omega) = \arg\{A(e^{j\omega})\}$ in order that the group delay $\tau(\omega)$ be well behaved. Now the unwrapped phase function $\theta_c(\omega)$ of a stable allpass function $A(z)$ is a monotonically decreasing function of ω so that $\tau(\omega)$ is everywhere positive in the range $0 < \omega < \pi$. Therefore, an Mth-order stable real allpass transfer function satisfies the property (Problem 4.71)

$$\int_0^\pi \tau(\omega)\, d\omega = M\pi \qquad (4.130)$$

Or in other words, the change in the phase of an Mth-order allpass function as ω goes from 0 to π is $M\pi$ radians.

4.4.3   A Simple Application

A simple but often used application of an allpass filter is as a *delay equalizer*. Let $G(z)$ be the transfer function of a digital filter that has been designed to meet a prescribed magnitude response. The nonlinear phase response of this filter can be corrected by cascading it with an allpass filter section $A(z)$ so that the overall cascade with a transfer function $G(z)A(z)$ has a constant group delay over the frequency domain of interest (see Figure 4.28). Since the allpass filter has a unity magnitude response, the magnitude response of the cascade is still equal to $|G(e^{j\omega})|$, while the overall delay is given by the sum of the group delays of $G(z)$ and $A(z)$. The allpass is designed so that the overall group delay is approximately a constant in the frequency region of interest.

Various other applications of the allpass filter are described in the latter parts of this book.

4.5   Minimum-Phase and Maximum-Phase Transfer Functions

Another useful classification of a transfer function is in terms of the behavior of its phase response. Consider the two first-order transfer functions $H_1(z)$ and $H_2(z)$:

$$H_1(z) = \frac{z+b}{z+a}, \quad H_2(z) = \frac{bz+1}{z+a}, \quad |a| < 1, \quad |b| < 1. \qquad (4.131)$$

As can be seen from the pole-zero plots given in Figure 4.29, both transfer functions have a pole inside the unit circle at the same location at $z = -a$ and are therefore stable. On the

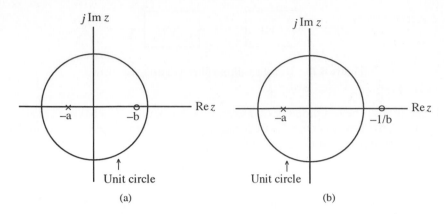

(a)   (b)

Figure 4.29   Pole-zero plots of the transfer functions of Eq. (4.131): (a) $H_1(z)$ and (b) $H_2(z)$.

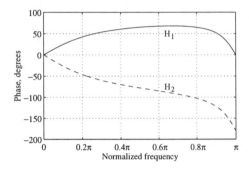

Figure 4.30   Unwrapped phase response of Eq. (4.131) for $a = 0.8$, $b = -0.5$.

other hand, the zero of $H_1(z)$ is inside the unit circle at $z = -b$, whereas the zero of $H_2(z)$ is outside the unit circle at $z = -1/b$ situated in a mirror-image symmetry with respect to the zero of $H_1(z)$. However, the two transfer functions have an identical magnitude function, since $H_1(z)H_1(z^{-1}) = H_2(z)H_2(z^{-1})$.

From Eq. (4.131),

$$\arg[H_1(e^{j\omega})] = \theta_1(\omega) = \tan^{-1}\frac{\sin\omega}{b + \cos\omega} - \tan^{-1}\frac{\sin\omega}{a + \cos\omega}, \qquad (4.132a)$$

$$\arg[H_2(e^{j\omega})] = \theta_2(\omega) = \tan^{-1}\frac{b\sin\omega}{1 + b\cos\omega} - \tan^{-1}\frac{\sin\omega}{a + \cos\omega}, \qquad (4.132b)$$

Figure 4.30 shows the unwrapped phase responses of the two transfer functions. From this figure it can be seen that $H_2(z)$ has an excess phase lag with respect to $H_1(z)$.

Generalizing the above result, a causal stable transfer function with all zeros outside the unit circle has an excess phase compared to a causal stable transfer function with identical magnitude but having all zeros inside the unit circle. As a result, a causal stable transfer function with all zeros inside the unit circle is called a *minimum-phase transfer function*, whereas a causal stable transfer function with all zeros outside the unit circle is called a *maximum-phase transfer function*.

It can be easily shown that any nonminimum-phase transfer function can be expressed as the product of a minimum-phase transfer function and a stable allpass transfer function (Problem 4.72).

4.6 Complementary Transfer Functions

A set of digital transfer functions with complementary chacteristics often finds useful applications in practice, such as efficient realizations of the transfer functions, low sensitivity realizations, filter bank design and implementation, etc. We describe next four useful complementary relations and indicate some of their applications.

4.6.1 Delay-Complementary Transfer Functions

A set of L transfer functions $\{H_0(z), H_1(z), \ldots, H_{L-1}(z)\}$ is defined to be *delay-complementary* of each other if the sum of their transfer functions is equal to some integer multiple of the unit delay [Vai93], i.e.,

$$\sum_{k=0}^{L-1} H_k(z) = \beta z^{-n_0}, \quad \beta \neq 0, \tag{4.133}$$

where n_0 is a nonnegative integer. A delay-complementary pair $\{H_0(z), H_1(z)\}$ can be readily designed if one of the pair is a known Type 1 linear-phase FIR transfer function of odd length. Let $H_0(z)$ be a Type 1 linear-phase frequency-selective FIR transfer function of length $M = 2K + 1$ with a magnitude response equal to $1 \pm \delta_p$ in the passband and less than or equal to δ_s in the stopband where δ_p and δ_s are very small numbers. From Eq. (4.94) its frequency response is of the form

$$H_0(e^{j\omega}) = e^{-jK\omega} \breve{H}_0(\omega), \tag{4.134}$$

where $\breve{H}_0(\omega)$ is the *amplitude response* of the FIR transfer function $H_0(z)$. Its delay-complementary transfer function $H_1(z)$ defined for $\beta = 1$ and $n_0 = K$ has a frequency response given by

$$H_1(e^{j\omega}) = e^{-jK\omega} \breve{H}_1(\omega) = e^{-jK\omega}[1 - \breve{H}_0(\omega)], \tag{4.135}$$

Now, in the passband, $1 - \delta_p \leq \breve{H}_0(\omega) \leq 1 + \delta_p$, and in the stopband, $-\delta_s \leq \breve{H}_0(\omega) \leq \delta_s$. It follows therefore from Eq. (4.135), in the stopband of $H_1(z)$, $-\delta_p \leq \breve{H}_0(\omega) \leq \delta_p$, and in the passband of $H_1(z)$, $1 - \delta_s \leq \breve{H}_0(\omega) \leq 1 + \delta_s$. As a result, $H_1(z)$ has a complementary magnitude response characteristic to that of $H_0(z)$ with a stopband exactly identical to the passband of $H_0(z)$ and a passband exactly equal to the stopband of $H_0(z)$. For example, if $H_0(z)$ is a lowpass filter, $H_1(z)$ will be a highpass filter, and vice-versa.

Example 4.17 Consider the Type 1 bandstop transfer function $H_{\mathrm{BS}}(z)$ given below [Aca83]:

$$H_{\mathrm{BS}}(z) = \tfrac{1}{64}(1 + z^{-2})^4(1 - 4z^{-2} + 5z^{-4} + 5z^{-8} - 4z^{-10} + z^{-12}). \tag{4.136a}$$

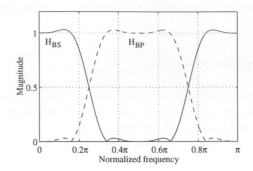

Figure 4.31   Delay-complementary linear-phase bandstop and bandpass FIR filters of Eqs. (4.136a) and (4.136b).

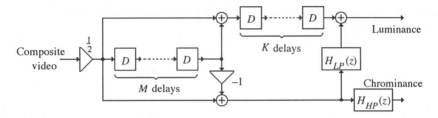

Figure 4.32   Luminance and chrominance components separation filter structure with improved vertical details.

Its delay-complementary bandpass transfer function is given by

$$H_{BP}(z) = z^{-10} - H_{BS}(z)$$
$$= \frac{1}{64}(1 - z^{-2})^4(1 + 4z^{-2} + 5z^{-4} + 5z^{-8} + 4z^{-10} + z^{-12}). \qquad (4.136b)$$

Figure 4.31 shows the plots of the magnitude responses of $H_{BS}(z)$ and $H_{BP}(z)$.

An interesting application of the delay-complementary FIR transfer function pair is in digital television receivers [Orf96]. It has been pointed out earlier that the structure of Figure 4.18 for the separation of the luminance and chrominance components of the composite video signal tends to blur the luminance output, resulting in the loss of vertical details. The low-frequency vertical details can be recovered from the output of the comb filter $G_1(z)$ of Figure 4.18 by a lowpass filter and added to the output of the comb filter $G_0(z)$. The vertical details can be removed from the output of $G_1(z)$ by filtering it through a filter with a delay-complementary frequency response to the filter employed to the lowpass filter. In practice the lowpass filter employed is a bandstop filter H_{BS} whose low-frequency passband coincides with the frequency range of the desired vertical details while its delay-complementary filter has a bandpass transfer function H_{BP}. The overall structure is thus as shown in Figure 4.32, where the delay-chain of length K in the top path is chosen to equalize the total delay in both the top and bottom paths.

One set of delay-complementary linear-phase bandpass/bandstop filters proposed for use in Figure 4.32 is the one given in Eqs. (4.136a) and (4.136b) for which $K = 10$. Other linear-phase bandstop FIR transfer functions suitable for vertical details recovery are given in Problem 4.76.

Delay-complementary filter sets can also be used as crossover filters for separating the digital audio input signals into two or three sub-signals occupying different frequency bands which are then used to drive the appropriate speakers of the loudspeaker system [Orf96]. Design of such delay-complementary crossover filters is considered in Problems M7.22 and M7.23. Another important application of the delay-complementary property is in the realization of low-sensitivity FIR digital filters discussed in Section 9.7.3.

4.6.2   Allpass-Complementary Filters

A set of M digital filters $\{H_i(z)\}, 0 \leq i \leq M - 1$, are defined to be *allpass-complementary* of each other, if the sum of their transfer functions is equal to an allpass function $A(z)$, [Gar80], [Neu84a], i.e.,

$$\sum_{i=0}^{M-1} H_i(z) = A(z). \tag{4.137}$$

4.6.3   Power-Complementary Filters

A set of M digital filters $\{H_i(z)\}, 0 \leq i \leq M - 1$, are defined to be *power-complementary* of each other, if the sum of the squares of their magnitude responses is equal to one [Neu84a], [Vai93], i.e.,[4]

$$\sum_{i=0}^{M-1} |H_i(e^{j\omega})|^2 = 1, \qquad \text{for all } \omega. \tag{4.138a}$$

By analytic continuation, the above property is equivalent to

$$\sum_{i=0}^{M-1} H_i(z^{-1}) H_i(z) = 1, \qquad \text{for all } z, \tag{4.138b}$$

for a real coefficient $H_i(z)$.

For a pair of power-complementary transfer functions, $H_0(z)$ and $H_1(z)$, the frequency ω_0 where $|H_0(e^{j\omega_0})|^2 = |H_1(e^{j\omega_0})|^2 = \frac{1}{2}$, is called the *cross-over frequency*. At this frequency the gain responses of both filters are 3 dB below their maximum values. As a result, ω_0 is also called the *3-dB cutoff frequency* of both filters.

Consider two transfer functions $H_0(z)$ and $H_1(z)$ described by

$$H_0(z) = \tfrac{1}{2}[A_0(z) + A_1(z)], \tag{4.139a}$$

$$H_1(z) = \tfrac{1}{2}[A_0(z) + A_1(z)], \tag{4.139b}$$

---

[4]A more general definition of the power-complementary property is given in [Vai93]

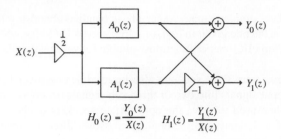

$$H_0(z) = \frac{Y_0(z)}{X(z)} \qquad H_1(z) = \frac{Y_1(z)}{X(z)}$$

Figure 4.33   Parallel allpass realization of a double-complementary IIR transfer functions.

where $A_0(z)$ and $A_1(z)$ are stable allpass transfer functions. It follows from the above that the sum of the two transfer functions is an allpass function $A_0(z)$, and hence, $H_0(z)$ and $H_1(z)$ of Eq. (4.139a) and (4.139b) are an *allpass-complementary* pair. It can be easily shown that the two filters described by Eq. (4.139a) and (4.139b) also form a *power-complementary* pair (Problem 4.78). It can also be easily shown that the the transfer functions $H_0(z)$ and $H_1(z)$ of Eq. (4.139a) and (4.139b) are also bounded real transfer functions (Problem 4.79).

A set of M transfer functions satisfying both the allpass-complementary property of Eq. (4.137) and the power-complementary property of Eq. (4.138a) is known as a *doubly complementary* set [Neu84a].

A pair of doubly complementary IIR transfer functions, $H_0(z)$ and $H_1(z)$, with a sum of allpass decomposition in the form of Eq. (4.139a) and (4.139b) can be simply realized by a parallel connection of the constituent allpass filters, as indicated in Figure 4.33. We shall demonstrate in Section 9.7.2 that such realizations also ensure low sensitivity in the passband with respect to the multiplier coefficients.

Example 4.18   The first-order lowpass transfer function of Eq. (4.70) can be expressed as:

$$H_{\mathrm{LP}}(z) = \frac{1}{2}\left[\frac{1 - \alpha + z^{-1} - \alpha z^{-1}}{1 - \alpha z^{-1}}\right]$$

$$= \frac{1}{2}\left[1 + \frac{-\alpha + z^{-1}}{1 - \alpha z^{-1}}\right] = \frac{1}{2}[A_0(z) + A_1(z)], \qquad (4.140)$$

where

$$A_0(z) = 1, \qquad A_1(z) = \frac{-\alpha + z^{-1}}{1 - \alpha z^{-1}}. \qquad (4.141)$$

From Eq. (4.139b) its power-complementary highpass transfer function is thus given by

$$H_{\mathrm{HP}}(z) = \frac{1}{2}[A_0(z) - A_1(z)] = \frac{1}{2}\left[1 - \frac{-\alpha + z^{-1}}{1 - \alpha z^{-1}}\right] = \frac{1 + \alpha}{2} \cdot \frac{1 - z^{-1}}{1 - \alpha z^{-1}}, \qquad (4.142)$$

which is precisely the first-order highpass transfer function of Eq. (4.73).

Figure 4.34 demonstrates the allpass complementary and the power-complementary properties of the first-order lowpass and highpass transfer functions of Eq. (4.140) and Eq. (4.142), respectively for $\alpha = 0.3$.

It can easily shown that the bandpass transfer function $H_{\mathrm{BP}}(z)$ of Eq. (4.74) and the band-stop transfer function $H_{\mathrm{BS}}(z)$ of Eq. (4.79) form a doubly complementary pair (Problem 4.81).

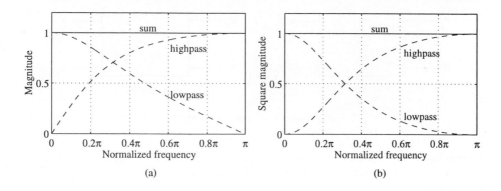

Figure 4.34 Illustration of the complementary properties of the first-order lowpass and highpass transfer functions with $\alpha = 0.3$. (a) Allpass complementary and (b) power-complementary.

4.6.4 Magnitude-Complementary Filters

A set of M digital filters $\{G_i(z)\}$, $i = 0, 1, \ldots, M - 1$, is defined to be *magnitude-complementary* of each other if the sum of their magnitude responses is equal to a constant [Reg87c], i.e.,

$$\sum_{i=0}^{M-1} |G_i(e^{j\omega})| = \beta \qquad \text{for all } \omega, \tag{4.143}$$

where β is a positive nonzero constant.

Consider two real coefficient double complementary transfer functions $H_0(z)$ and $H_1(z)$ that are related according to Eqs. (4.139a) and (4.139b). Define

$$G_0(z) = H_0^2(z) = \tfrac{1}{4}[A_0(z) + A_1(z)]^2, \tag{4.144a}$$

$$G_1(z) = -H_1^2(z) = -\tfrac{1}{4}[A_0(z) - A_1(z)]^2. \tag{4.144b}$$

It follows from above that $G_0(z) + G_1(z) = A_0(z)A_1(z)$, i.e., $G_0(z)$ and $G_1(z)$ are an allpass-complementary pair. It can be easily shown that (Problem 4.89)

$$|G_0(e^{j\omega})| + |G_1(e^{j\omega})| = |H_0(e^{j\omega})|^2 + |H_1(e^{j\omega})|^2 = 1, \tag{4.145}$$

i.e., $G_0(z)$ and $G_1(z)$ are a magnitude-complementary pair of transfer functions.

4.7 Digital Two-Pairs

The LTI discrete-time systems considered so far are single-input, single-output structures characterized by a transfer function. Often, such a system can be efficiently realized by interconnecting two-input, two-output structures, more commonly called *two-pairs* [Mit73b]. Figure 4.35 shows two commonly used block diagram representations of a two-pair, with Y_1 and Y_2 denoting the two outputs and X_1 and X_2 denoting the two inputs, where the dependence on the variable z has been omitted for simplicity. We consider here the transform-domain characterizations of such digital filter structures and discuss several

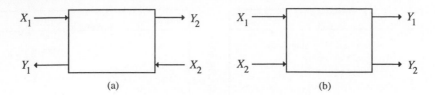

Figure 4.35   A digital two-pair.

two-pair interconnection schemes for the development of more complex structures. Later, we outline minimum-multiplier realizations of allpass transfer functions based on the two-pair representation.

4.7.1   Characterization

The input-output relation of a digital two-pair is given by

$$\begin{bmatrix} Y_1 \\ Y_2 \end{bmatrix} = \begin{bmatrix} t_{11} & t_{12} \\ t_{21} & t_{22} \end{bmatrix} \begin{bmatrix} X_1 \\ X_2 \end{bmatrix}, \tag{4.146}$$

In the above relation the matrix τ given by

$$\tau = \begin{bmatrix} t_{11} & t_{12} \\ t_{21} & t_{22} \end{bmatrix} \tag{4.147}$$

is called the *transfer matrix* of the two-pair. From Eq. (4.146) it follows that the transfer parameters can be found as follows:

$$t_{11} = \left. \frac{Y_1}{X_1} \right|_{X_2=0}, \quad t_{12} = \left. \frac{Y_1}{X_2} \right|_{X_1=0}, \quad t_{21} = \left. \frac{Y_2}{X_1} \right|_{X_2=0}, \quad t_{22} = \left. \frac{Y_2}{X_2} \right|_{X_1=0}. \tag{4.148}$$

An alternative characterization of the two-pair is in terms of its chain parameters as

$$\begin{bmatrix} X_1 \\ Y_1 \end{bmatrix} = \begin{bmatrix} A & B \\ C & D \end{bmatrix} \begin{bmatrix} Y_2 \\ X_2 \end{bmatrix}, \tag{4.149}$$

where the matrix Γ given by

$$\Gamma = \begin{bmatrix} A & B \\ C & D \end{bmatrix}, \tag{4.150}$$

is called the *chain matrix* of the two-pair.

The relations between the transfer parameters and the chain parameters can be easily derived and are given as

$$t_{11} = \frac{C}{A}, \quad t_{12} = \frac{AD - BC}{A}, \quad t_{21} = \frac{1}{A}, \quad t_{22} = -\frac{B}{A}, \tag{4.151a}$$

$$A = \frac{1}{t_{21}}, \quad B = -\frac{t_{22}}{t_{21}}, \quad C = \frac{t_{11}}{t_{21}}, \quad D = \frac{t_{12}t_{21} - t_{11}t_{22}}{t_{21}}. \tag{4.151b}$$

A two-pair is said to be *reciprocal* if $t_{12} = t_{21}$ [Mit73a]. From Eq. (4.151a) it follows that for a reciprocal two-pair, $AD - BC = 1$.

Figure 4.36   Γ-cascade connection of two-pairs.

Figure 4.37   τ-cascade connection of two-pairs.

4.7.2   Two-Pair Interconnection Schemes

Two or more two-pairs can be connected in cascade in two different ways. The cascade connection of two-pairs shown in Figure 4.36 is called a Γ-*cascade*. We next determine the characterization of the overall two-pair. Let the individual two-pairs be characterized by their chain parameters as:

$$\begin{bmatrix} X_1' \\ Y_1' \end{bmatrix} = \begin{bmatrix} A' & B' \\ C' & D' \end{bmatrix} \begin{bmatrix} Y_2' \\ X_2' \end{bmatrix}, \quad \begin{bmatrix} X_1'' \\ Y_1'' \end{bmatrix} = \begin{bmatrix} A'' & B'' \\ C'' & D'' \end{bmatrix} \begin{bmatrix} Y_2'' \\ X_2'' \end{bmatrix} \qquad (4.152)$$

But from Figure 4.36, $X_1'' = Y_2'$, and $Y_1'' = X_2'$. Substituting these relations in the first equation of Eq. (4.152) and combining the two equations, we arrive at

$$\begin{bmatrix} X_1' \\ Y_1' \end{bmatrix} = \begin{bmatrix} A' & B' \\ C' & D' \end{bmatrix} \begin{bmatrix} A'' & B'' \\ C'' & D'' \end{bmatrix} \begin{bmatrix} Y_2'' \\ X_2'' \end{bmatrix} \qquad (4.153)$$

Therefore, the chain matrix of the overall cascade is given by the product of the individual chain matrices, as indicated in Eq. (4.153).

A second type of cascade connection, called the τ-*cascade*, is shown in Figure 4.37. If the individual two-pairs are described by their transfer matrices

$$\begin{bmatrix} Y_1' \\ Y_2' \end{bmatrix} = \begin{bmatrix} t_{11}' & t_{12}' \\ t_{21}' & t_{22}' \end{bmatrix} \begin{bmatrix} X_1' \\ X_2' \end{bmatrix}, \quad \begin{bmatrix} Y_1'' \\ Y_2'' \end{bmatrix} = \begin{bmatrix} t_{11}'' & t_{12}'' \\ t_{21}'' & t_{22}'' \end{bmatrix} \begin{bmatrix} X_1'' \\ X_2'' \end{bmatrix}, \qquad (4.154)$$

then it follows that the overall cascade is characterized by a transfer matrix that is the product of the transfer matrix of the constituent two-pairs, i.e.

$$\begin{bmatrix} Y_1'' \\ Y_2'' \end{bmatrix} = \begin{bmatrix} t_{11}'' & t_{12}'' \\ t_{21}'' & t_{22}'' \end{bmatrix} \begin{bmatrix} t_{11}' & t_{12}' \\ t_{21}' & t_{22}' \end{bmatrix} \begin{bmatrix} X_1' \\ X_2' \end{bmatrix} \qquad (4.155)$$

Another useful interconnection is the constrained two-pair indicated in Figure 4.38. If we denote the transfer function Y_1/X_1 as $H(z)$, then it can shown that $H(z)$ can be

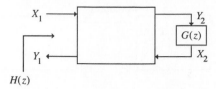

Figure 4.38   A constrained two-pair.

expressed in terms of the two-parameters and the constraining transfer function $G(z)$ as:

$$H(z) = \frac{Y_1}{X_1} = \frac{C + D \cdot G(z)}{A + B \cdot G(z)} \tag{4.156a}$$

$$= t_{11} + \frac{t_{12}t_{21}G(z)}{1 - t_{22}G(z)}. \tag{4.156b}$$

4.8   Stability Test

We have shown in Section 4.2.2 that the BIBO stability of a causal rational transfer function requires that all its poles be inside the unit circle. For very-high-order transfer functions it is difficult to determine the pole locations analytically, and the use of some type of root finding computer program is necessary. We outline here a simple algebraic stability test procedure that does not require the determination of the zero locations. The algorithm is based on the realization of an allpass transfer function with a denominator that is the same as that of the transfer function of interest.

4.8.1   The Stability Triangle

For a second-order transfer function the stability can be checked easily by examining its denominator coefficients. Let

$$D(z) = 1 + d_1 z^{-1} + d_2 z^{-2} \tag{4.157}$$

denote the denominator of the transfer function. In terms of its poles, $D(z)$ can be expressed as

$$D(z) = (1 - \lambda_1 z^{-1})(1 - \lambda_2 z^{-1}) = 1 - (\lambda_1 + \lambda_2)z^{-1} + \lambda_1\lambda_2 z^{-2}. \tag{4.158}$$

Comparing Eqs. (4.157) and (4.158) we obtain

$$d_1 = -(\lambda_1 + \lambda_2), \quad d_2 = \lambda_1\lambda_2. \tag{4.159}$$

Now for stability of the transfer function, its poles must be inside the unit circle, i.e.,

$$|\lambda_1| < 1, \quad |\lambda_2| < 1.$$

Since from Eq. (4.157) the coefficient d_2 is given by the product of the poles, we must have

$$|d_2| < 1. \tag{4.160}$$

Figure 4.39   Stability triangle for a second-order digital transfer function. Stability region is the cross-hatched area.

Now the roots of the polynomial $D(z)$ are given by

$$\lambda_1 = -\frac{d_1 + \sqrt{d_1^2 - 4d_2}}{2}, \qquad \lambda_2 = -\frac{d_1 - \sqrt{d_1^2 - 4d_2}}{2}, \qquad (4.161)$$

It can be shown (Problem 4.94) that a second coefficient condition is obtained using Eq. (4.161) and is given by

$$|d_1| < 1 + d_2. \qquad (4.162)$$

The region in the (d_1, d_2)-plane where the two coefficient conditions of Eqs. (4.160) and (4.162) are satisfied is a triangle, as sketched in Figure 4.39, and is known as the *stability triangle* for a second-order digital transfer function [Jac96].

Example 4.19   In the case of the second-order bandpass transfer function $H'_{\mathrm{BP}}(z)$ of Eq. (4.78a), note that $d_1 = -0.7343423986$ and $d_2 = 1.37638192$. Note that here $|d_2| > 1$ and hence, this transfer function is unstable.

For the second-order bandpass transfer function $H''_{\mathrm{BP}}(z)$ of Eq. (4.78b), we observe that $d_1 = -0.53353098$ and $d_2 = 0.726542528$. In this case, $|d_2| < 1$ and $|d_1| < 1 + d_2$, indicating that $H''_{\mathrm{BP}}(z)$ is a stable transfer function.

4.8.2   A Stability Test Procedure

It is impossible to develop simple conditions on the coefficients of the denominator polynomial of a higher order transfer function similar to that developed for a second-order function as given above. However, a number of methods have been proposed to determine the stability of an Mth-order transfer function $H(z)$ without factoring out the the roots of its denominator polynomial $D_M(z)$. We outline below one such method [Vai87e].

Let

$$D_M(z) = \sum_{i=0}^{M} d_i z^{-i}, \qquad (4.163)$$

where we assume for simplicity $d_0 = 1$. We first form an Mth-order allpass transfer function from $D_M(z)$ as

$$A_M(z) = \frac{\tilde{D}_M(z)}{D_M(z)}$$

$$= \frac{d_M + d_{M-1}z^{-1} + d_{M-2}z^{-2} + \cdots + d_1 z^{-(M-1)} + z^{-M}}{1 + d_1 z^{-1} + d_2 z^{-2} + \cdots + d_{M-1}z^{-(M-1)} + d_M z^{-M}}. \qquad (4.164)$$

If we express

$$D_M(z) = \prod_{i=1}^{M}(1 - \lambda_i z^{-1}),$$

it then follows that the coefficient d_M is the product of all roots, i.e.,

$$d_M = \prod_{i=1}^{M}\lambda_i.$$

Now for stability we must have $|\lambda_i| < 1$, which implies the condition $|d_M| < 1$. If we define

$$k_M = A_M(\infty) = d_M,$$

then a necessary condition for stability of $A_M(z)$, and hence, the original transfer function $H(z)$, is given by

$$k_M^2 < 1. \qquad (4.165)$$

Assume above condition holds. Next, we form a new function $A_{M-1}(z)$ according to

$$A_{M-1}(z) = z\left[\frac{A_M(z) - k_M}{1 - k_M A_M(z)}\right] = z\left[\frac{A_M(z) - d_M}{1 - d_M A_M(z)}\right]. \qquad (4.166)$$

Substituting in the above equation the expression for $A_M(z)$ from Eq. (4.164), we arrive at

$$A_{M-1}(z) = \frac{\begin{array}{c}(d_{M-1} - d_M d_1) + (d_{M-2} - d_M d_2)z^{-1} + \cdots \\ +(d_1 - d_M d_{M-1})z^{-(M-2)} + (1 - d_M^2)z^{-(M-1)}\end{array}}{\begin{array}{c}(1 - d_M^2) + (d_1 - d_M d_{M-1})z^{-1} + \cdots \\ +(d_{M-2} - d_M d_2)z^{-(M-2)} + (d_{M-1} - d_M d_1)z^{-(M-1)}\end{array}}, \qquad (4.167)$$

which is seen to be an $(M - 1)$th-order allpass function. $A_{M-1}(z)$ can be rewritten in the form

$$A_{M-1}(z) = \frac{d'_{M-1} + d'_{M-2}z^{-1} + \cdots + d'_1 z^{-(M-2)} + z^{-(M-1)}}{1 + d'_1 z^{-1} + \cdots + d'_{M-2}z^{-(M-2)} + d'_{M-1}z^{-(M-1)}}, \qquad (4.168)$$

where

$$d'_i = \frac{d_i - d_M d_{M-i}}{1 - d_M^2}, \qquad i = 1, 2, \ldots, M - 1. \qquad (4.169)$$

Now from Eq. (4.166) the poles λ_0 of $A_{M-1}(z)$ are given by the roots of the equation

$$A_M(\lambda_0) = \frac{1}{k_M}. \qquad (4.170)$$

By assumption Eq. (4.165) holds. Hence the above equation implies that

$$|A_M(\lambda_0)| > 1. \qquad (4.171)$$

If $A_M(z)$ is a stable allpass function, then according to Eq. (4.129), $|A_M(z)| < 1$ for $|z| > 1$, $|A_M(z)| = 1$ for $|z| = 1$, and $|A_M(z)| > 1$ for $|z| < 1$. Thus, if $A_M(z)$ is a stable allpass function then $|\lambda_0| < 1$, or in other words, $A_{M-1}(z)$ is a stable allpass function.

To prove the converse, we invert the relation of Eq. (4.167) to arrive at

$$A_M(z) = \frac{k_M + z^{-1}A_{M-1}(z)}{1 + k_M z^{-1}A_{M-1}(z)}. \qquad (4.172)$$

If ζ_0 is a pole of $A_M(z)$, then

$$\zeta_0^{-1}A_{M-1}(\zeta_0) = -\frac{1}{k_M}. \qquad (4.173)$$

By assumption Eq. (4.165) holds. Therefore, $|\zeta_0^{-1}A_{M-1}(\zeta_0)| > 1$, i.e.,

$$|A_{M-1}(\zeta_0)| > |\zeta_0|. \qquad (4.174)$$

Assume $A_{M-1}(z)$ is a stable allpass function, then according to Eq. (4.129), $|A_{M-1}(z)| \leq 1$ for $|z| \geq 1$. Now, if $|\zeta_0| \geq 1$, then from Eq. (4.129), $|A_{M-1}(\zeta_0)| \leq 1$, which contradicts Eq. (4.174). On the other hand, if $|\zeta_0| < 1$, then from Eq. (4.129), $|A_{M-1}(\zeta_0)| > 1$, satisfying the condition of Eq. (4.174). Thus, if Eq. (4.165) holds and if $A_{M-1}(z)$ is a stable allpass function, then $A_M(z)$ is also a stable allpass function.

Summarizing, a necessary and sufficient set of conditions for the allpass function $A_M(z)$ to be stable are therefore:

(a) $k_M^2 < 1$, and

(b) The allpass function $A_{M-1}(z)$ is stable.

Thus, once we have checked the condition $k_M^2 < 1$, we merely test for the stability of the lower order allpass function $A_{M-1}(z)$. The process can now be repeated, generating a set of coefficients

$$k_M, k_{M-1}, \ldots, k_2, k_1$$

and a set of allpass functions of decreasing orders,

$$A_M(z), A_{M-1}(z), \ldots, A_2(z), A_1(z), A_0(z) = 1.$$

The allpass function $A_M(z)$ is stable if and only if $k_i^2 < 1$ for all i.

Example 4.20 Let us test the stability of the transfer function

$$H(z) = \frac{1}{4z^4 + 3z^3 + 2z^2 + z + 1}. \tag{4.175}$$

From the above, we arrive at a fourth-order allpass function whose denominator is the same as that of $H(z)$ given above with the numerator being its mirror-image polynomial:

$$A_4(z) = \frac{\frac{1}{4}z^4 + \frac{1}{4}z^3 + \frac{1}{2}z^2 + \frac{3}{4}z + 1}{z^4 + \frac{3}{4}z^3 + \frac{1}{2}z^2 + \frac{1}{4}z + \frac{1}{4}} = \frac{d_4 z^4 + d_3 z^3 + d_2 z^2 + d_1 z + 1}{z^4 + d_1 z^3 + d_2 z^2 + d_3 z + d_4}. \tag{4.176}$$

Note that $k_4 = A_4(\infty) = d_4 = 1/4$. Hence $|k_4| < 1$.

Using Eq. (4.169), we next determine the coefficients $\{d_i'\}$ of the third-order allpass transfer function $A_3(z)$ derived from $A_4(z)$:

$$d_i' = \frac{d_i - d_4 d_{4-i}}{1 - d_4^2}, \qquad i = 1, 2, \ldots, 3. \tag{4.177}$$

Substituting the appropriate values of d_i from Eq. (4.176) in the above expression, we arrive at

$$A_3(z) = \frac{d_3' z^3 + d_2' z^2 + d_1' z + 1}{z^3 + d_1' z^2 + d_2' z + d_3'} = \frac{\frac{1}{15}z^3 + \frac{2}{5}z^2 + \frac{11}{15}z + 1}{z^3 + \frac{11}{15}z^2 + \frac{2}{5}z + \frac{1}{15}}.$$

It can be seen that $k_3 = A_3(\infty) = d_3' = 1/15$. Thus, $|k_3| < 1$.

Following the above procedure, we derive the next two lower order allpass functions:

$$A_2(z) = \frac{\frac{79}{224}z^2 + \frac{159}{224}z + 1}{z^2 + \frac{159}{224}z + \frac{79}{224}},$$

$$A_1(z) = \frac{\frac{53}{101}z + 1}{z + \frac{53}{101}}.$$

Note from the above that $k_2 = A_2(\infty) = 79/224$, $k_1 = A_1(\infty) = 53/101$ implying that $|k_2| < 1$ and $|k_1| < 1$.

It should be noted that is not necessary to derive $A_1(z)$ since $A_2(z)$ can be tested for stability using conditions of Eqs. (4.160) and (4.162). These two conditions are obviously satisfied for the $A_2(z)$ given above.

Since all of the stability test conditions are satisfied, $A_4(z)$, and hence, $H(z)$ of Eq. (4.175) is a stable transfer function.

Example 4.21 Determine whether all zeros of the denomintor polynomial of a causal transfer function

$$D_4(z) = z^4 + 3.25z^3 + 3.75z^2 + 2.75z + 0.5$$

are inside the unit circle.

We form the allpass function

$$A_4(z) = \frac{\tilde{D}_4(z)}{D_4(z)} = \frac{0.5z^4 + 2.75z^3 + 3.75z^2 + 3.25z + 1}{z^4 + 3.25z^3 + 3.75z^2 + 2.75z + 0.5}.$$

Note from the above that, $k_4 = A_4(\infty) = 0.5$ implying $|k_3| < 1$. Next, using Eq. (4.171) we derive the third-order allpass function $A_3(z)$, which can be shown to be equal to

$$A_3(z) = \frac{1.5z^3 + 2.5z^2 + 2.5z + 1}{z^3 + 2.5z^2 + 2.5z + 1.5}.$$

It can be seen from the above that $k_3 = A_3(\infty) = 1.5$, implying $|k_3| > 1$. Therefore, the allpass function $A_4(z)$ is unstable, or equivalently, not all zeros of $D_4(z)$ are inside the unit circle.

The M-function poly2rc in MATLAB can be utilized to determine the stability test arameters $\{k_i\}$ very efficiently on a computer. We illustrate its application below.

Example 4.22 We test the stability of the transfer function of Eq. (4.175) using MATLAB. To this end we make use of the following program:

```
% Program 4_5
% Stability Test of a Rational Transfer Function
%
% Read in the polynomial coefficients
den = input('Type in the denominator coefficients =');
% Generate the stability test parameters
k = poly2rc(den);
knew = fliplr(k);
disp('The stability test parameters are');disp(knew);
stable = all(abs(k) < 1)
```

The input data is the vector den of the coefficients of the denominator polynomial entered inside a square bracket in descending powers of z as indicated below:

```
den = [4    3    2    1    1]
```

The output data are the stability test parameters $\{k_i\}$. The program also has a logical output which is stable = 1 if the transfer function is stable, otherwise, stable = 0.
The output data generated for the transfer function of Eq. (4.175) are as follows:

```
The stability test parameters are
    0.2500    0.0667    0.3527    0.5248

stable =
    1
```

Note that the stability test parameters are identical to those computed in Example 4.20.

4.9 Discrete-time Processing of Random Signals

As we shall observe later in the text, there are occasions when we need to study the effect of processing a random discrete-time signal by a linear time-invariant discrete-time system. More precisely, we need to determine the statistical properties of the output signal $\{y[n]\}$ generated by a stable LTI system with an impulse response $\{h[n]\}$ when its input $x[n]$ is a particular realization of a wide-sense stationary (WSS) random process $\{X[n]\}$. For simplicity we assume the input sequence and the impulse response to be real-valued. Now, the output of an LTI system is given by the linear convolution of the input and the impulse response of the system, i.e.,

$$y[n] = \sum_{k=-\infty}^{\infty} h[k]x[n-k]. \tag{4.178}$$

Since the function of a random variable is also a random variable, it follows from the above that the output $y[n]$ is also a sample sequence of an output random process $\{Y[n]\}$.

4.9.1 Statistical Properties of the Output Signal

Since the input $x[n]$ is a sample sequence of a stationary random process, its mean m_x is a constant independent of the time index n.[5] The mean $E(y[n])$ of the output random process $y[n]$ is then given by

$$m_y = E(y[n]) = E\left(\sum_{k=-\infty}^{\infty} h[k]x[n-k]\right)$$

$$= \sum_{k=-\infty}^{\infty} h[k]E(x[n-k]) = m_x \sum_{k=-\infty}^{\infty} h[k] = m_x H(e^{j0}), \tag{4.179}$$

which is a constant independent of the time index n.

The autocorrelation function of the output of the LTI discrete-time system for a real-valued input is given by

$$\phi_{yy}[n+\ell, n] = E(y[n+\ell]y[n])$$

$$= E\left(\left\{\sum_{i=-\infty}^{\infty} h[i]x[n+\ell-i]\right\}\left\{\sum_{k=-\infty}^{\infty} h[k]x[n-k]\right\}\right)$$

$$= \sum_{i=-\infty}^{\infty} h[i] \sum_{k=-\infty}^{\infty} h[k]E(x[n+\ell-i]x[n-k])$$

$$= \sum_{i=-\infty}^{\infty} h[i] \sum_{k=-\infty}^{\infty} h[k]\phi_{xx}[n+\ell-i, n-k]. \tag{4.180}$$

Since the input is a sample sequence of a WSS random process, its autocorrelation sequence depends only on the difference $\ell + k - i$ of the time indices $n + \ell - i$ and $n - k$, i.e.,

$$\phi_{xx}[n+\ell-i, n-k] = \phi_{xx}[\ell+k-i]. \tag{4.181}$$

Substituting the above in Eq. (4.180), we arrive at

$$\phi_{yy}[n+\ell, n] = \sum_{i=-\infty}^{\infty} h[i] \sum_{k=-\infty}^{\infty} h[k]\phi_{xx}[\ell+k-i] = \phi_{yy}[\ell] \tag{4.182}$$

---

[5]For convenience, we drop the notational difference between the random process $\{X[n]\}$ and its particular realization $\{x[n]\}$.

indicating that the output autocorrelation sequence depends on the difference ℓ of the time indices $n + \ell$ and n. As a result of Eqs. (4.179) and (4.182), it follows that the output $y[n]$ is also a sample sequence of a WSS random process.

Substituting $r = i - k$ in Eq. (4.182) we arrive at

$$\phi_{yy}[\ell] = \sum_{r=-\infty}^{\infty} \phi_{xx}[\ell - r] \sum_{k=-\infty}^{\infty} h[k]h[r+k]$$

$$= \sum_{r=-\infty}^{\infty} \phi_{xx}[\ell - r]\psi[r], \qquad (4.183)$$

where

$$\psi[r] = \sum_{k=-\infty}^{\infty} h[k]h[r+k] = h[r] \circledast h[-r] \qquad (4.184)$$

is called the *aperiodic autocorrelation sequence* of the impulse response sequence $\{h[n]\}$. It should be noted that $\psi[r]$ is the autocorrelation of a finite energy deterministic sequence and is not the same as the autocorrelation of an infinite energy WSS random signal.

4.9.2   Transform Domain Representation

We now consider the z-transform representation of Eq. (4.182). As indicated in Section 3.10.2, the z-transform of $\phi_{xx}[\ell]$ may exist if the input random signal is of zero-mean. From Eq. (4.179), for a zero-mean random input, the output of an LTI system is also a zero-mean random signal. In this case we obtain from Eq. (4.182) by taking the z-transform of both sides

$$\Phi_{yy}(z) = \Psi(z)\Phi_{xx}(z), \qquad (4.185)$$

where $\Phi_{xx}(z)$, $\Phi_{yy}(z)$, and $\Psi(z)$ are, respectively, the z-transforms of $\phi_{xx}[\ell]$, $\phi_{yy}[\ell]$ and $\psi[r]$. But, from Eq. (4.184) $\Psi(z) = H(z)H(z^{-1})$, which when substituted in Eq. (4.185) yields

$$\Phi_{yy}(z) = H(z)H(z^{-1})\Phi_{xx}(z). \qquad (4.186)$$

On the unit circle, Eq. (4.186) reduces to

$$\Phi_{yy}(e^{j\omega}) = |H(e^{j\omega})|^2\Phi_{xx}(e^{j\omega}) \qquad (4.187)$$

Using the notations $P_{xx}(\omega)$ and $P_{yy}(\omega)$ to denote the input and output power spectral densities, $\Phi_{xx}(e^{j\omega})$ and $\Phi_{yy}(e^{j\omega})$, respectively, we can rewrite Eq. (4.187) as

$$P_{yy}(\omega) = |H(e^{j\omega})|^2 P_{xx}(\omega) \qquad (4.188)$$

Now, from Eq. (2.167), for a zero-mean WSS process $y[n]$, the total average power in the output is given by the mean-square value $E(y^2[n]) = \phi_{yy}[0]$. But, $\phi_{yy}[\ell]$ is given by the inverse Fourier transform of $\Phi_{yy}(e^{j\omega})$:

$$\phi_{yy}[\ell] = \frac{1}{2\pi} \int_{-\pi}^{\pi} \Phi_{yy}(e^{j\omega})e^{j\omega n} \, d\omega. \qquad (4.189)$$

Therefore, from Eqs. (4.187) and (4.188) we obtain

$$E(y^2[n]) = \phi_{yy}[0]$$

$$= \frac{1}{2\pi} \int_{-\pi}^{\pi} \Phi_{yy}(e^{j\omega}) \, d\omega$$

$$= \frac{1}{2\pi} \int_{-\pi}^{\pi} |H(e^{j\omega})|^2 P_{xx}(\omega) \, d\omega. \tag{4.190}$$

For a real-valued WSS random signal $x[n]$ the autocorrelation sequence $\phi_{xx}[\ell]$ is an even sequence, and hence, $\Phi_{xx}(e^{j\omega})$ is an even function of ω. Assume the LTI system $h[n]$ to be an ideal filter with a magnitude response

$$|H(e^{j\omega})|^2 = \begin{cases} 1, & \omega_{c1} \le |\omega| \le \omega_{c2}, \\ 0, & 0 \le |\omega| < \omega_{c1}, \omega_{c2} < |\omega| < \pi. \end{cases} \tag{4.191}$$

In this case, Eq. (4.190) reduces to

$$\phi_{yy}[0] = \frac{1}{\pi} \int_{-\omega_{c2}}^{-\omega_{c1}} P_{xx}(\omega) \, d\omega + \frac{1}{\pi} \int_{\omega_{c1}}^{\omega_{c2}} P_{xx}(\omega) \, d\omega$$

$$= \frac{2}{\pi} \int_{\omega_{c1}}^{\omega_{c2}} P_{xx}(\omega) \, d\omega. \tag{4.192}$$

Since the total average output power $\phi_{yy}[0]$ is always nonnegative independent of the bandwidth of the linear filter, it follows from the above that

$$P_{xx}(\omega) \ge 0, \tag{4.193}$$

proving that the power spectral density function of a real WSS random signal is also nonnegative in addition to being a real and even function.

Variance of the output signal

Now we develop the expression for the variance of the output random signal when the input to the LTI system is a real-valued white random process. From Eq. (3.129) we get

$$\sigma_y^2 = \gamma_{yy}[0] = \frac{1}{2\pi} \int_{-\pi}^{\pi} P_{yy}(\omega) \, d\omega - m_y^2. \tag{4.194}$$

Substituting Eq. (4.188) in the above equation and making use of Eq. (3.137), we arrive at

$$\sigma_y^2 = \frac{\sigma_x^2}{2\pi} \int_{-\pi}^{\pi} |H(e^{j\omega})|^2 \, d\omega, \tag{4.195}$$

which can be alternatively written as

$$\sigma_y^2 = \frac{\sigma_x^2}{2\pi j} \oint_C H(z) H(z^{-1}) z^{-1} \, dz, \tag{4.196}$$

where C is a counterclockwise closed contour in the ROC of $H(z)H(z^{-1})$.

Figure 4.40   A typical uniformly distributed random sequence (solid line) and the output of a 3-point running average filter (dashed line).

Using Parseval's relation in Eq. (4.195), we can also express the output variance as

$$\sigma_y^2 = \sigma_x^2 \sum_{n=-\infty}^{\infty} |h[n]|^2. \tag{4.197}$$

Example 4.23 Let $\{X[n]\}$ be an uniformly distributed random signal with a probability density function $p_{X[n]}(\alpha)$, as indicated in Figure 2.32(a). From Eq. (2.127) the mean and the variance of $X[n]$ are given by

$$m_x = 2, \qquad \sigma_x^2 = \tfrac{1}{3} \qquad \text{for all values of } n. \tag{4.198}$$

Figure 4.40 shows a typical realization (solid line) of the above random signal.

We process the above random signal by a causal 3-point moving-average filter with an impulse response

$$\{h[n]\} = \left\{ \tfrac{1}{3} \quad \tfrac{1}{3} \quad \tfrac{1}{3} \right\}.$$

The output $y[n]$ obtained for the input of Figure 4.40 is also shown in the same figure by the dotted lines. The first two samples of $y[n]$ are transients. Applying Eq. (4.179) we arrive at the mean value of the output of the 3-point moving-average filter given by

$$m_y = m_x(h[0] + h[1] + h[2]) = 2 \left(\tfrac{1}{3} + \tfrac{1}{3} + \tfrac{1}{3} \right) = 2,$$

and from Eq. (4.197) the variance of the output is obtained as

$$\sigma_y^2 = \sigma_x^2 (h^2[0] + h^2[1] + h^2[2]) = \tfrac{1}{3} \left(\tfrac{1}{9} + \tfrac{1}{9} + \tfrac{1}{9} \right) = \tfrac{1}{9}.$$

Example 4.24 A zero-mean WSS white noise sequence $x[n]$ with an autocorrelation sequence

$$\phi_{xx}[n] = \sigma_x^2 \delta[n],$$

is being filtered by a causal stable LTI discrete-system with a transfer function

$$H(z) = \frac{p_0}{1 + d_1 z^{-1}}, \qquad |d_1| < 1. \tag{4.199}$$

We determine the autocorrelation sequence $\phi_{yy}[n]$ of the output $y[n]$ of the linear system and the cross-correlation sequence $\phi_{yx}[n]$ between the output and the input.

Now the z-transform $\Phi_{yy}(z)$ of $\phi_{yy}[n]$ is related to the z-transform $\Phi_{xx}(z)$ of $\phi_{xx}[n]$ through Eq. (4.186). Substituting $H(z)$ from Eq. (4.199) and $\Phi_{xx}(z) = \sigma_x^2$ in Eq. (4.186) we arrive at

$$\Phi_{yy}(z) = \sigma_x^2 \left(\frac{p_0}{1 + d_1 z^{-1}} \right) \left(\frac{p_0}{1 + d_1 z} \right), \qquad |d_1| < |z| < \frac{1}{|d_1|}. \tag{4.200}$$

An inverse z-transform of the above is readily obtained by using the partial-fraction approach as described in Section 3.8.2 and is found to be

$$\phi_{yy}[n] = \frac{(p_0 \sigma_x)^2}{1 - (d_1)^2}(1 - d_1)^{|n|}.$$

4.10   Summary

The frequency-domain representations of discrete-time sequences, considered in the previous chapter, are applied in this chapter to develop the frequency-domain representations of an LTI discrete-time system. One such representation is the frequency response, obtained by applying the discrete-time Fourier transform (DTFT) to the impulse response sequence, which characterizes an LTI discrete-time system uniquely in the frequency-domain. The DTFT of the output sequence of an LTI discrete-time system is simply the product of its frequency response and the DTFT of the input sequence, which provides the input-output relation of the system in the frequency-domain.

A generalization of the frequency response concept is the transfer function defined by the z-transform of the impulse response sequence. An alternative input-output relation is thus given by the product of the transfer function and the z-transform of the input sequence that results in the z-transform of the output sequence. For a stable causal LTI discrete-time system, all poles of its transfer function must be strictly inside the unit circle. For a stable causal transfer function, the frequency response is given simply by its values on the unit circle in the z-plane.

The concept of filtering is introduced and several ideal filters are defined. Several simple approximations to the ideal filters are next introduced. In addition, various special types of transfer functions that are often encountered in practice are reviewed. The concept of complementary transfer functions relating a set of transfer functions is discussed, and several types of complementary conditions are introduced.

An important building block in the design of a single-input, single-output LTI discrete-time system is the digital two-pair, which is a two-input, two-output LTI discrete-time system. Characterizations of the digital two-pairs and their interconnections are discussed. A very simple algebraic procedure for testing the stability of a causal LTI transfer function is then introduced. Finally, the chapter concludes with a discussion on the statistical properties and transform-domain representation of the output signal of an LTI discrete-time system generated by a random input signal.

4.11   Problems

4.1 If the output of a system is the input multiplied by a complex constant, then the corresponding input is called an *eigenfunction*, of the system. For an LTI discrete-time system, $e^{j\omega n}$ is an eigenfunction and the corresponding output of an LTI discrete-time system is given by $H(e^{j\omega})e^{j\omega n}$, as indicated in Eq. (4.4). Show that the function $u[n] = z^n$, where z is a complex constant, is also an eigenfunction of an LTI discrete-time system. Is $v[n] = z^n \mu[n]$ with z a complex constant also an eigenfunction of an LTI discrete-time system?

4.2 Determine a closed-form expression for the frequency response $H(e^{j\omega})$ of the LTI discrete-time system characterized by an impulse response

$$h[n] = \delta[n] + \alpha\delta[n - R], \qquad (4.201)$$

where $|\alpha| < 1$. What are the maximum and the minimum values of its magnitude response? How many peaks and dips of the magnitude response occur in the range $0 \leq \omega < 2\pi$? What are the locations of the peaks and the dips? Sketch the magnitude and the phase responses for $R = 4$.

4.3 Determine a closed-form expression for the frequency response $G(e^{j\omega})$ of the LTI discrete-time system characterized by an impulse response

$$g[n] = h[n] \circledast h[n] \circledast h[n], \tag{4.202}$$

where $h[n]$ is given by Eq. (4.201).

4.4 Determine a closed-form expression for the frequency response $G(e^{j\omega})$ of an LTI discrete-time system with an impulse response given by

$$g[n] = \begin{cases} \alpha^n, & 0 \leq n \leq M - 1, \\ 0 & \text{otherwise,} \end{cases}$$

where $|\alpha| < 1$. What is the relation of $G(e^{j\omega})$ to $H(e^{j\omega})$ of Eq. (4.10)? Scale the impulse response by multiplying it with a suitable constant so that the dc value of the magnitude response is unity.

4.5 Show that the group delay $\tau(\omega)$ of an LTI discrete-time system characterized by a frequency response $H(e^{j\omega})$ can be expressed as

$$\tau(\omega) = \text{Re}\left\{ \frac{j\frac{dH(e^{j\omega})}{d\omega}}{H(e^{j\omega})} \right\}. \tag{4.203}$$

4.6 A noncausal LTI FIR discrete-time system is characterized by an impulse response $h[n] = a_1\delta[n-1] + a_2\delta[n] + a_3\delta[n+1]$. For what values of the impulse response samples will its frequency response $H(e^{j\omega})$ have a zero phase?

4.7 Repeat Problem 4.6 for an impulse response $h[n] = a_1\delta[n-2] + a_2\delta[n-1] + a_3\delta[n] + a_4\delta[n+1] + a_5\delta[n+2]$.

4.8 An FIR LTI discrete-time system is described by the difference equation

$$y[n] = a_1 x[n+k] + a_2 x[n+k-1] + a_3 x[n+k-2]$$
$$+ a_2 x[n+k-3] + a_1 x[n+k-4],$$

where $y[n]$ and $x[n]$ denote, respectively, the output and the input sequences. Determine the expression for its frequency response $H(e^{j\omega})$. For what values of the constant k will the system have a frequency response $H(e^{j\omega})$ that is a real function of ω?

4.9 Let $X(e^{j\omega})$ and $Y(e^{j\omega})$ denote the DTFTs of the input and output sequences of a nonlinear discrete-time system with $0 \leq |X(e^{j\omega})| \leq 1$. The relation between $X(e^{j\omega})$ and $Y(e^{j\omega})$ is given by

$$Y(e^{j\omega}) = |X(e^{j\omega})|^\alpha e^{j\arg X(e^{j\omega})}$$

where $0 < \alpha \leq 1$. Determine the expression for its frequency response $H(e^{j\omega}) = Y(e^{j\omega})/X(e^{j\omega})$ and show that it has zero phase.

4.10 Determine the expression for the frequency response $H(e^{j\omega})$ of a causal IIR LTI discrete-time system characterized by the input-output relation

$$y[n] = x[n] - \alpha y[n - R], \qquad |\alpha| < 1,$$

where $y[n]$ and $x[n]$ denote, respectively, the output and the input sequences. Determine the maximum and the minimum values of its magnitude response. How many peaks and dips of the magnitude response occur in the range $0 \le \omega < 2\pi$? What are the locations of the peaks and the dips? Sketch the magnitude and the phase responses for $R = 4$.

4.11 An IIR LTI discrete-time system is described by the difference equation

$$\alpha y[n] + \beta y[n - 1] = \gamma x[n] + \delta x[n - 1],$$

where $y[n]$ and $x[n]$ denote, respectively, the output and the input sequences. Determine the expression for its frequency response. For what values of the constants γ and δ, will the magnitude response be a constant for all values of ω?

4.12 The magnitude response of a digital filter with a real coefficient transfer function $H(z)$ is shown in Figure P4.1. Plot the magnitude response of the filter $H(z^3)$.

Figure P4.1

4.13 Consider an LTI discrete-time system with an impulse response $h[n] = (0.5)^n \mu[n]$. Determine the frequency response $H(e^{j\omega})$ of the system and evaluate its value at $\omega = \pm\pi/4$. What is the steady-state output $y[n]$ of the system for an input $x[n] = \cos(\pi n/4)\mu[n]$?

4.14 An FIR filter of length 3 is defined by a symmetric impulse response, i.e., $h[0] = h[2]$. Let the input to this filter be a sum of two cosine sequences of angular frequencies 0.1 rad/sec and 0.4 rad/sec, respectively. Determine the impulse response coefficients so that the filter passes only the low-frequency component of the input.

4.15 (a) Design a length-5 FIR lowpass filter with a symmetric impulse response $h[n]$, i.e., $h[n] = h[n - 4]$, $n = 0, 1, 2, 3, 4$, satisfying the following magnitude response values:

$$|H(e^{j0})| = 1, \quad |H(e^{j0.5\pi})| = 0.5, \quad |H(e^{j\pi})| = 0$$

   (b) Determine the exact expression for the frequency response of the filter designed and plot its magnitude and phase responses.

4.16   (a) Design a length-5 FIR bandpass filter with a symmetric impulse response $h[n]$, i.e., $h[n] = h[n-4]$, $n = 0, 1, 2, 3, 4$, satisfying the following magnitude response values:

$$|H(e^{j0})| = 0, \quad |H(e^{j0.5\pi})| = 1, \quad |H(e^{j\pi})| = 0$$

(b) Determine the exact expression for the frequency response of the filter designed and plot its magnitude and phase responses.

4.17   (a) Design a length-5 FIR lowpass filter with a symmetric impulse response $h[n]$, i.e., $h[n] = h[n-4]$, $n = 0, 1, 2, 3, 4$, satisfying the following conditions:

$$\left|H(e^{j0})\right| = 1, \quad \left.\frac{d\left|H(e^{j\omega})\right|}{d\omega}\right|_{\omega=0} = 0, \quad \left.\frac{d^2\left|H(e^{j\omega})\right|}{d\omega^2}\right|_{\omega=0} = 0.$$

(b) Determine the exact expression for the frequency response of the filter designed and plot its magnitude and phase responses.

4.18 An FIR filter of length 5 is defined by a symmetric impulse response, i.e., $h[0] = h[4]$ and $h[1] = h[3]$. Let the input to this filter be a sum of three cosine sequences of angular frequencies 0.1 rad/sec, 0.4 rad/sec, and 0.7 rad/sec. Determine the impulse response coefficients so that the filter passes only the mid-frequency component of the input.

4.19 The frequency response $H(e^{j\omega})$ of a length-4 FIR filter with real impulse response has the following specific values:

$$H(e^{j0}) = 10, \quad H(e^{j\pi/2}) = -2 + j2, \quad \text{and} \quad H(e^{j\pi}) = -2.$$

Determine $H(z)$.

4.20 The frequency response $H(e^{j\omega})$ of a length-4 FIR filter with a real and symmetric impulse response has the following specific values:

$$H(e^{j0}) = 6, \quad \text{and} \quad H(e^{j\pi/2}) = -1 - j1.$$

Determine $H(z)$.

4.21 Consider the two LTI causal digital filters with impulse responses given by

$$h_A[n] = 0.5\delta[n] - \delta[n-1] + 0.5\delta[n-2],$$
$$h_B[n] = 0.5\delta[n] + \delta[n-1] + 0.5\delta[n-2].$$

(a) Sketch the magnitude responses of the two filters and compare their characteristics.

(b) Let $h_A[n]$ be the impulse response of a causal digital filter with a frequency response $H_A(e^{j\omega})$. Define another digital filter whose impulse response $h_C[n]$ is given by

$$h_C[n] = (-1)^n h_A[n], \quad \text{for all } n.$$

What is the relation between the frequency response $H_C(e^{j\omega})$ of this new filter and the frequency response $H_A(e^{j\omega})$ of the parent filter?

4.22 As indicated in Example 2.31, the trapezoidal integration formula can be represented as an IIR digital filter represented by a difference equation given by

$$y[n] = y[n-1] + \tfrac{1}{2}\{x[n] + x[n-1]\},$$

with $y[-1] = 0$. Determine the frequency response of the above filter.

4.23 A recursive difference equation representation of the Simpson's numerical integration formula is given by [Ham89]

$$y[n] = y[n-2] + \tfrac{1}{3}\{x[n] + 4x[n-1] + x[n-2]\}.$$

Evaluate the frequency response of the above filter and compare it with that of the trapezoidal method of Problem 4.22.

4.24 Consider the digital filter structure of Figure P4.2, where $H_1(z)$, $H_2(z)$, and $H_3(z)$ are FIR digital filters with transfer functions given by

$$H_1(z) = \tfrac{1}{2} + \tfrac{3}{5}z^{-1} + \tfrac{2}{3}z^{-2}, \quad H_2(z) = \tfrac{1}{2} + \tfrac{7}{5}z^{-1} + \tfrac{7}{3}z^{-2}, \quad H_3(z) = 2 + 3z^{-1} + z^{-2}.$$

Determine the transfer function $H(z)$ of the composite filter.

Figure P4.2

4.25 Determine the transfer function of a causal LTI discrete-time system described by the following difference equation:

$$y[n] = 2x[n] + 2.6x[n-1] + 0.78x[n-2] + 0.054x[n-3]$$
$$+ 1.5y[n-1] - 0.47y[n-2] - 0.063y[n-3].$$

Express the transfer function in a factored form and sketch its pole-zero plot. Is the system **BIBO** stable?

4.26 Determine the transfer function of a causal LTI discrete-time system described by the following difference equation:

$$y[n] = 3x[n] + 1.2x[n-1] - 5.22x[n-2] - 1.548x[n-3]$$
$$+ 0.359x[n-4] - 0.6y[n-1] - 0.68y[n-2]$$
$$- 0.55y[n-3] + 0.477y[n-4].$$

Express the transfer function in a factored form and sketch its pole-zero plot. Is the system **BIBO** stable?

4.27 Determine the expression for the impulse response $\{h[n]\}$ of the following causal IIR transfer function $H(z)$:

$$H(z) = \frac{(2 + 0.4z^{-1})(0.5 + 2z^{-1})}{(1 - 2z^{-1} + 2z^{-2})(1 - 0.5z^{-1})}.$$

4.28 The transfer function of a causal LTI discrete-time system is given by

$$H(z) = \frac{2 - 4z^{-1}}{2 + z^{-1}} + \frac{1}{3 - z^{-1}}.$$

(a) Determine the impulse response $h[n]$ of the above system.

(b) Determine the output $y[n]$ of the above system for all values of n for an input

$$x[n] = (0.4)^n \mu[n] + (0.2)^{n-1} \mu[-n - 1].$$

4.29 Consider the discrete-time system of Figure P4.3. For $H_0(z) = 1 + z^{-1}$, find a suitable $F_0(z)$ so that the output $y[n]$ is a delayed and scaled replica of the input.

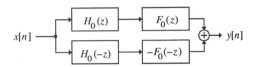

Figure P4.3

4.30 Let the transfer functions of Figure P4.3 be expressed in a polyphase form as

$$H_0(z) = E_0(z^2) + z^{-1}E_1(z^2), \quad F_0(z) = R_0(z^2) + z^{-1}R_1(z^2),$$

(a) Determine the condition so that the output $y[n]$ is a scaled and delayed replica of the input $x[n]$.

(b) Show that the transfer functions of Problem 4.29 satisfy this condition.

(c) Find at least another set of nontrivial realizable transfer functions $H_0(z)$ and $F_0(z)$ that satisfy this condition.

4.31 We have shown that a real coefficient FIR transfer function $H(z)$ with a symmetric impulse response has a linear phase response. As a result, the all-pole IIR transfer function $G(z) = 1/H(z)$ will also have a linear phase response. What are the practical difficulties in implementing $G(z)$? Justify your answer mathematically.

4.32 Consider the following causal IIR transfer function:

$$H(z) = \frac{2z^3 - 4z^2 + 9}{(z - 3)(z^2 + z + 0.5)}.$$

Is $H(z)$ a stable transfer function? If it is not stable, find a stable transfer function $G(z)$ such that $\left|G(e^{j\omega})\right| = \left|H(e^{j\omega})\right|$. Is there any other transfer function having the same magnitude response as $H(z)$?

4.33 Check the stability of the following causal IIR transfer function:

$$H(z) = \frac{(z^2 - 2z + 1)(z^2 + 3z + 4)}{(z^2 - 2.5z + 1.5)(z^2 - 0.5z + 0.25)}.$$

If it is unstable, find a stable transfer function $G(z)$ such that $\left|G(e^{j\omega})\right| = \left|H(e^{j\omega})\right|$. How many other transfer functions have the same magnitude response as $H(z)$?

4.34 Determine the transfer function of a causal LTI system characterized by the following state-space description:

$$\mathbf{A} = \begin{bmatrix} -0.4 & 0.2 \\ 1 & -0.3 \end{bmatrix}, \qquad \mathbf{B} = \begin{bmatrix} 1.5 \\ 0.4 \end{bmatrix}, \qquad \mathbf{C} = [-2 \;\; -1], \qquad D = 2.$$

4.35 Determine the state-space representation of the discrete-time system of Problem 4.32.

4.36 Determine the state-space representation of the discrete-time system of Problem 4.33.

4.37 Determine the impulse response $h_{HP}[n]$ of the ideal highpass filter of Figure 4.8(b).

4.38 Determine the impulse response $h_{BP}[n]$ of the ideal bandpass filter of Figure 4.8(c).

4.39 Determine the impulse response $h_{BS}[n]$ of the ideal bandstop filter of Figure 4.8(d).

4.40 Let $G_L(z)$ and $G_H(z)$ represent ideal lowpass and highpass filters with magnitude responses as sketched in Figure P4.4(a). Determine the transfer functions $H_k(z) = Y_k(z)/X(z)$ of the discrete-time system of Figure P4.4(b), $k = 0, 1, 2, 3$, and sketch their magnitude responses.

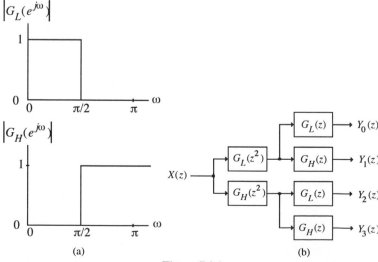

(a)   (b)

Figure P4.4

4.41 Let $H_{LP}(z)$ denote the transfer function of an ideal real coefficient lowpass filter with a cutoff frequency of ω_p. Sketch the magnitude response of $H_{LP}(-z)$ and show that it is a highpass filter. Determine the relation between the cutoff frequency of this highpass filter in terms of ω_p and its impulse response in terms of the impulse response $h_{LP}[n]$ of the parent lowpass filter.

4.42 Let $H_{LP}(z)$ denote the transfer function of an ideal real coefficient lowpass filter having a cutoff frequency of ω_p with $\omega_p < \pi/2$. Consider the complex coefficient transfer function $H_{LP}(e^{j\omega_0}z)$, where $\omega_p < \omega_0 < \pi - \omega_p$. Sketch its magnitude response for $-\pi \leq \omega \leq \pi$. What type of filter does it represent? Now consider the transfer function $G(z) = H_{LP}(e^{j\omega_0}z) + H_{LP}(e^{-j\omega_0}z)$. Sketch its magnitude response for $-\pi \leq \omega \leq \pi$. Show that $G(z)$ is a real coefficient bandpass filter with a passband centered at ω_0. Determine the width of its passband in terms of ω_p and its impulse response $g[n]$ in terms of the impulse response $h_{LP}[n]$ of the parent lowpass filter.

4.43 Let $H_{LP}(z)$ denote the transfer function of an ideal real coefficient lowpass filter with a cutoff frequency of ω_p with $0 < \omega_p < \pi/3$. Show that the transfer function $F(z) = H_{LP}(e^{j\omega_0}z) + H_{LP}(e^{-j\omega_0}z) + H_{LP}(z)$ where $\omega_0 = \pi - \omega_p$ is a real coefficient bandstop filter with the stopband centered at $\omega_0/2$. Determine the width of its stopband in terms of ω_p and its impulse response $f[n]$ in terms of the impulse response $h_{LP}[n]$ of the parent lowpass filter.

4.44 Show that the structure shown in Figure P4.5 implements the highpass filter of Problem 4.41.

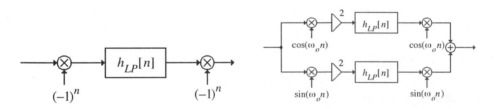

Figure P4.5   **Figure P4.6**

4.45 Show that the structure shown in Figure P4.6 implements the bandpass filter of Problem 4.42.

4.46 Let $H(z)$ be an ideal real coefficient lowpass filter with a cutoff at ω_c, where $\omega_c = \pi/M$. Figure P4.7 shows a single-input, M-output filter structure, called an M-band analysis filter bank, where $H_k(z) = H(ze^{-j2\pi k/M})$, $k = 0, 1, \ldots, M-1$. Sketch the magnitude response of each filter and describe the operation of the filter bank.

4.47 Consider a cascade of M sections of the first-order FIR lowpass filter of Eq. (4.64). Show that its 3-dB cutoff frequency is given by Eq. (4.66).

4.48 Consider a cascade of M sections of the first-order FIR highpass filter of Eq. (4.67). Develop the expression for its 3-dB cutoff frequency.

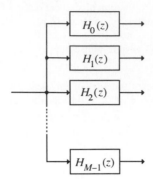

Figure P4.7

4.49 Verify that the value of α given by Eq. (4.72b) ensures that the transfer function $H_{\mathrm{LP}}(z)$ of Eq. (4.70) is stable.

4.50 Show by trigonometric manipulation that Eq. (4.72a) can be alternately expressed as

$$\tan\left(\frac{\omega_c}{2}\right) = \frac{1-\alpha}{1+\alpha}. \tag{4.204}$$

Next show that the transfer function $H_{\mathrm{LP}}(z)$ of Eq. (4.70) is stable for a value of α given by

$$\alpha = \frac{1 - \tan(\omega_c/2)}{1 + \tan(\omega_c/2)}. \tag{4.205}$$

4.51 Design a first-order lowpass IIR digital filter with a normalized 3-dB cutoff frequency at $\omega_c = 0.2$ rad/sec.

4.52 Show that the 3-dB cutoff frequency ω_c of the first-order highpass IIR digital filter of Eq. (4.73) is given by Eq. (4.72a).

4.53 The following first-order IIR transfer function has been proposed for clutter removal in MTI radars [Urk58]:

$$H(z) = \frac{1 - z^{-1}}{1 - kz^{-1}}.$$

Determine the magnitude response of the above transfer function and show that it has a highpass response. Scale the transfer function so that it has a 0-dB gain at $\omega = \pi$. Sketch the magnitude responses for $k = 0.95$, 0.9, and -0.5, respectively.

4.54 Show that the center frequency ω_0 and the 3-dB bandwidth $\Delta\omega_{3\mathrm{dB}}$ of the second-order IIR bandpass filter of Eq. (4.74) are given by Eqs. (4.76) and (4.77), respectively.

4.55 Show that the notch frequency ω_0 and the 3-dB notch bandwidth $\Delta\omega_{3\mathrm{dB}}$ of the second-order IIR bandstop filter of Eq. (4.79) are given by Eqs. (4.76) and (4.77), respectively.

4.56 Consider a cascade of K identical first-order lowpass digital filters with a transfer function given by Eq. (4.70). Show that the coefficient α of the first-order section is related to the 3-dB cutoff frequency ω_c of the cascade according to Eq. (4.83) with the parameter B given by Eq. (4.84).

4.57 Consider a cascade of K identical first-order highpass digital filters with a transfer function given by Eq. (4.73). Express the coefficient α of the first-order section in terms of the 3-dB cutoff frequency ω_c of the cascade.

4.58 If $H(z)$ is a lowpass filter, show that $H(-z)$ is a highpass filter. If ω_p and ω_s represent the passband and stopband edge frequencies, determine the locations of the passband and stopband edge frequencies of $H(-z)$. Determine the relations between the impulse response coefficients of the two filters.

4.59 Using the method of Problem 4.58, develop the transfer function $G_{HP}(z)$ of a first-order IIR highpass filter from the transfer function $H_{LP}(z)$ of the first-order IIR lowpass filter given by Eq. (4.70). Is it the same as that of the highpass transfer function of Eq. (4.73)? If not, determine the location of its 3-dB cutoff frequency as a function of the parameter α.

4.60 Let $H(z)$ be an ideal lowpass filter with a cutoff frequency at $\pi/2$. Sketch the magnitude responses of the following systems:
(a) $H(z^2)$, (b) $H(z)H(z^2)$, (c) $H(-z)H(z^2)$, and (d) $H(z)H(-z^2)$.

4.61 Let $H(z)$ be a lowpass filter with unity passband magnitude, a passband edge at ω_p, and a stopband edge at ω_s, as shown in Figure P4.8.

(a) Sketch the magnitude response of the digital filter $G_1(z) = H(z^M)F_1(z)$, where $F_1(z)$ is a lowpass filter with unity passband magnitude, a passband edge at ω_p/M and a stopband edge at $(2\pi - \omega_s)/M$. What are the band-edges of $G_1(z)$?

(b) Sketch the magnitude response of the digital filter $G_2(z) = H(z^M)F_2(z)$ where $F_2(z)$ is a bandpass filter with unity passband magnitude, and with passband edges at $(2\pi - \omega_p)/M$ and $(2\pi + \omega_p)/M$, and stopband edges at $(2\pi - \omega_s)/M$ and $(4\pi + \omega_s)/M$, respectively. What are the bandedges of $G_2(z)$?

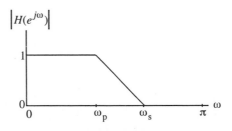

Figure P4.8

4.62 Let a causal LTI discrete-time system be characterized by a real impulse response $h[n]$ with a DTFT $H(e^{j\omega})$. Consider the system of Figure P4.9, where $x[n]$ is a finite length sequence. Determine the frequency response of the overall system $G(e^{j\omega})$ in terms of $H(e^{j\omega})$, and show that it has a zero-phase response.

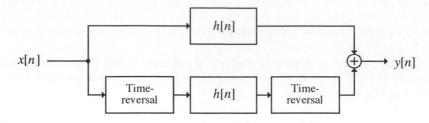

Figure P4.9

4.63 Show that the zero-phase response $\breve{H}(\omega)$ of Types 1, 2, 3, and 4 linear-phase FIR filters can be expressed as a product of the form of $P(\omega)Q(\omega)$ where $P(\omega)$ is a fixed function of ω and $Q(\omega)$ is a sum of cosines of the form of $\sum_{n=0}^{K} \alpha[n]\cos(\omega n)$. Determine the explicit expressions for $P(\omega)$ and $Q(\omega)$ for each case.

4.64 The zero locations of four FIR transfer functions of order 6 are sketched in Figure P4.10.

(a) Does any one of the FIR filters have a linear-phase response? If so, which one?

(b) Does any one of the FIR filters have a minimum-phase response? If so, which one?

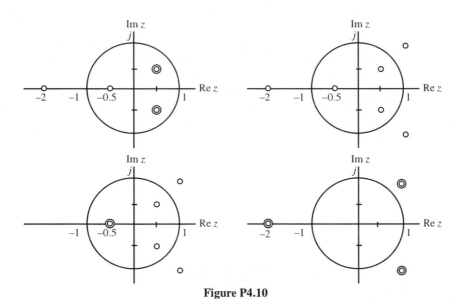

Figure P4.10

4.65 A third-order FIR filter has a transfer function $G_1(z)$ given by

$$G_1(z) = (6 - z^{-1} - 12z^{-2})(2 + 5z^{-1})$$

(a) Determine the transfer functions of all other FIR filters whose magnitude responses are identical to that of $G_1(z)$.

(b) Which one of these filters has a minimum-phase transfer function and which one has a maximum-phase transfer function?

(c) If $g_k[n]$ denotes the impulse response of the kth FIR filter determined in part (a), compute the partial energy of the impulse response given by

$$\mathcal{E}_k[n] = \sum_{m=0}^{n} g_k[m]^2, \qquad 0 \le n \le 3,$$

for all values of k, and show that

$$\sum_{m=0}^{n} |g_k[m]|^2 \le \sum_{m=0}^{n} |g_{min}[m]|^2,$$

and

$$\sum_{m=0}^{\infty} |g_k[m]|^2 = \sum_{m=0}^{\infty} |g_{min}[m]|^2$$

for all values of k, and where $g_{min}[n]$ is the impulse response of the minimum-phase FIR filter determined in part (a).

4.66 The z-transforms of five sequences of length 7 are given below:

$$H_1(z) = 0.0083653 - 0.001782726z^{-1} - 0.075506z^{-2} - 0.3413956532z^{-3}$$
$$+0.13529123z^{-4} - 1.50627293z^{-5} + 3.3207295z^{-6},$$

$$H_2(z) = 3.3207295 - 1.50627293z^{-1} + 0.13529123z^{-2} + 0.34139565z^{-3}$$
$$-0.07550603z^{-4} - 0.001782725z^{-5} + 0.0083652z^{06},$$

$$H_3(z) = 0.0269311 - 0.0840756z^{-1} + 0.02580603z^{-2} + 0.94049849z^{-3}$$
$$-2.2508765z^{-4} + 2.53245711z^{-5} + 1.03147943z^{-6},$$

$$H_4(z) = 1.03147943 + 2.5324571z^{-1} - 2.2508765z^{-2} + 0.9404985z^{-3}$$
$$+0.02580602z^{-4} - 0.0840756z^{-5} + 0.02693111z^{-6}$$

$$H_5(z) = 0.16667 - 0.05556z^{-1} - 0.75z^{-2} + 3.5z^{-3} - 0.75z^{-4}$$
$$-0.05556z^{-5} + 0.16667z^{-6}.$$

The magnitude of the DFT for each of the above sequences is the same. Which one of the above z-transforms has all its zeros outside the unit circle? Which one has all its zeros inside the unit circle? How many other real sequences of length 7 exist having the same DFT magnitude as those given above.

4.67 Let $h[n], 0 \le n \le N$, denote the impulse response of a Type 1 FIR filter of length $N + 1$. The frequency response is of the form $H(e^{j\omega}) = \breve{H}(\omega)e^{-j\omega N/2}$, where the amplitude response $\breve{H}(\omega)$ is as indicated in Figure P4.11(a).

(a) Construct another FIR filter of length $N + 1$ with an impulse response defined by

$$g[n] = \begin{cases} h[n], & \text{for all } n \text{ except } n = m, \\ \alpha, & n = m, \end{cases} \qquad 0 \le n \le N.$$

Determine a suitable m and α such that the frequency response of the new filter is of the form $G(e^{j\omega}) = \breve{G}(\omega)e^{-j\omega N/2}$, with its amplitude response $\breve{G}(\omega)$ as indicated in Figure P4.11(b) [Her70].

(b) Show that $\breve{G}(\omega)$ can be expressed as the square magnitude response of a minimum-phase FIR filter $F(z)$ of order $N/2$ and develop a method to construct $F(z)$.

(c) Can $\breve{G}(\omega)$ also be expressed as the square magnitude response of a minimum-phase FIR filter? If not, why not?

(a)   (b)

Figure P4.11

4.68 The time constant K of an LTI stable causal discrete-time system with an impulse response $h[n]$ is given by the value of the total time interval n at which the partial energy of the impulse response is within 95% of the total energy, i.e.,

$$\sum_{n=0}^{K} |h[n]|^2 = 0.95 \sum_{n=0}^{\infty} |h[n]|^2 .$$

Determine the time constant K of the first-order causal transfer function $H(z) = 1/(1 + \alpha z^{-1})$, $|\alpha| < 1$.

4.69 Let $F_1(z)$ denote one of the factors of a linear-phase FIR transfer function $H(z)$. Determine at least another factor $F_2(z)$ of $H(z)$ for the following choices of $F_1(z)$:

(a) $F_1(z) = 1 + 2z^{-1} + 3z^{-2}$,   (b) $F_1(z) = 3 + 5z^{-1} - 4z^{-2} - 2z^{-3}$.

4.70 Consider the first-order causal and stable allpass transfer function given by

$$A_1(z) = \frac{1 - d_1^* z}{z - d_1} .$$

Determine the expression for $\left(1 - |A_1(z)|^2 \right)$ and then show that

$$\left(1 - |A_1(z)|^2 \right) \begin{cases} < 0, & \text{for } |z|^2 < 1, \\ = 0, & \text{for } |z|^2 = 1, \\ > 0, & \text{for } |z|^2 > 1. \end{cases}$$

Now, using the above approach, show that property 2 given by Eq. (4.129) holds for any arbitrary causal stable allpass transfer function.

4.71 Derive property 3 of a stable allpass transfer function given by Eq. (4.130).

4.72 Let $G(z)$ be a causal stable nonminimum-phase transfer function, and let $H(z)$ denote another causal stable transfer function that is minimum-phase with $\left|G(e^{j\omega})\right| = \left|H(e^{j\omega})\right|$. Show that $G(z) = H(z)A(z)$, where $A(z)$ is a stable causal allpass transfer function.

4.73 The transfer function of a typical transmission channel is given by

$$H(z) = \frac{(z - 0.3)(z^2 - 2z + 2)}{(z - 0.9)(z - 0.5)}.$$

In order to correct for the magnitude distortion introduced by the channel on a signal passing through it, we wish to connect a stable digital filter characterized by a transfer function $G(z)$ at the receiving end. Determine $G(z)$.

4.74 Let $H(z)$ be a causal stable minimum-phase transfer function, and let $G(z)$ denote another causal stable transfer function which is nonminimum-phase with $\left|G(e^{j\omega})\right| = \left|H(e^{j\omega})\right|$. If $h[n]$ and $g[n]$ denote their respective impulse responses, show that

(a) $|g[0]| \leq |h[0]|$,

(b) $\sum_{\ell=0}^{n} |g[\ell]|^2 \leq \sum_{\ell=0}^{n} |h[\ell]|^2$.

4.75 Is the transfer function

$$H(z) = \frac{(z + 3)(z - 2)}{(z - 0.25)(z + 0.5)}$$

minimum-phase? If it is not minimum-phase, then construct a minimum-phase transfer function $G(z)$ such that $\left|G(e^{j\omega})\right| = \left|H(e^{j\omega})\right|$. Determine their corresponding unit sample responses, $g[n]$ and $h[n]$, for $n = 0, 1, 2, 3, 4$. For what values of m is $\sum_{n=0}^{m} |g[n]|^2$ bigger than $\sum_{n=0}^{m} |h[n]|^2$?

4.76 The following bandstop FIR transfer functions $H_{BS}(z)$ have also been proposed for the recovery of vertical details in the structure of Figure 4.18 employed for the separation of the luminance and the chrominance components [Aca83], [Pri80], [Ros75]:

(a) $H_{BS} = \frac{1}{4}(1 + z^{-2})^2$,

(b) $H_{BS} = \frac{1}{16}(1 + z^{-2})^2(-1 + 6z^{-2} - z^{-4})$,

(c) $H_{BS} = \frac{1}{32}(1 + z^{-2})^2(-3 + 14z^{-2} - 3z^{-4})$.

Develop their delay-complementary transfer functions $H_{BP}(z)$.

4.77 Let $A_0(z)$ and $A_1(z)$ be two causal stable allpass transfer functions. Define two causal stable IIR transfer functions as follows:

$$H_0(z) = A_0(z) + A_1(z), \qquad H_1(z) = A_0(z) - A_1(z).$$

Show that the numerators of $H_0(z)$ and $H_1(z)$ are, respectively, a symmetric and an antisymmetric polynomial.

4.78 Show that the two transfer functions of Eqs. (4.139a) and (4.139b) are a power-complementary pair.

4.79 Show that the two transfer functions of Eqs. (4.139a) and (4.139b) are each a BR function.

4.80 Consider the transfer function $H(z)$ given by

$$H(z) = \frac{1}{M} \sum_{k=0}^{M-1} A_k(z),$$

where $A_k(z)$ are stable real coefficient allpass functions. Show that $H(z)$ is a BR function.

4.81 Show that the bandpass transfer function $H_{BP}(z)$ of Eq. (4.74) and the band-stop transfer function $H_{BS}(z)$ of Eq. (4.79) form a doubly complementary pair.

4.82 Show that the following FIR transfer functions are BR functions:

(a) $H_1(z) = \dfrac{1}{1+\alpha}\left(1 + \alpha z^{-1}\right), \quad \alpha > 0,$

(b) $H_2(z) = \dfrac{1}{1+\beta}\left(1 - \beta z^{-1}\right), \quad \beta > 0,$

(c) $H_3(z) = \dfrac{1}{(1+\alpha)(1+\beta)}\left(1 + \alpha z^{-1}\right)\left(1 - \beta z^{-1}\right), \quad \alpha > 0, \quad \beta > 0,$

(d) $H_3 = \dfrac{1}{3.36}(1 + 0.4z^{-1})(1 + 0.5z^{-1})(1 + 0.6z^{-1}).$

4.83 Show that the following IIR transfer functions are BR functions:

(a) $H_1(z) = \dfrac{2 + 2z^{-1}}{3 + z^{-1}},$

(b) $H_2(z) = \dfrac{1 - z^{-1}}{4 + 2z^{-1}},$

(c) $H_3(z) = \dfrac{1 - z^{-2}}{4 + 2z^{-1} + 2z^{-2}},$

(d) $H_4(z) = \dfrac{3 + 6z^{-1} + 3z^{-2}}{6 + 5z^{-1} + z^{-2}},$

(e) $H_5(z) = \dfrac{3 + 2z^{-1} + 3z^{-2}}{4 + 2z^{-1} + 2z^{-2}},$

(f) $H_6(z) = \dfrac{3 + 9z^{-1} + 9z^{-2} + 3z^{-3}}{12 + 10z^{-1} + 2z^{-2}}.$

4.84 If $A_1(z)$ and $A_2(z)$ are two LBR functions, show that $A_1(1/A_2(z))$ is also an LBR function.

4.85 Let $G(z)$ be an LBR function of order N. Define

$$F(z) = z\left(\frac{G(z) + \alpha}{1 + \alpha G(z)}\right),$$

where $|\alpha| < 1$. Show that $F(z)$ is also LBR. What is the order of $F(z)$? Develop a realization of $G(z)$ in terms of $F(z)$.

4.86 If $G(z)$ is a BR function, show that $G(1/A(z))$ is a BR function, where $A(z)$ is an LBR function.

4.87 Show that each of the following pairs of transfer functions are doubly complementary:

(a) $H(z) = \dfrac{2 + 2z^{-1}}{3 + z^{-1}}$, $G(z) = \dfrac{1 - z^{-1}}{3 + z^{-1}}$,

(b) $H(z) = \dfrac{-1 + z^{-2}}{4 + 2z^{-1} + 2z^{-2}}$, $G(z) = \dfrac{3 + 2z^{-1} + 3z^{-2}}{4 + 2z^{-1} + 2z^{-2}}$,

4.88 Determine the power-complementary transfer function of each of the following BR transfer functions:

(a) $H_a = \dfrac{2(1 + z^{-1} + z^{-2})}{3 + 2z^{-1} + z^{-2}}$,

(b) $H_b = \dfrac{6(1 + z^{-1})^3}{18 + 21z^{-1} + 8z^{-2} + z^{-3}}$.

4.89 Verify Eq. (4.145).

4.90 Verify the relations between the transfer parameters and the chain parameters of a two-pair given in Eqs. (4.151a) and (4.151b).

4.91 Consider the Γ-cascade of Figure P4.12(a), where the two two-pairs are described by the transfer matrices

$$\tau_1 = \begin{bmatrix} k_1 & (1 - k_1^2)z^{-1} \\ 1 & -k_1 z^{-1} \end{bmatrix}, \qquad \tau_2 = \begin{bmatrix} k_2 & (1 - k_2^2)z^{-1} \\ 1 & -k_2 z^{-1} \end{bmatrix}.$$

Determine the transfer matrix of the cascade.

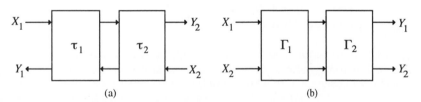

(a)   (b)

Figure P4.12

4.92 Consider the τ-cascade of Figure P4.12(b), where the two two-pairs are described by the chain matrices

$$\Gamma_1 = \begin{bmatrix} 1 & k_1 z^{-1} \\ k_1 & z^{-1} \end{bmatrix}, \qquad \Gamma_2 = \begin{bmatrix} 1 & k_2 z^{-1} \\ k_2 & z^{-1} \end{bmatrix}.$$

Determine the chain matrix of the cascade.

4.93 A transfer function $H(z)$ is realized in the form of Figure 4.38, where the constraining transfer function is given by $G(z)$. If the relation between $H(z)$ and $G(z)$ is of the form

$$H(z) = \frac{k_m + z^{-1}G(z)}{1 + k_m z^{-1}G(z)},$$

determine the transfer matrix and the chain matrix parameters of the two-pair of Figure 4.38.

4.94 Derive the inequality of Eq. (4.162).

4.95 Determine by inspection which one of the following second-order polynomials has roots inside the unit circle:

(a) $D_a(z) = 1 + 0.92z^{-1} + 0.1995z^{-2}$,

(b) $D_b(z) = 2 + 0.4z^{-1} - 2.86z^{-2}$,

(c) $D_c(z) = 1 + 1.4562z^{-1} + 0.81z^{-2}$,

(d) $D_d(z) = 1 + 2.1843z^{-1} + 0.81z^{-2}$.

4.96 Test analytically the BIBO stability of the following causal IIR transfer functions:

(a) $H_a(z) = \dfrac{z^2 + 0.3z - 99.17}{z^3 - \frac{1}{2}z^2 - \frac{1}{4}z + \frac{1}{12}}$,

(b) $H_b(z) = \dfrac{z^2 + 11.63z + 0.001}{z^3 + \frac{13}{6}z^2 + \frac{1}{6}z - \frac{1}{3}}$,

(c) $H_c(z) = \dfrac{9.25z^3 + 13.7z^2 + 4.04z + 0.3}{18z^4 + 30z^3 + 18.5z^2 + 5z + 0.5}$,

(d) $H_d(z) = \dfrac{9 + z^{-1} - 3z^{-2} + 7z^{-4} + z^{-5}}{1 + \frac{5}{2}z^{-1} + \frac{5}{2}z^{-2} + \frac{5}{4}z^{-3} + \frac{5}{16}z^{-4} + \frac{1}{32}z^{-5}}$,

(e) $H_e(z) = \dfrac{1}{6 + 5z^{-1} + 4z^{-2} + 3z^{-3} + z^{-4} + z^{-5}}$.

4.97 Determine analytically whether all roots of the following polynomials are inside the unit circle:

(a) $D_a(z) = 5 + 4z^{-1} + 3z^{-2} + 2z^{-3} + z^{-4}$,

(b) $D_b = z^3 + 0.2z^2 + 0.3z + 0.4$.

4.98 A polynomial $A(s)$ in the complex variable s is called a strictly Hurwitz polynomial if all zeros of $A(s)$ are in the left-half s-plane, i.e., if $s = s_k$ is a zero of $A(s)$, then $\text{Re}(s_k) < 0$. Let $D(z)$ be a polynomial in z of degree N with all roots inside the unit circle. If we replace z in $D(z)$ by the function $(1 + s)/(1 - s)$, we arrive at rational function of s given by $B(s)/(1 - s)^N$, where $B(s)$ is a polynomial in s of degree N. Show that $B(s)$ is a strictly Hurwitz polynomial.

4.99 Consider the structure of Figure P4.13, where $A(z) = (\alpha - z^{-1})/(1 - \alpha z^{-1})$ is a causal stable allpass filter with $|\alpha| < 1$. Let the input $x[n]$ be a stationary noise with power spectral density given by

$$P_{xx}(\omega) = \frac{1}{1 + d\cos\omega}.$$

(a) Calculate the power spectral density $P_{yy}(\omega)$ of the output $y[n]$.

(b) Does your answer depend upon α? If not, where did you use the information $|\alpha| < 1$?

4.100 A real, white zero-mean WSS random signal $x[n]$ is processed by an LTI digital filter with a real, causal, stable impulse response $h[n]$, as indicated in Figure P4.14. Let $P_{xy}(\omega)$ and $P_{xu}(\omega)$ denote the respective cross-power spectral density functions.

(a) Is $P_{xy}(\omega)$ a real function of ω?

(b) Is $P_{xu}(\omega)$ a real function of ω?

Justify your answers.

Figure P4.13

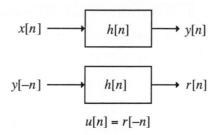

$$u[n] = r[-n]$$

Figure P4.14

4.12 MATLAB Exercises

M 4.1 Write a MATLAB program to compute the group delay using the expression of Problem 4.5 at a prescribed set of discrete frequencies.

M 4.2 Write a MATLAB program to simulate the filter designed in Problem 4.14 and verify its filtering operation.

M 4.3 Write a MATLAB program to simulate the filter designed in Problem 4.18 and verify its filtering operation.

M 4.4 A causal IIR digital filter is characterized by the following state-space description:

$$\mathbf{A} = \begin{bmatrix} 1 & -1 & 3 \\ 4 & 1 & 2 \\ 0 & 1 & 1 \end{bmatrix}, \quad \mathbf{B} = \begin{bmatrix} 1 \\ 2 \\ 1 \end{bmatrix}, \quad \mathbf{C} = [-1 \quad 0 \quad 0], \quad D = 2.$$

Using Program 4_3 determine its transfer function.

M 4.5 The following third-order IIR transfer function has been proposed for clutter rejection in MTI radar [Whi58]:

$$H(z) = \frac{z^{-1}(1 - z^{-1})^2}{(1 - 0.4z^{-1})(1 - 0.88z^{-1} + 0.61z^{-2})}.$$

Using MATLAB determine and plot its gain responses and show that it has a highpass response.

M 4.6 Show that for each case listed below, $H(z)$ and $H(-z)$ are power-complementary.

(a) $H(z) = \dfrac{0.75z^{-1} + 2z^{-2} + 2z^{-3} + z^{-4} + 0.75z^{-5}}{3 + 3.5z^{-2} + z^{-4}}$

(b) $H(z) = \dfrac{1 - 1.5z^{-1} + 3.75z^{-2} - 2.75z^{-3} + 2.75z^{-4} - 3.75z^{-5} + 1.55z^{-6} - z^{-7}}{6 + 6.5z^{-2} + 4.5z^{-4} + z^{-6}}.$

To verify the power-complementary property, write a MATLAB program to evaluate $H(z)H(z^{-1})+H(-z)H(-z^{-1})$ and show that this expression is equal to unity for each of the transfer functions given above.

M 4.7 Plot the magnitude and phase responses of the causal IIR digital transfer function

$$H(z) = \frac{0.0534(1+z^{-1})(1-1.0166z^{-1}+z^{-2})}{(1-0.683z^{-1})(1-1.4461z^{-1}+0.7957z^{-2})}.$$

What type of filter does this transfer function represent? Determine the difference equation representation of the above transfer function.

M 4.8 Plot the magnitude and phase responses of the causal IIR digital transfer function

$$H(z) = \frac{(1-z^{-1})^4}{(1-1.499z^{-1}+0.8482z^{-2})(1-1.5548z^{-1}+0.6493z^{-2})}.$$

What type of filter does this transfer function represent? Determine the difference equation representation of the above transfer function.

M 4.9 Design an FIR lowpass filter with a 3-dB cutoff frequency at 0.24π using a cascade of five first-order lowpass filters of Eq. (4.64). Plot its gain response.

M 4.10 Using the result of Problem 4.48 design a FIR highpass filter with a 3-dB cutoff frequency at 0.24π using a cascade of six first-order highpass filters of Eq. (4.67). Plot its gain response.

M 4.11 Design a first-order IIR lowpass and a first-order IIR highpass filter with a 3-dB cutoff frequency of 0.3π. Using MATLAB plot their magnitude responses on the same figure. Using MATLAB show that these filters are both allpass-complementary and power-complementary.

M 4.12 Design a second-order IIR bandpass and a second-order IIR notch filter with a center (notch) frequency $\omega_0 = 0.4\pi$ and a 3-dB bandwidth Ω_c (notch width) of 0.15π. Using MATLAB plot their magnitude responses on the same figure. Using MATLAB show that these filters are both allpass-complementary and power-complementary.

M 4.13 Design a stable second-order IIR bandpass filter with a center frequency at 0.7π and a 3-dB bandwidth of 0.15π. Plot its gain response.

M 4.14 Design a stable second-order IIR notch filter with a center frequency at 0.4π and a 3-dB bandwidth of 0.1π. Plot its gain response.

M 4.15 Using MATLAB show that the transfer function pairs of Problem 4.88 are both allpass-complementary and power-complementary.

M 4.16 Develop the pole-zero plots of the transfer functions of Problem 4.84 using the function zplane of MATLAB and show that they are stable. Next, plot the magnitude response of each transfer function using MATLAB and show that it satisfies the bounded real property.

M 4.17 Develop the pole-zero plots of the transfer functions of Problem 4.97 using the function zplane of MATLAB and then test their stability.

M 4.18 Using Program 4_5 test the stability of the transfer functions of Problem 4.97.

M 4.19 Using Program 4_5 determine whether the roots of the polynomials of Problem 4.98 are inside the unit circle or not.

M 4.20 The FIR digital filter structure of Figure P4.15 is used for aperture correction in television to compensate for high-frequency losses [Dre90]. A cascade of two such circuits is used, with one correcting the vertical aperture and the other correcting the horizontal aperture. In the former case, the delay z^{-1} is a line delay, whereas in the latter case it is 70 ns for the CCIR standard and the weighting factor k provides adjustable amount of correction. Determine the transfer function of this circuit and plot its magnitude response using MATLAB for two different values of k.

Figure P4.15

M 4.21 An improved aperture correction circuit for digital television is the FIR digital filter structure of Figure P4.16, where the delay z^{-1} is 70 ns for the CCIR standard, and the two weighting factors k_1 and k_2 provide adjustable amount of correction with $k_1 > 0$ and $k_2 < 0$ [Dre90]. Determine the transfer function of this circuit and plot its magnitude response using MATLAB for two different values of k_1 and k_2.

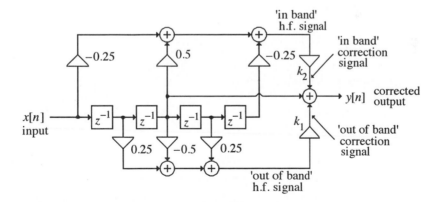

Figure P4.16

CHAPTER 5

Digital Processing of Continuous-Time Signals

5.1 Introduction

Even though this book is concerned primarily with the processing of discrete-time signals, most signals we encounter in the real world are continuous in time, such as speech, music, and images. Increasingly, discrete-time signal processing algorithms are being used to process such signals and are implemented employing discrete-time analog or digital systems. For processing by digital systems, the discrete-time signals are represented in digital form with each discrete-time sample as a binary word. Therefore, we need the analog-to-digital and digital-to-analog interface circuits to convert the continuous-time signals into discrete-time digital form, and vice-versa. As a result, it is necessary to develop the relations between the continuous-time signal and its discrete-time equivalent in the time-domain and also in the frequency-domain.[1] The latter relations are important in determining conditions under which the discrete-time processing of continuous-time signals can be done free of error under ideal situations.

The interface circuit performing the conversion of a continuous-time signal into a digital form is called the *analog-to-digital* (A/D) *converter*, often abbreviated as ADC. Likewise, the reverse operation of converting a digital signal into a continuous-time signal is implemented by the interface circuit called the *digital-to-analog* (D/A) *converter*, often abbreviated as DAC. In addition to these two devices, we need several additional circuits. Since the analog-to-digital conversion usually takes a finite amount of time, it is often necessary to ensure that the analog signal at the input of the A/D converter remains constant in amplitude until the conversion is complete to minimize the error in its representation. This is accomplished by a device called the *sample-and-hold* (S/H) circuit, which has dual purposes. It not only samples the input continuous-time signal at periodic intervals but also holds the analog sampled value constant at its output for sufficient time to permit accurate conversion by the A/D converter. In addition, the output of the D/A converter is a staircase-like waveform. It is therefore necessary to smooth the D/A converter output

---

[1]These relations also apply to discrete-time analog signal processing systems, such as the switched-capacitor networks.

Figure 5.1   Block diagram representation of the discrete-time digital processing of a continuous-time signal.

Figure 5.2   A simplified representation of Figure 5.1.

by means of an analog *reconstruction (smoothing) filter*. Finally, in most applications, the continuous-time signal to be processed has usually a larger bandwidth than the bandwidth of the available discrete-time processors. To prevent a detrimental effect called *aliasing*, an analog *anti-aliasing filter* is often placed before the S/H circuit. The complete block diagram illustrating the functional requirements for the discrete-time digital processing of a continuous-time signal is thus as indicated in Figure 5.1.

To understand the conditions under which the above system can work, we need to examine each of the interface circuits indicated in Figure 5.1. First, we assume a simpler mathematical equivalent of Figure 5.1, which enables us to derive the most fundamental condition that permits the discrete-time processing of continuous-time signals. To this end we assume that the A/D and D/A converters have infinite precision wordlengths, resulting in the simplified representation of Figure 5.2.[2] In this representation, the S/H circuit in cascade with an infinite precision A/D converter has been replaced with the ideal continuous-time-to-discrete-time (CT-DT) converter (i.e., an ideal sampler), which develops a discrete-time equivalent $x[n]$ of the continuous-time signal $x_a(t)$. Likewise, the infinite-precision D/A converter in cascade with the ideal reconstruction filter has been replaced with the ideal discrete-time-to-continuous-time (DT-CT) converter (i.e., an ideal interpolator), which develops a continuous-time equivalent $y_a(t)$ of the processed discrete-time signal $y[n]$.

In this chapter, we first derive the conditions for discrete-time representation of a band-limited continuous-time signal under ideal sampling and its exact recovery from the sampled version. If these conditions are not met, it is not possible to recover the original continuous-time signal from its sampled discrete-time equivalent, resulting in an undesirable distorted representation caused by an effect called aliasing. Since both the anti-aliasing filter and the reconstruction filter in Figure 5.1 are analog lowpass filters, we briefly review next the basic theory behind some commonly used analog filter design methods and illustrate them using MATLAB. It should be noted also that most widely used digital filter design methods are based on the conversion of an analog prototype, and a knowledge of analog filter design is thus useful in digital signal processing. We then examine the basic properties of the various interface circuits depicted in Figure 5.1.

---

[2]Effects of finite wordlength effects are considered in Chapter 9.

5.2 Sampling of Continuous-Time Signals

As indicated above, discrete-time signals in many applications are generated by sampling continuous-time signals. We also observed in Example 2.10 that identical discrete-time sequences may result from the sampling of more than one distinct continuous-time function. In fact, in general, there exists an infinite number of continuous-time signals, which when sampled lead to the same discrete-time signal. However, under certain conditions, it is possible to relate a unique continuous-time signal to a given discrete-time sequence, and it is possible to recover the original continuous-time signal from its sampled values. We develop this correspondence and the associated conditions next by considering the relation between the spectra of the original continuous-time signal and the discrete-time signal obtained by sampling.

5.2.1   Effect of Sampling in the Frequency-Domain

Let $g_a(t)$ be a continuous-time signal that is sampled uniformly at $t = nT$, generating the sequence $g[n]$ where

$$g[n] = g_a(nT), \qquad -\infty < n < \infty \tag{5.1}$$

with T being the *sampling period*. The reciprocal of T is called the *sampling frequency* F_T, i.e., $F_T = 1/T$. Now, the frequency-domain representation of $g_a(t)$ is given by its continuous-time Fourier transform $G_a(j\Omega)$,

$$G_a(j\Omega) = \int_{-\infty}^{\infty} g_a(t)e^{-j\Omega t}\, dt, \tag{5.2}$$

whereas the frequency-domain representation of $g[n]$ is given by its discrete-time Fourier transform $G(e^{j\omega})$,

$$G(e^{j\omega}) = \sum_{n=-\infty}^{\infty} g[n]\, e^{-j\omega n}. \tag{5.3}$$

To establish the relations between these two different types of Fourier spectra, $G_a(j\Omega)$ and $G(e^{j\omega})$, we treat the sampling operation mathematically as a multiplication of the continuous-time signal $g_a(t)$ by a periodic impulse train $p(t)$:

$$p(t) = \sum_{n=-\infty}^{\infty} \delta(t - nT), \tag{5.4}$$

consisting of a train of ideal impulses with a period T, as indicated in Figure 5.3. The multiplication operation yields an impulse train $g_p(t)$:

$$g_p(t) = g_a(t)p(t) = \sum_{n=-\infty}^{\infty} g_a(nT)\delta(t - nT). \tag{5.5}$$

which is seen to be a continuous-time signal consisting of a train of uniformly spaced impulses with the impulse at $t = nT$ weighted by the sampled value $g_a(nT)$ of $g_a(t)$ at that instant.

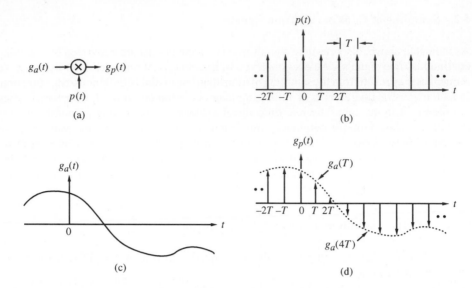

Figure 5.3   Mathematical rerpresentation of the uniform sampling process: (a) Ideal sampling model, (b) impulse train, (c) continuous-time signal, and (d) its sampled version.

There are two different forms of the continuous-time Fourier transform $G_p(j\Omega)$ of $g_p(t)$. One form is given by the weighted sum of the continuous-time Fourier transforms of $\delta(t - nT)$:

$$G_p(j\Omega) = \sum_{n=-\infty}^{\infty} g_a(nT)e^{-j\Omega nT}. \qquad (5.6)$$

To derive the second form, we note that the periodic impulse train $p(t)$ can be expressed as a Fourier series (Problem 5.1):

$$p(t) = \frac{1}{T} \sum_{k=-\infty}^{\infty} e^{j(2\pi/T)kt} = \frac{1}{T} \sum_{k=-\infty}^{\infty} e^{j\Omega_T kt}, \qquad (5.7)$$

where $\Omega_T = 2\pi/T$ denotes the sampling angular frequency. The impulse train $g_p(t)$ therefore can be expressed as

$$g_p(t) = \left(\frac{1}{T} \sum_{k=-\infty}^{\infty} e^{j\Omega_T kt} \right) \cdot g_a(t). \qquad (5.8)$$

Now from the frequency shifting property of the continuous-time Fourier transform, the continuous-time Fourier transform of $e^{j\Omega_T kt} g_a(t)$ is given by $G_a(j(\Omega - k\Omega_T))$. Hence, an alternative form of the continuous-time Fourier transform of $g_p(t)$ is given by

$$G_p(j\Omega) = \frac{1}{T} \sum_{k=-\infty}^{\infty} G_a(j(\Omega - k\Omega_T)). \qquad (5.9)$$

Therefore, $G_p(j\Omega)$ is a periodic function of frequency Ω consisting of a sum of shifted and scaled replicas of $G_a(j\Omega)$, shifted by integer multiples of Ω_T and scaled by $1/T$. The term on the right-hand side of Eq. (5.9) for $k = 0$ is the *baseband* portion of $G_p(j\Omega)$, and each of the remaining terms are the *frequency-translated* portions of $G_p(j\Omega)$. The frequency range $-\Omega_T/2 \le \Omega \le \Omega_T/2$ is called the *baseband* or *Nyquist band*.

Figure 5.4 illustrates the effect of time-domain sampling in the frequency-domain. Assume $g_a(t)$ is a bandlimited signal with a frequency response $G_a(j\Omega)$, as sketched in Figure 5.4(a) where Ω_m is the highest frequency contained in $g_a(t)$. The spectrum $P(j\Omega)$ of the periodic impulse train $p(t)$ with a sampling period $T = 2\pi/\Omega_T$ is indicated in Figures 5.4(b) and (d). Two possible spectra of $G_p(j\Omega)$ are shown in Figures 5.4(c) and (e). It is evident from Figure 5.4(c) that if $\Omega_T > 2\Omega_m$, there is no overlap between the shifted replicas of $G_a(j\Omega)$ generating $G_p(j\Omega)$. On the other hand, as indicated in Figure 5.4(e), if $\Omega_T < 2\Omega_m$, there is an overlap of the spectra of the shifted replicas of $G_a(j\Omega)$ generating $G_p(j\Omega)$. Consequently, if $\Omega_T > 2\Omega_m$, $g_a(t)$ can be recovered exactly from $g_p(t)$ by passing it through an ideal lowpass filter $H_r(j\Omega)$ with a gain T and a cutoff frequency Ω_c greater than Ω_m and less than $\Omega_T - \Omega_m$, as illustrated in Figure 5.5. However, if $\Omega_T < 2\Omega_m$, due to the overlap of the shifted replicas of $G_a(j\Omega)$, the spectrum $G_p(j\Omega)$ cannot be separated by filtering to recover $G_a(j\Omega)$ because of the distortion caused by a part of the replicas immediately outside the baseband folded back or *aliased* into the baseband. The frequency $\Omega_T/2$ is often referred to as the *folding frequency* or *Nyquist frequency*.

The above result is more commonly known as the *sampling theorem*,[3] which can be summarized as follows.

Let $g_a(t)$ be a bandlimited signal with $G_a(j\Omega) = 0$ for $|\Omega| > \Omega_m$. Then $g_a(t)$ is uniquely determined by its samples $g_a(nT)$, $n = 0, 1, 2, 3, \ldots$, if

$$\Omega_T \ge 2\Omega_m, \tag{5.10}$$

where

$$\Omega_T = 2\pi/T. \tag{5.11}$$

Given $\{g_a(nT)\}$, we can recover exactly $g_a(t)$ by generating an impulse train $g_p(t)$ of the form of Eq. (5.5) and then passing $g_p(t)$ through an ideal lowpass filter $H_r(j\Omega)$ with a gain T and a cutoff frequency Ω_c greater than Ω_m and less than $\Omega_T - \Omega_m$, i.e.,

$$\Omega_m < \Omega_c < (\Omega_T - \Omega_m). \tag{5.12}$$

The highest frequency Ω_m contained in $g_a(t)$ is usually called the *Nyquist frequency* since it determines the minimum sampling frequency $\Omega_T = 2\Omega_m$ that must be used to fully recover $g_a(t)$ from its sampled version. The frequency $2\Omega_m$ is called the *Nyquist rate*.

If the sampling rate is higher than the Nyquist rate, it is called *oversampling*. On the other hand, if the sampling rate is lower than the Nyquist rate, it is called *undersampling*. Finally, if the sampling rate is exactly equal to the Nyquist rate, it is called *critical sampling*.[4]

Typical sampling rates used in practice are, for example, 8 kHz in digital telephony and 44.1 kHz in compact disc (CD) music systems. In the former case, a 3.4-kHz signal

---

[3] Also called either the *Nyquist sampling theorem* or *Shannon sampling theorem*.

[4] It should be noted that a pure sinusoid may not be recoverable from its critically sampled version.

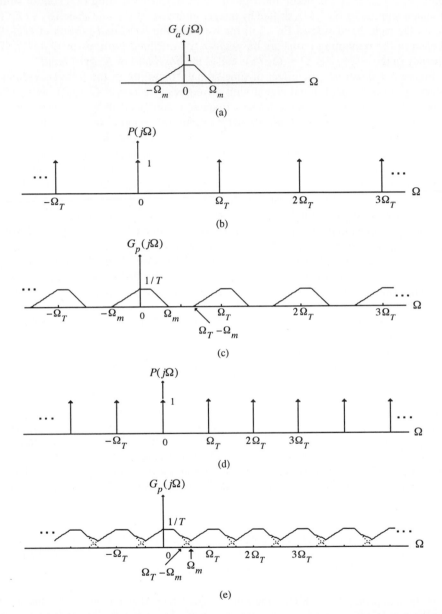

Figure 5.4 Illustration of the effect of time-domain sampling in the frequency-domain. (a) Spectrum of original continuous-time signal $g_a(t)$, (b) spectrum of the periodic impulse train $p(t)$, (c) spectrum of the sampled signal $g_p(t)$ with $\Omega_T > 2\omega_m$, (d) spectrum of the periodic impulse train $p(t)$ with a sampling periodic smaller than that shown in (b), and (e) spectrum of the sampled signal $g_p(t)$ with $\Omega_T < 2\Omega_m$. (Note: The spectrum of $g_a(t)$ is shown as not being an even function to emphasize the effect of sampling.

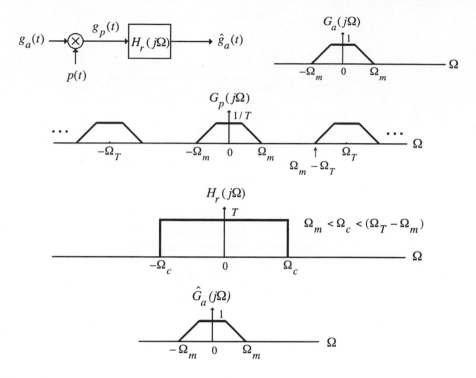

Figure 5.5 Reconstruction of the original continuous-time signal from its sampled version obtained by ideal sampling.

bandwidth is acceptable for telephone conversation. Hence, a sampling rate of 8 kHz, which is greater than 2 times that of the acceptable bandwidth of 6.8 kHz, is quite adequate. In high-quality analog music signal processing, on the other hand, a bandwidth of about 20 kHz has been determined to preserve the fidelity. As a result, here the analog music signal is sampled at a slightly higher rate of 44.1 kHz to ensure almost negligible aliasing distortion.

Example 5.1 Consider the three pure sinusoidal signals of Example 2.10: $g_1(t) = \cos(6\pi t)$, $g_2(t) = \cos(14\pi t)$, and $g_3(t) = \cos(26\pi t)$. The corresponding continuous-time Fourier transforms consist of two impulses each and are given by

$$G_1(j\Omega) = \pi[\delta(\Omega - 6\pi) + \delta(\Omega + 6\pi)],$$
$$G_2(j\Omega) = \pi[\delta(\Omega - 14\pi) + \delta(\Omega + 14\pi)],$$
$$G_3(j\Omega) = \pi[\delta(\Omega - 26\pi) + \delta(\Omega + 26\pi)].$$

These transforms are plotted in Figures 5.6(a) to (c), respectively.

The above signals are sampled at a rate of $T = 0.1$ sec, i.e., with a sampling frequency $\Omega_T = 20\pi$ rad/sec, generating the continuous-time impulse trains $g_{1p}(t)$, $g_{2p}(t)$, and $g_{3p}(t)$. From Eq. (5.9) their corresponding continuous-time Fourier transforms are

Figure 5.6 Effect of sampling on a pure cosine signal: (a) spectrum of $\cos(6\pi t)$, (b) spectrum of $\cos(14\pi t)$, (c) spectrum of $\cos(26\pi t)$ (d) spectrum of sampled version of $\cos(6\pi t)$ with $\Omega_T = 20\pi > 2\Omega_m = 12\pi$, (e) spectrum of sampled version of $\cos(14\pi t)$ with $\Omega_T = 20\pi < 2\Omega_m = 28\pi$, and (f) spectrum of sampled version of $\cos(2\pi t)$ with $\Omega_T = 20\pi < 2\Omega_m = 52\pi$.

given by

$$G_{\ell p}(j\Omega) = \frac{1}{T} \sum_{k=-\infty}^{\infty} G_\ell(j\Omega - jk\Omega_T), \qquad \ell = 1, 2, 3,$$

which have been plotted in Figure 5.6(d) to (f). These figures also indicate by dotted lines the frequency response of an ideal lowpass filter with a cutoff at $\Omega_c = \Omega_T/2 = 10\pi$ and a gain of $T = 0.1$ (not shown to scale). The Fourier transforms of the output of the lowpass filter in each case are also indicated in Figure 5.6(d) to (f). Note that in the case of $g_1(t)$, the sampling rate satisfies the Nyquist condition and there is no aliasing. The continuous-time reconstructed output is precisely the original continuous-time signal $g_1(t)$. However, in the other two cases, the sampling rate does not satisfy the Nyquist condition, resulting in aliasing, and the outputs here are all equal to $\cos(6\pi t)$. Note that in Figure 5.6(e), the impulse appearing at $\Omega = 6\pi$ in the positive frequency passband of the lowpass filter results from the aliasing of the impulse in $G_2(j\Omega)$ at $\Omega = -14\pi$. On the other hand, the impulse appearing at $\Omega = 6\pi$ in Figure 5.6(f) has resulted from the aliasing of the impulse in $G_3(j\Omega)$ at $\Omega = 26\pi$.

We now establish the relation between the discrete-time Fourier transform $G(e^{j\omega})$ of the sequence $g[n]$ and the continuous-time Fourier transform $G_a(j\Omega)$ of the analog signal $g_a(t)$. If we compare Eq. (5.3) with Eq. (5.6) and make use of Eq. (5.1), we observe that

$$G(e^{j\omega}) = G_p(j\Omega)\big|_{\Omega=\omega/T}, \tag{5.13a}$$

or equivalently,

$$G_p(j\Omega) = G(e^{j\omega})\big|_{\omega=\Omega T}. \tag{5.13b}$$

Therefore, from the above and Eq. (5.9), we arrive at the desired result given by

$$\begin{aligned} G(e^{j\omega}) &= \frac{1}{T} \sum_{k=-\infty}^{\infty} G_a(j\Omega - jk\Omega_T)\bigg|_{\Omega=\omega/T} \\ &= \frac{1}{T} \sum_{k=-\infty}^{\infty} G_a\left(j\frac{\omega}{T} - jk\Omega_T\right) \\ &= \frac{1}{T} \sum_{k=-\infty}^{\infty} G_a\left(j\frac{\omega}{T} - j\frac{2\pi k}{T}\right), \end{aligned} \tag{5.14a}$$

which can be alternatively expressed as

$$G(e^{j\Omega T}) = \frac{1}{T} \sum_{k=-\infty}^{\infty} G_a(j\Omega - jk\Omega_T). \tag{5.14b}$$

As can be seen from Eq. (5.13a) or (5.13b), $G(e^{j\omega})$ is obtained from $G_p(j\Omega)$ simply by scaling the frequency axis Ω according to the relation

$$\Omega = \frac{\omega}{T}. \tag{5.15}$$

Now, the continuous-time Fourier transform $G_p(j\Omega)$ is a periodic function of Ω with a period $\Omega_T = 2\pi/T$. Because of the above normalization, the discrete-time Fourier transform $G(e^{j\omega})$ is a periodic function of ω with a period 2π.

5.2.2   Recovery of the Analog Signal

We indicated earlier that if the discrete-time sequence $g[n]$ has been obtained by uniformly sampling a bandlimited continuous-time signal $g_a(t)$ with a highest frequency Ω_m at a rate $\Omega_T = 2\pi/T$ satisfying the condition of Eq. (5.10), then the original continuous-time signal $g_a(t)$ can be fully recovered by passing the equivalent impulse train $g_p(t)$ through an ideal lowpass filter $H_r(j\Omega)$ with a cutoff at Ω_c satisfying Eq. (5.12) and with a gain of T. We next derive the expression for the output $\hat{g}_a(t)$ of the ideal lowpass filter as a function of the samples $g[n]$.

Now, the impulse response $h_r(t)$ of the above ideal lowpass filter is obtained simply by taking the inverse Fourier transform of its frequency response $H_r(j\Omega)$:

$$H_r(j\Omega) = \begin{cases} T, & |\Omega| \le \Omega_c, \\ 0, & |\Omega| > \Omega_c, \end{cases} \tag{5.16}$$

and is given by

$$h_r(t) = \frac{1}{2\pi} \int_{-\infty}^{\infty} H_r(j\Omega)e^{j\Omega t}\, d\Omega = \frac{T}{2\pi} \int_{-\Omega_c}^{\Omega_c} e^{j\Omega}\, d\Omega$$

$$= \frac{\sin(\Omega_c t)}{\Omega_T t/2}, \quad -\infty \le t \le \infty. \tag{5.17}$$

Observe that the impulse train $g_p(t)$ is given by

$$g_p(t) = \sum_{n=-\infty}^{\infty} g[n]\delta(t - nT). \tag{5.18}$$

Therefore, the output $\hat{g}_a(t)$ of the ideal lowpass filter is given by the convolution of $g_p(t)$ with the impulse response $h_r(t)$ of the analog reconstruction filter:

$$\hat{g}_a(t) = \sum_{n=-\infty}^{\infty} g[n]h_r(t - nT). \tag{5.19}$$

Substituting $h_r(t)$ from Eq. (5.17) in Eq. (5.19) and assuming for simplicity $\Omega_c = \Omega_T/2 = \pi/T$, we arrive at

$$\hat{g}_a(t) = \sum_{n=-\infty}^{\infty} g[n]\frac{\sin[\pi(t - nT)/T]}{\pi(t - nT)/T}. \tag{5.20}$$

The above expression indicates that the reconstructed continuous-time signal $\hat{g}_a(t)$ is obtained by shifting in time the impulse response of the lowpass filter $h_r(t)$ by an amount nT and scaling it in amplitude by the factor $g[n]$ for all integer values of n in the range $-\infty < n < \infty$ and then summing up all shifted versions. The ideal bandlimited interpolation process is illustrated in Figure 5.7.

Now, it can be shown that when $\Omega_c = \Omega_T/2$ in Eq. (5.17), $h_r(0) = 1$ and $h_r(nT) = 0$ for $n \ne 0$ (Problem 5.7). As a result, from Eq. (5.20), $\hat{g}_a(rT) = g[r] = g_a(rT)$ for all integer values of r in the range $-\infty < r < \infty$, whether or not the condition of the

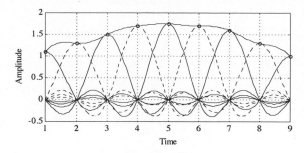

Figure 5.7   The ideal bandlimited reconstruction by interpolation.

sampling theorem has been satisfied. However, $\hat{g}_a(t) = g_a(t)$ for all values of t only if the sampling frequency Ω_T satisfies the condition of Eq. (5.10) of the sampling theorem.

It should be noted that the ideal analog lowpass filter of Eq. (5.17) has a doubly infinite-length impulse response and is thus noncausal, making it unrealizable. An analog lowpass filter is also needed to bandlimit the continuous-time signal before it is sampled to ensure that the conditions of the sampling theorem are satisfied. The magnitude response specification for the both the anti-aliasing lowpass filter and the analog reconstruction filter therefore must be modified to make them realizable. We review in Section 5.3 some commonly used methods for the determination of the transfer function of realizable and stable analog lowpass filters approximating the ideal magnitude characteristic of Eq. (5.17). In Sections 5.4 and 5.8 we consider specifically the issues concerning the design of an anti-aliasing filter and the reconstruction filter, respectively.

5.2.3   Implications of the Sampling Process

Consider again the sampling of three continuous-time signals of Example 5.1. From Figure 5.6(d) it should be apparent that from the sampled version $g_{1p}(t)$ of the continuous-time signal $g_1(t) = \cos(6\pi t)$, we can also recover any of its frequency translated versions $\cos[(20k \pm 6)\pi t]$ outside the baseband by passing $g_{1p}(t)$ through an ideal analog bandpass filter with a passband centered at $\Omega = (20k \pm 6)\pi$. For example, to recover the signal $\cos(34\pi t)$, it will be necessary to employ a bandpass filter with a frequency response

$$H_r(j\Omega) = \begin{cases} 0.1, & (34 - \Delta)\pi \leq |\Omega| \leq (34 + \Delta)\pi, \\ 0, & \text{otherwise,} \end{cases} \tag{5.21}$$

where Δ is a small number. Likewise, we can recover the aliased baseband component $\cos(6\pi t)$ from the sampled version of either $g_{2p}(t)$ or $g_{3p}(t)$ by passing it through an ideal lowpass filter:

$$H_r(j\Omega) = \begin{cases} 0.1, & (6 - \Delta)\pi \leq |\Omega| \leq (6 + \Delta)\pi, \\ 0, & \text{otherwise.} \end{cases} \tag{5.22}$$

There is no *aliasing distortion* unless the original continuous-time signal also contains the component $\cos(6\pi t)$. Similarly, from the sampled versions of either $g_{2p}(t)$ or $g_{3p}(t)$, we can recover any one of the frequency translated versions, including the parent continuous-time signal, $\cos(14\pi t)$ or $\cos(26\pi t)$ as the case may be, by employing suitable ideal bandpass filters.

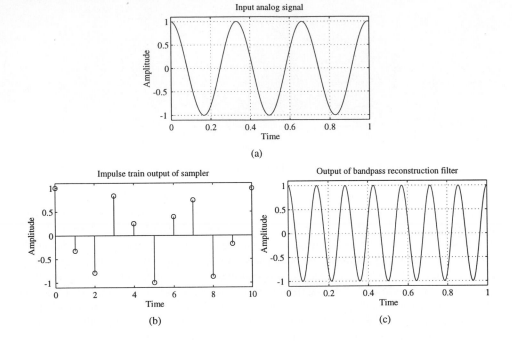

Figure 5.8   Illustration of frequency translation effect of sampling.

The frequency translation effect of sampling is further illustrated in the following two examples.

Example 5.2 Figure 5.8(a) shows the continuous-time sinusoidal signal $g_1(t) = \cos(6\pi t)$ of frequency 3 Hz. The impulse train $g_{1p}(t)$ generated from $g_1(t)$ by uniformly sampling it a rate of 10 Hz is depicted in Figure 5.8(b). The continuous-time output $y(t)$ by passing $g_{1p}(t)$ through an analog bandpass reconstruction filter with a passband from 10 Hz to 20 Hz is shown in Figure 5.8(c), which is seen to be a continuous-time sinusoidal signal $\cos(14\pi t)$ of frequency 7 Hz. In general, the continuous-time output of an analog bandpass reconstruction filter with a passband from $5k$ Hz to $5k + 15$ Hz, with k a positive integer, will be a sinusoidal signal of frequency of $10k + 6$ for k even and $10k - 6$ for k odd.

Example 5.3 The continuous-time sinusoidal signal $g_3(t) = \cos(26\pi t)$ of frequency 13 Hz shown in Figure 5.9(a) is sampled uniformly at 10-Hz rate, resulting in an impulse train $g_{3p}(t)$ sketched in Figure 5.9(b). The output generated from $g_{3p}(t)$ by passing it through an analog lowpass reconstruction filter with a passband edge at 5 Hz is a continuous-time sinusoidal signal of frequency 3 Hz, as shown in Figure 5.9(c). Note that the original high-frequency sinusoidal signal can be fully recovered from $g_{3p}(t)$ by passing it instead through a bandpass filter with a passband from 10 Hz to 15 Hz, as indicated in Figure 5.9(d).

Generalizing the above discussions, consider the continuous-time signals $g_1(t)$, $g_2(t)$, and $g_3(t)$ with bandlimited frequency spectrums $G_1(j\Omega)$, $G_2(j\Omega)$, and $G_3(j\Omega)$, as shown in Figures 5.10(a) to (c), respectively. Each of these continuous-time signals, when sampled

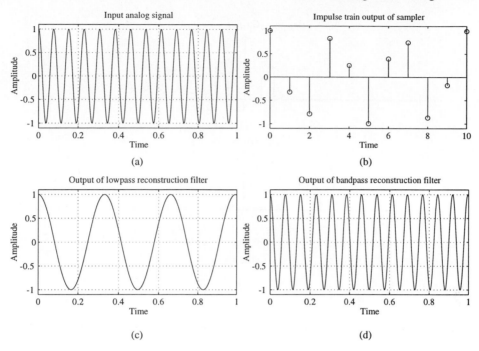

Figure 5.9   Illustration of effect of undersampling.

at a sampling frequency of Ω_T, develop a continuous-time signal $g_{\ell p}(t)$, $\ell = 1, 2, 3$, with an identical periodic frequency spectrum, as indicated in Figure 5.10(d). Therefore, by passing through an appropriate analog lowpass or bandpass filter of bandwidth greater than $\Omega_2 - \Omega_1$ but less than or equal to $\Omega_T/2$, we can recover either the original continuous-time signal or any one of its frequency-translated versions. Note that as long as the spectrum of the continuous-time signal being sampled at a sampling rate of Ω_T is bandlimited to the frequency range $k\Omega_T/2 \leq |\Omega| \leq (k+1)\Omega_T/2$, there is no aliasing distortion due to sampling, and it or its frequency-translated version can always be recovered from the sampled signal by appropriate filtering. There will be aliasing distortion only if there are frequency components in a wider frequency range than that indicated.

5.3   Analog Filter Design

There are a number of established approximation techniques for the design of analog lowpass filters [Vla69], [Dan74], [Tem73], [Tem77]. We describe four widely used design techniques here without their detailed derivations. Further details of these methods can be found in texts on analog filter design. Extensive tables for the design of analog lowpass filters are also available [Chr66], [Skw65], [Zve67]. As indicated earlier, a commonly used technique for the design of IIR digital filters is based on the conversion of an analog prototype transfer function designed employing one of the methods discussed here. The digital filter design techniques are the subject of discussion in Chapter 7.

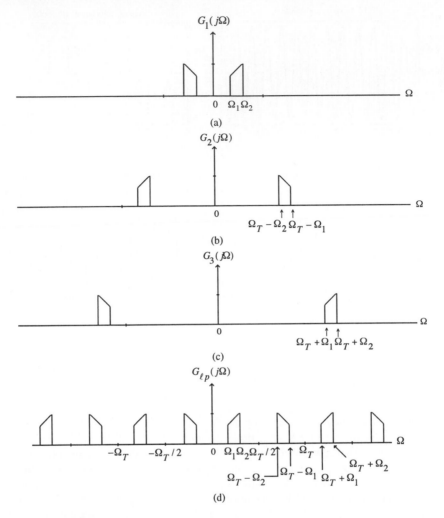

Figure 5.10   Further illustration of the effect of sampling.

5.3.1   Filter Specifications

Both the anti-aliasing filter and the reconstruction filter of Figure 5.1 are of the lowpass type, and ideally they should have a magnitude response of the form shown in Figure 5.5. In practice, the magnitude response characteristics in the passband and in the stopband cannot be constant and are therefore specified with some acceptable tolerances. Moreover, a transition band is specified between the passband and the stopband to permit the magnitude to drop off smoothly. For example, the magnitude $|H_a(j\Omega)|$ of a lowpass filter may be given as shown in Figure 5.11. As indicated in the figure, in the *passband* defined by $0 \le \Omega \le \Omega_p$, we require

$$1 - \delta_p \le |H_a(j\Omega)| \le 1 + \delta_p, \qquad \text{for } |\Omega| \le \Omega_p, \tag{5.23}$$

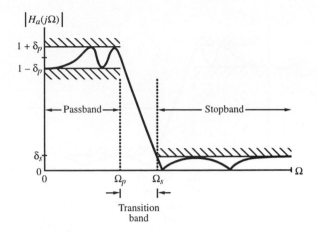

Figure 5.11 Typical magnitude specifications for an analog lowpass filter.

or in other words, the magnitude approximates unity within an error of $\pm\delta_p$. In the *stopband*, defined by $\Omega_s \leq \Omega \leq \infty$, we require

$$|H_a(j\Omega)| \leq \delta_s, \qquad \text{for } \Omega_s \leq |\Omega| \leq \infty, \tag{5.24}$$

implying that the magnitude approximate zero within an error of δ_s. The frequencies Ω_p and Ω_s are, respectively, called the *passband edge frequency* and the *stopband edge frequency*.

The limits of the tolerances in the passband and stopband, δ_p and δ_s, are called *ripples*. Usually these ripples are specified in dB in terms of the *peak passband ripple* α_p and the *minimum stopband attenuation* α_s, defined by:

$$\alpha_p = -20\log_{10}(1 - \delta_p)\,\text{dB}, \tag{5.25}$$
$$\alpha_s = -20\log_{10}(\delta_s)\,\text{dB}. \tag{5.26}$$

Often, the filter specifications are given in terms of the *loss function* or *attenuation function* $a(j\Omega)$ in dB, which is defined as the negative of the gain in dB, i.e., $-20\log_{10}|H_a(j\Omega)|$.

Example 5.4 Let the desired peak passband ripple of a lowpass filter be 0.01 dB, and the minimum attenuation in the stopband be 70 dB. Determine δ_p and δ_s.

Substituting $\alpha_p = 0.01$ in Eq. (5.25) and solving we get

$$\delta_p = 0.00115.$$

Likewise, substituting $\alpha_s = 70$ in Eq. (5.26) we obtain

$$\delta_s = 0.0003162.$$

The magnitude response specifications for an analog lowpass filter, in some applications, are given in a normalized form, as indicated in Figure 5.12. Here the maximum value of the magnitude in the passband is assumed to be unity and the passband ripple, denoted

Figure 5.12   Normalized magnitude specifications for an analog lowpass filter.

as $1/\sqrt{1+\varepsilon^2}$, is given by the minimum value of the magnitude in the passband. The maximum passband gain or the minimum passband loss is therefore 0 dB. The maximum stopband ripple is denoted by $1/A$, and the minimum stopband attenuation is therefore given by $-20\log_{10}(1/A)$.

In analog filter theory two additional parameters are defined. The first one, called the *transition ratio* or *selectivity parameter* is defined by the ratio of the passband edge frequency Ω_p and the stopband edge frequency Ω_s, and is usually denoted by k, i.e.,

$$k = \frac{\Omega_p}{\Omega_s}. \tag{5.27}$$

Note that for a lowpass filter, $k < 1$. The second one, called the *discrimination parameter* and denoted by k_1, is defined as

$$k_1 = \frac{\varepsilon}{\sqrt{A^2-1}}. \tag{5.28}$$

Usually, $k_1 \ll 1$.

5.3.2   Butterworth Approximation

The magnitude-squared response of an analog lowpass Butterworth filter $H_a(s)$ of Nth order is given by

$$|H_a(j\Omega)|^2 = \frac{1}{1+(\Omega/\Omega_c)^{2N}}. \tag{5.29}$$

It can be easily shown that the first $2N-1$ derivatives of $|H_a(j\Omega)|^2$ at $\Omega = 0$ are equal to zero, and as a result, the Butterworth filter is said to have a *maximally flat magnitude* at $\Omega = 0$. The gain of the Butterworth filter in dB is given by

$$\mathcal{G}(\Omega) = 10\log_{10}|H_a(j\Omega)|^2 \text{ dB}.$$

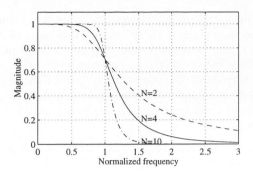

Figure 5.13 Typical Butterworth lowpass filter responses.

At dc, i.e., at $\Omega = 0$, the gain in dB is equal to zero, and at $\Omega = \Omega_c$, the gain is

$$\mathcal{G}(\Omega_c) = 10 \log_{10}\left(\tfrac{1}{2}\right) = -3.0103 \cong -3\,\text{dB}$$

and therefore, Ω_c is often called the *3-dB cutoff frequency*. Since the derivative of the squared-magnitude response, or equivalently, of the magnitude response is always negative for positive values of Ω, the magnitude response, is monotonically decreasing with increasing Ω. For $\Omega \gg \Omega_c$, the squared-magnitude function can be approximated by

$$|H_a(j\Omega)|^2 \approx \frac{1}{(\Omega/\Omega_c)^{2N}}.$$

The gain $\mathcal{G}(\Omega_2)$ in dB at $\Omega_2 = 2\Omega_1$ with $\Omega_1 \gg \Omega_c$ is given by

$$\mathcal{G}(\Omega_2) = -20\log_{10}\left(\frac{\Omega_2}{\Omega_c}\right)^{2N} = \mathcal{G}(\Omega_1) - 6N \; \text{dB},$$

where $\mathcal{G}(\Omega_1)$ is the gain in dB at Ω_1. As a result, the gain roll-off per octave in the stopband decreases by 6 dB, or equivalently, by 20 dB per decade for an increase of the filter order by one. In other words, the passband and the stopband behaviors of the magnitude response improve with a corresponding decrease in the transition band as the filter order N increases. A plot of the magnitude response of the normalized Butterworth lowpass filter with $\Omega_c = 1$ for some typical values of N is shown in Figure 5.13.

The two parameters completely characterizing a Butterworth filter are therefore the 3-dB cutoff frequency Ω_c and the order N. These are determined from the specified passband edge Ω_p, the minimum passband magnitude $1/\sqrt{1+\varepsilon^2}$, the stopband edge Ω_s, and the maximum stopband ripple $1/A$. From Eq. (5.29) we get

$$\left|H_a(j\Omega_p)\right|^2 = \frac{1}{1 + (\Omega_p/\Omega_c)^{2N}} = \frac{1}{1+\varepsilon^2}, \tag{5.30a}$$

$$\left|H_a(j\Omega_s)\right|^2 = \frac{1}{1 + (\Omega_s/\Omega_c)^{2N}} = \frac{1}{A^2}. \tag{5.30b}$$

Solving the above we arrive at the expression for the order N as

$$N = \frac{1}{2}\frac{\log_{10}\left[\left(A^2 - 1\right)/\varepsilon^2\right]}{\log_{10}\left(\Omega_s/\Omega_p\right)} = \frac{\log_{10}(1/k_1)}{\log_{10}(1/k)}. \tag{5.31}$$

Since the order N of the filter must be an integer, the value of N computed using the above expression is rounded up to the next higher integer. This value of N can be used next in either Eq. (5.30a) or (5.30b) to solve for the 3-dB cutoff frequency Ω_c. If it is used in Eq. (5.30a), the passband specification is met exactly, whereas the stopband specification is exceeded. On the other hand, if it is used in Eq. (5.30b), the stopband specification is met exactly, whereas the passband specification is exceeded.

The expression for the transfer function of the Butterworth lowpass filter is given by

$$H_a(s) = \frac{C}{D_N(s)} = \frac{\Omega_c^N}{s^N + \sum_{\ell=0}^{N-1} d_\ell s^\ell} = \frac{\Omega_c^N}{\prod_{\ell=1}^{N}(s - p_\ell)}, \tag{5.32}$$

where

$$p_\ell = \Omega_c e^{j[\pi(N+2\ell-1)/2N]}, \qquad \ell = 1, 2, \dots, N. \tag{5.33}$$

The denominator $D_N(s)$ of Eq. (5.32) is known as the *Butterworth polynomial* of order N and is easy to compute. These polynomials have been tabulated for easy design reference [Chr66], [Skw65], [Zve67]. The analog lowpass Butterworth filters can be readily designed using MATLAB (see Section 5.3.6).

Example 5.5 Determine the lowest order of a transfer function $H_a(s)$ having a maximally flat lowpass characteristic with a 1-dB cutoff frequency at 1 kHz and a minimum attenuation of 40 dB at 5 kHz.

We first determine ε and A. From Eq. (5.30a),

$$10 \log_{10}\left(\frac{1}{1+\varepsilon^2}\right) = -1,$$

which yields $\varepsilon^2 = 0.25895$. Likewise, from Eq. (5.30b),

$$10 \log_{10}\left(\frac{1}{A^2}\right) = -40,$$

which leads to $A^2 = 10,000$. Substituting the values in Eq. (5.28), we get $1/k_1 = 196.51334$.

The inverse transition ratio here is $1/k = 5000/1000 = 5$. Substituting these values in Eq. (5.31), we get

$$N = \frac{\log_{10}(1/k_1)}{\log_{10}(1/k)} = \frac{\log_{10}(196.51334)}{\log_{10}(5)} = 3.2811022. \tag{5.34}$$

Since the order of the transfer function must be an integer, we round the above up to the nearest integer $N = 4$.

5.3.3 Chebyshev Approximation

In this case, the approximation error, defined as the difference between the ideal brickwall characteristic and the actual response, is minimized over a prescribed band of frequencies. In fact, the magnitude error is equiripple in the band. There are two types of *Chebyshev transfer functions*. In the *Type 1* approximation, the magnitude characteristic is equiripple in the passband and monotonic in the stopband, whereas in the *Type 2* approximation, the magnitude response is monotonic in the passband and equiripple in the stopband.

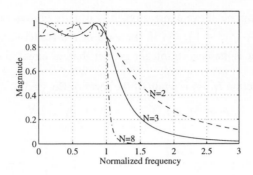

Figure 5.14   Typical Type 1 Chebyshev lowpass filter responses with 1 dB passband ripple.

Type 1 Chebyshev Approximation

The Type 1 Chebyshev transfer function $H_a(s)$ has a magnitude response given by

$$|H_a(j\Omega)|^2 = \frac{1}{1 + \varepsilon^2 T_N^2(\Omega/\Omega_p)}, \tag{5.35}$$

where $T_N(\Omega)$ is the Chebyshev polynomial of order N:

$$T_N(\Omega) = \begin{cases} \cos(N\cos^{-1}\Omega), & |\Omega| \leq 1, \\ \cosh(N\cosh^{-1}\Omega), & |\Omega| > 1. \end{cases} \tag{5.36}$$

The above polynomial can also be derived via a recurrence relation given by

$$T_r(\Omega) = 2\Omega T_{r-1}(\Omega) - T_{r-2}(\Omega), \qquad r \geq 2, \tag{5.37}$$

with $T_0(\Omega) = 1$ and $T_1(\Omega) = \Omega$.

Typical plots of the magnitude responses of the Type 1 Chebyshev lowpass filter are shown in Figure 5.14 for three different values of filter order N with the same passband ripple ε. From these plots it is seen that the square-magnitude response is equiripple between $\Omega = 0$ and $\Omega = 1$, and it decreases monotonically for all $\Omega > 1$.

The order N of the transfer function is determined from the attenuation specification in the stopband at a particular frequency. For example, if at $\Omega = \Omega_s$ the magnitude is equal to $1/A$, then from Eqs. (5.35) and (5.36),

$$|H(j\Omega_s)|^2 = \frac{1}{1 + \varepsilon^2 T_N^2(\Omega_s/\Omega_p)}$$

$$= \frac{1}{1 + \varepsilon^2 \left\{ \cosh[N\cosh^{-1}(\Omega_s/\Omega_p)] \right\}^2} = \frac{1}{A^2}. \tag{5.38}$$

Solving the above, we get

$$N = \frac{\cosh^{-1}(\sqrt{A^2 - 1}/\varepsilon)}{\cosh^{-1}(\Omega_s/\Omega_p)} = \frac{\cosh^{-1}(1/k_1)}{\cosh^{-1}(1/k)}. \tag{5.39}$$

In computing N using the above expression, it is usually convenient to use the identity $\cosh^{-1}(y) = \ln\left(y + \sqrt{y^2 - 1}\right)$. As in the case of the Butterworth filter, the order N of the filter is chosen as the nearest integer greater than or equal to the number given by Eq. (5.39).

The transfer function $H_a(s)$ is again of the form of Eq. (5.32), with

$$p_\ell = \sigma_\ell + j\,\Omega_\ell, \qquad \ell = 1, 2, \ldots, N, \tag{5.40}$$

where

$$\sigma_\ell = -\Omega_p \xi \sin\left[\frac{(2\ell - 1)\pi}{2N}\right], \qquad \Omega_\ell = \Omega_p \zeta \cos\left[\frac{(2\ell - 1)\pi}{2N}\right], \tag{5.41a}$$

$$\xi = \frac{\gamma^2 - 1}{2\gamma}, \qquad \zeta = \frac{\gamma^2 + 1}{2\gamma}, \qquad \gamma = \left(\frac{1 + \sqrt{1 + \varepsilon^2}}{\varepsilon}\right)^{1/N}. \tag{5.41b}$$

Type 2 Chebyshev Approximation

The Type 2 Chebyshev magnitude response, also known as the *inverse Chebyshev response*, exhibits a monotonic behavior in the passband with a maximally flat response at $\Omega = 0$ and an equiripple behavior in the stopband. The square-magnitude response expression here is given by

$$|H_a(j\Omega)|^2 = \frac{1}{1 + \varepsilon^2 \left[\dfrac{T_N(\Omega_s/\Omega_p)}{T_N(\Omega_s/\Omega)}\right]^2}. \tag{5.42}$$

Typical responses are as shown in Figure 5.15. The transfer function of a Type 2 Chebyshev lowpass filter is no longer an all-pole function and has both poles and zeros. If we write

$$H_a(s) = C_0 \frac{\prod_{\ell=1}^{N} (s - z_\ell)}{\prod_{\ell=1}^{N} (s - p_\ell)}, \tag{5.43}$$

the zeros are on the $j\Omega$-axis and are given by

$$z_\ell = j\frac{\Omega_s}{\cos\left[\frac{(2\ell-1)\pi}{2N}\right]}, \qquad \ell = 1, 2, \ldots, N. \tag{5.44}$$

If N is odd, then for $\ell = (N + 1)/2$, the zero is at $s = \infty$. The poles are located at

$$p_\ell = \sigma_\ell + j\Omega_\ell, \qquad \ell = 1, 2, \ldots, N, \tag{5.45}$$

where

$$\sigma_\ell = \frac{\Omega_s \alpha_\ell}{\alpha_\ell^2 + \beta_\ell^2}, \qquad \Omega_\ell = -\frac{\Omega_s \beta_\ell}{\alpha_\ell^2 + \beta_\ell^2}, \tag{5.46a}$$

$$\alpha_\ell = -\Omega_p \zeta \sin\left[\frac{(2\ell - 1)\pi}{2N}\right], \qquad \beta_\ell = \Omega_p \xi \cos\left[\frac{(2\ell - 1)\pi}{2N}\right], \tag{5.46b}$$

$$\zeta = \frac{\gamma^2 - 1}{2\gamma}, \qquad \xi = \frac{\gamma^2 + 1}{2\gamma}, \qquad \gamma = \left(A + \sqrt{A^2 - 1}\right)^{1/N}. \tag{5.46c}$$

The order N of the Type 2 Chebyshev lowpass filter is determined from given ε, Ω_s, and A using Eq. (5.39).

Figure 5.15 Typical Type 2 Chebyshev lowpass filter responses with 10 dB minimum stopband attenuation.

Example 5.6 We wish to determine the minimum order N required to design a lowpass filter with a Chebyshev or an inverse Chebyshev response with the specifications given in Example 5.5.

From Example 5.5, we have the following parameters: $1/k_1 = 196.51334$ and $1/k = 5$. Substituting these values in Eq. (5.39), we arrive at

$$N = \frac{\cosh^{-1}(1/k_1)}{\cosh^{-1}(1/k)} = \frac{\cosh^{-1}(196.51334)}{\cosh^{-1}(5)} = 2.60591. \qquad (5.47)$$

Since the order of the filter must be an integer, we choose the next highest integer value 3 for N. Note that the order of the Chebyshev lowpass filter, is lower than that of a Butterworth lowpass filter meeting the same specifications as given by Eq. (5.34). This is invariably the case for $N \geq 2$.

5.3.4   Elliptic Approximation

An *elliptic filter*, also known as a *Cauer filter*, has an equiripple passband and an equiripple stopband magnitude response, as indicated in Figure 5.16 for typical elliptic lowpass filters. The transfer function of an elliptic filter meets a given set of filter specifications, passband edge frequency Ω_p, stopband edge frequency Ω_s, passband ripple ε, and minimum stopband attenuation A, with the lowest filter order N. The theory of elliptic filter approximation is mathematically quite involved, and a detailed treatment of this topic is beyond the scope of this text. Interested readers are referred to the books by Antoniou [Ant93], Parks and Burrus [Par87], and Temes and LaPatra [Tem77].

The square-magnitude response of an elliptic lowpass filter is given by

$$|H_a(j\Omega)|^2 = \frac{1}{1 + \varepsilon^2 R_N^2(\Omega/\Omega_p)}, \qquad (5.48)$$

where $R_N(\Omega)$ is a rational function of order N satisfying the property $R_N(1/\Omega) = 1/R_N(\Omega)$, with the roots of its numerator lying within the interval $0 < \Omega < 1$ and the roots of its denominator lying in the interval $1 < \Omega < \infty$. For most applications, the filter order meeting a given set of specifications of passband edge frequency Ω_p, passband ripple ε, stopband edge frequency Ω_s, and the minimum stopband ripple A can be

Figure 5.16 Typical elliptic lowpass filter responses with 1 dB passband ripple and 10 dB minimum stopband attenuation.

estimated using the approximate formula [Ant93]:

$$N \cong \frac{2 \log_{10}(4/k_1)}{\log_{10}(1/\rho)},\tag{5.49}$$

where k_1 is the discrimination parameter defined in Eq. (5.28) and ρ is computed as follows:

$$k' = \sqrt{1 - k^2},\tag{5.50a}$$

$$\rho_0 = \frac{1 - \sqrt{k'}}{2(1 + \sqrt{k'})},\tag{5.50b}$$

$$\rho = \rho_0 + 2(\rho_0)^5 + 15(\rho_0)^9 + 150(\rho_0)^{13}.\tag{5.50c}$$

In Eq. (5.50a), k is the selectivity parameter defined in Eq. (5.27).

Example 5.7 We wish to determine the minimum order N required to design a lowpass elliptic filter with the specifications given in Example 5.5.

From Example 5.5, we note that $k = 1/5 = 0.2$ and $1/k_1 = 196.5134$. Substituting these values in Eq. (5.50a–5.50c), we arrive at

$$k' = 0.979796, \quad \rho_0 = 0.00255135, \quad \rho = 0.0025513525.$$

Using the above values in Eq. (5.49), we get

$$N = 2.23308.\tag{5.51}$$

Rounding the above value to the nearest higher integer, we obtain $N = 3$ as the appropriate order for the elliptic lowpass filter.

5.3.5 Linear-Phase Approximation

The previous three approximation techniques are for developing analog lowpass transfer functions meeting specified magnitude or gain response specifications without any concern for their phase responses. In a number of applications it is desirable that the analog lowpass

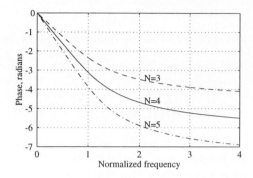

Figure 5.17   The phase responses of normalized fourth-order Bessel lowpass filter.

filter being designed have a linear-phase characteristic in the passband, in addition to approximating the magnitude specifications. One way to achieve this goal is to cascade an analog allpass filter with the filter designed to meet the magnitude specifications, so that the phase response of the overall cascade realization approximates linear-phase response in the passband. This approach increases the overall hardware complexity of the analog filter and may not be desirable for designing an analog anti-aliasing filter in some A/D conversion or designing an analog reconstruction filter in D/A conversion applications.

It is possible to design a lowpass filter that approximates a linear-phase characteristic in the passband but with a poorer magnitude response than that can be achieved by the previous three techniques. Such a filter has an all-pole transfer function of the form

$$H(s) = \frac{d_0}{B_N(s)} = \frac{d_0}{d_0 + d_1 s + \cdots + d_{N-1}s + s^N}, \tag{5.52}$$

and provides a maximally flat approximation to the linear-phase characteristic at $\Omega = 0$, i.e., has a maximally flat constant group delay at dc ($\Omega = 0$). For a normalized group delay of unity at dc, the denominator polynomial $B_N(s)$ of the transfer function, called the *Bessel polynomial*, can be derived via the recursion relation

$$B_N(s) = (2N - 1)B_{N-1}(s) + s^2 B_{N-2}(s), \tag{5.53}$$

starting with $B_1(s) = s + 1$ and $B_2(s) = s^2 + 3s + 3$. Alternatively, the coefficients of the Bessel polynomial $B_N(s)$ can be found from

$$d_\ell = \frac{(2N - \ell)!}{2^{N-\ell}\ell!(n - \ell)!}, \qquad \ell = 0, 1, \ldots, N - 1. \tag{5.54}$$

These filters are often referred to as *Bessel filters*. Figure 5.17 shows the phase responses of some typical Bessel filters.

5.3.6   Analog Filter Design Using MATLAB

The *Signal Processing Toolbox* in MATLAB includes a number of *M*-file functions to directly develop analog transfer functions for each one of the above approximation techniques. We next review these functions.

Butterworth filter

The *M*-file functions for the design of analog Butterworth filters are

```
[z,p,k]    = buttap(N),
[num,den]  = butter(N,Wn,'s')
[num,den]  = butter(N,Wn,'filtertype','s')
[N, Wn]    = buttord(Wp,Ws,Rp,Rs,'s')
```

The function `buttap(N)` computes the zeros, poles, and gain factor of the normalized analog Butterworth lowpass filter transfer function of order N with a 3-dB cutoff frequency of 1. The output files are the length *N* column vector p providing the locations the poles, a null vector z for the zero locations, and the gain factor k. The form of the transfer function obtained is given by

$$H_a(s) = \frac{P_a(s)}{D_a(s)} = \frac{k}{(s - p(1))\,(s - p(2)) \cdots (s - p(N))}. \qquad (5.55)$$

To determine the numerator and denominator coefficients of the transfer function from the zeros and poles computed, we need to use the *M*-file function `zp2tf(z,p,k)`.[5]

Alternatively, we can use the function `butter(N,Wn,'s')` to design an order N lowpass transfer function with a prescribed 3-dB cutoff frequency at Wn rad/sec, a nonzero number. The output data of this function are the numerator and the denominator coefficient vectors, num and den, respectively, in descending powers of s. If Wn is a two element vector [W1,W2] with W1 < W2, the function generates an order 2N bandpass transfer function with 3-dB bandedge frequencies at W1 and W2 with both being nonzero numbers. To design an order N highpass or an order 2N bandstop filter, the function `butter(N,Wn,'filtertype','s')` is employed where filtertype = high for a highpass filter with a 3-dB cutoff frequency at Wn or filtertype = stop for a bandstop filter with 3-dB stopband edges given by a two-element vector of nonzero numbers Wn = [W1,W2] with W1 < W2.

The function `buttord(Wp,Ws,Rp,Rs,'s')` computes the lowest order N of a Butterworth analog transfer function meeting the specifications given by the filter parameters, Wp, Ws, Rp, and Rs, where Wp is the passband edge angular frequency in rad/sec, Ws is the stopband edge angular frequency in rad/sec, Rp is the maximum passband attenuation in dB, and Rs is the minimum stopband attenuation in dB. The output data are the filter order N and the 3-dB cutoff angular frequency Wn in rad/sec. This function can also be used to calculate the order of any one of the four basic types of analog Butterworth filters. For lowpass design, Wp < Ws, whereas for highpass design, Wp > Ws. For the other two types, Wp and Ws are two-element vectors specifying the passband and stopband edge frequencies.

Type 1 Chebyshev Filter

The *M*-file functions for the design of analog Type 1 Chebyshev filters are as follows:

```
[z,p,k]    = cheb1ap(N,Rp)
```

---

[5]See Section 3.11.

```
[num,den]  = cheby1(N,Rp,Wn,'s')
[num,den]  = cheby1(N,Rp,Wn,'filtertype','s')
[N,Wn]     = cheb1ord(Wp,Ws,Rp,Rs,'s')
```

The function cheb1ap(N,Rp) computes the zeros, poles, and gain factor of the normalized analog Type 1 Chebyshev lowpass filter transfer function of order N with a passband ripple in dB given by Rp. The normalized passband edge frequency is set to 1. The output files are the column vector p providing the locations of the poles, a null vector z for the zero locations, and the gain factor k. The form of the transfer function is as in Eq. (5.55).

As in the previous case, the numerator and denominator coefficients of the transfer function can be determined using the *M*-file function zp2tf(z,p,k). The rational form of the Type 1 Chebyshev lowpass filter transfer function can also be determined directly using the function cheby1(N,Rp,Wn,'s'), where Wn is the passband edge angular frequency in rad/sec and Rp is the passband ripple in dB. The output data are the vectors, num and den, containing the numerator and the denominator coefficients of the transfer function in descending powers of s. If Wn is a two-element vector [W1,W2] with W1 < W2, the function computes the transfer function of an order 2*N* bandpass filter with passband edge angular frequencies in rad/sec given by W1 and W2. The function cheby1(N,Rp,Wn,'filtertype','s') is employed for the other two types of filter designs, where the filtertype = high for the highpass case and filtertype = stop for the bandstop case. Wn is a scalar representing the passband edge frequency for the highpass filter design, and is a two-element vector defining the stopband edge frequencies for a bandstop filter design.

The function cheb1ord(Wp,Ws,Rp,Rs,'s') determines the lowest order N of a Type 1 Chebyshev analog transfer function meeting the specifications given by the filter parameters, Wp, Ws, Rp, and Rs, where Wp is the passband edge angular frequency, Ws is the stopband angular edge frequency, Rp is the passband ripple in dB, and Rs is the minimum stopband attenuation in dB. The output data are the filter order N and the cutoff angular frequency Wn. This function can also be used to calculate the order of any one of the four basic types of analog Type 1 Chebyshev filters. For the lowpass design, Wp < Ws, whereas for the highpass design, Wp > Ws. For the other two types, Wp and Ws are two-element vectors specifying the passband and stopband edge frequencies. All bandedge frequencies are specified in rad/sec.

Type 2 Chebyshev Filter

The *M*-file functions for the design of analog Type 2 Chebyshev filters are

```
[z,p,k]    = cheb2ap(N,Rs)
[num,den]  = cheby2(N,Rs,Wn,'s')
[num,den]  = cheby2(N,Rs,Wn,'filtertype','s')
[N,Wn]     = cheb2ord(Wp,Ws,Rp,Rs,'s')
```

The function cheb2ap(N,Rs) returns the zeros, poles, and gain factor of a normalized analog Type 2 Chebyshev lowpass filter of order N with a minimum stopband attenuation of Rs in dB. The normalized stopband edge angular frequency is set to 1. The output files are the length-N column vectors z and p, providing the locations the zeros and the poles, respectively, and the gain factor k. If N is odd, z is of length N-1. The form of the transfer function obtained is given by

$$H_a(s) = \frac{P_a(s)}{D_a(s)} = k \frac{(s - z(1))\,(s - z(2))\cdots(s - z(N))}{(s - p(1))\,(s - p(2))\cdots(s - p(N))}. \tag{5.56}$$

The function cheby2(N,Rs,Wn,'s') can be employed to determine the transfer function of a Type 2 Chebyshev lowpass filter when Wn is a scalar defining the stopband edge angular frequency in rad/sec or a bandpass filter when Wn is a two-element vector defining the stopband edge angular frequencies in rad/sec. The function cheby2(N,Rs, Wn,'filtertype','s') provides the transfer function of a Type 2 Chebyshev high-pass when filtertype = high or a bandstop filter when filtertype = stop. In all cases, the specified minimum stopband attenuation is Rs in dB. The output files are the vectors, num and den, containing the numerator and denominator coefficients in descending powers of s.

The function cheb2ord(Wp,Ws,Rp,Rs,'s') determines the lowest order N of a Type 2 Chebyshev analog transfer function meeting the specifications given by the filter parameters, Wp, Ws, Rp, and Rs, as defined for the Type 1 Chebyshev filter.

Elliptic (Cauer) Filter

The *M*-file functions for the design of analog elliptic filters are

```
[z,p,k]     = ellipap(N,Rp,Rs)
[num,den]   = ellip(N,Rp,Rs,Wn,'s')
[num,den]   = ellip(N,Rp,Rs,Wn,'filtertype','s')
[N,Wn]      = ellipord(Wp,Ws,Rp,Rs,'s')
```

The function ellipap(N,Rp,Rs) determines the zeros, poles, and gain factor of a normalized analog elliptic lowpass filter of order N with a passband ripple of Rp dB and a minimum stopband attenuation of Rs dB. The normalized passband edge angular frequency is set to 1. The output files are the length-*N* column vectors z and p, providing the locations of the zeros and the poles, respectively, and the gain factor k. If N is odd, z is of length N-1. The form of the transfer function obtained is as given in Eq. (5.56).

The function ellip(N,Rp,Rs,Wn,'s') returns the transfer function of an elliptic analog lowpass filter when Wn is a scalar defining the passband edge angular frequency in rad/sec or a bandpass filter when Wn is a two-element vector defining the passband edge frequencies in rad/sec. The function ellip(N,Rp,Rs,Wn,'filtertype','s') is used to determine the transfer function of an elliptic highpass when filtertype = high, and Wn is a scalar defining the stopband edge angular frequency in rad/sec, or a bandstop filter when filtertype = stop, and Wn is a two-element vector defining the stopband edge angular frequencies in rad/sec. In all cases, the specified passband ripple is Rp dB and the minimum stopband attenuation is Rs in dB. The output files are the vectors, num and den, containing the numerator and denominator coefficients in descending powers of s.

The function ellipord(Wp,Ws,Rp,Rs,'s') determines the lowest order N of an elliptic analog transfer function meeting the specifications given by the filter parameters, Wp, Ws, Rp, and Rs, as defined for the Type 1 Chebyshev filter.

Bessel Filter

For the design of a Bessel filter , the available M-file functions are

```
[z,p,k]    = besselap(N)
[num,den]  = besself(N,Wn)
[num,den]  = besself(N,Wn,'filtertype')
```

The function `besselap(N)` is employed to compute the zeros, poles, and gain factor of an order-N Bessel lowpass filter prototype. The output files of these functions are the length N column vector p, providing the locations of the poles, and the gain factor k. Since there are no zeros, the output vector z is a null vector. The form of the transfer function is as in Eq. (5.55).

The function `besself(N,Wn)` is used to design an Nth order analog lowpass Bessel filter with a 3-dB cutoff angular frequency given by the scalar Wn in rad/sec. It generates the length-$(N + 1)$ vectors, num and den, containing the numerator and the denominator coefficients in descending powers of s. If Wn is a two-element vector, then it returns the transfer function of an order-$2N$ analog bandpass Bessel filter. For designing the other two types of Bessel filters, the function `besself(N,Wn,'filtertype')` is used. Here the `filtertype = high` with Wn representing the 3-dB stopband edge frequency in rad/sec for the highpass case, or `filtertype = stop` with Wn a two-element vector defining the 3-dB stopband edge frequencies in rad/sec for the bandstop case.

Limitations

The zero-pole-gain form is more accurate than the transfer function form for the design of Butterworth, Type 2 Chebyshev, elliptic, or Bessel filter design. It is recommended that the filter design function in these cases be used only for filter order less than 15 since numerical problems may arise for filter orders equal to or greater than 15.

Design examples

We provide below several examples illustrating the use of some of the above functions in the design of analog filters. In the first three examples, we repeat Examples 5.5 through 5.7 to determine the order of the transfer function using the respective M-file functions. In the remaining examples, we determine the corresponding transfer functions and then compute the frequency response using the M-file function `freqs(num,den,w)`, where num and den are the vectors of the numerator and denominator coefficients in descending powers of s, and w is a set of specified discrete angular frequencies. This function generates a complex vector of frequency response samples from which magnitude and/or phase response samples can be readily computed.

Example 5.8 To determine the order of the analog Butterworth lowpass filter meeting the frequency response specifications given in Example 5.5, we use the command `[N,Wn] = buttord (Wp,Ws,Rp,Rs,'s')` where Wp $= 2\pi(1000)$, Ws $= 2\pi(5000)$, Rp $= 1$, and Rs $= 40$. The outputs generated are given by N $= 4$ and Wn $= 7947.77$, where Wn is the 3-dB cutoff angular frequency. The computed order is same as that determined in Example 5.5 using the formula of Eq. (5.31).

Example 5.9 We next determine the order of analog Type 1 and Type 2 Chebyshev lowpass filter meeting the same specifications as above. To this end, we employ the commands `[N,Wn] = cheb1ord(Wp,Ws,Rp,Rs,'s')` and `[N,Wn] = cheb2ord(Wp,Ws,Rp,Rs,'s')`, respectively. For the Type 1 Chebyshev filter, the computed output data are `N = 3` and `Wn = 6283.18`, and for the Type 2 Chebyshev filter, the computed output data are `N = 3` and `Wn = 23440.97`. The computed `Wn` in the former case, is the 3-dB passband edge angular frequency in rad/sec, whereas in the second case, it is the 40-dB stopband edge angular frequency in rad/sec. The order determined is identical to that derived in Example 5.6 using the formula of Eq. (5.39).

Example 5.10 To determine the order of an analog elliptic lowpass filter meeting the above specifications, we use the command `[N,Wn] = ellipord(Wp,Ws,Rp,Rs,'s')`, resulting in the output data `N = 3` and `Wn = 6283.185` and verifying the order obtained in Example 5.7. Here `Wn` is the 1-dB passband edge angular frequency.

Example 5.11 We consider first the design of a fourth order maximally flat analog lowpass filter with a 3-dB cutoff frequency at $\Omega = 1$. The pertinent MATLAB program is given below.

```
% Program 5_1
% 4-th Order Analog Butterworth Lowpass Filter Design
%
format long
% Determine zeros and poles
[z,p,k] = buttap(4);
disp('Poles are at');disp(p);
% Determine transfer function coefficients
[pz, pp] = zp2tf(z, p, k);
% Print coefficients in ascending powers of s
disp('Numerator polynomial');
disp(pz)
disp('Denominator polynomial');
disp(real(pp))
omega = [0: 0.01: 5];
% Compute and plot frequency response
h = freqs(pz,pp,omega);
gain = 20*log10(abs(h));
plot (omega,gain);grid
xlabel('Normalized frequency');
ylabel('Gain, dB');
```

The output data generated by running this program are as follows:

```
Poles are at
 -0.38268343236509 + 0.92387953251129i
 -0.92387953251129 + 0.38268343236509i
 -0.92387953251129 - 0.38268343236509i
 -0.38268343236509 - 0.92387953251129i
```

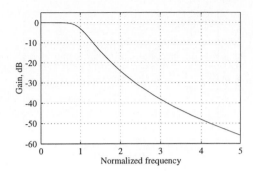

Figure 5.18   The gain response of a normalized fourth order Butterworth lowpass filter.

```
Numerator polynomial
       0      0      0      0      1
Denominator polynomial
   Columns 1 through 4

     1.00000000000000    2.61312592975275    3.41421356237310
     2.61312592975275


   Column 5
    1.00000000000000
```

The plot of the gain response generated by this program is sketched in Figure 5.18.

Example 5.12   We now use MATLAB to complete the design of the Butterworth lowpass filter of Example 5.5. To this end we modify Program 5_1 of Example 5.11 as indicated below in Program 5_2. During execution, the program asks for the order of the filter and the 3-dB cutoff angular frequency (determined in Example 5.8 to be equal to 4 and 7947.77, respectively) to be typed in. It then generates the gain response plot, as shown in Figure 5.19. As can be seen from this plot, the 1-dB passband edge is at 1 kHz, as desired, and at 5 kHz the attenuation is more than 40 dB, as expected.

```
% Program 5_2
% Program to Design Butterworth Lowpass Filter
%
% Type in the filter order and passband edge frequency
N = input('Type in filter order = ');
Fp = input('Passband edge frequency in Hz = ');
% Determine the transfer function
[num,den] = butter(N,Fp,'s');
%  Compute and plot the frequency response
omega = [0: 200: 12000*pi];
h = freqs(num,den,omega);
gain = 20*log10(abs(h));
plot (omega/(2*pi),gain);
```

Figure 5.19   The gain response of the Butterworth lowpass filter meeting the specifications of Example 5.5 designed using Program 5_2.

```
xlabel('Frequency in Hz');
ylabel('Gain, dB');
```

Example 5.13 We next illustrate the design of a Type 1 Chebyshev lowpass filter meeting the specifications of Example 5.5. The corresponding MATLAB program is given below in Program 5_3. As the program is run, it asks for the order of the filter, the passband edge angular frequency (determined in Example 5.9 to be equal to 3 and 6283.18, respectively), and the passband ripple (1 dB) to be typed in. It then generates the gain response plot as shown in Figure 5.20.

```
% Program 5_3
% Program to Design Type 1 Chebyshev Lowpass Filter
%
% Read in the filter order, passband edge frequency
% and passband ripple in dB
N = input('Order = ');
Fp = input('Passband edge frequency in Hz = ');
Rp = input('Passband ripple in dB = ');
% Determine the coefficients of the transfer function
[num,den] = cheby1(N,Rp,Fp,'s');
% Compute and plot the frequency response
omega = [0: 200: 12000*pi];
h = freqs(num,den,omega);
gain = 20*log10(abs(h));
plot (omega/(2*pi),gain);
xlabel('Frequency, Hz');
ylabel('Gain, dB');
```

Example 5.14 The design of an elliptic lowpass filter meeting the specifications of Example 5.5 follows a similar format. The pertinent MATLAB program is given below in Program 5_4. As the program is run, it asks for the order of the filter (determined in Example 5.10 to be equal to 3), the passband edge angular frequency ($2000\pi = 6283.185$), the passband ripple (1 dB), and the minimum stopband attenuation (40 dB) to be typed in. It then generates the gain response plot as shown in Figure 5.21.

Figure 5.20 The gain response of the Type 1 Chebyshev lowpass filter meeting the specifications of Example 5.5 designed using Program 5_3.

```
% Program 5_4
% Program to Design Elliptic Lowpass Filter
%
% Read in the filter order, passband edge frequency,
% passband ripple in dB and minimum stopband
% attenuation in dB
N = input('Order = ');
Fp = input('Passband edge frequency in Hz = ');
Rp = input('Passband ripple in dB = ');
Rs = input('Minimum stopband attenuation in dB = ');
% Determine the coefficients of the transfer function
[num,den] = ellip(N,Rp,Rs,Fp,'s');
% Compute and plot the frequency response
omega = [0: 200: 12000*pi];
h = freqs(num,den,omega);
gain = 20*log10(abs(h));
plot (omega/(2*pi),gain);
xlabel('Frequency in Hz');
ylabel('Gain, dB');
```

5.3.7 A Comparison of the Filter Types

In the previous four sections we have described four types of analog lowpass filter approximations, three of which have been developed primarily to meet the magnitude response specifications while the fourth has been developed primarily to provide a linear-phase approximation. In order to determine which filter type to choose to meet a given magnitude response specification, we need to compare the performances of the four types of approximations. To this end, we compare here the frequency responses of the normalized Butterworth, Chebyshev, and elliptic analog lowpass filters of same order. The passband ripple of the Type 1 Chebyshev and the equiripple filters are assumed to be the same, while the minimum stopband attenuation of the Type 2 Chebyshev and the equiripple filters are

Figure 5.21   The gain response of the elliptic lowpass filter meeting the specifications of Example 5.5 designed using Program 5_4.

assumed to be the same. The filter specifications used for comparison are as follows: filter order of 6, passband edge at $\Omega = 1$, maximum passband deviation of 1 dB, and minimum stopband attenuation of 40 dB. The frequency responses computed using MATLAB are plotted in Figure 5.22.

As can be seen from Figure 5.22, the Butterworth filter has the widest transition band, with a monotonically decreasing gain response. Both types of Chebyshev filters have a transition band of equal width that is smaller than that of the Butterworth filter but greater than that of the elliptic filter. The Type 1 Chebyshev filter provides a slightly faster roll-off in the transition band than the Type 2 Chebyshev filter. The magnitude response of the Type 2 Chebyshev filter in the passband is nearly identical to that of the Butterworth filter. The elliptic filter has the narrowest transition band, with an equiripple passband and an equiripple stopband response.

The Butterworth and Chebyshev filters have a nearly linear-phase response over about three-fourths of the passband, whereas the elliptic filter has a nearly linear-phase response over about one-half of the passband. One the other hand, the Bessel filter may be more attractive if the linearity of the phase response over a larger portion of the passband is desired at the expense of a poorer gain response. Figure 5.23 shows the gain and phase responses of a sixth order Bessel filter frequency scaled to have a passband edge at $\Omega = 1$ with a maximum passband deviation of 1 dB. However, the Bessel filter provides a minimum of 40 dB attenuation at approximately $\Omega = 9.4$ and as a result, has the largest transition band compared to the other three types.

Another way of comparing the performances of the Butterworth, Chebyshev, and elliptic filters would be to compare the order of these filters required to meet the filter specifications. For example, to meet these specifications: passband edge at $\Omega = 1$, maximum passband deviation of 1 dB, stopband edge at $\Omega = 1.2$, and minimum stopband attenuation of 40 dB, requires a Butterworth filter of order 29, a Chebyshev Type 1 or 2 filter of order 10, and an elliptic filter of order 6.

5.4   Anti-Aliasing Filter Design

According to the sampling theorem of Section 5.2.1, a bandlimited continuous-time signal $g_a(t)$ can be fully recovered from its uniformly sampled version if the condition of Eq. (5.10) is satisfied, i.e., if $g_a(t)$ is sampled at a sampling frequency Ω_T that is at least

Figure 5.22   A comparison of the frequency responses of the four types of analog lowpass filters: (a) gain responses, (b) passband details, and (c) phase responses.

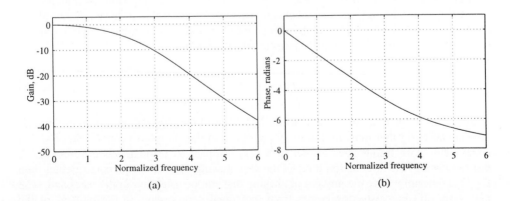

Figure 5.23   The frequency responses of a sixth-order analog Bessel filter: (a) gain response, and (b) unwrapped phase response.

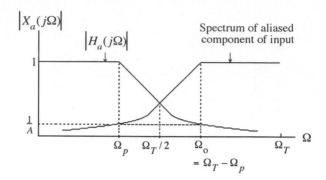

Figure 5.24   Anti-aliasing filter magnitude response and its effect in the signal band of interest.

twice the highest frequency Ω_m contained in $g_a(t)$. If this condition is not satisfied, the original continuous-time signal $g_a(t)$ cannot be recovered from its sampled version because of distortion caused by aliasing. In practice, $g_a(t)$ is passed through an analog anti-aliasing lowpass filter prior to sampling to enforce the condition of Eq. (5.10). This analog filter is the first circuit in the interface between the continuous-time and the discrete-time domains and is studied in this section.

Ideally, the anti-aliasing filter $H_a(s)$ should have a lowpass frequency response $H_a(j\Omega)$ given by

$$H_a(j\Omega) = \begin{cases} 1, & |\Omega| < \Omega_T/2, \\ 0, & |\Omega| \geq \Omega_T/2. \end{cases} \tag{5.57}$$

Such a "brickwall" type frequency response cannot be realized using practical analog circuit components and, hence, must be approximated. A practical anti-aliasing filter therefore should have a magnitude response approximating unity in the passband with an acceptable tolerance, a stopband magnitude response exceeding a minimum attenuation level, and an acceptable transition band separating the passband and the stopband, as indicated in Figure 5.12. In addition, in many applications, it is also desirable to have a linear-phase response in the passband. The passband edge frequency Ω_p, the stopband edge frequency Ω_s, and the sampling frequency Ω_T must satisfy the relation

$$\Omega_p < \Omega_s \leq \frac{\Omega_T}{2}. \tag{5.58}$$

The passband edge frequency Ω_p is determined by the highest frequency in the continuous-time signal $g_a(t)$ that must be faithfully preserved in the sampled version. Since signal components with frequencies greater than $\Omega_T/2$ appear as frequencies less than $\Omega_T/2$ due to aliasing, the attenuation level of the anti-aliasing filter at frequencies greater than $\Omega_T/2$ is determined by the amount of aliasing that can be tolerated in the passband. The maximum aliasing distortion comes from the signal components in the replicas of the input spectrum adjacent to the baseband. It follows from Figure 5.24, that the frequency $\Omega_0 = \Omega_T - \Omega_p$ is aliased into the frequency Ω_p, and if the acceptable amount of aliased spectrum at Ω_p is $a_p = -20\log_{10}(1/A)$ then the minimum attenuation of the anti-aliasing filter at Ω_0 must also be a_p [Jac96].

Table 5.1   Approximate minimum stopband attenuation of a Butterworth lowpass filter.

Ω_0	$2\Omega_p$	$3\Omega_p$	$4\Omega_p$
Attenuation (dB)	$6.02N$	$9.54N$	$12.04N$
Ω_T	$3\Omega_p$	$4\Omega_p$	$5\Omega_p$

Example 5.15 Consider the design of an anti-aliasing filter with a Butterworth lowpass magnitude response. From Eq. (5.29) the difference in dB in the attenuation levels at Ω_p and Ω_0 is given by

$$10\log_{10}\left[\frac{1+(\Omega_0/\Omega_c)^{2N}}{1+(\Omega_p/\Omega_c)^{2N}}\right] \cong 10\log_{10}\left(\frac{\Omega_0}{\Omega_p}\right)^{2N} \qquad (5.59)$$

This difference has been tabulated in Table 5.1 for several values of the ratio Ω_0/Ω_p as a function of the filter order N.

This table can be used to estimate the minimum sampling frequency Ω_T as a function of the passband edge Ω_p for a specified filter order and a specified minimum stopband attenuation level at a frequency $\Omega_0 = \Omega_T - \Omega_p$. For example, if the minimum stopband attenuation at $\Omega_T - \Omega_p$ is 60 dB, then from Table 5.1 we observe that for a sampling frequency $\Omega_T = 3\Omega_p$, the filter order should be $N = \lceil 60/6.02\rceil = \lceil 9.967\rceil = 10$.[6] If we increase the sampling frequency to $4\Omega_p$, then the filter order reduces to $N = \lceil 60/9.54\rceil = \lceil 6.29\rceil = 7$. For a still higher sampling frequency of $5\Omega_p$, the filter order becomes $N = \lceil 60/12.04\rceil = \lceil 4.98\rceil = 5$.

In practice, the sampling frequency chosen depends on the specific application. In applications requiring minimal aliasing, the sampling rate is typically chosen to be 3 to 4 times the passband edge Ω_p of the anti-aliasing analog filter. In noncritical applications, a sampling rate of twice the passband edge Ω_p of the anti-aliasing analog filter is more than adequate. For example, in pulse-code modulation (PCM) telephone systems, the voice signal is first bandlimited to 4 kHz by an anti-aliasing analog filter with a passband edge at 3.6 kHz and a stopband edge at 4 kHz. These specifications are typically met by a third-order elliptic lowpass filter. The output of the filter is then sampled at 8 kHz.

Requirements for the analog anti-aliasing filter can be relaxed by oversampling the analog signal and then decimating the high-sampling-rate digital signal to the desired low-rate digital signal. The decimation process can be implemented completely in the digital domain by first passing the high-rate digital signal through a digital anti-aliasing filter and then down-sampling its output. To understand the advantages of the oversampling approach, consider the sampling of an analog signal bandlimited to a frequency Ω_m. Figure 5.25 shows the spectra of the sampled version of this signal sampled at two different rates, Ω_T and $\Omega_T' = 2\Omega_T$, where Ω_T is slightly higher than the Nyquist rate of $2\Omega_m$. These figures also show the desired frequency response of the analog anti-aliasing filter in both cases. Note that the transition band of the analog anti-aliasing filter in the latter case is

---

[6] $\lceil x\rceil$ denotes the smallest integer greater than or equal to x.

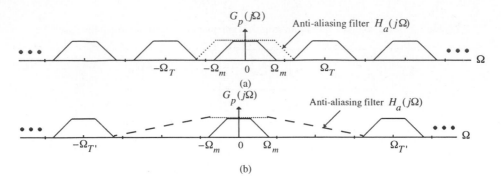

Figure 5.25   Analog anti-aliasing filter requirements for two different oversampling rates.

considerably more than 3 times that needed in the former situation. As a result, the filter specifications are met more easily with a much lower order analog filter.

For the design of the anti-aliasing filter, any one of the four approximations described in Section 5.3 can be employed. Of the four types, the Butterworth approximation provides a reasonable compromise between the desired magnitude response and a linear-phase response in the passband for a given filter order. For an improved phase response at the expense of a poorer magnitude response, the Bessel approximation can be used. On the other hand, for an improved magnitude response with a poorer phase response, either the Chebyshev or the elliptic approximation can be used, with the latter providing the smallest aliasing error for a given filter order.

Once the transfer function $H_a(s)$ for the anti-aliasing filter meeting the requirements has been determined, it can be implemented in a number of ways, such as a passive RLC filter, an active-RC filter, or a switched-capacitor filter [Lar93]. A detailed discussion of these implementations is beyond the scope of this book, and we refer the reader to the texts listed at the end of this book [Dar76], [Tem77], [Tem73].

5.5   Sample-and-Hold Circuit[7]

As indicated earlier, the bandlimited output of the analog anti-aliasing filter is fed into a sample-and-hold (S/H) circuit, which is the second circuit in the interface between the continuous-time and discrete-time domains. It samples the analog signal at uniform intervals and holds the sampled value after each sampling operation for sufficient time for accurate conversion by the A/D converter.

The basic S/H circuit, shown in Figure 5.26, operates as follows: During the sampling phase the periodically operated analog switch S remains closed allowing the capacitor C to track the input analog signal $x_a(t)$ and to charge up to a voltage equal to that of the input. During the hold period, the switch remains open, permitting the charged capacitor to hold the voltage across it until the next sampling phase begins. The operation of the switch is controlled by a digital clock signal. The voltage follower at the output of the S/H circuit acts as a buffer between the capacitor C and the input stage of the A/D converter.

---

[7]This section has been adapted from [Mit80] by permission of the author and the publisher.

Figure 5.26   The basic sample-and-hold circuit.

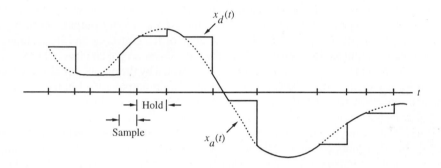

Figure 5.27   Input/output waveforms of a sample-and-hold circuit.

Typical input-output waveforms of an S/H circuit are as shown in Figure 5.27, where the dotted line represents the input continuous-time signal $x_a(t)$ and the solid line represents the output $x_d(t)$, assuming instantaneous change from the sample mode, to the hold mode and vice-versa.

Practical S/H circuits are often much more complex than the basic circuit of Figure 5.26. They typically include an additional operational amplifier at the input to provide better isolation between the source and the capacitor and better tracking of the input signal. They may contain additional circuit components to minimize the effect of the hold voltage decay, which otherwise may occur due to the leakage through the input resistance of the output buffer amplifier and through the finite OFF resistance of the switch. The major parameters characterizing the performance of a practical S/H circuit are acquisition time, aperture time, and droop. The total time needed to switch from the hold mode to the sample mode and acquire the input analog signal within a specified accuracy is defined as its *acquisition time*. This parameter depends on the switching delay time, the time constant of the RC circuit, and the dynamic performance of the output operational amplifier determined primarily by its slew time and settling time. The time taken by the switch to change from the sample mode to the hold mode is defined as the *aperture time*. The hold voltage drift per second due to leakage out of the holding capacitor is called the *droop*. In addition to these parameters, nonideal properties of the operational amplifiers used in the design of the S/H circuit should also be taken into account in determining the overall performance of the circuit.

Figure 5.28   Block diagram representation of an analog comparator.

5.6 Analog-to-Digital Converter

The next step in the digital processing of an analog signal is the conversion of the output
of the S/H circuit in its hold mode to a digital form by means of an analog-to-digital (A/D)
converter. For digital signal processing applications, the output of the A/D converter is
usually in binary code. The output is a sequence of *words*, with each word representing
a sample of the sequence. The *wordlength* of the A/D converter output, given by the
number of bits, limits the achievable dynamic range of the converter and its accuracy in
representing the input analog signal. The accuracy of conversion of an ideal A/D converter
is expressed in terms of its *resolution*, which is determined by the number of discrete levels
that can be assumed by the A/D converter output. For an output coded in natural binary
form with an N-bit wordlength, the number of available discrete levels is 2^N, and as a
result, the resolution or accuracy is 1 part in 2^N or $100/2^N$ percent.

There is a variety of A/D converters that are used in signal processing applications
[Lar93]. In all of these converters, the analog comparator is an important circuit component. The analog comparator is a device that compares two analog voltages at its input and
develops a binary output indicating which input voltage level is larger. With respect to its
circuit symbol shown in Figure 5.28, the input/output relation of an analog comparator is
thus as follows:

$$V_0 = \begin{cases} V^+ & \text{if } V_1 > V_2, \\ V^- & \text{if } V_1 < V_2, \end{cases} \tag{5.60}$$

where $V^+ > V^-$. Usually these circuits are designed such that their output voltage levels
are compatible with the logic levels of conventional digital circuits.

We briefly review next the operations of the following types of A/D converters: (1)
flash A/D converter, (2) serial-parallel A/D converter, (3) successive-approximation A/D
converter, (4) counting A/D converter, and (5) oversampling A/D converter. A detailed
discussion of their design and implementation is beyond the scope of this book.

5.6.1 Flash A/D Converters

In an N-bit flash converter, shown in block diagram form in Figure 5.29, the input analog
voltage V_A is compared simultaneously with a set of $2^N - 1$ uniformly separated reference
voltage levels by means of a set of $2^N - 1$ analog comparators, and the location of the
adjacent comparators for which the outputs changes from V^- to V^+ indicates the range
of the input voltage. A logic encoder circuit is then used to convert the location information into an N-bit binary code. The set of reference voltages is usually derived by a
potential-divider resistor string. In a flash A/D converter, all output bits are developed simultaneously, and as a result, it is the fastest converter with a conversion time given by the
comparator switching time plus the propagation delay of the encoder circuit. However, the
hardware requirements of this type of converter increase very rapidly (exponentially) with

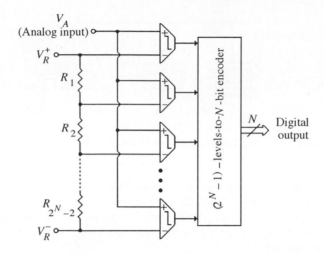

Figure 5.29   Block diagram representation of a flash unipolar A/D converter.

Figure 5.30   Block diagram representation of a subranging A/D converter.

an increase in the resolution. As a result, flash converters are employed for low-resolution (typically 8-bit or less) and high-speed conversion applications.

5.6.2   Serial-Parallel A/D Converter

Two $N/2$-bit flash converters in a serial-parallel configuration can be employed to reduce the hardware complexity of an N-bit flash converter at a slight increase in the conversion time [Lar93]. One such scheme, called a *subranging A/D converter*, is shown in Figure 5.30. Here, a coarse approximation of the input analog voltage V_A composed of the $N/2$ most significant bits (MSBs) is first generated by means of one of the $N/2$-bit flash converters. These MSBs are then fed into a D/A converter whose analog output is subtracted from the V_A. The difference voltage is scaled by an amplifier of gain $2^{N/2}$ and converted into digital form by the second (fine) $N/2$-bit flash converter, which provides the least significant bits (LSBs).

Figure 5.31   Block diagram representation of a ripple A/D converter.

The second scheme utilizing also two $N/2$-bit flash converters is called the *ripple A/D converter*, as shown in Figure 5.31. In this scheme, the coarse $N/2$-bit A/D converter has two functions. It generates the $N/2$ MSBs, and it controls a reference voltage generator that develops a reference voltage V'_R for the fine $N/2$-bit A/D converter generating the $N/2$ least significant bits (LSBs).

In both of the two serial-parallel A/D converters, while one of the $N/2$-bit A/D converters is operating, the other is idle, permitting the use of a single $N/2$-bit A/D converter twice in one conversion period. A generalization of the above two schemes employing N 1-bit A/D converters is called the *pipelined A/D converter* [Lar93].

5.6.3   Successive Approximation A/D Converter

In this type of converter, essentially a trial-and-error approach is successively used to obtain the digital word representing the input analog voltage V_A [Mit80]. The basic idea behind the operation of this converter can be explained with the aid of its block diagram representation given in Figure 5.32. The conversion procedure is an iterative process. At the kth step of the iteration, the digital approximation stored in the shift register is converted into an analog voltage V_D by the D/A converter in the system. If $V_D < V_A$, then the digital number is increased by setting to ONE the $(k+1)$th bit to the right of the kth bit, which is assumed to be a ONE. If, on the other hand, $V_D > V_A$, then the digital number is decreased by setting the kth bit to a ZERO, and the $(k+1)$th bit to a ONE. The above process is followed for all k from $k = 1, 2, \ldots, N$. After the Nth bit has been examined, the conversion process is terminated, with the contents of the shift register representing the digital equivalent of the input analog voltage. In practice, to round off the approximation to the nearest discrete level, an analog equivalent to one-half of the value of the LSB is added to the analog input signal before conversion begins.

The successive-approximation A/D converter can be designed with high resolution and reasonably high speed at a moderate cost and is therefore widely used in digital signal processing applications. The performance of this type of A/D converter depends primarily on the performance of its constituent D/A converter and the analog comparator.

5.6.4   Counting A/D Converter

In this type of converter, shown in block diagram form in Figure 5.33, an N-bit counter begins counting clock pulses when the conversion process is started [Mit80]. The analog equivalent V_D of the digital word of the counter formed by means of a D/A converter,

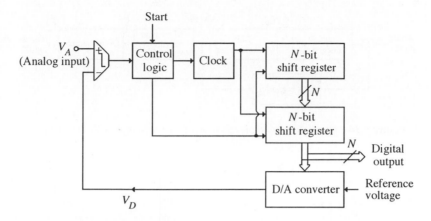

Figure 5.32 Block diagram representation of a successive-approximation A/D converter. (Adapted from [Mit80] by permission of the author and the publisher.)

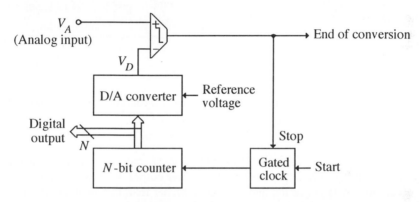

Figure 5.33 Block diagram representation of a counting A/D converter. (Adapted from [Mit80] by permission of the author and publisher.)

at each clock cycle, is compared with the input analog voltage level V_A. The conversion process is terminated as soon as $V_D > V_A$. Here, the conversion time depends on the value of V_A. It is maximum when the digital word in the counter becomes all ONEs, and thus, for an N-bit binary counter, the conversion time is equal to $(2^N - 1)T$, where T is the clock period. This is a long time for $N > 10$, and hence, this technique can be used only in low speed applications.

5.6.5 Oversampling A/D Converter

As the name implies, in this type of converter the analog signal is sampled at a rate much higher than the Nyquist rate, resulting in very closed spaced samples. As a consequence, the difference between the amplitudes of two consecutive samples is very small, permitting it to be represented in digital form using very few bits, usually by one bit. The sampling rate is then decreased by passing the digital signal first through an anti-aliasing lowpass

Figure 5.34   Block diagram representation of an oversampling A/D converter.

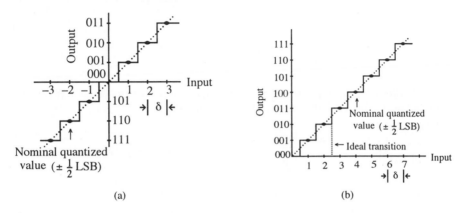

Figure 5.35   Input-output characteristics of an ideal 3-bit A/D converter: (a) bipolar converter, and (b) unipolar converter.

Mth band digital filter to reduce its bandwidth to π/M and then through a factor-of-M down-sampler. The wordlength of the down-sampler output determines the resolution of the oversampling A/D converter, and it is much higher than that of the high rate digital signal due to the effect of digital filtering. The basic block diagram representation of the oversampling A/D converter is shown in Figure 5.34.[8] This type of converter, often called a *sigma-delta A/D converter*, is discussed in more detail in Section 11.12.

5.6.6   Characteristics of a Practical A/D Converter [9]

A practical A/D converter is a nonideal device and exhibits a variety of errors that are usually determined in terms of the input analog values at which the transition in the digital output takes place, since these transitions can be measured accurately. In order to understand the effects of these errors on the performance of an A/D converter, consider first its operation as an ideal device. The input-output relation of an ideal 3-bit A/D converter is shown in Figure 5.35. The error introduced by an ideal A/D converter is simply the difference between the value of the analog input and the analog equivalent of the digital representation; this difference is called the *quantization error*. It follows from Figure 5.35

---

[8]The down-sampler shown in Figure 5.34 is employed to reduce the sampling rate and is reviewed in Section 10.1.

[9]This section has been adapted from [Mit80] by permission of the author and the publisher.

Figure 5.36 Linearity errors in a practical A/D converter.

that the quantization error $e[n]$ for an ideal converter satisfies

$$-\frac{\delta}{2} < e[n] \le \frac{\delta}{2}, \tag{5.61}$$

where δ is called the quantization step and is given by

$$\delta = 2^{-N} \tag{5.62}$$

for an N-bit wordlength. Note that δ is precisely the value of the least-significant bit (LSB).

A practical A/D converter exhibits *linearity error* if the difference between two consecutive transition values of the input, is not equal for the complete range of the input as illustrated in Figure 5.36. The maximum value of this difference value over the full range is called the *differential nonlinearity (DNL) error*. Note from Figure 5.36(b) that in some cases severe nonlinearity in the input-output relation may result in missing codes at the output.

The A/D converter exhibits *gain* or *scale-factor error* if the difference between the last and first transitions is not equal to the full scale value minus 1/2 LSB. An *offset error* occurs if all transitions are shifted by an equal amount from the ideal transition locations. These errors are illustrated in Figure 5.37.

The time needed by the A/D converter to generate the full digital equivalent of the input analog signal is called its *word-conversion time*. In the flash A/D converter, the word-conversion time is essentially equal to the *bit-conversion time*. On the other hand, in the successive approximation A/D converter, the word-conversion time is equal to the bit-conversion time multiplied by the wordlength.

It should be noted that an *overflow error* occurs if the input analog voltage V_A exceeds the dynamic range of the A/D converter. It is therefore important to ensure that V_A is properly scaled before it is fed into the A/D converter.

5.7 Digital-to-Analog Converter

The final step in the digital processing of analog signals is the conversion of the digital filter output into an analog form, which is accomplished by a digital-to-analog (D/A)

Figure 5.37   (a) Offset and (b) gain errors in a practical ideal A/D converter.

Figure 5.38   Block diagram representation of an N-bit D/A converter.

converter followed by a reconstruction filter. The basic idea behind most commonly used D/A converter can be explained by means of the simplified block diagram representation shown in Figure 5.38, where we have assumed, without any loss of generality, that the digital sample is positive and represented in a natural binary code. Here the ℓth switch S_ℓ is in its ON-position if the ℓth binary bit $a_\ell = 1$, and it is in the OFF-position if $a_\ell = 0$. The output V_0 of the D/A converter is then given by

$$V_0 = \sum_{\ell=1}^{N} 2^{\ell-1} a_\ell V_R. \tag{5.63}$$

There are a variety of D/A converters that are used in signal processing applications [Lar93]. We discuss below only the following types: (1) weighted-resistor D/A converter, (2) resistor-ladder D/A converter, and (3) oversampling D/A converter.

5.7.1   Weighted-Resistor D/A Converter

The schematic of an N-bit weighted-resistor D/A converter is shown in Figure 5.39. Here the operation of the switches are as in Figure 5.38. It can be shown that the output V_0 of

Figure 5.39 Schematic representation of an N-bit weighted resistor unipolar D/A converter.

the D/A converter is given simply by

$$V_0 = \sum_{\ell=1}^{N} 2^{\ell-1} a_\ell \left(\frac{R_L}{(2^N - 1)R_L + 1} \right) V_R. \tag{5.64}$$

The full-scale output voltage $V_{0,FS}$ is obtained when all a_ℓ's are ONEs. Then from Eq. (5.64),

$$V_{0,FS} = \left(\frac{(2^N - 1)R_L}{(2^N - 1)R_L + 1} \right) V_R.$$

In practice, usually $(2^N - 1)R_L \gg 1$ and, as a result, $V_{0,FS} \cong V_R$.

Usually a buffer amplifier is placed at the output to provide gain and prevent loading. For a D/A converter with a moderate to high resolution, the spread of the resistor values becomes very large, making this type of converter unsuitable for many applications.

Based on the same principle as discussed above, a weighted-capacitor D/A converter can be designed. Such circuits are more popular in IC design.

5.7.2 Resistor-Ladder D/A Converter

This type of converter is probably the most widely used in practice. From its schematic representation shown in Figure 5.40, it can be shown that the D/A converter output V_0 is given by

$$V_0 = \sum_{\ell=1}^{N} 2^{\ell-1} a_\ell \left(\frac{R_L}{2R_L + R} \right) \frac{V_R}{2^{N-1}}. \tag{5.65}$$

Because of the resistor values used and the ladder-like circuit connection, the structure is often referred to as the $R-2R$ ladder D/A converter. In practice, often $2R_L \gg R$ and, hence, the full-scale output voltage $V_{0,FS}$ is given by

$$V_{0,FS} \cong \left(\frac{2^N - 1}{2^N} \right) V_R.$$

As in the previous case, a buffer amplifier is also placed at the output to provide gain and prevent loading.

Figure 5.40   Schematic representation of an N-bit resistor-ladder (R−2R ladder) unipolar D/A converter.

Figure 5.41   Block diagram representation of an oversampling sigma-delta D/A converter.

5.7.3   Oversampling Sigma-Delta D/A Converter

The basic scheme employed in an oversampling sigma-delta D/A converter is shown in Figure 5.41. Here, the sampling rate of the input b-bit digital signal is first increased by a factor-of-L interpolator implemented by an up-sampler followed by an Lth band digital lowpass filter.[10] The output of the interpolator is fed into a digital sigma-delta quantizer, which creates a single bit output by extracting only the MSB. The MSB is then converted to analog form by a 1-bit D/A converter followed by an analog lowpass filter. The LSBs form the error signal that is subtracted from the interpolator output in the summer of the sigma-delta quantizer. A detailed analysis of the operation of the oversampling sigma-delta D/A converter is provided in Section 11.13.

5.7.4   Characteristics of a Practical D/A Converter[11]

A practical D/A converter is characterized by a number of parameters. The effects of these parameters on the performance of a D/A converter are best understood by first examining the input-output relation of an idealized device. Figure 5.42(a) shows the input-output relation of a 3-bit unipolar D/A converter. Here the analog outputs for each possible digital inputs are shown as vertical "bars."

The *resolution* of a D/A converter is defined in a manner identical to that of an A/D converter. For an N-bit D/A converter with the input digital word coded in natural binary form, the resolution is therefore 1 part in 2^{N-1}.

---

[10]The design of a factor-of-L interpolator is treated in Section 10.3.

[11]This section has been adapted from [Mit80] by permission of the author and publisher.

Figure 5.42 Input-output relation of a unipolar D/A converter: (a) ideal converter, and (b) linearity errors in a practical D/A converter.

In an ideal D/A converter, the analog outputs as a function of the input discrete levels will be on a straight line going through the origin, as shown in Figure 5.42(a), with the difference between the outputs for two consecutive input digital signals being 1 LSB. In a practical D/A converter, the actual outputs evaluated for each possible input digital signal may be unevenly distributed instead of being on a straight line as indicated in Figure 5.42(b). The *integral linearity (INL) error* is defined as the maximum deviation from the straight line. A measure of the variation in the difference of the analog outputs corresponding to two consecutive digital inputs is called the *differential nonlinearity*. If the analog outputs are on a straight line that is uniformly shifted by an equal amount for each input discrete value, as shown in Figure 5.43(a), the D/A converter is said to have an *offset error*. If the difference between the actual analog outputs and their corresponding ideal analog outputs increase linearly for increase in the digital inputs, as indicated in Figure 5.43(b), the D/A converter is said to exhibit a *gain error*.

The *accuracy* of a practical D/A converter is defined by the maximum deviation of its measured output from that of an ideal D/A converter. The *word-conversion time* of a D/A converter is given by the time taken to decode a digital word.

The finite turn-on and turn-off times of analog switches in the D/A converter, in general, are not equal, and as a result, they give rise to a dynamic error called a *glitch*. Consider the situation when at time $t = t_n$, the N-bit word is given as the MSB being ONE and the remaining bits as ZEROs. Assume that at time $t = t_{n+1}$, the MSB changes to a ZERO with all other bits becoming ONEs. If the turn-on time of the analog switch is greater than its turn-off time, then the Nth switch is turned off first and momentarily all switches in the D/A converter are in their OFF positions, resulting in a false analog output equal to zero before the switches settle to their correct positions. The temporary state of all switches in the off positions causes a narrow pulse or spike of half the height of full scale to appear at the converter output. These pulses appearing during the transition periods are called *glitches*. If such glitches are undesirable, they can be eliminated by placing an S/H circuit at the output of the D/A converter, which holds the previous D/A converter output

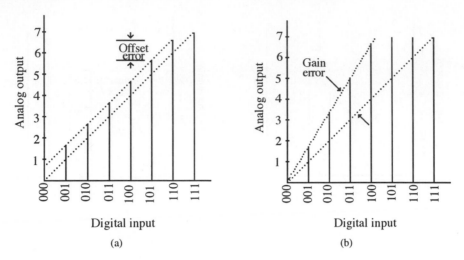

Figure 5.43   Input-output relation of a practical D/A converter: (a) offset error, and (b) gain error.

until the glitches disappear and then acquires and holds the new output.

5.8   Reconstruction Filter Design

The output of the D/A converter is finally passed through an analog reconstruction or smoothing filter to eliminate all the replicas of the spectrum outside the base band. As indicated earlier in Section 5.2.2, this filter ideally should have a frequency response such as given by Eq. (5.16). If the cutoff frequency Ω_c of the reconstruction filter is chosen as $\Omega_T/2$, where Ω_T is the sampling angular frequency, the corresponding frequency response is given by

$$H_r(j\Omega) = \begin{cases} T, & |\Omega| \leq \Omega_T/2, \\ 0, & |\Omega| > \Omega_T/2, \end{cases} \tag{5.66}$$

If we denote the input to the D/A converter as $y[n]$, then from Eq. (5.20) the reconstructed analog equivalent $y_a(t)$ is given by

$$y_a(t) = \sum_{n=-\infty}^{\infty} y[n] \frac{\sin[\pi(t-nT)T]}{\pi(t-nT)/T}. \tag{5.67}$$

Since the ideal reconstruction filter of Eq. (5.66) has a doubly infinite impulse response, it is noncausal and thus unrealizable. In practice, it is necessary to use filters that approximate the ideal lowpass frequency response.

   Almost always, a practical D/A converter unit contains a zero-order hold circuit at its output producing a staircase-like analog waveform $y_z(t)$, as shown in Figure 5.44(b). It is therefore important to analyze the effect of the zero-order hold circuit in order to determine the specifications for the smoothing lowpass filter that should follow the overall D/A converter structure.

   The zero-order hold operation can be modeled by an ideal impulse-train D/A output $y_p(t)$ followed by a linear, time-invariant analog circuit with an impulse response $h_z(t)$

(a)   (b)

Figure 5.44 Typical output waveforms: (a) ideal D/A converter, and (b) a practical D/A converter.

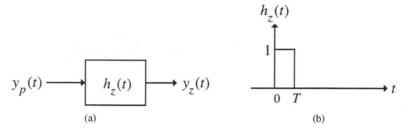

(a)   (b)

Figure 5.45 (a) Modeling of the zero-order hold operation and (b) impulse response of the zero-order hold circuit.

that is a rectangular pulse of width T and unity height, as indicated in Figure 5.45. It follows from this figure that if $Y_p(j\Omega)$ denotes the continuous-time Fourier transform of $y_p(t)$, the output of the ideal D/A converter, then the continuous-time Fourier transform $Y_z(j\Omega)$ of $y_z(t)$, the output of the zero-order hold circuit, is simply given by

$$Y_z(j\Omega) = H_z(j\Omega)Y_p(j\Omega), \tag{5.68}$$

where

$$H_z(j\Omega) = \int_0^T e^{-j\Omega t}\, dt = -\left.\frac{e^{-j\Omega t}}{j\Omega}\right|_0^T = \frac{1 - e^{-j\Omega T}}{j\Omega} = e^{-j\frac{\Omega T}{2}}\left[\frac{\sin(\Omega T/2)}{\Omega/2}\right]. \tag{5.69}$$

The magnitude response of the zero-order hold circuit, as indicated in Figure 5.46(a), has a lowpass characteristic with zeros at $\pm\Omega_T$, $\pm 2\Omega_T$, ..., where $\Omega_T = 1/T$ is the sampling angular frequency. Figure 5.46(b) shows the magnitude response $|Y_p(j\Omega)|$, which is a periodic function of Ω with a period Ω_T. Since the frequency response $Y_z(j\Omega)$ of the analog output $y_z(t)$ of the zero-order hold circuit is a product of $Y_p(j\Omega)$ and $H_z(j\Omega)$, the zero-order hold circuit somewhat attenuates the unwanted replicas centered at multiples of the sampling frequency Ω_T, as sketched in Figure 5.46(c). An *analog reconstruction filter*, also called a *smoothing filter*, $H_r(j\Omega)$ thus follows a practical D/A converter unit and is designed to further attenuate the residual portions of the signal spectrum centered at multiples of the sampling frequency Ω_T. Moreover, it should also compensate for the amplitude distortion, more commonly called *droop*, caused by the zero-order hold circuit in the band from dc to $\Omega_T/2$.

The general specifications for the analog reconstruction filter $H_r(j\Omega)$ can be easily determined if the effect of the droop is neglected. If Ω_c denotes the highest frequency of

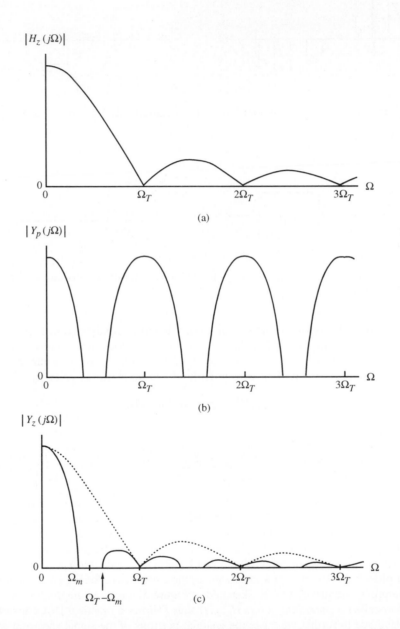

Figure 5.46   Magnitude responses of (a) the zero-order hold circuit, (b) the output of the ideal D/A converter, and (c) output of the practical D/A converter.

the signal $y_p(t)$ that should be preserved at the output of the reconstruction filter, then the lowest frequency component present in the residual images in the output of the zero-order hold circuit is of frequency $\Omega_0 = \Omega_T - \Omega_c$. The S/H circuit has a gain at frequency Ω_0 given by

$$20 \log_{10} |H_z(j\Omega_0)| = 20 \log_{10} \left[\frac{\sin(\Omega_0 T/2)}{\Omega_0/2} \right]. \tag{5.70}$$

Therefore, if the system specification calls for a minimum attenuation of A_s dB of all frequency components in the residual images, then the reconstruction filter should provide at least an attenuation of $A_s + 20 \log_{10} |H_z(j\Omega_0)|$ dB at Ω_0. For example, if the normalized value of Ω_c is 0.7π, then the gain of the zero-order hold circuit at 0.7π is -7.2 dB. Now, the lowest normalized frequency of the residual images is given by 1.3π. For a minimum attenuation of 50 dB of all signal components in the residual images at the output of the zero-order hold, the reconstruction filter must therefore provide at least an attenuation of 42.8 dB at frequency 1.3π.

The droop caused by the zero-order hold circuit can be compensated either before the D/A converter by means of a digital filter or after the zero-order hold circuit by the analog reconstruction filter. For the latter approach, we observe that the cascade of the zero-order hold circuit and the analog reconstruction filter must have a frequency response of an ideal reconstruction filter following an ideal D/A converter. If we denote the frequency response of the ideal reconstruction filter as $H_r(j\Omega)$ and that of the actual reconstruction filter as $\hat{H}_r(j\Omega)$, then we require

$$H_r(j\Omega) = H_z(j\Omega)\hat{H}_r(j\Omega), \tag{5.71}$$

where $H_r(j\Omega)$ is as given by Eq. (5.16). Therefore, from Eq. (5.71) the desired frequency response of the actual reconstruction filter is given by

$$\hat{H}_r(j\Omega) = \frac{H_r(j\Omega)}{\left[\frac{2\sin(\Omega T/2)}{\Omega} \right]} e^{j\frac{\Omega T}{2}}. \tag{5.72}$$

to ensure a faithful reconstruction of the original signal $g_a(t)$. The frequency response of the above filter for a cutoff frequency $\Omega_c = \Omega_T/2$ is sketched in Figure 5.47. However, the above frequency response cannot be physically implemented since the phase response has a positive slope, implying a negative group delay. In addition, there is a sharp transition from the passband to the stopband, indicating that the impulse response is of infinite length. As a result, an analog filter approximating the magnitude response of $\hat{H}_r(j\Omega)$ must therefore be designed.

Alternatively, the effect of the droop can be compensated by including a digital compensation filter prior to the D/A converter circuit with a modest increase in the digital hardware requirements. The digital compensation filter can be either an FIR or an IIR type. Two such very-low-order filters are as follows [Jac96]:

$$H_{\text{FIR}}(z) = -\tfrac{1}{16} + \tfrac{9}{8}z^{-1} - \tfrac{1}{16}z^{-2}, \tag{5.73a}$$

$$H_{\text{IIR}}(z) = \frac{9}{8 + z^{-1}}. \tag{5.73b}$$

Figure 5.48 shows the gain responses of the uncompensated and the droop-compensated D/A converters in the baseband. Since the above digital compensation filters have a

(a)   (b)

Figure 5.47   Desired frequency response of the filter following the zero-order hold circuit for exact reconstruction of the original continuous-time signal: (a) magnitude response, and (b) phase response.

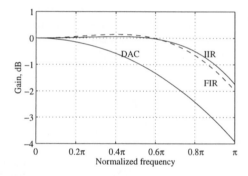

Figure 5.48   Gain responses of the uncompensated and compensated DAC in the base-band.

periodic frequency response of period Ω_T, the replicas of the baseband magnitude response outside the baseband need to be suppressed sufficiently to ensure minimal effect from aliasing. Even though the zero-order hold circuit in the D/A converter provides some attenuation of these unwanted replicas [(see Figure 5.46(c)], it may be necessary for the analog reconstruction filter following the D/A converter to provide additional attenuation.

5.9 Summary

Various issues concerned with the digital processing of continuous-time signals are studied in this chapter. A discrete-time signal is obtained by uniformly sampling a continuous-time signal. The discrete-time representation is unique if the sampling frequency is greater than twice the highest frequency contained in the continuous-time signal, and the latter can be fully recovered from its discrete-time equivalent by passing it through an ideal analog lowpass reconstruction filter with a cutoff frequency that is half the sampling frequency. If the sampling frequency is lower than twice the highest frequency contained in the continuous-time signal, the latter cannot be recovered from its discrete-time version due to aliasing. In practice, the continuous-time signal is first passed through an analog lowpass anti-aliasing filter with the cutoff frequency chosen as half of the sampling frequency whose output is sampled to prevent aliasing to occur.

A brief review of the theory behind some popular analog lowpass filter design techniques is included and their design using MATLAB is illustrated. The specifications of the analog filters are usually given in terms of the locations of the passband and stopband edge frequencies, and the passband and stopband ripples. Effects of these parameters on the performances of the anti-aliasing and reconstruction filters are examined.

Other interface devices involved in the digital processing of continuous-time signals are the sample-and-hold circuit, comparator, analog-to-digital converter, and digital-to-analog converter. A brief introduction to these devices is included for completeness.

5.10 Problems

5.1 Show that the periodic impulse train $p(t)$ defined in Eq. (5.4) can be expressed as a Fourier series as given by Eq. (5.7).

5.2 A continuous-time signal $x_a(t)$ is composed of a linear combination of sinusoidal signals of frequencies 300 Hz, 400 Hz, 1.3 kHz, 3.6 kHz, and 4.3 kHz. The signal $x_a(t)$ is sampled at a 2-kHz rate and the sampled sequence is passed through an ideal lowpass filter with a cutoff frequency of 900 Hz, generating a continuous-time signal $y_a(t)$. What are the frequency components present in the reconstructed signal $y_a(t)$?

5.3 A continuous-time signal $x_a(t)$ is composed of a linear combination of sinusoidal signals of frequencies F_1 Hz, F_2 Hz, F_3 Hz, and F_4 Hz. The signal $x_a(t)$ is sampled at a 3-kHz rate, and the sampled sequence is then passed through an ideal lowpass filter with a cutoff frequency of 1.4 kHz, generating a continuous-time signal $y_a(t)$ composed of three sinusoidal signals of frequencies 350 Hz, 425 Hz, and 918 Hz, respectively. What are the possible values of F_1, F_2, F_3, and F_4? Is your answer unique? If not, indicate another set of possible values of these frequencies.

5.4 The continuous-time signal $x_a(t) = 2\cos(600\pi t) + 4\sin(1400\pi t) + 3\cos(4400\pi t) + 7\sin(5400\pi t)$ is sampled at a 3-kHz rate generating the sequence $x[n]$. Determine the exact expression of $x[n]$.

5.5 A continuous-time signal $g_a(t) = \cos(10\pi t) + \cos(20\pi t) + \cos(40\pi t)$ is uniformly sampled to generate a discrete-time sequence $g[n]$. Plot the sampled sequence for the following sampling frequencies: (a) 45 Hz, (b) 40 Hz, and (c) 25 Hz.

5.6 The left and right channels of an analog stereo audio signal is sampled at a 44.1-kHz rate with each channel then converted into a digital bit stream using a 16-bit A/D converter. Determine the combined bit rate of the two channels after sampling and digitization.

5.7 Show that the impulse response $h_r(t)$ of an ideal lowpass filter as derived in Eq. (5.17) takes the value $h_r(nT) = \delta[n]$ for all n if the cutoff frequency $\Omega_c = \Omega_T/2$, where Ω_T is the sampling frequency.

5.8 Consider the system of Figure 5.2, where the input continuous-time signal $x_a(t)$ has a bandlimited spectrum $X_a(j\Omega)$ as sketched in Figure P5.1(a) and is being sampled at the Nyquist rate. The discrete-time processor is an ideal lowpass filter with a frequency response $H(e^{j\omega})$ as shown in Figure P5.1(b) and has a cutoff-frequency $\omega_c = \Omega_m T/3$, where T is the sampling period. Sketch as accurately as possible the spectrum $Y_a(j\Omega)$ of the output continuous-time signal $y_a(t)$.

(a)   (b)

Figure P5.1

5.9 A bandlimited continuous-time signal $x_a(t)$ has signal components with frequencies in the range from 30 Hz to 40 Hz. What is the smallest sampling rate F_T that can be employed to sample $x_a(t)$ so that it can be fully recovered from its sampled version $x[n]$? Sketch the Fourier transform of the sampled version $x[n]$ obtained by sampling $x_a(t)$ at the smallest sampling rate F_T if the Fourier transform of $x_a(t)$ is as shown in Figure P5.2. Sketch the frequency response of the ideal reconstruction filter needed to fully recover $x_a(t)$.

Figure P5.2

5.10 For each set of desired peak passband deviation A_p and the minimum stopband attenuation A_s of an analog lowpass filter given below, determine the corresponding passband and stopband ripples, δ_p and δ_s:

(a) $A_p = 0.05\,\text{dB}$, $A_s = 65\,\text{dB}$;

(b) $A_p = 0.02\,\text{dB}$, $A_s = 52\,\text{dB}$;

(c) $A_p = 0.5\,\text{dB}$, $A_s = 81\,\text{dB}$.

5.11 Show that the analog transfer function

$$H_1(s) = \frac{a}{s+a}, \qquad a > 0, \tag{5.74}$$

has a lowpass magnitude response with a monotonically decreasing magnitude response with $|H_a(j0)| = 1$ and $|H_a(j\infty)| = 0$. Determine the 3-dB cutoff frequency Ω_c at which the gain response is 3 dB below the maximum value of 0 dB at $\Omega = 0$.

5.12 Show that the analog transfer function

$$H_2(s) = \frac{s}{s+a}, \qquad a > 0, \tag{5.75}$$

has a highpass magnitude response with a monotonically increasing magnitude response with $|H_a(j0)| = 0$ and $|H_a(j\infty)| = 1$. Determine the 3-dB cutoff frequency Ω_c at which the gain response is 3 dB below the maximum value of 0 dB at $\Omega = \infty$.

5.13 The lowpass transfer function $H_1(s)$ of Eq. (5.74) and the highpass transfer function $H_2(s)$ of Eq. (5.75) can be expressed in the form

$$H_1(s) = \tfrac{1}{2}\{A_1(s) - A_2(s)\}, \qquad H_2(s) = \tfrac{1}{2}\{A_1(s) + A_2(s)\},$$

where $A_1(s)$ and $A_2(s)$ are stable analog allpass transfer functions. Determine $A_1(s)$ and $A_2(s)$.

5.14 Show that the analog transfer function

$$H_1(s) = \frac{bs}{s^2 + bs + \Omega_0^2}, \qquad b > 0, \tag{5.76}$$

has a bandpass magnitude response with $|H_a(j0)| = |H_a(j\infty)| = 0$ and $|H_a(j\Omega_0)| = 1$. Determine the frequencies Ω_1 and Ω_2 at which the gain is 3 dB below the maximum value of 0 dB at Ω_0. Show that $\Omega_1\Omega_2 = \Omega_0^2$. The difference $\Omega_2 - \Omega_1$ is called the 3-dB bandwidth of the bandpass transfer function. Show that $b = \Omega_2 - \Omega_1$.

5.15 Show that the analog transfer function

$$H_2(s) = \frac{s^2 + \Omega_0^2}{s^2 + bs + \Omega_0^2}, \qquad b > 0, \tag{5.77}$$

has a bandstop magnitude response with $|H_a(j0)| = |H_a(j\infty)| = 1$ and $|H_a(j\Omega_0)| = 0$. Since the magnitude is exactly zero at Ω_0, it is called the notch frequency, and $H_2(s)$ is often called notch transfer function. Determine the frequencies Ω_1 and Ω_2 at which the gain is 3 dB below the maximum value of 0 dB at $\Omega = 0$ and $\Omega = \infty$. Show that $\Omega_1\Omega_2 = \Omega_0^2$. The difference $\Omega_2 - \Omega_1$ is called the 3-dB notch bandwidth of the bandpass transfer function. Show that $b = \Omega_2 - \Omega_1$.

5.16 The bandpass transfer function $H_1(s)$ of Eq. (5.76) and the bandstop transfer function $H_2(s)$ of Eq. (5.77) can be expressed in the form

$$H_1(s) = \tfrac{1}{2}\{A_1(s) - A_2(s)\}, \qquad H_2(s) = \tfrac{1}{2}\{A_1(s) + A_2(s)\},$$

where $A_1(s)$ and $A_2(s)$ are stable analog allpass transfer functions. Determine $A_1(s)$ and $A_2(s)$.

5.17 Show that the first $2N - 1$ derivatives of the squared magnitude response $|H_a(j\Omega)|^2$ of a Butterworth filter of order N as given by Eq. (5.29) is equal to zero at $\Omega = 0$.

5.18 Using Eq. (5.31) determine the lowest order of a lowpass Butterworth filter with a 0.5-dB cutoff frequency at 1.5 kHz and a minimum attenuation of 60 dB at 6.1 kHz.

5.19 Using Eq. (5.33) determine the pole locations and the coefficients of a sixth order Butterworth polynomial with unity 3-dB cutoff frequency.

5.20 Using Eq. (5.39) determine the lowest order of a lowpass Chebyshev filter with a 0.5-dB cutoff frequency at 1.5 kHz and a minimum attenuation of 60 dB at 6.1 kHz.

5.21 Using Eq. (5.49) determine the lowest order of a lowpass elliptic filter with a 0.5-dB cutoff frequency at 1.5 kHz and a minimum attenuation of 60 dB at 6.1 kHz.

5.22 Determine the Bessel polynomials $B_N(s)$ for $N = 3$ and $N = 4$.

5.23 Verify Table 5.1.

5.24 Derive Eq. (5.64).

5.25 Derive Eq. (5.65).

5.11   MATLAB Exercises

M 5.1 Write a MATLAB program to compute the required order of a lowpass Butterworth analog filter according to Eq. (5.31). Using this program determine the lowest order of the lowpass filter of Problem 5.18.

M 5.2 Write a MATLAB program to compute the required order of a lowpass Chebyshev analog filter according to Eq. (5.34). Using this program determine the lowest order of the lowpass filter of Problem 5.20.

M 5.3 Write a MATLAB program to compute the required order of a lowpass elliptic analog filter according to Eq. (5.49). Using this program determine the lowest order of the lowpass filter of Problem 5.21.

M 5.4 Determine the transfer function of a lowpass Butterworth analog filter with specifications as given in Problem 5.18 using Program 5_2. Plot the gain response and verify that the filter designed meets the given specifications.

M 5.5 Determine the transfer function of a lowpass Type 1 Chebyshev analog filter with specifications as given in Problem 5.20 using Program 5_3. Plot the gain response and verify that the filter designed meets the given specifications.

M 5.6 Modify Program 5_3 to design lowpass Type 2 Chebyshev analog filters. Using this program, determine the transfer function of a lowpass Type 2 Chebyshev analog filter with specifications as given in Problem 5.20. Plot the gain response and verify that the filter designed meets the given specifications.

M 5.7 Determine the transfer function of a lowpass elliptic analog filter with specifications as given in Problem 5.21 using Program 5_4. Plot the gain response and verify that the filter designed meets the given specifications.

M 5.8 Write a MATLAB program to verify the plots of Figure 5.48.

CHAPTER 6

Digital Filter Structures

The description of the discrete-time system of Eq. (2.61a) or (2.61b) expresses the nth output sample as a convolution sum of the input with the impulse response of the system, and in some sense, it is the most fundamental characterization of an LTI digital filter. The convolution sum, in principle, can be employed to implement a digital filter with a known impulse response, and the implementation involves addition, multiplication, and delay, which are fairly simple operations. For an LTI system with an infinite-length impulse response, the approach is not practical. However, for the infinite-impulse response (IIR) LTI digital filter described by a constant constant coefficient difference equation of the form of Eq. (2.79), and for the finite-impulse response (FIR) LTI digital filter described by Eq. (2.82), the input-output relation involves a finite sum of products, and a direct implementation based on these equations is quite practical. In this text we deal only with these two types of LTI digital filters.

Now, the actual implementation of a digital filter could be either in software or hardware form, depending on applications. In both types of implementation, the signal variables and the filter coefficients cannot be represented with infinite precision. As a result, a direct implementation of a digital filter based on either Eq. (2.79) or Eq. (2.82) may not provide satisfactory performance due to the finite precision arithmetic. It is thus of interest to develop alternative realizations based on other types of time-domain representations with equivalent input-output relations to either Eq. (2.79) or Eq. (2.82), depending on the type of digital filter being implemented, and choose the realization that provides satisfactory performance under finite precision arithmetic.

In this chapter, we consider the realization problem of causal IIR and FIR transfer functions and outline realization methods based on both the time-domain and the transform-domain representations. In Chapter 9, we develop the methods for the analysis of such structures when implemented with finite precision arithmetic and present additional realizations that have been developed to minimize the effects of finite wordlength.

A structural representation using interconnected basic building blocks is the first step in the hardware or software implementation of an LTI digital filter. The structural representation provides the relations between some pertinent internal variables with the input and the output that in turn provide the keys to the implementation. There are various forms of the structural representation of a digital filter. We review in this chapter two such representations, and then describe some popular schemes for the realization of real causal IIR and FIR digital filters. In addition, we outline a method for the realization of IIR digital filter structures that can be used for the generation of a pair of orthogonal sinusoidal sequences.

6.1   Block Diagram Representation

As indicated earlier, the input-output relations of an LTI digital filter can be expressed in various ways. In the time-domain, it is given by the convolution sum

$$y[n] = \sum_{k=-\infty}^{\infty} h[k]x[n-k], \qquad (6.1)$$

or by the linear constant coefficient difference equation

$$y[n] = -\sum_{k=1}^{N} d_k y[n-k] + \sum_{k=0}^{M} p_k x[n-k], \qquad (6.2)$$

or by the state-space equations

$$s[n+1] = \mathbf{A}s[n] + \mathbf{B}x[n], \qquad (6.3a)$$
$$y[n] = \mathbf{C}s[n] + dx[n]. \qquad (6.3b)$$

A digital filter can be implemented on a general-purpose digital computer or with special-purpose hardware. To this end, it is necessary to describe the input-output relationship by means of a computational algorithm. To illustrate what we mean by a computational algorithm, consider a first-order causal LTI IIR digital filter described by

$$y[n] = -d_1 y[n-1] + p_0 x[n] + p_1 x[n-1]. \qquad (6.4)$$

Using Eq. (6.4) we can compute $y[n]$ for $n = 0, 1, 2, \ldots$, knowing the initial condition $y[-1]$ and the input $x[n]$ for $n = -1, 0, 1, 2, \ldots$:

$$y[0] = -d_1 y[-1] + p_0 x[0] + p_1 x[-1],$$
$$y[1] = -d_1 y[0] + p_0 x[1] + p_1 x[0],$$
$$y[2] = -d_1 y[1] + p_0 x[2] + p_1 x[1],$$
$$\vdots$$

We can continue this calculation for any value of n we desire. Each step of the calculation process requires a knowledge of the previously calculated value of the output sample (delayed value of the output), the present value of the input sample, and the previous value of the input sample (delayed value of the input). Knowing the data values, we multiply each of them appropriately with coefficients $-d_1$, p_0, and p_1 and sum the products to compute the present value of the output. As a result, the difference equation of Eq. (6.4) can be interpreted as a valid computational algorithm.

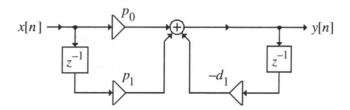

$x[n] \longrightarrow \bullet \longrightarrow x[n]$

$x[n]$

(a)

$x[n] \longrightarrow \oplus \longrightarrow x[n]+ y[n]$

$y[n]$

(b)

$x[n] \longrightarrow \triangleright \longrightarrow \alpha\, x[n]$

α

(c)

$x[n] \longrightarrow \boxed{z^{-1}} \longrightarrow x[n-1]$

(d)

Figure 6.1   Basic building blocks: (a) pick-off node, (b) adder, (c) multiplier, and (d) unit delay

Figure 6.2   A first-order LTI digital filter.

6.1.1   Basic Building Blocks

The computational algorithm of an LTI digital filter can be conveniently represented in block diagram form using the basic building blocks representing the unit delay, the multiplier, and the adder as shown earlier in Figure 2.12 and reproduced in Figure 6.1. In addition, we show in Figure 6.1 a pick-off node that splits a single incoming signal into multiple identical outgoing signals. Note also that the symbol for representing the unit delay is now the transfer function of the device and is the more commonly used z^{-1} instead of D used earlier in Figure 2.12. Thus, a block diagram representation of the first-order digital filter described by Eq. (6.4) is as indicated in Figure 6.2. The adder shown here is a three-input adder. Note also the pick-off nodes at the input and the output.

There are several advantages in representing the digital filter in block diagram form: (1) it is easy to write down the computational algorithm by inspection, (2) it is easy to analyze the block diagram to determine the explicit relation between the output and the input, (3) it is easy to manipulate a block diagram to derive other "equivalent" block diagrams yielding different computational algorithms, (4) it is easy to determine the hardware requirements, and finally (5) it is easier to develop block diagram representations from the transfer function directly leading to a variety of "equivalent" representations.

6.1.2   Analysis of Block Diagrams

Digital filter structures represented in block diagram form can often be analyzed by writing down the expressions for the output signals of each adder as a sum of its input signals, developing a set of equations relating the filter input and output signals in terms of all internal signals. Eliminating the unwanted internal variables then results in the expression

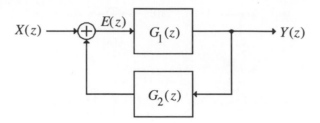

Figure 6.3   A single-loop digital filter structure.

for the output signal as a function of the input signal and the filter parameters that are the multiplier coefficients.

The following two examples illustrate the analysis approach.

Example 6.1 Consider the single-loop feedback structure of Figure 6.3. There is only one adder in this structure whose output signal $E(z)$ is given by

$$E(z) = X(z) + G_2(z)Y(z).$$

But from the figure,

$$Y(z) = G_1(z)E(z).$$

Eliminating $E(z)$ from the above two equations, we arrive at

$$\left[1 - G_1(z)G_2(z)\right]Y(z) = G_1(z)X(z),$$

which leads to the transfer function

$$H(z) = \frac{Y(z)}{X(z)} = \frac{G_1(z)}{1 - G_1(z)G_2(z)}. \tag{6.5}$$

Example 6.2 Analyze the cascaded lattice digital filter structure of Figure 6.4. The output signals of the four adders in Figure 6.4 are given by

$$W_1 = X - \alpha S_2, \tag{6.6a}$$
$$W_2 = W_1 - \delta S_1, \tag{6.6b}$$
$$W_3 = S_1 + \varepsilon W_2, \tag{6.6c}$$
$$Y = \beta W_1 + \gamma S_2. \tag{6.6d}$$

But, from the figure, $S_2 = z^{-1}W_3$ and $S_1 = z^{-1}W_2$. Substituting these relations in Eqs. (6.6a)–(6.6d), we obtain

$$W_1 = X - \alpha z^{-1}W_3, \tag{6.7a}$$
$$W_2 = W_1 - \delta z^{-1}W_2, \tag{6.7b}$$
$$W_3 = z^{-1}W_2 + \varepsilon W_2, \tag{6.7c}$$
$$Y = \beta W_1 + \gamma z^{-1}W_3. \tag{6.7d}$$

From Eq. (6.7b), $W_2 = W_1/(1+\delta z^{-1})$ and, from Eq. (6.7c), $W_3 = (\varepsilon + z^{-1})W_2$. Combining these two equations, we get

$$W_3 = \frac{\varepsilon + z^{-1}}{1 + \delta z^{-1}} W_1. \tag{6.8}$$

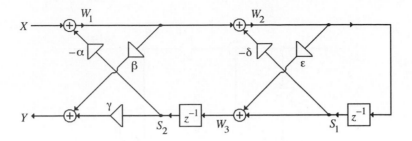

Figure 6.4 A cascaded lattice digital filter structure.

Figure 6.5 An example of a delay-free loop.

Substituting Eq. (6.8) in Eqs. (6.7a) and (6.7d), we finally obtain after some algebra the expression for the transfer function $H(z)$ of the digital filter structure as

$$H(z) = \frac{Y}{X} = \frac{\beta + (\beta\delta + \gamma\varepsilon)z^{-1} + \gamma z^{-2}}{1 + (\delta + \alpha\varepsilon)z^{-1} + \alpha z^{-2}}. \tag{6.9}$$

6.1.3 The Delay-Free Loop Problem

For physical realizability of the digital filter structure, it is necessary that the block diagram representation contains no delay-free loops, i.e., feedback loops without any delay elements. Figure 6.5 illustrates a typical delay-free loop that may appear unintentionally in a specific structure. Analysis of this structure yields

$$y[n] = B\{A(w[n] + y[n]) + v[n]\}.$$

The above expression implies that the determination of the current value of $y[n]$ requires the knowledge of the same value that is physically impossible to achieve due to the finite time required of all arithmetic operations in a digital machine.

A simple graph-theoretic-based method has been proposed to detect the presence of delay-free loops in an arbitrary digital filter structure, along with the methods to locate and remove these loops without altering the overall input-output relations [Szc75]. The removal is achieved by replacing the portion of the structure containing the delay-free loops by an equivalent realization with no delay-free loops. For example, Figure 6.6 shows an equivalent realization of the block diagram of Figure 6.5 without the delay-free loop.

Figure 6.6   An equivalent realization of the structure of Figure 6.5 with no delay-free loop.

6.1.4   Canonic and Noncanonic Structures

A digital filter structure is said to be *canonic* if the number of delays in the block diagram representation is equal to the order of the difference equation (i.e., the order of the transfer function). Otherwise, it is a *noncanonic* structure. For example, the structure of Figure 6.2 is a noncanonic realization since it employs two delays to realize the first-order difference equation of Eq. (6.4).

6.2   Signal Flow-Graph Representation

Another convenient representation of a digital filter is by means of a signal flow-graph which is primarily used for analyzing the digital filter structure. We describe this type of representation next.

6.2.1   Basic Concepts

In a signal flow-graph [Mas60], the dependent and the independent signal variables are represented by *nodes* v_k, whereas the multiplier and the delay units are represented by *directed branches*. In the latter case, the directed branch has an attached symbol denoting the *branch-gain* or the *transmittance*, which for a multiplier branch is the multiplier coefficient value and for a delay branch is simply z^{-1}. Often, a directed branch starting at a node v_1 and terminating at a node v_2 is denoted by the symbol (v_1, v_2). Examples of the basic elements in the signal flow-graph representation of an LTI digital filter are sketched in Figure 6.7. As shown in this figure, there are basically three types of nodes. In the example of Figure 6.7(b), the *sink node* v_4 has only incoming branches, whereas in Figure 6.7(c), the *source node* v_1 has only outgoing branches. However, in general, an internal node will have a combination of incoming and outgoing branches, as illustrated in Figure 6.7(d). A node with only one incoming branch and one outgoing branch is called a *chain node*.

For analysis purposes, often, we remove some of the branches and nodes of a given flow-graph. The reduced representation is called a *subgraph*. We now outline some of the basic rules for the analysis of a signal flow-graph.

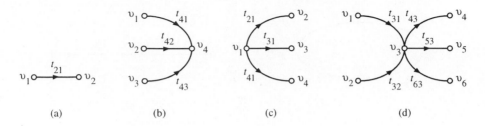

Figure 6.7   Basic elements of a signal flow-graph.

6.2.2   Analysis Rules

The branches and the nodes of a signal flow-graph obey some rules relating the node variables. In the case of Figure 6.7(a), the node variables v_1 and v_2 are related as

$$v_2 = t_{21}v_1, \tag{6.10}$$

where t_{21} is the transmittance of the directed branch connecting the starting node v_1 to the terminating node v_2. In the case of Figure 6.7(b), the node variable v_4 is related to the other three node variables by the relation

$$v_4 = t_{41}v_1 + t_{42}v_2 + t_{43}v_3. \tag{6.11}$$

Finally, in the case of Figure 6.7(c), the node variables v_2, v_3, and v_4 are related to the node variable v_1 through the relations

$$v_2 = t_{21}v_1, \quad v_3 = t_{31}v_1, \quad v_4 = t_{41}v_1. \tag{6.12}$$

Two branches are said to be in *parallel* if they have the same starting node v_1 and the same terminating node v_2 as illustrated in Figure 6.8(a). In this case, to simplify the analysis, the branches in parallel can be replaced by a single branch whose branch-gain is the sum of the branch-gains of the original branches, i.e.,

$$v_2 = t'_{21}v_1 + t''_{21}v_1 = (t'_{21} + t''_{21})v_1. \tag{6.13}$$

Note that this situation is equivalent to the parallel interconnection scheme of Figure 2.26.

Likewise, two branches are said to be in *series* if the termination node of the first branch acts as the starting node of the second branch, as shown in Figure 6.8(b). Again, these two branches can be replaced by a single branch with a branch-gain that is the product of the branch-gains of the original branches, i.e.,

$$v_3 = t_{32}v_2 = t_{32}(t_{21}v_1) = (t_{32}t_{21})v_1. \tag{6.14}$$

The series connection of branches is equivalent to the cascade combination scheme of Figure 2.25.

A particularly important subgraph is the *loop*, as shown in Figure 6.9, that is a connected graph containing only chain nodes. The *loop-gain* is defined as the product of all branch-gains involved in the loop. A special case of the loop is the *self-loop*, which

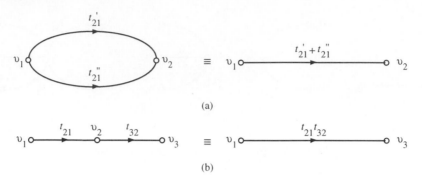

Figure 6.8   Basic flow-graph interconnection schemes: (a) parallel connection, and (b) series connection.

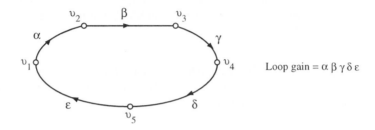

Figure 6.9   A loop.

contains only a single chain node, as sketched in Figure 6.10(a). The self-loop can be eliminated by absorbing it in the incoming branch. To develop the modified representation, consider the graph of Figure 6.10(b) containing a self-loop and described by the following two equations:

$$\upsilon_2 = t_{21}\upsilon_1 + t_{22}\upsilon_2,$$
$$\upsilon_3 = t_{32}\upsilon_2. \tag{6.15}$$

Rewriting Eq. (6.15), we arrive at

$$\upsilon_2 = \frac{t_{21}}{1 - t_{22}}\upsilon_1, \tag{6.16}$$

which leads to the equivalent representation shown in Figure 6.10(c).

Another important concept in the signal flow-graph is the connected subgraph called the *path*, which connects a source node to a sink node with only chain nodes. An example of a path is indicated in Figure 6.11. The product of all branch-gains in a path is defined as its *path-gain*.

Two loops or paths are defined to be *disjoint* if they do not have any node in common. A set of loops or paths is defined to be disjoint if every two of them in the set are pairwise disjoint.

Often it is possible to reduce a complicated signal flow-graph step-by-step to arrive at a single branch connecting the input source node to the output sink node. The branch-gain of

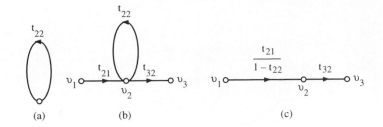

Figure 6.10   A self-loop and its elimination scheme.

Figure 6.11   A path.

this single branch is then the overall gain of the original signal flow-graph. Alternatively, it may be convenient to compute the overall gain of the original signal flow-graph by applying Mason's gain formula, described next.

6.2.3   Mason's Gain Formula

To determine the gain, we need to determine two expressions: the determinant of the signal flow-graph and the cofactor with respect to each path. The *determinant* Δ of a flow-graph is defined as

$$\Delta = 1 - \text{(sum of all loop-gains)}$$
$$+ \text{(sum of products of loop-gains of all pairs of disjoint loops)}$$
$$- \text{(sum of products of loop-gains of all triples of disjoint loops)}$$
$$+ \dots \tag{6.17}$$

For the kth path in the signal flow-graph with a path-gain g_k we define the *cofactor* Δ_k of the flow-graph with respect to this path as

$$\Delta_k = 1 - \text{sum of loop-gains of all loops that are disjoint from the path)}$$
$$+ \text{(sum of products of loop-gains of all pairs}$$
$$\text{of disjoint loops that are disjoint from the path)}$$
$$- \text{(sum of products of loop-gains of all triples of}$$
$$\text{disjoint loops that are disjoint from the path)}$$
$$+ \dots . \tag{6.18}$$

The overall *gain* (or the *transfer function*) H of the original signal flow-graph is then given by [Mas60]

$$H = \frac{1}{\Delta} \sum_k g_k \Delta_k . \tag{6.19}$$

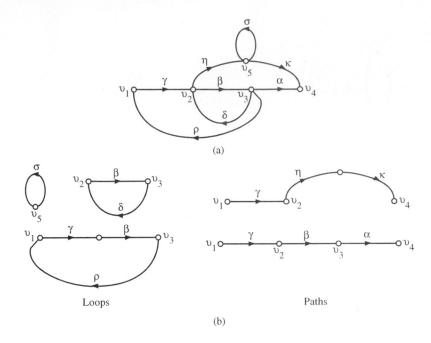

Figure 6.12   (a) A typical signal flow-graph, and (b) its pertinent sub-graphs.

Example 6.3 Consider the signal flow-graph of Figure 6.12(a). We determine its transfer function $H = v_4/v_1$ using Mason's gain formula.

The signal flow-graph of Figure 6.12(a) is seen to have three loops and two paths as shown in Figure 6.12(b). The loop-gains of the three loops in Figure 6.12(a) are given by

$$
\begin{array}{lll}
\text{Loop } L_1: & \text{self-loop } (v_5) & \text{loop-gain} = \sigma \\
\text{Loop } L_2: & (v_2, v_3) & \text{loop-gain} = \beta\delta \\
\text{Loop } L_3: & (v_1, v_2, v_3) & \text{loop-gain} = \gamma\beta\rho
\end{array}
$$

Note that loops L_1 and L_2 are disjoint, and similarly, loops L_1 and L_3 are disjoint. However, loops L_2 and L_3 have nodes v_2 and v_3 in common.

Likewise, the path-gains of the two paths in Figure 6.12(b) are given by

$$
\begin{array}{lll}
\text{Path } P_1: & (v_1, v_2, v_5, v_4) & \text{path-gain } g_1 = \gamma\eta\kappa \\
\text{Path } P_2: & (v_1, v_2, v_3, v_4) & \text{path-gain } g_2 = \gamma\beta\alpha.
\end{array}
$$

It should be noted also that none of the loops is disjoint with path P_1, whereas loop L_1 is disjoint with path P_2.

From Eq. (6.17), the determinant Δ of the above flow-graph is

$$\Delta = 1 - (\sigma + \beta\delta + \gamma\beta\rho) + (\sigma\beta\delta + \sigma\gamma\beta\rho).$$

The cofactor Δ_1 for the path P_1 is 1 since there are no disjoint loops. In the case of path P_2, the cofactor Δ_2 is given by $1 - \sigma$. Therefore, the overall gain $H = v_4/v_1$ is given by

$$
\begin{aligned}
H &= \frac{g_1\Delta_1 + g_2\Delta_2}{\Delta} \\
&= \frac{\gamma\kappa\eta + \gamma\alpha\beta(1-\sigma)}{1 - (\sigma + \beta\delta + \gamma\beta\rho) + (\sigma\beta\delta + \sigma\gamma\beta\rho)}.
\end{aligned}
\tag{6.20}
$$

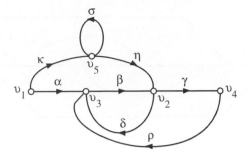

Figure 6.13 The transposed graph of Figure 6.12(a). Note that in the transposed graph, the input and the output node variables have been denoted as in the original graph of Figure 6.12(a).

6.2.4 Transposed Graph

Another interesting and useful operation is the transposition of a signal flow-graph which, is simply the reversal of all paths in a given graph [Jac70a], [Mas60]. The signal flow-graph $\mathcal{G}_t$ obtained by reversing the paths of a graph $\mathcal{G}$ is called its *transposed graph*. Note that the source nodes of $\mathcal{G}_t$ are the sink nodes of $\mathcal{G}$, and vice-versa. If $\mathcal{G}$ has a single source node and a single sink node with an overall gain H, then the gain of the transposed graph $\mathcal{G}_t$ is also H. This observation follows directly from the gain formula given in Eq. (6.19) since both the original and the transposed graphs have identical loops and paths leading to identical gain expressions.

Example 6.4 The transposed graph of the flow-graph of Figure 6.12(a) is shown in Figure 6.13. It is readily seen that the three loops (v_5), (v_3, v_2), and (v_3, v_2, v_4) are same as the loops L_1, L_2, and L_3, respectively, of the flow-graph of Figure 6.12(a) with identical loop-gains. Likewise, the two paths (v_1, v_5, v_2, v_4) and (v_1, v_3, v_2, v_4) are the same as the paths P_1 and P_2 with identical path-gains. These loops and paths have the same disjointness. As a result, the gain of the transposed graph is also given by the expression of Eq. (6.20).

6.2.5 Analysis of Digital Filter Structures

We now show with the aid of two examples how the signal flow-graph method can be applied to the analysis of digital filter structures.

Example 6.5 Consider the digital filter of Figure 6.4, which was analyzed earlier in Example 6.2. We analyze this structure again using the signal flow-graph methods. For ease in the signal flow-graph representation, we have labeled the adder outputs and the pick-off nodes as shown in the figure. The corresponding flow-graph in a redrawn form is sketched in Figure 6.14. In order to apply the gain formula, we identify the loops and the paths. The three loops and their loop-gains are as follows:

$$\text{Loop } L_1 : \quad (W_1, W_4, W_2, W_5, S_1, W_3, S_2) \qquad \text{loop-gain} = -\alpha z^{-2}$$
$$\text{Loop } L_2 : \quad (W_1, W_4, W_2, W_5, W_3, S_2) \qquad \text{loop-gain} = -\alpha \varepsilon z^{-1}$$
$$\text{Loop } L_3 : \quad (W_2, W_5, S_1) \qquad \text{loop-gain} = -\delta z^{-1}$$

Note that none of the loops are disjoint since they all have the node W_2 in common.

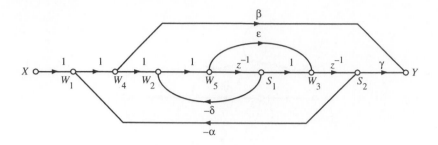

Figure 6.14   Signal flow-graph representation of the digital filter structure.

The three paths and their path-gains are given by

Path P_1 : $(X, W_1, W_4, W_2, W_5, S_1, W_3, S_2, Y)$ path-gain $g_1 = \gamma z^{-2}$
Path P_2 : $(X, W_1, W_4, W_2, W_5, W_3, S_2, Y)$ path-gain $g_2 = \gamma \varepsilon z^{-1}$
Path P_3 : (X, W_1, W_4, Y) path-gain $g_3 = \beta$

It should be noted that only path P_3 and loop L_3 are nontouching, i.e., disjoint.

Using Eq. (6.17) we obtain the determinant Δ of the flow-graph

$$\Delta = 1 + \alpha z^{-2} + \alpha \varepsilon z^{-1} + \delta z^{-1} = 1 + (\alpha \varepsilon + \delta)z^{-1} + \alpha z^{-2}.$$

To compute the numerator, we calculate the three cofactors corresponding to the three paths using Eq. (6.18):

$$\Delta_1 = 1, \qquad \Delta_2 = 1, \qquad \Delta_3 = 1 + \delta z^{-1}.$$

Substituting the above in Eq. (6.19) we finally arrive at the expression for the transfer function $H(z) = Y(z)/X(z)$ as

$$H(z) = \frac{g_1 \Delta_1 + g_2 \Delta_2 + g_3 \Delta_3}{\Delta}$$

$$= \frac{\gamma z^{-2} + \gamma \varepsilon z^{-1} + \beta(1 + \delta z^{-1})}{1 + (\alpha \varepsilon + \delta)z^{-1} + \alpha z^{-2}} = \frac{\beta + (\gamma \varepsilon + \beta \delta)z^{-1} + \gamma z^{-2}}{1 + (\alpha \varepsilon + \delta)z^{-1} + \alpha z^{-2}}. \qquad (6.21)$$

As expected, the above expression is identical to Eq. (6.9) derived in Example 6.2.

Example 6.6 In this example, we again consider the analysis of the signal flow-graph of Figure 6.14, but now using the flow-graph simplification approach.

Now, examination of the above flow-graph reveals that the node variable S_1 is precisely equal to $z^{-1}W_5$. Thus, if we split the node S_1 into two nodes as shown in Figure 6.15(a), the input-output relation of the flow-graph remains invariant since the signals coming into nodes W_2 and W_3 from node S_1 have not changed. But $W_5 = W_2$, and as a result, we can merge the two nodes into one, resulting in the signal flow-graph of Figure 6.15(b). The self-loop at node W_5 is next absorbed using formula of Eq. (6.16), leading to the flow-graph of Figure 6.15(c). Replacing the three series branches between nodes W_4 and S_2 by one and then splitting the node S_2, we arrive at the simplified flow-graph of Figure 6.15(d). We next absorb the self-loop at W_4 and replace the two parallel branches by one, resulting in the two-series-branch representation of Figure 6.15(e), from which the overall gain is seen to be given by,

$$\frac{\beta + (\gamma \varepsilon + \beta \delta)z^{-1} + \gamma z^{-2}}{1 + (\alpha \varepsilon + \delta)z^{-1} + \alpha z^{-2}}$$

which is same as the one given in Eq. (6.21), as expected.

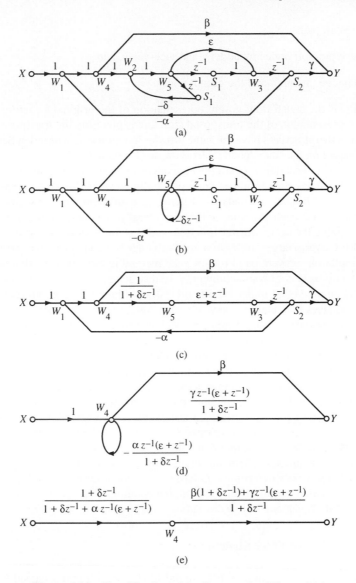

Figure 6.15   Simplification of the flow-graph of Figure 6.14.

6.3   Equivalent Structures

Our main objective in this chapter is to develop various realizations of a given transfer function. We define two digital filter structures to be *equivalent* if they have the same transfer function. We outline in this chapter and in Chapter 9 a number of methods for the generation of equivalent structures. However, a fairly simple way to generate an equivalent structure from a given realization is via the *transpose operation*, which is as follows [Jac70a]:

(i) Reverse all paths,

(ii) Replace pick-off nodes by adders, and vice-versa, and

(iii) Interchange the input and the output nodes.

Note that if $\mathcal{G}$ is the signal flow-graph representation of the original structure, the signal flow-graph representation of the transposed structure is precisely the transposed graph $\mathcal{G}_t$. As a result, the two structures have the same transfer function as discussed in Section 6.2.4. For a proof based on the state-space realization, see Problem 6.5.

   Another way to generate very simply equivalent structures is by applying linear transformation to the state-space representation of the digital filter. We discuss this approach in Section 6.6.3. All other methods for developing equivalent structures are based on a specific algorithm for each structure. There are literally an infinite number of equivalent structures realizing the same transfer function, and it is impossible to develop all such realizations. Moreover, a large variety of algorithms have been advanced by various authors, and space limitations prevent us to review each method in this text. We therefore restrict ourselves to a discussion of some commonly used structures.

   It should be noted that under infinite precision arithmetic any given realization of a digital filter behaves identically to any other equivalent structure. However, in practice, due to the finite wordlength limitations, a specific realization behaves totally differently from its other equivalent realizations. Hence, it is important to choose a structure that has the least quantization effects from the finite wordlength implementation. One way to arrive at such a structure is to determine a large number of equivalent structures, analyze the finite wordlength effects of each one, and then select the one showing the least effects. In certain cases, it is possible to develop a structure that by construction has the least quantization effects. The analysis of quantization effects is the subject of Chapter 9, which also describes additional structures specifically developed to minimize certain quantization effects. In this chapter, we discuss some simple realizations that in many applications are adequate. We do, however, compare each of the realizations discussed here with regard to their computational complexity determined in terms of the total number of multipliers and the total number of adders required for their implementation. This latter issue is important where the cost of implementation is critical.

6.4   Basic FIR Digital Filter Structures

We first consider the realization of FIR digital filters. Recall that a causal FIR filter of length M is characterized by a transfer function $H(z)$:

$$H(z) = \sum_{k=0}^{M-1} h[k]z^{-k}, \tag{6.22}$$

which is polynomial in z^{-1} of degree $M-1$. In the time-domain the input-output relation of the above FIR filter is given by

$$y[n] = \sum_{k=0}^{M-1} h[k]x[n-k], \tag{6.23}$$

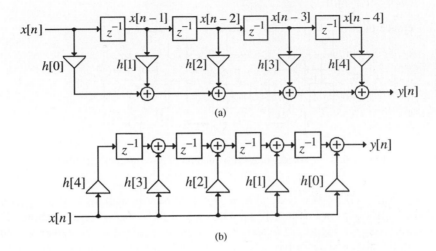

Figure 6.16 Direct form FIR structures.

where $y[n]$ and $x[n]$ are the output and input sequences, respectively. Since FIR filters can be designed to provide exact linear phase over the whole frequency range and are always BIBO stable independent of the filter coefficients, such filters are often preferred in many applications. We now outline several realization methods for such filters.

6.4.1 Direct Form

An FIR filter of length M is characterized by M coefficients, and in general, requires M multipliers and $M - 1$ two-input adders for implementation. Structures in which the multiplier coefficients are precisely the coefficients of the transfer function are called *direct form* structures. A direct form realization of an FIR filter can be readily developed from Eq. (6.23), as indicated in Figure 6.16(a) for $M = 5$. An analysis of this structure yields

$$y[n] = h[0]x[n] + h[1]x[n-1] + h[2]x[n-2] + h[3]x[n-3] + h[4]x[n-4],$$

which is precisely of the form of Eq. (6.23).

The structure of Figure 6.16(a) is also called a *tapped delay line* or a *transversal filter*. Its transpose as sketched in Figure 6.16(b) is the second direct form structure. Both direct form structures are canonic with respect to delays.

6.4.2 Cascade Form

A higher order FIR transfer function can also be realized as a cascade of FIR sections with each section characterized by either a first-order or a second-order transfer function. To this end, we factor the FIR transfer function $H(z)$ of Eq. (6.22) and write it in the form

$$H(z) = h[0] \prod_{k=1}^{K} \left(1 + \beta_{1k}z^{-1} + \beta_{2k}z^{-2} \right), \tag{6.24}$$

Figure 6.17   Cascade form FIR structure for a sixth-order FIR filter.

where $K = (M-1)/2$ if M is odd and $K = M/2$ if M is even, with $\beta_{2k} = 0$. A realization of Eq. (6.24) is shown in Figure 6.17 for a cascade of three second-order sections. Each second-order stage in Figure 6.17 can of course also be realized in the transposed direct form. Note that the cascade form is canonic with respect to delay, and also employs $(M-1)$ two-input adders and M multipliers for an $(M-1)$th order FIR transfer function.

The factorization of $H(z)$ in the form of Eq. (6.24) can be carried out on MATLAB using the function roots. For example, r = roots(h) will return the roots of the polynomial vector h containing the coefficients of $H(z)$ in ascending powers of z^{-1} in the output vector r. From the computed roots, the coefficients of the quadratic factors can be determined.

6.4.3 Polyphase Realization

Another interesting realization of an FIR filter is based on the polyphase decomposition of its transfer function and results in a parallel structure [Bel76]. To illustrate this approach, consider for simplicity a causal FIR transfer function $H(z)$ of length 9:

$$H(z) = h[0] + h[1]z^{-1} + h[2]z^{-2} + h[3]z^{-3} + h[4]z^{-4}$$
$$+ h[5]z^{-5} + h[6]z^{-6} + h[7]z^{-7} + h[8]z^{-8}. \tag{6.25}$$

The above transfer function can be expressed as a sum of two terms, with one term containing the even-indexed coefficients and the other containing the odd-indexed coefficients, as indicated below:

$$H(z) = \left(h[0] + h[2]z^{-2} + h[4]z^{-4} + h[6]z^{-6} + h[8]z^{-8} \right)$$
$$+ \left(h[1]z^{-1} + h[3]z^{-3} + h[5]z^{-5} + h[7]z^{-7} \right)$$
$$= \left(h[0] + h[2]z^{-2} + h[4]z^{-4} + h[6]z^{-6} + h[8]z^{-8} \right)$$
$$+ z^{-1} \left(h[1] + h[3]z^{-2} + h[5]z^{-4} + h[7]z^{-6} \right). \tag{6.26}$$

By using the notations

$$E_0(z) = h[0] + h[2]z^{-1} + h[4]z^{-2} + h[6]z^{-3} + h[8]z^{-4},$$
$$E_1(z) = h[1] + h[3]z^{-1} + h[5]z^{-2} + h[7]z^{-3}, \tag{6.27}$$

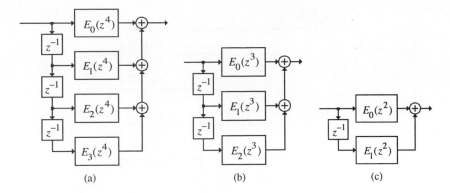

Figure 6.18   Polyphase realizations of an FIR transfer function.

we can rewrite Eq. (6.26) as

$$H(z) = E_0(z^2) + z^{-1}E_1(z^2). \qquad (6.28)$$

In a similar manner, by grouping the terms of Eq. (6.25) differently, we can reexpress it in the form

$$H(z) = E_0(z^3) + z^{-1}E_1(z^3) + z^{-2}E_2(z^3), \qquad (6.29)$$

where now

$$
\begin{aligned}
E_0(z) &= h[0] + h[3]z^{-1} + h[6]z^{-2}, \\
E_1(z) &= h[1] + h[4]z^{-1} + h[7]z^{-2}, \\
E_2(z) &= h[2] + h[5]z^{-1} + h[8]z^{-2}.
\end{aligned}
\qquad (6.30)
$$

The decomposition of $H(z)$ in the form of Eqs. (6.28) and (6.29) is more commonly known as the *polyphase decomposition*. In the general case, an L-branch polyphase decomposition of the transfer function of Eq. (6.22) of order $M-1$ is of the form

$$H(z) = \sum_{m=0}^{L-1} z^{-m} E_m(z^L), \qquad (6.31)$$

where[1]

$$E_m(z) = \sum_{n=0}^{\lfloor M/L \rfloor} h[Ln+m]z^{-n}, \qquad 0 \le m \le L-1, \qquad (6.32)$$

with $h[n] = 0$ for $n \ge M$. A realization of $H(z)$ based on the decomposition of Eq. (6.31) is called a *polyphase realization*. Figure 6.18 shows the four-branch, the three-branch, and the two-branch polyphase realizations of a transfer function. As indicated in Eqs. (6.27) and (6.30), the expression for the transfer function $E_0(z)$ is different for each structure and so are the expressions for $E_1(z)$ etc.

---

[1] $\lfloor x \rfloor$ is the integer part of x.

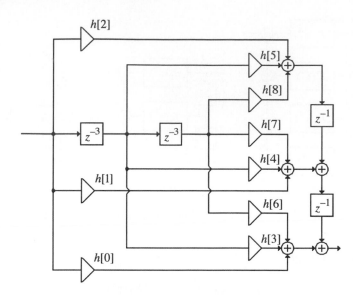

Figure 6.19   Canonic three-branch polyphase realization of a length-9 FIR filter.

The subfilters $E_k(z^L)$ in the polyphase realization of an FIR transfer function are also FIR filters and can be realized using any of the methods described earlier. However, to obtain a canonic realization of the overall structure, the delays in all subfilters must be shared. Figure 6.19 illustrates a canonic polyphase realization of a length-9 FIR transfer function obtained by delay-sharing. It should be noted that in developing this realization, we have used the transpose of the structure of Figure 6.18(b). Other canonic polyphase realizations can be similarly derived (Problems 6.17 to 6.19).

The polyphase structures are often used in multirate digital signal processing applications for computationally efficient realizations (see Chapter 10).

6.4.4   Linear-Phase FIR Structures

We showed in Section 4.3.6 that a linear-phase FIR filter of length M is characterized by

$$h[n] = h[M - 1 - n]. \tag{6.33}$$

We also showed that an antisymmetric impulse response condition

$$h[n] = -h[M - 1 - n] \tag{6.34}$$

is another way of achieving a constant group delay, but with an "almost linear-phase" property.

The symmetry (or antisymmetry) property of a linear-phase FIR filter can be exploited to reduce the total number of multipliers into almost half in the direct form implementations of the transfer function. To this end consider the realization of a length-7 Type 1 FIR transfer function with a symmetric impulse response:

$$H(z) = h[0] + h[1]z^{-1} + h[2]z^{-2} + h[3]z^{-3} + h[2]z^{-4} + h[1]z^{-5} + h[0]z^{-6},$$

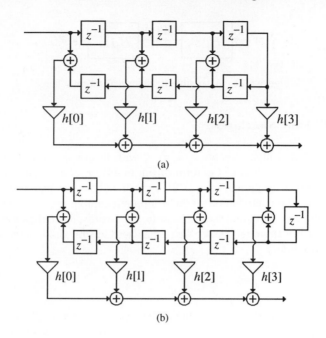

(a)

(b)

Figure 6.20   Linear-phase FIR structures: (a) Type 1, and (b) Type 2.

which can be rewritten in the form

$$H(z) = h[0]\left(1 + z^{-6}\right) + h[1]\left(z^{-1} + z^{-5}\right)$$
$$+ h[2]\left(z^{-2} + z^{-4}\right) + h[3]z^{-3}. \tag{6.35}$$

A realization of $H(z)$ based on the decomposition of Eq. (6.35) is shown in Figure 6.20(a). A similar decomposition can be applied for the realization of a Type 2 FIR transfer function. For example, for a length-8 Type 2 FIR transfer function, the pertinent decomposition is given by

$$H(z) = h[0]\left(1 + z^{-7}\right) + h[1]\left(z^{-1} + z^{-6}\right)$$
$$+ h[2]\left(z^{-2} + z^{-5}\right) + h[3]\left(z^{-3} + z^{-4}\right), \tag{6.36}$$

leading to the realization shown in Figure 6.20(b).

It should be noted that the structure of Figure 6.20(a) requires 4 multipliers, whereas, the original length-7 FIR filter structure of Figure 6.16(a) employs 7 multipliers. Likewise, the structure of Figure 6.20(b) requires 4 multipliers, compared to 8 multipliers in the direct realization of a length-8 FIR filter structure. A similar savings occur in the case of an FIR filter with an antisymmetric impulse response.

Figure 6.21   A possible IIR filter realization scheme.

6.5   Basic IIR Filter Structures

The causal IIR digital filters we are concerned with in this text are characterized by a real rational transfer function of the form of Eq. (4.44) or, equivalently, by the constant coefficient difference equation of Eq. (2.79). From the difference equation representation, it can be seen that the computation of the nth output sample requires the knowledge of several past samples of the output sequence, or in other words, the realization of a causal IIR filter requires some form of feedback. We outline here several simple and straightforward realizations of IIR filters.

6.5.1   Direct Forms

An Nth order IIR digital filter transfer function is characterized by $2N + 1$ unique coefficients and, in general, requires $2N + 1$ multipliers and $2N$ two-input adders for implementation. As in the case of FIR filter realization, IIR filter structures in which the multiplier coefficients are precisely the coefficients of the transfer function are called *direct form* structures. We now describe the development of these structures.

Consider for simplicity a third-order IIR filter characterized by a transfer function

$$H(z) = \frac{P(z)}{D(z)} = \frac{p_0 + p_1 z^{-1} + p_2 z^{-2} + p_3 z^{-3}}{1 + d_1 z^{-1} + d_2 z^{-2} + d_3 z^{-3}}, \qquad (6.37)$$

which we can implement as a cascade of two filter sections as shown in Figure 6.21 where

$$H_1(z) = \frac{W(z)}{X(z)} = P(z) = p_0 + p_1 z^{-1} + p_2 z^{-2} + p_3 z^{-3}, \qquad (6.38)$$

and

$$H_2(z) = \frac{Y(z)}{W(z)} = \frac{1}{D(z)} = \frac{1}{1 + d_1 z^{-1} + d_2 z^{-2} + d_3 z^{-3}}. \qquad (6.39)$$

The filter section $H_1(z)$ of Eq. (6.38) is seen to be an FIR filter and can be realized as shown in Figure 6.22(a). We next consider the realization of $H_2(z)$ given by Eq. (6.39). Note that a time-domain representation of this transfer function is given by

$$y[n] = w[n] - d_1 y[n-1] - d_2 y[n-2] - d_3 y[n-3], \qquad (6.40)$$

resulting in the realization indicated in Figure 6.22(b).

A cascade of the structures of Figures 6.22(a) and (b) as indicated in Figure 6.21 leads to a realization of the original IIR transfer function $H(z)$ of Eq. (6.37). The resulting structure is sketched in Figure 6.23(a) and is commonly known as the *direct form I* structure. Note that the overall realization is noncanonic since it employs six delays to implement a third-order transfer function. The transpose of this structure is sketched in

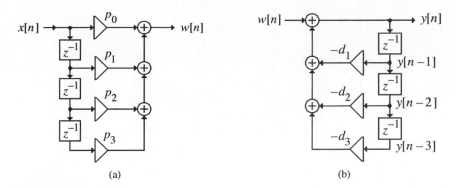

Figure 6.22   (a) Realization of the transfer function $H_1(z) = W(z)/X(z)$ and (b) realization of the transfer function $H_2(z) = Y(z)/W(z)$.

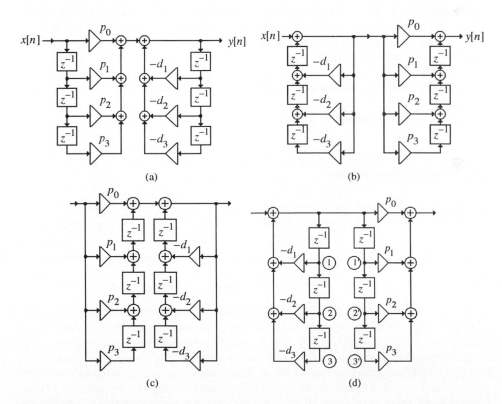

Figure 6.23   (a) Direct Form I, (b) direct form I$_t$ structures, (c) and (d) additional direct form structures.

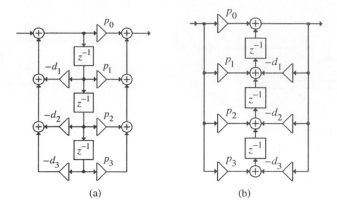

(a)   (b)

Figure 6.24   Direct form II and II$_t$ structures.

Figure 6.23(b). Various other noncanonic direct forms can be derived by simple block diagram manipulations. Two such realizations are shown in Figure 6.23(c) and (d).

To derive a canonic realization, we observe that in Figure 6.23(d) the signal variables at nodes ① and ①' are the same, and hence, the two top delays can be shared. Likewise, the signal variables at nodes ② and ②' are the same, which permits the sharing of the two middle delays. Following the same argument, we can share the two delays at the bottom leading to the final canonic structure shown in Figure 6.24(a), which is called the *direct form II* realization. The transpose of this is indicated in Figure 6.24(b).

The structures for direct form I and direct form II realizations of an Nth order IIR transfer function should be evident from the third-order structures of Figures 6.23 and 6.24.

6.5.2 Cascade Realizations

By expressing the numerator and the denominator polynomials of the transfer function $H(z)$ as a product of polynomials of lower degree, a digital filter is often realized as a cascade of low-order filter sections. Consider for example, $H(z) = P(z)/D(z)$ expressed as

$$H(z) = \frac{P_1(z)P_2(z)P_3(z)}{D_1(z)D_2(z)D_3(z)}. \tag{6.41}$$

Various different cascade realizations of $H(z)$ can be obtained by different pole-zero polynomial pairings. Some examples of such realizations are shown in Figure 6.25. Additional cascade realizations are obtained by simply changing the ordering of the sections.[2] Figure 6.26 illustrates examples of different structures obtained by changing the ordering of the sections. There are altogether a total of 36 cascade realizations for the factored form indicated in Eq. (6.41) based on pole-zero pairings and ordering (Problem 6.24). In practice due to the finite wordlength effects, each such cascade realization behaves differently from others.

---

[2]See Section 9.6.7.

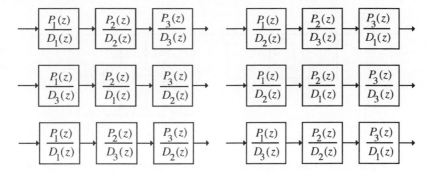

Figure 6.25   Examples of different equivalent cascade realizations obtained by different pole-zero pairings.

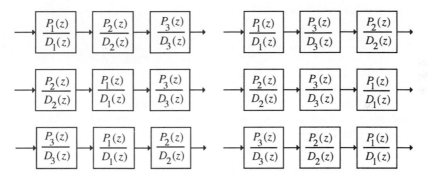

Figure 6.26   Different cascade realizations obtained by changing the ordering of the sections.

Usually, the polynomials are factored into a product of first-order and second-order polynomials. In this case, $H(z)$ is expressed as

$$H(z) = p_0 \prod_k \left(\frac{1 + \beta_{1k}z^{-1} + \beta_{2k}z^{-2}}{1 + \alpha_{1k}z^{-1} + \alpha_{2k}z^{-2}} \right). \qquad (6.42)$$

In the above, for a first-order factor, $\alpha_{2k} = \beta_{2k} = 0$. A possible realization of a third-order transfer function

$$H(z) = p_0 \left(\frac{1 + \beta_{11}z^{-1}}{1 + \alpha_{11}z^{-1}} \right) \left(\frac{1 + \beta_{12}z^{-1} + \beta_{22}z^{-2}}{1 + \alpha_{12}z^{-1} + \alpha_{22}z^{-2}} \right).$$

is shown in Figure 6.27.

As in the FIR filter case, the factors for the cascade realization of an IIR transfer function can also be determined by MATLAB using the function `roots`.

6.5.3   Parallel Realizations

An IIR transfer function can be realized in a parallel form by making use of the partial-fraction expansion of the transfer function. A partial-fraction expansion of the transfer

Figure 6.27   Cascade realization of a third-order IIR transfer function.

function in the form of Eq. (3.111) leads to the *parallel form I*. Thus, assuming simple poles, $H(z)$ is expressed in the form

$$H(z) = \gamma_0 + \sum_k \left(\frac{\gamma_{0k} + \gamma_{1k}z^{-1}}{1 + \alpha_{1k}z^{-1} + \alpha_{2k}z^{-2}} \right). \tag{6.43}$$

In the above, for a real pole, $\alpha_{2k} = \gamma_{1k} = 0$.

A direct partial-fraction expansion of the transfer function $H(z)$ expressed as a ratio of polynomials in z leads to the second basic form of the parallel structure, called the *parallel form II* [Mit77c]. Assuming simple poles, we arrive at

$$H(z) = \delta_0 + \sum_k \left(\frac{\delta_{1k}z^{-1} + \delta_{2k}z^{-2}}{1 + \alpha_{1k}z^{-1} + \alpha_{2k}z^{-2}} \right). \tag{6.44}$$

Here, for a real pole, $\alpha_{2k} = \delta_{2k} = 0$.

The two basic parallel realizations of a third-order IIR transfer function are sketched in Figure 6.28.

Example 6.7   We develop different realizations of the third-order IIR transfer function:

$$H(z) = \frac{0.44z^2 + 0.362z + 0.02}{z^3 + 0.4z^2 + 0.18z - 0.2} = \frac{0.44z^{-1} + 0.362z^{-2} + 0.02z^{-3}}{1 + 0.4z^{-1} + 0.18z^{-2} - 0.2z^{-3}}. \tag{6.45}$$

A direct form II realization of the above transfer function is shown in Figure 6.29(a).

By factoring the numerator and the denominator polynomials of $H(z)$ as given in Eq. (6.42), we obtain

$$H(z) = \frac{0.44z^2 + 0.362z + 0.02}{(z^2 + 0.8z + 0.5)(z - 0.4)}$$

$$= \left(\frac{0.44 + 0.362z^{-1} + 0.02z^{-2}}{1 + 0.8z^{-1} + 0.5z^{-2}} \right) \left(\frac{z^{-1}}{1 - 0.4z^{-1}} \right).$$

From the above, we arrive at a cascade realization of $H(z)$, as shown in Figure 6.29(b). Another cascade realization is obtained by using a different pole-zero pairing:

$$H(z) = \left(\frac{z^{-1}}{1 + 0.8z^{-1} + 0.5z^{-2}} \right) \left(\frac{0.44 + 0.362z^{-1} + 0.02z^{-2}}{1 - 0.4z^{-1}} \right),$$

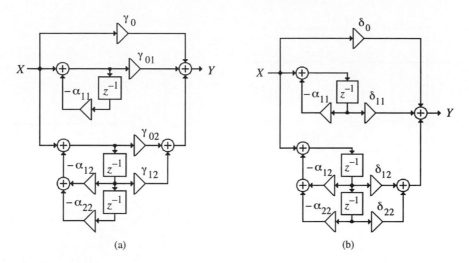

Figure 6.28   Parallel realizations of a third-order IIR transfer function: (a) parallel form I, and (b) parallel form II.

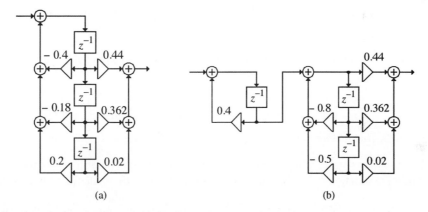

Figure 6.29   (a) Direct form II realization, and (b) cascade realization based on direct form II realization of each section.

whose realization is left as an exercise. However, it should be noted that the realization based on the above factored form will be noncanonic since it would require four delays instead of three delays employed in Figure 6.29(b). The total number of multipliers in both realizations remain the same.

Next, we make a partial-fraction expansion of $H(z)$ of the form of Eq. (3.111), resulting in

$$H(z) = -0.1 + \frac{0.6}{1 - 0.4z^{-1}} + \frac{-0.5 - 0.2z^{-1}}{1 + 0.8z^{-1} + 0.5z^{-2}},$$

which leads to the parallel form I realization indicated in Figure 6.30(a).

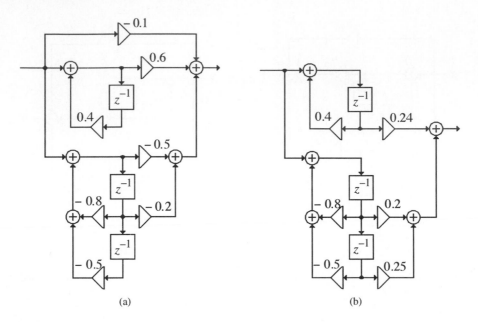

Figure 6.30   (a) Parallel form I realization, and (b) parallel form II realization.

Finally, a direct partial-fraction expansion of $H(z)$ expressed as a ratio of polynomials in z is given by

$$H(z) = \frac{0.24}{z - 0.4} + \frac{0.2z + 0.25}{z^2 + 0.8z + 0.5} = \frac{0.24z^{-1}}{1 - 0.4z^{-1}} + \frac{0.2z^{-1} + 0.25z^{-2}}{1 + 0.8z^{-1} + 0.5z^{-2}},$$

resulting in the parallel form II realization sketched in Figure 6.30(b).

The functions `residue` and `residuez` in MATLAB can be utilized to develop the two parallel realizations described above. We illustrate their use in the following example.

Example 6.8 We repeat the parallel realizations of the previous example using Program 6_1 given below.

```
% Program 6_1
% Parallel Realizations of an IIR Transfer Function
%
num = input('Numerator coefficient vector = ');
den = input('Denominator coefficient vector = ');
[r1,p1,k1] = residuez(num,den);
[r2,p2,k2] = residue(num,den);
disp('Parallel Form I')
disp('Residues are');disp(r1);
disp('Poles are at');disp(p1);
disp('Constant value');disp(k1);
disp('Parallel Form II')
```

```
disp('Residues are');disp(r2);
disp('Poles are at');disp(p2);
disp('Constant value');disp(k2);
```

The input data requested by the program are the vectors num and den, containing the numerator and denominator coefficients, respectively. For our example, these are given by

num = [0   0.44   0.362   0.02],   den = [1   0.4   0.18   -0.2].

The results generated by this program are:

```
Parallel Form I
Residues are
  -0.2500 - 0.0000i
  -0.2500 + 0.0000i
   0.6000
  -0.1000

Poles are at
  -0.4000 + 0.5831i
  -0.4000 - 0.5831i
   0.4000
        0

Constant value
  -0.1000

Parallel Form II
Residues are
   0.1000 - 0.1458i
   0.1000 + 0.1458i
   0.2400

Poles are at
  -0.4000 + 0.5831i
  -0.4000 - 0.5831i
   0.4000

Constant value
```

The complex conjugate pairs in each partial-fraction expansion given above can be combined in MATLAB using the statement [b,a] = residue[R, P, 0] where R is the two-element vector of the complex conjugate residues, and P is the two-element vector of the complex conjugate poles, verifying the correctness of the results obtained in the previous example.

6.6   State-Space Structures

The state-space description of a causal LTI digital filter as given by Eqs. (2.97) and (2.98) provides an alternative realization scheme that we consider in this section. Because of the straight forward method of generating an equivalent state-space description by a simple linear transformation of the state variables, the state-space-based realization lends itself to fast generation of equivalent digital filter structures. It should be noted that a state-space realization, in general, is computationally quite intensive, requiring considerably more multipliers compared to most other realizations and may not be attractive in many applications. However, the state-space-based formalism provides an elegant approach in the development of equivalent realization with improved performance under finite wordlength constraints from a given digital filter structure. For example, the quantization of the product at the output of a multiplier can be modeled as an additive noise source, which propagates to the output of the digital filter and appears as noise added to the desired signal. We outline in Section 9.8.2 a systematic method of determining an equivalent state-space structure with a minimum product roundoff noise. The quantization of arithmetic operations can also lead to undesirable oscillatory output called limit cycles. To this end we develop in Section 9.9.3 the conditions on the state-transition matrix to guarantee the absence of limit cycles. In this section we consider the development of state-space-based realization and its transformation to other equivalent structures.

6.6.1   General Form

Without any loss of generality, we consider the realization of a second-order IIR transfer function. Consider a second-order IIR digital filter described by the state-space equations

$$\begin{bmatrix} s_1[n+1] \\ s_2[n+1] \end{bmatrix} = \begin{bmatrix} a_{11} & a_{12} \\ a_{21} & a_{22} \end{bmatrix} \begin{bmatrix} s_1[n] \\ s_2[n] \end{bmatrix} + \begin{bmatrix} b_1 \\ b_2 \end{bmatrix} x[n], \tag{6.46}$$

$$y[n] = [\,c_1 \quad c_2\,] \begin{bmatrix} s_1[n] \\ s_2[n] \end{bmatrix} + dx[n]. \tag{6.47}$$

A direct realization of the above is sketched in Figure 6.31. Note that this realization requires a total of nine multipliers and six two-input adders, whereas a canonic direct form realization requires five multipliers and four two-input adders.

The form of the state-space structure for a higher order transfer function should be evident from this figure. In the general case, the state-space realization of an Nth order IIR transfer function requires $(N+1)^2$ multipliers and $N(N+1)$ two-input adders, whereas, a direct form realization requires $2N+1$ multipliers and $2N$ two-input adders.

6.6.2   Development of the State-Space Structure

It is quite straightforward to write down by inspection the state-space description of a digital filter from its structural representation. To this end, we observe that the outputs of the delay units are essentially the state-variables. Thus, to arrive at the state-space description, we express all other node variables and the output as a function of the variables at each delay output and the input. We illustrate this approach with the aid of an example.

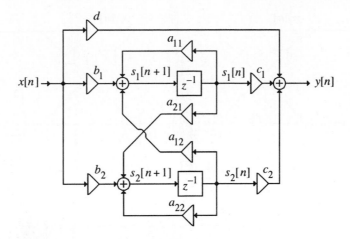

Figure 6.31 A second-order IIR state-space structure.

Example 6.9 Consider the digital filter of Figure 6.4, which we have redrawn in Figure 6.32 with node variables shown in the time-domain. Note that the two state-variables are $s_1[n]$ and $s_2[n]$ as shown in the figure. As carried out in Example 6.2, the output signals of the four adders are given by

$$w_1[n] = x[n] - \alpha s_2[n],$$
$$w_3[n] = w_1[n] - \delta s_1[n],$$
$$w_5[n] = s_1[n] + \varepsilon w_3[n],$$
$$y[n] = \gamma s_2[n] + \beta w_1[n].$$

But from the figure, $s_1[n] = w_3[n-1]$ and $s_2[n] = w_5[n-1]$. From these equations we therefore arrive at:

$$s_1[n+1] = w_3[n] = w_1[n] - \delta s_1[n] = x[n] - \alpha s_2[n] - \delta s_1[n],$$
$$s_2[n+1] = w_5[n] = s_1[n] + \varepsilon w_3[n] = s_1[n] + \varepsilon\{w_1[n] - \delta s_1[n]\}$$
$$= s_1[n] - \delta\varepsilon s_1[n] + \varepsilon\{x[n] - \alpha s_2[n]\},$$
$$y[n] = \gamma s_2[n] + \beta w_1[n] = \gamma s_2[n] + \beta\{x[n] - \alpha s_2[n]\}.$$

These equations can now be rewritten in a matrix form to yield

$$\begin{bmatrix} s_1[n+1] \\ s_2[n+1] \end{bmatrix} = \begin{bmatrix} -\delta & -\alpha \\ 1-\delta\varepsilon & -\alpha\varepsilon \end{bmatrix} \begin{bmatrix} s_1[n] \\ s_2[n] \end{bmatrix} + \begin{bmatrix} 1 \\ \varepsilon \end{bmatrix} x[n], \qquad (6.48)$$

$$y[n] = [0 \quad \gamma - \alpha\beta] \begin{bmatrix} s_1[n] \\ s_2[n] \end{bmatrix} + \beta x[n]. \qquad (6.49)$$

From the above, we thus identify the state-space parameters as

$$\mathbf{A} = \begin{bmatrix} -\delta & -\alpha \\ 1-\delta\varepsilon & -\alpha\varepsilon \end{bmatrix}, \quad \mathbf{B} = \begin{bmatrix} 1 \\ \varepsilon \end{bmatrix}, \quad \mathbf{C} = [0 \quad \gamma - \alpha\beta], \quad D = \beta. \qquad (6.50)$$

A realization of this filter based on the above state-space equations is shown in Figure 6.33. Note that the state-space realization requires 7 multipliers and 5 two-input adders, whereas the original realization of Figure 6.32 uses 5 multipliers and 4 two-input adders.

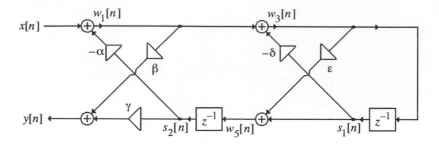

Figure 6.32 The cascaded lattice structure of Figure 6.4 with time-domain representations of all signal variables.

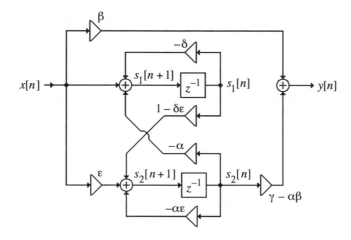

Figure 6.33 A second-order IIR state-space structure realizing the digital filter of Figure 6.32.

We next determine the transfer function using the formula of Eq. (4.56). To this end, we first compute $[z\mathbf{I} - \mathbf{A}]^{-1}$:

$$\begin{bmatrix} z+\delta & \alpha \\ \delta\varepsilon - 1 & z + \alpha\varepsilon \end{bmatrix}^{-1} = \frac{1}{\Delta}\begin{bmatrix} z + \alpha\varepsilon & -\alpha \\ 1 - \delta\varepsilon & z + \delta \end{bmatrix},$$

where Δ is the determinant of the $\mathbf{A}$ matrix, i.e.,

$$\Delta = (z+\delta)(z + \alpha\varepsilon) - \alpha(\delta\varepsilon - 1) = z^2 + (\alpha\varepsilon + \delta)z + \alpha.$$

Substituting the above and the state-space parameters of Eq. (6.50) in Eq. (4.56), we arrive at the transfer function of the digital filter structure of Figure 6.32 as

$$\begin{aligned} H(z) &= \frac{1}{\Delta}[0 \quad \gamma - \alpha\beta]\begin{bmatrix} z + \alpha\varepsilon & -\alpha \\ 1 - \delta\varepsilon & z + \delta \end{bmatrix}\begin{bmatrix} 1 \\ \varepsilon \end{bmatrix} + \beta \\ &= \frac{(\gamma - \alpha\beta)\varepsilon z + \beta z^2 + \beta(\alpha\varepsilon + \delta)z + \alpha\beta + \gamma - \alpha\beta}{z^2 + (\alpha\varepsilon + \delta)z + \alpha} \\ &= \frac{\beta z^2 + (\gamma\varepsilon + \beta\delta)z + \gamma}{z^2 + (\alpha\varepsilon + \delta)z + \alpha}, \end{aligned}$$

which is the same as that given by Eq. (6.9) if expressed as a ratio of polynomials in z^{-1}.

As indicated in earlier chapters, MATLAB includes a number of functions for state-space analysis of LTI discrete-time systems. For example, the functions ss2tf, tf2ss, ss2zp, and zp2ss can be utilized for various purposes.

6.6.3   Generation of Equivalent State-Space Structures

The nonuniqueness of the state-space description as discussed in Section 2.6.4 lends itself readily to the development of equivalent digital filter structures from a given state-space-based realization. As indicated there, for a given a state-space description as in Eqs. (2.97) and (2.98), we define a new state-vector $\hat{s}[n]$ as

$$\hat{s}[n] = \mathbf{Q}s[n]. \tag{6.51}$$

resulting in a new state-space description of the digital filter given by

$$\hat{s}[n+1] = \hat{\mathbf{A}}\hat{s}[n] + \hat{\mathbf{B}}x[n], \tag{6.52}$$

$$y[n] = \hat{\mathbf{C}}\hat{s}[n] + Dx[n], \tag{6.53}$$

where

$$\hat{\mathbf{A}} = \mathbf{Q}\mathbf{A}\mathbf{Q}^{-1}, \qquad \hat{\mathbf{B}} = \mathbf{Q}\mathbf{B}, \qquad \hat{\mathbf{C}} = \mathbf{C}\mathbf{Q}^{-1}, \tag{6.54}$$

are the new state-space parameters.

We showed in Section 2.6.4 that a nonsingular linear transformation of the state variables keeps the input-output relation of the digital filter invariant. This can also be demonstrated by computing the transfer function of the digital filter. For the new state-space characterization the transfer function is given by

$$\hat{H}(z) = \hat{\mathbf{C}}(z\mathbf{I} - \hat{\mathbf{A}})^{-1}\hat{\mathbf{B}} + D = \mathbf{C}\mathbf{Q}^{-1}(z\mathbf{I} - \mathbf{Q}\mathbf{A}\mathbf{Q}^{-1})^{-1}\mathbf{Q}\mathbf{B} + D$$
$$= \mathbf{C}\left\{\mathbf{Q}^{-1}(z\mathbf{I} - \mathbf{Q}\mathbf{A}\mathbf{Q}^{-1})\mathbf{Q}\right\}^{-1}\mathbf{B} + D$$
$$= \mathbf{C}\left\{z\mathbf{Q}^{-1}\mathbf{Q} - \mathbf{Q}^{-1}\mathbf{Q}\mathbf{A}\mathbf{Q}^{-1}\mathbf{Q}\right\}^{-1}\mathbf{B} + D$$
$$= \mathbf{C}(z\mathbf{I} - \mathbf{A})^{-1}\mathbf{B} + D = H(z). \tag{6.55}$$

indicating that the transfer function remains invariant under a nonsingular linear transformation of the state-vector.

Example 6.10 To illustrate the equivalence transformation, we determine an alternative state space realization of the digital filter of Figure 6.33. Let

$$\mathbf{Q} = \begin{bmatrix} 1 & -1 \\ 1 & 1 \end{bmatrix}. \tag{6.56}$$

The inverse of the matrix $\mathbf{Q}$ is given by

$$\mathbf{Q}^{-1} = \frac{1}{2}\begin{bmatrix} 1 & 1 \\ -1 & 1 \end{bmatrix}. \tag{6.57}$$

From Eqs. (6.50), (6.54), (6.56), and (6.57), we arrive at the new state-space parameters as

$$\hat{\mathbf{A}} = \frac{1}{2} \begin{bmatrix} 1 & -1 \\ 1 & 1 \end{bmatrix} \begin{bmatrix} -\delta & -\alpha \\ 1 - \delta\varepsilon & -\alpha\varepsilon \end{bmatrix} \begin{bmatrix} 1 & 1 \\ -1 & 1 \end{bmatrix}$$

$$= \frac{1}{2} \begin{bmatrix} (1-\varepsilon)(\alpha-\delta) - 1 & -(1-\varepsilon)(\alpha+\delta) - 1 \\ (1+\varepsilon)(\alpha-\delta) + 1 & -(1+\varepsilon)(\alpha+\delta) + 1 \end{bmatrix},$$

$$\hat{\mathbf{B}} = \begin{bmatrix} 1 & -1 \\ 1 & 1 \end{bmatrix} \begin{bmatrix} 1 \\ \varepsilon \end{bmatrix} = \begin{bmatrix} 1 - \varepsilon \\ 1 + \varepsilon \end{bmatrix},$$

$$\hat{\mathbf{C}} = \frac{1}{2} \begin{bmatrix} 0 & \gamma - \alpha\beta \end{bmatrix} \begin{bmatrix} 1 & 1 \\ -1 & 1 \end{bmatrix} = \frac{1}{2} \begin{bmatrix} \alpha\beta - \gamma & \gamma - \alpha\beta \end{bmatrix}.$$

Realization of the digital filter based on the new state-space description is left as an exercise (Problem 6.36).

6.7   Allpass Filters

We now turn our attention to the realization of a very special type of IIR transfer function, the allpass function, introduced earlier in Section 4.4. We recall from its definition in Eq. (4.123) that an IIR allpass transfer function $A(z)$ has a unity magnitude response for all frequencies, i.e., $|A(e^{j\omega})| = 1$ for all values of ω. The digital allpass filter is a versatile building block for signal processing applications. Some of its possible applications have already been pointed out earlier. For example, as indicated in Section 4.4.3, it is used often as a delay equalizer, in which case it is designed such that when cascaded with another discrete-time system, the overall cascade has a constant group delay in the frequency range of interest. Another application described in Section 4.6 is in the efficient implementation of a set of transfer functions satisfying certain complementary properties. For example, a pair of power-complementary first-order lowpass and highpass transfer functions can be implemented simultaneously employing only a single first-order allpass filter. Likewise, a pair of power-complementary second-order bandpass and bandstop transfer functions can be implemented simultaneously employing only a single second-order allpass filter. In fact, we shall demonstrate later in Section 6.10 that a large class of power-complementary transfer function pairs can be realized as a parallel connection of two allpass filters. It is thus of interest to develop techniques for the computationally efficient realization of allpass transfer functions, which is the subject of this section.

An Mth order real coefficient allpass transfer function is of the form of Eq. (4.124), with the numerator being the mirror-image polynomial of the denominator. A direct realization of an Mth order allpass transfer function requires $2M$ multipliers. Since an Mth-order allpass transfer function is characterized by M unique coefficients, our objective here is to develop realization methods requiring only M multipliers. We outline here two different approaches to the minimum multiplier realizations of allpass transfer functions.

6.7.1   Realization Based on the Multiplier Extraction Approach

Since an arbitrary allpass transfer function can be expressed as the product of second-order and/or first-order allpass transfer functions, we consider the realization of these lower order transfer functions here. Even though an allpass transfer function can be realized using any of the methods discussed in this chapter, our objective here is to develop structures

Figure 6.34 A multiplier-less two-pair constrained by a single multiplier.

Figure 6.35 Development of the Type 1A first-order allpass structure.

that remain allpass despite changes in the multiplier coefficients that may occur due to coefficient quantization [Mit74a].

Consider first the realization of a first-order allpass transfer function given by

$$A_1(z) = \frac{d_1 + z^{-1}}{1 + d_1 z^{-1}}. \tag{6.58}$$

Since the above transfer function is uniquely characterized by a single constant d_1, we attempt to realize it using a structure containing a single multiplier d_1 in the form of Figure 6.34. Substituting $G(z) = d_1$ in Eq. (4.156b), we express the input transfer function $A_1(z) = Y_1/X_1$ in terms of the transfer parameters of the two-pair as

$$A_1(z) = t_{11} + \frac{t_{12}t_{21}d_1}{1 - d_1 t_{22}} = \frac{t_{11} - d_1(t_{11}t_{22} - t_{12}t_{21})}{1 - d_1 t_{22}}. \tag{6.59}$$

A comparison of Eqs. (6.58) and (6.59) yields

$$t_{11} = z^{-1}, \qquad t_{22} = -z^{-1}, \tag{6.60}$$

$$t_{11}t_{22} - t_{12}t_{21} = -1. \tag{6.61}$$

Substituting Eq. (6.60) in Eq. (6.61), we arrive at

$$t_{12}t_{21} = 1 - z^{-2},$$

which leads to four possible solutions:

Type 1A : $t_{11} = z^{-1}$, $t_{22} = -z^{-1}$, $t_{12} = 1 - z^{-2}$, $t_{21} = 1$;   (6.62a)

Type 1B : $t_{11} = z^{-1}$, $t_{22} = -z^{-1}$, $t_{12} = 1 + z^{-1}$, $t_{21} = 1 - z^{-1}$;   (6.62b)

Type 1A$_t$: $t_{11} = z^{-1}$, $t_{22} = -z^{-1}$, $t_{12} = 1$, $t_{21} = 1 - z^{-2}$;   (6.62c)

Type 1B$_t$: $t_{11} = z^{-1}$, $t_{22} = -z^{-1}$, $t_{12} = 1 - z^{-1}$, $t_{21} = 1 + z^{-1}$.   (6.62d)

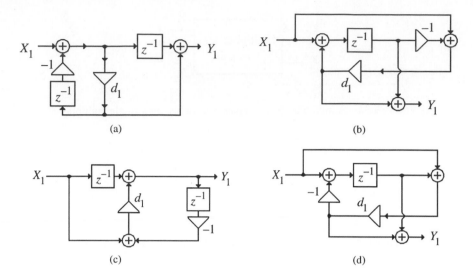

Figure 6.36   First-order single-multiplier allpass structures: (a) Type 1A, (b) Type 1B, (c) Type 1A$_t$, (d) Type B$_t$.

We now outline the development of the two-pair structure implementing the transfer parameters of Eq. (6.62a). From these equations we arrive at

$$Y_2 = X_1 - z^{-1}X_2,$$
$$Y_1 = z^{-1}X_1 + (1 - z^{-2})X_2 = z^{-1}Y_2 + X_2.$$

A realization of the above is sketched in Figure 6.35. By constraining the X_2, Y_2 terminal-pair with the multiplier d_1, we arrive at the single multiplier realization of the first-order allpass transfer function, as sketched in Figure 6.36(a). In a similar fashion, the other three single-multiplier allpass structures can be realized from their transfer parameter descriptions of Eqs. (6.62b), (6.62c) and (6.62d) (Problem 6.44). The final realizations are indicated in Figure 6.36(b) to (d). These structures have been called the *Type 1 allpass networks*. It should be noted that the structures of Figure 6.36(c) and (d) are, respectively, the transpose of the structures of Figure 6.36(a) and (b).

We next consider the realization of a second-order allpass transfer function. Such a function is characterized by two unique coefficients and, hence, can be realized by a structure with two multipliers with multiplier constants d_1 and d_2. Various forms of the second-order allpass transfer functions exist. Circuits realizing the transfer function of the form

$$A_2(z) = \frac{d_1 d_2 + d_1 z^{-1} + z^{-2}}{1 + d_1 z^{-1} + d_1 d_2 z^{-2}} \tag{6.63}$$

are called the *Type 2 allpass networks* and are sketched in Figure 6.37. Additional Type 2 allpass structures can be derived by transposing the structures of Figure 6.37.

Another form of the second-order allpass transfer function is given by

$$A_2(z) = \frac{d_2 + d_1 z^{-1} + z^{-2}}{1 + d_1 z^{-1} + d_2 z^{-2}}. \tag{6.64}$$

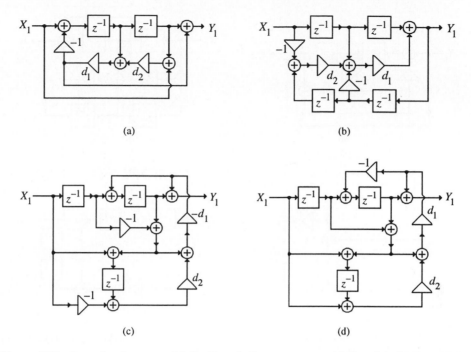

Figure 6.37   Second-order two-multiplier Type 2 allpass structures. (a) Type 2A, (b) Type 2D, (c) Type 2B, and (d) Type 2C.

The corresponding circuits are called the *Type 3 allpass structures* and are indicated in Figure 6.38.

Example 6.11   Realize the third-order allpass transfer function

$$A_3(z) = \frac{-0.2 + 0.18z^{-1} + 0.4z^{-2} + z^{-3}}{1 + 0.4z^{-1} + 0.18z^{-2} - 0.2z^{-3}}. \qquad (6.65)$$

By factoring the numerator and the denominator, we rewrite the above as

$$A_3(z) = \frac{(-0.4 + z^{-1})(0.5 + 0.8z^{-1} + z^{-2})}{(1 - 0.4z^{-1})(1 + 0.8z^{-1} + 0.5z^{-2})}. \qquad (6.66)$$

A three-multiplier realization of the above allpass transfer function is shown in Figure 6.39, where the first-order allpass section has been implemented using the structure of Figure 6.36(b) and the second-order allpass section has been implemented using the structure of Figure 6.38(a).

6.7.2   Realization Based on the Two-Pair Extraction Approach

The stability test algorithm described in Section 4.8.2 also leads to an elegant realization of an allpass transfer function in the form of a cascaded two-pair [Vai87e]. This algorithm is based on the development of a series of $(m-1)$th-order allpass transfer function $A_{m-1}(z)$

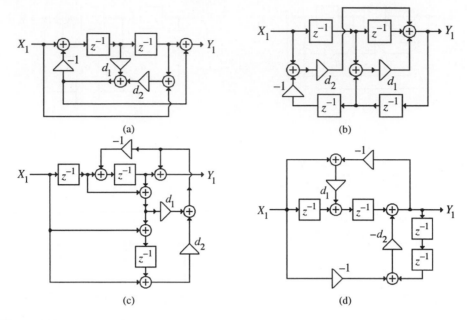

Figure 6.38   Second-order two-multiplier Type 3 allpass structures. (a) Type 3A, (b) Type 3D, (c) Type 3C, and (d) Type 3H.

Figure 6.39   A three-multiplier realization of the allpass transfer function of Eq. (6.66).

from an mth-order allpass transfer function $A_m(z)$:

$$A_m(z) = \frac{d_m + d_{m-1}z^{-1} + d_{m-2}z^{-2} + \cdots + d_1 z^{-(m-1)} + z^{-m}}{1 + d_1 z^{-1} + d_2 z^{-2} + \cdots + d_{m-1}z^{-(m-1)} + d_m z^{-m}} \qquad (6.67)$$

using the recursion

$$A_{m-1}(z) = z\left[\frac{A_M(z) - k_m}{1 - k_m A_m(z)}\right], \qquad m = M, M-1, \ldots 1, \qquad (6.68)$$

where $k_m = A_m(0) = d_m$. It has been shown in Section 4.8.2 that $A_m(z)$ is stable if and only if

$$k_m^2 < 1 \text{ for } m = M, M-1, \ldots, 1. \qquad (6.69)$$

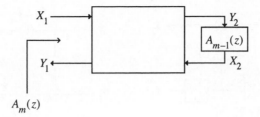

Figure 6.40   Realization of $A_m(z)$ by two-pair extraction.

If the allpass transfer function $A_{m-1}(z)$ is expressed in the form

$$A_{m-1}(z) = \frac{d'_{m-1} + d'_{m-2}z^{-1} + \cdots + d'_1 z^{-(m-2)} + z^{-(m-1)}}{1 + d'_1 z^{-1} + \cdots + d'_{m-2}z^{-(m-2)} + d'_{m-1}z^{-(m-1)}}, \qquad (6.70)$$

then the coefficients of $A_{m-1}(z)$ are simply related to the coefficients of $A_m(z)$ through the expression

$$d'_i = \frac{d_i - d_m d_{m-i}}{1 - d_m^2}, \qquad i = 1, 2, \ldots, m-1. \qquad (6.71)$$

To develop a realization of $A_m(z)$ using the above algorithm, we rewrite Eq. (6.68) as

$$A_m(z) = \frac{k_m + z^{-1}A_{m-1}(z)}{1 + k_m z^{-1}A_{m-1}(z)}. \qquad (6.72)$$

We realize $A_m(z)$ by extracting a two-pair constrained by $A_{m-1}(z)$ in the form of Figure 6.40.

Now from Eq. (4.156b), $A_m(z)$ can be expressed in terms of the transfer parameters of the two-pair as

$$A_m(z) = \frac{t_{11} - (t_{11}t_{22} - t_{12}t_{21})A_{m-1}(z)}{1 - t_{22}A_{m-1}(z)}. \qquad (6.73)$$

Comparing Eqs. (6.72) and (6.73), we readily obtain

$$t_{11} = k_m, \qquad t_{22} = -k_m z^{-1}, \qquad (6.74a)$$

$$t_{11}t_{22} - t_{12}t_{21} = -z^{-1}. \qquad (6.74b)$$

Substituting Eq. (6.74a) in Eq. (6.74b), we get

$$t_{12}t_{21} = (1 - k_m^2)z^{-1}. \qquad (6.75)$$

As can be seen from the above, there are a number of solutions for t_{12} and t_{21} leading to different realizations for the two-pair. Some possible solutions are as indicated below:

$$t_{11} = k_m, \; t_{22} = -k_m z^{-1}, \; t_{12} = (1 - k_m^2)z^{-1}, \; t_{21} = 1, \qquad (6.76a)$$

$$t_{11} = k_m, \; t_{22} = -k_m z^{-1}, \; t_{12} = (1 - k_m)z^{-1}, \; t_{21} = (1 + k_m), \qquad (6.76b)$$

$$t_{11} = k_m, \; t_{22} = -k_m z^{-1}, \; t_{12} = \sqrt{1 - k_m^2}\,z^{-1}, \; t_{21} = \sqrt{1 - k_m^2}, \qquad (6.76c)$$

$$t_{11} = k_m, \; t_{22} = -k_m z^{-1}, \; t_{12} = z^{-1}, \; t_{21} = (1 - k_m^2). \qquad (6.76d)$$

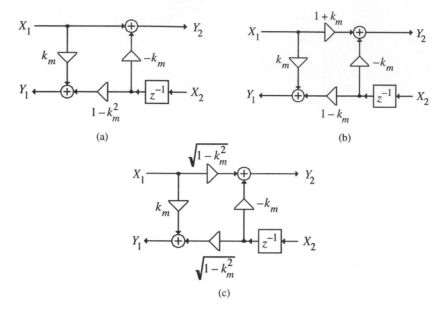

Figure 6.41   Direct realization of the two-pairs described by Eq. (6.76a) through (6.76c). (a) The two-pair described by Eq. (6.76a), (b) the two-pair described by Eq. (6.76b), and (c) the two-pair described by Eq. (6.76c).

The input-output relations of the two-pair described by Eq. (6.76a) are given by

$$Y_1 = k_m X_1 + (1 - k_m^2)z^{-1}X_2, \tag{6.77a}$$

$$Y_2 = X_1 - k_m z^{-1}X_2. \tag{6.77b}$$

A direct realization of the above equations leads to the 3-multiplier two-pair shown in Figure 6.41(a). In a similar manner, direct realizations based on Eqs. (6.76b) and (6.76c) result in the 4-multiplier structures of Figure 6.41(b) and (c), respectively. The structure of Figure 6.41(b) is called the *Kelly-Lochbaum form* [Kel62]. A direct realization of Eq. (6.76d) results in a 3-multiplier structure and is left as an exercise.

A 2-multiplier realization can be derived by manipulating the input-output relations of Eqs. (6.77a)and (6.77b). Using Eq. (6.77b), we can rewrite Eq. (6.77a) as

$$Y_1 = k_m Y_2 + z^{-1}X_2. \tag{6.78}$$

A realization of the two-pair based on Eqs. (6.77b) and (6.78) is given in Figure 6.42(a). The two-pair of Figure 6.42(a) is often referred to as a *lattice structure*.

The two-pair described by Eq. (6.76b) can be realized using a single multiplier. To this end, we first write its input-output relation from Eq. (6.76b) as

$$Y_1 = k_m X_1 + (1 - k_m)z^{-1}X_2, \tag{6.79a}$$

$$Y_2 = (1 + k_m)X_1 - k_m z^{-1}X_2. \tag{6.79b}$$

Defining

$$V_1 = k_m(X_1 - z^{-1}X_2), \tag{6.80}$$

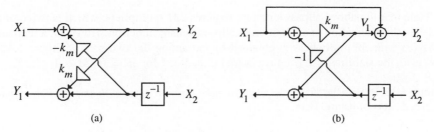

Figure 6.42   (a) A two-multiplier realization of the two-pair described by Eq. (6.76a), and (b) a one-multiplier realization of the two-pair described by Eq. (6.76b).

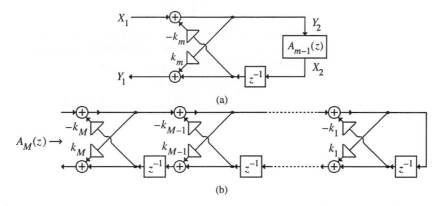

Figure 6.43   (a) Realization of $A_m(z)$ by extracting the lattice two-pair of Figure 6.42(a), and (b) cascaded lattice realization of $A_m(z)$.

we can rewrite Eqs. (6.79a) and (6.79b) as

$$Y_1 = V_1 + z^{-1} X_2, \tag{6.81a}$$
$$Y_2 = X_1 + V_1. \tag{6.81b}$$

A realization based on Eqs. (6.80), (6.81a) and (6.81b) leads to the single-multiplier two-pair of Figure 6.42(b). Note that the two-input adder with an incoming multiplier with a coefficient -1 is implemented as a subtractor.

The realization of the mth-order allpass transfer function $A_m(z)$ is therefore obtained by constraining any one of the two-pairs of Figure 6.41 and 6.42 by the $(m-1)$th-order allpass transfer function $A_{m-1}(z)$. For example, Figure 6.43(a) shows the realization of $A_m(z)$ by extracting the lattice two-pair of Figure 6.42(a).

Following the above algorithm, we can next realize $A_{m-1}(z)$ as a lattice two-pair constrained by the allpass transfer function $A_{m-2}(z)$. This process is repeated until the constraining transfer function is $A_0(z) = 1$. The complete realization of the original allpass transfer function $A_M(z)$ based on the extraction of the lattice two-pair of Figure 6.42(a) is therefore as indicated in Figure 6.43(b). It follows from our discussion in Section 4.8.2 that $A_M(z)$ is stable if the magnitudes of all multiplier coefficients in the realization are less than unity, i.e., $|k_m| < 1$ for $m = M, M-1, \ldots, 1$.

Note that the above allpass structure requires $2M$ multipliers, which is twice of that needed in the realization of an Mth-order allpass transfer function. However, a realization of $A_M(z)$ with M multipliers is obtained by extracting the two-pair of Figure 6.42(b). Here, also, the stability of $A_M(z)$ is ensured if $|k_m| < 1$ for $m = M, M - 1, \ldots, 1$.

Example 6.12 Realize the third-order allpass transfer function $A_3(z)$ of Example 6.10 by means of a cascaded lattice structure. Here,

$$
\begin{aligned}
A_3(z) &= \frac{d_3 + d_2 z^{-1} + d_1 z^{-2} + z^{-3}}{1 + d_1 z^{-1} + d_2 z^{-2} + d_3 z^{-3}} \\
&= \frac{-0.2 + 0.18 z^{-1} + 0.4 z^{-2} + z^{-3}}{1 + 0.4 z^{-1} + 0.18 z^{-2} - 0.2 z^{-3}}.
\end{aligned}
\tag{6.82}
$$

Using the method outlined above, we realize $A_3(z)$ first in the form of a lattice two-pair characterized by the multiplier coefficient $k_3 = d_3 = -0.2$ and constrained by an allpass $A_2(z)$ as indicated in Figure 6.44(a). The allpass $A_2(z)$ is of the form

$$
A_2(z) = \frac{d_2' + d_1' z^{-1} + z^{-2}}{1 + d_1' z^{-1} + d_2' z^{-2}},
\tag{6.83}
$$

whose coefficients are obtained using Eq. (6.71) and are given by

$$
d_1' = \frac{d_1 - d_3 d_2}{1 - d_3^2} = \frac{0.4 - (-0.2)(0.18)}{1 - (-0.2)^2} = 0.4541667,
\tag{6.84a}
$$

$$
d_2' = \frac{d_2 - d_3 d_1}{1 - d_3^2} = \frac{0.18 - (-0.2)(0.4)}{1 - (-0.2)^2} = 0.2708333.
\tag{6.84b}
$$

Next, the allpass $A_2(z)$ is realized as a lattice two-pair characterized by the multiplier coefficient $k_2 = d_2' = 0.2708333$ and constrained by an allpass $A_1(z)$, as indicated in Figure 6.44(b). The allpass $A_1(z)$ is of the form

$$
A_1(z) = \frac{d_1'' + z^{-1}}{1 + d_1'' z^{-1}},
\tag{6.85}
$$

where

$$
d_1'' = \frac{d_1' - d_2' d_1'}{1 - (d_2')^2} = \frac{d_1'}{1 + d_2'} = \frac{0.4541667}{1.2708333} = 0.3573771.
\tag{6.86}
$$

Finally, the allpass $A_1(z)$ is realized as a lattice structure characterized by the multiplier coefficient $k_1 = d_1'' = 0.3573771$, resulting in the complete realization of the allpass $A_3(z)$, as shown in Figure 6.44(c). Note from Figure 6.44(c) that all multiplier coefficients have magnitudes less than unity. Therefore, the allpass transfer function of Eq. (6.82) is stable.

A MATLAB program to realize an allpass transfer function in the cascaded lattice form is given below. It employs the function `poly2rc`. Note the similarity between this program and the Program 4_5 for the stability test.

```
% Program 6_2
% Cascaded Lattice Realization of an
% Allpass Transfer Function
%
```

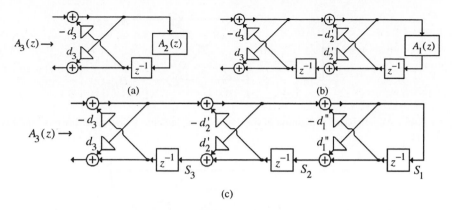

Figure 6.44 Cascaded lattice realization of the third-order allpass transfer function of Eqs. (6.79a) and (6.79b): $d_3 = -0.2$, $d_2' = 0.2708333$, and $d_1'' = 0.3573771$.

```
format long
den = input('Denominator coefficient vector = ');
k = poly2rc(den);
knew = fliplr(k);
disp('The lattice section multipliers are');disp(knew);
```

We illustrate its use in developing the realization of the allpass transfer function of Example 6.12 as given in Eq. (6.82). The program requests the input data, which are the vector of denominator coefficients in descending powers of z:

$$\text{den} = [1 \quad 0.4 \quad 0.18 \quad -0.2]$$

It then prints the lattice section multiplier coefficients as given below:

```
The lattice section multipliers are
   -0.2    0.27083333    0.35737705
```

These values are seen to be identical to those derived in Example 6.12.

6.8 Tunable IIR Digital Filters

In Section 4.3.3, we described two first-order and two second-order IIR digital transfer functions with tunable frequency response characteristics. As we show next, these transfer functions can be realized easily using allpass structures, and the resulting realizations provide independent tuning of the filter parameters such as the cutoff frequencies, bandwidth, etc. [Mit90a].

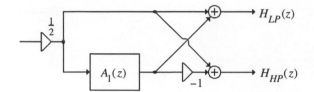

Figure 6.45   Allpass based realization of the doubly-complementary first-order lowpass and high-pass filters.

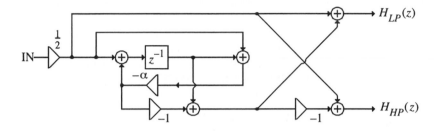

Figure 6.46   A tunable first-order lowpass/highpass filter structure.

6.8.1   Tunable Lowpass and Highpass First-Order Filters

In Example 4.18, we showed that the first-order lowpass transfer function $H_{\text{LP}}(z)$ of Eq. (4.70) and the first-order highpass transfer function $H_{\text{HP}}(z)$ of Eq. (4.73) are a doubly complementary pair and can be expressed as

$$H_{\text{LP}}(z) = \frac{1}{2}\left[\frac{1 - \alpha + z^{-1} - \alpha z^{-1}}{1 - \alpha z^{-1}}\right] = \frac{1}{2}\left[1 + \frac{-\alpha + z^{-1}}{1 - \alpha z^{-1}}\right]$$
$$= \tfrac{1}{2}\left[1 + A_1(z)\right], \tag{6.87}$$

$$H_{\text{HP}}(z) = \frac{1}{2}\left[\frac{1 + \alpha - z^{-1} - \alpha z^{-1}}{1 - \alpha z^{-1}}\right] = \frac{1}{2}\left[1 - \frac{-\alpha + z^{-1}}{1 - \alpha z^{-1}}\right]$$
$$= \tfrac{1}{2}\left[1 - A_1(z)\right], \tag{6.88}$$

where

$$A_1(z) = \frac{-\alpha + z^{-1}}{1 - \alpha z^{-1}}, \tag{6.89}$$

is a first-order allpass transfer function. A combined realization of $H_{\text{LP}}(z)$ and $H_{\text{HP}}(z)$ based on the decompositions of Eqs. (6.87) and (6.88) is as indicated in Figure 6.45, in which the allpass filter given by Eq. (6.89) can be realized using any one of the four single-multiplier allpass structures of Figure 6.36. Figure 6.46 shows one such realization in which the 3-dB cutoff frequency of both filters can be simultaneously varied by changing the multiplier coefficient α. Figure 6.47 shows the composite magnitude responses of the two filters for two different values of α.

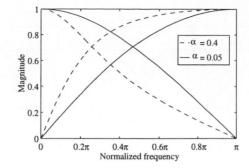

Figure 6.47 Magnitude responses of the lowpass/highpass filter structure of Figure 6.46 for two different values of the parameter α.

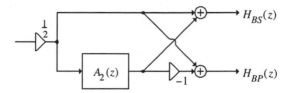

Figure 6.48 Allpass based realization of a doubly-complementary second-order bandpass and bandstop filters.

6.8.2   Tunable Bandpass and Bandstop Second-Order Filters

The second-order bandpass transfer function of Eq. (4.74) and the second-order band-stop transfer function of Eq. (4.79) also form a doubly complementary pair and can be expressed as

$$H_{\text{BP}}(z) = \tfrac{1}{2}\left[1 - A_2(z)\right], \tag{6.90}$$

$$H_{\text{BS}}(z) = \tfrac{1}{2}\left[1 + A_2(z)\right], \tag{6.91}$$

where $A_2(z)$ is a second-order allpass transfer function given by

$$A_2(z) = \frac{\alpha - \beta(1+\alpha)z^{-1} + z^{-2}}{1 - \beta(1+\alpha)z^{-1} + \alpha z^{-2}}. \tag{6.92}$$

Therefore, the bandpass transfer function of Eq. (4.74) and the bandstop transfer function of Eq. (4.79) can be realized together, as indicated in Figure 6.48, where the allpass filter $A_2(z)$ is given by Eq. (6.92). A tunable bandpass/bandstop filter structure is then obtained by realizing the allpass $A_2(z)$ by means of a cascaded lattice structure described in Section 6.7.2, with the lattice two-pair realized using its single-multiplier equivalent of Figure 6.42(a). The final structure is indicated in Figure 6.49. Note that in the structure of Figure 6.49, the multiplier β controls the center frequency and the multiplier α controls the 3-dB bandwidth. Figure 6.50 illustrates the parametric tuning property of the bandpass/bandstop filter structure of Figure 6.49.

Figure 6.49   Tunable second-order bandpass/bandstop filter structure.

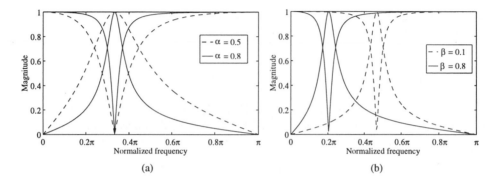

Figure 6.50   Magnitude responses of the bandpass/bandstops filter structure of Figure 6.49 for different values of the parameters α and β. (a) $\beta = 0.5$ and (b) $\alpha = 0.8$.

6.9   Cascaded Lattice Realization of IIR and FIR Filters

Several simple straightforward realizations of FIR and IIR transfer functions have been outlined in Sections 6.4 and 6.5, respectively. In most applications, these realizations work reasonably well even under finite wordlength constraints. However, in some cases, digital filters with more robust properties are needed to provide satisfactory performances. We describe in this section two such realizations.

6.9.1   Realization of Arbitrary IIR Transfer Function

The cascaded lattice structure of Figure 6.43 forms the basis of an often used method for the realization of an arbitrary Mth-order transfer function $H(z)$ originally proposed by Gray and Markel [Gra73]. In their method, $H(z) = P_M(z)/D_M(z)$ is realized in two steps. In the first step, an intermediate allpass transfer function $A_M(z) = z^{-M} D_M(z^{-1})/D_M(z) = \tilde{D}_M(z)/D_M(z)$ is realized in the form of a cascaded lattice structure.[3] The state variables of this structure are then summed in the second step, with appropriate weights to yield the desired numerator $P_M(z)$.

---

[3] $\tilde{D}_M(z)$ denotes the mirror-image polynomial of $D_M(z)$ of degree M, i.e. $\tilde{D}_M(z) = z^{-M} D_M(z^{-1})$.

Figure 6.51   The Gray-Markel structure for a third-order transfer function.

To illustrate the method of realizing the numerator, we consider for simplicity the implementation of a third-order IIR transfer function

$$H(z) = \frac{P_3(z)}{D_3(z)} = \frac{p_0 + p_1 z^{-1} + p_2 z^{-2} + p_3 z^{-3}}{1 + d_1 z^{-1} + d_2 z^{-2} + d_3 z^{-3}}. \tag{6.93}$$

As indicated earlier, in the first step, we form an intermediate allpass transfer function $A_3(z) = Y_1(z)/X_1(z) = \tilde{D}_3(z)/D_3(z)$. Realization of $A_3(z)$ has been illustrated in Example 6.12 resulting in the structure of Figure 6.44(c). Note that here the state variables are S_1, S_2, and S_3. Our objective is to sum the signal variables Y_1, S_1, S_2 and S_3 with weights $\{\alpha_i\}$ as shown in Figure 6.51 to arrive at the desired numerator $P_3(z)$. To this end, we need to analyze the digital filter structure of Figure 6.44(c) and determine the transfer functions $S_1(z)/X_1(z)$, $S_2(z)/X_1(z)$, and $S_3(z)/X_1(z)$.

Using any one of the analysis methods discussed earlier we obtain [4]

$$\frac{S_1(z)}{X_1(z)} = \frac{1}{D_3(z)}, \tag{6.94a}$$

$$\frac{S_2(z)}{X_1(z)} = \frac{d_1" + z^{-1}}{D_3(z)}, \tag{6.94b}$$

$$\frac{S_3(z)}{X_1(z)} = \frac{d_2 + d_1 z^{-1} + z^{-2}}{D_3(z)}. \tag{6.94c}$$

We now form

$$\frac{Y_0(z)}{X_1(z)} = \alpha_1 \frac{Y_1(z)}{X_1(z)} + \alpha_2 \frac{S_3(z)}{X_1(z)} + \alpha_3 \frac{S_2(z)}{X_1(z)} + \alpha_4 \frac{S_1(z)}{X_1(z)}. \tag{6.95}$$

Substituting Eqs. (6.94a)–(6.94c) and the expression for $Y_1(z)/X_1(z) = A_3(z)$ as given by Eq. (6.82) in Eq. (6.95), we arrive at

$$\frac{Y_0(z)}{X_1(z)} = \frac{\begin{matrix} \alpha_1(d_3 + d_2 z^{-1} + d_1 z^{-2} + z^{-3}) \\ + \alpha_2(d_2' + d_1' z^{-1} + z^{-2}) + \alpha_3(d_1" + z^{-1}) + \alpha_4 \end{matrix}}{D_3(z)}. \tag{6.96}$$

---

[4]It should be noted that the numerator of the transfer function $S_i(z)/X_1(z)$ is precisely the numerator of the allpass transfer function $S_i(z)/W_i(z)$.

384   Chapter 6: Digital Filter Structures

Comparing the numerators of the righthand sides of Eqs. (6.93) and (6.96), and by equating the coefficients of like powers of z^{-1}, we thus obtain

$$\alpha_1 d_3 + \alpha_2 d_2' + \alpha_3 d_1'' + \alpha_4 = p_0,$$
$$\alpha_1 d_2 + \alpha_2 d_1' + \alpha_3 = p_1,$$
$$\alpha_1 d_1 + \alpha_2 = p_2,$$
$$\alpha_1 = p_3.$$

Solving the above we get

$$\alpha_1 = p_3,$$
$$\alpha_2 = p_2 - \alpha_1 d_1,$$
$$\alpha_3 = p_1 - \alpha_1 d_2 - \alpha_2 d_1',$$
$$\alpha_4 = p_0 - \alpha_1 d_3 - \alpha_2 d_2' - \alpha_3 d_1'', \qquad (6.97)$$

which is in a form suitable for step-by-step calculation of the feed-forward tap coefficients α_i.

Example 6.13 Let us realize the transfer function of Eq. (6.45) in Example 6.7, repeated below for convenience:

$$H(z) = \frac{P_3(z)}{D_3(z)} = \frac{0.44z^{-1} + 0.36z^{-2} + 0.02z^{-3}}{1 + 0.4z^{-1} + 0.18z^{-2} - 0.2z^{-3}}, \qquad (6.98)$$

using the Gray-Markel method.

We form the intermediate allpass transfer function $A_3(z)$ having the same denominator as $H(z)$ in Eq. (6.98). Its realization has been carried out in Example 6.12 and is shown in Figure 6.44(c), where $d_3 = -0.2$, $d_2' = 0.2708333$, and $d_1'' = 0.3573771$.

To realize the numerator, we compute the tap coefficients $\{\alpha_i\}$ using Eq. (6.97). Substituting the values of $d_3, d_2', d_1'' = 0.4541667$, and the values of the numerator coefficients, $p_0 = 0$, $p_1 = 0.44$, $p_2 = 0.36$, and $p_3 = 0.02$ in Eq. (6.97), we arrive at

$$\alpha_1 = 0.02, \quad \alpha_2 = 0.352, \quad \alpha_3 = 0.2765333, \quad \alpha_4 = -0.19016.$$

The final realization is thus as indicated in Figure 6.51 with the multiplier values as given above.

Program 6.3 can be employed to develop the Gray-Markel cascaded lattice of an arbitrary IIR transfer function. The input data called by the program are the vectors num and den containing the coefficients of the numerator and the denominator polynomials in ascending powers of z^{-1}. The output data are the feed-forward coefficient vector alpha and the lattice parameter vector k.

```
% Program 6_3
% Computation of the Parameters of
% Gray-Markel Cascaded Lattice Structure
%
% den is the denominator coefficient vector
% num is the numerator coefficient vector
```

```
% k is the lattice parameter vector
% alpha is the vector of feedforward multipliers
%
format long
% Read in the transfer function coefficients
num = input('Numerator coefficient vector = ');
den = input('Denominator coefficient vector = ');
N = length(den)-1; % Order of denominator polynomial
k = ones(den(1:N));
a1 = den/den(1);
alpha = num(N+1:-1:1)/den(1);
for ii = N:-1:1,
    alpha(N+2-ii:N+1) = alpha(N+2-ii:N+1)-alpha(N-
ii+1)*a1(2:ii+1);
    k(ii) = a1(ii+1);
    a1(1:ii+1) = (a1(1:ii+1)-k(ii)*a1(ii+1:-1:1))/(1-
k(ii)*k(ii));
end
disp('Lattice parameters are');disp(k)
disp('Feedforward multipliers are');disp(alpha)
```

Applying the above program to the IIR transfer function of Example 6.13, we obtain the pertinent multiplier coefficients of the corresponding cascade lattice realization as:

```
Lattice parameters are
    0.35737704918033   0.27083333333333   -0.2

Feedforward multipliers are
    0.02   0.352   0.27653333333333   -0.19016
```

which are seen to be identical to those derived in Example 6.13.

6.9.2 Realization of an Arbitrary FIR Transfer Function

The cascaded lattice structure that can be used to realize an arbitrary FIR transfer function is shown in Figure 6.52 [Mak75]. To develop the algorithm for realizing a given Mth order FIR transfer function

$$H_M(z) = 1 + \sum_{n=1}^{M} p_n z^{-n}, \qquad (6.99)$$

in the form of this structure, we first analyze it to determine the relations between the input $X(z) = X_0(z)$ and the intermediate variables $X_i(z)$ and $X_i'(z)$, i.e., between the series of transfer functions $H_i(z) = X_i(z)/X_0(z)$ and $H_i'(z) = X_i'(z)/X_0(z)$, $i = 1, 2, \ldots, M$. Note that $X_M(z) = Y(z)$, is the output variable.

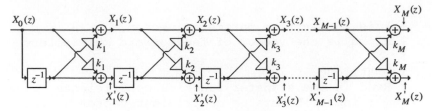

Figure 6.52   The FIR cascaded lattice structure.

From Figure 6.52 we obtain

$$X_1(z) = X_0(z) + k_1 z^{-1} X_0(z), \qquad (6.100a)$$
$$X_1'(z) = k_1 X_0(z) + z^{-1} X_0(z). \qquad (6.100b)$$

The corresponding transfer functions are given by

$$H_1(z) = \frac{X_1(z)}{X_0(z)} = 1 + k_1 z^{-1}, \qquad (6.101a)$$

$$H_1'(z) = \frac{X_1(z)}{X_0(z)} = k_1 + z^{-1}, \qquad (6.101b)$$

which are seen to be first-order FIR transfer functions. Moreover, it can be seen from Eqs. (6.101a) and (6.101b) that

$$\tilde{H}_1(z) = z^{-1} H_1(z^{-1}) = z^{-1} + k_1 = H_1'(z),$$

indicating that the FIR transfer function $H_1'(z)$ of Eq. (6.101b) is the mirror-image of the FIR transfer function of $H_1(z)$ Eq. (6.101a).

Next, we observe from Figure 6.52 that

$$X_2(z) = X_1(z) + k_2 z^{-1} X_1(z), \qquad (6.102a)$$
$$X_2'(z) = k_2 X_1(z) + z^{-1} X_1(z). \qquad (6.102b)$$

The corresponding transfer functions are given by

$$H_2(z) = \frac{X_2(z)}{X_0(z)} = H_1(z) + k_2 z^{-1} H_1'(z), \qquad (6.103a)$$

$$H_2'(z) = \frac{X_2(z)}{X_0(z)} = k_2 H_1(z) + z^{-1} H_1'(z). \qquad (6.103b)$$

Substituting Eqs. (6.101a) and (6.101b) in above we observe that the $H_2(z)$ and $H_2'(z)$ are second-order FIR transfer functions. Moreover, since $H_1'(z) = \tilde{H}_1(z)$, it follows from the above that $H_2'(z) = z^{-2} H_2(z^{-1}) = \tilde{H}_2(z)$, implying that $H_2'(z)$ is the mirror image of $H_2(z)$.

Now, the input-output relation of the ith lattice section is given by

$$X_i(z) = X_{i-1}(z) + k_i z^{-1} X_{i-1}'(z), \qquad (6.104a)$$
$$X_i'(z) = k_i X_{i-1}(z) + z^{-1} X_{i-1}'(z). \qquad (6.104b)$$

The corresponding transfer functions, $H_i(z) = X_i(z)/X_0(z)$ and $H_i'(z) = X_i'(z)/X_0(z)$, are related through

$$H_i(z) = H_{i-1}(z) + k_i z^{-1} H_{i-1}'(z), \tag{6.105a}$$

$$H_i'(z) = k_i H_{i-1}(z) + z^{-1} H_{i-1}'(z), \tag{6.105b}$$

where $H_{i-1}(z) = X_{i-1}(z)/X_0(z)$ and $H_{i-1}'(z) = X_{i-1}'(z)/X_0(z)$. It should be noted that the relations of Eqs. (6.105a) and (6.105b) again hold for all $i = 1, 2, \ldots, M$.

If we assume that $H_{i-1}(z)$ and $H_{i-1}'(z)$ are $(i-1)$th-order FIR transfer functions, then it follows from Eqs. (6.105a) and (6.105b) that $H_i(z)$ and $H_i'(z)$ are ith-order FIR transfer functions. Moreover, if $H_{i-1}'(z) = \tilde{H}_{i-1}(z)$, then it also follows from Eqs. (6.105a) and (6.105b) that $H_i'(z) = \tilde{H}_i(z)$. We have already shown that these two observations hold for $i = 1$ and $i = 2$. Therefore, by induction, they also hold for all $i = 1, 2, \ldots, M$.

To develop the realization algorithm, the above process should be reversed. That is, given the expressions for $H_i(z)$ and $\tilde{H}_i(z)$, we need to find the expressions for $H_{i-1}(z)$ and $\tilde{H}_{i-1}(z)$ for $i = M, M-1, \ldots, 2, 1$. We now develop a recursion relation to compute these intermediate transfer functions.

Solving Eqs. (6.105a) and (6.105b) for $i = M$, we obtain

$$H_{M-1}(z) = \frac{1}{(1 - k_M^2)z^{-1}} \left\{ z^{-1} H_M(z) - k_M z^{-1} \tilde{H}_M(z) \right\}, \tag{6.106}$$

$$\tilde{H}_{M-1}(z) = \frac{1}{(1 - k_M^2)z^{-1}} \left\{ -k_M H_M(z) + \tilde{H}_M(z) \right\}. \tag{6.107}$$

It can be easily verified that $\tilde{H}_{M-1}(z)$ of Eq. (6.107) is indeed the mirror-image of $H_{M-1}(z)$ of Eq. (6.106). Substituting the expression for $H_M(z)$ of Eq. (6.99) in Eq. (6.106) we get

$$H_{M-1}(z) = \frac{1}{(1 - k_M^2)} \left\{ (1 - k_M p_M) + (p_1 - k_M p_{M-1})z^{-1} + \ldots \right.$$
$$\left. + (p_{M-1} - k_M p_1)z^{-(M-1)} + (p_M - k_M)z^{-M} \right\}. \tag{6.108}$$

If we choose $k_M = p_M$, $H_{M-1}(z)$ reduces to an FIR transfer function of degree $M-1$ and can be written in the form

$$H_{M-1}(z) = 1 + \sum_{n=1}^{M-1} p_n' z^{-n}, \tag{6.109}$$

where

$$p_n' = \frac{p_n - k_M p_{M-n}}{1 - k_M^2}, \qquad n = 1, 2, \ldots, M-1. \tag{6.110}$$

Continuing the above recursion algorithm, all the multiplier coefficients of the structure of Figure 6.52 can be computed. The following example illustrates the procedure.

Example 6.14 Realize the fourth-order FIR transfer function

$$H_4(z) = 1 + 1.2z^{-1} + 1.12z^{-2} + 0.12z^{-3} - 0.08z^{-4}. \tag{6.111}$$

Figure 6.53   Cascaded lattice realization of the FIR transfer function of Eq. (6.111). $k_1 = 0.5$, $k_2 = 1.0$, $k_3 = 0.2173913$, and $k_4 = -0.08$.

From the above the multiplier coefficient of the rightmost lattice is given by $k_4 = p_4 = -0.08$. Using Eq. (6.110), we next determine the coefficients p_n' of the third-order transfer function $H_3(z)$ as,

$$p_3' = \frac{p_3 - k_4 p_1}{1 - k_4^2} = \frac{0.12 - (-0.08)(1.2)}{1 - (-0.08)^2} = 0.2173913,$$

$$p_2' = \frac{p_2 - k_4 p_2}{1 - k_4^2} = \frac{1.12}{1 + 0.08} = 1.2173913,$$

$$p_1' = \frac{p_1 - k_4 p_3}{1 - k_4^2} = \frac{1.2 - (-0.08)(0.12)}{1 - (-0.08)^2} = 1.2173913.$$

As a result,
$$H_3(z) = 1 + 1.2173913z^{-1} + 1.2173913z^{-2} + 0.2173913z^{-3}.$$

Thus, the multiplier coefficient of the third lattice two-pair is given by $k_3 = p_3' = 0.2173913$.

We continue the above recursion process to determine the coefficients p_n'' of the second-order transfer function $H_2(z)$ which are given by

$$p_2'' = \frac{p_2' - k_3 p_1'}{1 - k_3^2} = \frac{p_2'}{1 + k_3} = \frac{1.2173913}{1.2173913} = 1.0,$$

$$p_1'' = \frac{p_1' - k_3 p_2'}{1 - k_3^2} = \frac{p_1'}{1 + k_3} = 1.0,$$

leading to
$$H_2(z) = 1 + z^{-1} + z^{-2}.$$

Consequently, the multiplier coefficient of the second lattice two-pair is given by $k_2 = p_2'' = 1.0$.

The last recursion yields the multiplier coefficient of the first lattice two-pair:

$$k_1 = \frac{p_1''}{1 + k_2} = \frac{1.0}{2.0} = 0.5.$$

The final realization of the FIR transfer function of Eq. (6.111) is thus as shown in Figure 6.53.

The function `poly2rc` in MATLAB can again be employed to compute the multiplier coefficients of the cascade lattice structure of Figure 6.50 as illustrated in the following example.

Example 6.15 Using MATLAB develop the cascaded lattice realization of the FIR transfer function of Eq. (6.111). To this end, we make use of Program 6_4 given below.

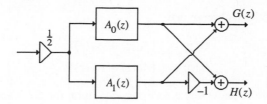

Figure 6.54 Realization of a doubly-complementary transfer function pair using a parallel allpass structure.

```
% Program 6_4
% FIR Cascaded Lattice Realization
%
format long
num = input('Transfer function coefficients = ');
k = poly2rc(num);
disp('Lattice coefficients are'); disp(k');
```

The input data called by the program is the vector num of transfer function coefficients entered in ascending powers of z^{-1}. For our example,

$$num = [1 \quad 1.2 \quad 1.12 \quad 0.12 \quad -0.08].$$

The output data is the vector of lattice coefficients as given below.

```
Lattice coefficients are
-0.08      0.2173913      1.0      0.5
```

which are seen to be identical to those computed in Example 6.14.

6.10 Parallel Allpass Realization of IIR Transfer Functions

In Section 6.8, we have demonstrated that a pair of doubly complementary first-order lowpass and highpass transfer functions can be realized as shown in Figure 6.45. Likewise, we have shown that a pair of doubly complementary second-order bandpass and bandstop transfer functions can be realized, as indicated in Figure 6.48. These two structures can be considered as special cases of a structure composed of a parallel connection of two stable allpass filters of the form of Figure 6.54, where one of the allpass section is a zeroth-order transfer function. We consider in this section the realization of an Nth-order transfer function $G(z)$ in the form of Figure 6.54, along with its power-complementary transfer function $H(z)$ [Vai86a]. As we shall point out later, such structures have a number of very attractive properties from the implementation point of view.

From Figure 6.54 we observe that

$$G(z) = \tfrac{1}{2}\{A_0(z) + A_1(z)\}, \qquad\qquad (6.112a)$$

$$H(z) = \tfrac{1}{2}\{A_0(z) - A_1(z)\}. \qquad\qquad (6.112b)$$

It is a simple exercise to show that a necessary condition for $G(z)$ and $H(z)$ to have the sum of allpass decomposition of the form of Eqs. (6.112a) and (6.112b) is that each be a bounded real (BR) transfer function (Problem 4.79).[5] The BR condition can be easily satisfied by any stable transfer function by a simple scaling. Let

$$G(z) = \frac{P(z)}{D(z)} = \frac{p_0 + p_1 z^{-1} + \ldots + p_N z^{-N}}{1 + d_1 z^{-1} + \ldots + d_N z^{-N}} \qquad\qquad (6.113)$$

be an Nth-order BR transfer function with its power-complementary transfer function given by

$$H(z) = \frac{Q(z)}{D(z)} = \frac{q_0 + q_1 z^{-1} + \ldots + q_N z^{-N}}{1 + d_1 z^{-1} + \ldots + d_N z^{-N}}. \qquad\qquad (6.114)$$

The power-complementary property implies that

$$\left| G(e^{j\omega}) \right|^2 + \left| H(e^{j\omega}) \right|^2 = 1. \qquad\qquad (6.115)$$

From Eq. (6.112a) it follows that $G(z)$ must have a symmetric numerator, i.e.,

$$p_n = p_{N-n}, \qquad\qquad (6.116)$$

and from Eq. (6.112b) it follows that $H(z)$ must have an antisymmetric numerator, i.e.,

$$q_n = -q_{N-n}. \qquad\qquad (6.117)$$

We develop next the procedure to identify the two allpass transfer functions from a specified transfer function $G(z)$. Now, the symmetric property of the numerator of $G(z)$ as given by Eq. (6.116) implies that

$$P(z^{-1}) = z^{-N} P(z). \qquad\qquad (6.118)$$

Likewise, the antisymmetric property of the numerator of $H(z)$ as given by Eq. (6.117) implies that

$$Q(z^{-1}) = -z^{-N} Q(z). \qquad\qquad (6.119)$$

By analytic continuation, we can rewrite Eq. (6.115) as

$$G(z^{-1})G(z) + H(z)H(z^{-1}) = 1. \qquad\qquad (6.120)$$

Substituting $G(z) = P(z)/D(z)$ and $H(z) = Q(z)/D(z)$ in the above equation, and making use of Eqs. (6.118) and (6.119), we arrive at

$$[P(z) + Q(z)][P(z) - Q(z)] = z^{-N} D(z^{-1})D(z). \qquad\qquad (6.121)$$

---

[5]See Section 4.3.8 for the definition of a bounded real transfer function.

From the relations of Eqs. (6.118) and (6.119), we can write

$$P(z) - Q(z) = z^N[P(z^{-1}) + Q(z^{-1})], \qquad (6.122)$$

If we denote the zeros of $[P(z) + Q(z)]$ as $z = \xi_k$, $1 \le k \le N$, then it follows from Eq. (6.121) that $z = 1/\xi_k$, $1 \le k \le N$, are the zeros of $[P(z) - Q(z)]$, i.e., $[P(z) - Q(z)]$ is the mirror-image polynomial of $[P(z) + Q(z)]$. From Eq. (6.122) it also follows that the zeros of $[P(z) + Q(z)]$ inside the unit circle are zeros of $D(z)$, whereas the zeros of $[P(z) + Q(z)]$ outside the unit circle are zeros of $D(z^{-1})$, since $G(z)$ and $H(z)$ have been assumed to be stable transfer functions. Let $z = \xi_k$, $1 \le k \le r$, be the r zeros of $[P(z) + Q(z)]$ inside the unit circle, and the remaining $N - r$ zeros, $z = \xi_k$, $r + 1 \le k \le N$, be outside the unit circle. Hence, it can be seen from Eq. (6.122) that the N zeros of $D(z)$ are given by

$$z = \xi_k, \quad 1 \le k \le r,$$
$$z = \frac{1}{\xi_k}, \quad r + 1 \le k \le N.$$

All we need to do now is to identify the above zeros with the appropriate allpass transfer functions $A_0(z)$ and $A_1(z)$. To this end, we obtain from Eqs. (6.112a) and (6.112b)

$$A_0(z) = G(z) + H(z) = \frac{P(z) + Q(z)}{D(z)}, \qquad (6.123a)$$

$$A_1(z) = G(z) - H(z) = \frac{P(z) - Q(z)}{D(z)}. \qquad (6.123b)$$

Therefore, the two allpass transfer functions can be expressed as

$$A_0(z) = \prod_{k=r+1}^{N} \frac{z^{-1} - (\xi_k^*)^{-1}}{1 - \xi_k^{-1} z^{-1}}, \qquad (6.124a)$$

$$A_1(z) = \prod_{k=1}^{r} \frac{z^{-1} - \xi_k^*}{1 - \xi_k z^{-1}}. \qquad (6.124b)$$

In order to arrive at the above expressions for the two allpass transfer functions, it is necessary to determine the transfer function $H(z)$ that is power-complementary to $G(z)$. Denoting the $2N$th-degree polynomial $P^2(z) - z^{-N} D(z^{-1}) D(z)$ as $U(z)$:

$$P^2(z) - z^{-N} D(z^{-1}) D(z) = U(z) = \sum_{n=0}^{2N} u_n z^{-n}, \qquad (6.125)$$

we can rewrite Eq. (6.121) as

$$Q^2(z) = \sum_{n=0}^{2N} u_n z^{-n}. \qquad (6.126)$$

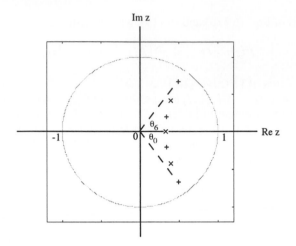

Figure 6.55   Pole interlacing property in the case of a seventh order digital Butterworth lowpass filter. The poles marked '+' belong to the allpass function $A_0(z)$ and the poles marked '×' belong to the allpass function $A_1(z)$.

Solving the above equation for the coefficients q_k of $Q(z)$ we finally arrive at

$$q_0 = \sqrt{u_0}, \quad q_1 = \frac{u_1}{2q_0}, \tag{6.127a}$$

$$q_k = -q_{N-k} = \frac{u_k - \sum_{\ell=1}^{k-1} q_\ell q_{n-\ell}}{2q_0}, \quad k \geq 2, \tag{6.127b}$$

where we have used the antisymmetric property of the coefficients. After $Q(z)$ has been determined, we form the polynomial $[P(z) + Q(z)]$, find its zeros $z = \xi_k$, and then determine the two allpass transfer functions using Eqs. (6.124a) and (6.124b).

It can be shown that IIR digital transfer functions derived from the analog Butter-worth, Chebyshev, and elliptic filters[6] via the bilinear transformation approach discussed in Section 7.3 can be decomposed in the form of Eqs. (6.112a) and (6.112b) [Vai86a]. Moreover, for lowpass-highpass filter pairs, the order N of the transfer function must be odd with the orders of $A_0(z)$ and $A_1(z)$ differing by 1. Likewise, for bandpass-bandstop filter pairs, the order N of the transfer function must be even with the orders of $A_0(z)$ and $A_1(z)$ differing by 2 [Vai86a].

In the case of odd-order digital Butterworth, Chebyshev, and elliptic lowpass or high-pass transfer functions, there is a simple approach to identify the poles of the allpass transfer functions $A_0(z)$ and $A_1(z)$ from the poles λ_k, $0 \leq k \leq N - 1$, of the parent lowpass transfer function $G(z)$ or $H(z)$. Let θ_k denote the angle of the pole λ_k. If we assume that the poles are numbered such that $\theta_k < \theta_{k+1}$, then the poles of $A_0(z)$ are given by λ_{2k} and the poles of $A_1(z)$ are given by λ_{2k+1} [Gas85]. Figure 6.55 illustrates this *pole interlacing property* of the two allpass transfer functions.

Example 6.16   Consider the transfer function of a seventh-order Butterworth lowpass digital filter

---

[6]See Section 5.3 for a review of these analog filter approximation techniques.

with a 3-dB cutoff frequency of 0.39:[7]

$$G(z) = \frac{0.000963(1 + 7z^{-1} + 21z^{-2} + 35z^{-3} + 35z^{-4} + 21z^{-5} + 7z^{-6} + z^{-7})}{\begin{array}{c}1 - 2.7825z^{-1} + 3.9668z^{-2} - 3.4051z^{-3} + 1.8757z^{-4}\\ -0.6509z^{-5} + 0.1308z^{-6} - 0.01166z^{-7}\end{array}}. \quad (6.128)$$

Its seven poles are located at (numbered according to increasing angles):

$$\xi_0 = 0.4981133 - j0.668405, \qquad \xi_1 = 0.3907071 - j0.4204395,$$
$$\xi_2 = 0.3399766 - j0.2030305, \qquad \xi_3 = 0.32491969623291,$$
$$\xi_4 = 0.3399766 + j0.2030305, \qquad \xi_5 = 0.3907071 + j0.4204395,$$
$$\xi_6 = 0.4981133 + j0.668405.$$

The corresponding z-plane pole-plot is indicated in Figure 6.55. Making use of the pole interlacing property, we identify the two allpass transfer functions as

$$A_0(z) = \frac{(z^{-1} - \xi_0)(z^{-1} - \xi_2)(z^{-1} - \xi_4)(z^{-1} - \xi_6)}{(1 - \xi_0 z^{-1})(1 - \xi_2 z^{-1})(1 - \xi_4 z^{-1})(1 - \xi_6 z^{-1})}$$

$$= \frac{(0.69488 - 0.99623z^{-1} + z^{-2})(0.15681 - 0.67995z^{-1} + z^{-2})}{(1 - 0.99623z^{-1} + 0.69488z^{-2})(1 - 0.67995z^{-1} + 0.15681z^{-2})}, \quad (6.129a)$$

$$A_1(z) = \frac{(z^{-1} - \xi_1)(z^{-1} - \xi_3)(z^{-1} - \xi_5)}{(1 - \xi_1 z^{-1})(1 - \xi_3 z^{-1})(1 - \xi_5 z^{-1})}$$

$$= \frac{(-0.32492 + z^{-1})(0.32942 - 0.781414z^{-1} + z^{-2})}{(1 - 0.32492z^{-1})(1 - 0.781414z^{-1} + 0.32942z^{-2})}. \quad (6.129b)$$

It can be easily verified that by substituting the above two allpass transfer functions in Eq. (6.112a), we indeed arrive at the transfer function G(z) of Eq. (6.128). In addition, making use of these in Eq. (6.112b), we also get its power-complementary highpass transfer function $H(z)$ given by

$$H(z) = \frac{0.108(1 - 7z^{-1} + 21z^{-2} - 35z^{-4} - 21z^{-5} + 7z^{-6} - z^{-7})}{\begin{array}{c}1 - 2.7825z^{-1} + 3.9668z^{-2} - 3.4051z^{-3} + 1.8757z^{-4}\\ -0.6509z^{-5} + 0.1308z^{-6} - 0.01166z^{-7}\end{array}}. \quad (6.130)$$

Figure 6.56 shows the magnitude response plots of $G(z)$ of Eq. (6.128) and $H(z)$ of Eq. (6.130) demonstrating their power-complementary property.

The parallel allpass structure has two other very attractive properties. As indicated in Section 6.7, an Mth-order allpass transfer function can be realized using only M multipliers. Therefore, the realization of an Nth-order IIR transfer function $G(z)$ based on the allpass decomposition of Eq. (6.112a) requires only $r + (N - r) = N$ multipliers. On the other hand, in general, a direct form realization of an Nth-order IIR transfer function uses $2N + 1$ multipliers. Moreover, with an additional adder and no additional multipliers, one can easily implement its power-complementary transfer function $H(z)$, also of Nth-order, as indicated in Figure 6.54. As a result, the parallel allpass realization of an IIR transfer function, if it exists, is computationally very efficient. We shall show later in Section 9.7.2, that the parallel allpass realization also has very low passband sensitivity with respect to multiplier coefficients if the allpass sections are realized in a structurally lossless form.

---

[7]This transfer function has been obtained using MATLAB. For details, see Section 7.8.3.

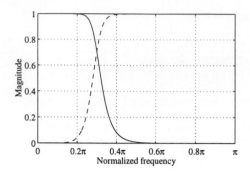

Figure 6.56 Magnitude responses of a pair of power-complementary seventh order Butterworth lowpass and highpass transfer functions with a 3-dB cutoff frequency at 0.3π.

6.11 Digital Sine-Cosine Generator

We now consider the realization of digital sinusoidal oscillators. In particular, we consider here the design of digital sine-cosine generators that produce two sinusoidal sequences that are exactly 90° out of phase with each other [Mit75]. These circuits have a number of applications, such as the computation of discrete Fourier transform and certain digital communication systems. In some applications, it is more convenient to use numerical algorithms for the computation of sine or cosine functions. These are discussed in Section 8.8.1.

Let $s_1[n]$ and $s_2[n]$ denote the two outputs of a digital sine-cosine generator given by

$$s_1[n] = \alpha \sin(n\theta), \tag{6.131a}$$

$$s_2[n] = \beta \cos(n\theta). \tag{6.131b}$$

From the above we arrive at

$$\begin{aligned} s_1[n+1] &= \alpha \sin\left((n+1)\,\theta\right) \\ &= \alpha \sin(n\theta)\cos\theta + \alpha \cos(n\theta)\sin\theta, \end{aligned} \tag{6.132a}$$

$$\begin{aligned} s_2[n+1] &= \beta \cos\left((n+1)\,\theta\right) \\ &= \beta \cos(n\theta)\cos\theta - \beta \sin(n\theta)\sin\theta. \end{aligned} \tag{6.132b}$$

Making use of Eqs. (6.131a) and (6.131b), we can rewrite Eqs. (6.132a) and (6.132b) in a matrix form as:

$$\begin{bmatrix} s_1[n+1] \\ s_2[n+1] \end{bmatrix} = \begin{bmatrix} \cos\theta & \frac{\alpha}{\beta}\sin\theta \\ -\frac{\beta}{\alpha}\sin\theta & \cos\theta \end{bmatrix} \begin{bmatrix} s_1[n] \\ s_2[n] \end{bmatrix}. \tag{6.133}$$

which is the state-space representation of a sine-cosine generator with a zero-input. Thus, such a structure must have at least two delays.

In order to arrive at a realization of the sine-cosine generator, we need to compare Eq. (6.133) with the equivalent expression of a general second-order structure with no delay-free loops. Such a structure is characterized by the following equations

Figure 6.57 A general second-order digital filter structure with no input node.

$$\begin{bmatrix} s_1[n+1] \\ s_2[n+1] \end{bmatrix} = \begin{bmatrix} 0 & A \\ 0 & 0 \end{bmatrix}\begin{bmatrix} s_1[n+1] \\ s_2[n+1] \end{bmatrix} + \begin{bmatrix} C & D \\ E & F \end{bmatrix}\begin{bmatrix} s_1[n] \\ s_2[n] \end{bmatrix} \tag{6.134}$$

and can be implemented using 5 multipliers, as indicated in Figure 6.57.

From Figure 6.57 we readily arrive at the state representation of this structure:

$$\begin{bmatrix} s_1[n+1] \\ s_2[n+1] \end{bmatrix} = \begin{bmatrix} AE+C & AF+D \\ E & F \end{bmatrix}\begin{bmatrix} s_1[n] \\ s_2[n] \end{bmatrix}. \tag{6.135}$$

Comparing Eqs. (6.133) and (6.135), we get

$$E = -\frac{\beta}{\alpha}\sin\theta, \qquad F = \cos\theta, \tag{6.136a}$$

$$AE + C = \cos\theta, \qquad AF + D = \frac{\alpha}{\beta}\sin\theta. \tag{6.136b}$$

From the above two equations, we obtain after some algebra

$$C\cos\theta + D\frac{\beta}{\alpha}\sin\theta = 1. \tag{6.136c}$$

Equations (6.136a) and (6.136c) ensure that the structure of Figure 6.57 will be a sine-cosine generator.

Expressing the multiplier constants A and D as a function of the multiplier constant C, we obtain after some simple manipulations, the 5-multiplier characterization of the sine-cosine generator as

$$\begin{bmatrix} s_1[n+1] \\ s_2[n+1] \end{bmatrix} = \begin{bmatrix} 0 & \frac{\alpha(C-\cos\theta)}{\beta\sin\theta} \\ 0 & 0 \end{bmatrix}\begin{bmatrix} s_1[n+1] \\ s_2[n+1] \end{bmatrix}$$

$$+ \begin{bmatrix} C & \frac{\alpha(1-C\cdot\cos\theta)}{\beta\sin\theta} \\ -\frac{\beta}{\alpha}\sin\theta & \cos\theta \end{bmatrix}\begin{bmatrix} s_1[n] \\ s_2[n] \end{bmatrix}. \tag{6.137}$$

To reduce the total number of multipliers to 4, we can choose the following specific values for the multiplier constant C: $\cos\theta, 0, +1$, and -1. For example, if $C = \cos\theta$, then Eq. (6.137) reduces to

$$\begin{bmatrix} s_1[n+1] \\ s_2[n+1] \end{bmatrix} = \begin{bmatrix} \cos\theta & \frac{\alpha}{\beta}\sin\theta \\ -\frac{\beta}{\alpha}\sin\theta & \cos\theta \end{bmatrix}\begin{bmatrix} s_1[n] \\ s_2[n] \end{bmatrix}. \tag{6.138}$$

Figure 6.58   A single multiplier sine-cosine generator.

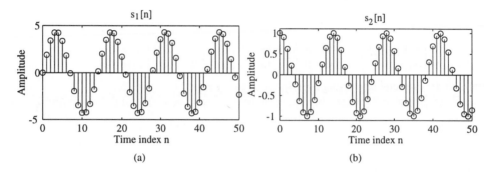

Figure 6.59   Sine-cosine sequences generated by the structure of Figure 6.58 for $\cos\theta = 0.9$.

which for $\beta = \pm\alpha$, leads to the 4-multiplier sine-cosine generator described by Gold and Rader [Gol69b]. On the other hand, if we set $\alpha\sin\theta = \pm\beta$, then Eq. (6.138) leads to a 3-multiplier realization (Problem 6.61). Another 3-multiplier sine-cosine generator is obtained if we set $\alpha = \pm\beta\sin\theta$ in Eq. (6.138) (Problem 6.62).

A single-multiplier structure can be derived by setting $\beta\sin\theta/\alpha = 1 - \cos\theta$, or equivalently, $\beta = \alpha\tan(\theta/2)$, in Eq. (6.138) resulting in

$$\begin{bmatrix} s_1[n+1] \\ s_2[n+1] \end{bmatrix} = \begin{bmatrix} \cos\theta & \cos\theta + 1 \\ \cos\theta - 1 & \cos\theta \end{bmatrix} \begin{bmatrix} s_1[n] \\ s_2[n] \end{bmatrix}. \tag{6.139}$$

leading to the realization of Figure 6.58. Another single-multiplier realization can be derived by setting $-\beta\sin\theta/\alpha = (1+\cos\theta)$, or equivalently, $\alpha = \beta\tan(\theta/2)$, in Eq. (6.138) (Problem 6.63). Various other single-multiplier sine-cosine generators can be developed for $C = 0$ and $C = \pm 1$ and are left as exercises (Problems 6.64 and 6.65).

Figure 6.59 shows the plots of the sequences $s_1[n]$ and $s_2[n]$ generated by simulating Eq. (6.139) on MATLAB for $\cos\theta = 0.9$. As can be seen from this figure, $s_1[n]$ and $s_2[n]$ are, respectively, cosinusoidal and sinusoidal sequences. Note that the maximum value of the amplitudes of the two sequences can be made equal by scaling one of the sequences appropriately.

The single-multiplier structures retain their characteristic roots on the unit circle under finite wordlength constraints. On the other hand, in other realizations of the sine-cosine generators, roots may go inside or outside the unit circle due to the quantization of the multiplier coefficients, causing the oscillations to decay or to build up as n increases.

Table 6.1 Computational complexity comparison of various realizations of a length M FIR filter.

Structure	No. of multipliers	No. of 2-input adders
Direct form	M	$M - 1$
Cascade form	M	$M - 1$
Polyphase	M	$M - 1$
Cascaded lattice	$2M$	$2M - 1$
Linear phase	$\lfloor \dfrac{M+1}{2} \rfloor$	$M - 1$

In addition, due to product roundoff errors, the sequences generated by the sine-cosine generator may not retain their sinusoidal behaviors, even in the case of a single-multiplier generator. Its is therefore advisable to reset the state variables after some iterations at prescribed points, so that the accumulated errors do not become unacceptable.

6.12   Computational Complexity of Digital Filter Structures

The computational complexity of a digital filter structure is given by the total number of multipliers and the total number of two-input adders required for its implementation which roughly provides an indication of its cost of implementation. We summarize this measure here for various realizations discussed in this chapter. It should be emphasized, though, that the computational complexity measure is not the only criterion that should be used in selecting a particular structure for implementation of a given transfer function. The performances of all equivalent realizations under finite wordlength constraints should also be considered together with the costs of implementation in selecting a structure.

Tables 6.1 and 6.2 show the computational complexity measures of all realizations discussed in this chapter. As can be seen from these tables, in general, the direct form realizations for the case of both the FIR and the IIR transfer functions require the least number of multipliers and two-input adders. However, in the case of the IIR transfer function, the parallel allpass realization is the most efficient if such a realization exists. Likewise, it is possible to realize certain special types of FIR transfer function with a lot fewer multipliers and adders than that indicated in this table.

6.13   Summary

This chapter considered the realization of a causal digital transfer function. Such realizations, called structures, are usually represented in block diagram forms that are formed by interconnecting adders, multipliers, and unit delays. A digital filter structure represented in block diagram form can be analyzed to develop its input-output relationship either in

Table 6.2   Computational complexity comparison of various realizations of an IIR filter
of order N.

Structure	No. of multipliers	No. of 2-input adders
Direct form II and II$_t$	$2N + 1$	$2N$
Cascade form	$2N + 1$	$2N$
Parallel	$2N + 1$	$2N$
State-space	$(N + 1)^2$	$N(N + 1)$
Gray-Markel	$3N + 1$	$3N$

the time-domain or in the transform-domain. Often, for analysis purposes, it is convenient to represent the digital filter structure as a signal flow-graph.

Several basic FIR and IIR digital filter structures are then reviewed. These structures, in most cases, can be developed from the transfer function description of the digital filter essentially by inspection.

The state-space representation of an IIR digital transfer function leads to an alternative structure that is computationally the most complex realization, requiring a very large number of multipliers and adders. In spite of its increased computational complexity, the state-space-based realization permits the development of an equivalent state-space structure that is more robust under finite wordlength constraints. The development of a robust state-space structure is considered in Section 9.8.2.

The digital allpass filter is a versatile building block and has a number of attractive digital signal processing applications. Since the numerator and the denominator polynomials of a digital transfer function exhibit mirror-image symmetry, an Mth-order digital allpass filter can be realized with only M distinct multipliers. Two different approaches to the minimum-multiplier realization of a digital allpass transfer function are described. One approach is based on a realization in the form of a cascade of first- and second-order allpass filters. The second approach results in a cascaded lattice realization. The final realizations in both cases remain allpass independent of the actual values of the multiplier coefficients and are thus less sensitive to multiplier coefficient quantization. An elgant application of the first- and second-order minimum multiplier allpass structures, considered here, is in the implementation of some simple transfer functions with parametrically tunable properties.

The Gray-Markel method to realize any arbitrary transfer function using the cascaded lattice form of allpass structure is outlined. The realization of a large class of arbitrary Nth order transfer functions using a parallel connection of two allpass filters is described. The final structure is shown to require only N multipliers. In Section 9.7.2 we demonstrate the low passband sensitivity property of these structures to small changes in the multiplier coefficients.

The cascaded lattice realization of an FIR transfer function is considered. The realization of a digital sine-cosine generator is then described and various oscillator structures are systematically developed. The chapter concludes with a comparison of the computational complexities of FIR and IIR digital filter structures.

The digital filter realization methods outlined in this chapter assume the existence of the corresponding transfer function. The following chapter considers the development of such transfer functions meeting the given frequency response specifications. The analysis of the finite wordlength effects on the performance of the digital filter structures is treated in Chapter 9.

6.14   Problems

6.1 The digital filter structure of Figure P6.1 has a delay-free loop and is therefore unrealizable. Determine a realizable equivalent structure with identical input-output relations and without any delay-free loop. (*Hint*: Express the output variables $y[n]$ and $w[n]$ in terms of the input variables $x[n]$ and $u[n]$ only and develop the correresponding block diagram representation from these input-output relations.)

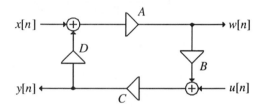

Figure P6.1

6.2 Determine by inspection whether or not the digital filter structures in Figure P6.2 have delay-free loops.   Identify these loops if they exist.   Develop equivalent structures without delay-free loops.

6.3   (a) Determine the gain Y/X of the signal flow-graphs of Figure P6.3 by simplifying each graph to a single branch connecting the input and output nodes.

(b) Verify the result of part (a) using Mason's gain formula.

6.4 Develop the transpose of the signal flow-graphs of Figure P6.3.

6.5 Consider the state-space representation of a causal IIR digital filter given by

$$s[n+1] = \mathbf{A}s[n] + \mathbf{B}x[n],$$
$$y[n] = \mathbf{C}s[n] + Dx[n], \tag{6.140}$$

where $y[n]$ and $x[n]$ denote, respectively, the output and the input sequences. Show that the transpose of the above state-space structure is characterized by the equations

$$\hat{s}[n+1] = \mathbf{A}^T \hat{s}[n] + \mathbf{C}^T x[n],$$
$$y[n] = \mathbf{B}^T \hat{s}[n] + Dx[n]. \tag{6.141}$$

Show that the transfer function $\hat{H}(z)$ of the transposed state-space structure is same as the transfer function $H(z)$ of the original state-space structure.

(a)

(b)

Figure P6.2

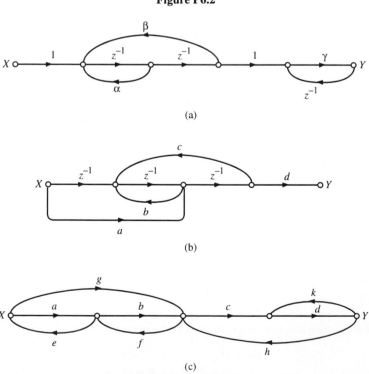

(a)

(b)

(c)

Figure P6.3

6.6 For the digital filter shown in Figure P6.4, find the digital angular frequency in rad/sec at which the magnitude response has a value of 0.9.

Figure P6.4

6.7 Analyze the digital filter structure of Figure P6.5 and determine its transfer function $H(z) = Y(z)/X(z)$.

(a) Is this a canonic structure?

(b) What should be the value of the multiplier coefficient K so that $H(z)$ has a unity gain at $\omega = 0$?

(c) What should be the value of the multiplier coefficient K so that $H(z)$ has a unity gain at $\omega = \pi$?

(d) Is there a difference between these two values of K? If not, why not?

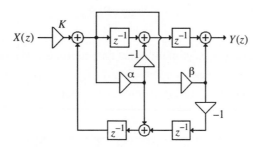

Figure P6.5

6.8 Consider the digital filter structure of Figure P6.6, where all multiplier coefficients are real.

(a) Develop its signal flow-graph representation and then determine the transfer function $H(z) = Y(z)/X(z)$ by using Mason's gain formula.

(b) What are the range of values of the multiplier coefficients for which the filter is BIBO stable?

Figure P6.6

6.9 Develop the signal flow-graph of the digital filter structure of Figure P6.7.

(a) Determine its transfer function $H(z) = Y(z)/X(z)$ by simplifying the signal flow-graph.

(b) Verify the result of part (a) using Mason's gain formula.

Figure P6.7

6.10 Develop the signal flow-graph of the digital filter structure of Figure P6.8.

(a) Determine its transfer function $H(z) = Y(z)/X(z)$ by simplifying the signal flow-graph.

(b) Verify the result of part (a) using Mason's gain formula.

Figure P6.8

6.11 Develop the signal flow-graph of the digital filter structure of Figure P6.9.

  (a) Determine its transfer function $H(z) = Y(z)/X(z)$ by simplifying the signal flow-graph.

  (b) Verify the result of part (a) using Mason's gain formula.

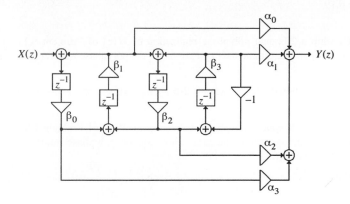

Figure P6.9

6.12 By using the block diagram analysis approach, determine the transfer function $H(z) = Y(z)/X(z)$ of the digital filter structure of Figure P6.6.

6.13 By using the block diagram analysis approach, determine the transfer function $H(z) = Y(z)/X(z)$ of the digital filter structure of Figure P6.7.

6.14 By using the block diagram analysis approach, determine the transfer function $H(z) = Y(z)/X(z)$ of the digital filter structure of Figure P6.8.

6.15 By using the block diagram analysis approach, determine the transfer function $H(z) = Y(z)/X(z)$ of the digital filter structure of Figure P6.9.

6.16 Realize the FIR transfer function

$$H(z) = (1 + 0.8z^{-1})^5 = 1 + 4z^{-1} + 6.4z^{-2} + 5.12z^{-3} + 2.048z^{-4} + 0.32768z^{-5}$$

in the following forms:

  (a) two different direct forms,

  (b) cascade of first-order sections,

  (c) cascade of one first-order and two second-order sections, and

  (d) cascade of one second-order and one third-order sections.

Compare the computational complexity of each of the above realizations.

6.17 Consider a length-9 FIR transfer function given by

$$H(z) = h[0] + h[1]z^{-1} + h[2]z^{-2} + h[3]z^{-3} + h[4]z^{-4} + h[5]z^{-5}$$
$$+ h[6]z^{-6} + h[7]z^{-7} + h[8]z^{-8}.$$

(a) Develop a 3-branch polyphase realization of $H(z)$ in the form of Figure 6.18(b) and determine the expressions for the polyphase transfer functions $E_0(z)$, $E_1(z)$, and $E_2(z)$. Compare these expressions with those given in Eq. (6.30).

(b) From this realization develop the canonic 3-branch polyphase realization shown in Figure 6.19.

6.18 (a) Develop a 2-branch polyphase realization of $H(z)$ of Problem 6.17 in the form of Figure 6.18(c) and determine the expressions for the polyphase transfer functions $E_0(z)$ and $E_1(z)$.

(b) From this realization develop a canonic 2-branch polyphase realization of $H(z)$.

6.19 (a) Develop a 4-branch polyphase realization of $H(z)$ of Problem 6.17 in the form of Figure 6.18(a) and determine the expressions for the polyphase transfer functions $E_0(z)$, $E_1(z)$, $E_2(z)$ and $E_3(z)$.

(b) From this realization develop a canonic 4-branch polyphase realization of $H(z)$.

6.20 Develop a minimum-multiplier realization of a length-9 Type 3 FIR transfer function.

6.21 Develop a minimum-multiplier realization of a length-10 Type 4 FIR transfer function.

6.22 Let $H(z)$ be a Type 1 linear-phase FIR filter of length M with $G(z)$ denoting its delay-complementary filter. Develop a realization of both filters using only $(M-1)$ delays and $(M+1)/2$ multipliers.

6.23 Show that a Type 1 linear-phase FIR transfer function $H(z)$ of length $N = 2M + 1$ can be expressed as

$$H(z) = z^{-M}\left[h[M] + \sum_{n=1}^{M} h[M-n]\left(z^n + z^{-n}\right) \right].$$ (6.142)

By using the relation

$$z^\ell + z^{-\ell} = 2T_\ell\left(\frac{z + z^{-1}}{2}\right),$$

where $T_\ell(x)$ is the ℓth-order Chebyshev polynomial[8] in x, express $H(z)$ in the form

$$H(z) = z^{-M} \sum_{n=0}^{M} a[n]\left(\frac{z + z^{-1}}{2}\right)^n.$$ (6.143)

Determine the relation between $a[n]$ and $h[n]$. Develop a realization of $H(z)$ based on Eq. (6.143) in the form of Figure P6.10, where $F_1(z^{-1})$ and $F_2(z^{-1})$ are causal structures. Determine the form of $F_1(z^{-1})$ and $F_2(z^{-1})$. The structure of Figure P6.10 is called the *Taylor structure* for linear-phase FIR filters [Sch72].

---

[8]For a definition of the Chebyshev polynomial and a recursive equation for generating such a polynomial, see Section 5.3.3.

Figure P6.10.   The Taylor structure shown for $M = 4$.

6.24 Show that there are 36 distinct cascade realizations of the transfer function $H(z)$ of Eq. (6.41) obtained by different pole-zero pairings and different orderings of the individual sections.

6.25 Consider a real coefficient IIR transfer function $H(z)$ with its numerator and denominator expressed as a product of polynomials:

$$H(z) = \prod_{i=1}^{K} \frac{P_i(z)}{D_i(z)},$$

where $P_i(z)$ and $D_i(z)$ are either first-order or second-order polyniomials with real coefficients. Determine the total number of distinct cascade realizations that can be obtained by different pole-zero pairings and different orderings of the individual sections.

6.26 Develop a canonic direct form realization of the transfer function

$$H(z) = \frac{2 + 4z^{-1} - 7z^{-2} + 3z^{-5}}{1 + 2z^{-1} + 5z^{-3}},$$

and then determine its transpose configuration.

6.27 Develop two different cascade canonic realizations of the following causal IIR transfer functions:

(a) $H_1(z) = \dfrac{(2 + 0.4z^{-1})(0.5 + 2z^{-1})}{(1 - 2z^{-1} + 2z^{-2})(1 - 0.5z^{-1})}$,

(b) $H_2(z) = \dfrac{3z\left(z + \frac{1}{2}\right)\left(z^2 + \frac{1}{3}\right)}{\left(z - \frac{1}{2}\right)\left(z - \frac{1}{3}\right)\left(z^2 - z + \frac{1}{2}\right)}$,

(c) $H_3(z) = \dfrac{(2 + 3z^{-1})(4 + 2z^{-1} + 3z^{-2})}{(1 + 0.6z^{-1})(1 + z^{-1} + 0.5z^{-2})}$.

6.28 Realize the transfer functions of Problem 6.27 in parallel forms I and II.

6.29 The structure shown in Figure P6.11 was developed in the course of a realization of the IIR digital transfer function

$$H(z) = \frac{3(2z^2 + 5z + 4)}{(2z + 1)(z + 2)}.$$

However, by a mistake in the labelling, two of the multiplier coefficients in this structure have incorrect values. Find these two multipliers and determine their correct values.

Figure P6.11

6.30 Figure P6.12 shows an incomplete realization of the causal IIR transfer function

$$H(z) = \frac{3z(5z - 2)}{(z + 0.5)(3z - 1)}.$$

Determine the values of the multiplier coefficients A and B.

Figure P6.12

6.31 Develop a two-multiplier canonic realization of the second-order transfer function

$$H(z) = \frac{1 + \alpha + \beta}{1 + \alpha z^{-1} + \beta z^{-2}},$$

with multiplier coefficients α and β.

6.32 A possible continued-fraction expansion of the digital transfer function

$$H(z) = \frac{p_0 z^N + p_1 z^{N-1} + \cdots + p_{N-1} z + p_N}{d_0 z^N + d_1 z^{N-1} + \cdots + d_{N-1} z + d_N} \qquad (6.144)$$

is given by

$$H(z) = A_0 + \cfrac{1}{B_1 z + \cfrac{1}{A_1 + \cfrac{1}{B_2 z + \cfrac{1}{\ddots \\ + \cfrac{1}{B_N z + \cfrac{1}{A_N}}}}}} \tag{6.145}$$

if $p_0 \neq 0$, $d_0 \neq 0$, and $d_N \neq 0$ [Mit72]. If $d_N = 0$, then A_N is infinite, and if $p_0 = 0$, then $A_0 = 0$. Develop a realization of $H(z)$ based on the continued-fraction expansion of Eq. (6.145).

6.33 Another possible continued-fraction expansion of the digital transfer function of Eq. (6.144) is of the form

$$H(z) = A_0 + \cfrac{1}{B_1 z + A_1 + \cfrac{1}{B_2 z + A_2 + \cfrac{1}{\ddots \\ + \cfrac{1}{B_N z + A_N}}}} \tag{6.146}$$

if $p_0 \neq 0$ and $d_0 \neq 0$ [Mit72]. As in Problem 6.32, if $p_0 = 0$, then the constant $A_0 = 0$. Develop a realization of $H(z)$ of Eq. (6.144) based on the continued-fraction expansion of Eq. (6.146) for N even and odd.

6.34 Show that a realization of the IIR transfer function of Eq. (6.144) based on a continued-fraction expansion of the form

$$H(z) = A_0 + \cfrac{1}{B_1 z^{-1} + A_1 + \cfrac{1}{B_2 z^{-1} + A_2 + \cfrac{1}{\ddots \\ + \cfrac{1}{B_N z^{-1} + A_N}}}} \tag{6.147}$$

is not realizable due to the presence of delay-free loops [Mit72].

6.35 Develop the continued-fraction expansion of the IIR transfer function [Mit72]

$$H(z) = \frac{720z^2 + 240z + 12}{720z^3 + 600z^2 + 72z + 1} \tag{6.148}$$

in the forms of Eqs. (6.145) and (6.146). From these expansions determine their corresponding realizations.

6.36 Develop the equivalent state-space structure from the transformed state-space parameters derived in Example 6.10. Use the value of D given in Eq. (6.50).

6.37 The state-space representation of a digital filter is given by:

$$A = \begin{bmatrix} -0.5 & 0.5 \\ 1 & 0 \end{bmatrix}, B = \begin{bmatrix} 1 \\ 0 \end{bmatrix}, C = [2 \quad 3], D = 2.$$

(a) Derive the transfer function of the filter from the state-space representation,

(b) Develop an equivalent state-space realization using the linear transformation

$$Q = \begin{bmatrix} 1 & 1 \\ 1 & -1 \end{bmatrix}.$$

6.38 The state-space representation of a digital filter is given by:

$$A = \begin{bmatrix} 2 & -1 \\ 3 & 1 \end{bmatrix}, B = \begin{bmatrix} 1 \\ 2 \end{bmatrix}, C = [1 \quad 3], D = 3.$$

(a) Derive the transfer function of the filter from the state-space representation,

(b) Develop an equivalent state-space realization using the linear transformation

$$Q = \begin{bmatrix} 2 & 1 \\ 2 & -1 \end{bmatrix}.$$

6.39   (a) Develop the state-space representations of the digital filter structure of Figure P6.13 and draw the block diagram representation of the this state-space structure.

(b) Derive the transfer function of the structure from the state-space representation.

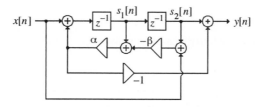

Figure P6.13

6.40 In this problem we develop an alternative cascaded lattice realization of an Nth-order IIR transfer function [Mit77b]

$$H_N(z) = \frac{p_0 + p_1 z^{-1} + \cdots + p_{N-1} z^{-N+1} + p_N z^{-N}}{1 + d_1 z^{-1} + \cdots + d_{N-1} z^{-N+1} + d_N z^{-N}}. \qquad (6.149)$$

The first stage of the realization process is shown in Figure P6.14.

(a) Show that if the two-pair chain parameters are chosen as

$$A = 1, \qquad B = d_N z^{-1}, \qquad C = p_0, \qquad D = p_N z^{-1}, \qquad (6.150)$$

then $H_{N-1}(z)$ is an $(N-1)$th-order IIR transfer function of the form

$$H_{N-1}(z) = \frac{p_0' + p_1' z^{-1} + \cdots + p_{N-1}' z^{-N+1}}{1 + d_1' z^{-1} + \cdots + d_{N-1}' z^{-N+1}}, \qquad (6.151)$$

with coefficients given by

$$p'_k = \frac{p_0 d_{k+1} - p_{k+1}}{p_0 d_N - p_N}, \quad k = 0, 1, \ldots, N - 1, \qquad (6.152a)$$

$$d'_k = \frac{p_k d_N - p_N d_k}{p_0 d_N - p_N}, \quad k = 1, 2, \ldots, N - 1. \qquad (6.152b)$$

(b) Develop a lattice realization of the two-pair.

(c) Continuing the above process, we can realize $H_N(z)$ as a cascade connection of N lattice sections constrained by a transfer function $H_0(z) = 1$. What are the total number of multipliers and two-input adders in the final realization of $H_N(z)$?

Figure P6.14

6.41 Realize the transfer function of Eq. (6.45) using the cascaded lattice realization method of Problem 6.40.

6.42 Realize the transfer functions of Problem 6.27 using the cascaded lattice realization method of Problem 6.40.

6.43 Show that the cascaded lattice realization method of Problem 6.40 results in the Gray-Markel cascaded lattice structure described in Section 6.7.2 when $H_N(z)$ is an allpass transfer function.

6.44 Develop the structures of Types 1B, 1A$_t$ and B$_t$ first-order allpass transfer functions shown in Figure 6.36(b), (c), and (d), from Eqs. (6.62b), (6.62c), and (6.62d), respectively.

6.45 (a) Develop a cascade realization of the third-order allpass transfer function

$$A(z) = \left(\frac{a + z^{-1}}{1 + az^{-1}}\right)\left(\frac{b + z^{-1}}{1 + bz^{-1}}\right)\left(\frac{c + z^{-1}}{1 + cz^{-1}}\right),$$

with each allpass section realized in Type 1A form. By sharing the delays between adjacent allpass sections, show that the total number of delays in the overall structure can be reduced from 6 to 4 [Mit74a].

(b) Repeat part (a) with each allpass section realized in Type 1A$_t$ form.

6.46 Develop formally the realizations of a second-order Type 2 allpass transfer function of Eq. (6.63) shown in Figure 6.37 using the multiplier extraction approach. Are there other Type 2 allpass structures?

6.47 Show that a cascade of two Type 2D second-order allpass structures can be realized with six delays by sharing the delays between adjacent sections. What is the minimum number of multipliers needed to implement a cascade of M Type 2D second-order allpass structures?

6.48 Develop formally the realizations of a second-order Type 3 allpass transfer function of Eq. (6.64) shown in Figure 6.38 using the multiplier extraction approach. Are there other Type 3 allpass structures?

6.49 Show that a cascade of two Type 3H second-order allpass structures can be realized with six delays by sharing the delays between adjacent sections. What is the minimum number of multipliers needed to implement a cascade of M Type 3H second-order allpass structures?

6.50 Realize each of the following IIR transfer functions in Gray-Markel's form and check the BIBO stability of each transfer function:

(a) $H_1(z) = \dfrac{2 + z^{-1} + 2z^{-2}}{1 - \frac{3}{4}z^{-1} + \frac{1}{8}z^{-2}}$,

(b) $H_2(z) = \dfrac{1 + 2z^{-1} + 3z^{-2}}{1 - z^{-1} + \frac{1}{4}z^{-2}}$,

(c) $H_3(z) = \dfrac{2 + 5z^{-1} + 8z^{-2} + 3z^{-3}}{1 + 0.75z^{-1} + 0.5z^{-2} + 0.25z^{-3}}$,

(d) $H_4(z) = \dfrac{1 + 1.6z^{-1} + 0.6z^{-2}}{1 - z^{-1} - 0.25z^{-2} + 0.25z^{-3}}$,

(e) $H_5(z) = \dfrac{3z\left(z + \frac{1}{2}\right)\left(z^2 + \frac{1}{3}\right)}{\left(z - \frac{1}{2}\right)\left(z - \frac{1}{3}\right)\left(z^2 - z + \frac{1}{2}\right)}$.

6.51 Realize each of the IIR transfer functions of Problem 6.27 in the Gray-Markel's form and check their BIBO stability.

6.52 Realize the IIR transfer function

$$H(z) = \frac{0.05634(1 + z^{-1})(1 - 1.0166z^{-1} + z^{-2})}{(1 - 0.683z^{-1})(1 - 1.4461z^{-1} + 0.7957z^{-2})}$$

in the following forms: (a) direct canonic form, (b) cascade form, (c) Gray-Markel form, and (d) cascaded lattice structure described in Problem 6.40. Compare their hardware requirements.

6.53 In this problem, the realization of a real coefficient transfer function using complex arithmetic is illustrated [Reg87a]. Let $G(z)$ be an Nth order real coefficient transfer function with simple complex conjugate poles and with numerator degree less than or equal to that of the denominator.

(a) Show that $G(z)$ can be expressed as a sum of two complex coefficient transfer functions of order $N/2$:

$$G(z) = H(z) + H^*(z^*), \tag{6.153}$$

where the coefficients of $H^*(z^*)$ are complex conjugate of their corresponding coefficients of $H(z)$.

(b) Generalize the above decomposition to the case when $G(z)$ has one or more simple real poles.

(c) Consider a realization of $H(z)$ that has one real input $x[n]$ and a complex output $y[n]$. Show that the transfer function from the input to the real part of the output is simply $G(z)$.

[*Hint*: Use partial-fraction expansion to obtain the decomposition of Eq. (6.153).]

6.54 Develop a realization of a first-order complex coefficient transfer function $H(z)$ given by

$$H(z) = \frac{A + jB}{1 + (\alpha + j\beta)z^{-1}},$$

where A, B, α, and β are real constants. Show the real and imaginary parts of all signal variables separately. Determine the transfer functions from the input to the real and imaginary parts of the output.

6.55 Develop a cascaded lattice realization of an Nth-order complex coefficient allpass transfer function $A_N(z)$.

6.56 Realize the following transfer functions in the form of a parallel connection of two allpass filters:

(a) $H_1(z) = \dfrac{2 + 2z^{-1}}{3 + z^{-1}}$,

(b) $H_2(z) = \dfrac{1 - z^{-1}}{4 + 2z^{-1}}$,

(c) $H_3(z) = \dfrac{1 - z^{-2}}{4 + 2z^{-1} + 2z^{-2}}$,

(d) $H_4(z) = \dfrac{3 + 9z^{-1} + 9z^{-2} + 3z^{-3}}{12 + 10z^{-1} + 2z^{-2}}$.

6.57 Consider a causal length-6 FIR filter described by the convolution sum:

$$y[n] = \sum_{k=0}^{5} h[k]x[n-k], \qquad n \geq 0,$$

where $y[n]$ and $x[n]$ denote, respectively, the output and the input sequences.

(a) Let the output and input sequences be blocked into length-2 vectors

$$\mathbf{Y}_\ell = \begin{bmatrix} y[2\ell] \\ y[2\ell + 1] \end{bmatrix}, \quad \mathbf{X}_\ell = \begin{bmatrix} x[2\ell] \\ x[2\ell + 1] \end{bmatrix}.$$

Show that the above FIR filter can be equivalently described by a block convolution sum given by

$$\mathbf{Y}_\ell = \sum_{r=0}^{3} \mathbf{H}_r \mathbf{X}_{\ell-r},$$

where $\mathbf{H}_r$ are 2×2 matrices composed of the impulse response coefficients. Determine the block convolution matrices $\mathbf{H}_r$. An implementation of the FIR filter based on the above block convolution sum is shown in Figure P6.15 where the block labeled "S/P" is a serial-to-parallel converter and the block marked "P/S" is a parallel-to-serial converter.

(b) Develop the block convolution sum description of the above FIR filter for an input and output block length of 3.

(c) Develop the block convolution sum description of the above FIR filter for an input and output block length of 4.

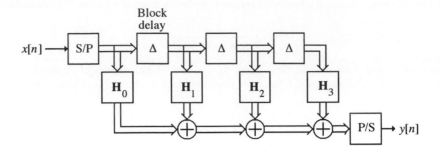

Figure P6.15

6.58 Consider a causal IIR filter described by a difference equation

$$\sum_{k=0}^{4} d_k y[n-k] = \sum_{k=0}^{4} p_k x[n-k], \qquad n \ge 0,$$

where $y[n]$ and $x[n]$ denote, respectively, the output and the input sequences.

(a) Let the output and input sequences be blocked into length-2 vectors

$$\mathbf{Y}_\ell = \begin{bmatrix} y[2\ell] \\ y[2\ell+1] \end{bmatrix}, \mathbf{X}_\ell = \begin{bmatrix} x[2\ell] \\ x[2\ell+1] \end{bmatrix}.$$

Show that the above IIR filter can be equivalently described by a block difference equation given by [Bur72]

$$\sum_{r=0}^{2} \mathbf{D}_r \mathbf{Y}_{\ell-r} = \sum_{r=0}^{2} \mathbf{P}_r \mathbf{X}_{\ell-r},$$

where $\mathbf{D}_r$ and $\mathbf{P}_r$ are 2×2 matrices composed of the difference equation coefficients $\{d_k\}$ and $\{p_k\}$, respectively. Determine the block difference equation matrices $\mathbf{D}_r$ and $\mathbf{P}_r$. An implementation of the IIR filter based on the above block difference equation is shown in Figure P6.16.

(b) Develop the block difference equation description of the above IIR filter for an input and output block length of 3.

(c) Develop the block difference equation description of the above IIR filter for an input and output block length of 4.

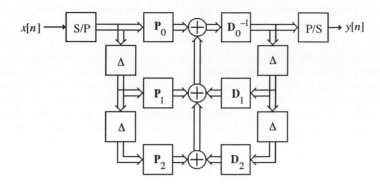

Figure P6.16

6.59 Develop a canonic realization of the block digital filter of Figure P6.16 employing only two block delays.

6.60 Consider the state-space representation of an Nth causal IIR transfer function given by

$$\mathbf{s}[n+1] = \mathbf{As}[n] + \mathbf{B}x[n],$$
$$y[n] = \mathbf{Cs}[n] + Dx[n],$$

where $y[n]$ and $x[n]$ are the output and input sequences, respectively, and $\mathbf{s}[n]$ is the state vector of order N. Define input and output block vectors as

$$\mathbf{Y}_k = [\, y[kL] \quad y[kL+1] \quad \cdots \quad y[kL+L-1]\,]^T,$$
$$\mathbf{X}_k = [\, x[kL] \quad x[kL+1] \quad \cdots \quad x[kL+L-1]\,]^T,$$

and the block state-vector as

$$\mathbf{S}_k = \mathbf{s}[kL].$$

The block state-space representation is of the form

$$\mathbf{S}_{k+1} = \boldsymbol{\mathcal{A}}\mathbf{S}_k + \boldsymbol{\mathcal{B}}\mathbf{X}_k,$$
$$\mathbf{Y}_k = \boldsymbol{\mathcal{C}}\mathbf{S}_k + \boldsymbol{\mathcal{D}}\mathbf{X}_k.$$

Express the block state-space parameter matrices $\boldsymbol{\mathcal{A}}$, $\boldsymbol{\mathcal{B}}$, $\boldsymbol{\mathcal{C}}$, and $\boldsymbol{\mathcal{D}}$, as a function of the state-space parameter matrices $\mathbf{A}$, $\mathbf{B}$, $\mathbf{C}$, and D.

6.61 Develop the 3-multiplier structure of a digital sine-cosine generator obtained by setting $\alpha \sin \theta = \pm\beta$ in Eq. (6.138).

6.62 Develop the 3-multiplier structure of a digital sine-cosine generator obtained by setting $\alpha = \pm\beta \sin \theta$ in Eq. (6.138).

6.63 Develop a one-multiplier structure of a digital sine-cosine generator obtained by setting $-\beta \sin \theta/\alpha = 1 + \cos \theta$ in Eq. (6.138).

6.64 Develop a one-multiplier structure of a digital sine-cosine generator obtained by setting $C = 0$ in Eq. (6.137) and then choosing α and β properly.

6.65 Develop a one-multiplier structure of a digital sine-cosine generator obtained by setting $C = -1$ in Eq. (6.137) and then choosing α and β properly. Show the final structure.

6.66 Signals generated by multple sources or multiple sensors, called a *multichannel signal*, are usually transmitted through independent channels in close proximity to each other. As a result, each component of the multichannel signal often get corrupted by signals from adjacent channels during transmission resulting in *cross-talk*. Separation of the multichannel signal at the receiver is thus of practical interest. A model representing the cross-talk between a pair of channels for a two-channel signal is depicted in Figure P6.17(a) and the corresponding discrete-time system for channel separation is as shown in Figure P6.17(b) [Yel96]. Determine two possible sets of conditions for perfect channel separation.

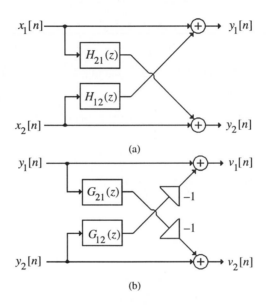

(a)

(b)

Figure P6.17

6.67 Develop a two-multiplier canonic realization of each of the following second-order transfer functions where the multiplier coefficients are α_1 and $-\alpha_2$, respectively [Hir73]:

(a) $H_1(z) = \dfrac{(1 - \alpha_1 + \alpha_2)(1 + z^{-1})^2}{1 - \alpha_1 z^{-1} + \alpha_2 z^{-2}},$

(b) $H_2(z) = \dfrac{(1 - \alpha_2)(1 - z^{-2})}{1 - \alpha_1 z^{-1} + \alpha_2 z^{-2}}.$

6.15   MATLAB Excercises

M 6.1 Using MATLAB develop a cascade realization of each of the following linear-phase FIR transfer functions:

(a) $H_1(z) = 0.01551933 - 0.157681075z^{-1} - 0.0558404z^{-2} + 0.1803771z^{-3}$
$+0.4136878z^{-4} + 0.4136878z^{-5} + 0.1803771z^{-6} - 0.0558404z^{-7}$
$-0.157681075z^{-8} + 0.01551933z^{-9}$,

(b) $H_2(z) = 0.0636786 + 0.0691228z^{-1} - 0.1010436z^{-2} - 0.2857499z^{-3}$
$+0.5892699z^{-4} - 0.2857499z^{-5} - 0.1010436z^{-6} + 0.0691228z^{-7} + 0.0636786z^{-8}$.

M 6.2 Consider the fourth-order IIR lowpass transfer function

$$G(z) = \frac{\begin{matrix}0.008363239 - 0.033452958z^{-1} + 0.05017944z^{-2} \\ -0.03345296z^{-3} + 0.008363239z^{-4}\end{matrix}}{1 + 2.37412317z^{-1} + 2.7056566z^{-2} + 1.5917092z^{-3} + 0.41031508z^{-4}}.$$

(a) Using MATLAB express $G(z)$ in factored form,

(b) Develop four different cascade realizations of $G(z)$,

(c) Develop two different parallel form realizations of $G(z)$.

M 6.3 Consider the fourth-order IIR highpass transfer function given below:

$$H(z) = \frac{1 - 2.0912246355z^{-1} + 2.840504z^{-2} - 2.0912246355z^{-3} + z^{-4}}{1 + 0.896172z^{-1} + 1.3142222z^{-2} + 0.6655849z^{-3} + 0.29383441z^{-4}}.$$

(a) Using MATLAB express $H(z)$ in factored form,

(b) Develop four different cascade realizations of $H(z)$,

(c) Realize $H(z)$ in parallel forms I and II.

Realize each second-order section in direct form II.

M 6.4   (a) Develop state-space realizations of the transfer functions of Problem 6.22 using MATLAB.

(b) Develop alternate state-space realizations of each realization obtained in part (a) by transforming the state-variables suitably.

M 6.5 Using Program 6_3 develop a Gray-Markel cascaded lattice realization of the IIR transfer function $H(z)$ of Problem M6.2.

M 6.6 Using Program 6_3 develop a Gray-Markel cascaded lattice realization of the IIR transfer function $H(z)$ of Problem M6.3.

M 6.7 Using Program 6_4 develop a cascade lattice realization of each of the FIR transfer functions of Problem M6.1.

M 6.8   (a) Realize the following IIR lowpass transfer function $G(z)$ in the form of a parallel allpass structure.

(b) From the allpass decomposition determine its power-complementary transfer function $H(z)$.

(c) Plot the square of the magnitude responses of the original transfer function $G(z)$ and its power-complementary transfer function $H(z)$ derived in part (b) and verify that their sum is equal to one at all frequencies.

$$G(z) = \frac{0.04934436(1 + 1.3z^{-1} + 2.21z^{-2} + 1.3z^{-3} + 2.21z^{-4} + z^{-5})}{1 - 1.9919z^{-1} + 2.819z^{-2} - 2.25648z^{-3} + 1.1866z^{-4} - 0.31214z^{-5}}.$$

M 6.9   (a) Realize the following IIR highpass transfer function $G(z)$ in the form of a parallel allpass structure.

(b) From the allpass decomposition determine its power-complementary transfer function $H(z)$.

(c) Plot the square of the magnitude responses of the original transfer function $G(z)$ and its power-complementary transfer function $H(z)$ derived in part (b) and verify that their sum is equal to one at all frequencies.

$$G(z) = \frac{0.1629245(1 - 3.09z^{-1} + 5.07z^{-2} - 5.07z^{-3} + 3.09z^{-4} + z^{-5})}{1 - 0.5192z^{-1} + 0.9241z^{-2} - 0.35113z^{-3} + 0.16492z^{-4} - 0.02515z^{-5}}.$$

M 6.10   (a) Realize the following IIR bandpass transfer function $G(z)$ in the form of a parallel allpass structure.

(b) From the allpass decomposition determine its power-complementary transfer function $H(z)$.

(c) Plot the square of the magnitude responses of the original transfer function $G(z)$ and its power-complementary transfer function $H(z)$ derived in part (b) and verify that their sum is equal to one at all frequencies.

$$G(z) = \frac{0.13110644(1 - z^{-2})^2}{1 + 0.48243z^{-1} + 0.81z^{-2} + 0.226876z^{-3} + 0.272215z^{-4}}.$$

M 6.11 Using MATLAB simulate the single multiplier sine-cosine generator of Problem 6.63 with $\cos\theta = 0.9$ and plot the first 50 samples of its two output sequences. Scale the outputs so that they both have a maximum amplitude of ± 1. What is the effect of initial values of the state-variables?

M 6.12 Using MATLAB simulate the single multiplier sine-cosine generator of Problem 6.64 with $\cos\theta = 0.9$ and plot the first 50 samples of its two output sequences. Scale the outputs so that they both have a maximum amplitude of ± 1. What is the effect of initial values of the state-variables?

M 6.13 Using MATLAB simulate the single multiplier sine-cosine generator of Problem 6.65 with $\cos\theta = 0.9$ and plot the first 50 samples of its two output sequences. Scale the outputs so that they both have a maximum amplitude of ± 1. What is the effect of initial values of the state-variables on the outputs?

CHAPTER 7

Digital Filter Design

An important step in the development of a digital filter is the determination of a realizable transfer function $G(z)$ approximating the given frequency response specifications. If an IIR filter is desired, it is also necessary to ensure that $G(z)$ is stable. The process of deriving the transfer function $G(z)$ is called *digital filter design*. After $G(z)$ has been obtained, the next step is to realize it in the form of a suitable filter structure. In the previous chapter, we outlined a variety of basic structures for the realization of FIR and IIR transfer functions. In this chapter, we consider the digital filter design problem.

First we review some of the issues associated with the filter design problem. Two popular approaches to IIR filter design based on the conversion of a prototype analog transfer function to a digital transfer function are discussed next. Typical design examples are included to illustrate both these approaches. We then consider the transformation of one type of filter transfer function into another type, which is achieved by replacing the complex variable z by a function of z. Four commonly used transformations are summarized. Two different approaches to FIR filter design are next described. Finally, we consider the computer-aided design of both IIR and FIR digital filters. To this end, we restrict our discussion to the use of MATLAB in determining the transfer functions.

7.1 Preliminary Considerations

There are two major issues that need to be answered before one can develop the digital transfer function $G(z)$. The first and foremost issue is the development of a reasonable filter frequency response specification from the requirements of the overall system in which the digital filter is to be employed. The second issue is to determine whether a finite-impulse response (FIR) or an infinite-impulse response (IIR) digital filter is to be designed. In this section, we examine these two issues. In addition, we review the basic approaches to the design of IIR and FIR digital filters and determination of the filter order to meet the prescibed specifications. We also discuss the scaling of the transfer function.

417

7.1.1   Digital Filter Specifications

As in the case of the analog filter, either the magnitude or the phase (delay) response is specified for the design of a digital filter for most applications. In some situations, the unit sample response or the step response may be specified. In most practical applications, the problem of interest is the development of a realizable approximation to a given magnitude response specification. We thus restrict our attention in this chapter to this problem only. As indicated in Section 4.3.1, there are four basic types of filters, whose magnitude responses are shown in Figure 4.8. Since the impulse response corresponding to each of these are noncausal and of infinite length, these ideal filters are not realizable. One way to develop a realizable approximation to these filters would be to truncate the impulse response as indicated in Eq. (4.63) for a lowpass filter. The magnitude response of the FIR lowpass filter obtained by truncating the impulse response of the ideal lowpass filter as sketched in Figure 4.22 does not have a sharp transition from passband to stopband but, rather, exhibits a gradual "roll-off".

Thus, as in the case of the analog filter design problem outlined in Section 5.3.1, the magnitude response specifications of a digital filter in the passband and in the stopband are given with some acceptable tolerances. In addition, a transition band is specified between the passband and the stopband to permit the magnitude to drop off smoothly. For example, the magnitude $\left|G(e^{j\omega})\right|$ of a lowpass filter may be given as shown in Figure 7.1. As indicated in the figure, in the *passband* defined by $0 \le \omega \le \omega_p$, we require that the magnitude approximates unity with an error of $\pm\delta_p$, i.e.,

$$1 - \delta_p \le \left|G(e^{j\omega})\right| \le 1 + \delta_p, \qquad \text{for } |\omega| \le \omega_p. \tag{7.1}$$

In the *stopband*, defined by $\omega_s \le \omega \le \pi$, we require that the magnitude approximate zero with an error of δ_s, i.e.,

$$\left|G(e^{j\omega})\right| \le \delta_s, \quad \text{for } \omega_s \le |\omega| \le \pi. \tag{7.2}$$

The frequencies ω_p and ω_s are, respectively, called the *passband edge frequency* and the *stopband edge frequency*. The limits of the tolerances in the passband and stopband, δ_p and δ_s, are usually called the *peak ripple values*. Note that the frequency response $G(e^{j\omega})$ of a digital filter is a periodic function of ω, and the magnitude response of a real coefficient digital filter is an even function of ω. As a result, the digital filter specifications are given only for the range $0 \le |\omega| \le \pi$.

Often, the specifications of the digital filter are given in terms of the loss function, $\mathcal{G}(\omega) = -20\log_{10}\left|G(e^{j\omega})\right|$, in dB. Here the *peak passband ripple* α_p and the *minimum stopband attenuation* α_s are given in dB, i.e., the loss specifications of a digital filter are given by

$$\alpha_p = -20\log_{10}(1 - \delta_p) \text{ dB}, \tag{7.3}$$
$$\alpha_s = -20\log_{10}(\delta_s) \text{ dB}. \tag{7.4}$$

As in the case of an analog lowpass filter, the specifications for a digital lowpass filter may alternatively be given in terms of its magnitude response, as in Figure 7.2. Here the maximum value of the magnitude in the passband is assumed to be unity, and the

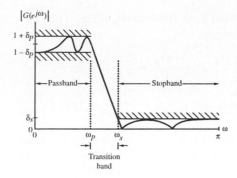

Figure 7.1 Typical magnitude specifications for a digital lowpass filter.

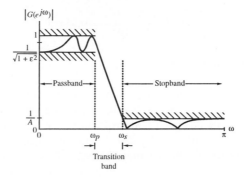

Figure 7.2 Alternate magnitude specifications for a digital lowpass filter.

maximum passband deviation, denoted as $1/\sqrt{1+\varepsilon^2}$, is given by the minimum value of the magnitude in the passband. The maximum stopband magnitude is denoted by $1/A$.[1]

For the normalized specification, the maximum value of the gain function or the minimum value of the loss function is therefore 0 dB. The quantity $\alpha_{\max}$ given by

$$\alpha_{\max} = 20\log_{10}\left(\sqrt{1+\varepsilon^2}\right)\ \text{dB} \tag{7.5}$$

is called the *maximum passband attenuation*. For $\delta_p \ll 1$, as is typically the case, it can be shown that

$$\alpha_{\max} \cong -20\log_{10}(1-2\delta_p) \cong 2\alpha_p. \tag{7.6}$$

The passband and stopband edge frequencies, in most applications, are specified in Hz, along with the sampling rate of the digital filter. Since all filter design techniques are developed in terms of normalized angular frequencies ω_p and ω_s, the specified critical frequencies need to be normalized first before a specific filter design algorithm can be applied. Let F_T denote the sampling frequency in Hz, and F_p and F_s denote, respectively,

---

[1]The minimum stopband attenuation is therefore $20\log_{10}(A)$.

the passband and stopband edge frequencies in Hz. Then the normalized angular edge frequencies in radians are given by

$$\omega_p = \frac{\Omega_p}{F_T} = \frac{2\pi F_p}{F_T} = 2\pi F_p T, \tag{7.7}$$

$$\omega_s = \frac{\Omega_s}{F_T} = \frac{2\pi F_s}{F_T} = 2\pi F_s T. \tag{7.8}$$

Example 7.1 Let the specified passband and stopband edge frequencies of a digital highpass filter operating at a sampling rate of 25 kHz be 7 kHz and 3 kHz, respectively. Using Eqs. (7.7) and (7.8) we determine the corresponding normalized angular bandedge frequencies as

$$\omega_p = \frac{2\pi(7 \times 10^3)}{25 \times 10^3} = 0.56\pi,$$

$$\omega_s = \frac{2\pi(3 \times 10^3)}{25 \times 10^3} = 0.24\pi.$$

7.1.2   Selection of the Filter Type

The second issue of interest is the selection of the digital filter type, i.e., whether an IIR or an FIR digital filter is to be employed. There are several advantages to using an FIR filter, since it can be designed with exact linear phase and the filter structure is always stable with quantized filter coefficients. However, in most cases, the order N_{FIR} of an FIR filter is considerably higher than the order N_{IIR} of an equivalent IIR filter meeting the same magnitude specifications. In general, the implementation of the FIR filter requires approximately N_{FIR} multiplications per output sample, whereas the IIR filter requires $2N_{\text{IIR}} + 1$ multiplications per output sample. In the former case, if the FIR filter is designed with a linear phase, then the number of multiplications per output sample reduces to approximately $(N_{\text{FIR}} + 1)/2$. Likewise, most IIR filter designs result in transfer functions with zeros on the unit circle, and the cascade realization of an IIR filter of order N_{IIR} with zeros on the unit circle requires $\lfloor(3N_{\text{IIR}} + 3)/2\rfloor$ multiplications per output sample.[2] It has been shown by Rabiner and Gold [Rab75] that for most practical filter specifications, the ratio $N_{\text{FIR}}/N_{\text{IIR}}$ is typically of the order of tens or more, and as a result, the IIR filter usually is computationally more efficient. However, if the group delay of the IIR filter is equalized by cascading it with an allpass equalizer, then the savings in computation may no longer be that significant [Rab75]. In many applications, the linearity of the phase response of the digital filter is not an issue, making the IIR filter preferable because of the lower computational requirements.

7.1.3   Basic Approaches to Digital Filter Design

In the case of an IIR filter design, the most common practice is to convert the digital filter specifications to analog lowpass prototype filter specifications, to determine the analog lowpass filter transfer function $H_a(s)$ meeting these specifications, and then to transform it into the desired digital filter transfer function $G(z)$. This approach has been widely used for many reasons:

---

[2]$\lfloor x \rfloor$ denotes the integer part of x.

(a) Analog approximation techniques are highly advanced

(b) They usually yield closed-form solutions

(c) Extensive tables are available for analog filter design

(d) Many applications require the digital simulation of analog filters.

In the sequel, we denote an analog transfer function as

$$H_a(s) = \frac{P_a(s)}{D_a(s)},$$
(7.9)

where the subscript "a" specifically indicates the analog domain. The digital transfer function derived from $H_a(s)$ is denoted by

$$G(z) = \frac{P(z)}{D(z)}.$$
(7.10)

The basic idea behind the conversion of an analog prototype transfer function $H_a(s)$ to a digital IIR transfer function $G(z)$ is to apply a mapping from the s-domain to the z-domain so that the essential properties of the analog frequency response are preserved. This implies that the mapping function should be such that:

(a) The imaginary ($j\Omega$) axis in the s-plane be mapped onto the unit circle of the z-plane

(b) A stable analog transfer function be transformed into a stable digital transfer function.

To this end, two popular transformations are the impulse invariance mapping and the bilinear transformation described in Sections 7.2 and 7.3, respectively.

Unlike the IIR digital filter design, the FIR filter design does not have any connection with the design of analog filters. The design of FIR filters is therefore based on a direct approximation of the specified magnitude response, with the often added requirement that the phase response be linear. A causal FIR transfer function $H(z)$ of length N is a polynomial in z^{-1} of degree $N-1$:

$$H(z) = \sum_{n=0}^{N-1} h[n]z^{-n}.$$
(7.11)

The corresponding frequency response is given by

$$H(e^{j\omega}) = \sum_{n=0}^{N-1} h[n]e^{-j\omega n}.$$
(7.12)

It has been shown in Section 3.2.2 that any finite duration sequence $x[n]$ of length N is completely characterized by N samples of its Fourier transform $X(e^{j\omega})$. As a result, the design of an FIR filter of length N may be accomplished by finding either the impulse response sequence $\{h[n]\}$ or N samples of its frequency response $H(e^{j\omega})$. Also, to ensure a linear phase design, the condition $h[N-1-n] = h[n]$ or $-h[n]$ must be satisfied. Two direct approaches to the design of FIR filters are the truncated Fourier series approach and the frequency sampling approach, described in Sections 7.7 and 7.8, respectively.

In Section 7.9 we outline computer-based digital filter design methods.

7.1.4   Estimation of the Filter Order

For the design of an IIR digital lowpass filter $G(z)$ based on the conversion of an analog lowpass filter $H_a(s)$, the filter order of $H_a(s)$ is first estimated from its specifications using the appropriate formulas given in Eqs. (5.31), (5.39), or (5.49), depending on whether a Butterworth, Chebyshev, or equiripple filter approximation is desired. The order of $G(z)$ is then determined automatically from the transformation being used to convert $H_a(s)$ to $G(z)$. There are several M-file functions in MATLAB that can be used to directly estimate the minimum order of an IIR digital transfer function meeting the filter specifications for the class of approximations discussed in Section 5.3. These are discussed in Section 7.9.2.

For the design of FIR lowpass digital filters, several authors have advanced formulas for estimating the minimum value of the filter length N directly from the digital filter specifications: normalized passband edge angular frequency ω_p, normalized stopband edge angular frequency ω_s, peak passband ripple δ_p, and peak stopband ripple δ_s [Her73], [Kai74], and [Rab73]. A rather simple approximate formula developed by Kaiser [Kai74] is given by

$$N \cong \frac{-20\log_{10}\left(\sqrt{\delta_p \delta_s}\right) - 13}{14.6(\omega_s - \omega_p)/2\pi}. \tag{7.13}$$

Note from the above formula that the filter length of an FIR filter is inversely proportional to the transition band width and does not depend on the actual location of the transition band. This implies that a sharp cutoff FIR filter with a narrow transition band would be of very long length, whereas an FIR filter with a wide transition band will have a very short length. Another interesting property of the Kaiser's formula is that the length depends on the product $\delta_p \delta_s$. This implies that if the values of δ_p and δ_s are interchanged, the length remains the same.

The formula of Eq. (7.13) provides a reasonable good estimate of the filter length in the case of FIR filters with a moderate passband width and may not work well in the case of very narrow passband or very wide passband FIR filters. In case of a narrow band filter, the stopband ripple essentially controls the length and an alternative approximate formula provides a more reasonable estimate of the filter length [Par87]:

$$N \cong \frac{-20\log_{10}(\delta_s) + 0.22}{(\omega_s - \omega_p)/2\pi}. \tag{7.14}$$

On the other hand, in the case of a very wideband filter, the passband ripple has more effect on the length, and a more reasonable estimate of the filter length can be obtained using the following formula [Par87]:

$$N \cong \frac{-20\log_{10}(\delta_p) + 5.94}{27(\omega_s - \omega_p)/2\pi}. \tag{7.15}$$

We illustrate the application of Kaiser's formula of Eq. (7.13) in the following example.

Example 7.2 Estimate the length of a linear-phase FIR transfer function with the following specifications: passband edge = 180 Hz, stopband edge = 200 Hz, passband ripple $\delta_p = 0.002$, and stopband ripple $\delta_s = 0.001$, and sampling rate of 12 kHz. To this end we employ the following MATLAB program.

```
% Program 7_1
% Computation of the length of a linear-phase
% FIR lowpass filter using Kaiser's formula
%
Rp = input('Type in the passband ripple = ');
Rs = input('Type in the stopband ripple = ');
Fp = input('Type in the passband edge in Hz = ');
Fs = input('Type in the stopband edge in Hz = ');
FT = input('Type in the sampling frequency in Hz = ');
num = -20*log10(sqrt(Rp*Rs))-13;
den = 14.6*(Fs - Fp)/FT;
N = ceil(num/den);
fprintf('Filter length is %f \n',N);
```

During execution, the program requests the relevant filter specifications and displays the filter order at the end of execution as indicated below:

```
Filter length is 1808.000000
```

The formula due to Hermann et al. [Her73] gives a slightly more accurate value for the length and is given by[3]

$$N \cong \frac{D_\infty(\delta_p, \delta_s) - F(\delta_p, \delta_s) \left[(\omega_s - \omega_p)2\pi \right]^2}{\left[(\omega_s - \omega_p)2\pi \right]}, \qquad (7.16)$$

where

$$D_\infty(\delta_p, \delta_s) = \left[a_1 (\log_{10} \delta_p)^2 + a_2(\log_{10} \delta_p) + a_3 \right] \log_{10} \delta_s$$
$$- \left[a_4 (\log_{10} \delta_p)^2 + a_5(\log_{10} \delta_p) + a_6 \right], \qquad (7.17a)$$

and

$$F(\delta_p, \delta_s) = b_1 + b_2 \left[\log_{10} \delta_p - \log_{10} \delta_s \right], \qquad (7.17b)$$

with

$$a_1 = 0.005309, \qquad a_2 = 0.07114, \qquad a_3 = -0.4761, \qquad (7.17c)$$

$$a_4 = 0.00266, \qquad a_5 = 0.5941, \qquad a_6 = 0.4278, \qquad (7.17d)$$

$$b_1 = 11.01217, \qquad b_2 = 0.51244. \qquad (7.17e)$$

The formula given in Eq. (7.16) is valid for $\delta_p > \delta_s$. If $\delta_p < \delta_s$, then the filter length formula to be used is obtained by interchanging δ_p and δ_s in Eq. (7.16).

For small values of δ_p and δ_s, both Eqs. (7.13) and (7.16) provide reasonably close and accurate results. On the other hand, when the values of δ_p and δ_s are large, Eq. (7.16) yields a more accurate value for the length. It should be noted that both formulas provide an estimate of the filter length. If the FIR filter designed using the estimated value of length

---

[3] It is obtained by following a least-squares approximation to N as a function of the bandedges and ripples.

does not meet the specifications, the value of the length should be increased gradually until the specifications are met.

The order of the FIR filter, which is the length minus one, can also be determined using the MATLAB function `remezord`. There are two different options available with this function:

```
[ord,f,m,wgt] = remezord(fedge,mag,dev)
[ord,f,m,wgt] = remezord(fedge,mag,dev,FT)
```

where the input data are the vector `fedge` of bandedges, the vector `mag` of desired magnitude values in each frequency band, and a vector `dev` specifying the maximum allowable deviation between the magnitude response of the designed filter and the desired magnitude. The length of `fedge` is two times that of `mag` minus 2, while the length of `dev` is the same as that of `mag`. The output data are the estimated value `ord` of the filter order, the normalized frequency bandedge vector `f`, the frequency band magnitude vector `m`, and the weight vector `wgt` meeting the filter specifications. The output data can then be directly used in filter design using the function `remez`.[4] The sampling frequency FT can be specified using the second version of this function. A default of 2 Hz is employed if FT is not explicitly indicated. `remezord` is based on the formula of Eq. (7.16).

Example 7.3 Estimate the length of a linear-phase FIR transfer function with the specifications given in Example 7.2 using `remezord`. The MATLAB code used for this example is given below:

```
% Program 7_2
% Estimation of FIR filter order using remezord
%
fedge = input('Type in the bandedges = ');
mag = input('Desired magnitude values in each band = ');
dev = input('Allowable deviation in each band = ');
FT = input('Type in the sampling frequency = ');
[ord, fo, mo, wgt] = remezord(fedge, mag, dev, FT);
fprintf('Filter order is %f \n',ord);
```

The input data called by this program are the vectors fedge, mag, and dev, and the scalar FT. For this example, these are given by `fedge = [180 200]`, `mag = [1 0]`, `dev = [0.002 0.001]`, and `FT = 12000`. The output data printed by this program is given by

```
Filter order is 1827.000000
```

Note that the filter length computed by this algorithm is slightly higher than that obtained using Kaiser's formula in the previous example.

---

[4]See Section 7.9.3.

7.1.5   Scaling the Digital Transfer Function

After a digital filter has been designed following any one of the techniques outlined in this chapter, the corresponding transfer function $G(z)$ has to be scaled in magnitude before it can be implemented. In magnitude scaling, the transfer function is multiplied by a scaling constant K so that the maximum magnitude of the scaled transfer function $G_t(z) = KG(z)$ in the passband is unity, i.e., the scaled transfer function has a maximum gain of 0 dB. For a stable transfer function $G(z)$ with real coefficients, the scaled transfer function $KG(z)$ is then a bounded real (BR) function.[5]

In the case of a lowpass transfer function, it is usual practice to use $K = 1/G(1)$, implying a dc gain of 0 dB for the scaled transfer function. Likewise, in the case of a highpass transfer function, K is selected equal to $1/G(-1)$ yielding a gain of 0 dB at $\omega = \pi$. For a bandpass transfer function, it is common to use K equal to $1/|G(e^{j\omega_c})|$, where ω_c is the center frequency. Finally, for a bandstop transfer function, K is typically chosen to be the reciprocal of $\max[G(-1), G(1)]$.

Example 7.4   Consider the fourth-order highpass transfer function

$$G(z) = \frac{p_0 + p_1 z^{-1} + p_2 z^{-2} + p_3 z^{-3} + p_4 z^{-4}}{1 + d_1 z^{-1} + d_2 z^{-2} + d_3 z^{-3} + d_4 z^{-4}}.$$

Its magnitude scaled version is then given by

$$G_t(z) = K \frac{p_0 + p_1 z^{-1} + p_2 z^{-2} + p_3 z^{-3} + p_4 z^{-4}}{1 + d_1 z^{-1} + d_2 z^{-2} + d_3 z^{-3} + d_4 z^{-4}},$$

where

$$K = \frac{1 - d_1 + d_2 - d_3 + d_4}{p_0 - p_1 + p_2 - p_3 + p_4}.$$

7.2   Impulse Invariance Method of IIR Filter Design

As indicated earlier, IIR filters are usually designed by transforming a prototype analog transfer function. Two different approaches to IIR filter design are considered in this chapter. The first approach discussed here is the impulse invariance method. The basic idea behind this method is to develop an IIR transfer function whose impulse response is exactly identical to the uniformly sampled version of the impulse response of the prototype analog transfer function. In the following section, we discuss the bilinear transformation approach. Design examples are included later that illustrate both approaches.

To develop the necessary design formulas for the impulse invariance method, consider a causal and stable analog transfer function $H_a(s)$. Its impulse response $h_a(t)$ is given by the inverse Laplace transform of $H_a(s)$, i.e.,

$$h_a(t) = \mathcal{L}^{-1}\{H_a(s)\}. \tag{7.18}$$

In the impulse invariance method, we require that the unit sample response $g[n]$ of the desired causal digital transfer function $G(z)$ be given by the sampled version of $h_a(t)$ sampled at uniform intervals of T seconds, i.e.,

---

[5] See Section 4.3.8 for a definition of a bounded real (BR) function.

$$g[n] = h_a(nT), \qquad n = 0, 1, 2, \ldots, \tag{7.19}$$

where T is the sampling period.[6]

It can be shown that the relation between the digital transfer function $G(z)$ and the parent analog transfer function $H_a(s)$ is given by

$$G(z) = \mathcal{Z}\{g[n]\} = \mathcal{Z}\{h_a(nT)\}$$

$$= \frac{1}{T} \sum_{k=-\infty}^{\infty} H_a\left(s + j\frac{2\pi k}{T}\right)\bigg|_{s=(1/T)\ln z}. \tag{7.20}$$

The relation between the corresponding frequency responses is obtained by letting $z = e^{j\omega}$ and $s = j\Omega$:

$$G(e^{j\omega}) = \frac{1}{T} \sum_{k=-\infty}^{\infty} H_a\left(j\frac{\omega}{T} + j\frac{2\pi k}{T}\right), \tag{7.21}$$

which is identical to the expression given in Eq. (5.14a).

Let us examine the transformation $s = (1/T)\ln z$ of Eq. (7.20), which can be written alternatively as

$$z = e^{sT}, \tag{7.22}$$

For $s = \sigma_0 + j\Omega_0$,

$$z = re^{j\omega} = e^{\sigma_0 T}e^{j\Omega_0 T}. \tag{7.23}$$

This then implies

$$|z| = r = e^{\sigma_0 T}. \tag{7.24}$$

Now, a point on the $j\omega$-axis in the s-plane is characterized by $\sigma_0 = 0$. Hence, from Eq. (7.24) it follows that this point is mapped to a point on the unit circle in the z-plane as $|z| = 1$. Likewise, a point in the left-half s-plane is characterized by $\sigma_0 < 0$. The corresponding point in the z-plane according to Eq. (7.24) has a magnitude $|z| < 1$, indicating that it is mapped inside the unit circle. Finally, a point in right-half s-plane is given by $\sigma_0 > 0$, which is now mapped onto a point outside the unit circle as $|z| > 1$. The transformation of Eq. (7.22) thus has the desired properties outlined earlier in Section 7.1.3. However, the transformation maps all points in the s-plane given by

$$s = \sigma_0 + j\Omega_0 + j\frac{2\pi k}{T}, \qquad k = 0, \pm 1, \pm 2, \ldots \tag{7.25}$$

onto a single point in the z-plane as

$$e^{\sigma_0 T}e^{j\left(\Omega_0 + \frac{2\pi k}{T}\right)T} = e^{\sigma_0 T}e^{j\Omega_0 T}. \tag{7.26}$$

The impulse invariance transformation is illustrated in Figure 7.3. Note that the strip of width $2\pi/T$ in the s-plane for values of s in the range $-\pi/T \leq \Omega \leq \pi/T$ is mapped into the entire z-plane. Likewise, the strip of width $2\pi/T$ in the s-plane for values of s in the range $\pi/T \leq \Omega \leq 3\pi/T$ is also mapped into the entire z-plane, and so is the strip of

---

[6]It has been assumed tacitly here that $h_a(t)$ is continuous at $t = nT$.

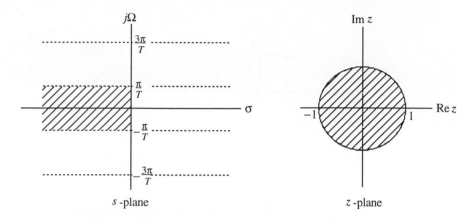

Figure 7.3 Impulse invariance mapping.

width $2\pi/T$ in the range $-3\pi/T \leq \Omega \leq -\pi/T$. In fact, the mapping is a many-to-one mapping with an infinite number of such strips of width $2\pi/T$ being mapped into the entire z-plane. Moreover, the portion of each strip in the left-half s-plane maps into the interior of the unit circle, whereas the right-half portion maps into the exterior of the unit circle in the z-plane.

Now, let us examine the effect of mapping on the frequency response. From Eq. (7.21) we observe that the frequency response of the derived digital transfer function is given by the sum of the frequency responses of the original analog transfer function and its shifted versions, shifted by $\pm 2\pi k/T$, with the overall sum scaled by the factor $1/T$. It thus follows from Eq. (7.21) and from our earlier discussions on sampling theorem (See Section 5.2.1), that if the analog frequency response $H_a(j\Omega)$ is bandlimited with

$$H_a(j\Omega) = 0 \qquad \text{for } |\Omega| \geq \frac{\pi}{T}, \tag{7.27}$$

then

$$G(e^{j\omega}) = \frac{1}{T} H_a\left(j\frac{\omega}{T}\right), \qquad \text{for } |\omega| \leq \pi, \tag{7.28}$$

and there is no aliasing. If the condition of Eq. (7.27) does not hold, then there will be aliasing. Figure 7.4 illustrates these conditions.

For rational analog transfer functions, the infinite series expression for $G(z)$ given by Eq. (7.20) can be replaced with a closed-form solution. We first develop the pertinent result for a (causal) first-order analog transfer function with a simple real pole,

$$H_a(s) = \frac{A}{s + \alpha}. \tag{7.29}$$

The corresponding impulse response $h_a(t)$ is given by

$$h_a(t) = Ae^{-\alpha t}\mu(t), \tag{7.30}$$

where $\mu(t)$ is the unit step function. From Eq. (7.19) we get

$$g[n] = h_a(nT) = Ae^{-\alpha nT}\mu[n]. \tag{7.31}$$

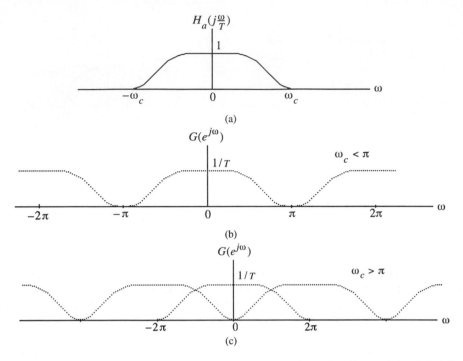

Figure 7.4   The relation between the frequency responses of the analog filter and the digital filter derived via impulse invariance method. (a) Frequency response of the analog filter, (b) and (c) frequency responses of the digital filter derived via the impulse invariance method. Note the absence of aliasing effect in (b) and the presence of aliasing in (c).

Therefore

$$G(z) = A \sum_{n=0}^{\infty} e^{-\alpha nT} z^{-n} = \frac{A}{1 - e^{-\alpha T} z^{-1}}, \tag{7.32}$$

provided $\left| e^{-\alpha T} \right| < 1$, which is always satisfied if $\alpha > 0$, indicating that $H_a(s)$ is a stable transfer function.

In the case of a causal second-order analog transfer function with a complex pole-pair, the corresponding result can be easily derived. For example, if $H_a(s)$ is of the form

$$H_a(s) = \frac{\lambda}{(s + \beta)^2 + \lambda^2}, \tag{7.33}$$

the digital transfer function $G(z)$ obtained by an impulse invariance transformation is given by (Problem 7.4)

$$G(z) = \frac{z e^{-\beta T} \sin \lambda T}{z^2 - 2z e^{-\beta T} \cos \lambda T + e^{-2\beta T}}. \tag{7.34}$$

Another form of the second-order analog transfer function is

$$H_a(s) = \frac{s + \beta}{(s + \beta)^2 + \lambda^2}, \tag{7.35}$$

whose transformation leads to (Problem 7.5)

$$G(z) = \frac{z^2 - ze^{-\beta T} \cos \lambda T}{z^2 - 2ze^{-\beta T} \cos \lambda T + e^{-2\beta T}}. \tag{7.36}$$

For a higher order analog transfer function $H_a(s)$ with simple poles, we can make a partial-fraction expansion and transform each term in the expansion according to the formulas derived above. For a first-order real pole, the term in the partial-fraction expansion is of the form of Eq. (7.29) and its z-domain equivalent obtained via the impulse-invariance transformation is given by Eq. (7.32). For a complex conjugate pole-pair, the term in the partial-fraction expansion is of the form

$$H_2(s) = \frac{Cs + D}{(s + \beta)^2 + \lambda^2}, \tag{7.37}$$

which can be rewritten as

$$H_2(s) = \frac{C\left(s + \frac{D}{C}\right)}{(s + \beta)^2 + \lambda^2} = \frac{C(s + \beta) + A\lambda}{(s + \beta)^2 + \lambda^2}, \tag{7.38}$$

where

$$A = \frac{(D - \beta C)}{\lambda}. \tag{7.39}$$

The corresponding z-domain equivalent obtained via the impulse-invariance transformation is then arrived at by applying Eqs. (7.34) and (7.36), and is given by

$$G_2(z) = \frac{Cz^2 + ze^{-\beta T}(A \sin \lambda T - C \cos \lambda T)}{z^2 - 2ze^{-\beta T} \cos \lambda T + e^{-2\beta T}}. \tag{7.40}$$

Example 7.5 Consider a causal analog transfer function given by

$$H_a(s) = \frac{16(s + 2)}{(s + 3)(s^2 + 2s + 5)}. \tag{7.41}$$

A partial-fraction expansion of the above yields

$$H_a(s) = \frac{-2}{s + 3} + \frac{2s + 14}{s^2 + 2s + 5}. \tag{7.42}$$

In order to make use of the formulas given earlier, we rewrite the second term in the right-hand side of Eq. (7.42) in the form of Eq. (7.37), resulting in

$$H_a(s) = \frac{-2}{s + 3} + \frac{2s + 14}{(s + 1)^2 + (2)^2}.$$

Making use of Eqs. (7.32) and (7.40), we obtain from the above

$$G(z) = \frac{-2z}{z - e^{-3T}} + \frac{2z^2 + ze^{-T}(6 \sin 2T - 2 \cos 2T)}{z^2 - 2ze^{-T} \cos 2T + e^{-2T}}. \tag{7.43}$$

Note that the impulse invariance ensures only that $g[n] = h_a[nT]$ and nothing more. If $|H_a(j\Omega)|$ is bandlimited according to Eq. (7.27), then the frequency response will also be preserved. Otherwise, there will be aliasing, and the frequency response $G(e^{j\omega})$ of the digital filter derived will no longer resemble that of the analog filter in the range

Table 7.1   Parameters of a first-order digital lowpass filter as a function of the 3-dB cutoff frequency α.

3 dB cutoff freq., Hz	α	d_1	$\left| H_a\left(j\frac{\pi}{T}\right) \right|$
1	6.2831853	-0.996863	0.001
5	31.415926	-0.984415	0.005
50	314.159265	-0.854636	0.05

$|\omega| \leq \pi$. In practice, no analog transfer function is strictly bandlimited. However, if $|H_a(j\Omega)| \leq 0.01 H_{\max}$ for $\Omega \geq \pi/T$, where $H_{\max}$ is the maximum value of $|H_a(j\Omega)|$ in the range $0 \leq \Omega \leq \pi/T$, then in most practical applications, $|H_a(j\Omega)|$ can be assumed to be sufficiently bandlimited and the effect of the aliasing can be considered to be minimal [Gol73].

Example 7.6   We consider here the design of a simple first-order digital transfer function using the impulse invariance method [Gol73]. Let $H_a(s)$ be a first-order analog lowpass transfer function,

$$H_a(s) = \frac{\alpha}{s + \alpha}, \tag{7.44}$$

with a 0-dB dc gain and a 3-dB cutoff frequency at $\Omega = \alpha$ rad/sec.

From Eq. (7.32), the digital transfer function derived via the impulse invariance method is

$$G(z) = \frac{\alpha}{1 - e^{-\alpha T} z^{-1}} = \frac{\alpha}{1 + d_1 z^{-1}}, \tag{7.45}$$

where $d_1 = -e^{-\alpha T}$. To make $|H_a(j\Omega)|$ sufficiently bandlimited, we must ensure that the 3-dB cutoff frequency α be considerably smaller than π/T, i.e., $\alpha \ll \pi/T$. In order to compare the magnitude responses of the digital filter of Eq. (7.45) with its parent analog filter of Eq. (7.44), we have computed the filter coefficient d_1 for several values of the cutoff frequency α for a sampling frequency of 2 kHz, i.e., for $T = 0.5$ msec, as shown in Table 7.1.

A plot of the digital filter gain response is shown in Figure 7.5 for each of the above three cases. In the plots shown, the dc value of the magnitude, $|G(e^{j0}|$, has been normalized to unity. Because of the 2-kHz sampling frequency, the usable frequency range of the digital filter is from 0 to 1 kHz. Note that only the digital filter derived from the analog filter of 3-dB cutoff frequency of 1 Hz has a gain response reasonably close to that of the parent analog filter. In the second case (cutoff at 5 Hz), the effect of aliasing on the digital filter response is visible. However, the digital filter still has a lowpass response with a reasonable stopband. In the last case (cutoff at 50 Hz), the frequency response has a very poor stopband response and the aliasing effect is quite severe.

Now, the frequency response of any analog filter is defined for the frequency range 0 to ∞, and as a result, there will be some aliasing in the frequency response of the digital filter designed by the impulse invariance approach. However, if the parent analog filter has a reasonably bandlimited frequency response, then the magnitude and the phase responses of the digital filter can be made to match those of the analog filter by making the sampling interval sufficiently small. Hence, the impulse invariance method is preferable when both magnitude and phase responses are to be preserved.

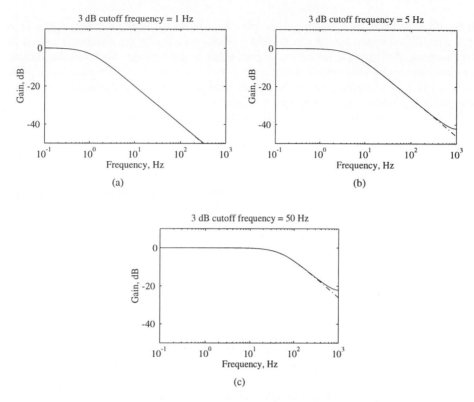

Figure 7.5 Gain responses of the first-order digital filter (solid line) and the parent analog filter (dashed line) for different values of the cutoff frequency.

The impulse invariance method is not applicable to the design of a frequency selective filter if the parent analog filter does not have a bandlimited frequency response, such as in the case of a highpass analog filter. However, if we wish to design a digital filter whose frequency response matches that of the analog filter up to $\Omega = \pi/T$, then we can use a suitably chosen *guard filter* $H_g(s)$ so that the product $H_a(s)H_g(s)$ is bandlimited with $|H_a(j\Omega)H_g(j\Omega)| \approx |H_a(j\Omega)|$ for $\Omega < \pi/T$ and $|H_a(j\Omega)H_g(j\Omega)| \approx 0$ for $\Omega \geq \pi/T$. Note that the guard filter thus should be a sharp cutoff lowpass filter with $|H_g(j\Omega)| \approx 1$ for $\Omega < \pi/T$ and $|H_g(j\Omega)| \approx 0$ for $\Omega \geq \pi/T$.

Finally, it should be noted that if a digital filter needs to be designed by the impulse invariance method meeting the frequency response specifications given in the digital domain, then the specifications of the analog prototype is obtained by applying the frequency transformation $\Omega = \omega/T$. After the analog filter has been designed it is transformed back into the desired digital filter using the transformation $z = e^{sT}$. Since the value of the sampling interval T is canceled because of the forward and reverse frequency transformations, it can be chosen arbitrarily and it is common practice to set $T = 1$ for convenience.

7.3   Bilinear Transform Method of IIR Filter Design

In the impulse invariance mapping, a strip in the s-plane between $\Omega = k\pi/T$ and $\Omega = (k+2)\pi/T$ is mapped completely into the z-plane. As a result, the aliasing is caused by the mapping of all such strips on top of each other in the z-plane. To avoid the aliasing problem, we need a transformation that is a one-to-one mapping, i.e., it maps a single point in the s-plane to a unique point in the z-plane, and vice-versa. The mapping with such a property is the bilinear transformation discussed next.

The bilinear transformation from the s-plane to the z-plane is given by [Kai66]

$$s = \frac{2}{T}\left(\frac{1-z^{-1}}{1+z^{-1}}\right). \tag{7.46}$$

The relation between the digital transfer function $G(z)$ and the parent analog transfer function $H_a(s)$ is then given by

$$G(z) = H_a(s)\big|_{s=\frac{2}{T}\left(\frac{1-z^{-1}}{1+z^{-1}}\right)}. \tag{7.47}$$

The bilinear transformation is derived by applying the trapezoidal numerical integration approach to the differential equation corresponding to $H_a(s)$ that leads to the difference equation corresponding to $G(z)$.[7] The parameter T represents the step size in the numerical integration. As we shall see later in this section, the digital filter design procedure consists of two steps: first, the inverse bilinear transformation is applied to the digital filter specifications to arrive at the specifications of the analog filter prototype; then the bilinear transformation of Eq. (7.46) is employed to obtain the desired digital transfer function $G(z)$ from the analog transfer function $H_a(s)$ designed to meet the analog filter specifications. As a result, the parameter T has no effect on the expression for $G(z)$, and we shall, for convenience, choose $T = 2$ to simplify the design procedure.

The corresponding inverse transformation for $T = 2$ is given by

$$z = \frac{1+s}{1-s}. \tag{7.48}$$

Let us now examine the above transformation. Note that for $s = j\Omega_0$,

$$z = \frac{1+j\Omega_0}{1-j\Omega_0}, \tag{7.49}$$

which has a unity magnitude. This implies that a point on the imaginary axis in the s-plane is mapped onto a point on the unit circle in the z-plane. In the general case, for $s = \sigma_0 + j\Omega_0$,

$$z = \frac{1+(\sigma_0+j\Omega_0)}{1-(\sigma_0+j\Omega_0)} = \frac{(1+\sigma_0)+j\Omega_0}{(1-\sigma_0)-j\Omega_0}. \tag{7.50}$$

Therefore,

$$|z|^2 = \frac{(1+\sigma_0)^2+(\Omega_0)^2}{(1-\sigma_0)^2+(\Omega_0)^2}. \tag{7.51}$$

---

[7]See Example 2.31.

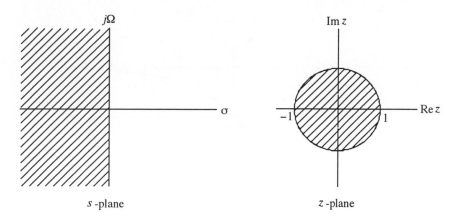

Figure 7.6   The bilinear transformation mapping.

Thus, a point in the left-half s-plane with $\sigma_0 < 0$ is mapped to a point inside the unit circle in the z-plane as $|z| < 1$. Likewise, a point in the right-half s-plane with $\sigma_0 > 0$ is mapped to a point outside the unit circle in the z-plane, as $|z| > 1$. Any point in the s-plane is mapped onto a unique point in the z-plane and vice-versa. The mapping of the s-plane into the z-plane via the bilinear transformation is illustrated in Figure 7.6 and is seen to have all the desired properties. Also, there is no aliasing due to the one-to-one mapping.

The exact relation between the imaginary axis in the s-plane and the unit circle in the z-plane is of interest. From Eq. (7.46) it follows that

$$j\Omega = \frac{1 - e^{-j\omega}}{1 + e^{-j\omega}} = j \tan\left(\frac{\omega}{2}\right),$$

or

$$\Omega = \tan\left(\frac{\omega}{2}\right), \tag{7.52}$$

which has been plotted in Figure 7.7. Note from this plot that the positive (negative) imaginary axis in the s-plane is mapped into the upper (lower) half of the unit circle in the z-plane. However, it is clear that the mapping is highly nonlinear since the complete negative imaginary axis in the s-plane from $\Omega = -\infty$ to $\Omega = 0$ is mapped to the lower half of the unit circle from $\omega = -\pi (z = -1)$ to $\omega = 0$ $(z = +1)$, and the complete positive imaginary axis in the s-plane from $\Omega = 0$ to $\Omega = +\infty$ is mapped to the upper half of the unit circle from $\omega = 0$ $(z = +1)$ to $\omega = +\pi (z = -1)$. This introduces a distortion in the frequency axis called *frequency warping*. The effect of warping is more evident in Figure 7.8, which shows the transformation of a typical analog filter magnitude response to a digital filter magnitude response derived via the bilinear transformation. Thus, to develop a digital filter meeting a specified magnitude response, we must first prewarp the critical bandedge frequencies (ω_p and ω_s) to find their analog equivalents (Ω_p and Ω_s) using the relation of Eq. (7.52), design the analog prototype $H_a(s)$ using the prewarped critical frequencies, and then transform $H_a(s)$ using the bilinear transformation to obtain the desired digital filter transfer function $G(z)$.

It should be noted that the bilinear transformation preserves the magnitude response of an analog filter only if the specification requires piecewise constant magnitude. However,

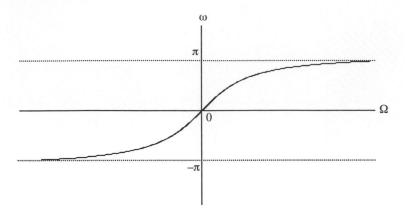

Figure 7.7   Mapping of the angular analog frequencies Ω to the angular digital frequencies ω via the bilinear transformation.

Figure 7.8   Illustration of frequency warping effect.

the phase response of the analog filter is not preserved after transformation. Hence, the transformation can be used only to design digital filters with prescribed magnitude response with piecewise constant values.

Example 7.7 It follows from Eqs. (5.32) and (5.33) that a first-order Butterworth lowpass transfer function with a 3-dB cutoff frequency at Ω_c is given by

$$H_a(s) = \frac{\Omega_c}{s + \Omega_c}. \tag{7.53}$$

Applying a bilinear transformation to Eq. (7.53), we arrive at the expression for a first-order digital lowpass transfer function $G(z)$ with a maximally flat magnitude response given by

$$G(z) = H_a(s)|_{s=(1-z^{-1})/(1+z^{-1})} = \frac{\Omega_c(1+z^{-1})}{(1-z^{-1}) + \Omega_c(1+z^{-1})}. \tag{7.54}$$

Rearranging terms, Eq. (7.54) can be written as

$$G(z) = \frac{1-\alpha}{2} \cdot \frac{1+z^{-1}}{1-\alpha z^{-1}}, \tag{7.55}$$

where

$$\alpha = \frac{1-\Omega_c}{1+\Omega_c}. \tag{7.56}$$

The 3-dB cutoff frequency ω_c of the digital transfer function of Eq. (7.55) is related to the 3-dB cutoff frequency Ω_c of the analog transfer function through Eq. (7.52). Using Eq. (7.52) in Eq. (7.56), we can express α as a function of ω_c, which is thus given by

$$\alpha = \frac{1-\tan(\omega_c/2)}{1+\tan(\omega_c/2)}. \tag{7.57}$$

It should be noted that the first-order lowpass transfer function of Eq. (7.55) is identical to that of Eq. (4.70), where it was introduced without any derivation. It can be easily shown that the relation between the 3-dB bandwidth and the transfer function parameter as given by Eq. (7.57) is the same as that given in Eq. (4.72b).

7.4 Design of Digital IIR Notch Filters

We consider next the design of a second-order IIR notch filter as an example of the application of the bilinear transformation method [Hir74]. Now, a second-order analog notch filter has a transfer function given by

$$H_a(s) = \frac{s^2 + \Omega_0^2}{s^2 + Bs + \Omega_0^2}. \tag{7.58}$$

Its magnitude response is then

$$|H_a(j\Omega)| = \frac{\Omega_0^2 - \Omega^2}{\sqrt{(\Omega_0^2 - \Omega^2)^2 + B^2\Omega^2}}, \tag{7.59}$$

which approaches unity values, i.e., a gain of 0 dB, at $\Omega = 0$ and ∞. The magnitude has a zero value at the notch frequency $\Omega = \Omega_0$. If Ω_1 and Ω_2, $\Omega_2 > \Omega_1$, denote the frequencies at which the gain is down by -3 dB, it can be shown that the 3-dB notch bandwidth defined by $(\Omega_2 - \Omega_1)$ is equal to B.

Applying a bilinear transformation to $H_a(s)$ of Eq. (7.58), we arrive at

$$\begin{aligned}
G(z) &= H_a(s)|_{s=(1-z^{-1})/(1+z^{-1})} \\
&= \frac{(1+\Omega_0^2) - 2(1-\Omega_0^2)z^{-1} + (1+\Omega_0^2)z^{-2}}{(1+\Omega_0^2+B) - 2(1-\Omega_0^2)z^{-1} + (1+\Omega_0^2-B)z^{-2}},
\end{aligned} \tag{7.60}$$

which can be rewritten as

$$G(z) = \frac{1}{2}\frac{(1+\alpha) - 2\beta(1+\alpha)z^{-1} + (1+\alpha)z^{-2}}{1 - \beta(1+\alpha)z^{-1} + \alpha z^{-2}}, \tag{7.61}$$

where

$$\alpha = \frac{1 + \Omega_0^2 - B}{1 + \Omega_0^2 + B}, \tag{7.62a}$$

$$\beta = \frac{1 - \Omega_0^2}{1 + \Omega_0^2}. \tag{7.62b}$$

It is a simple exercise to show that the notch frequency ω_0 and the 3-dB notch bandwidth $\Delta\omega_{3dB}$ of the digital notch filter of Eq. (7.61) are related to the constants α and β through

$$\alpha = \frac{1 - \tan(\Delta\omega_{3dB}/2)}{1 + \tan(\Delta\omega_{3dB}/2)}, \tag{7.63a}$$

$$\beta = \cos\omega_0. \tag{7.63b}$$

The above expressions are the desired design formulas to determine the constants α and β for a given notch frequency ω_0 and a 3-dB notch bandwidth $\Delta\omega_{3dB}$. It should be noted that Eq. (7.61) is precisely the transfer function of the second-order notch filter given in Eq. (4.79) and introduced earlier in Section 4.3.3 without any derivation.

Example 7.8 Design a second-order digital notch filter having a notch frequency at 60 Hz and a 3-dB notch bandwidth of 6 Hz. The sampling frequency employed is 400 Hz.

The normalized angular notch frequency ω_0 and the normalized angular 3-dB bandwidth $\Delta\omega_{3dB}$ are therefore given by

$$\omega_0 = 2\pi \left(\frac{60}{400} \right) = 0.3\pi,$$

$$\Delta\omega_{3dB} = 2\pi \left(\frac{6}{400} \right) = 0.03\pi.$$

Substituting these values in Eqs. (7.63a and 7.63b) we arrive at

$$\alpha = 0.8816186, \qquad \beta = 0.587785.$$

Then, substituting the above values in Eq. (7.61), we arrive at the desired transfer function

$$G(z) = \frac{0.940809 - 1.105987z^{-1} + 0.940809z^{-2}}{1 - 1.105987z^{-1} + 0.881618z^{-2}},$$

whose gain and phase responses are plotted in Figure 7.9. Observe the phase shift of π radians at 60 Hz, the notch frequency.

7.5   Lowpass IIR Digital Filter Design Examples

We illustrate in this section the development of a digital IIR transfer function meeting given specifications using both the impulse invariance and the bilinear transformation methods. To this end, we first obtain the specifications for a prototype analog filter from the specifications of the digital filter. We then determine the analog transfer function $H_a(s)$ meeting the specifications of the prototype analog filter. Finally, the analog transfer

(a)   (b)

Figure 7.9   Gain and phase responses of the notch transfer function of Example 7.8.

function $H_a(s)$ is transformed into a digital transfer function $G(z)$ using both the impulse invariance and bilinear transformation methods.

Specifically, consider the design of a lowpass IIR digital filter $G(z)$ with a maximally flat magnitude characteristic. The passband edge frequency ω_p is 0.25π, with a passband ripple not exceeding 0.5 dB. The minimum stopband attenuation at the stopband edge frequency ω_s of 0.55π is 15 dB. Thus, if $|G(e^{j0})| = 1$, then we require that

$$20 \log_{10} \left| G(e^{j0.25\pi}) \right| \geq -0.5 \text{ dB}, \qquad (7.64a)$$

$$20 \log_{10} \left| G(e^{j0.55\pi}) \right| \leq -15 \text{ dB}. \qquad (7.64b)$$

7.5.1   Design Based on Impulse Invariance Method

We assume first that no aliasing occurs. Then, the prototype analog transfer function $H_a(s)$ must satisfy the same specifications as the desired digital transfer function, i.e., the passband edge frequency is $\Omega_p = 0.25\pi$, the stopband edge frequency is $\Omega_s = 0.55\pi$, the passband ripple is 0.5 dB, and the minimum stopband attenuation is 15 dB. We choose the analog lowpass transfer function to have a Butterworth characteristic. Moreover, without any loss of generality, we assume $T = 1$. The requirements to be satisfied by the analog filter magnitude response $|H_a(j\Omega)|$ are then

$$20 \log_{10} |H_a(j0.25\pi)| \geq -0.5 \text{ dB}, \qquad (7.65a)$$
$$20 \log_{10} |H_a(j0.55\pi)| \leq -15 \text{ dB}. \qquad (7.65b)$$

From the specified passband ripple of 0.5 dB, we obtain $\varepsilon^2 = 0.1220185$, and from the minimum stopband attenuation of 15 dB, we obtain $A^2 = 31.622777$. Therefore, the inverse discrimination ratio is given by

$$\frac{1}{k_1} = \frac{\sqrt{A^2 - 1}}{\varepsilon} = 15.841979,$$

and the inverse transition ratio is given by

$$\frac{1}{k} = \frac{\Omega_s}{\Omega_p} = 2.2.$$

Substituting the above parameters in Eq. (5.31), we arrive at the value of the filter order as

$$N = \frac{\log_{10}(1/k_1)}{\log_{10}(1/k)} = \frac{\log_{10}(15.84194)}{\log_{10}(2.2)} = 3.503885.$$

By rounding the above value up to the nearest integer, we choose the desired filter order as $N = 4$.

From Eq. (5.30a), we have

$$\left(\frac{\Omega_p}{\Omega_c}\right)^{2N} = \varepsilon^2. \tag{7.66}$$

Substituting the values of ϵ^2, Ω_p and N in the above, we get

$$\Omega_c = 1.3007568(\Omega_p) = 0.3251892\pi = 1.021612.$$

Substituting this value of the 3-dB frequency in Eq. (5.30b), we next arrive at

$$10\log_{10}\left[1 + \left(\frac{0.55\pi}{0.3251892\pi}\right)^8\right] = 18.322506 \text{ dB}$$

which is greater than the specified minimum attenuation of 15 dB. However, the excess attenuation in the stopband also provides an additional margin of safety with regard to minimizing the effect of aliasing in the passband of the digital filter caused by the impulse invariance method and is therefore beneficial.

Now, using the function buttap in MATLAB,[8] we arrive at the the normalized Butterworth transfer function of order 4 given by

$$H_{an}(s) = \frac{1}{(s^2 + 0.7654s + 1)(s^2 + 1.8478s + 1)}.$$

For deriving the equivalent digital transfer function via the impulse invariance method, it is more convenient to express the above transfer function in a partial-fraction expansion form. To this end, we employ the function residue of MATLAB[9] and obtain

$$H_{an}(s) = \frac{-0.9238729s - 0.7071323}{s^2 + 0.7654s + 1} + \frac{0.9238729s + 1.7071323}{s^2 + 1.8478s + 1}.$$

Note that the normalized transfer function has a 3-dB frequency at $\Omega_c = 1$. We thus need to denormalize the transfer function $H_{an}(s)$ to move the 3-dB frequency to $\Omega_c = 1.021612$, leading to

$$H_a(s) = H_{an}\left(\frac{s}{1.021612}\right)$$
$$= \frac{-0.9438396s - 0.7380277}{s^2 + 0.7819418s + 1.0436911} + \frac{0.9438396s + 1.78171875}{s^2 + 1.88773465s + 1.0436911}.$$

[8]See Section 5.3.6.

[9]See Section 3.8.2.

(a)   (b)

Figure 7.10   Frequency response of the lowpass filter design based on the impulse invariance method. (a) Magnitude response and (b) gain response.

To convert the above into its digital equivalent, we rewrite it as

$$H_a(s) = \frac{-0.9438396s - 0.7380277}{(s + 0.3909709)^2 + (0.9438394)^2} + \frac{0.9438396s + 1.78171875}{(s + 0.9438673)^2 + (0.3909035)^2}.$$

Comparing each term in the above expression with Eq. (7.37) and applying the transformation of Eq. (7.40), we arrive at the desired expression for the digital transfer function $G(z)$ as

$$G(z) = \frac{-0.9438396 + 0.160387896z^{-1}}{1 - 0.7936648z^{-1} + 0.4575167z^{-2}}$$
$$+ \frac{0.9438396 - 0.0016672z^{-1}}{1 - 0.71953375z^{-1} + 0.1514144z^{-2}}.$$

The corresponding frequency response is shown in Figure 7.10. Note that since $H_a(s)$ was sufficiently bandlimited, $G(z)$ does meet the given specifications. If it did not, we should increase N and try again.

7.5.2   Design Based on Bilinear Transformation Method

In this approach it is necessary first to prewarp the digital frequency specifications to obtain the corresponding analog filter specifications. From Eq. (7.52) the pertinent analog bandedge frequencies Ω_p and Ω_s corresponding to the two digital frequencies ω_p and ω_s are given by

$$\Omega_p = \tan\left(\frac{\omega_p}{2}\right) = \tan\left(\frac{0.25\pi}{2}\right) = 0.4142135,$$

$$\Omega_s = \tan\left(\frac{\omega_s}{2}\right) = \tan\left(\frac{0.55\pi}{2}\right) = 1.1708496.$$

The inverse transition ratio is now

$$\frac{1}{k} = \frac{\Omega_s}{\Omega_p} = \frac{1.1708496}{0.4142135} = 2.8266814.$$

The inverse discrimination parameter $1/k_1$ remains the same as in the impulse invariance method and is given by

$$\frac{1}{k_1} = 15.841979.$$

Substituting these values in Eq. (5.31) we obtain the filter order N as

$$N = \frac{\log_{10}(1/k_1)}{\log_{10}(1/k)} = \frac{\log_{10}(15.84194)}{\log_{10}(2.8266814)} = 2.658697.$$

The nearest higher integer 3 is thus taken as the filter order, which is seen to be less than that of the analog filter designed using the impulse invariance method. In general, the filter order required for the bilinear transformation based design will be lower than that required for the impulse invariance method. This is due to the frequency compression as indicated by Eq. (7.52) and shown in Figure 7.7.

   Since there is no aliasing in the bilinear transformation method, the filter order can be used to determine the 3-dB cuttoff frequency Ω_c using either the Eq. (5.30a) or Eq. (5.30b). However, it is preferable to use the latter since this ensures the smallest ripple in the passband, or in other words, the smallest amplitude distortion to the signal being filtered in its band of interest. Substituting the values of ε^2, Ω_p, and N in Eq. (7.66), we arrive at

$$\Omega_c = 1.419915(\Omega_p) = 1.419915 \times 0.4142135 = 0.588148.$$

Using the function `buttap` of MATLAB,[10] we obtain the third-order normalized lowpass Butterworth transfer function as

$$H_{an}(s) = \frac{1}{(s+1)(s^2+s+1)},$$

which has a 3-dB frequency at $\Omega = 1$ and therefore has to be denormalized to bring the 3-dB frequency to $\Omega_c = 0.588148$. The denormalized transfer function is given by

$$H_a(s) = H_{an}\left(\frac{s}{0.588148}\right) = \frac{0.203451}{(s+0.588148)(s^2+0.588148s+0.345918)}.$$

Applying the bilinear transformation to the above, we finally arrive at the desired expression for the digital lowpass transfer function:

$$G(z) = H_a(s)|_{s=(1-z^{-1})/(1+z^{-1})}$$
$$= \frac{0.0662272(1+z^{-1})^3}{(1-0.2593284z^{-1})(1-0.6762858z^{-1}+0.3917468z^{-2})}. \qquad (7.67)$$

The corresponding magnitude and gain responses are plotted in Figure 7.11. It should be noted that, as expected, the magnitude and gain responses here fall much more rapidly (sharper transition band) than those of the filter designed via the impulse invariance method as indicated in Figure 7.10.

---

[10]See Section 5.3.6.

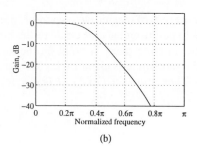

(a)   (b)

Figure 7.11   Magnitude and gain response of the lowpass filter design based on the bilinear transformation method.

7.6   Spectral Transformations of IIR Filters

All the analog filter approximations studied so far are for the design of lowpass filters. Hence, the digital filter obtained by the transformations of such filters are also of the lowpass type. Often, in practice, it may be necessary to modify the characteristics of a filter to meet the new specifications without repeating the filter design procedure. For example, after a lowpass filter with a passband edge at 2 kHz has been designed, it may be required to move the passband edge to 2.1 kHz. If a digital filter with a highpass or bandpass or bandstop characteristics is to be designed, it is usual practice to convert the specifications to that of a lowpass prototype, design the lowpass prototype, and then transform it back to the desired type by a spectral transformation. We describe here the spectral transformations that can be used to transform a given lowpass digital IIR transfer function $G_L(z)$ to another digital transfer function $G_D(z)$ that could be a lowpass, highpass, bandpass, or bandstop filter [Con70]. Figure 4.8 shows the magnitude responses of these four types of ideal filters.

To eliminate the confusion between the complex variable z of the lowpass transfer function $G_L(z)$ and that of the desired transfer function $G_D(z)$, we shall use different symbols. Thus we shall use z^{-1} to denote the unit delay in the prototype lowpass digital filter $G_L(z)$ and $\hat{z}^{-1}$ to denote the unit delay in the transformed filter $G_D(\hat{z})$. The unit circles in the z- and $\hat{z}$-planes are defined by

$$z = e^{j\omega}, \qquad \hat{z} = e^{j\hat{\omega}}.$$

We denote the transformation from the z-domain to the $\hat{z}$-domain as

$$z = F(\hat{z}). \tag{7.68}$$

Then, $G_L(z)$ is transformed to $G_D(\hat{z})$ through

$$G_D(\hat{z}) = G_L\{F(\hat{z})\}. \tag{7.69}$$

To transform a rational $G_L(z)$ into a rational $G_D(\hat{z})$, $F(\hat{z})$ must be a rational function of $\hat{z}$. In addition, to guarantee the stability of $G_D(\hat{z})$, the transformation should be such that the inside of the unit circle of the z-plane is mapped into the inside of the unit circle of the $\hat{z}$-plane. Finally, to ensure that a lowpass magnitude response is mapped into one of the four basic types of magnitude responses, points on the unit circle of the z-plane should be mapped to points on the unit circle of the $\hat{z}$-plane.

Now, in the z-plane, a point on the unit circle is characterized by $|z| = 1$, a point inside the unit circle is given by $|z| < 1$, and a point outside the unit circle is defined by $|z| > 1$. Thus, from Eq. (7.68), $|F(\hat{z})| = |z|$, and therefore,

$$|F(\hat{z})| = \begin{cases} > 1, & \text{if } |z| > 1, \\ 1, & \text{if } |z| = 1, \\ < 1, & \text{if } |z| < 1. \end{cases} \qquad (7.70)$$

Thus, from the above and Eq. (4.129), it follows that $F^{-1}(\hat{z})$ be a stable allpass function. From our earlier discussions on allpass function in Section 4.4, we note that the most general form of $F^{-1}(\hat{z})$ with real coefficients is given by

$$F^{-1}(\hat{z}) = \pm \prod_{\ell=1}^{L} \left(\frac{1 - \alpha_\ell^* \hat{z}}{\hat{z} - \alpha_\ell} \right), \qquad (7.71)$$

where $|\alpha_\ell|$ is either real or occurs in complex conjugate pairs with $|\alpha_\ell| < 1$ for stability.

7.6.1   Lowpass-to-Lowpass Transformation

To transform a prototype lowpass filter $G_L(z)$ with a cutoff frequency ω_c to another lowpass filter $G_D(\hat{z})$ with a cutoff frequency $\hat{\omega}_c$, we use the transformation

$$z^{-1} = F^{-1}(\hat{z}) = \frac{1 - \alpha \hat{z}}{\hat{z} - \alpha}, \qquad (7.72)$$

with α real. On the unit circle, the above transformation reduces to

$$e^{-j\omega} = \frac{e^{-j\hat{\omega}} - \alpha}{1 - \alpha e^{-j\hat{\omega}}},$$

from which we arrive at

$$\tan\left(\frac{\omega}{2}\right) = \left(\frac{1 + \alpha}{1 - \alpha}\right) \tan\left(\frac{\hat{\omega}}{2}\right). \qquad (7.73)$$

A plot of the relation between ω and $\hat{\omega}$ is given in Figure 7.12 for three values of α. Note that the mapping is nonlinear except for $\alpha = 0$, resulting in a warping of the frequency scale for nonzero values of α. However, if $G_L(z)$ is a piecewise constant lowpass magnitude response, then the transformed filter $G_D(\hat{z})$ will likewise have a similar piecewise constant lowpass magnitude response due to the monotonicity of the transformation of Eq. (7.73). The relation between the cutoff frequency ω_c of $G_L(z)$ with the cutoff frequency $\hat{\omega}_c$ of $G_D(\hat{z})$ follows from Eq. (7.73):

$$\tan\left(\frac{\omega_c}{2}\right) = \left(\frac{1 + \alpha}{1 - \alpha}\right) \tan\left(\frac{\hat{\omega}_c}{2}\right),$$

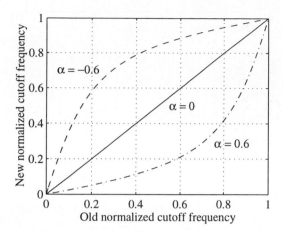

Figure 7.12 Mapping of the angular frequencies in the lowpass-to-lowpass transformation for three different values of the parameter α.

which can be solved for α yielding:

$$\alpha = \frac{\tan(\omega_c/2) - \tan(\hat{\omega}_c/2)}{\tan(\omega_c/2) + \tan(\hat{\omega}_c/2)} = \frac{\sin\left(\frac{\omega_c - \hat{\omega}_c}{2}\right)}{\sin\left(\frac{\omega_c + \hat{\omega}_c}{2}\right)}. \tag{7.74}$$

Example 7.9 Consider the third-order lowpass digital transfer function $G(z)$ of Eq. (7.67), which has a passband from dc to 0.25π with a 0.5 dB ripple, and a stopband from 0.55π to π with an attenuation greater than 15 dB.

  We wish to redesign this lowpass filter by applying the lowpass-to-lowpass transformation of Eq. (7.72) so that its passband edge moves from 0.25π to 0.35π. Here we have $\omega_c = 0.25\pi$ and $\hat{\omega}_c = 0.35\pi$. Substituting these values in Eq. (7.74), we get

$$\alpha = \frac{\sin\left(\frac{0.25\pi - 0.35\pi}{2}\right)}{\sin\left(\frac{0.25\pi + 0.35\pi}{2}\right)} = -\frac{\sin(0.05\pi)}{\sin(0.3\pi)} = -0.1933636.$$

Thus the desired lowpass transfer function $G_D(\hat{z})$ is given by

$$G_D(\hat{z}) = G(z)\Big|_{z^{-1} = \frac{\hat{z}^{-1} + 0.1933636}{1 + 0.1933636\hat{z}^{-1}}}$$

$$= \frac{0.2172235(1 + \hat{z}^{-1})^3}{(1 - 0.0694472\hat{z}^{-1})(1 - 0.1848053\hat{z}^{-1} + 0.337566\hat{z}^{-2})}. \tag{7.75}$$

  A plot of the gain responses of the original lowpass filter $G(z)$ of Eq. (7.66) and the new lowpass filter $G_D(\hat{z})$ of Eq. (7.75) is shown in Figure 7.13. Note that the transformed transfer function $G_D(\hat{z})$ of Eq. (7.75) has been scaled in the plot to have a 0-dB dc gain.

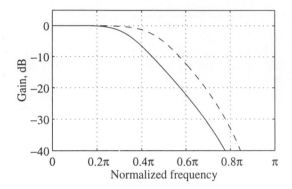

Figure 7.13   Gain responses of the prototype lowpass filter (solid line) and the transformed lowpass filter (dashed line).

7.6.2   Other Transformations

Table 7.2 lists other useful transformations such as the lowpass-to-highpass, lowpass-to-bandpass, and lowpass-to-bandstop transformations, in addition to the lowpass-to-lowpass transformation discussed above. It should be noted that these spectral transformations can be used only to map one frequency point ω_c in the magnitude response of the lowpass prototype filter into a new position $\hat{\omega}_c$, with the same magnitude response value for the transformed lowpass and highpass filters, or into two new positions, $\hat{\omega}_{c1}$ and $\hat{\omega}_{c2}$, with the same magnitude response values for the transformed bandpass and bandstop filters. Hence, it is possible only to map either the passband edge or the stopband edge of the lowpass prototype filter to the desired position(s), but not both.

Example 7.10   We now consider the design of a highpass filter by applying a spectral transformation to the third-order lowpass digital transfer function $G(z)$ of Eq. (7.67). The desired passband edge $\hat{\omega}_c$ of the highpass filter is 0.55π, whereas the passband edge of the prototype lowpass filter is at $\omega_c = 0.25\pi$. Substituting these values in the lowpass-to-highpass transformation given in Table 7.2, we arrive at

$$\alpha = -\frac{\cos\left(\frac{0.55\pi + 0.25\pi}{2}\right)}{\cos\left(\frac{0.55\pi - 0.25\pi}{2}\right)} = -\frac{\cos(0.4\pi)}{\cos(0.15\pi)} = -0.3468179.$$

Thus, from Table 7.2, the desired lowpass-to-highpass transformation is given by

$$z^{-1} = -\frac{\hat{z}^{-1} - 0.3468179}{1 - 0.3468179\hat{z}^{-1}}.$$

Using the above transformation, we obtain from Eq. (7.67) and (7.69) the desired highpass transfer function

$$G_D(\hat{z}) = G(z)\big|_{z^{-1} = -\frac{\hat{z}^{-1} - 0.3468179}{1 - 0.3468179\hat{z}^{-1}}}$$

$$= \frac{0.2187917(1 - \hat{z}^{-1})^3}{(1 - 0.091006602\hat{z}^{-1})(1 + 0.7460066\hat{z}^{-1} + 0.3414851\hat{z}^{-2})}.$$

The gain response of the above transfer function has been plotted in Figure 7.14.

Table 7.2 Spectral transformations of a lowpass filter with a cutoff frequency ω_c.

Filter type	Spectral transformation	Design parameters
Lowpass	$z^{-1} = \frac{\hat{z}^{-1} - \alpha}{1 - \alpha \hat{z}^{-1}}$	$\alpha = \dfrac{\sin\left(\frac{\omega_c - \hat{\omega}_c}{2}\right)}{\sin\left(\frac{\omega_c + \hat{\omega}_c}{2}\right)}$ $\hat{\omega}_c = $ desired cutoff frequency
Highpass	$z^{-1} = -\frac{\hat{z}^{-1} + \alpha}{1 + \alpha \hat{z}^{-1}}$	$\alpha = -\dfrac{\cos\left(\frac{\omega_c + \hat{\omega}_c}{2}\right)}{\cos\left(\frac{\omega_c - \hat{\omega}_c}{2}\right)}$ $\hat{\omega}_c = $ desired cutoff frequency
Bandpass	$z^{-1} = -\dfrac{\hat{z}^{-2} - \frac{2\alpha\beta}{\beta+1}\hat{z}^{-1} + \frac{\beta-1}{\beta+1}}{\frac{\beta-1}{\beta+1}\hat{z}^{-2} - \frac{2\alpha\beta}{\beta+1}\hat{z}^{-1} + 1}$	$\alpha = \dfrac{\cos\left(\frac{\hat{\omega}_{c2} + \hat{\omega}_{c1}}{2}\right)}{\cos\left(\frac{\hat{\omega}_{c2} - \hat{\omega}_{c1}}{2}\right)}$ $\beta = \cot\left(\dfrac{\hat{\omega}_{c2} - \hat{\omega}_{c1}}{2}\right)\tan\left(\dfrac{\omega_c}{2}\right)$ $\hat{\omega}_{c2}, \hat{\omega}_{c1} = $ desired upper and lower cutoff frequencies
Bandstop	$z^{-1} = \dfrac{\hat{z}^{-2} - \frac{2\alpha}{1+\beta}\hat{z}^{-1} + \frac{1-\beta}{1+\beta}}{\frac{1-\beta}{1+\beta}\hat{z}^{-2} - \frac{2\alpha}{1+\beta}\hat{z}^{-1} + 1}$	$\alpha = \dfrac{\cos\left(\frac{\hat{\omega}_{c2} + \hat{\omega}_{c1}}{2}\right)}{\cos\left(\frac{\hat{\omega}_{c2} - \hat{\omega}_{c1}}{2}\right)}$ $\beta = \tan\left(\dfrac{\hat{\omega}_{c2} - \hat{\omega}_{c1}}{2}\right)\tan\left(\dfrac{\omega_c}{2}\right)$ $\hat{\omega}_{c2}, \hat{\omega}_{c1} = $ desired upper and lower cutoff frequencies

The lowpass-to-bandpass transformation given in Table 7.2 can be simplified if we consider the case when the bandwidth of the passband for the prototype lowpass filter is same as that of the transformed bandpass filter, i.e., $\omega_c = \hat{\omega}_{c2} - \hat{\omega}_{c1}$. Applying this constraint to the respective spectral transformation in Table 7.2, we arrive at the modified spectral transformation given by

$$z^{-1} = -\hat{z}^{-1}\frac{\hat{z}^{-1} - \alpha}{1 - \alpha\hat{z}^{-1}}. \tag{7.76}$$

The parameter α is determined from the desired location of the center frequency $\hat{\omega}_0$ of the bandpass filter derived via the transformation of Eq. (7.76), which maps the zero frequency

Figure 7.14   Gain response of the highpass filter.

of the lowpass filter, i.e., $\omega = 0$, to $\hat{\omega}_0$. From Eq. (7.76) we get

$$e^{-j\omega} = -e^{-j\hat{\omega}} \frac{e^{-j\hat{\omega}} - \alpha}{1 - \alpha e^{-j\hat{\omega}}}. \qquad (7.77)$$

Substituting $\omega = 0$ and $\hat{\omega} = \hat{\omega}_0$ in above, we arrive at

$$\alpha = \cos \hat{\omega}_0. \qquad (7.78)$$

Example 7.11   We develop the transfer function of a second-order bandpass filter by applying the transformation of Eq. (7.76) to the first-order lowpass transfer function of Eq. (7.55). The desired transfer function is thus given by

$$
\begin{aligned}
G_D(\hat{z}) &= \left. G(z) \right|_{z^{-1} = -\hat{z}^{-1}(\hat{z}^{-1} - \beta)/(1 - \beta \hat{z}^{-1})} \\
&= \frac{1-\alpha}{2} \left[\frac{1 - \hat{z}^{-2}}{1 - \beta(1-\alpha)\hat{z}^{-1} + \alpha \hat{z}^{-2}} \right],
\end{aligned} \qquad (7.79)
$$

whose passband center frequency is given by $\beta = \cos \hat{\omega}_0$, and the 3-dB passband bandwidth $\Delta\hat{\omega}_{3\text{dB}}$ is given by $\alpha = [1 - \tan(\Delta\hat{\omega}_{3\text{dB}})]/[1 + \tan(\Delta\hat{\omega}_{3\text{dB}})]$. It should be noted that the above bandpass transfer function is precisely the same as in Eq. (4.74), where it was introduced without any derivation. The relation between the transfer function parameters β and α with the passband center frequency $\hat{\omega}_0$ and the 3-dB bandwidth $\Delta\hat{\omega}_{3\text{dB}}$ given above are identical to those given in Eqs. (4.76) and (4.77), respectively.

7.7   FIR Filter Design Based on Truncated Fourier Series

So far we have considered only the design of real coefficient IIR digital filters that are described by a real rational transfer function that is a ratio of polynomials in z^{-1} with real coefficients. Since the transfer function of an analog filter is also a real rational transfer function in the complex frequency variable s, it is more convenient to design IIR digital filters by the conversion of a prototype analog transfer function, and two commonly used methods based on this approach were discussed. In addition, we outlined a method to transform one type of digital filter to another type. We now turn our attention to the design of real coefficient FIR filters. These filters are described by a transfer function that is a polynomial in z^{-1} and therefore require different approaches for their design.

A variety of approaches have been proposed for the design of FIR digital filters. We discuss in this chapter two direct design methods. The direct and straightforward method is based on truncating the Fourier series representation of the prescribed frequency response and is discussed in this section. The second method discussed in the following section is based on the observation that, for a length-N FIR digital filter, N distinct equally spaced frequency samples of its frequency response constitute the N-point DFT of its impulse-response, and hence, the impulse-response sequence can be readily computed by applying an inverse DFT to these frequency samples.

7.7.1   Least Integral-Squared Error Design of FIR Filters

Let $H_d(e^{j\omega})$ denote the desired frequency response function. Since $H_d(e^{j\omega})$ is a periodic function of ω with a period 2π, it can be expressed as a Fourier series:

$$H_d(e^{j\omega}) = \sum_{n=-\infty}^{\infty} h_d[n]e^{-j\omega n}, \tag{7.80}$$

where the Fourier coefficients $\{h_d[n]\}$ are precisely the corresponding impulse-response samples and are given by

$$h_d[n] = \frac{1}{2\pi} \int_{-\pi}^{\pi} H_d(e^{j\omega})e^{j\omega n}d\omega, \qquad -\infty \leq n \leq \infty. \tag{7.81}$$

Thus, given a frequency response specification $H_d(e^{j\omega})$, we can compute $h_d[n]$ using Eq. (7.81) and, hence, determine the transfer function $H_d(z)$. However, for most practical applications, the desired frequency response is piecewise constant with sharp transitions between bands, in which case, the corresponding impulse response sequence $\{h_d[n]\}$ is of infinite length and noncausal.

Our objective is to find a finite-duration impulse response sequence $\{h_t[n]\}$ of length $2M + 1$ whose DTFT $H_t(e^{j\omega})$ approximates the desired DTFT $H_d(e^{j\omega})$ in some sense. One commonly used approximation criteria is to minimize the integral squared-error

$$\Phi = \frac{1}{2\pi} \int_{-\pi}^{\pi} \left| H_t(e^{j\omega}) - H_d(e^{j\omega}) \right|^2 d\omega, \tag{7.82}$$

where

$$H_t(e^{j\omega}) = \sum_{n=-M}^{M} h_t[n]e^{-j\omega n}. \tag{7.83}$$

Using Parseval's relation (Table 3.2) we can rewrite Eq. (7.82) as

$$\Phi = \sum_{n=-\infty}^{\infty} |h_t[n] - h_d[n]|^2$$

$$= \sum_{n=-M}^{M} |h_t[n] - h_d[n]|^2 + \sum_{n=-\infty}^{-M-1} h_d^2[n] + \sum_{n=M+1}^{\infty} h_d^2[n]. \tag{7.84}$$

It is evident from Eq. (7.84) that the integral-squared error is a minimum when $h_t[n] = h_d[n]$ for $-M \leq n \leq M$, or in other words, the best finite-length approximation to the ideal infinite-length impulse response in the mean-square error sense is simply obtained by truncation.

A causal FIR filter with an impulse response $h[n]$ can be derived from $h_t[n]$ by delaying the latter sequence by M samples, i.e., by forming

$$h[n] = h_t[n - M]. \tag{7.85}$$

Note that the causal filter $h[n]$ has the same magnitude response as that of the noncausal filter $h_t[n]$ and its phase response has a linear phase shift of ωM radians with respect that of the noncausal filter.

7.7.2   Impulse Responses of Ideal Filters

Four commonly used frequency selective filters are the lowpass, highpass, bandpass, and bandstop filters whose ideal frequency responses are shown in Figure 4.8. It is straightforward to develop their corresponding impulse responses. For example, the ideal *lowpass filter* of Figure 4.8(a) has a zero-phase frequency response

$$H_{LP}(e^{j\omega}) = \begin{cases} 1, & |\omega| \leq \omega_c, \\ 0, & \omega_c < \omega \leq \pi. \end{cases} \tag{7.86}$$

The corresponding impulse response coefficients were determined in Example 4.11 and are given by

$$h_{\mathrm{LP}}[n] = \frac{\sin \omega_c n}{\pi n}, \quad -\infty \leq n \leq \infty. \tag{7.87}$$

As can be seen from the above, the impulse response of an ideal lowpass filter is doubly infinite, not absolutely summable, and therefore unrealizable. By setting all impulse response coefficients outside the range $-M \leq n \leq M$ equal to zero, we arrive at a finite-length noncausal approximation of length $N = 2M + 1$, which when shifted to the right yields the coefficients of a causal FIR lowpass filter:

$$\hat{h}_{\mathrm{LP}}[n] = \begin{cases} \frac{\sin(\omega_c(n-M))}{\pi(n-M)}, & 0 \leq n \leq N - 1, \\ 0, & \text{otherwise.} \end{cases} \tag{7.88}$$

It should be noted that the above expression also holds for N even, in which case M is a fraction.

Likewise, the impulse response coefficients $h_{\mathrm{HP}}[n]$ of the ideal *highpass filter* of Figure 4.8(b) are given by

$$h_{\mathrm{HP}}[n] = \begin{cases} 1 - \frac{\omega_c}{\pi}, & \text{for } n = 0, \\ -\frac{\sin(\omega_c n)}{\pi n}, & \text{for } |n| > 0. \end{cases} \tag{7.89}$$

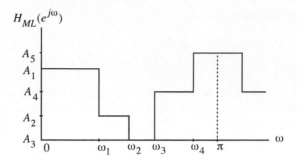

Figure 7.15   A typical zero-phase multilevel frequency response.

Correspondingly, the impulse-response coefficients $h_{BP}[n]$ of an ideal *bandpass filter* of Figure 4.8(c) with cutoffs at ω_{c1} and ω_{c2} is given by

$$h_{BP}[n] = \frac{\sin(\omega_{c2}n)}{\pi n} - \frac{\sin(\omega_{c1}n)}{\pi n}, \quad |n| \geq 0, \tag{7.90}$$

and that of an ideal *bandstop filter* of Figure 4.8(d) with cutoffs at ω_{c1} and ω_{c2} is given by

$$h_{BS}[n] = \begin{cases} 1 - \frac{(\omega_{c2}-\omega_{c1})}{\pi}, & \text{for } n = 0, \\ \frac{\sin(\omega_{c1}n)}{\pi n} - \frac{\sin(\omega_{c2}n)}{\pi n}, & \text{for } |n| > 0. \end{cases} \tag{7.91}$$

All the above design methods are for single passband or single stopband filters with two magnitude levels. However, it is quite straightforward to generalize the method to the design of multilevel FIR filters and obtain the expression for the impulse-response coefficients. The zero-phase frequency response of an ideal L-band digital filter $H_{ML}(z)$ is given by

$$H_{ML}(e^{j\omega}) = A_k, \quad \text{for } \omega_{k-1} \leq \omega \leq \omega_k, \quad k = 1, 2, \ldots, L, \tag{7.92}$$

where $\omega_0 = 0$ and $\omega_L = \pi$. Figure 7.15 shows the zero-phase frequency response of a typical multilevel filter. Its impulse response $h_{ML}[n]$ is given by

$$h_{ML}[n] = \sum_{\ell=1}^{L} (A_\ell - A_{\ell+1}) \cdot \frac{\sin(\omega_\ell n)}{\pi n}, \tag{7.93}$$

with $A_{L+1} = 0$.

Two other types of FIR digital filters that find applications are the discrete-time *Hilbert transformer* and the *differentiator*. The ideal Hilbert transformer, also called a *90-degree phase shifter*, is characterized by a frequency response

$$H_{HT}(e^{j\omega}) = \begin{cases} j, & -\pi < \omega < 0, \\ -j & 0 < \omega < \pi. \end{cases} \tag{7.94}$$

It finds application in the generation of analytic signals (see Section 11.6). The impulse response $h_{HT}[n]$ of the Hilbert transformer is obtained by computing the inverse Fourier transform of Eq. (7.94) and is given by

$$h_{\text{HT}}[n] = \begin{cases} 0, & n = 0, \\ \frac{2\sin^2(\pi n/2)}{\pi n}, & n \neq 0. \end{cases} \qquad (7.95)$$

The ideal discrete-time differentiator is employed to perform the differentiation operation in discrete-time on the sampled version of a continuous-time signal. It is characterized by a frequency response given by

$$H_{\text{DIF}}(e^{j\omega}) = j\omega, \quad |\omega| < \pi. \qquad (7.96)$$

The impulse response $h_{\text{DIF}}[n]$ of the ideal discrete-time differentiator is determined by an inverse Fourier transform of Eq. (7.96) and is given by

$$h_{\text{DIF}}[n] = \begin{cases} 0, & n = 0, \\ \frac{\cos \pi n}{n}, & |n| > 0. \end{cases} \qquad (7.97)$$

Like the ideal lowpass filter, all the above five ideal filters are also characterized by doubly infinite impulse responses that are not absolutely summable, making them unrealizable. They can be made realizable by truncating the impulse response sequences to finite lengths and shifting the truncated coefficients to the right appropriately.

7.7.3   Gibbs Phenomenon

The causal FIR filters obtained by simply truncating the impulse response coefficients of the ideal filters given in the previous section exhibit an oscillatory behavior in their respective magnitude responses, which is more commonly referred to as the *Gibbs phenomenon*. We illustrate here the occurrence of the Gibbs phenomenon by considering the design of lowpass filters. Figure 7.16 shows the magnitude responses of a lowpass filter with a cutoff at $\omega_c = 0.4\pi$ designed using the formula of Eq. (7.88) for two different values of filter lengths. The oscillatory behavior of the magnitude response on both sides of the cutoff frequency is clearly visible in both cases. Moreover, as the length of the filter is increased, the number of ripples in both passband and stopband increases, with a corresponding decrease in the widths of the ripples. However, the height of the largest ripples, which occur on both sides of the cutoff frequency, remain the same independent of the filter length and is approximately 11% of the difference between the passband and stopband magnitudes of the ideal filter [Par87].

A similar oscillatory behavior is also observed in the frequency responses of the truncated versions of the impulse responses of other types of ideal filters described in the previous section (Problems M7.10 to M7.12). For example, Figure 7.17 shows the magnitude response of a length-51 differentiator designed by truncating the impulse response coefficients of Eq. (7.97).

The reason behind the Gibbs phenomenon can be explained by considering the truncation operation as multiplication by a finite-length window sequence $w[n]$ and by examining the windowing process in the frequency domain. Thus, the FIR filter obtained by truncation can be alternatively expressed as

$$h_t[n] = h_d[n] \cdot w[n]. \qquad (7.98)$$

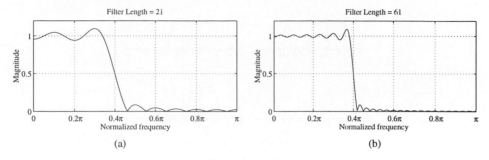

Figure 7.16 Magnitude responses of lowpass filters designed using the truncated impulse response of Eq. (7.88): (a) length $N = 21$, and (b) length $N = 61$.

Figure 7.17 Magnitude response of a length 51 differentiator designed by truncating the impulse response of Eq. (7.96).

From the modulation theorem of Table 3.2, the Fourier transform of Eq. (7.98) is given by

$$H_t(e^{j\omega}) = \frac{1}{2\pi} \int_{-\pi}^{\pi} H_d(e^{j\varphi}) \Psi(e^{j(\omega-\varphi)}) \, d\varphi. \tag{7.99}$$

where $H_t(e^{j\omega})$ and $\Psi(e^{j\omega})$ are the Fourier transforms of $h_t[n]$ and $w[n]$, respectively. Eq. (7.99) implies that $H_t(e^{j\omega})$ is obtained by a periodic continuous convolution of the desired frequency response $H_d(e^{j\omega})$ with the Fourier transform $\Psi(e^{j\omega})$ of the window. The process is illustrated in Figure 7.18 with all Fourier transforms shown as real functions for convenience.

From Eq. (7.99) it follows that if $\Psi(e^{j\omega})$ is a very narrow pulse centered at $\omega = 0$ (ideally a delta function) compared to variations in $H_d(e^{j\omega})$, then $H_t(e^{j\omega})$ will approximate $H_d(e^{j\omega})$ very closely. This implies that the length $2M + 1$ of the window function $w[n]$ should be very large. On the other hand, the length $2M + 1$ of $h_t[n]$, and hence that of $w[n]$, should be as small as possible to make the computational complexity of the filtering process small.

Now, the window used to achieve simple truncation of the ideal infinite-length impulse response is called a *rectangular window* and is given by

$$w_R[n] = \begin{cases} 1, & 0 \le |n| \le M, \\ 0, & \text{otherwise.} \end{cases} \tag{7.100}$$

Figure 7.18   Illustration of the effect of windowing in the frequency domain.

The presence of the oscillatory behavior in the Fourier transform of a truncated Fourier series representation of an ideal filter is basically due to two reasons. First, the impulse response of an ideal filter is infinitely long and not absolutely summable, and as a result, the filter is unstable. Second, the rectangular window has an abrupt transition to zero. The oscillatory behavior can be easily explained by examining the Fourier transform $\Psi_R(e^{j\omega})$ of the rectangular window function of Eq. (7.100):

$$\Psi_R(e^{j\omega}) = \sum_{n=-M}^{M} e^{-j\omega n} = \frac{\sin([2M+1]\omega/2)}{\sin(\omega/2)}. \tag{7.101}$$

A plot of the above is sketched in Figure 7.19 for $M = 4$ and 10. The frequency response $\Psi_R(e^{j\omega})$ has a narrow "main lobe" centered at $\omega = 0$. All the other ripples in the frequency response are called the "sidelobes." The main lobe is characterized by its width $4\pi/(2M+1)$ defined by the first zero crossings on both sides of $\omega = 0$. Thus, as M increases, the width of the main lobe decreases as desired. However, the area under each lobe remains constant, while the width of each lobe decreases with an increase in M. This implies that with increasing M, ripples in $H_t(e^{j\omega})$ around the point of discontinuity occur more closely but with no decrease in amplitude.

Recall also that ideally the Fourier transform of the window function should closely resemble an impulse function centered at $\omega = 0$, with its length $2M + 1$ being as small as possible to reduce the computational complexity of the FIR filter. An increase in the length of the rectangular window function reduces the main lobe width but unfortunately increases the computational complexity.

The rectangular window has an abrupt transition to zero outside the range $-M \leq n \leq M$, which is the reason behind the appearance of the Gibbs phenomenon in the magnitude response of the windowed ideal filter impulse-response sequence. The Gibbs phenomenon can be reduced by either using a window that tapers smoothly to zero at each end or by providing a smooth transition from the passband to the stopband. Use of a tapered window causes the height of the sidelobes to diminish, with a corresponding increase in the main

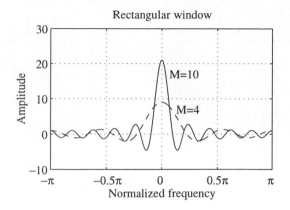

Figure 7.19   Frequency responses of the rectangular window for $M = 4$ and $M = 10$.

lobe width resulting in a wider transition at the discontinuity. We review in the next two sections a few of these windows and study their properties. Elimination of the Gibbs phenomenon by introducing a smooth transition in the filter specifications is considered in Section 7.7.5.

7.7.4   Fixed Window Functions

A large number of tapered windows have been proposed by various authors. A discussion of all these suggested windows is beyond the scope of this text. We restrict our discussion to three commonly used tapered windows of length $2M + 1$, which are listed below [Sar93][11]:

$$Hann : ^{12} \quad w[n] = \frac{1}{2}\left[1 + \cos\left(\frac{2\pi n}{2M + 1}\right)\right], \qquad -M \leq n \leq M, \quad (7.102)$$

$$Hamming : \quad w[n] = 0.54 + 0.46\cos\left(\frac{2\pi n}{2M + 1}\right), \qquad -M \leq n \leq M, \quad (7.103)$$

$$Blackman : \quad w[n] = 0.42 + 0.5\cos\left(\frac{2\pi n}{2M + 1}\right)$$
$$+ 0.08\cos\left(\frac{4\pi n}{2M + 1}\right), \qquad -M \leq n \leq M, \quad (7.104)$$

A plot of the magnitude of the Fourier transform of each of the above windows in the dB scale is shown in Figure 7.20 for $M = 25$. As can be seen from these plots, the magnitude spectrum of each window is characterized by a large main lobe centered at $\omega = 0$ followed by a series of sidelobes with decreasing amplitudes. Two parameters that somewhat predict the performance of a window in FIR filter design are its *main lobe width* and the *relative sidelobe level*. The main lobe width Δ_{ML} is the distance between the nearest zero crossings

---

[11] The expressions for the window functions given here are slightly different from that given in the literature.

[12] In the literature, this window is often called the Hanning window or the von Hann window.

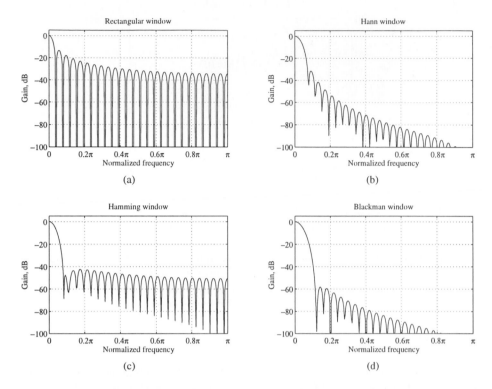

Figure 7.20   Gain response of the fixed window functions.

on both sides of the main lobe, and the relative sidelobe level $A_{s\ell}$ is the difference in dB between the amplitudes of the largest side lobe and the main lobe.

To understand the effect of the window function on FIR filter design, we show in Figure 7.21 a typical relation between $H_t(e^{j\omega})$, $\Psi(e^{j\omega})$, and $H_d(e^{j\omega})$, the frequency responses of the windowed lowpass filter, the window function, and the desired ideal lowpass filter, respectively [Sar93]. Since the corresponding impulse responses are symmetric with respect to $n = 0$, the frequency responses are of zero-phase. From this figure, we observe that for the windowed filter, $H_t(e^{j(\omega_c+\omega)}) + H_t(e^{j(\omega_c-\omega)}) \cong 1$, around the cutoff frequency ω_c. As a result, $H_t(e^{j\omega_c}) \cong 0.5$. Moreover, the passband and stopband ripples are the same. In addition, the distance between the maximum passband deviation and the minimum stopband value is approximately equal to the width Δ_{ML} of the main lobe of the window, with the center at ω_c. The width of the transition band, defined by $\Delta\omega = \omega_p - \omega_s$, is less than Δ_{ML}. Therefore, to ensure a fast transition from the passband to the stopband, the window should have a very small main lobe width. On the other hand, to reduce the passband and stopband ripple δ, the area under the sidelobes should be very small. Unfortunately, these two requirements are contradictory.

In the case of the window functions of Eqs. (7.100) and (7.102) to (7.104), the value of the ripple δ does not depend on the filter length, and the cutoff frequency ω_c, and is essentially constant. In addition, the transition bandwidth is approximately given by

Figure 7.21 Relations between the frequency responses of an ideal lowpass filter, a typical window and the windowed filter.

Table 7.3 Properties of Some Fixed Window Functions.[13]

Type of window	Main lobe width Δ_{ML}	Relative side-lobe level $A_{s\ell}$	Minimum stopband attenuation	Transition bandwidth $\Delta\omega$
Rectangular	$4\pi/(2M+1)$	13.3 dB	20.9 dB	$0.92\pi/M$
Hann	$8\pi/(2M+1)$	31.5 dB	43.9 dB	$3.11\pi/M$
Hamming	$8\pi/(2M+1)$	42.7 dB	54.5 dB	$3.32\pi/M$
Blackman	$12\pi/(2M+1)$	58.1 db	75.3 dB	$5.56\pi/M$

$$\Delta\omega \approx \frac{c}{M}, \qquad (7.105)$$

where c is a constant for most practical purposes [Sar93].

Table 7.3 summarizes the essential properties of the above window functions.

For designing an FIR filter using one of the above windows, first the cutoff frequency ω_c is determined from the specified passband and stopband edge frequencies, ω_p and ω_s, by setting $\omega_c = (\omega_p + \omega_s)/2$. Next, M is estimated using Eq. (7.105), where the value of the constant c is obtained from Table 7.3 for the window chosen. The following example illustrates the effect of each of the above windows on the frequency response of an FIR lowpass filter designed using the windowed Fourier series approach.

Example 7.12 The impulse response $h_{\mathrm{LP}}[n]$ of the ideal lowpass filter is given in Eq. (7.87). To create a finite-duration zero-phase FIR filter of length N, we form $h_t[n] = h_{\mathrm{LP}}[n] \cdot w_w[n]$, where $M = (N-1)/2$. Figure 7.22 shows the impulse response samples $h_t[n]$ for $N = 51$ and $\omega_c = \pi/2$.

---

[13]This table has been adapted from [Sar93] with the values shown in the table for $\omega_c = 0.4\pi$ and $M = 128$.

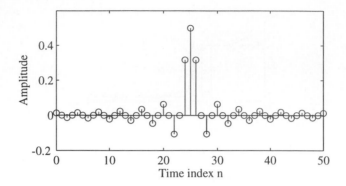

Figure 7.22   Impulse response of the truncated ideal lowpass FIR filter of length 51 with a cutoff at $\pi/2$.

The gain response of each of the filters obtained by windowing the above impulse-response samples with various fixed window functions is sketched in Figure 7.23.[14] It should be noted from this figure that the increase in the main lobe width of the window functions is clearly associated with an increase in the transition width. Likewise, a decrease in the sidelobe amplitude results in an increase in the stopband attenuation as expected.

7.7.5   Adjustable Window Functions

As indicated above, the ripple δ of the filter designed using any one of the fixed window functions is fixed. Several windows have been developed that provide control over δ by means of an additional parameter characterizing the window. We describe here two such windows.

The *Dolph-Chebyshev window* of length $2M + 1$ is defined by [Hel68]

$$w[n] = \frac{1}{2M+1}\left[\frac{1}{\gamma} + 2\sum_{k=1}^{M} T_k\left(\beta\cos\frac{k\pi}{2M+1}\right)\cos\frac{2nk\pi}{2M+1}\right], \quad -M \le n \le M,$$

(7.106)

where γ is the relative side lobe amplitude expressed as a fraction,

$$\gamma = \frac{\text{amplitude of sidelobe}}{\text{main lobe amplitude}},$$

(7.107)

$$\beta = \cosh\left(\frac{1}{2M}\cosh^{-1}\frac{1}{\gamma}\right),$$

(7.108)

and T_ℓ is the ℓth order Chebyshev polynomial defined by

$$T_\ell(x) = \begin{cases} \cos(\ell\cos^{-1}x), & \text{for } |x| \le 1, \\ \cosh(\ell\cosh^{-1}x), & \text{for } |x| > 1. \end{cases}$$

(7.109)

---

[14]These window functions have been generated using MATLAB *M*-file functions (see Section 7.9.3).

Figure 7.23 Gain responses of a lowpass FIR filter design using the fixed window functions.

The above window can be designed with any specified relative sidelobe level, and as in the case of the other windows, its main lobe width can be adjusted by choosing the length appropriately.

Figure 7.24 shows the gain response of a Dolph-Chebyshev window for $M = 25$, i.e., a window length of 51, and a relative sidelobe level of 50 dB.[15] As can be seen from this plot, all sidelobes are of equal height. As a result, the stopband approximation error of filters designed using this window have essentially an equiripple behavior. Another interesting property of this window is that for a given window length it has the smallest main lobe width compared to other windows, resulting in filters with the smallest transition band.

The most widely used adjustable window is the *Kaiser window* given by [Kai74]:

$$w[n] = \frac{I_0\left\{\beta\sqrt{1-(n/M)^2}\right\}}{I_0(\beta)}, \qquad -M \leq n \leq M, \qquad (7.110)$$

where β is an adjustable parameter and $I_0(u)$ is the modified zeroth-order Bessel function,

---

[15]The coefficients of the window have been computed using the MATLAB function `chebwin` (see Section 7.9.3).

Figure 7.24   Gain response of a length-51 Dolph-Chebyshev window with a relative sidelobe level of 50 dB.

which can be expressed in a power series form:

$$I_0(u) = 1 + \sum_{r=1}^{\infty} \left[\frac{(u/2)^r}{r!} \right]^2, \qquad (7.111)$$

which is seen to be positive for all real values of u. In practice, it is sufficient to keep only the first 20 terms in the summation of Eq. (7.111) to arrive at a reasonably accurate value of $I_0(u)$.

The parameter β controls the minimum attenuation α_s, i.e., the ripple δ_s, in the stopband of the windowed filter response. Formulas for estimating β and the filter length $N = 2M + 1$, for specified α_s and transition bandwidth Δf, have been developed by Kaiser [Kai74]. The parameter β is computed from

$$\beta = \begin{cases} 0.1102(\alpha_s - 8.7), & \text{for } \alpha_s > 50, \\ 0.5842(\alpha_s - 21)^{0.4} + 0.07886(\alpha_s - 21), & \text{for } 21 \le \alpha_s \le 50, \\ 0, & \text{for } \alpha_s < 21. \end{cases} \qquad (7.112)$$

The filter length N is estimated using the formula

$$N \approx \begin{cases} \frac{\alpha_s - 7.95}{14.36 \Delta f} + 1, & \text{for } \alpha_s > 21, \\ \frac{0.9222}{\Delta f} + 1 & \text{for } \alpha_s \le 21. \end{cases} \qquad (7.113)$$

It should be noted that the Kaiser window provides no independent control over the passband ripple δ_p. However, in practice, δ_p is approximately equal to δ_s.

We illustrate the design of a linear-phase lowpass filter using the Kaiser window in the following example.

Example 7.13 The desired specifications for the lowpass filter are $\omega_p = 0.3\pi$, $\omega_s = 0.5\pi$, and $\alpha_s = 40$ dB. The cutoff frequency is therefore given by $\omega_c = (\omega_p + \omega_s)/2 = 0.4\pi$.

Figure 7.25 (a) Gain response of the Kaiser window of Example 7.13, and (b) gain response of the lowpass filter designed using this window.

For the specified minimum stopband attenuation, the peak stopband ripple δ_s is computed using Eq. (7.4) and is found to be $\delta_s = 0.01$. Using Eq. (7.112), the parameter β is next calculated and is determined to be $\beta = 3.3953$. The transition bandwidth Δf is given by

$$\Delta f = \frac{\omega_s - \omega_p}{2\pi} = \frac{0.5\pi - 0.3\pi}{2\pi} = 0.1.$$

Applying Eq. (7.113), we next determine the filter length N to be 23.3189. The next higher odd integer value of 25 is chosen as the filter length, implying $M = 12$.

From Eqs. (7.87) and (7.110), we then arrive at the impulse-response coefficients of the FIR filter obtained by windowing as

$$h_t[n] = \frac{\sin \omega_c n}{\pi n} \cdot w[n], \quad -M \leq n \leq M,$$

where $M = 12$, $\omega_c = 0.4\pi$, and $w[n]$ is the nth coefficient of a length-25 Kaiser window. It should be noted that the above filter is noncausal and can be converted into a causal filter by delaying the filter coefficients by M samples. Since M is even for our example, the delayed filter here is a Type 1 FIR linear-phase filter.

The coefficients of the truncated ideal lowpass filter, the Kaiser window, and the windowed lowpass filter are listed in Table 7.4.[16] The gain responses of the Kaiser window and the windowed filter are sketched in Figure 7.25.

7.7.6 Impulse Responses of FIR Filters with a Smooth Transition

We showed earlier that the FIR filter obtained by truncating the infinite-length impulse response of a digital filter developed from a frequency response specification with sharp discontinuities exhibits oscillatory behavior, called the Gibbs phenomenon, in its frequency response. One way to reduce the heights of the ripples to acceptable values is to truncate the infinite-length impulse response by tapered window functions. Another approach to eliminate the Gibbs phenonomenon is by modifying the frequency response specification of the digital filter to have a transition band between the passband and the stopband and to

---

[16]The coefficients of the window have been computed using the MATLAB function kaiser (see Section 7.9.3).

Figure 7.26 (a) Lowpass filter specification with a transition region and (b) the specification of its derivative function.

provide a smooth transition between the bands. We now discuss this second approach for lowpass filter design. Similar modifications can be carried out for the design of the other types of filters discussed in Section 7.7.2.

The simplest modification to the zero-phase lowpass filter specification is to provide a transition band between the passband and the stopband responses and to connect these two with a first-order spline function (straight line), as indicated in Figure 7.26(a). An inverse Fourier transform of the modified frequency response $H_{\mathrm{LP}}(e^{j\omega})$ leads to the expression for its corresponding impulse-response coefficients $h_{\mathrm{LP}}[n]$. However, as illustrated below, a simpler method is to compute $h_{\mathrm{LP}}[n]$ from the inverse Fourier transform of the derivative of the specified frequency response $H_{\mathrm{LP}}(e^{j\omega})$.

Let $G(e^{j\omega})$ denote $dH_{\mathrm{LP}}(e^{j\omega})/d\omega$. Its specification will be thus as indicated in Figure 7.26(b). It follows from the differentiation-in-frequency property of the DTFT given in Table 3.2, $h_{\mathrm{LP}}[n] = jg[n]/n$. From the inverse DTFT $g[n]$ of $G(e^{j\omega})$ given in Figure 7.26(b), we thus arrive at the impulse response of the modified lowpass filter [Par87]:

$$
h_{\mathrm{LP}}[n] = \begin{cases} \dfrac{\omega_c}{\pi}, & n = 0, \\[2mm] \dfrac{2\sin(\Delta\omega n/2)}{\Delta\omega n} \cdot \dfrac{\sin(\omega_c n)}{\pi n}, & |n| > 0, \end{cases} \tag{7.114}
$$

where $\Delta\omega = \omega_s - \omega_p$ and $\omega_c = (\omega_p + \omega_s)/2$.

Still smoother transition between the passband and the stopband of the lowpass filter can be provided by specifying the transition function by a higher order spline. The corresponding impulse response for a Pth order spline as the transition function is given by [Bur92], [Par87]:

$$
h_{\mathrm{LP}}[n] = \begin{cases} \dfrac{\omega_c}{\pi}, & n = 0, \\[2mm] \left(\dfrac{\sin(\Delta\omega n/2P)}{\Delta\omega n/2P}\right)^P \cdot \dfrac{\sin(\omega_c n)}{\pi n}, & |n| > 0, \end{cases} \tag{7.115}
$$

Example 7.14 Design a lowpass filter with a passband edge at 0.35π and a stopband edge at 0.45π using a first-order and a second-order spline as the transition functions. The magnitude response of a length-41 filter obtained by truncating the impulse response of Eq. (7.114) is shown by solid line in Figure 7.27. The magnitude response of a length-61 FIR filter derived by truncating the impulse response of Eq. (7.115) for $P = 2$ (second-order spline) is shown in Figure 7.27 by dashed line.

As the above example points out, the effect of P on the frequency response of the truncated filter is not that obvious. For a given filter length N and transition bandwidth $\Delta\omega$, the optimum value of P minimizing the integral-squared approximation error has been

Table 7.4 Coefficients of the truncated ideal lowpass filter, the Kaiser window, and the windowed lowpass filter of Example 7.13.

n	Truncated Ideal Filter	Kaiser Window	Windowed Filter
0	0.4	1.0	0.4
1	0.30273069145626	0.99018600306076	0.22975969337690
2	0.09354892837886	0.096116367723306	0.08991583299184
3	−0.06236595225258	0.91416906485180	−0.05701302424933
4	−0.07568267286407	0.85118713849723	−0.06442011774899
5	0	0.77484399922791	0
6	0.05045511524271	0.68826531740711	0.03472650590734
7	0.02672826525110	0.59490963898165	0.01590090263114
8	−0.02338723209472	0.49838664295215	−0.01165588409163
9	−0.02752097195057	0.40227124342555	−0.01353109463058
10	0	0.30992453405416	0
11	0.02752097195057	0.22433197338245	0.00617383394707
12	0.01559148806314	0.14796795346662	0.00230704058020

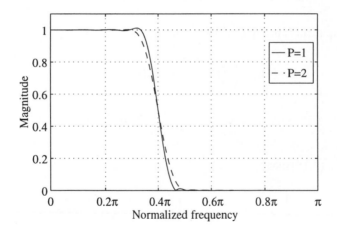

Figure 7.27 Magnitude responses of FIR lowpass filters with smooth transition between bands obtained using spline transition functions.

shown to be given by [Bur92]

$$P = 0.624(\Delta\omega)N. \tag{7.116}$$

Various other transition functions have been investigated (Problem 7.25) [Bur92], [Par87]. These transition functions can also be employed for the design of other types of filters. So far, no specific guidelines have been advanced to select the optimum transition function for a given filter design problem. As a result, it should be selected by a trial-and-error procedure.

7.8   FIR Filter Design Based on Frequency Sampling Approach

The second direct method of FIR filter design is the frequency sampling approach. In this approach, the specified frequency response $H_d(e^{j\omega})$ is first uniformly sampled at N equally spaced points $\omega_k = 2\pi k/N, k = 0, 1, \ldots, N-1$, providing N *frequency samples*. These N frequency samples constitute an N-point DFT $H[k]$ whose N-point inverse-DFT thus yields the impulse response coefficients $h[n]$ of the FIR filter of length N [Gol69a]. The basic assumption here is that the specified frequency response is uniquely characterized by the N frequency samples and, hence, can be fully recovered from these samples.

Now, $H_d(e^{j\omega})$ is a periodic function of ω with a Fourier series representation given by Eq. (7.80). Its Fourier coefficients $h_d[n]$ are given by Eq. (7.81). It is instructive to develop the relation between $h_d[n]$ and $h[n]$.

From Eq. (7.80),

$$H[k] = H_d(e^{j\omega_k}) = H_d(e^{j(2\pi k/N)}) = \sum_{\ell=-\infty}^{\infty} h_d[\ell] W_N^{k\ell}, \tag{7.117}$$

where $W_N = e^{-j(2\pi/N)}$. An inverse-DFT of $H[k]$ yields

$$h[n] = \frac{1}{N} \sum_{k=0}^{N-1} H[k] W_N^{-kn}. \tag{7.118}$$

Substituting Eq. (7.117) in Eq. (7.118), we get

$$h[n] = \frac{1}{N} \sum_{k=0}^{N-1} \sum_{\ell=-\infty}^{\infty} h_d[\ell] W_N^{k\ell} W_N^{-kn}$$

$$= \sum_{\ell=-\infty}^{\infty} h_d[\ell] \left[\frac{1}{N} \sum_{k=0}^{N-1} W_N^{-k(n-\ell)} \right]. \tag{7.119}$$

Recall from Eq. (3.18) that

$$\frac{1}{N} \sum_{k=0}^{N-1} W_N^{-k(n-r)} = \begin{cases} 1, & \text{for } r = n + mN, \\ 0, & \text{otherwise.} \end{cases} \tag{7.120}$$

Making use of the above identity in Eq. (7.119), we finally arrive at the desired relation

$$h[n] = \sum_{m=-\infty}^{\infty} h_d(n + mN), \quad 0 \le n \le N - 1. \tag{7.121}$$

The above relation indicates that $h[n]$ is obtained from $h_d[n]$ by adding an infinite number of shifted replicas of $h_d[n]$ to $h_d[n]$, with each replica shifted by an integer multiple of N sampling instants, and observing the sum only for the interval $0 \le n \le N - 1$. Thus, if $h_d[n]$ is a finite-length sequence of length less than or equal to N, then $h[n] = h_d[n]$ for $0 \le n \le N - 1$, otherwise there is a time-domain aliasing of samples with $h[n]$ bearing no resemblance to $h_d[n]$, as illustrated in the following example.

Example 7.15 Let $h_d[n]$ be a length-6 sequence given by

$$h_d[n] = \{0 \quad 1 \quad 2 \quad 3 \quad 4 \quad 5\}.$$
$$\uparrow$$

By sampling the Fourier transform $H_d(e^{j\omega})$ of $h_d[n]$ at 4 equally spaced points given by $\omega_k = \pi k/4$, $k = 0, 1, 2, 3$, and then applying a 4-point inverse-DFT to these samples, we arrive at a sequence $h[n]$ given by[17]

$$h[n] = h_d[n] + h_d[\langle n+4 \rangle_5] + h_d[\langle n-4 \rangle_5], \quad n = 0, 1, 2, 3, 4, 5,$$

i.e.,

$$h[n] = \{4 \quad 6 \quad 2 \quad 3 \quad 4 \quad 6\}.$$
$$\uparrow$$

We derive next the expression for the transfer function $H(z)$ of the FIR filter designed using the frequency sampling approach outlined above in terms of the frequency samples $H[k]$. Now,

$$H(z) = \sum_{n=0}^{N-1} h[n] z^{-n}. \tag{7.122}$$

Substituting Eq. (7.118) in Eq. (7.122) we arrive at

$$H(z) = \sum_{n=0}^{N-1} \frac{1}{N} \sum_{k=0}^{N-1} H[k] W_N^{-nk} z^{-n}, \tag{7.123}$$

which after some algebra results in

$$H(z) = \frac{1 - z^{-N}}{N} \sum_{k=0}^{N-1} \frac{H[k]}{1 - W_N^{-k} z^{-1}}. \tag{7.124}$$

A possible realization of $H(z)$ based on the above expression is indicated in Figure 7.28. Note that the structure in this figure makes use of recursive resonators in realizing an FIR transfer function. The pole of each of the resonator on the unit circle is exactly canceled by a zero at the corresponding point on the unit circle in the FIR stage in cascade. Such cancellation may not occur in practice due to the finite wordlength representations of the filter coefficients and may result in an unstable operation.

On the unit circle, Eq. (7.124) reduces to

$$H(e^{j\omega}) = \frac{e^{-j\omega[(N-1)/2]}}{N} \sum_{k=0}^{N-1} H[k] \frac{\sin\left(\frac{\omega N - 2\pi k}{2}\right)}{\sin\left(\frac{\omega N - 2\pi k}{2N}\right)} e^{j\pi k[(N-1)/N]}. \tag{7.125}$$

It can be shown that

$$H(e^{j\omega})\Big|_{\omega = 2\pi k/N} = H[k], \quad k = 0, 1, \ldots, N - 1. \tag{7.126}$$

---

[17] $\langle n \rangle_M$ denotes n modulo M.

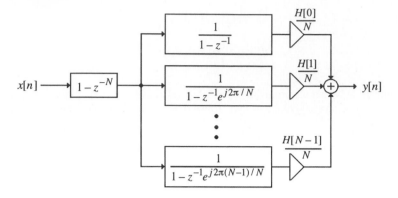

Figure 7.28   Frequency sampling-based FIR digital filter realization.

That is, the FIR filter designed via the frequency sampling approach has exactly the specified frequency samples $H[k] = H_d(e^{j\omega})$, where $\omega_k = 2\pi k/N$, $k = 0, 1, \ldots, N - 1$, whether or not the length of $h_d[n]$ is less than or equal to N.

Example 7.16   Consider the design of a linear-phase lowpass FIR filter with a passband edge at $\omega_p = 0.3\pi$ using the frequency sampling approach. We choose the length $N = 37$. For a linear phase design, the delay is therefore equal to $(N - 1)/2 = 18$.

If we assume the desired magnitude response is an ideal brickwall characteristic of Figure 4.8(a), the specified frequency samples are then given by

$$H[k] = \begin{cases} e^{-j(2\pi k/37)18}, & k = 0, 1, \ldots, 5, 32, \ldots, 36, \\ 0, & k = 6, \ldots, 31 \end{cases}$$

A 37-point inverse-DFT of the above yields the filter's impulse-response coefficients.

A plot of the magnitude response of the above filter is sketched in Figure 7.29 along with the specified frequency samples.

It can be seen that even though the response in the passband of the above FIR filter is acceptable, the stopband response is not that great, since here the minimum attenuation is merely 17.1886 dB, which in most applications is not satisfactory. In addition, the maximum passband ripple is 1.1525 dB. The reason behind the overall unsatisfactory magnitude response is that the impulse response $h_d[n]$ corresponding to the ideal filter is of infinite length, and as a result of the relation of Eq. (7.121), there is a severe time-domain aliasing in determining the impulse response coefficients of the FIR filter $h[n]$. One of the earliest approaches to alleviate this drawback was to change the frequency response specifications and provide a transition band containing more than one frequency sample. Further improvement in the filter's response has been obtained by adjusting the values of the transition band frequency samples by using a linear programming approach [Rab70].

A more effective approach is to specify the desired frequency response as a simple analytic function with a few adjustable parameters that are then chosen to permit almost an exact realization by an FIR filter [Lig94]. Separate functions are used to represent the frequency responses in the passband and stopband, respectively, with a weighted sum of these functions defining the transition band. The analytic function is sampled to produce

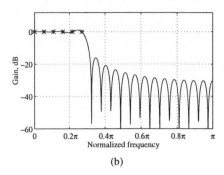

(a)   (b)

Figure 7.29   Magnitude and gain responses of the linear-phase lowpass FIR filter of Example 7.16. The original frequency samples in the passband are marked with "*". Note that the zero-valued samples in the stopband have infinite attenuation and therefore do not show up in the gain plot.

the finite number of frequency samples $H[k]$ whose inverse DFT yields the filter's impulse-response coefficients.

We consider the design of a linear-phase FIR lowpass filter with passband edge at ω_p, stopband edge at ω_s, and passband ripple of δ_p and stopband ripple of δ_s. Let $\breve{H}(\omega)$ denote the desired amplitude response of the linear-phase FIR filter. We represent the specified amplitude response $\breve{H}(\omega)$ by the function

$$\breve{H}_p(\omega) = 1 - \delta_p T_P(X_p(\omega)) \tag{7.127}$$

in the passband ($0 \leq \omega \leq \omega_p$) and by the function

$$\breve{H}_s(\omega) = \delta_s T_S(X_s(\omega)), \tag{7.128}$$

in the stopband ($\omega_s \leq \omega \leq \pi$). In the transition band, $\breve{H}(\omega)$ is expressed as a linear combination of the above two functions:

$$\breve{H}_t(\omega) = W_p(\omega)X_p(\omega) + W_s(\omega)X_s(\omega), \tag{7.129}$$

where $W_p(\omega)$ and $W_s(\omega)$ are appropriately chosen weighting functions to ensure a smooth transition between $\breve{H}_p(\omega)$ and $\breve{H}_s(\omega)$. In Eqs. (7.127) and (7.128), $T_L(\omega)$ is an Lth-order Chebyshev polynomial. The functions $X_p(\omega)$ and $X_s(\omega)$ are chosen as

$$X_p(\omega) = A\cos(a\omega + b) + B, \tag{7.130a}$$
$$X_s(\omega) = C\cos(c\omega + d) + D, \tag{7.130b}$$

with the additional requirements

$$X_p(\omega_p) = 1, \qquad X_s(\omega_s) = 1, \tag{7.131a}$$
$$X_p(0) = -1, \qquad X_s(\pi) = -1. \tag{7.131b}$$

The two constraints in Eq. (7.131a) ensure better approximation of $\breve{H}(\omega)$ to the amplitude response of an optimal equiripple lowpass filter. It should be pointed out that most lowpass filter specifications can be met by satisfying these two constraints by slightly increasing the filter length and by a smaller change in the ripple ratio δ_p/δ_s than that obtained with the optimal filters, if necessary.

From Eqs. (7.131b), (7.130a), and (7.130b), we arrive at the conditions

$$B = A - 1, \tag{7.132a}$$

$$D = -C - 1. \tag{7.132b}$$

Additional restrictions need to be placed on the functions $X_p(\omega)$ and $X_s(\omega)$ so that $\breve{H}(\omega)$ approximates the amplitude response of an optimal equiripple lowpass filter. To this end we set

$$d = 0 \tag{7.133a}$$

to force $\breve{H}_p(\omega)$ to be symmetric around $\omega = 0$ and set

$$b = \pi(1 - a) \tag{7.133b}$$

so that $\breve{H}_s(\omega)$ is symmetric around $\omega = \pi$.

Because of the above constraints, only two of the eight parameters of Eqs. (7.130a) and (7.130b) can be chosen independently. Further restrictions on the design parameters are obtained by requiring that the functions $\breve{H}_p(\omega)$ and $\breve{H}_s(\omega)$ remain between $-\delta_s$ and $1 + \delta_p$ at all frequencies, which are equivalent to the conditions

$$\min_{\omega} \left[\breve{H}_p(\omega) \right] = -\delta_s, \tag{7.134a}$$

$$\max_{\omega} \left[\breve{H}_s(\omega) \right] = 1 + \delta_p. \tag{7.134b}$$

Eqs. (7.133a) and (7.133b) leads to

$$1 - \delta_p T_P(-2C - 1) = -\delta_s, \tag{7.135a}$$

$$\delta_s T_S(2A - 1) = 1 + \delta_s. \tag{7.135b}$$

Applying all the above constraints, closed form expressions for all eight parameters of Eqs. (7.130a) and (7.130b) have been derived and are given by

$$A = \frac{1}{2}\left[T_S^{-1}\left(\frac{1 + \delta_p}{\delta_s} \right) + 1 \right], \tag{7.136a}$$

$$B = A - 1, \tag{7.136b}$$

$$C = -\frac{1}{2}\left[T_P^{-1}\left(\frac{1 + \delta_p}{\delta_s} \right) + 1 \right], \tag{7.136c}$$

$$D = -C - 1, \tag{7.136d}$$

$$a = \frac{1}{\omega_s - \pi}\left[\cos^{-1}\left(\frac{2}{A} - 1 \right) - \pi \right], \tag{7.136e}$$

$$b = (1 - a)\pi, \tag{7.136f}$$

$$c = \frac{1}{\omega_p}\left[\cos^{-1}\left(\frac{2}{C} + 1 \right) \right], \tag{7.136g}$$

$$d = 0. \tag{7.136h}$$

(a)   (b)

Figure 7.30 (a) Magnitude response with the desired frequency samples marked with "*" and (b) the gain response of the linear-phase lowpass FIR filter of Example 7.17.

The required order P of the Chebyshev polynomial $T_P(\omega)$ and the order S of the Chebyshev polynomial $T_S(\omega)$ are obtained as follows. First, for the given filter specifications, ω_p, ω_s, δ_s, and δ_s, the necessary filter length N is estimated using Eq. (7.13) or (7.16). The polynomial order P and S are determined from filter length N by weighting them proportionally to the passband width and stopband width such that $P + S = N/2$ when N is even and $P + S = (N - 1)/2$ when N is odd. For a smoother transition from $\breve{H}_p(\omega)$ to $\breve{H}_s(\omega)$, the analytic function in the transition band is calculated from a weighted average of these two functions with the weights given by

$$W_p(\omega) = \left[\frac{1}{2}\cos\left(\frac{(\omega - \omega_p)\pi}{\omega_s - \omega_p}\right) + \frac{1}{2}\right]^{\delta_p/\delta_s}, \tag{7.137a}$$

$$W_s(\omega) = 1 - W_p(\omega). \tag{7.137b}$$

The analytic functions $\breve{H}_p(\omega)$, $\breve{H}_t(\omega)$, and $\breve{H}_s(\omega)$ are then sampled at equally spaced intervals, and an inverse DFT is applied to these frequency samples to determine the FIR filter coefficients.

Example 7.17 We consider again the design of an FIR lowpass filter with a passband edge at $\omega_p = 0.3\pi$ and length $N = 37$ that are the same as those in Example 7.16. However, we now fix the stopband edge at $\omega_s = 0.4\pi$ to provide a smooth transition of the gain response. The maximum passband ripple δ_p is set at 0.1807 dB, while the minimum stopband attenuation is set at $\alpha_s = 25.0974$ dB. Using these specifications, we then determine the analytic functions $\breve{H}_p(\omega)$, $\breve{H}_t(\omega)$, and $\breve{H}_s(\omega)$ following the method outlined above and sample these functions uniformly at 37 points on the unit circle, as indicated in Figure 7.30(a). An inverse DFT of these samples lead to an FIR filter with a gain response as indicated in Figure 7.30(b). Note the improvement in the gain response compared to that obtained in Example 7.16 by applying the frequency sampling approach to an ideal filter response.

7.9   Computer-Aided Design of Digital Filters

The filter design algorithms described in Sections 7.2, 7.3, 7.7, and 7.8 can be easily implemented on a computer. In addition, a number of filter design algorithms have been advanced that rely on some type of iterative optimization techniques that are used to minimize the error between the desired frequency response and that of the computer-generated filter. In this section, we consider the computer-aided design of FIR and IIR digital filters. First, we describe two specific design approaches based on iterative optimization techniques. A variety of software packages are presently commercially available that have made the design of digital filters rather simple to implement on a computer. In this section, we consider only the application of MATLAB.

The basic idea behind the computer-based iterative technique is as follows. Let $|H(e^{j\omega})|$ denote the magnitude response of the digital transfer function $H(z)$ to be designed to approximate the desired magnitude response $D(\omega)$ in some sense. The objective is to determine iteratively the transfer function coefficients so that the difference between $|H(e^{j\omega})|$ and $D(\omega)$ for all values of ω in the range $0 \leq \omega \leq \pi$ is minimized. Usually, this difference is specified as an weighted error function $\mathcal{E}(\omega)$ given by

$$\mathcal{E}(\omega) = P(\omega)\left[\left|H(e^{j\omega})\right| - D(\omega)\right], \tag{7.138}$$

where $P(\omega)$ is some user-specified weighting function. A commonly used approximation measure is the *Chebyshev* or *minimax criterion* in which the peak absolute value of the weighted error $\mathcal{E}(\omega)$,

$$\varepsilon = \max_{0 \leq \omega \leq \pi} |\mathcal{E}(\omega)|, \tag{7.139}$$

is minimized. A second approximation measure is the *least-p criterion*, which minimizes the function

$$\varepsilon = \sum_{i=1}^{L} \left\{ P(\omega_i)\left(\left|H(e^{j\omega_i})\right| - D(\omega_i)\right)\right\}^p, \tag{7.140}$$

where ω_i, $i = 1, \ldots, L$, are suitably chosen digital angular frequencies, and p is an even positive integer. It can be shown that for a large value of L, as $p \to \infty$, the least pth solution approaches the minimax solution. The *least-squares criterion* obtained from Eq. (7.140) with $p = 2$ is used often in practice.

7.9.1   Design of Optimal Linear-Phase FIR Filters

The linear-phase FIR filter obtained by minimizing the peak absolute value of the weighted error ε given by Eq. (7.139) is usually called the *optimal FIR filter*, since here, after ε has been minimized, the weighted error function $\mathcal{E}(\omega)$ exhibits an equiripple behavior in the frequency range of interest. We briefly outline below the basic idea behind the most widely used highly efficient algorithm, more commonly known as the *Parks-McClellan algorithm,* for designing the optimum linear-phase FIR filter [Par72].

In Section 4.3.6, we defined the four types of linear-phase FIR filters. The general form of the frequency response $H(e^{j\omega})$ of a linear-phase FIR filter of length N is given by

$$H(e^{j\omega}) = e^{-j(N-1)\omega/2}e^{j\beta}\breve{H}(\omega), \qquad \beta = 0 \text{ or } \frac{\pi}{2}, \tag{7.141}$$

where the amplitude response $\breve{H}(\omega)$ is a real function of ω. Without any loss of generality, we consider here the design of optimal Type 1 linear-phase FIR filters. However, the algorithm can be easily applied to the design of the other three types of linear-phase FIR filters. The Type 1 linear-phase FIR filter has a symmetric impulse response of odd length $N = 2M + 1$. From Eq. (4.98), we note that the amplitude response of this type of FIR filter is of the form

$$\breve{H}(\omega) = \sum_{n=0}^{M} a[n] \cos(\omega n), \tag{7.142}$$

where the coefficients $a[n]$ are related to the filter's impulse response coefficients $h[n]$ through

$$a[0] = h[M], \qquad a[n] = 2\,h[M - n], \qquad 1 \leq n \leq M. \tag{7.143}$$

The algorithm is based on the following theorem [Par72]:

The amplitude function $\breve{H}(\omega)$ of Eq. (7.142) is the best unique approximation minimizing ε of Eq. (7.139) if and only if there exist at least $M + 2$ extremal angular frequencies, $\omega_1, \omega_2, \ldots, \omega_{M+2}$, in a closed subset S (specified passbands and stopbands) of the frequency range $0 \leq \omega \leq \pi$ such that $\omega_1 < \omega_2 < \cdots < \omega_{M+1} < \omega_{M+2}$ and $\mathcal{E}(\omega_i) = -\mathcal{E}(\omega_{i+1})$ with $1 \leq i \leq M + 1$ and $|\mathcal{E}(\omega_i)| = \varepsilon$ with $1 \leq i \leq M + 2$.

The optimum solution can thus be obtained by solving the set of $M + 2$ equations

$$P(\omega_i)[\breve{H}(\omega_i) - D(\omega_i)] = (-1)^i \varepsilon, \qquad 1 \leq i \leq M + 2, \tag{7.144}$$

for the unknowns $a[i]$ and ε, provided the $M + 2$ extremal angular frequencies are known. To solve for the unknowns, Eq. (7.144) is rewritten in a matrix form as

$$
\begin{bmatrix}
1 & \cos(\omega_1) & \cdots & \cos(M\omega_1) & 1/P(\omega_1) \\
1 & \cos(\omega_2) & \cdots & \cos(M\omega_2) & -1/P(\omega_2) \\
\vdots & \vdots & \ddots & \vdots & \vdots \\
1 & \cos(\omega_{M+1}) & \cdots & \cos(M\omega_{M+1}) & (-1)^M/P(\omega_{M+1}) \\
1 & \cos(\omega_{M+2}) & \cdots & \cos(M\omega_{M+2}) & (-1)^{M+1}/P(\omega_{M+2})
\end{bmatrix}
\begin{bmatrix}
a[0] \\
a[1] \\
\vdots \\
a[M] \\
\varepsilon
\end{bmatrix}
$$

$$
=
\begin{bmatrix}
D(\omega_1) \\
D(\omega_2) \\
\vdots \\
D(\omega_{M+1}) \\
D(\omega_{M+2})
\end{bmatrix}, \tag{7.145}
$$

which can, in principle, be solved for the unknowns if the locations of the $M + 2$ extremal frequencies are known a priori. The Remez algorithm, a highly efficient iterative procedure, is used to determine the locations of the extremal frequencies and consists of the following steps at each iteration stage:

Step 1: A set of initial values for the extremal frequencies are either chosen or are available from the completion of the previous stage.

Step 2: The value ε is then computed by solving Eq. (7.145), resulting in the expression

$$\varepsilon = \frac{c_1 D(\omega_1) + c_2 D(\omega_2) + \cdots + c_{M+2} D(\omega_{M+2})}{\frac{c_1}{P(\omega_1)} - \frac{c_2}{P(\omega_2)} + \cdots + \frac{(-1)^{M+1} c_{M+2}}{P(\omega_{M+2})}}, \tag{7.146}$$

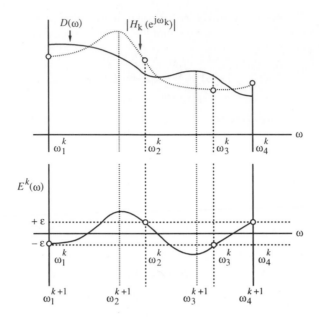

Figure 7.31   Illustration of one stage of the iteration process.

where the constant c_n is given by

$$c_n = \prod_{\substack{i=1 \\ i \neq n}}^{M+2} \frac{1}{\cos(\omega_n) - \cos(\omega_i)} \tag{7.147}$$

Step 3: The values of the weighted error function over the closed subset S of the frequency range $0 \leq \omega \leq \pi$ are determined by interpolating the values at the $M + 2$ extremal frequencies using the Lagrange interpolation formulae.[18]

Step 4: The $M + 2$ new extremal frequencies are next determined from the weighted error function obtained by interpolation in Step 3.

Figure 7.31 demonstrates how the weighted error function changes from one iteration stage to the next and how the new extremal frequencies are determined. Finally, the iteration process is stopped after the difference between the value of the peak error ε calculated at any stage and that at the previous stage is below a preset threshold value, such as 10^{-6}. In practice, the process converges after very few iterations.

The M-file function `remez` in the *Signal Processing Toolbox* of MATLAB used for the design of optimal FIR filters is based on this algorithm (see Section 7.9.3).

The basic principle of the Remez exchange algorithm is illustrated in the following example [Par87].

---

[18] see Section 11.11.3 for a review of the Lagrange interpolation method.

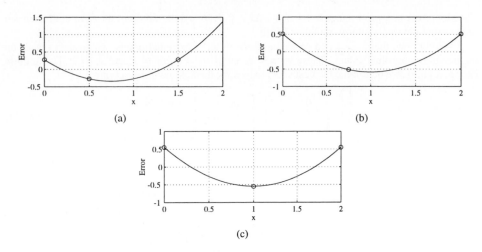

Figure 7.32   Illustration of the Remez algorithm.

Example 7.18 Let the desired function be a quadratic function $D(x) = 1.1x^2 - 0.1$ defined for the range $0 \le x \le 2$. We wish to approximate $D(x)$ by a linear function $a_1x + a_0$ by minimizing the peak value of the absolute error , i.e.,

$$\max_{x \in [0,2]} \left| 1.1x^2 - 0.1 - a_0 - a_1x \right|. \tag{7.148}$$

Since there are three unknowns, a_0, a_1, and ε, we need three extremal points on x which we arbitrarily choose as $x_1 = 0$, $x_2 = 0.5$, and $x_3 = 1.5$. We then solve the three linear equations

$$a_0 + a_1x_\ell - (-1)^\ell \varepsilon = D(x_\ell) \qquad \ell = 1, 2, 3. \tag{7.149}$$

which leads to

$$\begin{bmatrix} 1 & 0 & 1 \\ 1 & 0.5 & -1 \\ 1 & 1.5 & 1 \end{bmatrix} \begin{bmatrix} a_0 \\ a_1 \\ \varepsilon \end{bmatrix} = \begin{bmatrix} -0.1 \\ 0.175 \\ 2.375 \end{bmatrix}, \tag{7.150}$$

for the given extremal points. The above matrix equation, when solved, yields

$$a_0 = -0.375, \qquad a_1 = 1.65, \qquad \epsilon = 0.275. \tag{7.151}$$

Figure 7.32(a) shows the plot of the corresponding error $\mathcal{E}_1(x) = 1.1x^2 - 1.65x + 0.275$, along with the values of error at the chosen extremal points. As expected, at the extremal points, the errors are equal in magnitude and alternate in sign.

The next set of extremal points are those points where $\mathcal{E}_1(x)$ assumes its maximum absolute values. These extremal ponts are given by $x_1 = 0$, $x_2 = 0.75$, and $x_3 = 2$. The new values of the unknowns are obtained by solving

$$\begin{bmatrix} 1 & 0 & 1 \\ 1 & 0.75 & -1 \\ 1 & 2 & 1 \end{bmatrix} \begin{bmatrix} a_0 \\ a_1 \\ \varepsilon \end{bmatrix} = \begin{bmatrix} -0.1 \\ 0.5188 \\ 4.3 \end{bmatrix}, \tag{7.152}$$

which yields

$$a_0 = -0.6156, \qquad a_1 = 2.2, \qquad \varepsilon = 0.5156. \tag{7.153}$$

The plot of the new error $\mathcal{E}_2(x) = 1.1x^2 - 2.2x + 0.5156$ and the values of the error at the new set of extremal points are shown in Figure 7.32(b).

We continue the process and determine the new set of extremal points, where $\mathcal{E}_2(x)$ takes on the maximum absolute values. These are now given by $x_1 = 0$, $x_2 = 1$, and $x_3 = 2$. Using these extremal points we determine the next set of values for the unknowns by solving

$$\begin{bmatrix} 1 & 0 & 1 \\ 1 & 1 & -1 \\ 1 & 2 & 1 \end{bmatrix} \begin{bmatrix} a_0 \\ a_1 \\ \varepsilon \end{bmatrix} = \begin{bmatrix} -0.1 \\ 1.0 \\ 4.3 \end{bmatrix}, \tag{7.154}$$

which results in

$$a_0 = -0.65, \quad a_1 = 2.2, \quad \varepsilon = 0.55. \tag{7.155}$$

Figure 7.32(c) depicts the new error $\mathcal{E}_3(x) = 1.1x^2 - 2.2x + 0.55$ and the values of the error at the new extremal points. This time the algorithm has converged as ε is also the maximum value of the absolute error.

7.9.2   IIR Digital Filter Design Using MATLAB

As in the case of analog filter design described in Section 5.3.5, the *Signal Processing Toolbox* in MATLAB includes a large number of built-in functions to develop both IIR and FIR digital transfer functions for meeting given frequency response specifications.

For IIR filter design based on the bilinear transformation, it offers functions for each one of the four magnitude approximation techniques, i.e., Butterworth, Type 1 and 2 Chebyshev, and elliptic approximations. Specifically, the following *M*-file functions are available:

```
[b, a] = butter(N,Wn)
[b, a] = cheby1(N,Rp,Wn)
[b, a] = cheby2(N,Rs,Wn)
[b, a] = ellip(N,Rp,Rs,Wn)
```

These functions directly determine the digital lowpass filter transfer function of order N with a normalized passband edge frequency Wn that must be a number between 0 and 1 with a sampling frequency assumed to be equal to 2 Hz. The additional parameters for the Chebyshev and the elliptic filters are Rp, specifying the passband ripple in dB, and Rs, specifying the minimum stopband attenuation in dB. The output files of these functions are the length N+1 column vectors b and a providing, respectively, the numerator and denominator coefficients in ascending powers of z^{-1}. The form of the transfer function obtained is given by

$$G(z) = \frac{B(z)}{A(z)} = \frac{b(1) + b(2)z^{-1} + \cdots + b(N+1)z^{-N}}{1 + a(2)z^{-1} + \cdots + a(N+1)z^{-N}}. \tag{7.156}$$

After these coefficients have been computed, the frequency response can be computed using the *M*-file function freqz(b,a,w), where w is a set of specified angular frequencies.[19] The function freqz(b,a,w) generates a complex vector of frequency response samples from which magnitude and/or phase response samples can be readily computed.

The following four IIR lowpass filter design functions can be employed to determine the zeros and poles of the transfer function:

```
[z,p,k] = butter(N,Wn)
[z,p,k] = cheby1(N,Rp,Wn)
[z,p,k] = cheby2(N,Rs,Wn)
[z,p,k] = ellip(N,Rp,Rs,Wn)
```

The output files of these functions are the zeros and poles of the transfer function given as the N-length, vectors z and p, and k is the scalar gain factor. The numerator and denominator coefficients of the transfer function can then determined using the function poly(r), where r is the vector of roots of the polynomial.

We illustrate the design of a digital lowpass filter using MATLAB in the next example.

Example 7.19 Determine the transfer function and plot the gain response of a 5th-order elliptic IIR lowpass filter with a passband edge at 0.4, a passband ripple of 0.5 dB, and a minimum stopband attenuation of 40 dB.

The following MATLAB program can be used to design such a filter. As the program is run it asks for the following filter parameters to be typed in: filter order N, passband ripple Rp, minimum stopband attenuation Rs, and passband edge Wn.

```
% Program 7_3
% Elliptic IIR Lowpass Filter Design
%
N = input('Type in the order = ');
Rp = input('Type in the passband ripple = ');
Rs = input('Type in the min. stopband attenuation = ');
Wn = input('Type in the passband edge frequency = ');
[b,a] = ellip(N,Rp,Rs,Wn);
disp('Numerator polynomial');
disp(b)
disp('Denominator polynomial');
disp(a)
w = 0:0.01/pi:pi;
h = freqz(b,a,w);
gain = 20*log10(abs(h));
plot (w/pi,gain);grid;
xlabel('Normalized frequency');
ylabel('Gain, dB');
```

The output generated by the program are the numerator and the denominator coefficients in ascending powers of z^{-1}, as given below, and the gain response plot, as shown in Figure 7.33.

---

[19]The range of frequencies w in freqz(b,a,w) should be between 0 and π, whereas, the range of frequencies in any of the filter functions are between 0 and 1.

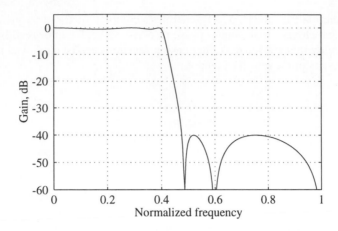

Figure 7.33   Gain response of the IIR elliptic lowpass filter of Example 7.19.

```
Numerator polynomial
  Columns 1 through 4

    0.05281676474336        0.07970816344795
    0.12949752895344        0.12949752895343


  Columns 5 through 6

    0.07970816344795        0.05281676474336

Denominator polynomial
  Columns 1 through 4

    1.00000000000000       -1.81069194743518
    2.49467590243473       -1.88009328220852


  Columns 5 through 6

    0.95373953873315       -0.23358529723470
```

The transfer function determined by this program is therefore given by

$$G(z) = \frac{\begin{aligned}0.05281676 + 0.07970816z^{-1} + 0.12949753z^{-2} + 0.12949753z^{-3} \\ + 0.07970816z^{-4} + 0.05281676z^{-5}\end{aligned}}{\begin{aligned}1 - 1.81069195z^{-1} + 2.4946759z^{-2} - 1.88009328z^{-3} \\ + 0.95373954z^{-4} - 0.23358529z^{-5}\end{aligned}}.$$

Other types of digital filters can be designed by simple modifications of the filter function commands given above. Bandpass digital filters of order 2N are designed by using the above functions if Wn is a two-element vector:

```
Wn = [w1 w2]
```

with a passband defined by w1 $< \omega <$ w2. For designing highpass digital filters, one uses the string *'high'* as the last argument for the filter function, e.g.,

```
[b,a] = cheby1(N,Rp,Wn,'high')
```

Similarly, the string *'stop'* is included as a final argument for designing bandstop digital filters, e.g.,

```
[b,a] = ellip(N,Rp,Rs,Wn,'stop')
```

The following two examples illustrate the design of a highpass and a bandpass digital filter, respectively.

Example 7.20 Determine the transfer function and plot the gain response of a 4th-order Type 1 Chebyshev IIR highpass filter with a passband edge at 0.7 and a passband ripple of 1 dB.

Program 7_4 below provides the MATLAB program that can be used to design such a filter. During execution the program asks for the following filter parameters to be typed in: filter order N, passband ripple Rp, and passband edge Wn.

```
% Program 7_4
% Type 1 Chebyshev IIR Highpass Filter Design
%
N = input('Type in the order = ');
Rp = input('Type in the passband ripple = ');
Wn = input('Type in the passband edge frequency = ');
[b,a] = cheby1(N,Rp,Wn,'high');
disp('Numerator polynomial');
disp(b)
disp('Denominator polynomial');
disp(a)
w = 0:0.01/pi:pi;
h = freqz(b,a,w);
gain = 20*log10(abs(h));
plot (w/pi,gain);grid;
xlabel('Normalized frequency');
ylabel('Gain, dB');
```

The output of the above program is as follows:

```
Numerator polynomial
  Columns 1 through 4
```

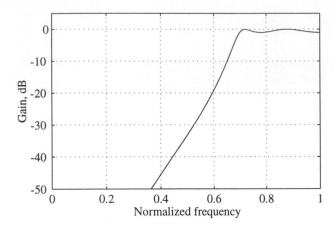

Figure 7.34   Gain response of the IIR Type 1 Chebyshev highpass filter of Example 7.20.

```
0.00836323955556   -0.03345295822222
0.05017943733333   -0.03345295822222

Column 5

0.00836323955556

Denominator polynomial
  Columns 1 through 4

  1.00000000000000    2.37412317472661
  2.70565666020506    1.59170922154748

Column 5

0.41031508197432
```

The desired transfer function is thus given by

$$G(z) = \frac{\begin{array}{c}0.008363239 - 0.033452958z^{-1} + 0.05017944z^{-2} \\ -0.03345296z^{-3} + 0.00833239z^{-4}\end{array}}{\begin{array}{c}1 + 2.37412317z^{-1} + 2.7056566z^{-2} + 1.5917092z^{-3} \\ +0.41031508z^{-4}\end{array}}.$$

and the corresponding gain response is as shown in Figure 7.34.

Example 7.21 Determine the transfer function and plot the gain response of an 8th-order Butterworth IIR bandpass filter with passband edges at 0.4 and 0.7, respectively.

The MATLAB program for designing such a filter is given in Program 7_5. As the program is executed it asks for the following filter parameters to be typed in: filter order N and lower and upper passband edges W1 and W2.

```
% Program 7_5
% Design of IIR Butterworth Bandpass Filter
%
format long
N = input('Type in the filter order = ');
M = N/2;
W1 = input('Lower passband edge frequency = ');
W2 = input('Upper passband edge frequency = ');
Wn = [W1 W2];
[b,a] = butter(M,Wn);
disp('Numerator polynomial');
disp(b)
disp('Denominator polynomial');
disp(a)
w = 0:0.01/pi:pi;
h = freqz(b,a,w);
gain = 20*log10(abs(h));
plot (w/pi,gain);grid;
xlabel('Normalized frequency');
ylabel('Gain, dB');
```

The output data consist of the numerator and the denominator coefficients of the transfer function in ascending powers of z^{-1}:

```
Numerator polynomial
  Columns 1 through 4

   0.01856301062690   0.00000000000000
  -0.07425204250759   0.00000000000000

  Columns 5 through 8

   0.11137806376138   0.00000000000000
  -0.07425204250759   0.00000000000000

  Column 9

   0.01856301062690

Denominator polynomial
  Columns 1 through 4

   1.00000000000000   0.97799773704953
   1.93989229067322   1.33864067564165

  Columns 5 through 8

   1.62710138513352   0.73488127769299
   0.58261200429173   0.13855877212735
```

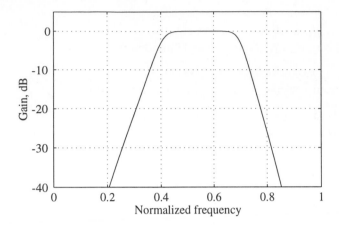

Figure 7.35   Gain response of the IIR Butterworth bandpass filter of Example 7.21.

```
Column 9
  0.07619706461033
```

The computed gain response is plotted in Figure 7.35.

For the design of higher order filters, the functions computing the zeros and poles of the transfer function are more accurate than the functions computing the transfer function coefficients. Moreover, numerical accuracy could be a problem in designing filters of order 15 or above [Kra94].

The *Signal Processing Toolbox* of MATLAB also includes functions for estimating the order of the four types of digital filters designed via the bilinear transformation approach. These functions are

```
[N,Wn] = buttord(Wp,Ws,Rp,Rs)
[N,Wn] = cheb1ord(Wp,Ws,Rp,Rs)
[N,Wn] = cheb2ord(Wp,Ws,Rp,Rs)
[N,Wn] = ellipord(Wp,Ws,Rp,Rs)
```

For lowpass filters, Wp and Ws are the normalized passband and stopband edge frequencies with Wp < Ws, respectively. These frequency points must be between 0 and 1, where the sampling frequency is equal to 2. The other two parameters, Rp and Rs, are the passband ripple and the minimum stopband attenuation specified in dB, respectively. The outputs of these functions are the filter order N and the frequency scaling factor Wn. The latter parameter is needed in the MATLAB filter design functions, butter, cheby1, cheby2, and ellip, to meet the desired filter response specifications.

For highpass filters Wp > Ws. For bandpass and bandstop digital filters, Wp and Ws are vectors of length-2 specifying the transition bandedges, with the lower frequency edge being the first element of the vectors. Here, Wn is also a length 2 vector and is used as input to the MATLAB functions for IIR filter design.

The use of the order estimating function is illustrated in the next example.

Example 7.22 Determine the minimum order of the transfer function of a Type 2 Chebyshev digital highpass filter operating at a sampling rate of 4 kHz with the following specifications: passband edge at 1000 Hz, stopband edge at 600 Hz, passband ripple of 1 dB, and minimum stopband attenuation of 40 dB.

To this end we can use Program 7.6 given below.

```
% Program 7_6
% Estimation of the Order of a Type 2 Chebyshev Filter
%
FT = input('Type in the sampling frequency = ');
Fp = input('Type in the passband edge frequency = ');
Fs = input('Type in the stopband edge frequency = ');
Rp = input('Type in the passband ripple = ');
Rs = input('Type in the stopband ripple = ');
Wp = Fp*2/FT; Ws = Fs*2/FT;
[N,Wn] = cheb2ord(Wp,Ws,Rp,Rs);
fprintf('Order of filter is %f \n',N);
```

The input data called by the program are the sampling frequency FT in Hz, passband edge frequency Fp in Hz, stopband edge frequency Fs in Hz, passband ripple Rp in dB, and stopband ripple Rs in dB. The output data is the order N of the IIR digital filter given by

```
        Order of filter is 5.000000
```

For designing IIR filters using the impulse invariance approach, MATLAB offers the function impinvar. There are two versions of this function:

```
[b,a] = impinvar(num,den,FT),[b,a] = impinvar(num,den)
```

These functions determine the IIR digital transfer function $G(z)$ whose impulse response coefficients are equal to the sampled values of the impulse response of the analog transfer function $H_a(s)$ when sampled at a frequency FT Hz. The input data are the coefficients of the numerator and the denominator polynomials of $H_a(s)$ entered as vectors num and den. The output files are the vectors b and a of the numerator and denominator polynomials, respectively, of $G(z)$. In the second version, where FT is not specified, a default value of FT = 1 Hz is used.

In the following example, we illustrate the use of the above function in designing IIR digital filters via the impulse invariance approach.

Example 7.23 We consider the design of an IIR lowpass filter of Section 7.5 using the impulse invariance method as discussed in Section 7.5.1. To this end, we can employ the MATLAB program given below, which makes use of the function impinvar.

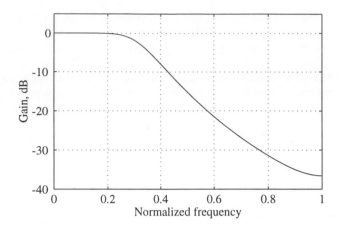

Figure 7.36   Gain response of the IIR Butterworth lowpass filter of Example 7.23.

```
% Program 7_7
% Design of Butterworth Lowpass Filter Using the
% Impulse Invariance Method
N = input('Type in filter order = ');
Wn = input('Type in 3-dB cutoff frequency = ');
[num,den] = butter(N, Wn, 's');
FT = input('Type in sampling frequency = ');
[b,a] = impinvar(num,den,FT);
[h,omega] = freqz(b,a,512);
mag = 20*log10(abs(h));
plot(omega/pi,mag);grid;
xlabel('Normalized Frequency');
ylabel('Gain, dB');
```

As the program is run, it asks for the order N of the analog Butterworth filter, its 3-dB cutoff frequency Wn, and the value of the sampling frequency FT. The gain response of the digital filter designed by this program for N = 4, Wn = 1.021612, and FT = 1 is as indicated in Figure 7.36. Note that this gain plot is identical to that shown in Figure 7.10(b).

7.9.3   FIR Digital Filter Design Using MATLAB

MATLAB offers linear-phase FIR filter design programs based on both the Parks-McClellan algorithm (Section 7.9.1) and the windowed Fourier series method (Section 7.7). FIR filters based on the former method can be designed using the function remez. There are various versions of this function:

```
b = remez(N,f,m)
b = remez(N,f,m,wt)
```

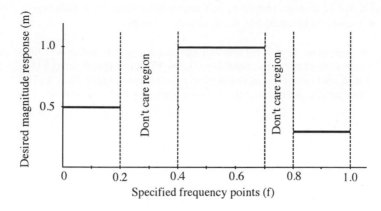

Figure 7.37   Illustration of relationship between vectors f and m.

```
b = remez(N,f,m,'ftype')
b = remez(N,f,m,wt,'ftype')
```

It can design any type of multiband linear-phase filters. Its output is a vector b of length N+1 containing the impulse response coefficients of the linear-phase FIR filter with bandedges specified by the vector f. The frequency points of the vector f must be in the range between 0 and 1, with sampling frequency being equal to 2, and must be specified in an increasing order. The first element of f must be 0 and the last element must be 1. The bandedges between a passband and its adjacent stopband must be separated by at least 0.1. If same values are used for these frequency points, the program automatically separates them by 0.1. The desired magnitudes of the FIR filter frequency response at the specified bandedges are given by the vector m, with the elements given in equal-valued pairs. The desired magnitudes between two specified consecutive frequency points f(k) and f(k+1) are determined according to the following rules. For k odd, the magnitude is a line segment joining the points {m(k), f(k)} and {m(k+1), f(k+1)}, whereas for k even, it is unspecified, with the frequency range [f(k), f(k+1)] being a *transition* or *"don't care"* region. The vectors f and m must be of same length with the length being even. Figure 7.37 illustrates the relationship between the vectors f and m given by

```
f = [0     0.2    0.4    0.7    0.8    1.0],
m = [0.5   0.5    1.0    1.0    0.3    0.3].
```

The desired magnitude responses in the passband(s) and the stopband(s) can be weighted by an additional vector wt included as the argument of the function remez. The function can design equiripple Types 1, 2, 3, and 4 linear-phase FIR filters. Types 1 and 2 are the default designs for order N even and odd, respectively. Types 3 (N even) and 4 (N odd) are used for specialized filter designs, Hilbert transformer and differentiator. To design these two types of FIR filters the flags 'hilbert' and 'differentiator' are used for 'ftype' in the last two versions of remez.

Incorrect use of `remez` results in self-explanatory diagnostic messages. The use of the function `remez` is illustrated in the following example.

Example 7.24 Determine the transfer function and plot the gain response of a linear-phase FIR bandpass filter of order 27 with a passband from 0.3 to 0.5 and stopbands from 0 to 0.25 and from 0.55 to 1.0. The desired passband and stopband magnitudes are, respectively, 1 and 0.

The program for designing an equiripple linear-phase FIR filter is as follows:

```
% Program 7_8
% Design of Equiripple Linear-Phase FIR Filters
%
format long
N = input('Type in the filter order = ');
f = input('Type in the band edges = ');
m = input('Type in the magnitude values = ');
b = remez(N,f,m);
disp('FIR Filter Coefficients');
disp(b)
[h,w] = freqz(b,1,256);
mag = 20*log10(abs(h));
plot(w/pi,mag);grid;
xlabel('Normalized Frequency');
ylabel('Gain, dB')
```

During execution, the program asks for the following data to be typed in: filter order, specified bandedges of passband(s) and stopband(s), and desired magnitude values at these bandedges. Note that since the set of bandedges and the set of magnitude values are vectors, the corresponding data must be entered within square brackets.

The output data generated by this program are given below.

```
       FIR Filter Coefficients
         Columns 1 through 4

      0.00115007769608    0.07967706858212
      0.01194683012627   -0.04117777843979


         Columns 5 through 8

     -0.02976148300198   -0.00006846550026
     -0.01796059325940   -0.02007716296898


         Columns 9 through 12

      0.07315361857354    0.10629839147908
     -0.05696315694865   -0.21362025959366


         Columns 13 through 16

     -0.07543778032552    0.19780526636535
      0.19780526636535   -0.07543778032552
```

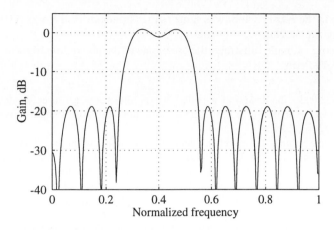

Figure 7.38   Gain response of the FIR equiripple bandpass filter of Example 7.24.

```
Columns 17 through 20

-0.21362025959366   -0.05696315694865
 0.10629839147908    0.07315361857354

Columns 21 through 24

-0.02007716296898   -0.01796059325940
-0.00006846550026   -0.02976148300198

Columns 25 through 28

-0.04117777843979    0.01194683012627
 0.07967706858212    0.00115007769608
```

In the above list, the first number is the filter coefficient $h[0]$, the second is the filter coefficient $h[1]$, and so on. As can be seen from above, the filter coefficients satisfy the symmetry constraint $h[k] = h[N - k]$, as expected. The computed gain response is plotted in Figure 7.38.

For designing linear-phase FIR filters based on the windowed Fourier series approach, MATLAB includes the following functions for generating the windows discussed earlier in Section 7.7:

```
w = blackman(N),    w = hamming(N),     w = hanning(N)
w = chebwin(N,Rs),  w = kaiser (N,beta)
```

The above functions generate a vector w of window coefficients of odd length N.[20] The

---

[20]It should be noted that the Hann and Blackman windows generated by MATLAB have zero-valued first and last coefficients. As a result, to keep the length the same for all windows, these two windows should be generated using a value of N that is greater by 2 than the value of N used for the Hamming and the Kaiser windows.

parameter *beta* in the Kaiser window is same as the parameter β of Eq. (7.110) and can be computed using Eq. (7.112).

The following example illustrates the use of one of the above functions in determining the window coefficients.

Example 7.25 Consider the design of an FIR lowpass filter using a Kaiser window. The filter specifications are $\omega_p = 0.3\pi$, $\omega_s = 0.4\pi$, and $\alpha_s = 50$ dB. The cutoff frequency is thus given by $\omega_c = (\omega_p + \omega_s)/2 = 0.35\pi$.

Substituting the value of the minimum stopband attenuation α_s in Eq. (7.4), we arrive at the value of the peak stopband ripple $\delta_s = 0.003162$. We next calculate the value of the parameter β by substituting in Eq. (7.112), which yields $\beta = 4.533514$. The transition bandwidth is given by $\Delta f = (\omega_s - \omega_p)/2\pi = 0.05$. Using the values of Δf and α_s in Eq. (7.113), we find the estimated value of the filter length to be $N = 59.5655$. The next higher odd integer value of 61 is then chosen as the filter length.

We use the following MATLAB program to determine the window coefficients.

```
% Program 7_9
% Kaiser Window Generation Program
%
N = input('Type in the window length = ');
beta = input('Type in the value of beta = ');
w = kaiser(N,beta);
disp('Window Coefficients'); disp(w)
[h,omega] = freqz(w,1,256);
mag = 20*log10(abs(h));
plot(omega/pi,mag);grid;
xlabel('Normalized Frequency');
ylabel('Gain, dB);
```

The program requests the window length N to be typed in followed by the value of β. It displays the window coefficients generated by the program and plots the gain response of the window as indicated in Figure 7.39.

Two additional functions available in MATLAB aid in the design of FIR filters using the windowed Fourier series approach. These are `fir1` and `fir2`.

The function `fir1` can be used to design conventional lowpass, highpass, bandpass, and bandstop FIR filters. The various forms of this function are as follows:

```
b = fir1(L,Wn)
b = fir1(L,Wn,'filtertype')
b = fir1(L,Wn,window)
b = fir1(L,Wn,'filtertype',window)
```

The basic form, `b = fir1(L,Wn)`, generates the length `N = L+1` vector `b` containing the impulse-response coefficients of a lowpass FIR filter of order `L` with a normalized cutoff frequency of `Wn` between 0 and 1. In the design of highpass filters, the function `b = fir1(L,Wn,'high')` should be employed, with filter order `L` an even number. The function `b = fir1(L,Wn)` is used for the design of bandpass filters with `Wn` being a length-2 vector containing the passband edge frequencies. And for the design of bandstop

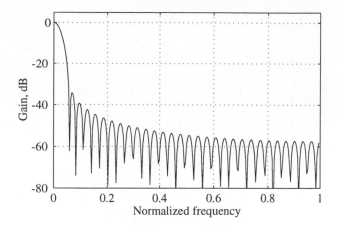

Figure 7.39 Gain response of the Kaiser window generated in Example 7.25.

filters, the function b = fir1(L,Wn,'stop') can be employed, with Wn being a length-2 vector containing the stopband edge frequencies and the filter order L being an even number. The last two forms are utilized if a specific window is to be used for the filter design with the window coefficients entered through the vector window. The Hamming window for filter design is used if no window is specified in the argument of fir1.

The function fir2 is employed to design FIR filters with arbitrarily shaped magnitude response. The various forms of this function are as follows:

```
b = fir2(L,f,m)
b = fir2(L,f,m,window)
```

As in the previous case, the basic form b = fir2(L,f,m) is used to design a lowpass FIR filter of order L whose magnitude response samples at the specified frequency points given by the vector f match the specified magnitude response samples defined by the vector m. The output vector b is of length N = L+1 and contains the FIR filter coefficients ordered in ascending powers of z^{-1}. The vectors f and m must be of the same length. Elements of f must be in the range from 0 to 1, where 1 corresponds to one-half of normalized sampling frequency. Moreover, the first and last element of f must be 0 and 1, respectively. Duplicate frequency points in f are permissible for approximating magnitude response specifications with jumps. Any one of the window functions available in the MATLAB *Signal Processing Toolbox* can be used for designing FIR filters by making use of its corresponding *M*-function to generate the window coefficients and entering them into the function fir2 by means of the length L+1 vector window in the argument. As before, if the window coefficients are not explicitly included, the Hamming window is used as default.

We illustrate the use of the above two functions in FIR filter design in the next two examples.

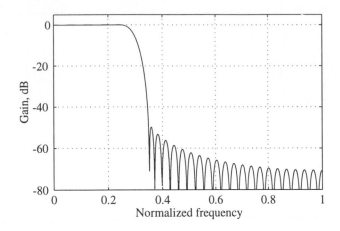

Figure 7.40   Gain response of the lowpass filter of Example 7.26.

Example 7.26 We continue here with the remaining steps in the design of the lowpass FIR filter of Example 7.25 using the Kaiser window. The desired specifications for the filter are $\omega_p = 0.3\pi$, $\omega_s = 0.4\pi$, and $A_s = 50$ dB. From Example 7.25, we observe $\beta =$ 4.533514 and N = 2M + 1 = 61. The MATLAB program used for this filter design problem is indicated below.

```
% Program 7_10
% Design of Lowpass Filter Using Kaiser Window
% Length = 61, Beta = 4.533514
format long
kw = kaiser(61,4.533514);
b = fir1(60,0.3,kw);
[h,omega] = freqz(b,1,512);
mag = 20*log10(abs(h));
plot(omega/pi,mag);grid;
xlabel('Normalized Frequency');
ylabel('Gain, dB');
```

The gain response of the resulting lowpass filter is sketched in Figure 7.40. The filter coefficients generated by the above program are contained in the vector b and can be printed if desired.

   In the following example we consider the application of the function fir2 of MATLAB in designing an FIR filter with passbands having different gain levels.

Example 7.27 Consider the design of an FIR filter of order 100 with three different constant magnitude levels: 0.3 in the frequency range 0 to 0.28, 1.0 in the frequency range 0.3 to 0.5, and 0.7 in the frequency range 0.52 to 1.0. To this end we use the following MATLAB program:

```
% Program 7_11
% Design of Multiband Filter Using Hamming Window
fpts = [0 0.28 0.3 0.5 0.52 1];
```

Figure 7.41   Magnitude response of the multilevel filter of Example 7.27.

```
mval = [0.3 0.3 1.0 1.0 0.7 0.7];
b = fir2(100,fpts,mval);
[h,omega] = freqz(b,1,512);
plot(omega/pi,abs(h));grid;
xlabel('Normalized Frequency');
ylabel('Magnitude');
```

The filter coefficients are contained in the vector b. The resulting magnitude response is as indicated in Figure 7.41.

7.10   Summary

This chapter considered the design of both infinite impulse-response (IIR) and finite impulse-response (FIR) causal digital filters. The digital filter design problem is concerned with the development of a suitable transfer function meeting the frequency response specifications, which in this chapter, is restricted to magnitude (or, equivalently gain) response specifications. These specifications are usually given in terms of the desired passband edge and stopband edge frequencies and the allowable deviations from the desired passband and stopband magnitude (gain) levels.

IIR filter design is usually carried out by transforming a prototype analog transfer function by means of a suitable mapping of the complex frequency variable s to the complex variable z. The impulse invariance method and the bilinear transform method, discussed in this chapter, are based on this approach. Of these two, the latter is less restricted and is more widely used than the former.

FIR filter design, on the other hand, is carried out directly from the digital filter specifications. Two such design methods are outlined. The first method outlined is based on the truncated Fourier series expansion of the desired frequency response. Formulas for computing the Fourier series coefficients of some ideal frequency response specifications are included. To reduce the effect of the Gibbs phenomenon, the truncation can be carried out by applying a suitable window function. Some commonly used window functions are reviewed along with their properties. The effect of the Gibbs phenomenon can also be reduced by providing a smooth transition between the passband and the stopband. The second method is based on uniformly sampling the desired frequency response and obtaining the impulse-response coefficients by an inverse DFT of the frequency samples.

Finally, the chapter considers the computer-aided design of digital filters. To this end it primarily discusses design algorithms that are available in the *Signal Processing Toolbox* of MATLAB as functions. For the design of IIR digital filters, MATLAB provides functions for designing digital Butterworth, Chebyshev, and elliptic filters. For the design of FIR filters, it includes functions for the windowed Fourier series approach and the Parks-McClellan algorithm. The latter develops a transfer function with an equiripple passband and stopband magnitude responses and makes use of the Remez optimization algorithm.

7.11   Problems

7.1 Let $H(z)$ be the transfer function of a lowpass digital filter with a passband edge at ω_p, stopband edge at ω_s, a passband ripple of δ_p, and a stopband ripple of δ_s, as indicated in Figure 7.1. Consider a cascade of two identical filters with a transfer function $H(z)$. What are the passband and stopband ripples of the cascade at ω_p and ω_s, respectively? Generalize the results for a cascade of M identical sections.

7.2 Let $H_{\mathrm{LP}}(z)$ denote the transfer function of a real coefficient lowpass filter with a passband edge at ω_p, stopband edge at ω_s, passband ripple of δ_p, and stopband ripple of δ_s as indicated in Figure 7.1. Sketch the magnitude response of the highpass transfer function $H_{\mathrm{LP}}(-z)$ for $-\pi \leq \omega \leq \pi$ and determine its passband and stopband edges in terms of ω_p and ω_s.

7.3 Consider the transfer function $G(z) = H_{\mathrm{LP}}(e^{j\omega_0}z)$, where $H_{\mathrm{LP}}(z)$ is the lowpass transfer function of Problem 7.2. Sktech its magnitude response for $-\pi \leq \omega \leq \pi$, and determine its passband and stopband edge frequencies in terms of ω_p, ω_s, and ω_0.

7.4 Show that the digital transfer function $G(z)$ of Eq. (7.34) is obtained by an impulse invariance transformation of Eq. (7.33) by making use of the relation between Eqs. (7.32) and (7.29).

7.5 Show that the digital transfer function $G(z)$ of Eq. (7.36) is obtained by an impulse invariance transformation of Eq. (7.35) by making use of the relation between Eqs. (7.32) and (7.29).

7.6 Show that the digital transfer function $G(z)$ obtained from an arbitrary rational analog transfer function $H_a(s)$ with simple poles via the impulse invariance method is given by

$$G(z) = \sum_{\substack{\text{all poles} \\ \text{of } H_a(s)}} \text{Residues} \left[\frac{H_a(s)}{1 - e^{sT}z^{-1}} \right]. \tag{7.157}$$

7.7 Verify the relation between Eqs. (7.32) and (7.29) using the above formula.

7.8 Develop a digital transfer function $G(z)$ from the causal analog transfer function

$$H_a(s) = \frac{3s^3 + 7s^2 + 10s + 7}{(s^2 + s + 1)(s^2 + 2s + 3)} \tag{7.158}$$

using the impulse-invariance method. Choose $T = 0.1$ sec.

7.9 Develop a digital transfer function $G(z)$ from the causal analog transfer function

$$H_a(s) = \frac{4s^2 + 10s + 8}{(s^2 + 2s + 3)(s + 1)} \tag{7.159}$$

using the impulse-invariance method. Choose $T = 0.2$ sec.

7.10 This problem illustrates how aliasing can be suitably exploited in order to realize interesting frequency response characteristics. An ideal causal analog lowpass filter with an impulse response $h_a(t)$ has a frequency response given by

$$H_a(j\Omega) = \begin{cases} 1, & |\Omega| < \Omega_c, \\ 0, & \text{otherwise.} \end{cases}$$

Let $H_1(e^{j\omega})$ and $H_2(e^{j\omega})$ be the frequency responses of digital filters obtained by sampling $h_a(t)$ at $t = nT$, where $T = 3\pi/2\Omega_c$ and π/Ω_c, respectively. Assume the transfer functions are later normalized so that $H_1(e^{j0}) = H_2(e^{j0}) = 1$.

(a) Sketch the frequency responses $G_1(e^{j\omega})$ and $G_2(e^{j\omega})$ of the two digital filter structures shown in Figure P7.1.

(b) What type of filters are $G_1(z)$ and $G_2(z)$ (lowpass, highpass, etc.)?

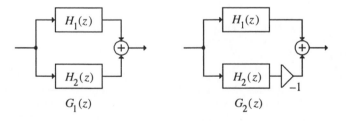

Figure P7.1

7.11 This problem illustrates the method of digital filter design by *step-response invariance method* [Hay70], [Whi71]. Let $H_a(s)$ be a real coefficient causal and stable analog transfer function. Its unit step response $h_{a,\mu}(t)$ is given by the inverse Laplace transform of $H_a(s)/s$. Let $G(z)$ be a causal digital transfer function with a unit step response $g_\mu[n]$ such that

$$g_\mu[n] = h_{a,\mu}(nT), \qquad n = 0, 1, 2, \ldots \tag{7.160}$$

Determine the expression for $G(z)$ if

$$H_a(s) = \sum_{k=0}^{R} \frac{A_k}{s - \alpha_k}, \tag{7.161}$$

and show that if $\omega \ll 1/T$ and

$$\frac{H_a(j\omega)}{j\omega} \approx 0 \qquad \text{for } |\omega| \geq \frac{1}{2T}, \tag{7.162}$$

then

$$G(e^{j\omega}) \cong H_a(j\omega) \qquad \text{for } |\omega| < \frac{1}{2T} \tag{7.163}$$

7.12 Let $H_a(s)$ be a real coefficient causal and stable analog transfer function with a magnitude response bounded above by unity. Show that the digital transfer function $G(z)$ obtained by a bilinear transformation of $H_a(s)$ is a BR function.

7.13 Verify the expression for the impulse-response coefficients $h_{ML}[n]$ given in Eq. (7.93) for the zero-phase multiband filter with a frequency response $H_{ML}(e^{j\omega})$ defined in Eq. (7.92) and shown in Figure 7.15.

7.14 Show that the ideal Hilbert transformer with a frequency response $H_{HT}(^{j\omega})$ defined in Eq. (7.94) has an impulse response $h_{HT}[n]$ as given in Eq. (7.95). Since the impulse response is doubly infinite, the ideal discrete-time Hilbert transformer is not realizable. To make it realizable, the impulse response has to be truncated to $|n| \leq M$. What type of linear-phase FIR filter is the truncated impulse response? Plot the frequency response of the truncated approximation for various values of M. Comment on your results.

7.15 Let $\mathcal{H}\{\cdot\}$ denote the ideal operation of Hilbert transformation defined by

$$\mathcal{H}\{x[n]\} = \sum_{\ell=-\infty}^{\infty} h_{HT}[n - \ell]x[\ell],$$

where $h_{HT}[n]$ as given in Eq. (7.95). Evaluate the following quantities:

(a) $\mathcal{H}\{\mathcal{H}\{\mathcal{H}\{\mathcal{H}\{x[n]\}\}\}\}$,   (b) $\displaystyle\sum_{\ell=-\infty}^{\infty} x[\ell]\mathcal{H}\{x[\ell]\}$.

7.16 Show that the ideal differentiator with a frequency response $H_{DIF}(e^{j\omega})$ defined in Eq. (7.96) has an impulse response $h_{DIF}[n]$ as given in Eq. (7.97). Since the impulse response is doubly-infinite, the ideal discrete-time differentiator is not realizable. To make it realizable, the impulse response has to be truncated to $|n| \leq M$. What type of linear-phase FIR filter is the truncated impulse response? Plot the frequency response of the truncated approximation for various values of M. Comment on your results.

7.17 Develop the expression for the impulse response $\hat{h}_{HP}[n]$ of a causal highpass FIR filter of length $N = 2M + 1$ obtained by truncating and shifting the impulse response $h_{HP}[n]$ of the ideal highpass filter given by Eq. (7.89). Show that the causal lowpass FIR filter $\hat{h}_{LP}[n]$ of Eq. (7.88) and $\hat{h}_{HP}[n]$ are delay-complementary pair.

7.18 Determine the impulse response $h_{LLP}[n]$ of a zero-phase ideal linear passband lowpass filter characterized by a frequency response shown in Figure P7.2(a).

7.19 Determine the impulse response $h_{BLDIF}[n]$ of a zero-phase ideal bandlimited differentiator characterized by a frequency response shown in Figure P7.2(b).

$$\text{Figure P7.2}$$

7.20 By computing the inverse discrete-time Fourier transform of the frequency response $H_{LP}(e^{j\omega})$ of the zero-phase modified lowpass filter of Figure 7.26(a) with a first-order spline as the transition function, verify the expression for its impulse response $h_{LP}[n]$ as given in Eq. (7.114). Show that $h_{LP}[n]$ of Eq. (7.114) can also be derived by computing the inverse discrete-time Fourier transform of the derivative function $G(e^{j\omega})$ of Figure 7.26(b) and then using the differentiation-in-frequency property of the discrete-time Fourier transform given in Table 3.2.

7.21 Show that the impulse response $h_{LP}[n]$ of the zero-phase modified lowpass filter with a Pth-order spline as the transition function is given by Eq. (7.115).

7.22 The frequency response of a zero-phase lowpass filter with a passband edge at ω_p, a stopband edge at ω_s, and a raised cosine transition function is given by [Bur92], [Par87].

$$H_{LP}(e^{j\omega}) = \begin{cases} 1, & 0 \le |\omega| < \omega_p, \\ \frac{1}{2}\left(1 + \cos\left(\frac{\pi(\omega-\omega_p)}{\omega_s-\omega_p}\right)\right), & \omega_p \le |\omega| \le \omega_s, \\ 0, & \omega_s < |\omega| < \pi. \end{cases} \quad (7.164)$$

Show that its impulse response is of the form

$$h_{LP}[n] = \left[\frac{\cos(\Delta\omega n/2)}{1-(\Delta\omega/\pi)^2 n^2}\right] \cdot \frac{\sin(\omega_c n)}{\pi n}, \quad (7.165)$$

where $\Delta\omega = \omega_s - \omega_p$ and $\omega_c = (\omega_p + \omega_s)/2$.

7.23 Show that the length-$(2M + 1)$ *Bartlett window* sequence given by

$$w[n] = 1 - \frac{|n|}{M+1}, \quad -M \le n \le M, \quad (7.166)$$

can be obtained by a linear convolution of two scaled length-N rectangular windows. Determine N and the scale factor. From this relation, determine the expression for the frequency response of the length-$(2M + 1)$ Bartlett window. Determine the main lobe width Δ_{ML} and the relative sidelobe level $A_{s\ell}$ of the Bartlett window sequence.

7.24 The length-$(2M+1)$ Hann, Hamming and Blackman window sequences given in Eqs. (7.102) to (7.104) are all of the form of raised cosine windows and can be expressed as

$$w_{GC}[n] = \left[\alpha + \beta \cos\left(\frac{2\pi n}{2M+1}\right) + \gamma \cos\left(\frac{4\pi n}{2M+1}\right)\right] w_R[n], \quad (7.167)$$

where $w_R[n]$ is a length-$(2M + 1)$ rectangular window sequence. Express the Fourier transform of the above generalized cosine window in terms of the Fourier transform of the rectangular window $\Psi_R(e^{j\omega})$. From this expression, determine the Fourier transform of the Hann, Hamming and Blackman window sequences.

7.25 Many applications require the fitting of a set of $2L + 1$ equally spaced data samples $x[n]$ by a smooth polynomial $x_a(t)$ of degree N where $N < 2L$. In the least-squares fitting approach the polynomial coefficients α_i, $i = 0, 1, \ldots < N$, are determined so that the mean-square error

$$\varepsilon(\alpha_i) = \sum_{k=-L}^{L} \{x[k] - x_a(k)\}^2 \qquad (7.168)$$

is a minimum [Ham89]. In smoothing a very long data sequence $x[n]$ based on the least-squares fitting approach, the central sample in a set of consecutive $2L + 1$ data samples is replaced by the polynomial coefficient minimizing the corresponding mean-square error.

(a) Develop the smoothing algorithm for $N = 1$ and $L = 5$, and show that it is a moving average FIR filter of length 5,

(b) Develop the smoothing algorithm for $N = 2$ and $L = 5$. What type of digital filter is represented by this algorithm?

(c) By comparing the frequency responses of the previous two FIR smoothing filters, select the filter that provides better smoothing.

7.26 An improved smoothing algorithm is *Spencer's* 15-point smoothing formula given by [Ham89]

$$y[n] = \frac{1}{320} \{-3x[n-7] - 6x[n-6] - 5x[n-5] + 3x[n-4]$$
$$+21x[n-3] + 46x[n-2] + 67x[n-1] + 74x[n]$$
$$+67x[n+1] + 46x[n+2] + 21x[n+3] + 3x[n+4]$$
$$-5x[n+5] - 6x[n+6] - 3x[n+7]\}. \qquad (7.169)$$

Evaluate its frequency response and, comparing it with that of the two smoothing filters of Problem 7.25, show why Spencer's formula yields the better result.

7.27 In Problem 7.1, we considered filtering by a cascade of a number of identical filters. While the cascade provides more stopband attenuation than that obtained by a single filter section, it also increases the passband ripple or in effect decreases the passband width for a given maximum passband deivation. In the case of an FIR filter $H(z)$ with a symmetric impulse response, improved passband and stopband performances can be achieved by employing the *filter sharpening approach* [Kai77] in which the overall system $G(z)$ is implemented as

$$G(z) = \sum_{\ell=1}^{L} \alpha_\ell [H(z)]^\ell, \qquad (7.170)$$

where $\{\alpha_\ell\}$ are real constants. In this problem, we outline the method of selecting the weighting coefficients $\{\alpha_\ell\}$ for a given L. It follows from above that $G(z)$ is also an FIR filter with a symmetric impulse response. Let x denote a specific value of the amplitude response of $H(z)$ at a given angular frequency ω. If we denote the value of the amplitude response of $G(z)$ at this value of ω as $P(x)$, then it is related to x through

$$P(x) = \sum_{\ell=1}^{L} \alpha_\ell x^\ell. \qquad (7.171)$$

$P(x)$ is called the *amplitude change function*. For a BR $H(z)$, $0 \le x \le 1$, where $x = 0$ is in the stopband and $x = 1$ is in the passband. If we further desire $G(z)$ to be BR, then the amplitude change function must satisfy the two basic properties $P(0) = 0$ and $P(1) = 1$. Additional conditions on the amplitide change function are obtained by constraining the behavior of its slope at $x = 0$ and $x = 1$. To improve the performance of $G(z)$ in the stopband, we need to ensure

$$\left. \frac{d^k P(x)}{dx^k} \right|_{x=0} = 0, \quad k = 1, 2, \ldots, n, \tag{7.172}$$

and to improve the performance of $G(z)$ in the passband, we need to ensure

$$\left. \frac{d^k P(x)}{dx^k} \right|_{x=1} = 0, \quad k = 1, 2, \ldots, m, \tag{7.173}$$

where $m + n = L - 1$. Determine the coefficients $\{\alpha_\ell\}$ for $L = 3, 4$, and 5.

7.28 In this problem, we consider the design of computationally efficient linear-phase narrowband FIR filters using the *interpolated finite-impulse response* (IFIR) approach [Neu84b]. Let $H(z)$ be a lowpass FIR transfer function with a passband edge at ω_p and a stopband edge at ω_s, as shown in Figure P7.3(a) The magnitude response of the transfer function $H(z^M)$ is then as shown in P7.3(b), where M is a positive integer satisfying the condition $M\omega_s \leq \pi$. If we cascade $H(z^M)$ with another lowpass filter $F(z)$ with passband edge at ω_p and stopband edge at $(2\pi - \omega_s)/M$, as shown in Figure P7.3(c), the cascade $G(z) = H(z^M)F(z)$ is a sharp cutoff lowpass filter as indicated in Figure P7.3(d). By choosing $F(z)$ appropriately, any one of the M passbands of $H(z^M)$ can be retained and the remaining $(M - 1)$ passbands attenuated resulting in either a narrowband lowpass or highpass or bandpass filter.

By reversing the above process, we can develop the specifications of $H(z)$ and $F(z)$ from the desired specifications of the overall narrowband filter $G(z)$ obtained by the cascade realization. Since the gain in dB of a cascade is the sum of the gains in dB of the individual sections, the passband ripples of $H(z)$ and $F(z)$ can be made one-half of the desired passband ripple δ_p of the narrowband filter $G(z)$. On the other hand, the stopband ripples of $H(z)$ and $F(z)$ can be made equal to the desired passband ripple δ_s of the narrow-band filter $G(z)$ for simplicity. The computational complexity of the cascade $G(z)$ is thus equal to that of $H(z)$ together with that of $F(z)$.

Using the above approach, design a linear-phase lowpass FIR filter $G(z)$ in the form of $H(z^M)F(z)$, where $H(z)$ and $F(z)$ are linear-phase FIR filters. The specifications for $G(z)$ are as follows: $\omega_p = 0.15\pi$, $\omega_s = 0.2\pi$, $\delta_p = 0.002$, $\delta_s = 0.001$. Choose the largest possible value for M. Compare the computational complexities of $G(z)$ designed as a single stage and designed as a cascade $H(z^M)F(z)$.

7.29 Let $\left| H_d(e^{j\omega}) \right|$ denote the desired magnitude response of a real linear-phase FIR filter of length N.

(a) For N odd (Type 1 FIR filter), show that the DFT samples $H[k]$ needed for a frequency sampling based design are given by

$$H[k] = \begin{cases} \left| H_d(e^{j2\pi k/N}) \right| e^{-j(2\pi/N)k[(N-1)/2]}, & k = 0, 1, \ldots, \frac{N-1}{2}, \\ \left| H_d(e^{j2\pi k/N}) \right| e^{j(2\pi/N)(N-k)[(N-1)/2]}, & k = \frac{N+1}{2}, \ldots, N - 1. \end{cases} \tag{7.174}$$

(b) Develop the corresponding expressions for N even (Type 2 FIR filter).

(c) Show that the values of the frequency samples used in Example 7.16 satisfy Eq. (7.174).

7.30 By solving Eq. (7.145), derive the value of ε given by Eq. (7.146).

(a)

(b)

(c)

(d)

Figure P7.3

7.12 MATLAB Exercises

M 7.1 Design a digital filter by an impulse invariant transformation of a fourth-order analog Bessel transfer function for the following values of sampling frequencies: (a) $F_T = 1.25$ Hz, and (b) $F_T = 2.5$ Hz. Plot the gain and the group delay responses of both designs using MATLAB, and compare these responses with that of the original Bessel transfer function. Comment on your results.

M 7.2 Design a digital Butterworth lowpass filter operating at sampling rate of 44.1 kHz with a 0.5-dB cutoff frequency at 2 kHz and a minimum stopband attenuation of 50 dB at 10 kHz using the impulse invariance method and the bilinear transformation method. Assume the sampling interval for the impulse invariance design to be equal to 1. Determine the order of the analog filter prototype using the formula given in Eq. (5.34) and then design the analog prototype filter using the function `buttap` of MATLAB. Plot the gain and phase responses of both designs using MATLAB. Compare the performances of the two filters. Show all steps used in the design. Does the sampling interval has any effect on the design of the digital filter design based on the impulse invariance method?

M 7.3 Modify Program 7_5 to design a digital Butterworth lowpass filter using the bilinear transformation method. The input data required by the modified program should be the desired passband and stopband edges, and maximum passband deviation and the minimum stopband attenuation in dBs. Using the modified program, design the digital Butterworth lowpass filter of Problem M7.2.

M 7.4 Using Program 7_7 design the digital Butterworth lowpass filter of Problem M7.2. Use the analog prototype filter order determined using the formula given in Eq. (5.34).

M 7.5 Design a digital Type 1 Chebyshev lowpass filter operating at a sampling rate of 44.1 kHz with a passband edge frequency at 2 kHz, a passband ripple of 0.5 dB, and a minimum stopband attenuation of 50 dB at 10 kHz using the impulse invariance method and the bilinear transformation method. Determine the order of the analog filter prototype using the formula given in Eq. (5.39) and then design the analog prototype filter using the function `cheb1ap` of MATLAB. Plot the gain and phase responses of both designs using MATLAB. Compare the performances of the two filters. Show all steps used in the design.

M 7.6 Modify Program 7_4 to design a digital Type 1 Chebyshev lowpass filter using the bilinear transformation method. The input data required by the modified program should be the desired passband and stopband edges, and maximum passband deviation and the minimum stopband attenuation in dBs. Using the modified program, design the digital Type 1 Chebyshev lowpass filter of Problem M7.5.

M 7.7 Write a MATLAB program to design a digital Type 1 Chebyshev lowpass filter using the impulse invariance method. The input data required by the modified program should be the desired passband and stopband edges, and maximum passband deviation and the minimum stopband attenuation in dBs. Using your program, design the digital Type 1 Chebyshev lowpass filter of Problem M7.5.

M 7.8 Design a digital elliptic lowpass filter operating at sampling rate of 44.1 kHz with a passband edge frequency at 2 kHz, a stopband edge frequency at 10 kHz, passband ripple of 0.5 dB, and a stopband ripple of 50 dB using the impulse invariance method and the bilinear transformation method. Determine the order of the analog filter prototype using the formula given in Eq. (5.49) and then design the analog prototype filter using the function `ellipap` of MATLAB. Plot the gain and phase responses of both designs using MATLAB. Compare the performances of the two filters. Show all steps used in the design.

M 7.9 Modify Program 7_3 to design a digital elliptic lowpass filter using the bilinear transformation method. The input data required by the modified program should be the desired passband and stopband edges, and maximum passband deviation and the minimum stopband attenuation in dBs. Using the modified program, design the digital elliptic lowpass filter of Problem M7.8.

M 7.10 Plot the magnitude response of a linear-phase FIR highpass filter by truncating the impulse response $h_{HP}[n]$ of the ideal highpass filter of Eq. (7.89) to length $N = 2M + 1$ for two different values of M and show that the truncated filter exhbits oscillatory behavior on both sides of the cutoff frequency.

M 7.11 Plot the magnitude response of a linear-phase FIR bandpass filter by truncating the impulse response $h_{BP}[n]$ of the ideal bandpass filter of Eq. (7.90) to length $N = 2M + 1$ for two different values of M and show that the truncated filter exhbits oscillatory behavior on both sides of the cutoff frequency.

M 7.12 Plot the magnitude response of a linear-phase FIR Hilbert transformer by truncating the impulse response $h_{HT}[n]$ of the ideal Hilbert transformer of Eq. (7.95) to length $N = 2M + 1$ for two different values of M and show that the truncated filter exhbits oscillatory behavior near $\omega = 0$ and $\omega = \pi$.

M 7.13 Determine a quadratic approximation $a_0 x + a_1 x + a_2 x^2$ to the cubic function $D(x) = 1.1x^3 - 9x^2 + 1$ defined for the range $-1 \le x \le 6$ by minimizing the peak value of the absolute error $\left| D(x) - a_0 - a_1 x - a_2 x^2 \right|$, i.e.,

$$\max_{-1 \le x \le 6} \left| D(x) - a_0 - a_1 x - a_1 x^2 \right|,$$

using the Remez algorithm. Plot the error function after convergence of the algorithm.

M 7.14 Design using the windowed Fourier series approach a linear-phase FIR lowpass filter with the following specifications: passband edge at 1 rad/s, stopband edge at 2 rad/s, maximum passband attenuation of 0.1 dB, minimum stopband attenuation of 45 dB, and a sampling frequency of 10 rad/s. Use each of the following windows for the design: Hamming, Hann, and Blackman. Show the impulse-response coefficients and plot the gain response of the designed filters for each case. Comment on your results. Do not use the function `firl` of MATLAB.

M 7.15 Repeat Problem M7.14 using the Kaiser window. Do not use the function `firl` of MATLAB.

M 7.16 Design a linear-phase highpass FIR filter of length-40 with a passband edge at $\omega_p = 0.6\pi$ using the frequency sampling approach. Show the impulse response coefficients and plot the magnitude response of the designed filter using MATLAB.

M 7.17 Design a linear-phase bandpass FIR filter of length-45 with a passband edges at $\omega_{p1} = 0.3\pi$ and $\omega_{p2} = 0.5\pi$ using the frequency sampling approach. Show the impulse response coefficients and plot the magnitude response of the designed filter using MATLAB.

M 7.18 Using the frequency sampling approach redesign the length-37 linear-phase lowpass filter of Example 7.16 by including a transition band with one frequency sample of magnitude 1/2. Plot the magnitude response of the new filter using MATLAB and compare it with that shown in Figure 7.29.

M 7.19 Repeat Problem M7.18 by including a transition band with two frequency samples of magnitude 2/3 and 1/3, respectively.

M 7.20 Design the linear-phase FIR lowpass filter of Problem M7.14 using the function `fir1` of MATLAB. Use each of the following windows for the design: Hamming, Hann, Blackman and Kaiser. Show the impulse response coefficients and plot the gain response of the designed filters for each case. Compare your results with those obtained in Problems M7.14 and M7.15.

M 7.21 Using the function `fir1` of MATLAB design a linear-phase FIR highpass filter with the following specifications: stopband edge at 0.5π, passband edge at 0.65π, maximum passband attenuation of 0.2 dB, and a minimum stopband attenuation of 50 dB. Use each of the following windows for the design: Hamming, Hann, Blackman and Kaiser. Show the impulse response coefficients and plot the gain response of the designed filters for each case.

M 7.22 Using the function `fir1` of MATLAB design a linear-phase FIR bandpass filter with the following specifications: stopband edges at 0.4π and 0.85π, passband edges at 0.5π and 0.7π, maximum passband attenuation of 0.15 dB, and a minimum stopband attenuation of 40 dB. Use each of the following windows for the design: Hamming, Hann, Blackman and Kaiser. Show the impulse response coefficients and plot the gain response of the designed filters for each case.

M 7.23 Design a two-channel crossover FIR lowpass and highpass filter pair for digital audio applications. The lowpass and the highpass filters are of length 31 and have a crossover frequency of 2.5 kHz operating at a sampling rate of 44.1 kHz. Use the function `fir1` with a Hamming window to design the lowpass filter while the highpass filter is derived from the lowpass filter using the delay-complementary property. Plot the gain responses of both filters on the same figure. What is the minimum number of delays and multipliers needed to implement the crossover network?

M 7.24 Design a three-channel crossover FIR filter system for digital audio applications. All filters are of length 35 and operate at a sampling rate of 44.1 kHz. The two crossover frequencies are at 2.3 kHz and 6.2 kHz, respectively. Use the function `fir1` with a Hann window to design the lowpass and the highpass filters while the bandpass filter is derived from the lowpass and highpass filters using the delay-complementary property. Plot the gain responses of all filters on the same figure. What is the minimum number of delays and multipliers needed to implement the crossover network?

M 7.25 The function `fir2` of MATLAB is employed to design FIR filters with arbitrarily shaped magnitude response. Using this function design an FIR filter of order 90 with three different constant magnitude levels: 0.6 in the frequency range 0 to 0.5, 0.2 in the frequency range 0.52 to 0.79, and 1.0 in the frequency range 0.81 to 1.0. Plot the gain response of the designed filter.

M 7.26 Design the linear-phase FIR lowpass filter of Problem M7.14 using the `remez` function of MATLAB and plot its magnitude response.

M 7.27 Design the linear-phase FIR highpass filter of Problem M7.21 using the `remez` function of MATLAB and plot its magnitude response.

M 7.28 Design the linear-phase FIR bandpass filter of Problem M7.22 using the `remez` function of MATLAB and plot its magnitude response.

M 7.29 Design a length 27 discrete-time FIR differentiator using the `remez` function of MATLAB and plot its magnitude response.

M 7.30 Design a 26th-order FIR Hilbert transformer using the `remez` function of MATLAB. The passband is from 0.1π to 0.9π. The two stopbands are from 0.01π to 0.08π, and from 0.92π to π. Plot its magnitude response.

CHAPTER 8

DSP Algorithm Implementation Considerations

There are basically two types of digital signal processing (DSP) algorithms: filtering algorithms and signal analysis algorithms. These algorithms can be based on either difference equations, both recursive and nonrecursive, or DFT and can be implemented in any one of the following forms: hardware, firmware, and software. In the hardware approach, the algorithm may be implemented using digital circuitry, such as the shift register to provide the delaying operation, the digital multiplier, and the digital adder. Alternatively, a special purpose VLSI chip may be designed and fabricated to implement a specific filtering algorithm. In the firmware approach, the algorithm is implemented on a read-only-memory (ROM) chip. Additional control circuitry, and storage registers, are usually needed in the final hardware or firmware realization. Finally, in the software approach, the algorithm is implemented as a computer program on a general purpose computer such as a workstation, a minicomputer, a personal computer, or a programmable DSP chip. This chapter is concerned with the implementation aspects of DSP algorithms. We first examine the two major issues concerning all the above types of approaches to implementation. We then discuss the software implementation of digital filtering and DFT algorithms on a computer using MATLAB to illustrate the main points. It is followed by a review of various schemes for the representation of number and signal variables on a digital machine. The number representation scheme is basic to the development of methods for the analysis of finite word-length effects considered in the following chapter. Next, we review algorithms that are employed to implement addition and multiplication, the two key arithmetic operations in digital signal processing. We then briefly review operations developed to handle overflow. Finally, two general methods for the design and implementation of tunable digital filters are outlined, followed by a discussion of algorithms for the approximation of certain special functions. A detailed discussion of hardware, firmware, and DSP chip implementations is beyond the scope of this book. Information on programming the DSP chips can be found in the books and application notes published by the manufacturers of these chips. Discussion on selected DSP chips can also be found in the following books and chapters [Cha90], [Fad93], [Had91], [Ife93], [Lin87], [Pap90], [Bab95], [Mar92].

8.1   Basic Issues

We examine first two specific problems that may be encountered before a digital filter is actually implemented. The first problem is concerned with the computability of the equations describing the structure, and the second problem is concerned with the verification of the structure developed to realize a prescribed transfer function.

8.1.1   Matrix Representation of the Digital Filter Structure

As indicated in Chapter 6, a digital filter can be described in the time-domain by a set of equations relating the output sequence to the input sequence, and in some cases, one or more internally generated sequences. The ordering of these equations in computing the output samples is important, as discussed next.

Consider the digital filter structure of Figure 8.1. We can describe this structure by the following set of equations relating the signal variables $W_k(z)$, the output $Y(z)$, and the input $X(z)$:

$$W_1(z) = X(z) - \alpha W_5(z), \tag{8.1a}$$
$$W_2(z) = W_1(z) - \delta W_3(z), \tag{8.1b}$$
$$W_3(z) = z^{-1} W_2(z), \tag{8.1c}$$
$$W_4(z) = W_3(z) + \varepsilon W_2(z), \tag{8.1d}$$
$$W_5(z) = z^{-1} W_4(z), \tag{8.1e}$$
$$Y(z) = \beta W_1(z) + \gamma W_5(z). \tag{8.1f}$$

In the time-domain, the above set of equations is equivalent to

$$w_1[n] = x[n] - \alpha w_5[n], \tag{8.2a}$$
$$w_2[n] = w_1[n] - \delta w_3[n], \tag{8.2b}$$
$$w_3[n] = w_2[n-1], \tag{8.2c}$$
$$w_4[n] = w_3[n] + \varepsilon w_2[n], \tag{8.2d}$$
$$w_5[n] = w_4[n-1], \tag{8.2e}$$
$$y[n] = \beta w_1[n] + \gamma w_5[n]. \tag{8.2f}$$

The above set of equations do not describe a valid computational algorithm since the equations cannot be implemented in the order shown with each variable on the left side computed before the variable below is computed. For example, computation of $w_1[n]$ in the first step requires the knowledge of $w_5[n]$ that is computed in the fifth step. Likewise, the computation of $w_2[n]$ in the second step requires the knowledge of $w_3[n]$ that is computed in the following step. We call the ordered set of equations of Eqs. (8.2a)–(8.2f) as *noncomputable*.

Suppose we reorder the above equations and write them in the form

$$w_3[n] = w_2[n-1], \tag{8.3a}$$
$$w_5[n] = w_4[n-1], \tag{8.3b}$$
$$w_1[n] = x[n] - \alpha w_5[n], \tag{8.3c}$$

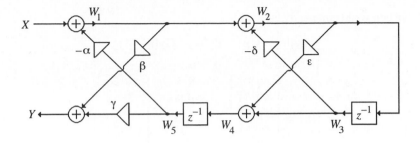

Figure 8.1 A cascaded lattice digital filter structure.

$$w_2[n] = w_1[n] - \delta w_3[n], \tag{8.3d}$$

$$y[n] = \beta w_1[n] + \gamma w_5[n], \tag{8.3e}$$

$$w_4[n] = w_3[n] + \varepsilon w_2[n]. \tag{8.3f}$$

It can be seen that the above ordered set of equations now describe a valid computational algorithm since the equations can be implemented in the sequential order shown, with each variable on the left side computed before the variable below is computed.

In most practical applications, the equations characterizing the digital filter can be put into a computable order by inspection. It is, however, instructive to examine the computability of the equations describing a digital filter in a more formal fashion, which is described next [Cro75].

To this end, we write the equations of the digital filter in matrix form. Thus, a matrix representation of Eqs. (8.2a)–(8.2f) is given by

$$\begin{bmatrix} w_1[n] \\ w_2[n] \\ w_3[n] \\ w_4[n] \\ w_5[n] \\ y[n] \end{bmatrix} = \begin{bmatrix} x[n] \\ 0 \\ 0 \\ 0 \\ 0 \\ 0 \end{bmatrix} + \begin{bmatrix} 0 & 0 & 0 & 0 & -\alpha & 0 \\ 1 & 0 & -\delta & 0 & 0 & 0 \\ 0 & 0 & 0 & 0 & 0 & 0 \\ 0 & \varepsilon & 1 & 0 & 0 & 0 \\ 0 & 0 & 0 & 0 & 0 & 0 \\ \beta & 0 & 0 & 0 & \gamma & 0 \end{bmatrix} \begin{bmatrix} w_1[n] \\ w_2[n] \\ w_3[n] \\ w_4[n] \\ w_5[n] \\ y[n] \end{bmatrix}$$

$$+ \begin{bmatrix} 0 & 0 & 0 & 0 & 0 & 0 \\ 0 & 0 & 0 & 0 & 0 & 0 \\ 0 & 1 & 0 & 0 & 0 & 0 \\ 0 & 0 & 0 & 0 & 0 & 0 \\ 0 & 0 & 0 & 1 & 0 & 0 \\ 0 & 0 & 0 & 0 & 0 & 0 \end{bmatrix} \begin{bmatrix} w_1[n-1] \\ w_2[n-1] \\ w_3[n-1] \\ w_4[n-1] \\ w_5[n-1] \\ y[n-1] \end{bmatrix}, \tag{8.4}$$

which we can write compactly as

$$\mathbf{y}[n] = \mathbf{x}[n] + \mathbf{F}\mathbf{y}[n] + \mathbf{G}\mathbf{y}[n-1] \tag{8.5}$$

where

$$\mathbf{y}[n] = \begin{bmatrix} w_1[n] \\ w_2[n] \\ w_3[n] \\ w_4[n] \\ w_5[n] \\ y[n] \end{bmatrix}, \qquad \mathbf{x}[n] = \begin{bmatrix} x[n] \\ 0 \\ 0 \\ 0 \\ 0 \\ 0 \end{bmatrix}, \tag{8.6a}$$

$$\mathbf{F} = \begin{bmatrix} 0 & 0 & 0 & 0 & -\alpha & 0 \\ 1 & 0 & -\delta & 0 & 0 & 0 \\ 0 & 0 & 0 & 0 & 0 & 0 \\ 0 & \varepsilon & 1 & 0 & 0 & 0 \\ 0 & 0 & 0 & 0 & 0 & 0 \\ \beta & 0 & 0 & 0 & \gamma & 0 \end{bmatrix}, \qquad \mathbf{G} = \begin{bmatrix} 0 & 0 & 0 & 0 & 0 & 0 \\ 0 & 0 & 0 & 0 & 0 & 0 \\ 0 & 1 & 0 & 0 & 0 & 0 \\ 0 & 0 & 0 & 0 & 0 & 0 \\ 0 & 0 & 0 & 1 & 0 & 0 \\ 0 & 0 & 0 & 0 & 0 & 0 \end{bmatrix}. \tag{8.6b}$$

If we examine Eq. (8.4), we observe that, for the computation of the present value of a particular signal variable, the nonzero entries in the corresponding rows of the matrices **F** and **G** determine the variables whose present and previous values are needed. If the diagonal element in **F** is nonzero, then the computation of the present value, of the corresponding variable requires the knowledge of its present value indicating the presence of a delay-free loop making the structure totally noncomputable. Any nonzero entries in the same row above the diagonal of **F** imply that the computation of the present value of the corresponding variable requires the present values of other variables that have not yet been computed, thus making the set of equations noncomputable.

In the **F** matrix of this example, the diagonal elements are all zeros, indicating that there are no delay-free loops in the structure. However, there are non-zero entries above the diagonal in the first and second rows of **F**, indicating that the set of equations of Eqs. (8.2a)–(8.2f) is not in proper order for computation.

On the other hand, the matrix representation of Eqs. (8.3a)–(8.3f) results in

$$\begin{bmatrix} w_3[n] \\ w_5[n] \\ w_1[n] \\ w_2[n] \\ y[n] \\ w_4[n] \end{bmatrix} = \begin{bmatrix} 0 \\ 0 \\ x[n] \\ 0 \\ 0 \\ 0 \end{bmatrix} + \begin{bmatrix} 0 & 0 & 0 & 0 & 0 & 0 \\ 0 & 0 & 0 & 0 & 0 & 0 \\ 0 & -\alpha & 0 & 0 & 0 & 0 \\ -\delta & 0 & 1 & 0 & 0 & 0 \\ 0 & \gamma & \beta & 0 & 0 & 0 \\ 1 & 0 & 0 & \varepsilon & 0 & 0 \end{bmatrix} \begin{bmatrix} w_3[n] \\ w_5[n] \\ w_1[n] \\ w_2[n] \\ y[n] \\ w_4[n] \end{bmatrix}$$

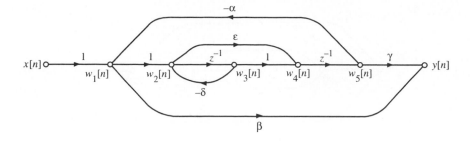

Figure 8.2   Signal flow-graph representation of the digital filter structure of Figure 8.1.

$$+ \begin{bmatrix} 0 & 0 & 0 & 1 & 0 & 0 \\ 0 & 0 & 0 & 0 & 0 & 1 \\ 0 & 0 & 0 & 0 & 0 & 0 \\ 0 & 0 & 0 & 0 & 0 & 0 \\ 0 & 0 & 0 & 0 & 0 & 0 \\ 0 & 0 & 0 & 0 & 0 & 0 \end{bmatrix} \begin{bmatrix} w_3[n-1] \\ w_5[n-1] \\ w_1[n-1] \\ w_2[n-1] \\ y[n-1] \\ w_4[n-1] \end{bmatrix}, \tag{8.7}$$

for which the $\mathbf{F}$ matrix can be seen to satisfy the computability condition, and thus, the equations describing the filter are in proper order.

8.1.2   Precedence Graph

We now describe a simple algorithm for testing the computability of a set digital filter structure and for developing the proper ordering sequence for a set of equations describing a computable structure. To this end, we form a signal flow-graph representation of the digital filter structure with each delay and multiplier element represented as single branches whose gain is either a unit delay z^{-1} or its multiplier coefficient. Accordingly, the signal flow-graph representation of the structure of Figure 8.1 is as shown in Figure 8.2.

As the output of the delay branches can always be computed at any instant since they are the delayed values of their respective input signals computed at the previous instant, we remove all delay branches from the complete signal flow-graph of the digital filter structure. Similarly, all branches coming out of the input node are also removed since the input variables are always available at each instant. For our example, the resulting reduced signal flow-graph is as shown in Figure 8.3.

We now group the remaining nodes in the reduced signal flow-graph as follows. All nodes with only outgoing branches are grouped into one set labeled $\{\mathcal{N}_1\}$. Next, we form the set $\{\mathcal{N}_2\}$ containing nodes that have branches coming in from one or more of the nodes in the set $\{\mathcal{N}_1\}$ and have outgoing branches to the other nodes. We then form a set $\{\mathcal{N}_3\}$ containing nodes that have branches coming in from one or more of the nodes in the sets $\{\mathcal{N}_1\}$ and $\{\mathcal{N}_2\}$ and have outgoing branches to the other nodes. This process is continued until we have a set $\{\mathcal{N}_f\}$ containing only nodes with only incoming branches.

Since the signal variables belonging to the set $\{\mathcal{N}_1\}$ do not depend on the present values of the other signal variables, these variables should be computed first. Next, the signal variables belonging to the set $\{\mathcal{N}_2\}$ can be computed since they depend on the present values of the signal variables contained in the set $\{\mathcal{N}_1\}$ that have already been computed.

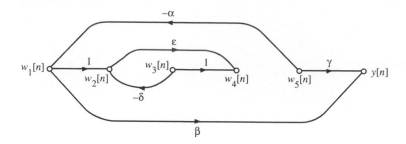

Figure 8.3   Reduced signal flow-graph obtained by removing the branches going out of the input node and the delay branches from the signal flow-graph of Figure 8.2.

This is followed by the computation of the signal variables in the sets $\{\mathcal{N}_3\}$, $\{\mathcal{N}_4\}$, etc. Finally, in the last step, the signal variables belonging to the set $\{\mathcal{N}_f\}$ are computed. This process of sequential computation ensures the development of a valid computational algorithm. However, if there is no final set $\{\mathcal{N}_f\}$ containing only incoming branches, the signal flow-graph is noncomputable. The rearranged signal flow-graph without the delay branches and with nodes grouped as indicated above is called a *precedence graph* [Cro75].

For our example, the pertinent groupings of node variables according to their precedence relations are as follows:

$$\{\mathcal{N}_1\} = \{w_3[n], \ w_5[n]\},$$
$$\{\mathcal{N}_2\} = \{w_1[n]\},$$
$$\{\mathcal{N}_3\} = \{w_2[n]\},$$
$$\{\mathcal{N}_4\} = \{w_4[n], \ y[n]\}.$$

The precedence graph redrawn according to the above groupings is indicated in Figure 8.4. Since the final node set $\{\mathcal{N}_4\}$ has only incoming nodes, the structure of Figure 8.1 has no delay-free loops. Therefore, for our example structure, we can compute first the signal variables $w_3[n]$ and $w_5[n]$ in any order, then compute the signal variable $w_1[n]$, followed by the computation of the signal variable $w_2[n]$, and finally, compute the signal variables $w_4[n]$ and $y[n]$ in any order to arrive at a valid computational algorithm.

8.1.3   Structure Verification

An important step that needs to be considered in the hardware or software implementation of a digital transfer function is to ensure that no computational and/or other errors have taken place during the course of the realization process and that the structure obtained is indeed characterized by the prescribed transfer function $H(z)$. A simple technique to verify the structure is outlined next [Mit77a].

Without any loss of generality, consider a causal LTI digital filter structure characterized by a fourth-order transfer function

$$H(z) = \frac{P(z)}{D(z)} = \frac{p_0 + p_1 z^{-1} + p_2 z^{-2} + p_3 z^{-3} + p_4 z^{-4}}{1 + d_1 z^{-1} + d_2 z^{-2} + d_3 z^{-3} + d_4 z^{-4}}. \tag{8.8}$$

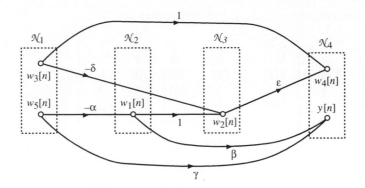

Figure 8.4   Precedence graph of Figure 8.3 redrawn with signal variables grouped according to their precedence relations.

If $\{h[n]\}$ denotes its unit sample response, then

$$H(z) = \sum_{n=0}^{\infty} h[n]z^{-n}. \tag{8.9}$$

From Eqs. (8.8) and (8.9) it follows that

$$P(z) = H(z)D(z), \tag{8.10}$$

or equivalently, in the time-domain by the convolution sum,

$$p_n = h[n] \circledast d_n. \tag{8.11}$$

Writing out Eq. (8.11) explicitly for $n = 0, 1, 2, \ldots, 9$, we obtain

$$
\begin{aligned}
p_0 &= h[0], \\
p_1 &= h[1] + h[0]d_1, \\
p_2 &= h[2] + h[1]d_1 + h[0]d_2, \\
p_3 &= h[3] + h[2]d_1 + h[1]d_2 + h[0]d_3, \\
p_4 &= h[4] + h[3]d_1 + h[2]d_2 + h[1]d_3 + h[0]d_4, \\
0 &= h[5] + h[4]d_1 + h[3]d_2 + h[2]d_3 + h[1]d_4, \\
0 &= h[6] + h[5]d_1 + h[4]d_2 + h[3]d_3 + h[2]d_4, \\
0 &= h[7] + h[6]d_1 + h[5]d_2 + h[4]d_3 + h[3]d_4, \\
0 &= h[8] + h[7]d_1 + h[6]d_2 + h[5]d_3 + h[4]d_4.
\end{aligned}
$$

In matrix form the above equations can be rewritten as

$$
\begin{bmatrix} p_0 \\ p_1 \\ p_2 \\ p_3 \\ p_4 \\ 0 \\ 0 \\ 0 \\ 0 \end{bmatrix} =
\begin{bmatrix}
h[0] & 0 & 0 & 0 & 0 \\
h[1] & h[0] & 0 & 0 & 0 \\
h[2] & h[1] & h[0] & 0 & 0 \\
h[3] & h[2] & h[1] & h[0] & 0 \\
h[4] & h[3] & h[2] & h[1] & h[0] \\
\cdots & \cdots & \cdots & \cdots & \cdots & \cdots \\
h[5] & \vdots & h[4] & h[3] & h[2] & h[1] \\
h[6] & \vdots & h[5] & h[4] & h[3] & h[2] \\
h[7] & \vdots & h[6] & h[5] & h[4] & h[3] \\
h[8] & \vdots & h[7] & h[6] & h[5] & h[4]
\end{bmatrix}
\begin{bmatrix} 1 \\ d_1 \\ d_2 \\ d_3 \\ d_4 \end{bmatrix}.
\tag{8.12}
$$

In partitioned form, Eq. (8.12) can be reexpressed as

$$
\begin{bmatrix} \mathbf{p} \\ \cdots \\ \mathbf{0} \end{bmatrix} =
\begin{bmatrix} & \mathbf{H}_1 & \\ \cdots & \cdots & \cdots \\ \mathbf{h} & \vdots & \mathbf{H}_2 \end{bmatrix}
\begin{bmatrix} 1 \\ \cdots \\ \mathbf{d} \end{bmatrix}
\tag{8.13}
$$

where

$$
\mathbf{p} = \begin{bmatrix} p_0 \\ p_1 \\ p_2 \\ p_3 \\ p_4 \end{bmatrix}, \quad
\mathbf{d} = \begin{bmatrix} d_1 \\ d_2 \\ d_3 \\ d_4 \end{bmatrix}, \quad
\mathbf{0} = \begin{bmatrix} 0 \\ 0 \\ 0 \\ 0 \end{bmatrix},
\tag{8.14a}
$$

$$
\mathbf{H}_1 = \begin{bmatrix}
h[0] & 0 & 0 & 0 & 0 \\
h[1] & h[0] & 0 & 0 & 0 \\
h[2] & h[1] & h[0] & 0 & 0 \\
h[3] & h[2] & h[1] & h[0] & 0 \\
h[4] & h[3] & h[2] & h[1] & h[0]
\end{bmatrix}, \quad
\mathbf{h} = \begin{bmatrix} h[5] \\ h[6] \\ h[7] \\ h[8] \end{bmatrix},
$$

$$
\mathbf{H}_2 = \begin{bmatrix}
h[4] & h[3] & h[2] & h[1] \\
h[5] & h[4] & h[3] & h[2] \\
h[6] & h[5] & h[4] & h[3] \\
h[7] & h[6] & h[5] & h[4]
\end{bmatrix}.
\tag{8.14b}
$$

Eq. (8.12) therefore can be written as two matrix equations:

$$
\mathbf{p} = \mathbf{H}_1 \begin{bmatrix} 1 \\ \mathbf{d} \end{bmatrix},
\tag{8.15}
$$

$$
\mathbf{0} = [\mathbf{h} \quad \mathbf{H}_2] \begin{bmatrix} 1 \\ \mathbf{d} \end{bmatrix}.
\tag{8.16}
$$

Solving Eq. (8.16), we obtain the vector $\mathbf{d}$ composed of the denominator coefficients:

$$
\mathbf{d} = -\mathbf{H}_2^{-1}\mathbf{h}.
\tag{8.17}
$$

Substituting Eq. (8.17) in Eq. (8.15), we arrive at the vector $\mathbf{p}$ containing the numeraor coefficients:

$$\mathbf{p} = \mathbf{H}_1 \begin{bmatrix} 1 \\ -\mathbf{H}_2^{-1}\mathbf{h} \end{bmatrix}. \tag{8.18}$$

We illustrate the above approach for the reconstruction of the transfer function of a causal IIR filter from its impulse response coefficients in the following example.

Example 8.1 Consider the causal IIR transfer function

$$H(z) = \frac{2 + 6z^{-1} + 3z^{-2}}{1 + z^{-1} + 2z^{-2}}.$$

Using long division, we can determine the first five terms in the series expansion of $H(z)$:

$$H(z) = 2 + 4z^{-1} - 5z^{-2} - 3z^{-3} + 13z^{-4} + \ldots.$$

Therefore, here,

$$h[0] = 2, \quad h[1] = 4, \quad h[2] = -5, \quad h[3] = -3, \quad h[4] = 13.$$

The equation corresponding to Eq. (8.12) here is given by

$$\begin{bmatrix} p_0 \\ p_1 \\ p_2 \\ 0 \\ 0 \end{bmatrix} = \begin{bmatrix} 2 & 0 & 0 \\ 4 & 2 & 0 \\ -5 & 4 & 2 \\ -3 & -5 & 4 \\ 13 & -3 & -5 \end{bmatrix} \begin{bmatrix} 1 \\ d_1 \\ d_2 \end{bmatrix}.$$

Therefore, from Eq. (8.17),

$$\begin{bmatrix} d_1 \\ d_2 \end{bmatrix} = -\begin{bmatrix} -5 & 4 \\ -3 & -5 \end{bmatrix}^{-1} \begin{bmatrix} -3 \\ 13 \end{bmatrix} = -\frac{1}{37}\begin{bmatrix} -5 & -4 \\ 3 & -5 \end{bmatrix}\begin{bmatrix} -3 \\ 13 \end{bmatrix} = \begin{bmatrix} 1 \\ 2 \end{bmatrix}.$$

Next, from Eq. (8.18),

$$\begin{bmatrix} p_0 \\ p_1 \\ p_2 \end{bmatrix} = \begin{bmatrix} 2 & 0 & 0 \\ 4 & 2 & 0 \\ -5 & 4 & 2 \end{bmatrix}\begin{bmatrix} 1 \\ 1 \\ 2 \end{bmatrix} = \begin{bmatrix} 2 \\ 6 \\ 3 \end{bmatrix}.$$

This gives us a straightforward means of finding the transfer function of any discrete-time structure knowing the first $2M + 1$ samples of $\{h[n]\}$, where M is the order of the transfer function $H(z)$. This approach to structure verification is illustrated later in Examples 8.8 and 8.9. It can also be used to determine the effect of coefficient quantization by computing the transfer function realized with the multiplier coefficients quantized to the desired number of bits. This actual transfer function is then used to compute the frequency response, to determine the actual pole locations, etc. Another application of the above method is in the determination of noise transfer functions for computing the output noise power due to product roundoffs in fixed-point digital filter implementations considered in Section 9.5, and in the determination of scaling transfer functions needed for dynamic range scaling discussed in Section 9.6. In most cases, we can assume that the denominator coefficients are known, and therefore, the solution of Eq. (8.17) that requires a matrix inversion can be avoided. The numerator coefficients are easily found from the first $M + 1$ samples of $\{h[n]\}$ and using Eq. (8.18).

Figure 8.5 Dirrect form IIR structure.

8.2 Software Implementation

As indicated earlier, we concentrate in this book only on software implementation of DSP algorithms. In this section, we consider only the implementation of digital filtering algorithms. The following section is devoted to the implementation of discrete Fourier transform algorithms.

The software implementation of a digital filtering algorithm on a general purpose computer is often carried out first to verify that the algorithm chosen does indeed meet the goals of the application on hand before the algorithm is implemented in a hardware form. Moreover, such an implementation is adequate if the application under consideration does not require real-time signal processing. In either situation, the first issue to be resolved is the selection of the programming language. To this end, the three most preferred languages are BASIC, FORTRAN and C, with the last one being used in the majority of cases.

For simulation purposes, it is quite convenient to make use of a number of powerful software packages that have been developed specifically for signal processing. We illustrate in this section the use of one such package, the MATLAB. It includes two functions, `filter` and `filtfilt`. The basic forms of these functions are as follows:

```
y = filter(num,den,x)
[y,sf] = filter(num,den,x,si)
y = filtfilt(num,den,x)
```

The numerator and the denominator coefficients are contained in the vectors `num` and `den`, respectively. These vectors do not have to be same size. The input vector is `x` while the output vector generated by the filtering algorithm is `y`. The function `filter` basically implements the digital filter in the direct form II_t structure shown in Figure 8.5 for a third-order structure.[1] As indicated in this figure, `d(1)` has been assumed to be equal to 1. If `d(1)` $\neq$ 1, the program automatically normalizes all filter coefficients in `p` and `d` to make `d(1)` = 1.[2]

As indicated in Figure 8.5, the function `filter` simulates the filtering operation in the time-domain in accordance with the following state-variable representation of the digital filter:

---

[1] See Section 6.5.1 for the development of IIR direct form structures.

[2] It should be noted that in Figure 8.5 we have used the MATLAB notations for vector elements instead of the notations used elsewhere in the text for representing filter coefficient and signal variables.

```
s3(n+1) = p(4)x(n) - d(4)y(n),
s2(n+1) = p(3)x(n) - d(3)y(n) + s3(n),
s1(n+1) = p(2)x(n) - d(2)y(n) + s2(n),
   y(n) = p(1)x(n) + s1(n).
```

In the second form of the function filter, the initial conditions of the delay (state) variables, sk(n), $k = 1, 2, \ldots$, can be specified through the argument si. Moreover, the function filter can return the final values of the delay (state) variables through the output vector sf. The size of the initial (final) condition vector si (sf) is one less than the maximum of the sizes of the filter coefficient vectors num and den. The final values of the state variables given as vector sf are useful if the input vector to be processed is very long and needs to be segmented into small blocks of data for processing in stages. In such a situation, after the ith block of input data has been processed, the final state vector sf is fed as the initial state vector si for the processing of the $(i + 1)$th block of input data, and so on.

The following example illustrate the application of the M-function filter in generating the impulse response coefficients of a causal IIR digital filter.

Example 8.2 Determine and plot the first 25 samples of the impulse response sequence, from $n = 0$ to $n = 24$, of the IIR lowpass digital filter described by the transfer function of Eq. (7.67) repeated below in a form suitable for direct form implementation:

$$G(z) = \frac{0.0662272 + 0.1986816z^{-1} + 0.1986816z^{-2} + 0.0662272z^{-3}}{1 - 0.93561z^{-1} + 0.567127z^{-2} - 0.101591z^{-3}}. \tag{8.19}$$

The MATLAB program given below was used to compute the impulse response.

```
% Program 8_1
% Impulse Response Computation
%
n = input('Impulse response length desired = ');
num = input('Numerator coefficients = ');
den = input('Denominator coefficients = ');
x = [1 zeros(1,n-1)];
order = max(length(num),length(den))-1;
si = [zeros(1,order)];
y = filter(num,den,x,si);
stem(n-1,y)
xlabel('Time index n'); ylabel('Amplitude');
title('Impulse response samples')
```

During execution, the program first requests the length of the impulse response to be computed. The program then requests the numerator and denominator vectors to be typed in. These vectors must be entered using the square brackets. After execution, it plots the impulse response as shown in Figure 8.6 for the above IIR filter.

The digital filtering application of the M-function filter is considered in the next five examples.

Figure 8.6   Impulse response samples of the IIR digital filter of Eq. (8.19).

Example 8.3 We illustrate in this example the filtering of a signal composed of the sum of two sinusoids of normalized angular frequencies, 0.1π and 0.8π using the lowpass IIR digital filter of Eq. (8.19). MATLAB Program 8_2 given below was utilized for this purpose.

```
% Program 8_2
% Illustration of Filtering by a Lowpass IIR Filter
%
% Generate the input sequence
k = 1:51;
w1 = 0.8*pi;w2 = 0.1*pi;
A = 1.5;B = 2.0;
x1 = A*cos(w1*(k-1)); x2 = B*cos(w2*(k-1));
x = x1+x2;
% Generate the output sequence by filtering the input
si = [0 0 0];
num = 0.0662272*[1 3 3 1];
den = [1 -0.93561864 0.56712991716567 -0.10159281020491];
y = filter(num,den,x,si);
% Plot the input and the output sequences
subplot(2,1,1);
stem(k-1,x); axis([0 50 -4 4]);
xlabel('Time index n'); ylabel('Amplitude');
title('Input sequence');
subplot(2,1,2);
stem(k-1,y); axis([0 50 -4 4]);
xlabel('Time index n'); ylabel('Amplitude');
title('Output sequence');
```

Figure 8.7 shows the plots generated by this program, verifying the blocking of the high-frequency sinusoidal sequence by the lowpass filter. Note also the transients generated by the zero initial conditions at the beginning of the output sequence.

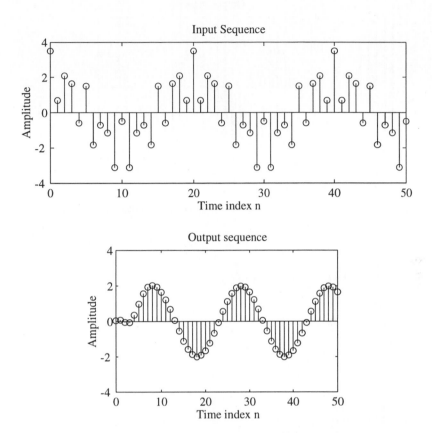

Figure 8.7   Illustration of lowpass filtering.

Example 8.4 We repeat here Example 8.3, using instead a cascade realization of the digital filter of Eq. (8.19). From Eq. (7.67) we arrive at one possible realization using a cascade of a first-order transfer function $G_1(z)$ and a second-order transfer function $G_2(z)$ given by

$$G_1(z) = \frac{0.0662272(1 + z^{-1})}{1 - 0.2593284z^{-1}}, \tag{8.20a}$$

and

$$G_2(z) = \frac{1 + 2z^{-1} + z^{-2}}{1 - 0.6762858z^{-1} + 0.3917468z^{-2}} \tag{8.20b}$$

The MATLAB program used for simulation is as indicated below

```
% Program 8_3
% Illustration of Cascade Realization of IIR Filters
%
k = 1:51;
```

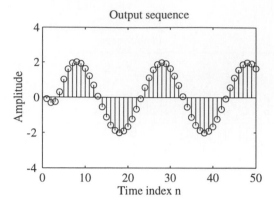

Figure 8.8   Output of the cascade realization of the lowpass filter of Eq. (8.19).

```
w1 = 0.8*pi;w2 = 0.1*pi;
A = 1.5;B = 2.0;
x1 = A*cos(w1*(k-1)); x2 = B*cos(w2*(k-1));
x = x1+x2;
b1 = 0.0662272*[1 1];
a1 = [1 -0.2593284];
y1 = filter(b1,a1,x,0);
si = [0 0];
b2 = [1 2 1];
a2 = [1 -0.6762858 0.3917468];
y2 = filter(b2,a2,y1,si);
stem(k-1,y2);axis([0 50 -4 4]);
xlabel('Time index n'); ylabel('Amplitude');
title('Output sequence');
```

The output generated by the above program is shown in Figure 8.8.

Example 8.5 As indicated in Section 3.5.2, a long input sequence can be filtered using the overlap-add method in which the input sequence is segmented into a set of contiguous short input blocks, each block is then filtered separately, and the overlaps in the output blocks are added appropriately to generate the long output sequence. This method of filtering can be easily implemented on MATLAB using the second form of the function `filter`. Here, the final values of the state variable vector `sf` at any stage of filtering is fed back in the following stage of filtering as the initial condition vector `si`. We use the following program to illustrate the method by repeating the above lowpass filtering example.

```
% Program 8_4
% Illustration of Overlap-Add Filtering Approach
%
k = 1:51;
w1 = 0.8*pi; w2 = 0.1*pi;
```

```
A = 1.5;B = 2.0;
x1 = A*cos(w1*(k-1)); x2 = B*cos(w2*(k-1));
x = x1+x2;
xs = zeros(10,5);
xs(:) = x;
si = [0 0 0];
b = 0.0662272*[1 3 3 1];
a = [1 -0.93561864 0.56712991716567 -0.10159281020491];
for n = 1:5
    for m = 1:10
        [ys(m,n),sf] = filter(b,a,xs(m,n),si);
        si = sf;
    end
end
y = ys(:);
stem(k-1,y);
xlabel('Time index n'); ylabel('Amplitude');
title('Output Sequence');
```

The output plot generated by the above program is identical to the one shown in Figure 8.8.

The function filter can also be used to implement an FIR digital filter as illustrated in the following example.

Example 8.6 We consider in this example the filtering of a signal formed by the sum of two sinsoidal sequences of normalized angular frequencies, 0.2π and 0.75π, by an equiripple FIR lowpass filter. The coefficients are generated using the function remez. For simplicity, we consider the design of a ninth order Type 1 FIR filter. Program 8_5 given below was utilized to illustrate the filtering process.

```
% Program 8_5
% Determine the FIR filter coefficients
%
f = [0   0.3   0.5 1];
m = [1 1 0 0];
num = remez(9,f,m);
% Generate the input sequence
k = 1:51;
w1 = 0.75*pi;w2 = 0.2*pi;
A = 1.5;B = 2.0;
x1 = A*cos(w1*(k-1)); x2 = B*cos(w2*(k-1));
x = x1+x2;
% Filter the input sequence
si = [zeros(1,9)];
y = filter(num,1,x,si);
% Plot the input and the output sequences
subplot(2,1,1)
stem(k-1,x);
xlabel('Time index n'); ylabel('Amplitude');
title('Input Sequence');
```

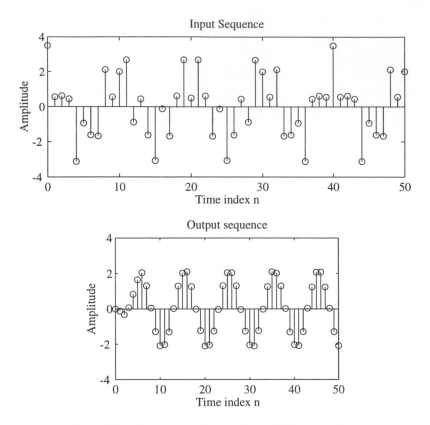

Figure 8.9   Illustration of filtering by an FIR lowpass filter.

```
subplot(2,1,2)
stem(k-1,y);
xlabel('Time index n'); ylabel('Amplitude');
title('Output Sequence');
```

Figure 8.9 shows the input and the output sequences generated by this program. These plots verify the lowpass filtering operation. The transients generated by the zero initial conditions at the beginning of the output sequence are again clearly visible.

The function `filtfilt` can be used to implement the zero-phase filtering scheme discussed in Section 4.3.5. In its basic form, `y = filtfilt(p,d,x)` implements zero-phase filtering by performing both forward and time-reversed processing operations. The resulting filter thus has double the order of the filter characterized by the coefficients p and d. If $G(z)$ denotes the transfer function of the original filter, `filtfilt` then realizes a filter with a zero-phase frequency response given by $|G(e^{j\omega})|^2$ and therefore has a passband ripple in dB and a minimum stopband attenuation in dB that are twice those of the original filter, respectively. We illustrate below one possible application of the function `filtfilt`.

Example 8.7 We have indicated in Section 7.1.2 that IIR digital filters are, in general, computation-ally more efficient than equivalent FIR digital filters meeting the same magnitude specifications. On the other hand, IIR filters cannot be designed with linear phase, whereas, it is much easier to design FIR filters with linear phase. One way to implement an IIR filter with linear phase is to use both forward and time-reversed filtering operations as discussed in Section 4.3.5. To this end we repeat the lowpass filtering problem of Example 8.3 and compare the results of the functions `filter` and `filtfilt` using the following MATLAB program.

```
% Program 8_6
% Illustration of Lowpass Digital Filtering
%
clf
% Generate the input sequence
k = 1:51;
w1 = 0.75*pi;w2 = 0.2*pi;
A = 1.5;B = 2.0;
x1 = A*cos(w1*(k-1)); x2 = B*cos(w2*(k-1));
x = x1+x2;
si = [0 0 0];
b = 0.0662272*[1 3 3 1];
a = [1 -0.93561864 0.56712991716567 -0.10159281020491];
y = filter(b,a,x,si);
stem(k-1,x2);axis([0 50 -2 2]);
xlabel('Time index n'); ylabel('Amplitude');
title('Desired component of the input sequence');
pause
stem(k-1,y);
xlabel('Time index n'); ylabel('Amplitude');
title('Output sequence generated by function filter ');
pause
si = [0 0 0];
y2 = filtfilt(b,a,x);
stem(k-1,y2);
xlabel('Time index n'); ylabel('Amplitude');
title('Output sequence generated by function filtfilt');
```

Figure 8.10 depicts the low-frequency component of the input, the output generated by forward-only filtering, and the output generated by both forward and time-reversed filtering. Several com-ments are in order here. Both outputs show the effect of the transients. In the case of forward-only filtering, as shown in Figure 8.10(b), only the first few output samples are corrupted by the transients generated by the initial conditions. On the other hand, in the case of both forward and time-reversed filtering indicated in Figure 8.10(c), effects of transients can be seen at both the beginning and the end. Next, as expected, both outputs are nearly identical to the input except for the transient part, but the output in the first case is delayed with respect to the input, whereas the output in the second case is not.

It is quite straight forward to write a MATLAB program to simulate any digital filter structure. We illustrate below the simulation of the IIR and FIR cascaded lattice structures of Section 6.9.

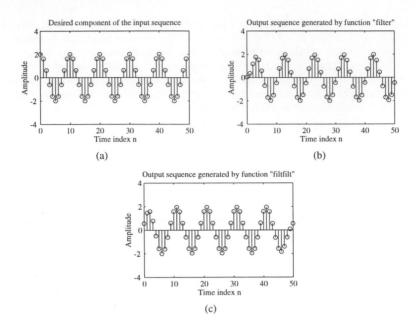

Figure 8.10 (a) Low-frequency component of the input, (b) the output generated by forward only filtering, and (c) the output generated by both forward and time-reversed filtering.

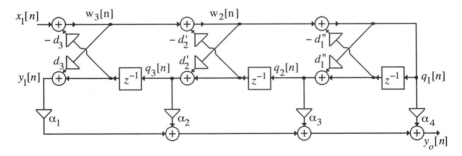

Figure 8.11 Cascaded lattice realization of the IIR transfer function of Eq. (8.21).

Example 8.8 Consider the cascaded lattice realization of the third-order IIR transfer function of Example 6.13 repeated below for convenience.

$$H(z) = \frac{0.44z^{-1} + 0.36z^{-2} + 0.02z^{-3}}{1 + 0.4z^{-1} + 0.18z^{-2} - 0.2z^{-3}}. \tag{8.21}$$

The realization derived in this example is depicted in Figure 8.11. The multiplier coefficients as derived in Examples 6.12 and 6.13 are given by

$$d_3 = -0.2, \qquad d_2' = 0.270833, \qquad d_1'' = 0.357377,$$
$$\alpha_1 = 0.02, \qquad \alpha_2 = 0.352, \qquad \alpha_3 = 0.276533, \qquad \alpha_4 = -0.19016$$

From Figure 8.11, we obtain by inspection the equations describing this structure as:

$$w_3[n] = x_1[n] - d_3 q_3[n-1],$$

$$w_2[n] = w_3[n] - d_2'q_2[n-1],$$
$$q_1[n] = w_2[n] - d_1''q_1[n-1],$$
$$q_2[n] = d_1''q_1[n] + q_1[n-1],$$
$$q_3[n] = d_2'w_2[n] + q_2[n-1],$$
$$y_1[n] = d_3w_3[n] + q_3[n-1],$$
$$y_0[n] = \alpha_1 y_1[n] + \alpha_2 q_3[n] + \alpha_3 q_2[n] + \alpha_4 q_1[n].$$

The MATLAB program for the simulation of this structure is given below. The input vector x1 consists of the first 7 coefficients of an unit sample sequence. The output vector yo contains the first 7 impulse response coefficients which are then used to determine the numerator and the denominator coefficients of the actual transfer function implemented using the method described in Section 8.1.3.

```
% Program 8_7
% Simulation of IIR Cascaded Lattice Structure
%
x1 = [1 zeros(1,6)];
q3old = 0; q2old = 0; q1old = 0;
% Enter filter coefficients
D1 = 0.357377; D2 = 0.27083; D3 =-0.2;
alpha1 = 0.02; alpha2 = 0.352; alpha3 = 0.276533;
alpha4 = -0.19016;
% Compute first 7 impulse response samples
for n = 1:7
w3 = x1(n) - D3*q3old;
w2 = w3 - D2*q2old;
q1new = w2 - D1*q1old;
q2new = D1*q1new + q1old;
q3new = D2*w2 + q2old;
y1 = D3*w3 + q3old;
yo(n) = alpha1*y1 + alpha2*q3new + alpha3*q2new + alpha4*q1new;
q3old = q3new; q2old = q2new; q1old = q1new;
end
% Form the matrices required for structure verification
H1 = [yo(1) 0 0 0
      yo(2) yo(1) 0 0
      yo(3) yo(2) yo(1) 0
      yo(4) yo(3) yo(2) yo(1)];
H2 = [yo(4) yo(3) yo(2)
      yo(5) yo(4) yo(3)
      yo(6) yo(5) yo(4)];
h = [yo(5) yo(6) yo(7)]';
% Compute the denominator coefficients
d = -(inv(H2))*h;
disp('Denominator Coefficients')
disp(d);
% Compute the numerator coefficients
p = H1*[1;d];
```

```
disp('Numerator Coefficients')
disp(p);
```

The resulting transfer function coefficients generated by this program are as follows:

```
Denominator Coefficients
    0.4000      0.1800     -0.2000

Numerator Coefficients
   -0.0000      0.4400      0.3600      0.0200
```

which are seen to be identical to those given in Eq. (8.21), thus verifying the structure of Figure 8.11.

Example 8.9 Consider the cascaded lattice realization of the fourth-order FIR transfer function of Example 6.14 repeated below for convenience:

$$H_4(z) = 1 + 1.2z^{-1} + 1.12z^{-2} + 0.12z^{-3} - 0.08z^{-4}. \tag{8.22}$$

The realization derived in this example is shown in Figure 8.12. The multiplier coefficients as determined in Example 6.14 are given by:

$$k_1 = 0.5, \qquad k_2 = 1, \qquad k_3 = 0.2173913, \qquad k_4 = -0.08$$

From Figure 8.12, we obtain by inspection the equations describing this structure as

$$x_1[n] = x_0[n] + k_1 x_0[n-1],$$
$$w_1[n] = k_1 x_0[n] + x_0[n-1],$$
$$x_2[n] = x_1[n] + k_2 w_1[n-1],$$
$$w_2[n] = k_2 x_1[n] + w_1[n-1],$$
$$x_3[n] = x_2[n] + k_3 w_2[n-1],$$
$$w_3[n] = k_3 x_2[n] + w_2[n-1],$$
$$x_4[n] = x_3[n] + k_4 w_3[n-1],$$
$$w_5[n] = k_4 x_3[n] + w_3[n-1].$$

The MATLAB program for the simulation of the structure of Figure 8.12 is given below. The input x is an unit sample sequence, while the output vector x4 contains the first five samples of the impulse response sequence. The program also computes the mirror image output vector w4.

```
% Program 8_8
% Simulation of FIR Cascaded Lattice Structure
%
% Enter Multiplier Values
k1 = 0.5; k2 = 1; k3 = 0.2173913; k4 = -0.08;
% Set initial conditions
w1old = 0; w2old = 0; w3old = 0; xold = 0;x = 1;
x4 = [zeros(1,5)]; w4 = x4;
```

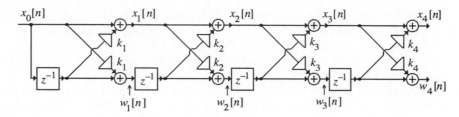

Figure 8.12   Cascaded lattice realization of the FIR transfer function of Eq. (8.22): $k_1 = 0.5$, $k_2 = 1.0$, $k_3 = 0.2173913$, and $k_4 = -0.08$.

```
% Compute the first 5 impulse response samples
for n = 1:5;
        x1 = x + k1*xold; w1 = k1*x + xold;
        x2 = x1 + k2*w1old; w2 = k2*x1 + w1old;
        x3 = x2 + k3*w2old; w3 = k3*x2 + w2old;
        x4(n) = x3 + k4*w3old; w4(n) = k4*x3 + w3old;
        xold = x; x = 0;
        w1old = w1; w2old = w2; w3old = w3;
end
disp('Filter Coefficients');disp(x4);
disp('Coefficients of Mirror Image Filter');disp(w4)
```

The output data displayed are the coefficients of the transfer function $H_4(z)$ and the coefficients of its mirror-image transfer function $\tilde{H}_4(z)$ as given below:

```
Filter Coefficients
        1.0000    1.2000    1.1200    0.1200    -0.0800

Coefficients of Mirror Image Filter
        -0.0800    0.1200    1.1200    1.2000    1.0000
```

The coefficients of $H_4(z)$ are seen to be identical to that given in Eq. (8.22).

8.3   Computation of the Discrete Fourier Transform

The discrete Fourier transform (DFT) is another widely used DSP algorithm. As indicated earlier, it can be employed to implement linear convolution of two sequences, the key digital filtering operation and is also used for the frequency-domain analysis of signals, which is discussed in Sections 11.2 and 11.3. Because of the widespread use of the DFT, it is of interest to investigate its efficient implementation methods which are considered in this section.

In Section 3.2 we introduced the concept of the N-point DFT $X[k]$ of a sequence $x[n]$ of length N as the N samples of its Fourier transform, $X(e^{j\omega})$, evaluated uniformly on the ω-axis at $\omega_k = 2\pi k/N$, $0 \leq k \leq N - 1$:

$$X[k] = X(e^{j\omega})\Big|_{\omega=2\pi k/N} = \sum_{n=0}^{N-1} x[n]e^{-j2\pi kn/N}, \qquad k = 0, 1, \ldots, N - 1. \quad (8.23)$$

Since a finite-length sequence is always absolutely summable, the DFT is thus the samples of its z-transform $X(z)$ evaluated on the unit circle at N equally spaced points:

$$X[k] = X(z)|_{z=e^{j2\pi k/N}}, \qquad k = 0, 1, \ldots, N-1. \tag{8.24}$$

As can be seen from Eq. (8.23), the computation of each sample of the DFT sequence requires N complex multplications and $N-1$ complex additions. Hence, the computation of the N-point DFT sequence requires N^2 complex multplications and $(N-1)N$ complex additions. In the case of a sequence of length N, it can be shown that the computation of its N-point DFT sequence requires $4N^2$ real multplications and $(4N-2)N$ real additions (Problem 8.12). As a result, the total number of computations to compute an N-point DFT increases very rapidly as N increases. For large N, the number of complex multiplications and additions is approximately equal to N^2. Hence, it is of practical interest to develop more efficient or fast algorithms for computing the DFT.

8.3.1   Goertzel's Algorithm

An elegant approach to computing the DFT is to use a recursive computation scheme. To this end, the most popular approach is the *Goertzel's algorithm* derived next [Goe58]. This algorithm makes use of the identity

$$W_N^{-kN} = 1, \tag{8.25}$$

obtained using the periodicity of W_N^{-kn}. Using the above identity, we can rewrite Eq. (8.23) as

$$X[k] = \sum_{\ell=0}^{N-1} x[\ell]W_N^{k\ell} = W_N^{-kN} \sum_{\ell=0}^{N-1} x[\ell]W_N^{k\ell} = \sum_{\ell=0}^{N-1} x[\ell]W_N^{-k(N-\ell)}. \tag{8.26}$$

The above expression can be expressed in the form of a convolution. To this end, we define a new sequence

$$y_k[n] = \sum_{\ell=0}^{n} x_e[\ell]W_N^{-k(n-\ell)}, \tag{8.27}$$

which is a direct convolution of the causal sequence $x_e[n]$ defined by

$$x_e[n] = \begin{cases} x[n], & 0 \le n \le N-1, \\ 0, & n < 0, n \ge N, \end{cases} \tag{8.28a}$$

with a causal infinite-length sequence

$$h_k[n] = \begin{cases} W_N^{-kn}, & n \ge 0, \\ 0, & n < 0. \end{cases} \tag{8.28b}$$

It follows from Eqs. (8.26) and (8.27) that

$$X[k] = y_k[n]|_{n=N} .$$

Figure 8.13   Recursive computation of the kth DFT sample.

By taking the z-transform of both sides of Eq. (8.27) we arrive at

$$Y_k(z) = \frac{X_e(z)}{1 - W_N^{-k}z^{-1}}, \tag{8.29}$$

where $1/(1 - W_N^{-k}z^{-1})$ is the z-transform of the causal sequence $h_k[n]$ and $X_e(z)$ is the z-transform of $x_e[n]$. The above equation implies that $y_k[n]$ is the output of an initially relaxed LTI digital filter with a transfer function

$$H_k(z) = \frac{1}{1 - W_N^{-k}z^{-1}}, \tag{8.30}$$

with an input $x[n]$ as indicated in Figure 8.13. When $n = N$, the output of the filter $y_k[N]$ is precisely $X[k]$.

From the above figure, the DFT computation algorithm is given by

$$y_k[n] = x[n] + W_N^{-k}y_k[n-1] \qquad n = 0, 1, \ldots, N, \tag{8.31}$$

with $y_k[-1] = 0$ and $x[N] = 0$. Since a complex multiplication can be implemented using 4 real multiplications and 2 real additions, computation of each new value of $y_k[n]$ thus, in general, requires 4 real multplications and 4 real additions.[3] As a result, the computation of $X[k] = y_k[N]$ involves $4N$ real multplications and $4N$ real additions, resulting in $4N^2$ real multiplications and $4N^2$ real additions for the computation of all N DFT samples.

Hence, the above algorithm, in comparison to the direct DFT computation, requires the same number of real multiplications but $2N$ more real additions. As a result, it is computationally slightly more inefficient than the direct approach. The advantage of the recursive algorithm, however, is that the N complex coefficients W_N^{kn} required to compute $X[k]$ do not have to be either computed or stored in advance, but are computed recursively as needed.

The algorithm can be made more efficient by observing that $H_k(z)$ of Eq. (8.30) can be rewritten as

$$H_k(z) = \frac{1}{1 - W_N^{-k}z^{-1}} = \frac{1 - W_N^k z^{-1}}{(1 - W_N^{-k}z^{-1})(1 - W_N^k z^{-1})}$$

$$= \frac{1 - W_N^k z^{-1}}{1 - 2\cos\left(\frac{2\pi}{N}k\right)z^{-1} + z^{-2}}, \tag{8.32}$$

resulting in the new realization shown in Figure 8.14.

---

[3] It can be shown that a simple modification of the complex multiplication algorithm can reduce the number of real multiplications to 3 while increasing the number of real additions to 5 (see Problem 8.13).

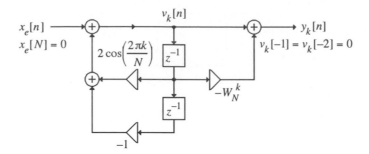

Figure 8.14   Modified approach to the recursive computation of the kth DFT sample.

The DFT computation equations are now given by

$$v_k[n] = x[n] + 2\cos\left(\frac{2\pi k}{N}\right) v_k[n-1] - v_k[n-2], \quad 0 \le n \le N, \quad (8.33\text{a})$$

$$X[k] = y_k[N] = v_k[N] - W_N^k v_k[N-1]. \quad (8.33\text{b})$$

Note that the computation of each sample of the intermediate variable $v_k[n]$ involves only 2 real multiplications and 4 real additions.[4] The complex multiplication by the constant W_N^k needs to be performed only once at $n = N$. Thus, to compute one sample $X[k]$ of the N-point DFT, we need $(2N+4)$ real multiplications and $(4N+4)$ real additions. As a result, the modified Goertzel's algorithm for computing the N-point DFT requires $2(N+2)N$ real multiplications and $4(N+1)N$ real additions.

Further savings in computational requirements can be obtained by comparing the realization of $H_{N-k}(z)$ with that of $H_k(z)$ shown in Figure 8.14. In the case of the former, the multiplier in the feedback path is $2\cos[2\pi(N-k)/N] = 2\cos(2\pi k/N)$ which is same as in Figure 8.14. Hence, $v_{N-k}[n] = v_k[n]$, indicating that the intermediate variables computed for determining $X[k]$ need no longer be computed again for determining $X[N-k]$. The only difference between these two structures is in the feedforward path, where the multiplier is instead $W_N^{N-k} = W_N^{-k}$, which is the complex conjugate of the coefficient W_N^k used in Figure 8.14. Thus, the computation of the two samples of the N-point DFT, $X[k]$ and $X[N-k]$, requires $2(N+4)$ real multiplications and $4(N+2)$ real additions. Or in other words, all N samples of the DFT can be determined using approximately N^2 real multiplications and approximately $2N^2$ real additions. The number of real multiplications is thus about one-fourth and the number of real additions is about one-half of those needed in the direct DFT computation.

The MATLAB function gfft(x,N,k) given below implements the modified Goertzel's algorithm to compute the kth DFT sample of an N-point DFT of the sequence x. The computed DFT sample is XF. The length of the input sequence must be less than or equal to N. If the length is less than N, the sequence length is increased to N by zero-padding.

```
% Function to Compute a DFT Sample
% Using Goertzel's Algorithm
% XF = gfft(x,N,k)
```

---

[4]In general, $x[n]$ and $v_k[n]$ are complex sequences.

```
% x is the input sequence of length <= N
% N is the DFT length
% k is the specified bin number
% XF is the desired DFT sample
%
function XF = gfft(x,N,k)
if length(x) < N
        xe = [x zeros(1,N-length(x))];
else
        xe = x;
end
x1 = [xe 0];
d1 = 2*cos(2*pi*k/N);W = exp(-i*2*pi*k/N);
y = filter(1,[1 -d1 1],x1);
XF = y(N+1) - W*y(N);
```

Goertzel's algorithm is attractive in applications requiring the computation of a few samples of the DFT. One such example is the *dual-tone multifrequency* (DTMF) signal detection in the TOUCH-TONE$^{\circledR}$ telephone dialing system discussed in Section 11.1. In such an application, the input $x[n]$ is a real sequence and the square magnitude of the DFT sample, $|X[k]|^2$, is of interest. Since $x[n]$ is a real sequence, the intermediate sequence $v_k[n]$ generated in the modified Goertzel's algorithm is also a real sequence. As a result, we obtain from Eq. (8.33b):

$$|X[k]|^2 = |y_k[N]|^2 = v_k^2[N] + v_k^2[N-1] - 2\cos(2\pi k/N)v_k[N]v_k[N-1]. \quad (8.34)$$

The above scheme uses only real multiplications, avoiding the complex multiplication required for the computation of $y_k[N]$ as indicated in Eq. (8.33b).

We next describe a fast algorithm for the computation of the DFT when the length N is a composite number. As we shall demonstrate, in the new algorithm in the case of N a power of 2, the total number of computations can be made proportional to $N \log_2(N)$ and is highly preferable in applications requiring the computation of all DFT samples.

8.3.2 Fast Fourier Transform Algorithms

The basic idea behind all fast algorithms for computing the discrete Fourier transform (DFT), commonly called the *fast Fourier transform* (FFT) algorithms, is to decompose successively the N-point DFT computation into computations of smaller size DFTs and to take advantage of the periodicity and symmetry of the complex number W_N^{kn}. Such decompositions, if properly carried out, can result in a significant savings in the computational complexity. There are various versions of the FFT algorithms. We review here the main concepts behind the two most basic FFT algorithms [Coo65].

Decimation-In-Time FFT Algorithm

Consider a sequence $x[n]$ of length N that is assumed to be a power of 2. Using a 2-band polyphase decomposition[5] of $x[n]$ we can express its z-transform $X(z)$ as

$$X(z) = X_0(z^2) + z^{-1}X_1(z^2), \tag{8.35}$$

where

$$X_0(z) = \sum_{n=0}^{\frac{N}{2}-1} x_0[n]z^{-n} = \sum_{n=0}^{\frac{N}{2}-1} x[2n]z^{-n}, \tag{8.36a}$$

$$X_1(z) = \sum_{n=0}^{\frac{N}{2}-1} x_1[n]z^{-n} = \sum_{n=0}^{\frac{N}{2}-1} x[2n+1]z^{-n}. \tag{8.36b}$$

Thus, $X_0(z)$ is the z-transform of the sequence $x_0[n] = x[2n]$ of length N/2 formed from the even-indexed samples of $x[n]$, while $X_1(z)$ is the z-transform of the sequence $x_1[n] = x[2n+1]$ of length $N/2$ formed from the odd-indexed samples of $x[n]$.

Evaluating $X(z)$ on the unit circle at N equally spaced points, $z = W_N^{-k}$, we arrive at the N-point DFT of $x[n]$ given by

$$X[k] = X_0[\langle k \rangle_{N/2}] + W_N^k X_1[\langle k \rangle_{N/2}], \qquad k = 0, 1, \ldots, N-1, \tag{8.37}$$

where $X_0[k]$ and $X_1[k]$ are the $(N/2)$-point DFTs of the $(N/2)$-length sequences $x_0[n]$ and $x_1[n]$, respectively, i.e.,

$$X_0[k] = \sum_{r=0}^{\frac{N}{2}-1} x_0[r]W_{N/2}^{rk}$$
$$= \sum_{r=0}^{\frac{N}{2}-1} x[2r]W_{N/2}^{rk}, \quad k = 0, 1, \ldots, \frac{N}{2} - 1, \tag{8.38a}$$

$$X_1[k] = \sum_{r=0}^{\frac{N}{2}-1} x_1[r]W_{N/2}^{rk}$$
$$= \sum_{r=0}^{\frac{N}{2}-1} x[2r+1]W_{N/2}^{rk}, \quad k = 0, 1, \ldots, \frac{N}{2} - 1. \tag{8.38b}$$

It is instructive at this point to examine a block diagram interpretation of the modified DFT computation scheme of Eq. (8.37) that computes an N-point DFT of the original length-N sequence $x[n]$ by forming a weighted sum of two $(N/2)$-point DFTs of two $(N/2)$-length subsequences formed from the even-indexed samples $x_0[n] = x[2n]$ and the odd-indexed samples $x_1[n] = x[2n+1]$. To this end we need a device, called a *down-sampler,* that can be used to develop the two subsequences from $x[n]$. It should be noted

---

[5]See Section 6.4.3 for a discussion on the polyphase decomposition.

$$x[n] \longrightarrow \boxed{\downarrow M} \longrightarrow y[n]$$

Figure 8.15   Representation of a factor-of-M down-sampler.

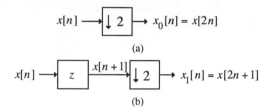

(a)

(b)

Figure 8.16   (a) Generation of subsequence containing even-indexed input samples, and (b) generation of subseqence containing odd-indexed input samples.

that the down-sampler is an important component in multirate digital signal processing and is treated in detail in Chapter 10. In its general form, a factor-of-M down-sampler retains at its output every Mth sample of the input with an input-output relation defined by

$$y[n] = x[Mn]. \tag{8.39}$$

The block diagram representation of a factor-of-M down-sampler is as shown in Figure 8.15.

It follows from the definition of Eq. (8.39), if the input $x[n]$ to a factor-of-2 down-sampler is a length N sequence defined for $n = 0, 1, \ldots, N - 1$, its output $x_0[n]$ is a length $N/2$ sequence defined for $n = 0, 1, \ldots, (N/2) - 1$ and is composed of the even-indexed samples of $x[n]$, i.e., $x_0[n] = x[2n]$, as shown in Figure 8.16(a). To generate the subsequence $x_1[n]$ composed of the odd-indexed samples of $x[n]$, we need to pass $x[n + 1]$ through a factor-of-2 down-sampler. The sequence $x[n + 1]$ can be developed from the sequence $x[n]$ by means of an advance operation. The process is illustrated in Figure 8.16(b).

From the above discussion it follows then that a block diagram interpretation of the DFT computation scheme of Eq. (8.37) is as indicated in Figure 8.17. Figure 8.18 shows its flow-graph representation for the case of $N = 8$.

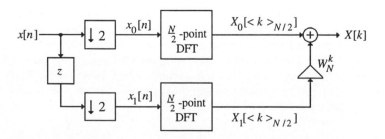

Figure 8.17   Structural interpretation of the DFT decomposition scheme of Eq. (8.37).

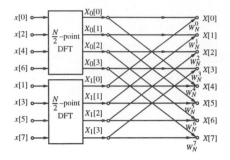

Figure 8.18   Flow-graph of the first stage in the decimation-in-time FFT algorithm for $N = 8$.

Before proceeding further, let us evaluate the computational requirements for computing an N-point DFT using two $(N/2)$-point DFTs based on the decomposition of Eq. (8.37). Now a direct computation of an N-point DFT requires N^2 complex multiplications and $N^2 - N \cong N^2$ complex additions. On the other hand, computation of an N-point DFT using the decomposition of Eq. (8.37) requires the computations of two $(N/2)$-point DFTs that need to be combined with N complex multiplications and N complex additions, resulting in a total of $N + (N^2/2)$ complex multiplications and approximately $N + (N^2/2)$ complex additions. It can easily be verified that for $N \geq 3$, $N + (N^2/2) < N^2$.

We can continue the above process by expressing each of the two $(N/2)$-point DFTs, $G[k]$ and $H[k]$, as a weighted combination of two $(N/4)$-point DFTs since by assumption $N/2$ is even. For example, we can express $X_0[k]$ as

$$X_0[k] = X_{00}[\langle k \rangle_{N/4}] + W_{N/2}^k X_{01}[\langle k \rangle_{N/4}], \quad k = 0, 1, \ldots, \frac{N}{2} - 1, \qquad (8.40)$$

where $X_{00}[k]$ and $X_{01}[k]$ are the $(N/4)$-point DFTs of the $(N/4)$-length sequences, $x_{00}[n]$ and $x_{01}[n]$, generated from the even and odd samples of $x_0[n]$, respectively. Likewise, we can express $X_1[k]$ as

$$X_1[k] = X_{10}[\langle k \rangle_{N/4}] + W_{N/2}^k X_{11}[\langle k \rangle_{N/4}], \quad k = 0, 1, \ldots, \frac{N}{2} - 1, \qquad (8.41)$$

where $X_{10}[k]$ and $X_{11}[k]$ are the $(N/4)$-point DFTs of the $(N/4)$-length sequences, $x_{10}[n]$ and $x_{11}[n]$, generated from the even and odd samples of $x_1[n]$, respectively.

Substituting Eqs. (8.40) and (8.41) in Eq. (8.37) and making use of the identity $W_{N/2}^k = W_N^{2k}$, we then arrive at the two-stage decomposition of an N-point DFT computation in terms of four $(N/4)$-point DFTs, as indicated by the block diagram of Figure 8.19. The corresponding flow-graph representation is shown in Figure 8.20 for $N = 8$. In the case of the 8-point DFT computation illustrated in Figure 8.20, the $(N/4)$-point DFT is a 2-point DFT and no further decomposition is possible. The 2-point DFTs, $X_{00}[k]$, $X_{01}[k]$, $X_{10}[k]$, and $X_{11}[k]$, can be easily computed. For example, for the computation of $X_{00}[k]$ the pertinent expressions are

$$X_{00}[k] = \sum_{n=0}^{1} x_{00}[n] W_2^{nk} = x[0] + W_2^k x[4], \quad k = 0, 1. \qquad (8.42)$$

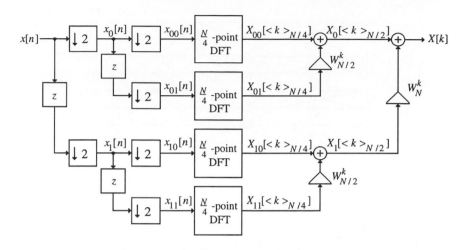

Figure 8.19   Structural interpretation of the two-stage DFT decomposition scheme.

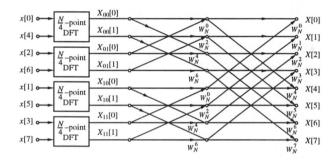

Figure 8.20   Flow-graph of the second stage in the decimation-in-time FFT algorithm for $N = 8$.

The corresponding flow-graph is indicated in Figure 8.21 where we have used the identity $W_2^k = W_N^{(N/2)k}$. Replacing each of the 2-point DFTs in Figure 8.20 with their respective flow-graph representations, we finally arrive at the complete flow-graph of the basic decimation-in-time DFT algorithm as shown in Figure 8.22.

If we examine the flow-graph of Figure 8.22, we note that it consists of three stages. The first stage computes the four 2-point DFTs, the second stage computes the two 4-point DFTs, and finally the last stage computes the desired 8-point DFT. Moreover, the number of complex multiplications and additions performed at each stage is equal to 8, the size of

Figure 8.21   Flow-graph of the 2-point DFT.

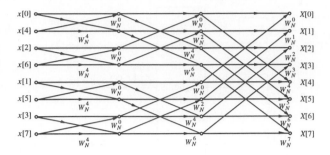

Figure 8.22 Complete flow-graph of the basic decimation-in-time FFT algorithm for $N = 8$.

the transform. As a result, the total number of complex multiplications and additions in computing all 8 DFT samples is equal to $3 \times 8 = 24$.

It follows from the above observation that in the general case when $N = 2^\mu$, the number of stages of computation of the (2^μ)-point DFT in the fast algorithm will be $\mu = \log_2 N$. Therefore, the total number of complex multiplications and additions in computing all N DFT samples is equal to $N(\log_2 N)$. In developing this count, we have for the present considered multiplying with $W_N^0 = 1$ and $W_n^{N/2} = -1$ as complex multiplications. In addition, we have not taken advantage of the symmetry property $W_N^{(N/2)+k} = -W_N^k$. These properties can be made use of in reducing the computational complexity further.

Computational Considerations

Examination of the flow-graph of Figure 8.22 also reveals that each stage of the DFT computation process employs the same basic computational module in which two output variables are generated from a weighted combination of two input variables. To see this aspect more clearly, let us label the N-input and the N-output variables in the rth stage of the DFT computation as $\Psi_r[m]$ and $\Psi_{r+1}[m]$, respectively, with $r = 1, 2, \ldots, \mu$, and $m = 0, 1, \ldots, N - 1$. According to this labeling, in the case of Figure 8.22 for the 8-point DFT computation, $\Psi_1[0] = x[0]$, $\Psi_1[1] = x[4]$, and so on. Similarly, here, $\Psi_4[0] = X[0]$, $\Psi_4[1] = X[1]$, and so on. Based on this labeling scheme, it can be easily verified that the basic computational module is represented by the flow-graph of Figure 8.23 and described by the following input-output relations:

$$\Psi_{r+1}[\alpha] = \Psi_r[\alpha] + W_N^\ell \Psi_r[\beta], \tag{8.43a}$$

$$\Psi_{r+1}[\beta] = \Psi_r[\alpha] + W_N^{\ell+(N/2)} \Psi_r[\beta]. \tag{8.43b}$$

Because of its shape, the basic computational module of Figure 8.23 is referred to in the literature as the *butterfly computation*.

Substituting $W_N^{\ell+(N/2)} = -W_N^\ell$ in Eq. (8.43b) we can rewrite it as

$$\Psi_{r+1}[\beta] = \Psi_r[\alpha] - W_N^\ell \Psi_r[\beta]. \tag{8.43c}$$

The modified butterfly computation is thus as indicated in Figure 8.24 and requires only one complex multiplication. Use of this modified butterfly computational module in the

Figure 8.23 Flow-graph of the basic computational module in the DIT FFT method.

Figure 8.24 Flow-graph of the modified butterfly computational module.

FFT computation leads to a reduction in the total number of complex multiplications by 50%, as can be seen from the new flow-graph for the $N = 8$ case illustrated in Figure 8.25. Further savings in the computational complexity arise by taking into consideration that multiplications by $W_N^0 = 1$, $W_N^{N/2} = -1$, $W_N^{N/4} = j$, and $W_N^{3N/4} = -j$ can be avoided in the DFT computation process.

Another attractive feature of the FFT algorithm described above is with regard to memory requirements. Since each stage employs the same butterfly computation to compute the two output variables $\Psi_{r+1}[\alpha]$ and $\Psi_{r+1}[\beta]$ from the input variables $\Psi_r[\alpha]$ and $\Psi_r[\beta]$, after $\Psi_{r+1}[\alpha]$ and $\Psi_{r+1}[\beta]$ have been determined, they can be stored in the same memory locations where $\Psi_r[\alpha]$ and $\Psi_r[\beta]$ were previously stored. Thus, at the end of the computation at any stage, the output variables $\Psi_{r+1}[m]$ can be stored in the same registers previously occupied by the corresponding input variables $\Psi_r[m]$. This type of memory location sharing feature is commonly known as the *in-place computation*, resulting in a significant saving in the overall memory requirements.

It should be noted, however, from Figure 8.25 that while the DFT samples $X[k]$ appear at the output in a sequential order, the input time-domain samples $x[n]$ appear in a different order. Thus, a sequentially ordered input $x[n]$ must be reordered appropriately before the FFT algorithm described by the above structure can begin. To understand the basic scheme, in the input reordering scheme consider the 8-point DFT computation illustrated in Figure 8.25. If we represent the arguments of the input samples $x[n]$ and their sequentially order

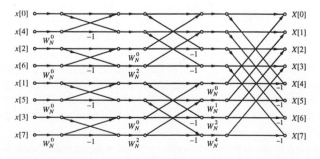

Figure 8.25 Flow-graph of the modified DIT FFT algorithm.

new representations $\Psi_1[m]$ in binary forms, we arrive at the following relations between m and n:

m	n
000	000
001	100
010	010
011	110
100	001
101	101
110	011
111	111

It follows from the above that if $(b_2b_1b_0)$ represents the index n of $x[n]$ in binary form, then the sample $x[b_2b_1b_0]$ appears in the location $m = b_0b_1b_2$ as $\Psi[b_0b_1b_2]$ before the DFT computation is started, or in other words, the location of $\Psi_1[m]$ is in *bit-reversed order* from that of the original input array $x[n]$.

   Various alternative forms of the FFT computation can be easily obtained by reordering the order of computation, such as the input in normal order and the output in bit-reversed order (Problem 8.15), both input and output in normal order (Problem 8.16), etc.

   The FFT algorithm outlined above assumed the input sequence $x[n]$ to be a power of 2. If it is not, we can extend the length of the sequence $x[n]$ by zero-padding (see Section 3.5.1) to make its length to be a power of 2.[6] Even after zero-padding, the DFT computation based on the fast algorithm derived above may be computationally more efficient than the direct DFT computation of the original shorter sequence. Alternatively, we can develop fast algorithms that make use of polyphase decomposition with more than two subsequences. To illustrate this modification to the basic FFT algorithm, consider a sequence $x[n]$ of a length that is a power of 3. Here, the DFT samples in the first stage are computed using a 3-band polyphase decomposition of $X(z)$,

$$X(z) = X_0(z^3) + z^{-1}X_1(z^3) + z^{-2}X_2(z^3), \qquad (8.44)$$

resulting in

$$X[k] = X_0[\langle k \rangle_{\frac{N}{3}}] + W_N^k X_1[\langle k \rangle_{\frac{N}{3}}] + W_N^{2k} X_2[\langle k \rangle_{\frac{N}{3}}], \quad k = 0, 1, \ldots, N-1, \quad (8.45)$$

where $X_0[k]$, $X_1[k]$, and $X_2[k]$ are now $(N/3)$-point DFTs. This process can be repeated to compute each of the $(N/3)$-point DFTs in terms of three $(N/9)$-point DFTs, and so on, until the smallest computational module is a 3-point DFT and no further decomposition is possible.

   The FFT computation schemes described above are called *decimation-in-time (DIT) FFT algorithms* since here the input sequence $x[n]$ is first decimated to form a set of subsequences before the DFT is computed. For example, the relation between the input sequence $x[n]$ and its even and odd parts, $x_0[n]$ and $x_1[n]$, respectively, generated by the

---

[6]It should be noted that zero-padding increases the effective length of the original sequence, and hence, the longer length DFT samples are different frequency samples of the frequency response and are more closely spaced on the unit circle than that of the original shorter length DFT samples.

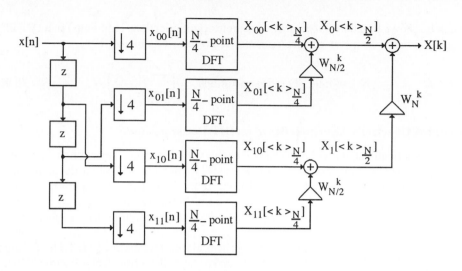

Figure 8.26   First stage in a radix-4 decimation-in-time FFT algorithm.

first stage of the DIT algorithm shown in Figure 8.17 is as follows:

$$
\begin{array}{llllllll}
x[n]: & x[0] & x[1] & x[2] & x[3] & x[4] & x[5] & x[6] & x[7] \\
x_0[n]: & x[0] & & x[2] & & x[4] & & x[6] & \\
x_1[n]: & x[1] & & x[3] & & x[5] & & x[7] &
\end{array}
$$

Likewise, the relation between the input sequence $x[n]$ and the sequences $x_{00}[n]$, $x_{01}[n]$, $x_{10}[n]$, and $x_{11}[n]$, generated by the two-stage decomposition of the DIT algorithm and illustrated in Figure 8.19, is given by

$$
\begin{array}{lllllllll}
x[n]: & x[0] & x[1] & x[2] & x[3] & x[4] & x[5] & x[6] & x[7] \\
x_{00}[n]: & x[0] & & & & x[4] & & & \\
x_{01}[n]: & x[2] & & & & x[6] & & & \\
x_{10}[n]: & x[1] & & & & x[5] & & & \\
x_{11}[n]: & x[3] & & & & x[7] & & &
\end{array}
$$

Or in other words, the subsequences $x_{00}[n]$, $x_{01}[n]$, $x_{10}[n]$ and $x_{11}[n]$ can be generated directly by a factor-of-4 decimation process, leading to the single stage decomposition given in Figure 8.26.

   If at each stage the decimation is by a factor of R, the resulting FFT algorithm is called a *radix-R* FFT algorithm. Thus, Figure 8.25 shows a radix-2 DIT FFT algorithm. Likewise, Figure 8.26 illustrates the first stage of a radix-4 DIT FFT algorithm. It also follows from the above discussion that, depending on the value of N, various combinations of decompositions of $X[k]$ can be used to develop different types of DIT FFT algorithms. If the scheme uses a mixture of decimations by different factors, it is called a *mixed radix* FFT algorithm.

   For N which is a composite number expressible in the form of a product of integers:

$$ N = r_1 \cdot r_2 \cdots r_\nu $$

it can be shown that the total number of complex multiplications (additions) in a DIT FFT algorithm based on a ν-stage decomposition is given by (Problem 8.17)

$$\text{No. of multiply (add) operations} = \left(\sum_{i=1}^{\nu} r_i - \nu \right) N. \tag{8.46}$$

Modified Goertzel's Algorithm Based on DIT Decomposition

In some applications, it may be convenient to apply the Goertzel's algorithm in computing the smaller size DFTs after one or two or more stages of decimation-in-time decompositions of the input sequence $x[n]$. For example, the four $(N/4)$-DFTs in Figure 8.19 can be computed using Goertzel's algorithm. This approach reduces the overall computational complexity of the direct Goertzel's algorithm while still permitting the computation of a few DFT samples.

It should be noted, however, that the structural interpretations of the DIT FFT algorithms given in Figures 8.17, 8.19, and 8.26 make use of the advance block with a transfer function z that is not physically realizable. If a hardware implementation of these structures is desired, we can insert an appropriate amount of delays at the input and then move them through the chain of advance blocks to make the overall structure physically realizable. Figure 8.27 shows the realizable version of Figure 8.26. Moreover, the input sequence $x[n]$ should be delayed by 3 sample periods before the down-sampling operations are carried out. The relations between the sub-sequences $x_{00}[n]$, $x_{01}[n]$, $x_{10}[n]$, and $x_{11}[n]$ and the original sequence $x[n]$ are now

$$
\begin{array}{llllllllll}
x[n]: & x[0] & x[1] & x[2] & x[3] & x[4] & x[5] & x[6] & x[7] \\
x_{00}[n]: & & & & x[0] & & & & x[4] \\
x_{01}[n]: & & & & x[2] & & & & x[6] \\
x_{10}[n]: & & & & x[1] & & & & x[5] \\
x_{11}[n]: & & & & x[3] & & & & x[7].
\end{array}
$$

Decimation-In-Frequency FFT Algorithm

The basic idea behind the decimation-in-time FFT algorithm is to decompose sequentially the N-point sequence $x[n]$ into sets of smaller and smaller subsequences and then form a weighted combinations of the DFTs of these subsequences. The same idea can be applied to the N-point DFT sequence $X[k]$ and decompose it sequentially into sets of smaller and smaller subsequences. This approach leads to another class of DFT computation schemes collectively known as the *decimation-in-frequency (DIF) FFT algorithm.*

To illustrate the basic difference between the above two decomposition schemes, we develop below the first stage of the DIF FFT algorithm for the case when N is a power of 2. We first express the z-transform $X(z)$ of $x[n]$ as

$$X(z) = X_0(z) + z^{-N/2} X_1(z), \tag{8.47}$$

where

$$X_0(z) = \sum_{n=0}^{(N/2)-1} x[n]z^{-n}, \quad X_1(z) = \sum_{n=0}^{(N/2)-1} x[\frac{N}{2} + n]z^{-n}. \tag{8.48}$$

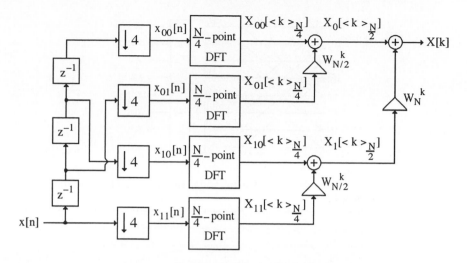

Figure 8.27 Modified structure for the first stage of a radix-4 decimation-in-time FFT algorithm.

Evaluating $X(z)$ on the unit circle at $z = W_N^{-k}$, we get from Eqs. (8.47) and (8.48),

$$X[k] = \sum_{n=0}^{(N/2)-1} x[n]W_N^{nk} + W_N^{(N/2)k} \sum_{n=0}^{(N/2)-1} x[\frac{N}{2}+n]W_N^{nk}. \qquad (8.49)$$

The above equation can be rewritten as

$$X[k] = \sum_{n=0}^{(N/2)-1} \left[x[n] + (-1)^k x[\frac{N}{2}+n] \right] W_N^{nk}, \qquad (8.50)$$

where we have used the identity $W_N^{(N/2)k} = (-1)^k$. Two different forms of Eq. (8.50) are obtained, depending on whether k is even or odd:

$$X[2\ell] = \sum_{n=0}^{(N/2)-1} \left[x[n] + x[\frac{N}{2}+n] \right] W_N^{2n\ell}$$

$$= \sum_{n=0}^{(N/2)-1} \left[x[n] + x[\frac{N}{2}+n] \right] W_{N/2}^{n\ell} \quad 0 \le \ell \le \frac{N}{2} - 1, \qquad (8.51a)$$

$$X[2\ell+1] = \sum_{n=0}^{(N/2)-1} \left[x[n] - x[\frac{N}{2}+n] \right] W_N^{n(2\ell+1)}$$

$$= \sum_{n=0}^{(N/2)-1} \left[x[n] - x[\frac{N}{2}+n] \right] W_N^{n} W_{N/2}^{n\ell} \quad 0 \le \ell \le \frac{N}{2} - 1. \qquad (8.51b)$$

The above two expressions represent the $(N/2)$-point DFTs of the following two $(N/2)$-

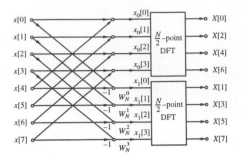

Figure 8.28   Flow-graph of the first stage of the decimation-in-frequency FFT algorithm for $N = 8$.

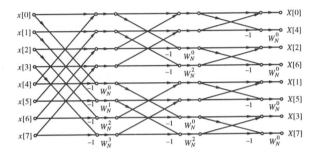

Figure 8.29   Complete flow-graph of the decimation-in-frequency FFT algorithm for $N = 8$.

point sequences:

$$x_0[n] = \left\{ x[n] + x\left[\frac{N}{2} + n\right] \right\}, \quad x_1[n] = \left\{ x[n] - x\left[\frac{N}{2} + n\right] \right\} W_N^n,$$

$$n = 0, 1, \ldots, \frac{N}{2} - 1, \tag{8.52}$$

respectively. The flow-graph of the first stage of the DFT computation scheme defined by Eqs. (8.51a) and (8.51b) is shown in Figure 8.28 for $N = 8$. As can be seen from this figure, here the input samples are in a sequential order, while the output DFT samples appear in a decimated form with the even-indexed samples appearing as the outputs of one $(N/2)$-point DFT and the odd-indexed samples appearing as the outputs of the other $(N/2)$-point DFT.

We can continue the above decomposition process by expressing the even- and odd-indexed samples of each one of the two $(N/2)$-point DFTs as a sum of two $(N/4)$-point DFTs. This process can continue until the smallest DFTs are 2-point DFTs. The complete flow-graph of the decimation-in-frequency DFT computation scheme for $N = 8$ case is shown in Figure 8.29.

It can be seen from Figure 8.29 that in the DIF FFT algorithm, the input $x[n]$ appears in the normal order while the output $X[k]$ appears in the bit-reversed order. Just as in the case of the radix-2 DIT FFT algorithm, the total number of complex multiplications per stage in the radix-2 DIF FFT algorithm is $N/2$. Hence, the total number of complex multiplications for computing the N-point DFTs samples here is also equal to $(N/2)\log_2 N$ ignoring the

Figure 8.30   Inverse DFT computation via DFT.

fact that multiplications by $W_N^{(N/4)\ell}$, $\ell = 0, 1, 2, 3, 4$ can be avoided. As before, various forms of the DIF FFT algorithm can be generated.

An examination of the flow-graph of the first stage of the radix-2 DIF FFT algorithm given in Figure 8.28 reveals that the even- and odd-indexed samples of $X[k]$ can be computed independent of each other. It has been shown by Duhamel [Duh86] that significant reduction in the computational complexity can be obtained by using a radix-2 DIF algorithm to compute the even-indexed DFT samples and a radix-4 DIF FFT algorithm to compute the odd-indexed DFT samples. This type of computational scheme has been called a *split-radix FFT algorithm* (Problem 8.28).

Inverse DFT Computation

An FFT algorithm for computing the DFT samples can also be used to calculate efficiently the inverse DFT (IDFT). To show this, consider an N-point sequence $x[n]$ with an N-point DFT $X[k]$. The sequence $x[n]$ is related to the samples $X[k]$ through

$$x[n] = \frac{1}{N} \sum_{k=0}^{N-1} X[k] W_N^{-nk}.$$ (8.53)

If we multiply both sides of the above equation by N and take the complex conjugate, we arrive at

$$N x^*[n] = \sum_{k=0}^{N-1} X^*[k] W_N^{nk}.$$ (8.54)

The right-hand side of the above expression can be recognized as the N-point DFT of a sequence $X^*[k]$ and can be computed using any one of the FFT algorithms discussed earlier. The desired IDFT $x[n]$ is then obtained as

$$x[n] = \frac{1}{N} \left\{ \sum_{k=0}^{N-1} X^*[k] W_N^{nk} \right\}^*.$$ (8.55)

In summary, given an N-point DFT $X[k]$, we first form its complex conjugate sequence $X^*[k]$, then compute the N-point DFT of $X^*[k]$, form the complex conjugate of the DFT computed, and finally, divide each sample by N. The inverse DFT computation process is illustrated in Figure 8.30. Two other approaches to inverse DFT computation are described in Problems 8.22 and 8.23.

8.3.3   DFT and IDFT Computation Using MATLAB

The following functions are included in the MATLAB package for the computation of the DFT and the IDFT:

```
fft(x),          fft(x,N)
ifft(X),         ifft(X,N)
```

These programs employ efficient FFT algorithms for the computation. The function fft(x) computes the DFT of a vector x of the same length as that of x. For computing the DFT of a specific length N, the function fft(x,N) is used. Here, if the length of x is greater than N, it is truncated to the first N samples, whereas if the length of x is less than N, the vector x is zero-padded at the end to make it into a length N-sequence. Likewise, the function ifft(X) computes the IDFT of a vector X of the same length as that of X, whereas the function ifft(X,N) can be used to compute the IDFT of a specified length N. In the latter case, the restrictions on N are the same as that in the function fft(x,N). It should be noted that the function ifft in MATLAB employs the IDFT computation scheme outlined in the previous section.

MATLAB uses a high-speed radix-2 algorithm when the sequence length of x or X is a power of 2. Moreover, the radix-2 FFT program has been optimized specifically to compute the DFT of a real input sequence faster than the DFT of a complex sequence. If the sequence length is not a power of 2, it employs a mixed-radix FFT algorithm and usually takes a much longer time to compute the DFT or the IDFT of a sequence with a non power of 2 length N than those of a sequence of a power of 2 length that is closest to N.

Since vectors in MATLAB are indexed from 1 to N instead of 0 to $N - 1$, the DFT and the IDFT computed in the above MATLAB functions make use of the expressions

$$X[k] = \sum_{n=1}^{N} x[n] W_N^{(n-1)(k-1)}, \qquad k = 1, 2, \ldots, N, \tag{8.56a}$$

$$x[n] = \frac{1}{N} \sum_{k=1}^{N} X[k] W_N^{-(n-1)(k-1)}, \quad n = 1, 2, \ldots, N, \tag{8.56b}$$

Example 8.10 We consider here the computation of the DFT of a sinusoidal sequence of finite length. The MATLAB program used for this purpose is given below:

```
% Program 8_9
% Spectral Analysis Using the DFT
%
N = input('Type in the length of DFT = ');
T = input('Type in the sampling period = ');
freq = input('Type in the sinusoid frequency = ');
k = 0:N-1;
f = sin(2*pi*freq*k*(1/T));
```

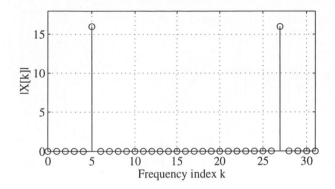

Figure 8.31   The magnitudes of a 32-point DFT of a sinusoid frequency 10 Hz sampled at a 64 Hz rate.

```
F = fft(f);
stem(k,abs(F));grid
xlabel('k')
ylabel('|X[k]|')
```

During execution, the program requests the following input data: length N of the DFT, the sampling frequency F_T in Hz, and frequency of the sinusoid in Hz. We compute here the 32-point DFT $X[k]$ of a length-32 sinusoidal sequence $x[n]$ of frequency 10 Hz with a sampling rate of 64 Hz. Note that the sampling frequency of 64 Hz is considerably higher than 20 Hz, the Nyquist frequency, and as result, no aliasing distortion will occur after sampling. The magnitudes of the DFT samples generated by this program are plotted in Figure 8.31.

Several comments are in order here. Since the time-domain sequence $x[n]$ is a pure sinusoid, its Fourier transform $X(e^{j2\pi f})$ contains two impulses at $f = \pm 10$ Hz and is zero everywhere else. Its 32-point DFT is obtained by sampling $X(e^{j2\pi f})$ at $f = 2k$ Hz, $k = 0, 1, \ldots, 31$. As a result, the impulse at $f = 10$ Hz appears as $X[5]$ at the DFT frequency bin location $k = 5$, while the impulse at $f = -10$ Hz appears as $X[27]$ at the bin location $k = 32 - 5 = 27$. Thus, the DFT has correctly identified the frequency of the sinusoid in this case. Note also that the first half of the DFT samples for $k = 0$ to $k = (N/2) - 1$ correspond to the positive frequency axis from $f = 0$ to $f = F_T/2$ excluding the point $F_T/2$, while the second half for $k = N/2$ to $k = N - 1$ corresponds to the negative frequency axis from $f = -F_T/2$ to $f = 0$ excluding the point $f = 0$.

We next compute the 32-point DFTs $X[k]$ of a sinusoid $x[n]$ of frequency 11 Hz with a sampling frequency of 64 Hz. Since the time-domain sequence $x[n]$ is a pure sinusoid, its Fourier transform $X(e^{j2\pi f})$ contains two impulses at $f = \pm 11$ Hz and is zero everywhere else. However, these frequency locations are between the bins $k = 5$ and $k = 6$, and $k = 26$ and $k = 27$. From the computed magnitudes of the DFT samples computed, shown in Figure 8.32, it can be seen that the spectrum contains frequency components at all bins, with two strong components at bins $k = 5$ and $k = 6$ and with two strong components at bins $k = 26$ and $k = 27$. This type of phenomenon is called the *leakage*. It is further discussed in detail in Section 11.2.

Example 8.11   In this example we illustrate the application of the FFT and IFFT functions of MAT-LAB in performing the linear convolution of two finite-length sequences. The MATLAB program that can be employed for this purpose is as follows:

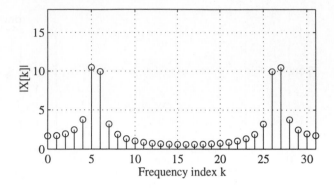

Figure 8.32   The magnitudes of a 32-point DFT of a sinusoid of frequency 11 Hz sampled at a 64 Hz rate.

```
% Program 8_10
% Computation of Linear Convolution Using DFT
%
g = input('Type in first sequence = ');
h = input('Type in second sequence = ');
ga = [g zeros(1,length(h)-1)];
ha = [h zeros(1,length(g)-1)];
G = fft(ga);H = fft(ha);
Y = G.*H;
y = ifft(Y);
disp('New Sequence')
disp(real(y))
```

We illustrate its use in developing the convolution of the two length-4 sequences of Example 3.11:

$$g[n] \quad = \quad \{1 \quad 2 \quad 0 \quad 1\}, \qquad h[n] \quad = \quad \{2 \quad 2 \quad 1 \quad 1\}.$$
$$\uparrow \qquad\qquad\qquad\qquad \uparrow$$

The input data requested by this program are the two sequences entered as vectors with square brackets. The result of the linear convolution using Program 8_10 for the above two sequences is given by

```
New Sequence
    2.0000    6.0000    5.0000    5.0000    4.0000
    1.0000    1.0000
```

which is seen to be identical to that determined in Example 3.11.

8.4   Number Representation

The binary representation is used to represent numbers (and signal variables) in most digital computers and special-purpose digital signal processors used for implementing digital filtering algorithms. In this form, the number is represented using the symbols 0 and 1, called *bits*,[7] with the *binary point* separating the integer part from the fractional part. For example, the binary representation of the decimal number 11.625 is given by

$$1011_\triangle 101$$

where $\triangle$ denotes the binary point. The four bits, 1011, to the left of the binary point form the integer part and the three bits, 101, to the right of the binary point represent the fractional part. In general, the decimal equivalent of a binary number η consisting of B integer bits and b fractional bits,

$$a_{B-1}a_{B-2}\cdots a_1 a_0{}_\triangle a_{-1}a_{-2}\cdots a_{-b}$$

is given by

$$\sum_{i=-b}^{B-1} a_i 2^i,$$

where each bit a_i is either a 0 or a 1. The leftmost bit a_{B-1} is called the *most significant bit*, abbreviated *MSB*, and the rightmost bit a_{-b} is called the *least significant bit*, abbreviated *LSB*.

To avoid the confusion between a decimal number containing the digits 1 and 0, and a binary number containing the bits 1 and 0 we shall include a subscript 10 to the right of the least significant digit to indicate a decimal number and a subscript 2 to the right of the least significant bit to indicate a binary number. Thus, for example, 1101_{10} represents a decimal number, whereas 1101_2 represents a binary number whose decimal equivalent is 13_{10}. If there is no ambiguity in the representation, then the subscript is dropped.

The block of bits representing a number is called a *word*, and the number of bits in the word is called the *wordlength* or *word size*. The wordlength is typically a positive integer power of 2, such as 8, 16, 32, and so on. The word size is often expressed in units of 8 bits called a *byte*. For example, a 4-byte word is equivalent to a 32-bit-size word.

Digital circuits implementing the arithmetic operations, addition and multiplication of two binary numbers, are specifically designed to develop the results, the sum and the product, respectively, in binary form with their binary points in the assumed locations. There are two basic types of binary representations of numbers, fixed-point and floating-point, as discussed below.

8.4.1 Fixed-Point Representation

In this type of representation, the binary point is assumed to be fixed at a specific location, and the hardware implementation of the arithmetic circuits take into account the fixed location in performing the arithmetic operations. Implementation of the addition operation carried out by a digital adder circuit is independent of the location of the binary points in the two numbers being added as long as they are in the same location for both numbers. On the other hand, it is not simple to locate the binary point in the product of two binary numbers unless they are both integers or are both fractions. In the case of the multiplication of two integers by the multiplier circuit, the result is also an integer. Likewise, multiplying

---

[7] bit is an abbreviated form of binary digit.

two fractions results in a fraction. In digital signal processing applications, therefore, fixed-point numbers are always represented as fractions.

The range of positive integers η that can be represented by B bits in a fixed-point representation is given by

$$0 \leq \eta \leq 2^B - 1. \tag{8.57}$$

Similarly, the range of positive fractions η that can be represented by B bits in a fixed-point representation is given by

$$0 \leq \eta \leq 1 - 2^{-B}. \tag{8.58}$$

In either case, the range is fixed. If η_{max} and η_{min} denote, respectively, the maximum and the minimum value of the numbers that can be represented in a B-bit fixed-point representation, then the dynamic range R of the numbers that can be represented with B bits is given by as $R = \eta_{max} - \eta_{min}$, and the *resolution* of the representation is defined by

$$\delta = \frac{R}{2^B - 1}, \tag{8.59}$$

where δ is also known as the *quantization level*.

8.4.2 Floating-Point Representation

In the normalized floating-point representation, a positive number η is represented using two parameters, the mantissa M and the exponent or the characteristic E in the form

$$\eta = M \cdot 2^E, \tag{8.60}$$

where the mantissa M is a binary fraction restricted to lie in the range

$$\tfrac{1}{2} \leq M < 1, \tag{8.61}$$

and the exponent E is either a positive or a negative binary integer.

The floating-point system provides a variable resolution for the range of numbers being represented with the resolution increasing exponentially with an increase in the value of the number being represented. Floating-point numbers are stored in a register by assigning B_E bits of the register to the exponent and the remaining B_M bits to the mantissa. If the same number of total bits are used in both floating-point and fixed-point representations, i.e., $B = B_M + B_E$, then the former provides a larger dynamic range than the latter (Problem 8.41).

The most widely followed floating-point formats for 32-bit and 64-bit representations are that given by the ANSI/IEEE Standard 754-1985 [IEEE85]. In this format, a 32-bit number is divided into fields. The exponent field is of 8-bit length, the mantissa[8] field is of 23-bit length, and 1 bit is assigned for the sign, as indicated in Figure 8.33. The exponent is coded in a biased form as $E - 127$. Thus, a floating-point number η under this scheme is represented as

$$\eta = (-1)^S \cdot 2^{E-127}(M), \tag{8.62}$$

with the mantissa in the range

$$0 \leq M < 1. \tag{8.63}$$

The following conventions are followed in interpreting the representation of Eq. (8.62):

---

[8] In the IEEE floating-point standard, the mantissa is called the *significand*.

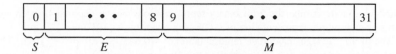

Figure 8.33   IEEE 32-bit floating-point format.

Figure 8.34   Representation of a general signed b-bit fixed-point fraction.

1. If $E = 255$ and $M \neq 0$, then η is not a number (abbreviated as *NaN*),

2. If $E = 255$ and $M = 0$, then $\eta = (-1)^S \cdot \infty$,

3. If $0 < E < 255$, then $\eta = (-1)^S \cdot 2^{E-127}(1_\Delta M)$,

4. If $E = 0$ and $M \neq 0$, then $\eta = (-1)^S \cdot 2^{-126}(0_\Delta M)$, and

5. If $E = 0$ and $M = 0$, then $\eta = (-1)^S \cdot 0$,

where $1_\Delta M$ is a number with one integer bit and 23 fractional bits and $0_\Delta M$ is a fraction. The range of a 32-bit floating point number in the above format is from 1.18×10^{-38} to 3.4×10^{38} (Problem 8.42). The IEEE 32-bit standard has been adopted for number representation in almost all commercial floating-point digital signal processor chips.

8.4.3   Representation of Negative Numbers

To accommodate the representation of both positive and negative b-bit fractions, an additional bit, called the *sign bit*, is placed at the leading position of the register to indicate the sign of the number (Figure 8.34). Independent of the scheme being used to represent the negative number, the *sign bit* is 0 for a positive number and 1 for a negative number.

A fixed-point negative number is represented in one of three different forms. In the *sign-magnitude* format, if the sign bit $s = 0$, the b-bit fraction is a positive number with a magnitude $\sum_{i=1}^{b} a_{-i} 2^{-i}$, and if $s = 1$, the b-bit fraction is a negative number with a magnitude $\sum_{i=1}^{b} a_{-i} 2^{-i}$.

In the *ones'-complement* form, a positive fraction is represented as in the sign-magnitude form, while its negative is represented by complementing each bit of the binary representation of the positive fraction. In this representation, the decimal equivalent of a positive or a negative fraction is thus given by $-s(1 - 2^{-b}) + \sum_{i=1}^{b} a_{-i} 2^{-i}$.

Finally, in the *two's-complement* representation, the positive fraction is represented as in the sign-magnitude form, while its negative is represented by complementing each bit (i.e., by replacing each 0 with a 1, and vice-versa) of the binary representation of the positive fraction, and adding a 1 to the LSB, the bth bit. In this form, the decimal equivalent of a positive or a negative fraction is thus given by $-s + \sum_{i=1}^{b} a_{-i} 2^{-i}$.

Example 8.12 We determine the 5-bit binary representations of the decimal fraction -0.625_{10}. Now, the sign-magnitude representation of 0.625_{10} is given by $0_\Delta 1010$ ($= 2^{-1} + 2^{-3} = 0.5 + 0.125$). Therefore, the sign-magnitude representation of -0.625_{10} is simply $1_\Delta 1010$. The ones'-complement representation -0.625_{10} is obtained by complementing each bit of $0_\Delta 1010$ individually, leading to $1_\Delta 0101$. Finally, the two's-complement representation of -0.625_{10} is obtained by adding a 1 to the LSB of the ones'-complement representation, resulting in $1_\Delta 0110$.

Table 8.1 illustrates the above three representations for a 4-bit number (3-bit fraction and a sign bit).

8.4.4 Offset Binary Representation

The offset binary representation, used primarily in bipolar digital-to-analog conversion, a b-bit fraction with an additional sign bit is considered as a $(b+1)$-bit number representing 2^{b+1} decimal numbers. About half of these numbers represent the negative fractions and the remaining half represent the positive fractions, as illustrated in Table 8.1 for $b = 3$. It should be noted that the two's-complement representation can be converted to the offset binary representation by simply complementing the sign bit.

8.4.5 Signed Digit Representation

The radix-2 signed digit (SD) format is a 3-valued representation of a radix-2 number and employs three digit values, 0, 1, and $\bar{1}$ (with the last symbol representing -1). In many instances, the SD representation of a binary number requires fewer nonzero digits and has been exploited in developing an algorithm for a faster hardware implementation of the multiplication operation (see Section 8.5.2). A simple algorithm for the conversion of a radix-2 binary number into an equivalent SD representation is as follows [Boo51]. Let $a_{-1}a_{-2}\ldots a_b$ denote a binary number with its equivalent SD representation denoted by $c_{-1}c_{-2}\ldots c_b$. The bits c_{-i} of the SD number are determined through the relation

$$c_{-i} = a_{-i-1} - a_{-i}, \qquad i = b, b-1, \ldots, 1,$$

where $a_{-b-1} = 0$.

The following example illustrates the radix-2 SD representation.

Example 8.13 Consider the binary number $\eta = 0_\Delta 0111_2$ whose decimal equivalent is 0.4375_{10}. Its various equivalent SD representations are indicated below, along with their decimal equivalents:

$$0_\Delta 100\bar{1} = (0.5)_{10} - (0.0625)_{10}$$
$$0_\Delta 10\bar{1}1 = (0.5625)_{10} - (0.125)_{10}$$
$$0_\Delta 1\bar{1}11 = (0.6875)_{10} - (0.25)_{10}$$

As can be seen from the above, the SD representation is not unique, and a representation with the fewest number of nonzero digits is called a *minimal* SD representation. A minimal SD representation containing no adjacent nonzero digits is called a *canonic signed-digit (CSD)* representation. An algorithm to derive the CSD representation of a binary number is given in Hwang [Hwa79].

Table 8.1 Binary number representations.

Decimal Equivalent	Sign-Magnitude	One's-Complement	Two's-Complement	Offset Binary
7/8	$0_\triangle 111$	$0_\triangle 111$	$0_\triangle 111$	$1_\triangle 111$
6/8	$0_\triangle 110$	$0_\triangle 110$	$0_\triangle 110$	$1_\triangle 110$
5/8	$0_\triangle 101$	$0_\triangle 101$	$0_\triangle 101$	$1_\triangle 101$
4/8	$0_\triangle 100$	$0_\triangle 100$	$0_\triangle 100$	$1_\triangle 100$
3/8	$0_\triangle 011$	$0_\triangle 011$	$0_\triangle 011$	$1_\triangle 011$
2/8	$0_\triangle 010$	$0_\triangle 010$	$0_\triangle 010$	$1_\triangle 010$
1/8	$0_\triangle 001$	$0_\triangle 001$	$0_\triangle 001$	$1_\triangle 001$
0	$0_\triangle 000$	$0_\triangle 000$	$0_\triangle 000$	$1_\triangle 000$
−0	$1_\triangle 000$	$1_\triangle 111$	N/A	N/A
−1/8	$1_\triangle 001$	$1_\triangle 110$	$1_\triangle 111$	$0_\triangle 111$
−2/8	$1_\triangle 010$	$1_\triangle 101$	$1_\triangle 110$	$0_\triangle 110$
−3/8	$1_\triangle 011$	$1_\triangle 100$	$1_\triangle 101$	$0_\triangle 101$
−4/8	$1_\triangle 100$	$1_\triangle 011$	$1_\triangle 100$	$0_\triangle 100$
−5/8	$1_\triangle 101$	$1_\triangle 010$	$1_\triangle 011$	$0_\triangle 011$
−6/8	$1_\triangle 110$	$1_\triangle 001$	$1_\triangle 010$	$0_\triangle 010$
−7/8	$1_\triangle 111$	$1_\triangle 000$	$1_\triangle 001$	$0_\triangle 001$
−8/8	N/A	N/A	$1_\triangle 000$	$0_\triangle 000$

8.4.6 Hexadecimal Representation

In writing the codes for programmable DSP chips, we often prefer the hexadecimal representation of a binary number for compactness. In this format, each binary number is divided into groups of 4 bits beginning at the binary point. The decimal equivalent of each 4-bit group is then represented by one of 16 symbols formed by the 10 decimal digits 0 through 9 and the six letters of the alphabet A through F, with A representing the decimal 10, B representing decimal 11, etc. The conversion process is illustrated next.

Example 8.14 Consider the 16-bit binary number 1101001011000101. We group this number into 4 groups of 4-bit each as indicated below. The decimal equivalent of the first group of 4 bits is 13, the decimal equivalent of the second group of 4 bits is 2, decimal equivalent of the third group of 4 bits is 12, and finally the decimal equivalent of the last group of 4 bits is 5:

$$\underbrace{1101}_{D}\ \underbrace{0010}_{2}\ \underbrace{1100}_{C}\ \underbrace{0101}_{5} \quad \leftarrow \text{binary}$$
$$\leftarrow \text{hexadecimal}$$

Since the hexadecimal digits denoting 13 and 12 are, respectively, D and C, the hexadecimal representation of the 16-bit binary number 1101001011000101 is D2C5.

8.5   Arithmetic Operations

The two basic arithmetic operations in the implementation of digital filtering algorithms are addition (subtraction) and multiplication. We illustrate the corresponding implementation algorithms for binary numbers.

8.5.1   Fixed-Point Addition

The procedure for the addition of two binary numbers represented in the signed-magnitude representation, is similar to that used in the addition of two decimal numbers. The serial addition of two positive binary fractions $0_\Delta a_{-1} a_{-2} \ldots a_{-b}$ and $0_\Delta d_{-1} d_{-2} \ldots d_{-b}$ is carried out in b steps. At the first step, we add the two LSBs a_{-b} and d_{-b} whose sum will consist of two bits, the first bit is a carry bit c_{-b} and the second bit is the LSB s_{-b} of the final sum. At all other steps, we add three bits, a_{-i}, d_{-i}, and c_{-i-1}, where c_{-i-1} is the carry bit generated at the previous step. Their sum again will consist of a carry bit c_i and a sum bit s_i. If the addition of the two positive binary fractions is still a binary fraction, the above process yields a correct result. This happens if at the bth step, the addition of a_{-1}, d_{-1}, and c_{-2} does not create a carry bit 1. However, if their sum is no longer a fraction, an *overflow* is said to have occurred, and the carry bit $c_i = 1$ generated is used to indicate the overflow.

We illustrate the addition process in the following example.

Example 8.15 Consider the binary addition of the two positive binary fractions $0_\Delta 1001$ and $0_\Delta 0101$, whose decimal equivalents are, respectively, 0.5625_{10} and 0.3125_{10}. The serial addition of these two numbers is shown below and is carried out in four steps:

Note that the sum of these two numbers is the positive binary fraction $0_\Delta 1110$ whose decimal equivalent is 0.875_{10} and no overflow has occurred.

Now, consider the binary addition of the two positive binary fractions $0_\Delta 1101$ and $0_\Delta 0111$, whose decimal equivalents are, respectively, 0.8175_{10} and 0.4375_{10}. The serial addition of these two numbers is illustrated below:

In this case, the addition of two positive fractions has resulted in a sum whose decimal equivalent, 1.25_{10}, is no longer a fraction and thus appears as a negative fraction. The appearance of a sign bit of 1 in the sum of two positive fractions indicates an overflow.

The subtraction of a positive binary number from another positive binary number, or equivalently, the addition of a positive binary number to a negative binary number in signed-magnitude form, is slightly more complicated since here we may have to introduce a borrowing process. The following example illustrates the method.

Example 8.16 Consider the subtraction of $0_\triangle 0111$ (called the subtrahend) from $0_\triangle 1101$ (called the minuend). The subtraction is carried out sequentially in four steps as shown below.

$$
\begin{array}{rcccccl}
 & -1 & -1 & & & & \leftarrow \text{ borrow} \\
0_\triangle & 1 & 11 & 10 & 1 & & \leftarrow \text{ minuend} \\
- \; 0_\triangle & 0 & 1 & 1 & 1 & & \leftarrow \text{ subtrahend} \\
\hline
0_\triangle & 0 & 1 & 1 & 0 & & \leftarrow \text{ difference}
\end{array}
$$

In the first step, the LSB of the subtrahend is subtracted from the LSB of the minuend. Since both LSBs are 1, their difference is 0. Next, the difference of the two bits to the left of the LSB is formed. Note that here the bit in the minuend is 0, whereas the corresponding bit in the subtrahend is 1. Thus, to form the difference, we borrow a 1 from the bit to its left in the minuend, making its effective value 10. Subtracting a 1 from it results in a difference of 1. This process is continued until the difference of the MSBs is computed.

Note that in the above example, the subtrahend is smaller in value than of the minuend, resulting in the correct difference. However, in the opposite case, the arithmetic operation will cause an *overflow*.

As illustrated above, the algorithms for the addition and subtraction operations for binary numbers in the signed-magnitude form are different, and therefore, separate circuits are needed in their hardware implementations. On the other hand, both operations can be carried out using essentially the same algorithm for binary numbers in either the ones'-complement or the two's-complement form. The basic difference in the arithmetic operations between the two representations is in the handling of the carry bit resulting in the last step. The following two examples illustrate the respective algorithms.

Example 8.17 Consider the computation of the difference $0_\triangle 1101 - 0_\triangle 0111$. We can treat the subtraction operation as an addition of a positive number $0_\triangle 1101$ to a negative number $-0_\triangle 0111$. The two's-complement representation of $-0_\triangle 0111$ is simply $1_\triangle 1001$. In the addition of the positive binary fraction $0_\triangle 1101$ and the two's-complement negative binary fraction $1_\triangle 1001$, we treat the sign bits as the rest of the number as indicated below.

$$
\begin{array}{rcccccc}
 & 0_\triangle & 1 & 1 & 0 & 1 \\
+ & 1_\triangle & 1 & 0 & 0 & 1 \\
\hline
1 & 0_\triangle & 0 & 1 & 1 & 0 \\
\uparrow & & & & & \\
\text{drop} & & & & &
\end{array}
$$

By dropping the carry bit generated at the fifth step, we arrive at the correct result given by $0_\triangle 0110$.

Example 8.18 We repeat the problem of the previous example by representing the negative number $-0_\triangle 0111$ in its ones'-complement form, which is given by $1_\triangle 1000$. The addition of $0_\triangle 1101$ and $1_\triangle 1000$ is shown below.

$$
\begin{array}{cccccccc}
 & 0 & _\Delta 1 & 1 & 0 & 1 \\
+ & 1 & _\Delta 1 & 0 & 0 & 0 \\
\hline
1 & 0 & _\Delta 0 & 1 & 0 & 1 \\
\end{array}
$$

$+\qquad\qquad\qquad\qquad\qquad 1 \leftarrow$ end around carry

$$
\begin{array}{cccccc}
\hline
0 & _\Delta 0 & 1 & 1 & 0 \\
\end{array}
$$

Here also, the addition operation has created an extra bit 1 to the left of the sign bit due to the carry bit 1 generated at the last step. This bit is next added to the LSB position of the intermediate sum resulting in the correct sum $0_\Delta 0110$.

8.5.2   Fixed-Point Multiplication

The multiplication of two b-bit binary fractions, $A = a_{s\Delta} a_{-1} a_{-2} \ldots a_{-b}$ (called the multiplicand) and $D = d_{s\Delta} d_{-1} d_{-2} \ldots d_{-b}$ (called the multiplier), in sign-magnitude form is carried out by forming the product $P^{(b)}$ of their respective magnitudes first, and then assigning the appropriate sign to the product from the signs of the multiplier and the multiplicand. The product of the magnitudes can be implemented serially in b steps where at the ith step the partial product $P^{(i)}$ is determined as follows:

$$
P^{(i)} = \left(P^{(i-1)} + d_{-b+i-1} \cdot A \right) \cdot 2^{-1}, \qquad i = 1, 2, \cdots, b, \tag{8.64}
$$

with $P^{(0)} = 0$. It follows from above that if $d_{-b+i-1} = 0$, the new partial product is obtained by simply shifting the previous partial product to the right by one bit position. On the other hand, if $d_{-b+i-1} = 1$, the multiplicand A is added to the previous partial product and then shifted to the right by one bit position to arrive at the new partial product. The final product is a $2b$-bit fraction. The following example illustrates the algorithm.

Example 8.19 Consider the multiplication of the two 3-bit binary fractions, $a_{s\Delta}110$ and $d_{s\Delta}101$. We first form the product of their magnitudes using the recursive algorithm of Eq. (8.64) as illustrated below.

$$
\begin{array}{rccccccl}
  & 1 & 1 & 0 &   &   & = & A \\
\times & 1 & 0 & 1 &   &   & = & D \\
\hline
  & 0 & 0 & 0 &   &   & = & P^{(0)} \\
+ & 1 & 1 & 0 &   &   &   & \\
\hline
  & 1 & 1 & 0 &   &   &   & \\
\hline
  & 0 & 1 & 1 & 0 &   &   & P^{(1)} \\
\hline
  & 0 & 0 & 1 & 1 & 0 &   & P^{(2)} \\
+ & 1 & 1 & 0 &   &   &   & \\
\hline
  & 1 & 1 & 1 & 1 & 0 &   & \\
\hline
  & 0 & 1 & 1 & 1 & 1 & = & P^{(3)} \\
\end{array}
$$

The final answer in sign-magnitude form is $p_{s\Delta}011110$ where the sign of the product $p_s = 0$ if $a_s = d_s$, otherwise it is 1. Note that the product is a 6-bit fraction.

If the multiplicand is a negative fraction in either two's-complement form or in ones'-complement form, and the multiplier is a positive fraction, the algorithm of Eq. (8.64) can be followed without any change, except that the addition of the negative multiplicand is carried out according to the method outlined in Section 8.5.1, and after shifting the sum, the sign bit of the sum is left as is. On the other hand, if the multiplier is a negative fraction, a correction step is needed at the end to arrive at the correct result [Kor93].

In Booth's multiplication algorithm, employed in most DSP chips, the serial multiplication process described above is implemented, with the multiplier recoded in the SD form resulting, in general, in a faster operation [Boo51]. For example, a multiplier $0_\Delta 011001111_2$ requires 6 additions, whereas its SD representation would require 4 add/subtract operations. Booth's algorithm generates the correct product if both the multiplicand and the multiplier are represented in the two's-complement form provided the sign bit of the multiplicand is employed in determining whether to perform an add or a subtract operation at the last step.

Example 8.20 Let $A = 1_\Delta 101 = -0.375_{10}$ and $D = 1_\Delta 011 = -0.625_{10}$ represented in two's-complement form. SD representation of D is given by $0_\Delta \bar{1}0\bar{1}$. The multiplication of $1_\Delta 101$ with $0_\Delta \bar{1}0\bar{1}$ using Booth's algorithm is illustrated below:

```
sign bit ⌐
            1   1   0   1              = A = − 0.325₁₀
   ×    0   1̄   0   1̄
        ─────────────────
            0   0   0   0              = P⁽⁰⁾
   +    0   0   1   1                  Add − A
        ─────────────────
            0   0   1   1
        ─────────────────
            0   0   0   1   1          = P⁽¹⁾
            0   0   0   0   1   1       = P⁽²⁾
   +    0   0   1   1                  Add − A
        ─────────────────
            0   0   1   1   1   1
        ─────────────────
        0 ₐ 0   0   1   1   1   1      = P⁽³⁾
```

The final product is a positive binary fraction $0_\Delta 001111$ of value 0.234375_{10}, as expected.

8.5.3 Floating-Point Arithmetic

Addition of two floating-point binary numbers can be carried out easily, if their exponents are equal. Thus, in a floating-point adder, the mantissa of the smaller number is shifted to the right by an appropriate number of bits to make its exponent equal to that of the larger number, and then the two mantissas are added. If the sum of the two mantissas is within the range of Eq. (8.60), then no additional normalization is necessary; otherwise, a normalization is carried out to bring it to the proper range with a corresponding adjustment to the exponent.

Example 8.21 Consider the addition of the two floating-point numbers

$$\eta_1 = (0_\Delta 1110)2^{01}$$
$$\eta_2 = (0_\Delta 1101)2^{10}$$

whose decimal equivalents are, respectively, 1.75_{10} and 3.25_{10}. We first rewrite the floating-point representation of the smaller number η_1 so that its characteristic is equal to that of the larger number η_2:

$$\eta_1 = (0_\Delta 0111)2^{10}.$$

Then the sum of these two numbers is given by

$$\eta_1 + \eta_2 = (0_\Delta 1101 + 0_\Delta 0111)2^{10} = (1_\Delta 0100)2^{10}.$$

Since the mantissa of the sum is greater than 1, its characteristic is therefore increased by one unit and the mantissa is shifted to the right by one bit, resulting in

$$\eta_1 + \eta_2 = (0_\Delta 1010)2^{11},$$

whose decimal equivalent is 5_{10} as expected.

In a floating-point multiplier, on the other hand, the multiplication is carried out by multiplying the two mantissas and adding their corresponding exponents. Since the range of the product of the mantissas is now between $\frac{1}{4}$ and 1, a normalization of the product is carried out if it is less than $\frac{1}{2}$ along with a corresponding adjustment to the new exponent.

Example 8.22 Consider the multiplication of the two floating-point numbers

$$\eta_1 = (0_\Delta 1010)2^{01},$$
$$\eta_2 = (0_\Delta 1001)2^{10},$$

whose decimal equivalents are, respectively, 1.25_{10} and 2.25_{10}. Their product is thus given by

$$\eta_1 \eta_2 = (0_\Delta 1010)(0_\Delta 1001)2^{01+10} = (0_\Delta 0101101)2^{11}.$$

As the mantissa of the product is less than $\frac{1}{2}$, the characteristic is reduced by one unit and the mantissa is shifted to the left by one bit resulting in the correct floating-point representation

$$\eta_1 \eta_2 = (0_\Delta 1011010)2^{10},$$

whose decimal equivalent is 2.8125_{10} as expected.

8.6   Handling of Overflow

In Section 8.5, we pointed out that the results of the addition of two fixed-point fractions can result in a sum exceeding the dynamic range of the register storing the result of the addition, thus resulting in an overflow. Occurrence of overflow leads to severe output distortion and may often result in large-amplitude oscillations at the filter output (see Section 9.8). Therefore, the sum should be substituted with another number that is within the dynamic range. Two widely used schemes for handling the overflow are described next.

Let η denote the sum, then in either of the two schemes, if η exceeds the dynamic range $[-1, 1)$, it is substituted with a number ξ which is within the range. In the *saturation overflow* scheme shown in Figure 8.35(a), if $\eta \geq 1$, it is replaced with $1 - 2^{-b}$, where the number is assumed to be a b-bit fraction with an additional bit for the sign, and if $\eta \leq -1$, it is replaced with -1. On the other hand, in the *two's-complement overflow* scheme shown in Figure 8.35(b), whenever η is outside the range $[-1, 1)$, it is replaced with $\xi = \langle \eta \rangle_2$.

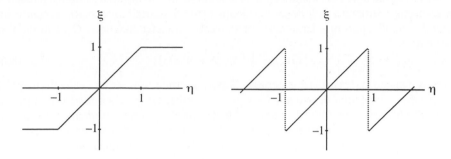

Figure 8.35   (a) Saturation overflow, and (b) two's-complement overflow.

Basically, here, when η is outside the range, the bits to the left of the sign bit (overflow bits) are ignored. The second scheme is usually implemented in nonrecursive digital filters employing two's-complement arithmetic. In most applications, the first scheme is usually preferred.

We shall discuss in Section 9.6 the dynamic range scaling of the digital filter structure to either eliminate completely or reduce the probability of overflow. The effects of the two overflow handling schemes on the performance of the digital filter are also considered in this section.

8.7   Tunable Digital Filters

Many applications require the use of digital filters with easily tunable characteristics. In Section 6.8, we outlined the design of tunable first-order and second-order digital filters that may provide adequate solutions in some applications. There are other applications where use of higher order tunable digital filters may be necessary. The design of such tunable filters is the subject of this section.

8.7.1   Tunable IIR Digital Filters

The basis for the design of tunable digital filters is the spectral transformation discussed in Section 7.6, which can be used to tune a given digital filter realization with a specified cutoff frequency to another realization with a different cutoff frequency . Thus, if $G_{\text{old}}(z)$ is the transfer function of the original realization, the transfer function of the new structure is $G_{\text{new}}(z)$ where

$$G_{\text{new}}(z) = G_{\text{old}}(z)|_{z^{-1}=F^{-1}(z)}, \tag{8.65}$$

in which $F^{-1}(z)$ is a stable allpass function of the form given in Table 7.2 with the parameters of the transformation being the tuning parameters. One straightforward way to implement this transformation would be to replace each delay block in the realization of $G_{\text{old}}(z)$ with an allpass structure realizing $F^{-1}(z)$. However, such an approach leads, in general, to a structure realizing $G_{\text{new}}(z)$ with delay-free-loops and cannot be implemented, as explained in Section 6.1.3.

We describe now a very simple practical modification to the above approach that does not result in a structure with delay-free-loops [Mit90b]. In Section 6.10, we outlined a method of realization of a large class of stable IIR transfer functions $G(z)$ in the form [Vai86a]:

$$G(z) = \tfrac{1}{2} \{A_0(z) + A_1(z)\}, \tag{8.66}$$

where $A_0(z)$ and $A_1(z)$ are stable allpass filters. The conditions for the realization of $G(z)$ as a parallel connection of two allpass sections are that $G(z)$ be a bounded real transfer function with a symmetric numerator and that it have a power complementary transfer function $H(z)$ with an antisymmetric numerator. These conditions are satisfied by all odd-order lowpass Butterworth, Chebyshev, and elliptic transfer functions.

The allpass filters $A_0(z)$ and $A_1(z)$ can be realized using any one of the approaches discussed in Section 6.7. We consider here their realizations as a cascade of first-order and second-order sections. As indicated in Section 6.7, there are a large variety of structurally lossless canonic structures realizing the first-order and second-order allpass transfer functions [Mit74a]; [Szc88]. These structures use only one multiplier and one delay for the realization of a first-order allpass function, and two multipliers and two delays for the realization of a second-order allpass function.

Consider first the tuning of the cutoff frequency of a lowpass IIR filter realized by a parallel allpass structure. From Section 7.6, we note that the lowpass-to-lowpass transformation is given by

$$z^{-1} \rightarrow F^{-1}(z^{-1}) = \frac{z^{-1} - \alpha}{1 - \alpha z^{-1}}, \tag{8.67}$$

where the parameter α is related to the old and new cutoff frequencies, ω_c and $\hat{\omega}_c$, respectively, through

$$\alpha = \frac{\sin[(\omega_c - \hat{\omega}_c)/2]}{\sin[(\omega_c + \hat{\omega}_c)/2]}. \tag{8.68}$$

Substituting the transformation of Eq. (8.67) on a Type 1 first-order allpass transfer function

$$a_1(z) = \frac{d_1 + z^{-1}}{1 + d_1 z^{-1}}, \tag{8.69}$$

we obtain the new first-order allpass transfer function

$$\hat{a}_1(z) = a_1(z)|_{z^{-1}=(z^{-1}-\alpha)/(1-\alpha z^{-1})} = \frac{d_1 + \left(\frac{z^{-1}-\alpha}{1-\alpha z^{-1}}\right)}{1 + d_1 \left(\frac{z^{-1}-\alpha}{1-\alpha z^{-1}}\right)}$$

$$= \frac{(d_1 - \alpha) + (1 - \alpha d_1)z^{-1}}{(1 - \alpha d_1) + (d_1 - \alpha)z^{-1}} = \frac{\left(\frac{d_1 - \alpha}{1 - \alpha d_1}\right) + z^{-1}}{1 + \left(\frac{d_1 - \alpha}{1 - \alpha d_1}\right)z^{-1}}. \tag{8.70}$$

If α is very small, we can make a Taylor series expansion of the coefficient $(d_1 - \alpha)/(1 - \alpha d_1)$ of the allpass function $\hat{a}_1(z)$ of Eq. (8.70) and arrive at an approximation

$$\hat{a}_1(z) \cong \frac{[d_1 + \alpha(d_1^2 - 1)] + z^{-1}}{1 + [d_1 + \alpha(d_1^2 - 1)]z^{-1}}, \tag{8.71}$$

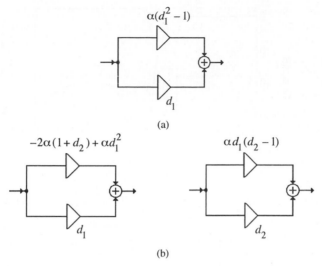

Figure 8.36   Multiplier replacement schemes in the constituent allpass sections for designing tunable IIR filters: (a) Type 1 allpass network and (b) Type 3 allpass network.

which is seen to be a Type 1 first-order allpass transfer function with a coefficient that is now a linear function of α. The approximated allpass section $\hat{a}_1(z)$ can be simply implemented by replacing each multiplier d_1 in Figure 6.36 with a parallel connection of two multipliers, as indicated in Figure 8.36(a). Note that for $\alpha = 0$, $\hat{a}_1(z)$ of Eq. (8.71) reduces to $a_1(z)$ of Eq. (8.69), as expected.

Applying a similar procedure to the Type 3 second-order allpass transfer function

$$a_2(z) = \frac{d_2 + d_1 z^{-1} + z^{-2}}{1 + d_1 z^{-1} + d_2 z^{-2}}, \tag{8.72}$$

we arrive at

$$\hat{a}_2(z) = a_2(z)\big|_{z^{-1}=(z^{-1}-\alpha)/(1-\alpha z^{-1})} = \frac{d_2 + d_1 \left(\frac{z^{-1}-\alpha}{1-\alpha z^{-1}}\right) + \left(\frac{z^{-1}-\alpha}{1-\alpha z^{-1}}\right)^2}{1 + d_1 \left(\frac{z^{-1}-\alpha}{1-\alpha z^{-1}}\right) + d_2 \left(\frac{z^{-1}-\alpha}{1-\alpha z^{-1}}\right)^2}. \tag{8.73}$$

Again, if we assume α to be very small, we can rewrite the above expression as

$$\begin{aligned}
\hat{a}_2(z) &\cong \frac{(d_2 - \alpha d_1) + (d_1 - 2\alpha[1 + d_2]) z^{-1} + (1 - \alpha d_1)z^{-2}}{(1 - \alpha d_1) + (d_1 - 2\alpha[1 + d_2]) z^{-1} + (d_2 - \alpha d_1)z^{-2}} \\[2mm]
&= \frac{\left(\frac{d_2-\alpha d_1}{1-\alpha d_1}\right) + \left(\frac{d_1-2\alpha[1+d_2]}{1-\alpha d_1}\right) z^{-1} + z^{-2}}{1 + \left(\frac{d_1-2\alpha[1+d_2]}{1-\alpha d_1}\right) z^{-1} + \left(\frac{d_2-\alpha d_1}{1-\alpha d_1}\right) z^{-2}},
\end{aligned} \tag{8.74}$$

by neglecting coefficients containing α^2. Next, an approximation based on the Taylor's series expansion of the last term in the above equation results in

$$\hat{a}_2(z) \cong \frac{[d_2 + \alpha d_1 (d_2 - 1)] + [d_1 - 2\alpha(1 + d_2) + \alpha d_1^2] z^{-1} + z^{-2}}{1 + [d_1 - 2\alpha(1 + d_2) + \alpha d_1^2] z^{-1} + [d_2 + \alpha d_1(d_2 - 1)] z^{-2}}, \tag{8.75}$$

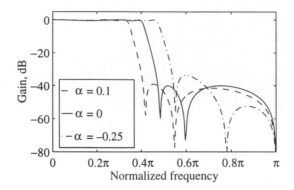

Figure 8.37   Gain responses of a fifth-order tunable elliptic lowpass filter for various values α.

Figure 8.38   Lowpass-to-bandpass transformation.

which is seen to be a Type 3 second-order allpass transfer function with coefficients that are a linear function of α. The approximated allpass section $\hat{a}_2(z)$ can be simply implemented by replacing the multipliers d_1 and d_2 in Figure 6.38 with a parallel connection of two multipliers, as indicated in Figure 8.36(b). Note that for $\alpha = 0$, $\hat{a}_2(z)$ of Eq. (8.75) reduces to $a_2(z)$ of Eq. (8.72) as expected.

Note that in the case of both the first-order and the second-order allpass filters, the tuning rule is now a linear function of α. Even though this tuning algorithm is approximate and has been derived for small values of α, in practice, tuning ranges of several octaves have been observed to hold in the case of narrowband lowpass elliptic filters [Mit90b]. Figure 8.37 shows the gain responses of a fifth-order tunable elliptic lowpass filter for three values of the tuning parameter α. The prototype filter ($\alpha = 0$) has a passband edge at $\omega_p = 0.4\pi$, passband ripple of 0.5 dB, and minimum stopband attenuation of 40 dB.

By applying a lowpass-to-bandpass transformation

$$z^{-1} \rightarrow F^{-1}(z^{-1}) = -z^{-1}\frac{z^{-1}+\beta}{1+\beta z^{-1}} \tag{8.76}$$

to a tunable lowpass IIR filter, we can design a tunable bandpass filter whose center frequency ω_0 is tuned by adjusting the parameter $\beta = \cos\omega_0$, and the bandwidth is tuned by changing α [Mit90b]. Unlike the lowpass-to-lowpass transformation of Eq. (8.67), the transformation of Eq. (8.76) can be directly implemented on the structure realizing the tunable lowpass filter by replacing each delay with the structure shown in Figure 8.38 to arrive at a tunable bandpass filter structure.

8.7.2 Tunable FIR Digital Filters

The spectral transformation approach of Section 7.6 can also be applied to an FIR filter to develop a filter structure with tunable characteristics. However, the resulting structure is no longer FIR since the replacement of the delays in the prototype FIR structure by allpass sections implementing the spectral transformation makes it an IIR filter. We outline below a straightforward method for designing tunable linear-phase lowpass FIR filter and later show how to modify the method to the design of other types of tunable FIR filters [Jar88]. The method discussed preserves the FIR structure and permits easy tuning of the cutoff frequency.

The basic idea behind the tuning procedure is the observation that, for an ideal lowpass FIR filter with a zero-phase response given by

$$H_d(e^{j\omega}) = \begin{cases} 1, & \text{for } 0 \le |\omega| \le \omega_c, \\ 0, & \text{for } \omega_c < |\omega| \le \pi. \end{cases} \qquad (8.77)$$

the impulse response coefficients are given by

$$h_d[n] = \frac{\sin(\omega_c n)}{\pi n}, \quad 0 \le |n| < \infty, \qquad (8.78)$$

as derived in Example 4.11. We can truncate the above expression and obtain the coefficients of a realizable approximation given by

$$h_{\text{LP}}[n] = \begin{cases} c[n]\omega_c, & \text{for } n = 0, \\ c[n]\sin(\omega_c n), & \text{for } 1 \le |n| \le N, \\ 0, & \text{otherwise}, \end{cases} \qquad (8.79)$$

where ω_c is the 6-dB cutoff frequency, and

$$c[n] = \begin{cases} 1/\pi, & \text{for } n = 0, \\ 1/\pi n, & \text{for } 1 \le |n| \le N. \end{cases} \qquad (8.80)$$

It follows from the above that, once an FIR lowpass filter has been designed for a given cutoff frequency, it can be tuned simply by changing ω_c and recomputing the filter coefficients according to the above expression. It can be shown that Eq. (8.79) can also be used to design a tunable FIR lowpass filter by equating the coefficients of a prototype filter developed using any of the FIR filter design methods outlined in Chapter 7 with those of Eq. (8.79) and solving for $c[n]$ [Jar88]. Thus, if $h_{\text{LP}}[n]$ denotes the coefficients of the prototype lowpass filter designed for a cutoff frequency ω_c, from Eq. (8.79) the constants $c[n]$ are given by

$$c[0] = \frac{h_{LP}[0]}{\omega_c}, \qquad (8.81a)$$

$$c[n] = \frac{h_{LP}[n]}{\sin(\omega_c n)}, \quad 1 \le |n| \le N. \qquad (8.81b)$$

Then, the coefficients $\hat{h}_{\text{LP}}[n]$ of the transformed FIR filter with a cutoff frequency $\hat{\omega}_c$ are given by

$$\hat{h}_{\text{LP}}[0] = c[0]\hat{\omega}_c = \left(\frac{\hat{\omega}_c}{\omega_c}\right) h_{LP}[0], \qquad (8.82a)$$

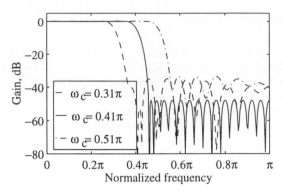

Figure 8.39   Tunable lowpass FIR filter of length 51.

$$\hat{h}_{\text{LP}}[n] = c[n]\sin(\hat{\omega}_c n) = \left(\frac{\sin(\hat{\omega}_c n)}{\sin(\omega_c n)}\right) h_{LP}[n], \quad 1 \le |n| \le N. \qquad (8.82b)$$

This tuning procedure has been recommended for filters with equal passband and stopband ripples. It has also been recommended that the prototype filter be designed such that its coefficients have values not too close to zero.

We illustrate the design of a tunable FIR filter in the following example.

Example 8.23 The prototype filter is a linear-phase lowpass FIR filter of length 51 with a passband edge ω_p at 0.36π and a stopband edge ω_s at 0.46π and a transition bandwidth of 0.1π. The cutoff frequency ω_c is thus at 0.41π. It has been designed using the function remez of MATLAB, assuming equal weights to the passband and the stopband. Using the method discussed above, the filter coefficients for a cutoff at 0.31π and at 0.51π have been computed. Figure 8.39 shows the plots of the gain responses of the three FIR filters.

For the design of tunable FIR filters with unequal passband and stopband ripples, the following modification is used. The impulse response coefficients of the tunable filter of odd length are now given by

$$h_{\text{LP}}[n] = \begin{cases} c[n]\omega_c + d[n], & \text{for } n = 0, \\ c[n]\sin(\omega_c n) + d[n]\cos(\omega_c n), & \text{for } 1 \le |n| \le N. \end{cases} \qquad (8.83)$$

The constants $c[n]$ and $d[n]$ are determined from two different optimal prototype filters, substituting them in Eq. (8.83), and then solving for these constants. For example, if the passband weight W_p is greater than the stopband weight W_s, the cutoff frequencies ω_{ca} and ω_{cb} of the two filters are chosen as

$$\omega_{ca} = 0.8 - \frac{0.25}{N}, \quad \omega_{cb} = 0.8 + \frac{0.25}{N}.$$

On the other hand, if the passband weight W_p is less than the stopband weight W_s, the cutoff frequencies ω_{ca} and ω_{cb} of the two filters are chosen as

$$\omega_{ca} = 0.2 - \frac{0.25}{N}, \omega_{cb} = 0.2 + \frac{0.25}{N}.$$

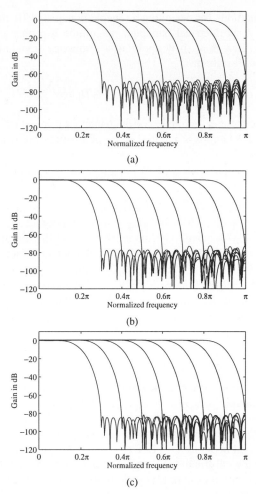

Figure 8.40   Length 41, transition bandwidth of 0.1π, and passband/stopband weights of (a) 1/1, (b) 1/10, and (c) 1/50.

The two optimal filters $h_a[n]$ and $h_b[n]$ are then designed using the Parks-McClellan algorithm. As before, the prototype filters should be designed with nonzero coefficients.

The above modification is not recommended for the design of tunable wideband or very narrowband filters. Figure 8.40 shows the gain responses of a lowpass FIR filter of length 41 and a transition width of 0.1π with a variable 6-dB cutoff frequency for various values of passband and stopband weights. The cutoff frequencies of the filters vary from from 0.2 to 0.9, as indicated in Figure 8.40.

The above methods can be directly applied to the design of a tunable highpass FIR filter from a highpass FIR prototype filter. Alternatively, the latter can be designed as a delay-complementary tunable Type 1 FIR lowpass filter.[9]

———————————

[9]See Section 4.6.1.

To develop the methods for designing a tunable bandpass FIR filter, we observe that a filter $H_{BP}(z)$ with a symmetrical bandpass magnitude response can be derived from a lowpass prototype filter $H_{LP}(z)$ by applying a frequency translation to its frequency response. This process results in

$$H_{BP}(e^{j\omega}) = H_{LP}(e^{j(\omega+\omega_0)}) + H_{LP}(e^{j(\omega-\omega_0)}), \qquad (8.84)$$

where ω_0 is the desired center frequency of the bandpass filter. The relation between the impulse responses of the bandpass and the prototype lowpass filters is given by

$$h_{BP}[n] = \left(e^{-j\omega_0 n} + e^{j\omega_0 n}\right) h_{LP}[n] = 2\cos(\omega_0 n) h_{LP}[n]. \qquad (8.85)$$

If δ_p and δ_s denote the passband and stopband ripples of the lowpass prototype, the corresponding ripples of the bandpass filter are $\delta_p + \delta_s$ for the passband and $2\delta_s$ for the stopband. Eq. (8.85) forms the basis for designing a tunable bandpass FIR filter with adjustable center frequency. A similar approach can be followed for the design of a tunable bandstop FIR filter, in which case the prototype is a highpass FIR filter.

It should be noted that the tunable FIR filters designed using the methods suggested above have the same hardware requirements as that of their prototypes. They also have linear phase if the prototype filter has linear phase.

8.8   Function Approximation

Often there is a need to use transcendental functions and random numbers in certain DSP applications. For example, the FFT computations require the generation of the complex exponential sequences. Certain digital communication systems require sinusoidal sequences. Both of these sequences can of course be generated by second-order digital filter structures described in Section 6.11. We outline below an alternative approach to the generation of transcendental functions based on truncated polynomial expansions. These types of expansions are often used in DSP chip implementations [Mar92].

8.8.1   Trigonometric Function Approximation

The sine of a number x can be approximated using the expansion [Abr72]:

$$\sin(x) \cong x - 0.166667x^3 + 0.008333x^5 - 0.0001984x^7 + 0.0000027x^9. \qquad (8.86)$$

where the argument x is in radians, and its range is restricted to the first quadrant, i.e., from 0 to $\pi/2$. If x is outside this range, its sine can be computed by making use of the identities $\sin(-x) = -\sin(x)$, and $\sin[(\pi/2)+x] = \sin[(\pi/2)-x]$. Figure 8.41 shows the plots of the sine aprroximation computed using Eq. (8.86) and the error due to the approximation.

For computing the arctangent of a number x where $-1 \leq x \leq 1$, a recommended expansion is given by [Abr72]

$$\tan^{-1}(x) \cong 0.999866x - 0.3302995x^3 + 0.180141x^5$$
$$-0.085133x^7 + 0.0208351x^9. \qquad (8.87)$$

Figure 8.41   (a) Plot of the sine value computed using Eq. (8.86), and (b) plot of the error between the actual sine value and the approximation.

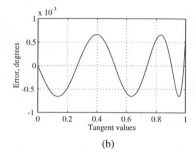

Figure 8.42   (a) Plot of the arctangent value computed using Eq. (8.87), and (b) plot of the error between the actual arctangent value and the approximation.

Figure 8.42 shows the plots of the arctangent aprroximation computed using Eq. (8.87) and the error due to the approximation. If $x \geq 1$, then its arcangent can be computed by making use of the identity $\tan^{-1}(x) = (\pi/2) - \tan^{-1}(1/x)$.

Problems M8.12 and M8.13 list several polynomial approximations for computing these trigonometric functions. Various trigonometric functions can also be computed from Eq. (8.86) using trigonometric identities.

8.8.2   Square Root and Logarithm Approximation

The square root of a positive number x in the range $0.5 \leq x \leq 1$ can be evaluated using the truncated polynomial approximation [Mar92]:

$$\sqrt{x} \cong 0.2075806 + 1.454895x - 1.34491x^2 + 1.106812x^3$$
$$- 0.536499x^4 + 0.1121216x^5. \tag{8.88}$$

Figure 8.43 shows the plot of the error due to the above approximation. If x is outside the range from 0.5 to 1, it can be multiplied by a binary constant K^2 to bring the product $x' = K^2 x$ into the desirable range, compute $\sqrt{x'}$ using Eq. (8.84), and then determine $\sqrt{x} = \sqrt{x'}/K$.

Polynomial expansions have also been advanced for the approximate computation of both the logarithm (base 10) and natural logarithm (base e) of any number x between one and two. Two such expressions are given by [Bur73], [Mar92]:

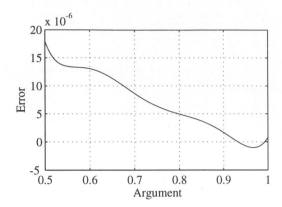

Figure 8.43 Plot of the error between the actual square root value and the approximation given by Eq. (8.88).

$$\log_{10}(x) \cong 0.4339142(x-1) - 0.21278385(x-1)^2 + 0.1240692(x-1)^3$$
$$-0.05778505(x-1)^4 + 0.0136261(x-1)^5, \qquad (8.89)$$
$$\log_e(x) \cong 0.999115(x-1) - 0.4899597(x-1)^2 + 0.2856751(x-1)^3$$
$$-0.13305665(x-1)^4 + 0.031372071(x-1)^5. \qquad (8.90)$$

For the calculation of the logarithm of a number x that is outside the range from 1 to 2, the number x must be scaled by an appropriate factor K with known logarithm to bring the product $x' = Kx$ to this range, then compute the logarithm of the product x'. From the logarithm of the product x', the logarithm of K is subtracted to get the logarithm of x.

8.8.3   Random Number Generation

Random number generation is used in a number of applications. For example, it is used as a training signal for the adaptive equalizer in high-speed modems [Mar92]. Another application is in the elimination of limit cycles in IIR digital filter structures using the random rounding method (Section 9.9.4). A variety of approaches has been proposed for the generation of uniformly distributed random numbers [Knu69]. One rather simple method is based on the linear congruence method and is given by the recursive equation

$$x[n+1] = \langle \alpha x[n] + \beta \rangle_M, \qquad (8.91)$$

where the modulus M is a positive integer. The rules for the selection of the constants α and β are given in [Knu69]. The above equation generates a periodic pseudo-random sequence having a period M with proper choice of α and β fairly independent of the *seed value* $x[0]$. The samples of the sequence generated by Eq. (8.91) within a period are uniformly distributed integers from 0 to $M-1$. A family of such random sequences can be generated by choosing different seed values for each sequence. To ensure the randomness of each sequence generated, M should be chosen as large as possible.

The M-function `rand` can be used to generate random numbers and matrices with elements uniformly distributed in the interval $(0, 1)$. Various versions of these functions are available as indicated below:

```
rand(N),          rand(M,N),          rand(size(A)),
rand('seed',N),   rand('seed')
```

The output of `rand(N)` is an $N \times N$ matrix and that of `rand(M,N)` is an $M \times N$ matrix with random entries. `rand(size(A))` generates an output of the same size as A. If `rand` is used without any argument, its output is a scalar with a value that changes each time it is called. The output of `rand('seed')` is the current value of the seed employed by the random number generator. Finally, `rand('seed',N)` uses the value N for the seed.

A similar set of M-functions for the generation of random numbers and matrices with elements that are normally (Gaussian) distributed with zero mean and unity variance are given by:

```
randn(N),          randn(M,N),          randn(size(A)),
randn('seed',N),   randn('seed')
```

The function `rand` was used in Program 2.4 of Example 2.14 to generate a random sequence.

8.9 Summary

Some of the common factors in the implementation of DSP algorithms, which are independent of the type of implementation being carried out, are discussed in this chapter. The implementation is usually carried out by a sequential implementation of the set of equations describing the algorithm. These equations can be derived directly from the structure realizing the algorithm. The computability condition of these equations is derived and an algorithm for testing the computability is described. A simple algebraic technique to verify the structure from the input and output samples is also outlined.

MATLAB-based software implementation of digital filtering algorithm and the discrete Fourier transform (DFT) are then considered. The basic ideas behind the fast Fourier transform (FFT) used for faster computation of the DFT samples are explained and several commonly used FFT algorithms are derived.

Various schemes for the binary representation of the numbers and the signal variables that are employed in the digital computers and special purpose DSP chips are reviewed, followed by a discussion of the algorithms used for the implementation of the addition and the multiplication operation. In certain cases, the result of an addition of numbers may cause an overflow of the dynamic range of the register storing the sum. Two methods of preventing physically the overflow are outlined.

Methods for the design of IIR and FIR digital filters with tunable characteristics are introduced next. The chapter concludes with a discussion on the approximation of certain functions that are needed in a number of applications requiring implementation using DSP chips. In particular, the approximation of trigonometric functions, square roots, logarithms, and the generation of random numbers are considered.

8.10   Problems

8.1 Develop a set of time-domain equations describing the digital filter structure of Figure P8.1 in terms of the input $x[n]$, output $y[n]$, and of equations describe a valid computational algorithm? Justify your answer by developing a matrix representation of the digital filter structure and by examining the matrix **F**.

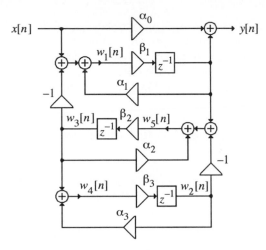

Figure P8.1

8.2 Develop a computable set of time-domain equations describing the digital filter structure of Figure P8.1. Verify the computability condition by forming an equivalent matrix representation and by examining the matrix **F**.

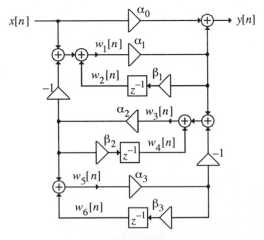

Figure P8.2

8.3 Develop a set of time-domain equations describing the digital filter structure of Figure P8.2 in terms of the input $x[n]$, output $y[n]$, and the intermediate variables $w_k[n]$ in a sequential order. Does this set of equations describe a valid computational algorithm? Justify your answer by developing a matrix representation of the digital filter structure and by examining the matrix **F**.

8.4 Develop the precedence graph of the digital filter structure of Figure P8.1 and investigate its realizability. If the structure is found to be realizable, then from the precendence graph, determine a valid computational algorithm describing the structure.

8.5 Develop the precedence graph of the digital filter structure of Figure P8.2 and investigate its realizability. If the structure is found to be realizable, then from the precendence graph, determine a valid computational algorithm describing the structure.

8.6   (a) Write down the time-domain equations relating the node variables $w_i[n]$, $y[n]$ and the input $x[n]$ of the digital filter structure of Figure P8.3. Check formally the computability of the set of equations if the equations are ordered sequentially with increasing values of the node indices.

  (b) Develop a signal flow-graph representation of this digital filter structure and then determine its precedence graph. From the precedence graph, develop a set of computable equations describing the structure and show formally that these equations are indeed computable.

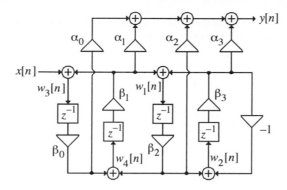

Figure P8.3

8.7 The measured impulse-response samples of a causal second-order IIR digital filter with a transfer function $H(z) = P(z)/(1 + 0.1z^{-1} - 0.2z^{-2})$ are given by

$$h[0] = 3.0, \ h[1] = 1.72, \ h[2] = 4.43, \ h[3] = -0.103, \ h[4] = 0.8963, \ldots$$

Determine the numerator $P(z)$ of the transfer function $H(z)$.

8.8 The first 5 impulse response samples of a causal second-order IIR digital filter are given by

$$h[0] = 5, \ h[1] = -1, \ h[2] = -8, \ h[3] = 19, \ h[4] = -14.$$

Determine the transfer function $H(z)$.

8.9 Determine the transfer function of a third-order causal IIR digital filter whose first 10 impulse response samples are given by

$$\{h[n]\} = \{3 \quad -2 \quad -3 \quad 1 \quad 15 \quad -21 \quad -7 \quad 17 \quad 71 \quad -165\}.$$

8.10 The first 4 impulse response samples of a causal third-order IIR digital filter with a transfer function $G(z) = P(z)/(1 + 3z^{-1} + 2z^{-2} + 4z^{-3})$ are given by

$$g[0] = 2, \ g[1] = -5, \ g[2] = 14, \ g[3] = 14.$$

Determine the numerator polynomial $P(z)$ of the transfer function.

8.11 The first 11 impulse response samples of a causal IIR transfer function

$$H(z) = \frac{p_0 + p_1 z^{-1} + p_2 z^{-2} + p_3 z^{-3} + p_4 z^{-4}}{1 + 2z^{-1} + 2z^{-2} + 3z^{-3} + 3z^{-4}},$$

are given by

$$\{h[n]\} = \{5 \quad -11 \quad 15 \quad -25 \quad 42 \quad -46 \quad 38 \quad -35 \quad 6 \quad 82 \quad -185\}.$$

Determine its numerator coefficients $\{p_i\}$.

8.12 Show that the direct computation of the N-point DFT of a length-N sequence requires $4N^2$ real multiplications and $(4N - 2)N$ real additions.

8.13 Develop an algorithm for the complex multiplication of two complex numbers using only 3 real multiplications and 5 real additions.

8.14 Let $H_k(z)$ denote the transfer function of the kth filter used in the Goertzel's algorithm to calculate the N-point DFT $X[k]$ of a length-N sequence $x[n]$. Consider the following input sequence applied to $H_k(z)$:

$$x[n] = \{1, \ 0, \ 0, \cdots, 1, \ 0, \ 0, \cdots\},$$
$$\qquad\quad \uparrow \qquad\qquad \uparrow$$
$$\qquad\quad 0 \qquad\qquad N/2$$

where $N/2$ is assumed to be divisible by k. What is the output sequence $y[n]$ for $k = 1$? What is the output sequence $y[n]$ for $k = N/2$? Give a qualitative sketch.

8.15 Develop the flow-graph for the FFT algorithm from the radix-2 DIT FFT algorithm of Figure 8.25 for the $N = 8$ case in which the input is in normal order and the output is in the bit-reversed order.

8.16 Develop the flow-graph for the FFT algorithm from the radix-2 DIT FFT algorithm of Figure 8.25 for the $N = 8$ case in which both the input and the output are in normal order.

8.17 Verify Eq. (8.46).

8.18 Develop the structural interpretation of the first stage of radix-3 DIT FFT algorithm.

8.19 Develop the flow-graph for the radix-3 DIT FFT algorithm for the $N = 9$ case in which the input is in digit-reversed order and the output is in the normal order.

8.20 Develop the flow-graph for a mixed-radix DIT FFT algorithm for the $N = 15$ case in which the input is in digit-reversed order and the output is in the normal order.

8.21 Form the transposed graph of the flow-graph of Figure 8.25. Replace each complex multiplicand W_N^r in the transposed flow-graph with $\frac{1}{2}W_N^{-r}$. Show that the final flow-graph implements an inverse DFT if the input is $X[k]$.

8.22 A second approach to the inverse DFT computation using a DFT algorithm is illustrated in Figure P8.4. Let $X[k]$ be the N-point DFT of a length-N sequence $x[n]$. Define a length-N time-domain sequence $q[n]$ as

$$\mathrm{Re}\{q[n]\} = \mathrm{Im}\, X[k]|_{k=n}\,, \quad \mathrm{Im}\{q[n]\} = \mathrm{Re}\, X[k]|_{k=n}\,,$$

with $Q[k]$ denoting its N-point DFT. Show that

$$\mathrm{Re}\{x[n]\} = \frac{1}{N}\cdot \mathrm{Im}\, Q[k]\bigg|_{k=n}\,, \quad \mathrm{Im}\{x[n]\} = \frac{1}{N}\cdot \mathrm{Re}\, Q[k]\bigg|_{k=n}\,.$$

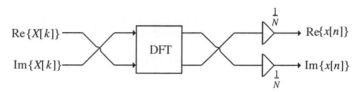

Figure P8.4

8.23 A third approach to the inverse DFT computation using a DFT algorithm is described below. Let $X[k]$ be the N-point DFT of a length-N sequence $x[n]$. Define a length-N time-domain sequence $r[n]$ as

$$r[n] = X[k]|_{k=\langle -n\rangle_N}\,,$$

with $R[k]$ denoting its N-point DFT. Show that

$$x[n] = \frac{1}{N}\cdot R[k]|_{k=n}\,.$$

8.24 We wish to determine the sequence $y[n]$ generated by a linear convolution of a length-4 real sequence $x[n]$ and a length-3 real sequence $h[n]$. To this end, we can follow one of the following methods:

Method # 1. Direct computation of the linear convolution.

Method # 2. Computation of the linear convolution via a single circular convolution.

Method # 3. Computation of the linear convolution using radix-2 FFT algorithms.

Determine the least number of real multiplications needed in each of the above methods. For the radix-2 FFT algorithm, do not include in the count multiplications by ± 1, $\pm j$ and W_N^0.

8.25 Repeat Problem 8.24 for the computation of the linear convolution of a length-8 sequence $x[n]$ with a length-6 sequence $h[n]$.

8.26 An input sequence $x[n]$ of length 1024 is to be filtered using a linear phase FIR filter $h[n]$ of length 65. This filtering process involves linear convolution of two finite-length sequences and can be computed using the overlap-add algorithm discussed in Section 3.5.2 where the short linear convolutions are performed using the DFT-based approach of Figure 3.12 with the DFTs implemented by the Cooley-Tukey FFT algorithm.

(a) Determine the appropriate transform length of power-of-2 that would result in a minimum number of multiplications and calculate the total number of multiplications that would be required.

(b) What would be the total number of multiplications if the direct convolution method is used?

8.27 We pointed out in Eq. (3.30) that the vector $\mathbf{X}$ of DFT samples can be expressed as the product of the DFT matrix $\mathbf{D}_N$ and the vector $\mathbf{x}$ of input samples, where $\mathbf{D}_N$ is given by Eq. (3.33).

(a) Show that the 8-point DIT FFT algorithm shown in Figure 8.18 is equivalent to expressing the DFT matrix as a product of 4 matrices as indicated below:

$$\mathbf{D}_N = \mathbf{V}_8 \mathbf{V}_4 \mathbf{V}_2 \mathbf{E}. \tag{8.92}$$

Determine the matrices given above and show that multiplication by each matrix $\mathbf{V}_k$, $k = 8, 4, 2$, requires at most 8 complex multiplications.

(b) Since the DFT matrix $\mathbf{D}_N$ is its own transpose, i.e., $\mathbf{D}_N = \mathbf{D}_N^T$ another FFT algorithm is readily obtained by forming the transpose of the right-hand side of Eq. (8.92), resulting in a factorization of $\mathbf{D}_N$ given by

$$\mathbf{D}_N = \mathbf{E}^T \mathbf{V}_2^T \mathbf{V}_4^T \mathbf{V}_8^T. \tag{8.93}$$

Show that the flow-graph representation of the above factorization is precisely the 8-point DIF FFT algorithm of Figure 8.27.

8.28 The basic idea behind the split-radix FFT (SRFFT) is explored in this problem [Duh86]. As in the case of the decimation-in-frequency FFT algorithm, in the SRFFT algorithm, the even-indexed and the odd-indexed samples of the DFT are computed separately. For the computation of the even-indexed samples we write, as in the DIF FFT approach,

$$X[2\ell] = \sum_{n=0}^{N-1} x[n] W_N^{2\ell n}$$
$$= \sum_{n=0}^{(N/2)-1} x[n] W_N^{2\ell n} + \sum_{n=N/2}^{N-1} x[n] W_N^{2\ell n}, \quad \ell = 0, 1, \ldots, \frac{N}{2} - 1. \tag{8.94}$$

Show that the above can be reexpressed as an $(N/2)$-point DFT in the form

$$X[2\ell] = \sum_{n=0}^{(N/2)-1} \left(x[n] + x[n + \frac{N}{2}] \right) W_{N/2}^{\ell n}, \quad \ell = 0, 1, \ldots, \frac{N}{2} - 1. \tag{8.95}$$

For the computation of the odd-indexed samples, we write two different expressions, depending on whether the frequency index k can be expressed as $4\ell + 1$ or $4\ell + 3$:

$$X[4\ell + 1] = \sum_{n=0}^{(N/4)-1} x[n] W_N^{(4\ell+1)n} + \sum_{n=\frac{N}{4}}^{(N/2)-1} x[n] W_N^{(4\ell+1)n}$$
$$+ \sum_{n=N/2}^{(3N/4)-1} x[n] W_N^{(4\ell+1)n} + \sum_{n=3N/4}^{N-1} x[n] W_N^{(4\ell+1)n},$$
$$\ell = 0, 1, \ldots, \frac{N}{4} - 1, \tag{8.96a}$$

$$X[4\ell + 3] = \sum_{n=0}^{(N/4)-1} x[n]W_N^{(4\ell+3)n} + \sum_{n=N/4}^{(N/2)-1} x[n]W_N^{(4\ell+3)n}$$

$$+ \sum_{n=N/2}^{(3N/4)-1} x[n]W_N^{(4\ell+3)n} + \sum_{n=3N/4}^{N-1} x[n]W_N^{(4\ell+3)n},$$

$$\ell = 0, 1, \ldots, \frac{N}{4} - 1. \tag{8.96b}$$

Show that Eqs. (8.96a) and (8.96b) can be rewritten as two ($N/4$)-point DFTs of the form

$$X[4\ell + 1] = \sum_{m=0}^{(N/4)-1} \left\{ \left(x[m] - x\left[m + \frac{N}{2}\right] \right) \right.$$

$$\left. -j\left(x\left[m + \frac{N}{4}\right] - x\left[m + \frac{3N}{4}\right] \right) \right\} W_N^m W_{N/4}^{\ell m}, \tag{8.97a}$$

$$X[4\ell + 3] = \sum_{m=0}^{(N/4)-1} \left\{ \left(x[m] - x\left[m + \frac{N}{2}\right] \right) \right.$$

$$\left. +j\left(x\left[m + \frac{N}{4}\right] - x\left[m + \frac{3N}{4}\right] \right) \right\} W_N^{3m} W_{N/4}^{\ell m}, \tag{8.97b}$$

where $\ell = 0, 1, \ldots, \frac{N}{4} - 1$. Sketch the flow-graph of a typical butterfly in the above algorithm for computing two even-numbered and two-odd numbered points, and show that the split-radix FFT algorithm requires only two complex multiplications per butterfly.

8.29 Develop the flow-graph for the computation of an 8-point DFT based on the split-radix algorithm described in Problem 8.28. What is the total number of real multiplications needed to implement this algorithm? How does this number compare with that required in a radix-2 DIF FFT algorithm? Ignore multiplications by ± 1 and $\pm j$.

8.30 Develop the flow-graph for the computation of a 16-point DFT based on the split-radix algorithm described in Problem 8.28. What is the total number of real multiplications needed to implement this algorithm? How does this number compare with that required in a radix-2 DIF FFT algorithm? Ignore multiplications by ± 1 and $\pm j$.

8.31 An alternative approach to the development of FFT algorithms is via index mapping [Coo65], which is studied in this problem. Consider a length-N sequence $x[n]$ with $X[k]$ denoting its N-point DFT where $N = N_1 N_2$. Define the index mappings

$$n = n_1 + N_1 n_2 \qquad \begin{cases} 0 \leq n_1 \leq N_1 - 1, \\ 0 \leq n_2 \leq N_2 - 1, \end{cases} \tag{8.98a}$$

$$k = N_2 k_1 + k_2 \qquad \begin{cases} 0 \leq k_1 \leq N_1 - 1, \\ 0 \leq k_2 \leq N_2 - 1. \end{cases} \tag{8.98b}$$

(a) Using the above mappings, show that $X[k]$ can be expressed as

$$X[k] = X[N_2 k_1 + k_2]$$

$$= \sum_{n_1=0}^{N_1-1} \left[\left(\sum_{n_2=0}^{N_2-1} x[n_1 + N_1 n_2]W_{N_2}^{k_2 n_2} \right) W_N^{k_2 n_1} \right] W_{N_1}^{k_1 n_1},$$

$$0 \leq k_1 \leq N_1 - 1, \quad 0 \leq k_2 \leq N_2 - 1. \tag{8.99}$$

As indicated above, the computation of the N-point DFT $X[k]$ is now carried out in three steps: (1) compute N_2-point DFT $H[n_1, k_2]$ of the set of N_1 sequences $x[n_1 + N_1 n_2]$ of length N_2, (2) multiply these DFTs with the twiddle factors $W_N^{k_2 n_1}$ to form $\hat{H}[n_1, k_2]$, and (3) compute N_2-point DFT $X[N_2 k_1 + k_2]$ of the set of N_2 sequences $\hat{H}[n_1, k_2]$ of length N_1.

(b) Show that for $N_1 = 2$ and $N_2 = N/2$, the above decomposition scheme leads to the DIT FFT algorithm and for $N_1 = N/2$ and $N_2 = 2$, the above decomposition scheme leads to the DIF FFT algorithm .

(c) If $\mathcal{R}(N)$ denotes the total number of multiplications needed to compute an N-point DFT, show that for the DFT computation algorithm based on the above decomposition scheme outlined above,

$$\mathcal{R}(N) = N_1 \mathcal{R}(N_2) + N_2 \mathcal{R}(N_1) + N$$
$$= N \left[\frac{1}{N_1} \mathcal{R}(N_1) + \frac{1}{N_2} \mathcal{R}(N_2) + 1 \right]. \qquad (8.100)$$

(d) Determine the total number of multiplications needed for the FFT algorithm based on the above decomposition scheme for $N = 2^{\nu}$.

8.32 Develop the index mapping for implementing an N-point DFT $X[k]$ of a length-N sequence $x[n]$ using the Cooley-Tukey FFT algorithm for (a) $N = 10$, (b) $N = 12$, (c) $N = 15$, and (d) $N = 20$.

8.33   (a) The twiddle factors needed in the DFT computation scheme outlined in Problem 8.31 can be eliminated, resulting in a more computationally efficient FFT algorithm for the case when the factors N_1 and N_2 are relatively prime by using the following index mappings [Bur77]:

$$n = \langle A n_1 + B n_2 \rangle_N, \qquad \begin{cases} 0 \le n_1 \le N_1 - 1, \\ 0 \le n_2 \le N_2 - 1, \end{cases} \qquad (8.101a)$$

$$k = \langle C k_1 + D k_2 \rangle_N, \qquad \begin{cases} 0 \le k_1 \le N_1 - 1, \\ 0 \le k_2 \le N_2 - 1. \end{cases} \qquad (8.101b)$$

Show that the twiddle factors are eliminated totally if the constants A, B, C and D in the above mappings are chosen satisfying the following conditions:

$$\langle AC \rangle_N = N_2, \quad \langle BD \rangle_N = N_1, \quad \text{and} \quad \langle AD \rangle_N = \langle BC \rangle_N = 0.$$

(b) Show that the following set of constants satisfy the above condition

$$A = N_2, \ B = N_1, \ C = N_2 \langle N_2^{-1} \rangle_{N_1}, \ \text{and} \ D = N_1 \langle N_1^{-1} \rangle_{N_2},$$

where $\langle N_1^{-1} \rangle_{N_2}$ denotes the multiplicative inverse of N_1 evaluated modulo N_2. Based on this choice of constants, develop the algorithm for the computation of $X[k]$. FFT computation schemes based on this approach are called the *prime factor algorithms* [Bur81], [Kol77].

8.34 Develop the index mapping for implementing an N-point DFT $X[k]$ of a length-N sequence $x[n]$ using the prime factor algorithm for (a) $N = 10$, (b) $N = 12$, (c) $N = 15$, and (d) $N = 20$.

8.35 Consider the computation of a 12-point DFT $X[k]$ of a length-N sequence $x[n]$. The index mapping to be used can be either

$$n = \langle An_1 + Bn_2 \rangle_{12}, \quad k = \langle Ck_1 + Dk_2 \rangle_{12},$$

or

$$n = \langle Cn_1 + Dn_2 \rangle_{12}, \quad k = \langle Ak_1 + Bk_2 \rangle_{12},$$

where A, B, C, and D are appropriately chosen constants. Let the mapping used be given by $g[n_1, n_2] = x[\langle An_1 + Bn_2 \rangle_{12}]$ resulting in

$$G[k_1, k_2] = \sum_{n_1=0}^{2} \sum_{n_2=0}^{3} g[n_1, n_2] W_4^{n_2 k_2} W_3^{n_1 k_1}, \tag{8.102}$$

$$X[\langle Ck_1 + Dk_2 \rangle_{12}] = G[k_1, k_2]. \tag{8.103}$$

If, on the other hand, the index mapping used is $h[n_1, n_2] = x[\langle Cn_1 + Dn_2 \rangle_{12}]$, we obtain

$$H[k_1, k_2] = \sum_{n_1=0}^{2} \sum_{n_2=0}^{3} h[n_1, n_2] W_4^{n_2 k_2} W_3^{n_1 k_1}, \tag{8.104}$$

$$Y[\langle Ak_1 + Bk_2 \rangle_{12}] = H[k_1, k_2]. \tag{8.105}$$

What is the relation between $X[k]$ and $Y[k]$?

8.36 Develop the flow-graph for the computation of a 10-point DFT based on the prime factor algorithm.

8.37 Develop the flow-graph for the computation of a 12-point DFT based on the prime factor algorithm.

8.38 Let $H(z) = \sum_{n=0}^{N-1} h[n]z^{-n}$ and $X(z) = \sum_{n=0}^{N-1} x[n]z^{-n}$ be two real polynomials of degree $(N-1)$. Their product $Y(z)$ is then a polynomial of degree $(2N-2)$ and a direct evaluation of $Y(z)$ requires N^2 multiplications and $N+1$ additions. Note also that the coefficients $y[n]$ of $Y(z)$ are the same as that obtained by a linear convolution of the two length-N sequences $h[n]$ and $x[n]$. The number of multiplications in computing the product $H(z)X(z)$ can be reduced but with an increase in the number of additions by making use of the Cook-Toom algorithm [Aga77], [Knu69] which is studied in this problem.

Let z_k, $k = 0, 1, \ldots, 2N-1$, be $2N-1$ distinct points in the z-plane. Then $\hat{Y}[z_k] = H(z_k)X(z_k)$ represent the $(2N-1)$-point NDFT of $y[n]$, and from these $2N-1$ samples we can uniquely determine $Y(z)$ using the Lagrange interpolation formula as discussed in Problem 3.7. By choosing the values of z_k appropriately, $\hat{Y}[z_k]$ can be evaluated with only $2N-1$ multiplications if we ignore multiplications (or divisions) by a power-of-2 integer that can be implemented using simple shifts in a binary representation.

(a) We first develop the Cook-Toom algoritm for $N = 2$. Here,

$$Y(z) = y[0] + y[1]z^{-1} + y[2]z^{-2} = \left(h[0] + h[1]z^{-1} \right) \left(x[0] + x[1]z^{-1} \right).$$

Using the Lagrange interpolation formula express $Y(z)$ in terms of its 3-point NDFT samples evaluated at $z_0 = -1$, $z_1 = \infty$, and $z_2 = +1$. Develop the expression for $y[0]$, $y[1]$, and $y[2]$ in terms of the parameters $X(z_k)$ and $H(z_k)$ and show that the computations of $\{y[n]\}$ require only a total of three multiplications. Determine the total number of additions required by this algorithm. Note that in many applications $h[n]$ represents a fixed FIR filter and, hence, the multiplications by integers needed to evaluate $y[n]$ can be included in the constants $H(z_k)$ to eliminate them from future computations.

(b) Develop the Cook-Toom algorithm for the linear convolution of two length-3 sequences.

8.39 Let $H(z) = h[0] + h[1]z^{-1}$ and $X(z) = x[0] + x[1]z^{-1}$. Show that $Y(z) = H(z)X(z) = y[0] + y[1]z^{-1} + y[2]z^{-2}$ can be written as [Jen91]

$$Y(z) = h[0]x[0] + [(h[0] + h[1])(x[0] + x[1]) \\ - (h[0]x[0] + h[1]x[1])]z^{-1} + h[1]x[1]z^{-2}.$$

The evaluation of the product of the two first-order real polynomials thus can be carried out using 3 multiplications instead of 4 as required in the direct product. Equivalently, a linear convolution of two length-2 sequences can thus be implemented using only 3 multiplications.

8.40 Develop a multistage algorithm to compute the linear convolution of two length-N real sequences based on the scheme outlined in Problem 8.39 for the case when N is a power-of-2 [Jen91]. What is the least number of multiplications required to compute the convolution using the multistage algorithm?

8.41 Let a 32-bit register be used to represent a floating-point number with E bits assigned for the exponent and M bits plus a sign bit for the mantissa. Determine the approximate dynamic range of this floating-point representation for the following pairs of bit assignments for the exponent and the mantissa by evaluating the values of the smallest and the largest numbers that can be represented by the floating-point representation: (a) $E = 6$ and $M = 25$, (b) $E = 7$ and $M = 24$, and (c) $E = 8$ and $M = 23$. Determine the dynamic range of a 32-bit fixed-point representation of a signed integer. Show that the floating-point representation provides a larger dynamic range than the fixed-point representation.

8.42 Show that the range of a 32-bit floating-point number in the IEEE standard is from 1.18×10^{-38} to 3.4×10^{38}.

8.43 Show that the decimal equivalent of a positive or a negative binary fraction given by $s_\triangle a_{-1} a_{-2} \ldots a_{-b}$ in two's-complement form is $-s + \sum_{i=1}^{b} a_{-i} 2^{-i}$.

8.44 Show that the decimal equivalent of a positive or a negative binary fraction given by $s_\triangle a_{-1} a_{-2} \ldots a_{-b}$ in ones'-complement form is $-s(1 - 2^{-b}) + \sum_{i=1}^{b} a_{-i} 2^{-i}$.

8.45 Determine the 9-bit sign-magnitude, ones'-complement, and two's-complement representations of the following negative decimal fractions:
(a) -0.625_{10}, (b) -0.7734375_{10}, (c) -0.36328125_{10}, (d) -0.94921875_{10}.

8.46 Determine the 7-bit offset binary representations of the following decimal numbers:
(a) 0.625_{10},   (b) -0.625_{10},   (c) 0.359375_{10},   (d) -0.359375_{10},
(e) 0.90625_{10},   (f) -0.90625_{10}.

8.47 Develop the signed-digit (SD) representation of the following binary numbers:
(a) $0_\triangle 11101101$,   (b) $0_\triangle 01111101$,   (c) $0_\triangle 10101111$.

8.48 Develop the hexadecimal representation of the following binary numbers:
(a) 1101011011000111,   (b) 0101111110101001,   (c) 1011010000101110.

8.49 Perform the following binary additions of positive binary fractions and comment on your results:
(a) $0_\triangle 10101 + 0_\triangle 01111$, (b) $0_\triangle 01011 + 0_\triangle 10001$.

8.50 Compute the following differences of positive binary fractions by performing binary additions of a positive fraction and a negative number represented in two's-complement form:
(a) $0_\triangle 10101 - 0_\triangle 01111$, (b) $0_\triangle 10001 - 0_\triangle 01011$.

8.51 Repeat Problem 8.50 by representing the negative number in ones'-complement form.

8.52 Consider the binary addition of the following three numbers represented in two's-complement form with 5-bits: $\eta_1 = 0.6875_{10}$, $\eta_2 = 0.8125_{10}$, and $\eta_3 = -0.5625_{10}$. The addition is carried out in two steps: first we form the sum $\eta_1 + \eta_2$ and then we form the sum $(\eta_1 + \eta_2) + \eta_3$. Since the magnitudes of η_1 and η_2 are both greater than 0.5, their sum will lead to overflow indicated by a 1 in the sign bit of the sum. Ignore this overflow by keeping all bits in the partial sum and add η_3. Show that the final sum is correct inspite of the overflow generated by the first addition.

8.53 Develop the products of the following binary fractions, where the negative numbers are in two's-complement forms:
(a) $(0_\triangle 11101) \times (1_\triangle 10111)$, (b) $(1_\triangle 10101) \times (0_\triangle 10111)$.

8.54 Repeat Problem 8.53 with the negative numbers are considered to be in ones'-complement form.

8.55 The Taylor structure of Figure P8.5(a) developed for the realization of a Type 1 linear-phase FIR transfer function in Problem 6.23 has been proposed for designing tunable FIR filters [Cro76b], [Opp76]. From Eq. (6.143) observe that the zero-phase frequency response, also called the amplitude response, of a Type 1 FIR transfer function of length $2M + 1$ is given by

$$\breve{H}(\omega) = \sum_{n=0}^{M} a[n] (\cos \omega)^n. \tag{8.106}$$

Let $\hat{\omega}$ denote the angular frequency variable of the transformed FIR filter $\hat{H}(z)$, show that a lowpass-to-lowpass transformation can be achieved by substituing

$$\cos \omega = \alpha + \beta \cos \hat{\omega}. \tag{8.107}$$

in Eq. (8.106). Show that this transformation can be implemented by replacing each block with a transfer function $(1 + z^{-2})/2$ in Figure P8.5 by a block with a transfer function

$$\alpha z^{-1} + \frac{\beta}{2}(1 + z^{-2}). \tag{8.108}$$

Let ω_c and $\hat{\omega}_c$ denote, respectively, the cutoff frequency of the prototype filter and the desired cutoff frequency of the transformed filter. Show that if $\hat{\omega}_c < \omega_c$, it is convenient to choose $\beta = 1 - \alpha$, with $0 \le \alpha < 1$. Sketch the mapping from $\cos \omega_c$ to $\cos \hat{\omega}_c$ for this case. On the other hand, if $\hat{\omega}_c > \omega_c$, show that it is convenient to choose $\beta = 1 + \alpha$, with $-1 < \alpha \le 0$. Sketch the mapping from $\cos \omega_c$ to $\cos \hat{\omega}_c$ for this second case.

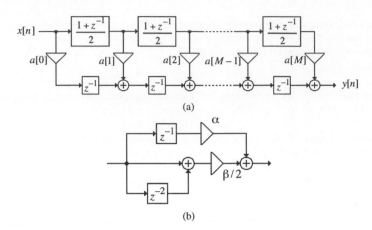

(a)

(b)

Figure P8.5

8.11 MATLAB Exercises

M 8.1 Using Program 8_1 determine the first 30 samples of the impulse response coefficients of the fifth-order elliptic lowpass filter developed in Example 7.19.

M 8.2 Using Program 8_1 determine the first 35 samples of the impulse response coefficients of the fourth-order Type 1 Chebyshev highpass filter developed in Example 7.20.

M 8.3 Using Program 8_1 determine the first 30 samples of the impulse response coefficients of the eighth-order Butterworth bandpass filter developed in Example 7.21.

M 8.4 Modify Program 8_2 to demonstrate filtering of a sum of two sinusoidal sequences by an arbitrary IIR causal digital filter. The input data to the modified program should be the angular frequencies of the sinusoidal sequences, the numerator and the denominator coefficients of the transfer function of the IIR digital filter. Apply the modified program to filter a sum of two sinusoidal sequences of angular frequencies 0.3π and 0.6π by the filter developed in Example 7.19 and verify its lowpass filtering property.

M 8.5 Apply the modified program developed in Problem M8.4 to filter a sum of two sinusoidal sequences of angular frequencies, 0.3π and 0.6π by the filter developed in Example 7.20 and verify its highpass filtering property.

M 8.6 Modify Program 8_3 to demonstrate filtering of a sum of two sinusoidal sequences by an arbitrary IIR causal digital filter implemented in a cascade form. The individual sections in the cascade are either second-order or first-order with real coefficients. The input data to the modified program should be the angular frequencies of the sinusoidal sequences, the numerator and the denominator coefficients of the transfer function of the individual sections in the cascade. Apply the modified program to filter a sum of two sinusoidal sequences of angular frequencies 0.3π and 0.6π by the filter developed in Example 7.19 and verify its lowpass filtering property. The output plot generated by the modified Program 8_3 should be identical to that generated in Problem M8.4.

M 8.7 Apply the modified program developed in Problem M8.6 to filter a sum of two sinusoidal sequences of angular frequencies 0.3π and 0.6π by the filter developed in Example 7.20 and verify its highpass filtering property. The output plot generated by the modified Program 8_3 should be identical to that generated in Problem M8.5.

M 8.8 Write an M-file function `filter2` to implement an IIR causal digital filter in direct form II.[10] Write a MATLAB program using this new function to demonstrate filtering of input sequences. Apply this program to filter a sum of two sinusoidal sequences of angular frequencies 0.3π and 0.6π by the filter developed in Example 7.19 and verify its lowpass filtering property. The output plot generated by your new program should be identical to that generated in Problem M8.4.

M 8.9 Using the M-file function `gfft` described in Section 8.3.1 for computing a single DFT sample by Goertzel's algorithm, write a MATLAB program to compute the N-point DFT of an arbitrary length-N sequence and compare the DFT samples generated with those obtained using the M-file function `fft`. Verify the DFT computation of several sequences of lengths $N = 8, 12,$ and 16.

M 8.10 Write a MATLAB program to verify the plots given in Figure 8.37.

M 8.11 Write a MATLAB program to verify the plots given in Figure 8.39.

M 8.12 A polynomial expansion suggested for the approximation of the sine of a number x is given by [Mar92]:

$$\sin(x) \cong 3.140625x + 0.02026367x^2 - 5.325196x^3$$
$$+0.5446778x^4 + 1.800293x^4, \tag{8.109}$$

where x is in radians, normalized by $\pi/2$, and its range is restricted to the first quadrant given by 0 to 0.5. (*Note*: 0.5 is the normalized value of $\pi/2$.) Using MATLAB compute and plot the values of $\sin(x)$ given by Eq. (8.109) and the error due to the above approximation. Compare this approximation with that given by Eq. (8.86).

M 8.13 A polynomial expansion suggested for the approximation of the arctangent of a number x in the range $-1 \le x \le 1$ is given by [Mar92]

$$\arctan(x) \cong 0.318253x + 0.003314x^2 - 0.130908x^3$$
$$+0.068542x^4 - 0.009159x^5. \tag{8.110}$$

Using MATLAB compute and plot the values of $\arctan(x)$ given by Eq. (8.110) and the error due to the above approximation. Compare this approximation with that given by Eq. (8.87).

---
[10]See Section 6.5.1.

CHAPTER 9

Analysis of Finite Word-Length Effects

So far, we have assumed that we are dealing with discrete-time systems characterized by linear difference equations with constant coefficients, where both the coefficients and the signal variables have infinite precision taking any value between $-\infty$ and ∞. However, when implemented in either software form on a general purpose computer or in special purpose hardware form, the system parameters along with the signal variables can take only discrete values within a specified range since the registers of the digital machine where they are stored are of finite length. The discretization process results in nonlinear difference equations characterizing the discrete-time systems. These nonlinear equations, in principle, are almost impossible to analyze and deal with exactly. Fortunately, if the quantization amounts are small compared to the values of the signal variables and filter constants, a simpler approximate theory based on a statistical model can be applied, and it is possible to derive the effects of discretization and develop results that can be verified experimentally.

To illustrate the various sources of errors arising from the discretization process in the implementation of a digital filter, consider for simplicity, the first-order IIR digital filter of Figure 9.1 defined by the linear constant coefficient difference equation

$$y[n] = \alpha y[n-1] + x[n], \qquad (9.1)$$

where $y[n]$ and $x[n]$ are the output and the input signal variables, respectively. The corresponding transfer function describing the above digital filter is given by

$$H(z) = \frac{1}{1 - \alpha z^{-1}} = \frac{z}{z - \alpha}. \qquad (9.2)$$

When implemented on a digital machine, the filter coefficient α can assume only certain discrete values $\hat{\alpha}$, and in general, can only approximate the original design value of α. As a result, the actual transfer function implemented is given by

$$\hat{H}(z) = \frac{z}{z - \hat{\alpha}}, \qquad (9.3)$$

Figure 9.1 A first-order IIR digital filter.

which is different from the desired transfer function $H(z)$ of Eq. (9.2). Therefore, the actual frequency response may be quite different from the desired frequency response. This coefficient quantization problem is similar to the sensitivity problem encountered in analog filter implementation.

If we assume that the input sequence $x[n]$ has been obtained by sampling an analog signal $x_a(t)$, it is discretized by the A/D converter being employed to convert the output of the sample-and-hold to digital samples. If we represent the output of the A/D converter as $\hat{x}[n]$, then the actual input to the digital filter of Figure. 9.1 is given by

$$\hat{x}[n] = x[n] + e[n], \tag{9.4}$$

where $e[n]$ is the A/D conversion error generated by the input quantization process.

The quantization of arithmetic operations leads to another source of errors. In the case of our simple digital filter of Eq. (9.1), the output of the multiplier $v[n]$ generated by multiplying the signal $y[n-1]$ with α,

$$v[n] = \alpha y[n-1], \tag{9.5}$$

is quantized to fit the register containing the product. The quantized signal $\hat{v}[n]$ can be represented as

$$\hat{v}[n] = v[n] + e_\alpha[n], \tag{9.6}$$

where $e_\alpha[n]$ is the error sequence generated by the product quantization process. The properties of this type of roundoff error are somewhat similar to those of the A/D conversion error.

In addition to the above sources of errors, another type of error occurs in digital filters due to the nonlinearity caused by the quantization of arithmetic operations. These errors manifest themselves in the form of oscillations, called limit cycles, at the output of the filter, usually in the absence of input or sometimes in the presence of constant input signals or sinusoidal input signals. In this chapter, we analyze the effects of the above sources of quantization errors and then describe structures that are less sensitive to these effects.

9.1   The Quantization Process and Errors

There are two basic types of the binary representations of data, fixed-point and floating-point formats, as described in Section 8.4. In each of these formats, a negative number can be represented in one of three different forms. The arithmetic operations involved in digital signal processing are the addition (subtraction) and the multiplication operations, discussed in Section 8.5. Various problems can arise in the digital implementation of the arithmetic operations involving the binary data due to the finite wordlength limitations of the registers

Figure 9.2   A general $(b + 1)$-bit fixed-point fraction.

Figure 9.3   The quantization process model.

storing the numbers and the results of the arithmetic operations. For example, in fixed-point arithmetic, as demonstrated in Example 8.19, the product of two b-bit numbers is $2b$ bits long, which has to be quantized to b bits to fit the prescribed wordlength of the registers. Moreover, in fixed-point arithmetic, the addition operation can result in a sum exceeding the register wordlength, causing an overflow as illustrated in Example 8.15. On the other hand, there is essentially no overflow in a floating-point addition. However, the results of both addition and multiplication may have to be quantized to fit the prescribed wordlength of the registers.

An analysis of the various quantization effects on the performance of a digital filter in practice depends on whether the numbers are in fixed-point or floating-point format, the type of representation for the negative numbers being used, the quantization method being employed to quantize the data, and the digital filter structure being used for implementation. Since the number of all possible combinations of the type of arithmetic, type of quantization method, and digital filter structure (of which there are literally thousands) is very large, we consider in this chapter analysis of quantization effects in some selected practical cases. However, the analysis presented can be easily extended to other cases without too much difficulty.

We now describe the three different types of quantization that can be employed. As indicated in Section 8.4, it is a common practice in digital signal processing applications, to represent data in a digital machine either as a fixed-point fraction or as a floating-point binary number with the mantissa as a binary fraction. We assume the available wordlength is $(b + 1)$ bits with the *most-significant bit (MSB)* representing the sign of the number. Consider first the data to be a $(b + 1)$-bit fixed-point fraction with the binary point just to the right of the sign bit, as indicated in Figure 9.2. The smallest positive number that can be represented in this format will have a *least-significant bit (LSB)* of 1, with the remaining bits being all 0's. Its decimal equivalent is 2^{-b}. Numbers represented with $(b + 1)$ bits are thus quantized in steps of 2^{-b}, called the *quantization step* or the *width of quantization*.

Before quantization, the wordlength is much larger than that indicated above. Assume that the original data x is represented as a $(\beta + 1)$-bit fraction with $\beta >> b$. To convert it into a $(b + 1)$-bit fraction, to be denoted as $\mathcal{Q}(x)$, we can employ either *truncation* or *rounding*. In either case, the quantization process can be modeled as shown in Figure 9.3. Since the representation of a positive binary fraction is the same independent of the format being used to represent the negative binary fraction, the effect of quantization of a positive

Figure 9.4   Illustration of the truncation operation.

fraction remains unchanged. However, the effect on negative fractions is not the same for the three different types of representations.

9.1.1   Quantization of Fixed-Point Numbers

To truncate a fixed-point number from $(\beta + 1)$ bits to $(b + 1)$ bits, we simply discard the least significant $(\beta - b)$ bits, as indicated in Figure 9.4. Let ε_t denote the truncation error defined by

$$\varepsilon_t = \mathcal{Q}(x) - x. \tag{9.7}$$

For a positive number x, the magnitude of the number $\mathcal{Q}(x)$ obtained after truncation is less than or equal to the magnitude of x. Therefore, $\varepsilon_t \leq 0$ for a positive x. The error ε_t is equal to zero if all bits being discarded are 0's and is largest if all bits being discarded are 1's. In the latter case, the decimal equivalent of the portion being discarded is equal to $2^{-b} - 2^{-\beta}$. Hence the range of the error ε_t in the case of truncation of a positive number x is given by

$$-(2^{-b} - 2^{-\beta}) \leq \varepsilon_t \leq 0. \tag{9.8}$$

For a negative number x, each one of the three different representations needs to be examined individually. For a negative fraction in sign-magnitude form, the magnitude of the truncated number $\mathcal{Q}(x)$ is smaller than that of the unquantized negative number x. Thus, it follows from the definition of the quantization error ε_t given by Eq. (9.7) that here

$$0 \leq \varepsilon_t \leq 2^{-b} - 2^{-\beta}. \tag{9.9}$$

For a negative fraction x in ones'-complement form $1_\Delta a_{-1}a_{-2}\ldots a_{-\beta}$, the numerical value is $-(1 - 2^{-\beta}) + \sum_{i=1}^{b} a_{-i}2^{-i}$. The numerical value of its quantized version $\mathcal{Q}(x)$ is likewise

$$\varepsilon_t = \mathcal{Q}(x) - x = -(1 - 2^{-b}) + \sum_{i=1}^{b} a_{-i}2^{-i} + (1 - 2^{-\beta}) - \sum_{i=1}^{\beta} a_{-i}2^{-i}$$

$$= (2^{-b} - 2^{-\beta}) - \sum_{i=b+1}^{\beta} a_{-i}2^{-i}. \tag{9.10}$$

The truncation error for this representation is always positive and has a range

$$0 \leq \varepsilon_t \leq 2^{-b} - 2^{-\beta}. \tag{9.11}$$

Now, consider a negative fraction x given in the two's-complement format $1_\triangle a_{-1}a_{-2}\ldots a_{-\beta}$. Its numerical value is given by $(-1 + \sum_{i=1}^{\beta} a_{-i}2^{-1})$. The representation of $Q(x)$, obtained after truncating x, is given by $1_\triangle a_{-1}a_{-2}\ldots a_{-b}$, with a numerical value $(-1 + \sum_{i=1}^{b} a_{-i}2^{-i})$. Therefore,

$$\varepsilon_t = Q(x) - x = \left(-1 + \sum_{i=1}^{b} a_{-i}2^{-i}\right) - \left(-1 + \sum_{i=1}^{\beta} a_{-i}2^{-i}\right) = -\sum_{i=b+1}^{\beta} a_{-i}2^{-i}. \tag{9.12}$$

The truncation error is here always negative and has a range

$$-(2^{-b} - 2^{-\beta}) \le \varepsilon_t \le 0. \tag{9.13}$$

In the case of rounding, the number is quantized to the nearest quantization level. We assume that a number exactly halfway between two quantization levels is rounded up to the nearest higher level. Therefore, if the bit $a_{-(b+1)}$ is 0, rounding is equivalent to truncation, and if this bit is 1, 1 is added to the LSB position of the truncated number. It should be noted that the rounding error ε_r does not depend on the format being used to represent the negative fraction since the operation is solely based on the magnitude of the number. To determine the range of ε_r, we observe that the quantization step after rounding has a value 2^{-b}. The maximum rounding error ε_r therefore has a magnitude $(2^{-b})/2$. As a result, the range of ε_r is given by

$$-\tfrac{1}{2}(2^{-b} - 2^{-\beta}) < \varepsilon_r \le \tfrac{1}{2}(2^{-b} - 2^{-\beta}). \tag{9.14}$$

In practice, $\beta >> b$. For example, the wordlength of a product is typically twice that of the numbers being multiplied. Hence, we can set $2^{-\beta} \cong 0$ in the inequalities of Eqs. (9.9), (9.11), (9.13), and (9.14), leading to the simpler inequalities shown in Table 9.1, where we have set the quantization step as $\delta = 2^{-b}$. Plots of the input-output characteristics of the quantizer for the three different representations and the two different quantization methods are shown in Figure 9.5.

9.1.2 Quantization of Floating-Point Numbers

In the case of floating-point numbers, quantization is carried out only on the mantissa. As a result, here it is more relevant to consider the relative error caused by the quantization process. To this end, we define the relative error e in terms of the numerical values of quantized floating-point number $Q(x) = 2^E Q(M)$ and the unquantized number $x = 2^E M$ as

$$e = \frac{Q(x) - x}{x} = \frac{Q(M) - M}{M}. \tag{9.15}$$

It can be shown (Problem 9.1) that the range of the relative errors for the different representations of a binary number are as indicated in Table 9.2, assuming $2^{-\beta} << 2^{-b}$.

A discussion of the analysis of quantization effects of digital filters implemented using floating-point arithmetic is beyond the scope of this text. Interested readers are referred to the publications listed at the end of this book [Kan71], [Liu69], [Opp75], [San67], [Wei69b], [Wei69a]. We consider here the fixed-point implementation case.

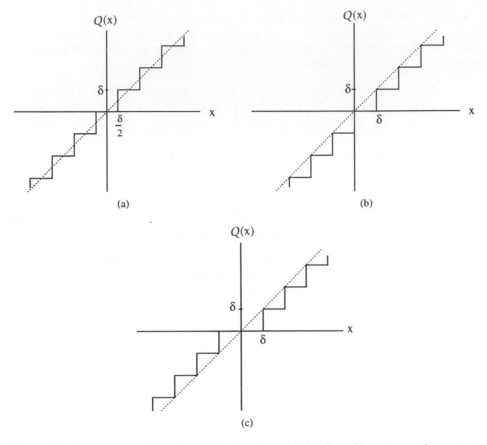

Figure 9.5   Input-output relationships of the quantizer: (a) rounding, (b) two's complement truncation, and (c) ones'-complement and sign-magnitude truncation. Here $\delta = 2^{-b}$.

9.2   Analysis of Coefficient Quantization Effects

The effect of multiplier coefficient quantization on digital filters is similar to that observed in analog filters. The transfer function $\hat{H}(z)$ of the digital filter implemented in hardware or software form with quantized coefficients is different from the desired transfer function $H(z)$. The main effect of the coefficient quantization is therefore on the poles and zeros that move to different locations than the original desired locations. As a result, the actual frequency response $\hat{H}(e^{j\omega})$ is different from the desired frequency response $H(e^{j\omega})$ and may not be acceptable to the user. Moreover, the poles may move outside the unit circle causing the implemented digital filter to become unstable even though the original transfer function with unquantized coefficients is stable, as illustrated earlier in Example 4.9. However, as demonstrated in Example 4.10, the same transfer function when implemented in a cascade form remained stable after coefficient quantization.

To illustrate the effect of the coefficient quantization on the frequency response of a digital filter, we show in Figure 9.6 the gain response of an ideal fifth-order IIR elliptic

Table 9.1   Range of quantization error.

Type of quantization	Number representation	Range of Error $\mathcal{Q}(x) - x$
Truncation	Positive number Two's-complement negative number	$-\delta < \varepsilon_t \leq 0$
Truncation	Sign-magnitude negative number Ones'-complement negative number	$0 \leq \varepsilon_t < \delta$
Rounding	All positive and negative numbers	$-\frac{\delta}{2} < \varepsilon_r \leq \frac{\delta}{2}$

Note: $\delta = 2^{-b}$.

Table 9.2   Range of relative error $e = (\mathcal{Q}(x) - x)/x$.

Type quantization	Number representation	Range of relative error
Truncation	Two's-complement	$-2\delta < e_t \leq 0, \quad x > 0$ $0 \leq e_t < 2\delta, \quad x < 0$
Truncation	Sign-magnitude Ones'-complement	$-2\delta < e_t \leq 0$
Rounding	All numbers	$-\delta < e_r \leq \delta$

Note: $\delta = 2^{-b}$.

filter with a cutoff at 0.4π, a passband ripple of 0.5 dB, a minimum stopband attenuation of 40 dB (shown with solid line), and the gain response obtained when the coefficients of the numerator and denominator are truncated to 8-bit length (shown with dotted line).

Figure 9.7 shows the locations of the poles and zeros of the original elliptic lowpass filter transfer function with unquantized coefficients and of the transfer function of the elliptic filter implemented with quantized coefficients. As can be seen from this plot, coefficient quantization can cause substantial displacement of the poles and zeros from their desired nominal locations. In this example, the zero closest to the pole has moved the farthest from its original location and has moved closer to the new location of its nearest pole, which is now much closer to the unit circle. These displacements do indicate how sensitive are the poles and zeros of an IIR filter to coefficient quantization. We next develop approximate expressions for estimating the pole-zero displacements due to coefficient quantization [Mit74a].

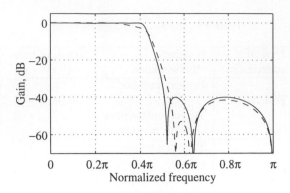

Figure 9.6   Gain responses of a fifth-order IIR elliptic lowpass filter with unquantized (shown with solid line) and quantized coefficients (shown with dotted line).

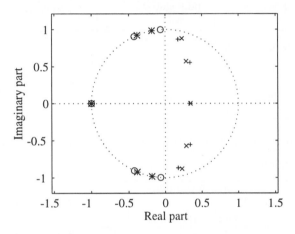

Figure 9.7   Pole-zero movements of a fifth-order IIR elliptic filter due to coefficient quantization: Pole and zero locations of the filter with quantized coefficients denoted by 'x' and 'o', respectively, and pole and zero locations of the filter with unquantized coefficients denoted by '+' and '*', respectively.

Consider an Nth degree polynomial $B(z)$ with simple roots:

$$B(z) = \sum_{i=0}^{N} b_i z^i = \prod_{k=1}^{N} (z - z_k), \qquad (9.16)$$

where $b_N = 1$. The roots z_k of $B(z)$ are given by

$$z_k = r_k e^{j\theta_k}. \qquad (9.17)$$

Note that $B(z)$ can be either the denominator polynomial or the numerator polynomial of the digital transfer function. The effect of coefficient quantization is manifested by the change of the polynomial coefficients from b_i to $b_i + \Delta b_i$, and as result, the polynomial

$B(z)$ changes to a new polynomial $\hat{B}(z)$ given by

$$\hat{B}(z) = \sum_{i=0}^{N} (b_i + \Delta b_i)z^i = B(z) + \sum_{i=0}^{N-1} (\Delta b_i)z^i = \prod_{k=1}^{N}(z - \hat{z}_k), \qquad (9.18)$$

with $\hat{z}_k$ denoting the roots of the polynomial $\hat{B}(z)$. Note that $\hat{z}_k$ are the new locations to which the roots z_k of $B(z)$ have moved. For small changes, $\hat{z}_k$ will be close to z_k and can be expressed as

$$\hat{z}_k = (r_k + \Delta r_k)e^{j(\theta_k + \Delta\theta_k)}, \qquad (9.19)$$

where Δr_k and $\Delta\theta_k$ represent the changes in the radius and the angle of the kth root due to coefficient quantization. Our aim is to develop simple expressions for estimating Δr_k and $\Delta\theta_k$ knowing the changes Δb_i in the coefficients of the polynomial $B(z)$. If we assume the change Δb_i to be very small, the changes in the radius and the angle, Δr_k and $\Delta\theta_k$, of the kth root can be also considered to be very small, and we can rewrite the expression for $\hat{z}_k$ in Eq. (9.19) as

$$\hat{z}_k = (r_k + \Delta r_k)e^{j\Delta\theta_k}e^{j\theta_k} \cong (r_k + \Delta r_k)(1 + j\Delta\theta_k)e^{j\theta_k}$$
$$\cong r_k e^{j\theta_k} + (\Delta r_k + jr_k\Delta\theta_k)e^{j\theta_k}, \qquad (9.20)$$

neglecting higher order terms. The root displacement can now be expressed as

$$\hat{z}_k - z_k = \Delta z_k \cong (\Delta r_k + jr_k\Delta\theta_k)e^{j\theta_k}. \qquad (9.21)$$

Now consider the rational function $1/B(z)$. Its partial-fraction expansion is given by

$$\frac{1}{B(z)} = \sum_{k=1}^{N} \frac{\rho_k}{z - z_k}, \qquad (9.22)$$

where ρ_k is the residue of $1/B(z)$ at the pole $z = z_k$, i.e.,

$$\rho_k = \left.\frac{(z - z_k)}{B(z)}\right|_{z=z_k} = R_k + jX_k. \qquad (9.23)$$

If we assume that $\hat{z}_k$ is very close to z_k, then we can write

$$\frac{1}{B(\hat{z}_k)} \cong \frac{\rho_k}{\hat{z}_k - z_k}, \qquad (9.24)$$

or

$$\Delta z_k = \rho_k \cdot B(\hat{z}_k). \qquad (9.25)$$

But

$$\hat{B}(\hat{z}_k) = 0 = B(\hat{z}_k) + \sum_{i=0}^{N-1} (\Delta b_i)(\hat{z}_k)^i. \qquad (9.26)$$

Therefore, from Eqs. (9.25) and (9.26), we arrive at

$$\Delta z_k = -\rho_k \left\{ \sum_{i=0}^{N-1} (\Delta b_i)(\hat{z}_k)^i \right\} \cong -\rho_k \left\{ \sum_{i=0}^{N-1} (\Delta b_i)(z_k)^i \right\}, \qquad (9.27)$$

assuming that $\hat{z}_k$ is very close to z_k. Rewriting Eq. (9.27) we obtain

$$(\Delta r_k + jr_k\Delta\theta_k)e^{j\theta_k} = -(R_k + jX_k)\left\{\sum_{i=0}^{N-1}(\Delta b_i)(r_k e^{j\theta_k})^i\right\}. \qquad (9.28)$$

Equating real and imaginary parts of the above, we arrive at, after some algebra,

$$\Delta r_k = (-R_k \mathbf{P}_k + X_k \mathbf{Q}_k)\cdot\Delta\mathbf{B} = \mathbf{S}_b^{r_k}\cdot\Delta\mathbf{B}, \qquad (9.29a)$$

$$\Delta\theta_k = -\frac{1}{r_k}(X_k \mathbf{P}_k + R_k \mathbf{Q}_k)\cdot\Delta\mathbf{B} = \mathbf{S}_b^{\theta_k}\cdot\Delta\mathbf{B}, \qquad (9.29b)$$

where

$$\mathbf{P}_k = \left[\cos\theta_k \quad r_k \quad r_k^2\cos\theta_k \cdots r_k^{N-1}\cos(N-2)\theta_k\right], \qquad (9.30a)$$

$$\mathbf{Q}_k = \left[-\sin\theta_k \quad 0 \quad r_k^2\sin\theta_k \cdots r_k^{N-1}\sin(N-2)\theta_k\right], \qquad (9.30b)$$

$$\Delta\mathbf{B} = \left[\Delta b_0 \; \Delta b_1 \; \Delta b_2 \cdots \Delta b_{N-1}\right]^T. \qquad (9.30c)$$

It should be noted that the sensitivity vectors $\mathbf{S}_b^{r_k}$ and $\mathbf{S}_b^{\theta_k}$ depend on only $B(z)$ and are independent of $\Delta\mathbf{B}$. Hence, once this vector has been calculated, pole-zero displacements for any sets of $\Delta\mathbf{B}$ can be rapidly calculated using Eqs. (9.29a) and (9.29b). Moreover, $\Delta\mathbf{B}$ are multiplier coefficient changes for only the direct form realization.

We illustrate in the following example the application of Eq. (9.29a) and Eq. (9.29b) in computing the pole displacements of a second-order direct form IIR digital filter structure due to coefficient quantization.

Example 9.1 Consider the direct form II realization of

$$H(z) = \frac{z^2}{z^2 - Kz + L} = \frac{z^2}{B(z)}. \qquad (9.31)$$

as indicated in Figure 9.8. Here,

$$B(z) = z^2 - 2r\cos\theta z + r^2 = (z - z_1)(z - z_2), \qquad (9.32)$$

where

$$z_1 = re^{j\theta}, \qquad z_2 = re^{-j\theta}. \qquad (9.33)$$

From Eqs. (9.31) and (9.32), we get

$$K = 2r\cos\theta, \qquad L = r^2. \qquad (9.34)$$

The residue ρ_1 of $1/B(z)$ at $z = z_1$ is given by

$$\rho_1 = \left.\frac{z - z_1}{B(z)}\right|_{z=z_1} = -\frac{j}{2r\sin\theta}. \qquad (9.35)$$

Therefore, for the IIR filter of Figure 9.8, we have

$$\Delta\mathbf{B} = [\Delta L \quad -\Delta K]^T, \qquad (9.36a)$$

$$\mathbf{Q}_1 = [-\sin\theta \quad 0], \qquad (9.36b)$$

$$\mathbf{P}_1 = [\cos\theta \quad r]. \qquad (9.36c)$$

Figure 9.8   Direct form II realization of the second-order IIR transfer function of Eq. (9.31).

Substituting Eqs. (9.36a)–(9.36c) in Eqs. (9.29a) and (9.29b), we arrive at

$$\Delta r = X_1 \mathbf{Q}_1 \Delta \mathbf{B} = \frac{1}{2r} \Delta L, \tag{9.37a}$$

$$\Delta \theta = -\frac{1}{r}(X_1 \mathbf{P}_1 \Delta \mathbf{B}) = \frac{\Delta L}{2r^2 \tan \theta} - \frac{\Delta K}{2r \sin \theta}. \tag{9.37b}$$

It is evident from above expressions that the second-order direct form IIR structure is highly sensitive to coefficient quantizations for transfer functions with poles close to $\theta = 0$ or π, i.e., for narrow bandwidth lowpass and highpass filters.

We now extend the results of Eqs. (9.29a) and (9.29b) to an arbitrary structure with R multipliers given by $\alpha_k, k = 1, 2, \ldots, R$. Due to coefficient quantization these coefficients change to $\alpha_k + \Delta \alpha_k$. Since the multiplier coefficients α_i are multilinear functions of the coefficients b_i of the polynomial $B(z)$, we can express the change Δb_i in the transfer function coefficient b_i to the change $\Delta \alpha_k$ in the multiplier coefficient α_k through

$$\Delta b_i = \sum_{k=1}^{R} \frac{\partial b_i}{\partial \alpha_k} \Delta \alpha_k, \qquad i = 0, 1, \ldots, N-1. \tag{9.38}$$

In matrix form the above can be written as

$$\Delta \mathbf{B} = \mathbf{C} \cdot \Delta \boldsymbol{\alpha}, \tag{9.39}$$

where

$$\mathbf{C} = \begin{bmatrix} \dfrac{\partial b_0}{\partial \alpha_1} & \dfrac{\partial b_0}{\partial \alpha_2} & \cdots & \dfrac{\partial b_0}{\partial \alpha_R} \\ \dfrac{\partial b_1}{\partial \alpha_1} & \dfrac{\partial b_1}{\partial \alpha_2} & \cdots & \dfrac{\partial b_1}{\partial \alpha_R} \\ \vdots & \vdots & \ddots & \vdots \\ \dfrac{\partial b_{N-1}}{\partial \alpha_1} & \dfrac{\partial b_{N-1}}{\partial \alpha_2} & \cdots & \dfrac{\partial b_{N-1}}{\partial \alpha_R} \end{bmatrix}, \tag{9.40a}$$

$$\Delta \boldsymbol{\alpha} = [\Delta \alpha_1 \quad \Delta \alpha_2 \cdots \Delta \alpha_R]^T. \tag{9.40b}$$

Substituting Eq. (9.39) in Eq. (9.29a) and Eq. (9.29b), we arrive at the desired result:

$$\Delta r_k = \mathbf{S}_b^{r_k} \cdot \mathbf{C} \cdot \Delta \boldsymbol{\alpha}, \tag{9.41a}$$

$$\Delta \theta_k = \mathbf{S}_b^{\theta_k} \cdot \mathbf{C} \cdot \Delta \boldsymbol{\alpha}, \tag{9.41b}$$

where the sensitivity vectors are as given in Eq. (9.29a) and Eq. (9.29b). It should be noted that **C** depends on the structure but has to be computed only once.

The application of Eq. (9.41a) and Eq. (9.41b) in computing the pole displacements of a second-order IIR digital filter structure is treated in the following example.

Example 9.2 Consider the coupled-form structure [Gol69b] of Figure 9.9 for which the transfer function is given by

$$H(z) = \frac{\gamma z^2}{z^2 - (\alpha + \delta)z + (\alpha\delta - \beta\gamma)}. \tag{9.42}$$

If $\alpha = \delta = r\cos\theta$ and $\beta = -\gamma = r\sin\theta$, then the transfer function of Eq. (9.42) becomes

$$H(z) = \frac{\gamma z^2}{z^2 - 2r\cos\theta z + r^2}. \tag{9.43}$$

Comparing the denominator polynomial of Eq. (9.42) with that of Eq. (9.31). we observe

$$K = \alpha + \delta = 2\alpha, \tag{9.44a}$$
$$L = \alpha\delta - \beta\gamma = \alpha^2 + \beta^2, \tag{9.44b}$$

from which we arrive at

$$\begin{bmatrix} \Delta L \\ \Delta K \end{bmatrix} = \begin{bmatrix} 2r\cos\theta & 2r\sin\theta \\ 2 & 0 \end{bmatrix} \begin{bmatrix} \Delta\alpha \\ \Delta\beta \end{bmatrix}. \tag{9.45}$$

Making use of Eq. (9.37a) and Eq. (9.37b) in the above we then get

$$\begin{bmatrix} \Delta r \\ \Delta\theta \end{bmatrix} = \begin{bmatrix} \frac{1}{2r} & 0 \\ \frac{1}{2r^2\tan\theta} & -\frac{1}{2r\sin\theta} \end{bmatrix} \begin{bmatrix} 2r\cos\theta & 2r\sin\theta \\ 2 & 0 \end{bmatrix} \begin{bmatrix} \Delta\alpha \\ \Delta\beta \end{bmatrix}$$
$$= \begin{bmatrix} \cos\theta & \sin\theta \\ -\frac{1}{r}\sin\theta & \frac{1}{r}\cos\theta \end{bmatrix} \begin{bmatrix} \Delta\alpha \\ \Delta\beta \end{bmatrix}. \tag{9.46}$$

As can be seen from the above, the coupled-form structure is less sensitive to multiplier coefficient quantization than the direct-form structure. Several other low-sensitivity second-order structures are considered in Problems 9.2 and 9.3.

9.3   Analysis of Coefficient Quantization Effects in FIR Filters

The analysis of the displacement of the roots of a polynomial due to coefficient quantization outlined in the previous section can of course be applied to FIR transfer function to determine the sensitivity of its zeros to changes in the coefficients. A more meaningful analysis is obtained by examining the changes in the frequency response due to coefficient quantization as described next.

Consider an $(M - 1)$th-order FIR transfer function:

$$H(z) = \sum_{n=0}^{M-1} h[n]z^{-n}. \tag{9.47}$$

Quantization of the filter coefficients results in a new transfer function

$$\hat{H}(z) = \sum_{n=0}^{M-1} \hat{h}[n]z^{-n} = \sum_{n=0}^{M-1} (h[n] + e[n])z^{-n}, \tag{9.48}$$

Figure 9.9   A second-order coupled form structure.

Figure 9.10   Model of the FIR filter with quantized coefficients.

which can be rewritten as

$$\hat{H}(z) = H(z) + E(z), \tag{9.49}$$

where

$$E(z) = \sum_{n=0}^{M-1} e[n]z^{-n}. \tag{9.50}$$

Thus, the FIR filter with quantized coefficients can be modeled as a parallel connection of two FIR filters, $H(z)$ and $E(z)$, as shown in Figure 9.10, where $H(z)$ represents the desired FIR filter with unquantized coefficients, and $E(z)$ is the FIR filter representing the error in the transfer function due to coefficient quantization.

Without any loss of generality, assume the FIR filter $H(z)$ to be a Type 1 linear-phase filter of length M with impulse response $h[n]$. Hence, $E(z)$ is also a Type 1 linear-phase FIR transfer function. The frequency response of $H(z)$ can be expressed as

$$H(e^{j\omega}) = e^{-j\omega(M-1)/2} \left(h\left[\frac{M-1}{2}\right] + \sum_{n=0}^{(M-3)/2} 2h[n] \cos\left[\left(\frac{M-1}{2} - n\right)\omega\right] \right).$$

$$\tag{9.51}$$

The frequency response of the actual FIR filter with quantized coefficients $\hat{h}[n]$ can be expressed as

$$\hat{H}(e^{j\omega}) = H(e^{j\omega}) + E(e^{j\omega}), \tag{9.52}$$

where $E(e^{j\omega})$ represents the error in the desired frequency response $H(e^{j\omega})$:

$$E(e^{j\omega}) = \sum_{n=0}^{M-1} e[n]e^{-j\omega n}, \tag{9.53}$$

with $e[n] = \hat{h}[n] - h[n]$. The error in the frequency response is thus bounded by

$$\left| E(e^{j\omega}) \right| = \left| \sum_{n=0}^{M-1} e[n]e^{-j\omega n} \right| \leq \sum_{n=0}^{M-1} |e[n]| \left| e^{-j\omega n} \right| \leq \sum_{n=0}^{M-1} |e[n]|. \tag{9.54}$$

Assume each impulse response coefficient $h[n]$ is a $(b+1)$-bit signed fraction. In this case, the range of the coefficient quantization error $e[n]$ is exactly same as that indicated in Table 9.1. Using the data given in this table, an upper bound on $\left| E(e^{j\omega}) \right|$ can be derived from Eq. (9.54). For example, for rounding, $|e[n]| \leq \delta/2$, where $\delta = 2^{-b}$ is the quantization step. As a result

$$\left| E(e^{j\omega}) \right| \leq \frac{M\delta}{2}. \tag{9.55}$$

The above bound is rather conservative and can be reached only if all errors in Eq. (9.54) are of same sign and have the maximum value in the range. A more realistic bound can be derived assuming $e[n]$ are statistically independent random variables[Cha73]. From Eqs. (9.51) to (9.53), we obtain

$$E(e^{j\omega}) = e^{-j\omega(M-1)/2} \left(e\left[\frac{M-1}{2} \right] + \sum_{n=0}^{(M-3)/2} 2e[n] \cos\left[\left(\frac{M-1}{2} - n \right) \omega \right] \right). \tag{9.56}$$

From the above, we observe that $E(e^{j\omega})$ is a sum of independent random variables. If we denote the variance of $e[n]$ as σ_e^2, then the variance of $E(e^{j\omega})$ is simply given by

$$\sigma_E^2(\omega) = \sigma_e^2 \left(1 + 4 \sum_{n=1}^{(M-1)/2} \cos^2(\omega n) \right) = \sigma_e^2 \left(M - 1 + \frac{\sin M\omega}{\sin \omega} \right). \tag{9.57}$$

Using the notation

$$W_M(\omega) = \left[\frac{1}{2M-1} \left(M - 1 + \frac{\sin M\omega}{\sin \omega} \right) \right]^{1/2}, \tag{9.58}$$

the standard deviation of $E(e^{j\omega})$ can be expressed as

$$\sigma_E(\omega) = \sigma_e \left(\sqrt{2M-1} \right) W_M(\omega). \tag{9.59}$$

For uniformly distributed $e[n]$, $\sigma_e = \delta/\sqrt{12} = 2^{-b-1}/\sqrt{3}$.

A plot of a typical weighting function $W_M(\omega)$ is sketched in Figure 9.11. In fact, it can be shown that $W_M(\omega)$ is in the range $(0, 1)$, and hence, the standard deviation $\sigma_E(\omega)$ is bounded by

$$\sigma_E(\omega) \leq \delta \sqrt{\frac{2M-1}{12}}. \tag{9.60}$$

Based on the above bound, Chan and Rabiner [Cha73] have advanced a method to estimate the wordlength of the FIR filter coefficients to meet the prescribed filter specifications.

Figure 9.11   Plot of a typical $W_M(\omega)$.

9.4   A/D Conversion Noise Analysis

In many applications, digital signal processing techniques are employed to process continuous-time (analog) bandlimited signals that are either voltage or current waveforms. These analog signals must be converted into digital form before they can be processed digitally. According to the sampling theorem given in Section 5.2.1, an analog band-limited signal $x_a(t)$ can be represented uniquely by its sampled version $x_a(nT)$ if the sampling frequency $\Omega_T = 1/T$ is greater than twice the highest frequency Ω_m contained in $x_a(t)$. In order to ensure that the sampling frequency chosen does satisfy this condition, an anti-aliasing filter is used to bandlimit the analog signal to half of the sampling frequency. The discrete-time sequence $x_a(nT) = x[n]$ is then converted into a digital sequence for digital signal processing. As indicated by Figure 5.1, the conversion of an analog signal into a digital sequence is implemented in practice by a cascade of two devices, a sample-and-hold (S/H) circuit followed by an analog-to-digital (A/D) converter.

The digital samples produced by the A/D converter are usually represented in a binary form. As indicated in Section 8.4, there are several different forms of binary representations of which the two's-complement representation is usually employed in digital signal processing for convenience in the implementation of the arithmetic operations. Consequently, the A/D converters used for the digital signal processing of analog signals in general are those that employ the two's-complement fixed-point representation to represent the digital equivalent of the input analog signal. Moreover, for the processing of bipolar analog signals, the A/D converter generates a bipolar output represented as a fixed-point signed binary fraction.

9.4.1   Quantization Noise Model

The digital sample generated by the A/D converter is the binary representation of the quantized version of that produced by an ideal sampler with infinite precision. Because of the finite wordlength of the output register, the digital equivalent can take a value from a finite set of discrete values within the dynamic range of the register. For example, if the output word is of length $(b+1)$ bits including the sign bit, the total number of discrete levels

Figure 9.12   Model of a practical A/D conversion system.

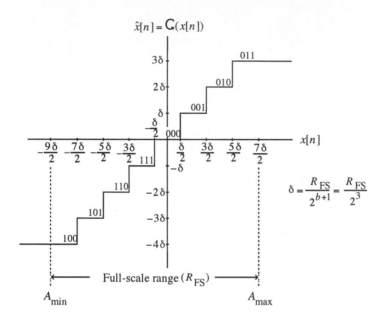

Figure 9.13   Input-output characteristic of a 3-bit bipolar A/D converter with two's-complement representation.

available for representation is 2^{b+1}. The dynamic range of the output register depends on the binary number representation selected for the A/D converter. The operation of a practical analog-to-digital conversion system consisting of a sample-and-hold circuit followed by an A/D converter therefore can be modeled as shown in Figure 9.12. The quantizer maps the input analog sample $x[n]$ into $\hat{x}[n]$, one of a set of discrete values, and the coder determines its binary equivalent $\hat{x}_{eq}[n]$ based on the binary representation scheme adopted by the A/D converter.

The quantization process employed by the quantizer in the A/D converter can be either rounding or truncation. Assuming rounding is used, the input-output characteristic of a 3-bit A/D converter with the output in two's-complement form is as shown in Figure 9.13. This figure also shows the binary equivalents of the quantized samples.

As indicated earlier and shown explicitly in Figure 9.13, for a two's-complement binary representation, the binary equivalent $\hat{x}_{eq}[n]$ of the quantized input analog sample $\hat{x}[n]$ is a binary fraction in the range

$$-1 \le \hat{x}_{eq}[n] < 1, \tag{9.61}$$

It is related to the quantized sample $\hat{x}[n]$ through

$$\hat{x}_{eq}[n] = \frac{2\hat{x}[n]}{R_{FS}},$$ (9.62)

where R_{FS} denotes the *full-scale range* of the A/D converter. We shall assume that the input signal has been scaled to be in the range of ± 1 by dividing its amplitude by $R_{FS}/2$, as is usually the case. Then the decimal equivalent of $\hat{x}_{eq}[n]$ is equal to $\hat{x}[n]$, and we shall not differentiate between these two numbers.

For a $(b+1)$-bit bipolar A/D converter, the total number of quantization levels is 2^{b+1} and the full-scale range R_{FS}, usually given in volts or amperes, is given by

$$R_{FS} = 2^{b+1}\delta,$$ (9.63)

where δ is the quantization step size, also called the quantization width, in volts or amperes. For the 3-bit A/D converter depicted in Figure 9.13, the total number of levels is $2^3 = 8$ and the full-scale range is $R_{FS} = 8\delta$, with a maximum value of $A_{\max} = 7\delta/2$ and a minimum value of $A_{\min} = -9\delta/2$. If the input analog sample $x_a(nT)$ is within the full-scale range,

$$-\frac{9\delta}{2} < x_a(nT) \le \frac{7\delta}{2},$$ (9.64)

it is quantized to one of the 8 discrete values indicated in Figure 9.13. In general, for a $(b+1)$-bit wordlength A/D converter employing two's-complement representation, the full-scale range is given by

$$-(2^{b+1}+1)\frac{\delta}{2} < x_a(nT) \le (2^{b+1}-1)\frac{\delta}{2}.$$ (9.65)

If we denote the difference between the quantized value $\mathcal{Q}(x[n]) = \hat{x}[n]$ and the input sample $x_a(nT) = x[n]$ as the quantization error $e[n]$:

$$e[n] = \mathcal{Q}(x[n]) - x[n] = \hat{x}[n] - x[n],$$ (9.66)

it follows from Figure 9.13 that $e[n]$ is within the range

$$-\frac{\delta}{2} < e[n] \le \frac{\delta}{2},$$ (9.67)

assuming that a sample exactly halfway between two levels is rounded up to the nearest higher level and assuming that the analog input is within the A/D converter full-scale range as given by Eq. (9.61). In this case, the quantization error $e[n]$, called the *granular noise*, is bounded in magnitude according to Eq. (9.67). A plot of the quantization error $e[n]$ of the above 3-bit A/D converter as a function of the input sample $x[n]$ is given in Figure 9.14.

As can be seen from this figure, when the input analog sample is outside the A/D converter full-scale range, the magnitude of the error $e[n]$ increases linearly with an increase in the magnitude of the input. In the latter situation, the A/D converter error $e[n]$ is called the *saturation error* or the *overload noise* as the A/D converter output is "clipped" to the maximum value $(1 - 2^{-b})$ if the analog input is positive or the minimum value -1 if the analog input is negative independent of the actual value of the input. A clipping of the A/D converter output causes signal distortion with highly undesirable effects and must be

Figure 9.14   Quantization error as a function of the input for a 3-bit bipolar A/D converter.

avoided by scaling down the analog input $x_a(nT)$ to ensure that it remains within the A/D converter full-scale range.

In order to develop the necessary mathematical model for analyzing the effect of the finite wordlength of the A/D converter output, we assume that analog input samples are within the full-scale range, and as a result, there is no saturation error at the converter output. Since the input-output characteristic of an A/D converter is nonlinear and the analog input signal, in most practical cases, is not known a priori, it is reasonable to assume for analysis purposes that the quantization error $e[n]$ is a random signal and to use a statistical model of the quantizer operation as indicated in Figure 9.15. Furthermore, for simplified analysis, we make the following assumptions:

(a) The error sequence $\{e[n]\}$ is a sample sequence of a wide-sense stationary (WSS) white noise process, with each sample $e[n]$ being uniformly distributed over the range of the quantization error as indicated in Figure 9.16, where δ is the quantization step.

(b) The error sequence is uncorrelated with its corresponding input sequence $\{x[n]\}$.

(c) The input sequence is a sample sequence of a stationary random process.

These assumptions hold in most practical situations for input signals whose samples are large and change in amplitude very rapidly in time relative to the quantization step in a somewhat random fashion, and have been verified to be valid assumptions experimentally [Ben48], [Wid56], [Wid61] and by computer simulations [DeF88]. The statistical model also makes the analysis of A/D conversion noise more tractable, and results derived have been found to be useful for most applications. It should be pointed out that in the case of an A/D converter employing ones'-complement or sign-magnitude truncation, the quantization error is correlated to the input signal since here the sign of each error sample $e[n]$ is exactly opposite to the sign of the corresponding input sample $x[n]$. As a result, practical A/D converters use either rounding or two's-complement truncation. The mean and the variance of the uniformly distributed random variables were computed in Example 2.36. Using the results of this example, we observe that mean and the variance of the error sample in the case of rounding are given by

$$m_e = \frac{(\delta/2) - (\delta/2)}{2} = 0, \tag{9.68}$$

$$\sigma_e^2 = \frac{((\delta/2) - (-\delta/2))^2}{12} = \frac{\delta^2}{12}. \tag{9.69}$$

Figure 9.15   A statistical model of the A/D quantizer.

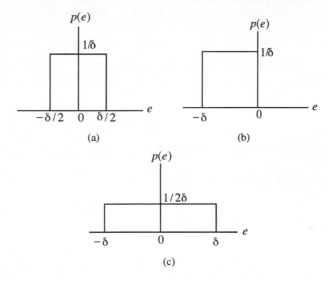

Figure 9.16   Quantization error probability density functions: (a) rounding, (b) two's-complement truncation, and (c) ones'-complement truncation.

The corresponding parameters for the two's-complement truncation are as follows:

$$m_e = \frac{0 - \delta}{2} = -\frac{\delta}{2}, \tag{9.70}$$

$$\sigma_e^2 = \frac{(0 - \delta)^2}{12} = \frac{\delta^2}{12}. \tag{9.71}$$

9.4.2   Signal-to-Quantization Noise Ratio

Based on the model of Figure 9.15, we can evaluate the effect of the additive quantization noise $e[n]$ on the input signal $x[n]$ by computing the *signal-to-quantization noise ratio* (SNR$_{A/D}$) in dB defined by

$$\text{SNR}_{A/D} = 10 \log_{10} \left(\frac{\sigma_x^2}{\sigma_e^2} \right) \text{dB}, \tag{9.72}$$

where σ_x^2 is the input signal variance representing the signal power and σ_e^2 is the noise variance representing the quantization noise power. For rounding, the quantization error is uniformly distributed in the range $(-\delta/2, \delta/2)$ and for two's-complement truncation,

Table 9.3 Signal-to-quantization ratio of an A/D converter as a function of wordlength and full-scale range.

	$b = 7$	$b = 9$	$b = 11$	$b = 13$	$b = 15$
$K = 4$	46.91	58.95	70.99	83.04	95.08
$K = 6$	43.39	55.43	67.47	79.51	91.56
$K = 8$	40.89	52.93	64.97	77.01	89.05

the quantization error is uniformly distributed in the range $(-\delta, 0)$ as indicated in Figure 9.16(a) and (b), respectively. In the case of a bipolar $(b + 1)$-bit A/D converter, $\delta = 2^{-(b+1)} R_{\text{FS}}$, and hence,

$$\sigma_e^2 = \frac{2^{-2b}(R_{\text{FS}})^2}{48}. \tag{9.73}$$

Substituting Eq. (9.73) in Eq. (9.72), we arrive at

$$\text{SNR}_{\text{A/D}} = 10 \log_{10} \left(\frac{48\sigma_x^2}{2^{-2b}(R_{\text{FS}})^2} \right)$$

$$= 6.02b + 16.81 - 20 \log_{10} \left(\frac{R_{\text{FS}}}{\sigma_x} \right) \text{dB}. \tag{9.74}$$

The above expression is used to determine the minimum A/D converter wordlength needed to meet a specified signal-to-quantization noise ratio. As can be seen from this expression, the SNR increases by approximately 6 dB for each bit added to the wordlength. For a given wordlength, the actual SNR depends on the last term in Eq. (9.74), which in turn depends on the σ_x, the rms value of the input signal amplitude, and the full-scale range R_{FS} of the converter, as illustrated by the following example.

Example 9.3 Determine the signal-to-quantization noise ratio in the digital equivalent of an analog sample $x[n]$ with a zero-mean Gaussian distribution using a $(b + 1)$-bit A/D converter having a full-scale range $R_{\text{FS}} = K\sigma_x$.
  From Eq. (9.74) we obtain

$$\text{SNR}_{\text{A/D}} = 6.0206b + 16.81 - 20 \log 10(K). \tag{9.75}$$

Table 9.3 shows the computed values of the SNR for various values of b and K.

Now, the probability of a particular input analog sample with a zero-mean Gaussian distribution staying within the full-scale range $K\sigma_x$ is given by [Par60]

$$2\Phi(K) - 1 = \sqrt{\frac{2}{\pi}} \int_0^K e^{-y^2/2} \, dy. \tag{9.76}$$

For example, for $K = 4$, the probability of an input analog sample staying within the full-scale range of $4\sigma_x$ is 0.9544. This implies that on average about 456 samples out of 10,000 samples will fall outside the range and be clipped. If we increase the full-scale range to $6\sigma_x$, the probability of an input analog sample staying within the expanded full-scale range increases to 0.9974, in which case on average about 26 samples out of 10,000 samples will now be outside the range. In most applications, a full-scale range of $8\sigma_x$ is more than adequate to ensure no clipping in conversion.

9.4.3 Effect of Input Scaling on SNR

Now, consider the effect of scaling of the input on the SNR. Let the input scaling factor be A with $A > 0$. Since the variance of the scaled input $Ax[n]$ is $A^2\sigma_x^2$, the signal-to-quantization noise expression of Eq. (9.75) changes to

$$\text{SNR}_{\text{A/D}} = 6.0206b + 16.81 - 20\log_{10}(K) + 20\log_{10}(A). \qquad (9.77)$$

For a given b, the SNR can also be increased by scaling up the input analog signal by making $A > 1$. However, this process also increases the probability of some of the input analog samples to be outside the full-scale range R_{FS}, and as a result, Eq. (9.77) no longer holds. Furthermore, the output is clipped, causing severe distortion in the digital representation of the analog input. On the other hand, a scaling down of the input analog signal by making $A < 1$ decreases the SNR. It is therefore necessary to ensure that the analog sample range matches as close as possible to the full-scale range of the A/D converter to get the maximum possible SNR without any signal distortion.

It should be noted here that the above analysis assumed an ideal A/D conversion. However, as pointed out in Section 5.6.6, a practical A/D converter is a nonideal device exhibiting a variety of errors, resulting in the actual signal-to-quantization noise ratio being smaller than that predicted by Eq. (9.74). Hence, the effective wordlength of the A/D converter, in general, is less than the wordlength computed using Eq. (9.74) by 1 to 2 bits. This factor should be taken into consideration in selecting an appropriate A/D converter for a given application.

9.4.4 Propagation of Input Quantization Noise to Digital Filter Output

In most applications, the quantized signal $\hat{x}[n]$ generated by the A/D converter is processed by a linear time-invariant discrete-time system $H(z)$. It is thus of interest to determine how the input quantization noise propagates to the output of the digital filter. In determining the noise at the filter output generated by the input noise, we can assume that the digital filter is implemented using infinite precision. However, as we shall point out later, in practice, the quantization of the arithmetic operations generates errors inside the digital filter structure, which also propagate to the output and appear as noise. These noise sources are assumed to be independent of the input quantization noise, and their effects can be analyzed separately and added to that due to the input noise.

As indicated in Figure 9.15, the quantized signal $\hat{x}[n]$ can be considered as a sum of two sequences: the unquantized input $x[n]$ and the input quantization noise $e[n]$. Because of the linearity property and the assumption that $x[n]$ and $e[n]$ are uncorrelated, the output $\hat{y}[n]$ of the LTI system can thus be expressed as a sum of two sequences: $y[n]$ generated by the unquantized input $x[n]$ and $v[n]$ generated by the error sequence $e[n]$, as shown in Figure 9.17. As a result, we can compute the output noise component $v[n]$ as a linear convolution of $e[n]$ with the impulse response sequence $h[n]$ of the LTI system:

$$v[n] = \sum_{m=-\infty}^{\infty} e[m]h[n-m]. \qquad (9.78)$$

From Eq. (4.179), the mean m_v of the output noise $v[n]$ is given by

$$m_v = m_e H(e^{j0}), \qquad (9.79)$$

Figure 9.17   Model for the analysis of the effect of processing a quantized input by an LTI discrete-time system.

and, from Eq. (4.195) its variance σ_v^2 is given by

$$\sigma_v^2 = \frac{\sigma_e^2}{2\pi} \int_{-\pi}^{\pi} \left| H(e^{j\omega}) \right|^2 d\omega. \tag{9.80}$$

The output noise power spectrum is given by

$$P_{vv}(\omega) = \sigma_e^2 \left| H(e^{j\omega}) \right|^2. \tag{9.81}$$

The normalized output noise variance is given by

$$\sigma_{v,n}^2 = \frac{\sigma_v^2}{\sigma_e^2} = \frac{1}{2\pi} \int_{-\pi}^{\pi} \left| H(e^{j\omega}) \right|^2 d\omega, \tag{9.82a}$$

which can be alternately expressed as

$$\sigma_{v,n}^2 = \frac{1}{2\pi j} \oint_C H(z) H(z^{-1}) z^{-1} \, dz, \tag{9.82b}$$

where C is a counterclockwise contour in the ROC of $H(z)H(z^{-1})$. An equivalent expression for Eq. (9.82b) obtained using Eq. (4.197) is given by

$$\sigma_{v,n}^2 = \sum_{n=-\infty}^{\infty} |h[n]|^2. \tag{9.82c}$$

9.4.5   Algebraic Computation of Output Noise Variance

We now outline a simple algebraic approach for computing the normalized output noise variance using Eq. (9.82b) [Mit74b]. Several other algebraic approaches have been suggested for computing integrals of the form of Eq. (9.82b) [Dug80], [Dug82].

In general, $H(z)$ is a causal stable real rational function with all poles inside the unit circle in the z-plane. It can be expressed in a partial-fraction form as

$$H(z) = \sum_{i=1}^{R} H_i(z), \tag{9.83}$$

where $H_i(z)$ is a low-order real rational causal stable transfer function. Substituting the above in Eq. (9.82b), we arrive at

$$\sigma_{v,n}^2 = \frac{1}{2\pi j} \sum_{k=1}^{R} \sum_{\ell=1}^{R} \oint_C H_k(z) H_\ell(z^{-1}) z^{-1} \, dz. \tag{9.84}$$

Since $H_k(z)$ and $H_\ell(z)$ are stable transfer functions, it can be shown that

$$\oint_C H_k(z) H_\ell(z^{-1}) z^{-1} \, dz = \oint_C H_\ell(z) H_k(z^{-1}) z^{-1} \, dz. \tag{9.85}$$

As a result, Eq. (9.84) can be rewritten as

$$\sigma_{v,n}^2 = \frac{1}{2\pi j} \left\{ \sum_{k=1}^{R} \oint_C H_k(z) H_k(z^{-1}) z^{-1} \, dz \right.$$
$$\left. + 2 \sum_{k=1}^{R-1} \sum_{\ell=k+1}^{R} \oint_C H_k(z) H_\ell(z^{-1}) z^{-1} \, dz \right\}. \tag{9.86}$$

In most practical cases, $H(z)$ has only simple poles with $H_k(z)$ being either a first-order or a second-order transfer function. As a result, each of the above contour integrations is much simpler to perform using the Cauchy's residue theorem, and the results are tabulated for easy reference. Typical terms in the partial-fraction expansion of $H(z)$ are as follows:[1]

$$A, \qquad \frac{B_k}{z - a_k}, \qquad \frac{C_k z + D_k}{z^2 + b_k z + d_k}. \tag{9.87}$$

Let us denote a typical contour integral in Eq. (9.86) as

$$I_i = \frac{1}{2\pi j} \oint_C H_k(z) H_\ell(z^{-1}) z^{-1} \, dz. \tag{9.88}$$

The expressions obtained after performing the contour integrations for different I_i are listed in Table 9.4.

Example 9.4 Let us determine the variance of the noise generated by the A/D quantization noise at the output of a first-order digital filter with a transfer function

$$H(z) = \frac{1}{1 - \alpha z^{-1}}. \tag{9.89}$$

We rewrite $H(z)$ as

$$H(z) = \frac{z}{z - \alpha},$$

whose partial-fraction expansion is given by

$$H(z) = 1 + \frac{\alpha}{z - \alpha}. \tag{9.90}$$

---

[1] It should be noted that $H(z)$ is expressed as the ratio of polynomials in z before a direct partial-fraction expansion is carried out to arrive at the terms given in Eq. (9.87)

Table 9.4 Expressions for typical contour integrals in the output noise variance calculation.

$H_k(z)$	\multicolumn{3}{c}{$H_\ell(z^{-1})$}		
	A	$\dfrac{B_\ell}{z^{-1}-a_\ell}$	$\dfrac{C_\ell z^{-1}+D_\ell}{z^{-2}+b_\ell z^{-1}+d_\ell}$
A	I_1	0	0
$\dfrac{B_k}{z-a_k}$	0	I_2	I_4'
$\dfrac{C_k z+D_k}{z^2+b_k z+d_k}$	0	I_4	I_3

$$I_1 = A^2$$

$$I_2 = \frac{B_k B_\ell}{1-a_k a_\ell}$$

$$I_3 = \frac{(C_k C_\ell + D_k D_\ell)(1-d_k d_\ell)-(D_k C_\ell - C_k D_\ell d_k)b_\ell -(C_k D_\ell - D_k C_\ell d_\ell)b_k}{(1-d_k d_\ell)^2 + d_k b_\ell^2 + d_\ell b_k^2 - (1+d_k d_\ell)b_k b_\ell}$$

$$I_4 = \frac{B_\ell(C_k + D_k a_\ell)}{1+b_k a_\ell + d_k a_\ell^2}$$

$$I_4' = \frac{B_k(C_\ell + D_\ell a_k)}{1+b_\ell a_k + d_\ell a_k^2}$$

The two terms in the partial-fraction expansion are

$$H_1(z)=1, \qquad H_2(z)=\frac{\alpha}{z-\alpha}.$$

Therefore, using the results of Table 9.4, we finally obtain the expression for the normalized output noise variance as

$$\sigma_{v,n}^2 = 1 + \frac{\alpha^2}{1-\alpha^2} = \frac{1}{1-\alpha^2}. \tag{9.91}$$

If the pole is close to the unit circle, we can write $\alpha = 1-\varepsilon$, where $\varepsilon \cong 0$, in which case, we can rewrite Eq. (9.91) as

$$\sigma_{v,n}^2 = \frac{1}{1-(1-\varepsilon)^2} \cong \frac{1}{2\varepsilon}. \tag{9.92}$$

indicating that as the pole gets closer to the unit circle, the output noise increases rapidly to very high values approaching infinity. Thus for high Q realizations, the wordlengths of the registers storing the signal variables should be of longer length to keep the output roundoff noise below a prescribed level.

The exact value of the output noise variance can be obtained from Eq. (9.91) by multiplying it with the variance of the A/D conversion noise σ_e^2.

9.4.6 Computation of Output Noise Variance Using MATLAB

The algebraic method of calculating the output noise variance developed in the previous section can be carried out much more easily using MATLAB. To this end, it is more convenient to develop a partial-fraction expansion of real coefficient transfer function $H(z)$ using the function $residue$, which results in terms of the form A and $B_k/(z - a_k)$ only, where the residues B_k and the poles a_k are either real or complex numbers. Thus, for the variance calculation, only the terms I_1 and I_2 of Table 9.4 are employed. Program 9_1 given below is based on the approach of Section 9.4.5. The input data called by the program are the numerator and denominator coefficients in ascending powers of z^{-1} entered as vectors inside square brackets. The program first determines the partial-fraction expansion of the transfer function and then computes the normalized output round-off noise variance using Eq. (9.86) and the tabulated integrals I_1 and I_2 in Table 9.4. The output data is the desired noise variance.

```
% Program 9_1
%
%   Computation of the output noise variance due
%   to input quantization of a digital filter
%   based on a partial-fraction approach
%
num = input('Type in the numerator = ');
den = input('Type in the denominator = ');
[r,p,K] = residue(num,den); % partial fraction expansion
R = size(r,1);
R2 = size(K,1);
    if R2 > 1
        disp('Cannot continue...');
        return;
    end

    if R2 == 1
        nvar = K^2;
    else
        nvar = 0;
    end
% Compute output noise variance
for k = 1:R,
    for m = 1:R,
        integral = r(k)*conj(r(m))/(1-p(k)*conj(p(m)));
        nvar = nvar + integral;
    end
end
disp('Output Noise Variance = ');disp(real(nvar))
```

We illustrate its use in the following example.

Example 9.5 Determine using Program 9_1 the output noise variance due to the input quantization of a causal fourth-order elliptic lowpass digital filter with a transfer function given by

$$H(z) = \frac{\begin{array}{l} 0.06891875 + 0.13808186z^{-1} + 0.18636107z^{-2} \\ +0.13808186z^{-3} + 0.06891875z^{-4} \end{array}}{\begin{array}{l} 1 - 1.30613249z^{-1} + 1.48301305z^{-2} \\ -0.77709026z^{-3} + 0.2361457z^{-4} \end{array}}. \tag{9.93}$$

The input data are the numerator and the denominator coefficients entered as vectors:

```
num = [0.06891875   0.13808186   0.18636107   0.13808186
       0.06891875]
den = [1.0    -1.30613249    1.48301305   -0.77709026
       0.2361457].
```

The output generated by the program is

```
Output Noise Variance = 0.40263012511156.
```

An alternative, fairly simple and straightforward computer-based method for the computation of the approximate value of the normalized output noise makes use of its equivalent expression given by Eq. (9.82c). For a causal and stable digital filter, the impulse response decays rapidly to zero values, and hence, Eq. (9.82c) can be approximated as a finite sum

$$\sigma_{v,n}^2 = S_L \cong \sum_{n=0}^{L} |h[n]|^2. \tag{9.94}$$

To determine $\sigma_{v,n}^2$, we can iteratively compute the above partial sum for $L = 1, 2, \ldots$, and stop the computation when the difference $S_L - S_{L-1}$ becomes smaller than a specified value κ, which is typically chosen as 10^{-6}.

The following example illustrates this approach.

Example 9.6 We determine again the normalized round-off noise variance of the fourth-order digital filter of Example 9.5 due to input quantization. To this end, we make use of the following MATLAB program.

```
% Program 9_2
% Computation of Approximate Output Roundoff Noise
%
num = input('Numerator coefficients = ');
den = input('Denominator coefficients = ');
x = 1;
order = max(length(num),length(den))-1;
si = [zeros(1,order)];
varnew = 0; k = 1;
while k > 0.000001
```

```
        [y,sf] = filter(num,den,x,si);
        si = sf;
        varold = varnew;
        varnew = varnew + abs(y)*abs(y);
        k = varnew - varold;
        x = 0;
    end
    disp('Output Noise Variance = ');disp(varnew)
```

During execution the program calls for the numerator and the denominator polynomial coefficients in ascending powers of z^{-1}. These data are entered as vectors inside square brackets as in Example 9.5. The program computes the impulse response coefficients of the filter and then evaluates the partial sum using Eq. (9.94) and displays the normalized round-off noise variance when the difference between the two successive partial sums falls below 10^{-6}. For the transfer function of Eq. (9.93), the output is given as indicated below:

```
        Output Noise Variance = 0.40254346992755
```

The approximate value computed using method of Eq. (9.94) is seen to be quite close to the actual value computed in the previous example.

9.5   Analysis of Arithmetic Round-Off Errors

As illustrated earlier in Section 8.5.2, in the fixed-point implementation of a digital filter the result of only the multiplication operation is quantized. In this section we develop the tools for the analysis of product round-off errors. Figure 9.18(a) shows the representation of a practical multiplier with the quantizer at its output. Its statistical model is indicated in Figure 9.18(b), which will be used to develop the error analysis methods. Here, the output $v[n]$ of the ideal multiplier is quantized to a value $\hat{v}[n]$, where $\hat{v}[n] = v[n] + e_\alpha[n]$. For analysis purposes we again make assumptions similar to those made for the A/D conversion error analysis, i.e.,

(a) The error sequence $\{e_\alpha[n]\}$ is a sample sequence of a stationary white noise process, with each sample $e_\alpha[n]$ being uniformly distributed over the range of the quantization error.

(b) The quantization error sequence $\{e_\alpha[n]\}$ is uncorrelated with the sequence $\{v[n]\}$, the input sequence $\{x[n]\}$ to the digital filter, and all other quantization noise sources.

Recall that the assumption of $\{e_\alpha[n]\}$ being uncorrelated with $\{v[n]\}$ holds only for rounding and two's-complement truncation. The range of the error sample $e_\alpha[n]$ for these two cases are as given in Table 9.1. The mean and variance of the error sample for rounding are given by Eqs. (9.68) and (9.69), respectively, while those for the two's-complement truncation are given by Eqs. (9.70) and (9.71), respectively.

Using the above model for each multiplier, the representation of a digital filter to determine the effect of product quantizations at the output of the digital filter is as indicated in Figure 9.19(a), which explicitly shows the ℓth adder with an output $v_\ell[n]$ summing the

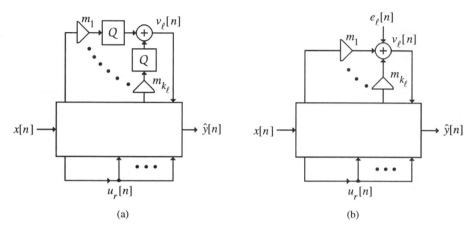

Figure 9.18   (a) Product quantization process and (b) its statistical model for the product round-off error analysis.

Figure 9.19   (a) Representation of a digital filter structure with product round-off before summation, and (b) its statistical model.

quantized outputs of the k_ℓ multipliers at its input. This figure also shows the internal rth branch node associated with the signal variable $u_r[n]$ that needs to be scaled to prevent overflows at these nodes. These nodes are typically the inputs to the multipliers, as indicated in Figure 9.20. In digital filters employing two's-complement arithmetic, these nodes are outputs of adders forming sums of products, since here the sums will have still the correct values even though some of the products and/or the partial sums overflow (Problem 8.52). If we assume the error sources are statistically independent of each other, then each error source develops a round-off noise at the output of the digital filter. An equivalent statistical model is then as shown in Figure 9.19(b).

Let the impulse response from the digital filter input to the rth branch node be denoted as $f_r[n]$ and the impulse response from the input of the ℓth adder to the digital filter output be denoted as $g_\ell[n]$, with their corresponding z-transforms denoted by $F_r(z)$ and $G_\ell(z)$. $F_r(z)$ is called the *scaling transfer function* and plays a role in Section 9.6 in the dynamic

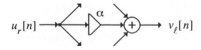

Figure 9.20   A typical multiplier branch with input as a branch node and output feeding into an adder.

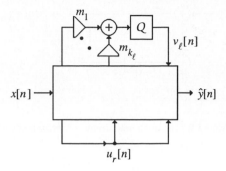

Figure 9.21   Digital filter structure with product round-off after summation.

range scaling schemes employed in fixed-point digital filter structures. $G_\ell(z)$ is called the *noise transfer function*, which is used in computing the noise power at the filter output due to product round-off as described next.

If σ_0^2 denotes the variance of each individual noise source at the output of each multiplier, the variance of $e_\ell[n]$ in Figure 9.19(b) is simply $k_\ell \sigma_0^2$ since we have assumed each noise source to be statistically independent of all others. The variance of the output noise caused by $e_\ell[n]$ is then given by

$$\sigma_0^2 \left[k_\ell \left(\frac{1}{2\pi j} \oint_C G_\ell(z) G_\ell(z^{-1}) z^{-1} dz \right) \right] = \sigma_0^2 \left[k_\ell \left(\frac{1}{2\pi} \int_{-\pi}^{\pi} \left| G_\ell(e^{j\omega}) \right|^2 d\omega \right) \right].$$
(9.95)

If there are L such adders in the digital filter structure, the total output noise power due to all product round-offs is given by

$$\sigma_\gamma^2 = \sigma_o^2 \sum_{\ell=1}^{L} k_\ell \left(\frac{1}{2\pi j} \oint_C G_\ell(z) G_\ell(z^{-1}) z^{-1} dz \right).$$
(9.96)

In many hardware implementation schemes such as those employing DSP chips, the multiplication operation is carried out as a multiply-add operation with the result stored in a double-precision register. In such cases, if the signal variable being generated is by a sum of product operations, the quantization operation can be carried out after all the multiply-add operations have been completed, reducing the number of quantization error sources to one for each such sum of product operations, as indicated in Figure 9.21. In such a case, the statistical model of Figure 9.19(b) can still be used except the variance of the noise source $e_\ell[n]$ is now σ_0^2, thus resulting in considerably lower noise at the digital filter output.

We illustrate the product round-off error analysis method for two simple digital filter structures.

Example 9.7 Consider first the filter of Figure 9.1. The model for the arithmetic round-off error analysis of this structure is as given in Figure 9.22, where the output is now given by a sum of two sequences, the desired output $y[n]$ generated by the input sequence $x[n]$ and the round-off noise $\gamma_\alpha[n]$ generated by the error sequence $e_\alpha[n]$. It follows from the figure that the noise transfer function

Figure 9.22   Model for the arithmetic round-off error analysis of Figure 9.1.

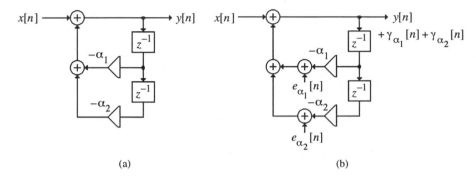

(a)   (b)

Figure 9.23   (a) A second-order digital filter structure, and (b) its model for product round-off error analysis.

$G_\alpha(z)$ is the same as the filter transfer function $H(z)$ as given by Eq. (9.2):

$$G_\alpha(z) = \frac{z}{z - \alpha} \qquad (9.97)$$

and thus the output noise variance due to product round-off is the same as that due to input quantization given in Eq. (9.91) and is given by

$$\sigma_\gamma^2 = \frac{\sigma_0^2}{1 - \alpha^2}, \qquad (9.98)$$

where σ_0^2 is the variance of the error sequence $e_\alpha[n]$. The quantity $\sigma_\gamma^2/\sigma_0^2$ is called the *noise gain* or the *normalized round-off noise variance* denoted as $\sigma_{\gamma,n}^2$.

Example 9.8 Consider next the causal second-order digital filter structure of Figure 9.23(a). Its transfer function is given by

$$H(z) = \frac{Y(z)}{X(z)} = \frac{z^2}{z^2 + \alpha_1 z + \alpha_2}. \qquad (9.99)$$

A model for the analysis of the product round-off error is as indicated in Figure 9.23(b), where we have assumed that each product is rounded first before addition. It follows from this figure that the two noise transfer functions $G_{\alpha_1}(z)$ and $G_{\alpha_2}(z)$ are same as the transfer function $H(z)$ of the digital filter given by Eq. (9.99). Thus, the total normalized variance $\sigma^2_{\gamma,n}$ of the output noise generated by the error sources $e_{\alpha_1}[n]$ and $e_{\alpha_2}[n]$ is given by

$$\sigma^2_{\gamma,n} = 2\left[\frac{1}{2\pi j}\oint_C H(z)H(z^{-1})z^{-1}\,dz\right], \tag{9.100}$$

where $H(z)$ is as given in Eq. (9.99).

Now a direct partial-fraction expansion of $H(z)$ is given by

$$H(z) = 1 + \frac{-\alpha_1 z - \alpha_2}{z^2 + \alpha_1 z + \alpha_2}. \tag{9.101}$$

Applying the technique of Section 9.4.5, we arrive at the expression for the normalized output noise variance:

$$\sigma^2_{\gamma,n} = 2\left[1 + \frac{(\alpha_1^2 + \alpha_2^2)(1 - \alpha_2^2) - 2(\alpha_1\alpha_2 - \alpha_1\alpha_2^2)\alpha_1}{(1 - \alpha_2^2)^2 + 2\alpha_2\alpha_1^2 - (1 + \alpha_2^2)\alpha_1^2}\right], \tag{9.102}$$

which, after some algebra, simplifies to

$$\sigma^2_{\gamma,n} = 2\left(\frac{1 + \alpha_2}{1 - \alpha_2}\right)\left(\frac{1}{1 + \alpha_2^2 + 2\alpha_2 - \alpha_1^2}\right). \tag{9.103}$$

It is instructive to express the multiplier coefficients in terms of the pole positions $re^{\pm j\theta}$ in the above expression. In this case, $\alpha_1 = -2r\cos\theta$ and $\alpha_2 = r^2$. Substituting these values in Eq. (9.103), we arrive after some simple manipulations at

$$\sigma^2_{\gamma,n} = 2\left(\frac{1 + r^2}{1 - r^2}\right)\left(\frac{1}{1 + r^4 - 2r^2\cos 2\theta}\right). \tag{9.104}$$

If the pole is close to the unit circle, we can replace r in Eq. (9.104) with $1 - \varepsilon$, where ε is a very small positive number. This leads to

$$\sigma^2_{\gamma,n} \cong \frac{1}{2\sin^2\theta}\left(\frac{1 - \varepsilon}{\varepsilon}\right)\left(\frac{1}{1 - 2\varepsilon}\right). \tag{9.105}$$

It can be seen from the above that as the poles get closer to the unit circle, $\varepsilon \to 0$, and as a result, the total output noise variance increases rapidly.

9.6 Dynamic Range Scaling

In a digital filter implemented using fixed-point arithmetic, overflow may occur at certain internal nodes such as inputs to multipliers and/or the adder outputs that may lead to large amplitude oscillations at the filter output causing unsatisfactory operation as discussed in Section 9.9.2. The probability of overflow can be minimized significantly by properly scaling the internal signal levels with the aid of scaling multipliers inserted at appropriate points in the digital filter structure. In many cases, most of these scaling multipliers can be absorbed with existing multipliers in the structure, thus reducing the total number needed to implement the scaled filter.

To understand the basic concepts involved in scaling, consider again the digital filter structure of Figure 9.19 showing explicitly the rth node variable $u_r[n]$ that needs to be scaled. We assume that all fixed-point numbers are represented as binary fractions and the input sequence of the filter is bounded by unity, i.e.,

$$|x[n]| \leq 1, \qquad \text{for all values of } n. \tag{9.106}$$

The objective of scaling is to ensure that

$$|u_r[n]| \leq 1, \qquad \text{for all } r \text{ and for all values of } n. \tag{9.107}$$

We now derive three different conditions to ensure that $u_r[n]$ satisfies the above bound.

9.6.1   An Absolute Bound

The inverse z-transform of the scaling transfer function $F_r(z)$ in Figure 9.19 is the impulse response $f_r[n]$ from the filter input to the rth node. Therefore, $u_r[n]$ can be expressed as the linear convolution of $f_r[n]$ and the input $x[n]$:

$$u_r[n] = \sum_{k=-\infty}^{\infty} f_r[k]x[n-k]. \tag{9.108}$$

From Eq. (9.108) it follows then that

$$|u_r[n]| = \left| \sum_{k=-\infty}^{\infty} f_r[k]x[n-k] \right| \leq \sum_{k=-\infty}^{\infty} |f_r[k]|$$

Thus, Eq. (9.107) is satisfied if

$$\sum_{k=-\infty}^{\infty} |f_r[k]| \leq 1 \qquad \text{for all } r. \tag{9.109}$$

The above condition is both necessary and sufficient to guarantee no overflow [Jac70a].[2] If it is not satisfied in the unscaled realization, we scale the input signal with a multiplier K of value

$$K = \frac{1}{\max\limits_{r} \sum_{k=-\infty}^{\infty} |f_r[k]|}. \tag{9.110}$$

It should be noted that the above scaling rule is based on a worst case bound and does not fully utilize the dynamic range of all adder output registers and, as a result, reduces the SNR significantly. Even though it is difficult to compute analytically the value of K, an approximate value can be computed on a computer using an approach similar to that given by Eq. (9.94).

More practical and easy to use scaling rules can be derived in the frequency domain if some information about the input signals is known a priori [Jac70a]. To this end, we define the $\mathcal{L}_p$-norm ($p \geq 1$) of a Fourier transform $F(e^{j\omega})$ as

$$\|F\|_p \triangleq \left(\frac{1}{2\pi} \int_{-\pi}^{\pi} \left| F(e^{j\omega}) \right|^p d\omega \right)^{1/p}. \tag{9.111}$$

---

[2]It also follows from the BIBO stability requirements discussed in Section 2.5.3 and Eq. (2.70).

From the above definition it follows that the $\mathcal{L}_2$-norm, $\|F\|_2$, is the root-mean-squared (rms) value of $F(e^{j\omega})$, and the $\mathcal{L}_1$-norm, $\|F\|_1$, is the mean absolute value of $F(e^{j\omega})$ over ω. Moreover $\lim_{p \to \infty} \|F\|_p$ exists for a continuous $F(e^{j\omega})$ and is given by its peak absolute value:

$$\|F\|_\infty = \max_{-\pi \le \omega \le \pi} \left| F(e^{j\omega}) \right|. \tag{9.112}$$

The bounds to be derived assume that the input $x[n]$ to the digital filter is a deterministic signal with a Fourier transform $X(e^{j\omega})$.

9.6.2 $\mathcal{L}_\infty$-Bound

Now from Eq. (9.108) we get

$$U_r(e^{j\omega}) = F_r(e^{j\omega})X(e^{j\omega}), \tag{9.113}$$

where $U_r(e^{j\omega})$ and $F_r(e^{j\omega})$ are the Fourier transforms of $u_r[n]$ and $f_r[n]$, respectively. An inverse Fourier transform of Eq. (9.113) yields

$$u_r[n] = \frac{1}{2\pi} \int_{-\pi}^{\pi} F_r(e^{j\omega})X(e^{j\omega})e^{j\omega n} d\omega. \tag{9.114}$$

As a result,

$$
\begin{aligned}
|u_r[n]| &\le \frac{1}{2\pi} \int_{-\pi}^{\pi} \left| F_r(e^{j\omega}) \right| \cdot \left| X(e^{j\omega}) \right| d\omega \\
&\le \left\| F_r(e^{j\omega}) \right\|_\infty \left[\frac{1}{2\pi} \int_{-\pi}^{\pi} \left| X(e^{j\omega}) \right| d\omega \right] \\
&\le \|F_r\|_\infty \cdot \|X\|_1 .
\end{aligned} \tag{9.115}
$$

If $\|X\|_1 \le 1$, then the dynamic range constraint of Eq. (9.107) is satisfied if

$$\left\| F_r(e^{j\omega}) \right\|_\infty \le 1. \tag{9.116}$$

Thus, if the mean absolute value of the input spectrum is bounded by unity, then there will be no adder overflow if the peak gains from the filter input to all adder outputs are scaled satisfying the bound of Eq. (9.116). In general, this scaling rule is rarely used since, with most input signals encountered in practice, $\|X\|_1 \le 1$ does not hold!

9.6.3 $\mathcal{L}_2$-Bound

Applying Schwarz inequality to Eq. (9.114), we arrive at [Jac70a]

$$|u_r[n]|^2 \le \left(\frac{1}{2\pi} \int_{-\pi}^{\pi} \left| F_r(e^{j\omega}) \right|^2 d\omega \right) \cdot \left(\frac{1}{2\pi} \int_{-\pi}^{\pi} \left| X(e^{j\omega}) \right|^2 d\omega \right) \tag{9.117}$$

or equivalently,

$$|u_r[n]| \le \|F_r\|_2 \cdot \|X\|_2 . \tag{9.118}$$

Thus, if the input to the filter has finite energy bounded by unity, i.e., $\|X\|_2 \le 1$, then the adder overflow can be prevented by scaling the filter such that the rms values of the transfer functions from the input to all adder outputs are bounded by unity:

$$\|F_r\|_2 \le 1, \qquad r = 1, 2, \ldots, R. \tag{9.119}$$

9.6.4   A General Scaling Rule

A more general scaling rule obtained using Holder's inequality is given by [Jac70a]

$$|u_r[n]| \leq \|F_r\|_p \cdot \|X\|_q, \qquad (9.120)$$

for all $p, q \geq 1$ satisfying $(1/p) + (1/q) = 1$. Note that for the $\mathcal{L}_\infty$-bound of Eq. (9.115), $p = \infty$ and $q = 1$, and for the $\mathcal{L}_2$-bound of Eq. (9.118), $p = q = 2$. Another useful scaling rule, $\mathcal{L}_1$-bound, is obtained for $p = 1$ and $q = \infty$.

After scaling, the scaling transfer functions become $\|\bar{F}_r\|_p$ and the scaling constants should be chosen such that

$$\|\bar{F}_r\|_p \leq 1, \qquad r = 1, 2, \ldots, R. \qquad (9.121)$$

In many structures, all scaling multipliers can be absorbed into the existing feed-forward multipliers without any increase in the total number of multipliers, and hence, noise sources. However, in some cases, the scaling process may introduce additional multipliers in the system. If all scaling multipliers are regular b-bit units, then the Eq. (9.121) can be satisfied with an equality sign, providing a full utilization of the dynamic range of each adder output, and thus, yielding a maximum SNR. An attractive option from a hardware point of view and preferred in cases where scaling introduces new multipliers, is to make as many scaling multiplier coefficients as possible in the scaled structure take a value that is a power-of-2 [Jac70a]. In which case, these multipliers can be implemented simply by a shift operation. Now, the norm of the scaling transfer function satisfies

$$\tfrac{1}{2} < \|\bar{F}_r\|_p \leq 1 \qquad (9.122)$$

with a slight decrease in the SNR.

It should be pointed out here that the output round-off noise should be always computed after the digital filter structure has been scaled. For the scaled structure, the expression for the output round-off noise of Eq. (9.95) thus changes to

$$\sigma_\gamma^2 = \sigma_0^2 \sum_{\ell=1}^{L} \bar{k}_\ell \left(\frac{1}{2\pi j} \oint_C \bar{G}_\ell(z) \bar{G}_\ell(z^{-1}) z^{-1} \, dz \right), \qquad (9.123)$$

where $\bar{k}_\ell$ is the total number of multipliers feeding the ℓth adder with $\bar{k}_\ell \geq k_\ell$ and $\bar{G}_\ell(z)$ is the modified noise transfer function from the input of the ℓth adder to the filter output.

We illustrate next the application of the above method in the scaling of a cascade realization of an IIR transfer function [Jac70b].

9.6.5   Scaling of a Cascade Form IIR Digital Filter Structure

Consider the unscaled structure of Figure 9.24 consisting of a cascade of R second-order IIR sections realized in direct form II. Its transfer function is given by

$$H(z) = K \prod_{i=1}^{R} H_i(z), \qquad (9.124)$$

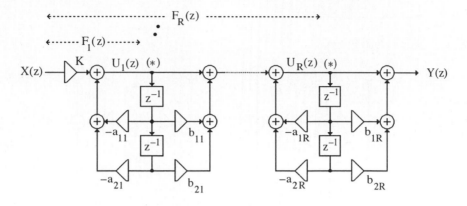

Figure 9.24   An unscaled cascade realization of second-order IIR sections.

where

$$H_i(z) = \frac{B_i(z)}{A_i(z)} = \frac{1 + b_{1i}z^{-1} + b_{2i}z^{-2}}{1 + a_{1i}z^{-1} + a_{2i}z^{-2}}. \tag{9.125}$$

The branch nodes to be scaled are marked by $(*)$ in the figure, which are seen to be the inputs to the multipliers in each second-order section. The transfer functions from the input to these branch nodes are the scaling transfer functions, which are given by

$$F_r(z) = \frac{K}{A_r(z)} \prod_{\ell=1}^{r-1} H_\ell(z), \qquad r = 1, 2, \ldots, R. \tag{9.126}$$

The scaled version of the above structure is shown in Figure 9.25 with new values for the feed-forward multipliers. Note that the scaling process has introduced a new multiplier $\bar{b}_{0\ell}$ in each second-order section. If the zeros of the transfer function are on the unit circle, as is usually the case in practice, then $b_{2\ell} = \pm 1$ in which case we can choose $\bar{b}_{0\ell} = \bar{b}_{2\ell} = 2^{-\beta}$ to reduce the total number of actual multiplications in the final realization.

It can be seen from Figure 9.25 that

$$\bar{F}_r(z) = \frac{\bar{K}}{A_r(z)} \prod_{\ell=1}^{r-1} \bar{H}_\ell(z), \tag{9.127a}$$

$$\bar{H}(z) = \bar{K} \prod_{\ell=1}^{R} \bar{H}_\ell(z), \tag{9.127b}$$

where

$$\bar{H}_\ell(z) = \frac{\bar{b}_{0\ell} + \bar{b}_{1\ell}z^{-1} + \bar{b}_{2\ell}z^{-2}}{1 + a_{1\ell}z^{-1} + a_{2\ell}z^{-2}}. \tag{9.128}$$

Denote

$$\|F_r\|_p \overset{\Delta}{=} \alpha_r, \qquad r = 1, 2, \ldots, R, \tag{9.129a}$$

$$\|H\|_p \overset{\Delta}{=} \alpha_{R+1}, \tag{9.129b}$$

Figure 9.25   The scaled cascade realization.

and choose the scaling constants as

$$\bar{K} = \beta_0 K; \qquad \bar{b}_{\ell r} = \beta_r b_{\ell r}, \qquad \ell = 1, 2, 3; \quad r = 1, 2, \ldots, R. \tag{9.130}$$

Now, it follows from Eqs. (9.130), (9.127a) and (9.127b) that

$$\bar{F}_r(z) = \left(\prod_{i=0}^{r-1} \beta_i \right) F_r(z), \qquad r = 1, 2, \ldots, R, \tag{9.131a}$$

$$\bar{H}(z) = \left(\prod_{i=0}^{R} \beta_i \right) H(z). \tag{9.131b}$$

After scaling we require

$$\| \bar{F}_r \|_p = \left(\prod_{i=0}^{r-1} \beta_i \right) \| F_r \|_p = \alpha_r \left(\prod_{i=0}^{r-1} \beta_i \right) = 1, \quad r = 1, 2, \ldots, R, \tag{9.132a}$$

$$\| \bar{H} \|_p = \left(\prod_{i=0}^{R} \beta_i \right) \| H \|_p = \alpha_{R+1} \left(\prod_{i=0}^{R} \beta_i \right) = 1. \tag{9.132b}$$

Solving the above for the scaling constants, we arrive at

$$\beta_0 = \frac{1}{\alpha_1}, \tag{9.133a}$$

$$\beta_r = \frac{\alpha_r}{\alpha_{r+1}}, \qquad r = 1, 2, \ldots, R. \tag{9.133b}$$

9.6.6   Dynamic Range Scaling Using MATLAB

The dynamic range scaling using the $\mathcal{L}_2$-norm rule can be easily performed using MATLAB by actual simulation of the digital filter structure. If we denote the impulse response from the input of the digital filter to the output of the rth branch node as $\{f_r[n]\}$ and assume that the branch nodes have been ordered in accordance to their precedence relations with increasing i (see Section 8.1.2), then we can compute the $\mathcal{L}_2$-norm $\|F_1\|_2$ of $\{f_1[n]\}$ first, and divide the input by a multiplier $k_1 = \|F_1\|_2$. Next, we compute the $\mathcal{L}_2$-norm $\|F_2\|_2$ of $\{f_2[n]\}$ and scale the multipliers feeding into the second adder by dividing with a constant $k_2 = \|F_2\|_2$. This process is continued until the output node has been scaled to yield an $\mathcal{L}_2$-norm of unity.

We illustrate the method by means of the following example.

Example 9.9 Consider the cascade realization of the third-order lowpass digital filter of Eq. (8.19). The transfer functions of the two sections in the cascade are given by Eqs. (8.20a,b), repeated below for convenience:

$$G_1(z) = \frac{0.0662272(1 + z^{-1})}{1 - 0.2593284z^{-1}}, \tag{9.134a}$$

$$G_2(z) = \frac{1 + 2z^{-1} + z^{-2}}{1 - 0.6762858z^{-1} + 0.3917468z^{-2}}. \tag{9.134b}$$

Their realization in direct form II structure is as shown in Figure 9.26. The MATLAB program simulating this structure is given by Program 9_3 below. The program is first run with all scaling constants set to unity, i.e., $\texttt{k1} = \texttt{k2} = \texttt{k3} = 1$. In the line indicating the approximate $\mathcal{L}_2$-norm calculation, the output variable is chosen as $\texttt{y1}$. The program computes the square of the $\mathcal{L}_2$-norm of the impulse norm at node $\texttt{y1}$ as 1.07210002757252, which is used to set $\texttt{k1} = \sqrt{1.07210002757252}$ with other scaling constants still set to unity. A second run of the program shows the $\mathcal{L}_2$-norm of the impulse norm at node $\texttt{y1}$ as 1.00000000000000, verifying the success of scaling the input. In the second step, in the line indicating the approximate $\mathcal{L}_2$-norm calculation, the output variable is chosen as $\texttt{y2}$. The program then yields the square of the $\mathcal{L}_2$-norm of the impulse norm at node $\texttt{y2}$ as 0.02679820762398, which is used to set $\texttt{k2} = \sqrt{0.02679820762398}$ with $\texttt{k3}$ still set to unity. This process is repeated for node $\texttt{y3}$, resulting in $\texttt{k3} = \sqrt{11.96975400608943}$ with the final value of the $\mathcal{L}_2$-norm of the impulse norm at node $\texttt{y3}$ as 0.99999683131540. The program listing shown below is after all adder outputs have been scaled.

```
% Program 9_3
% Illustration of Scaling & Round-off Noise Calculation
%
format long
k1 = sqrt(1.07210002757252);
k2 = sqrt(0.0334831780179);
k3 = sqrt(13.60517805970979);
x1 = 1/k1;
si1 = 0;
si2 = [0 0];
b1 = 1;
varnew = 0; k = 1;
while k > 0.000001
y1 = 0.2593284*si1 + x1;
x2 = (0.0662272/k2)*(y1 + si1);
si1 = y1;
```

Figure 9.26   Cascade realization of the transfer functions of Eqs. (9.134) in direct form II.

```
y2 = 0.6762858*si2(1) - 0.3917468*si2(2) + x2;
y3 = (y2 + 2*si2(1) + si2(2))/k3;
si2(2) = si2(1); si2(1) = y2;
varold = varnew;
varnew = varnew + abs(y3)*abs(y3);
% Compute approximate L2 norm square
k = varnew - varold;
x1 = 0;
end
disp('L2 norm square = ');disp(varnew);
```

The above program can be easily modified to compute the product round-off noise variance at the output of the scaled structure. To this end, we set the digital filter input to zero and apply an impulse at the input of the first adder. This is equivalent to setting $x1 = 1$ in Program 9_3. The normalized output noise variance due to a single error source is computed as 1.07209663042567. Next, we apply an impulse at the input of the second adder with the digital filter input set to zero. This is achieved by replacing $x2$ in the calculation of $y2$ as $x1$. The program yields the normalized output noise variance due to a single error source at the second adder as 1.26109014071707. The total normalized output round-off noise, assuming all products to be quantized before addition, is then equal to

$$2 \times 1.07209663042567 + 4 \times 1.26109014071707 + 3 = 10.18855382371962.$$

On the other hand, if we assume quantization after addition of all products at each adder, the total normalized output round-off noise becomes

$$1.07209663042567 + 1.26109014071707 + 1 = 3.33318677114274.$$

Example 9.10 We repeat the problem of Example 9.9 by interchanging the locations of the two sections $G_1(z)$ and $G_2(z)$ in the cascade connection, as indicated in Figure 9.27. By modifying Program 9_3, we arrive at the following values for the scaling constants:

$$k1 = \sqrt{1.54643736440090}, \qquad k2 = \sqrt{14.44567893169408},$$
$$k3 = \sqrt{0.01539326750782}$$

The total normalized output round-off noise assuming all products to be quantized before addition in this case is equal to

$$3 \times 1.54652209477041 + 4 \times 0.76938948942268 + 2 = 9.717124200195,$$

Figure 9.27   Alternate cascade realization of the transfer functions of Eqs. (9.134) in direct form II.

whereas the total normalized output round-off noise assuming quantization after addition is given by

$$0.76938948942268 + 1.54652209477041 + 2 = 3.31591158419309.$$

Note from above that the second cascade connection of Figure 9.27 yields lower round-off noise than the first in Figure 9.26.

9.6.7 Optimum Ordering and Pole-Zero Pairing of a Cascade Form IIR Digital Filter

As indicated in Section 6.5.2, and illustrated in Figures 6.25 and 6.26, there are many possible cascade realizations of a higher-order IIR transfer function obtained by different pole-zero pairings and ordering. In fact, for a cascade of R second-order sections, there are $(R!)^2$ different possible realizations (Problem 9.25). Each one of these realizations will have different output noise power, as illustrated in the previous two examples, and hence, it is of interest to determine the cascade realization with the lowest output noise power.

A fairly simple heuristic set of rules for determining an optimum pole-zero pairing and ordering of sections in a cascade realization has been advanced by Jackson [Jac70b]. To understand the reasoning behind these rules, we first develop the expression for the output noise variance due to product round-off in a cascade IIR structure implemented in fixed-point arithmetic. To this end we make use of the noise model of the scaled cascade structure of Figure 9.25 as shown in Figure 9.28.

It follows from Figure 9.28 and Eq. (9.130) that the scaled noise transfer functions are given by

$$\bar{G}_\ell(z) = \prod_{i=\ell}^{R} \bar{H}_i(z) = \left(\prod_{i=\ell}^{R} \beta_i \right) G_\ell(z), \qquad \ell = 1, 2, \ldots, R, \qquad (9.135a)$$

$$\bar{G}_{R+1}(z) = 1, \qquad (9.135b)$$

where $\bar{H}_i(z)$ is as given in Eq. (9.128) and the unscaled noise transfer function is given by

$$G_\ell(z) = \prod_{i=\ell}^{R} H_i(z). \qquad (9.136)$$

Figure 9.28   The noise model of the scaled cascade realization of Figure 9.24.

Hence, the output noise power spectrum due to product round-off is given by

$$P_{\gamma\gamma}(\omega) = \sigma_0^2 \left[\sum_{\ell=1}^{R+1} k_\ell \left| \bar{G}_\ell(e^{j\omega}) \right|^2 \right]. \qquad (9.137)$$

and the output noise variance is thus

$$\sigma_\gamma^2 = \sigma_0^2 \left[\sum_{\ell=1}^{R+1} k_\ell \left(\frac{1}{2\pi} \int_{-\pi}^{\pi} \left| \bar{G}_\ell(e^{j\omega}) \right|^2 d\omega \right) \right]$$

$$= \sigma_0^2 \left[\sum_{\ell=1}^{R+1} k_\ell \left\| \bar{G}_\ell \right\|_2^2 \right], \qquad (9.138)$$

where we have used the fact that the quantity inside the parentheses in the middle expression is the square of the $\mathcal{L}_2$-norm of $\bar{G}_\ell(e^{j\omega})$ as defined in Eq. (9.111). In Eqs. (9.137) and (9.138), k_ℓ is the total number of multipliers connected to the ℓth adder. If all products are rounded before summation, then

$$k_1 = k_{R+1} = 3, \qquad (9.139a)$$

$$k_\ell = 5, \qquad \text{for } \ell = 2, 3, \ldots, R. \qquad (9.139b)$$

On the other hand, if all products are rounded after summation, then

$$k_\ell = 1, \qquad \text{for } \ell = 1, 2, \ldots, R+1. \qquad (9.140)$$

From Eqs. (9.133a) and (9.133b),

$$\prod_{i=\ell}^{R} \beta_i = \frac{\alpha_\ell}{\alpha_{R+1}} = \frac{\|F_\ell\|_p}{\|H\|_p}. \qquad (9.141)$$

Substituting Eq. (9.141) in Eq. (9.137), we obtain for the output noise power spectrum

$$P_{\gamma\gamma}(\omega) = \frac{\sigma_0^2}{\|H\|_p^2} \left[k_{R+1} \|H\|_p^2 + \sum_{\ell=1}^{R} k_\ell \|F_\ell\|_p^2 \left| G_\ell(e^{j\omega}) \right|^2 \right]. \qquad (9.142)$$

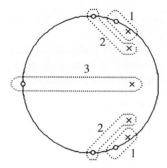

Figure 9.29   Illustration of the pole-zero pairing and ordering rules.

Correspondingly, the output noise variance is given by

$$\sigma_\gamma^2 = \frac{\sigma_0^2}{\|H\|_p^2} \left[k_{R+1} \|H\|_p^2 + \sum_{\ell=1}^{R} k_\ell \|F_\ell\|_p^2 \|G_\ell\|_2^2 \right]. \tag{9.143}$$

Now, the scaling transfer function $F_\ell(z)$ contains a product of section transfer functions, $H_i(z)$, $i = 1, 2, \ldots, \ell - 1$, whereas the noise transfer function $G_\ell(z)$ contains the product of section transfer functions, $H_i(z)$, $i = \ell, \ell + 1, \ldots, R$. Thus, every term in the sum in Eqs. (9.142) and (9.143) includes the transfer function of all R sections in the cascade realization. This implies that to minimize the output noise power, the norms of $H_i(z)$ should be minimized for all values of i by appropriately pairing the poles and zeros. To this end, Jackson has proposed the following rule [Jac70b]: First, the poles closest to the unit circle should be paired with the nearest zeros. Next, the poles that are closest to the previous set of poles should be matched with their nearest zeros. This process should be continued until all poles and zeros have been paired. This type of pairing of poles and zeros is likely to lower the peak gain of the section characterized by the paired pole and zeros. Lowering of the peak gain in turn reduces the possibility of overflow and attenuates the roundoff noise.

The above pole-zero pairing rule is illustrated in Figure 9.29 for the fifth-order elliptic lowpass IIR filter considered earlier in Section 9.2.

Once the appropriate pole-zero pairings have been made, the next question that needs to be answered is how to order the sections in the cascade structure [Jac70b]. A section in the front part of the cascade has its transfer function $H_i(z)$ appearing more frequently in the scaling transfer function expressions in Eqs. (9.142) and (9.143), whereas a section near the output end of the cascade has its transfer function $H_i(z)$ appearing more frequently in the noise transfer function expressions. The best location for $H_i(z)$ obviously depends on the type of norms being applied to the scaling and noise transfer functions.

A careful examination of Eqs. (9.142) and (9.143) reveals that if the $\mathcal{L}_2$-scaling is used, then ordering of paired sections does not influence too much the output noise power since all norms in the expressions are $\mathcal{L}_2$-norms. If, however, an $\mathcal{L}_\infty$-scaling is being employed, the sections with the poles closest to the unit circle exhibit a peaking magnitude response and should be placed closer to the output end. The ordering rule in this case is therefore to place the sections from least-peaked to most-peaked. On the other hand, the ordering scheme is exactly opposite if the objective is to minimize the peak noise

$\|P_{\gamma\gamma}(\omega)\|_{\infty}$ and an $\mathcal{L}_2$-scaling is used. However, the ordering has no effect on the peak noise if an $\mathcal{L}_{\infty}$-scaling is used.

MATLAB *Signal Processing Toolbox* includes two functions, `ss2sos` and `zp2sos`, that can be employed to determine the optimum pole-zero pairing and ordering according to the above discussed rules. The basic forms of these functions are:

```
sos = ss2sos(A,B,C,D),        sos = zp2sos(z,p,k)
```

Both functions generate a matrix `sos` containing the coefficients of each second order section of the equivalent transfer function $H(z)$ determined from the specified state-space representation, **A**, **B**, **C**, and D, or its zero-pole form. `sos` is an $L \times 6$ matrix of the form

$$\text{sos} = \begin{bmatrix} b_{01} & b_{11} & b_{21} & a_{01} & a_{11} & a_{21} \\ b_{02} & b_{12} & b_{22} & a_{02} & a_{12} & a_{22} \\ \vdots & \vdots & \vdots & \vdots & \vdots & \vdots \\ b_{0L} & b_{1L} & b_{2L} & a_{0L} & a_{1L} & a_{2L} \end{bmatrix},$$

whose ith row contains the coefficients, $\{b_{i\ell}\}$ and $\{a_{i\ell}\}$, of the numerator and denominator polynomials of the ith second-order section with L denoting the total number of sections in the cascade. The form of the overall transfer function is thus given by

$$H(z) = \prod_{i=1}^{L} H_i(z) = \prod_{i=1}^{L} \frac{b_{0i} + b_{1i}z^{-1} + b_{2i}z^{-2}}{a_{0i} + a_{1i}z^{-1} + a_{2i}z^{-2}}.$$

The rows are ordered so that the first row ($i = 1$) contains the coefficients of the pole pair farthest from the unit circle and its nearest zero pair. If a reverse ordering is desired with the first row containing the coefficients of the pole pair closest to the unit circle and its nearest zero pair, then the function to be used are

```
sos = ss2sos(A,B,C,D,'down'), sos = zp2sos(z,p,k,'down').
```

Example 9.11 We develop the optimum pairing and ordering of the fifth-order elliptic lowpass filter of Section 9.2. Its specifications are: passband ripple of 0.5 dB, stopband ripple of 40 dB, and normalized passband edge frequency of 0.3. We first determine the state-space matrices of the desired lowpass filter using the command `[A,B,C,D] = ellip(5,0.5,40,0.3)` and then employ the command `sos = ss2sos(A,B,C,D)` to determine the coefficients of the second-order sections. The matrix `sos` obtained is given by

```
sos =

    Columns 1 through 3

    0.19321569747185      0.19321569747185      0
    0.23971259321206     -0.02063964629270      0.23971259321206
    0.64775994647835     -0.49477798804715      0.64775994647835

    Columns 4 through 6

    1.00000000000000     -0.61356860505629      0
    1.00000000000000     -1.15286286138340      0.61164840151481
    1.00000000000000     -1.09833407780228      0.89907598271183
```

The transfer functions of the sections are thus given by

$$H_1(z) = \left[\frac{0.1932157 + 0.1932157z^{-1}}{1 - 0.6135686z^{-1}} \right],$$

$$H_2(z) = \left[\frac{0.2397126 - 0.0206396z^{-1} + 0.2397126z^{-2}}{1 - 1.15286294z^{-1} + 0.6116484z^{-2}} \right],$$

$$H_3(z) = \left[\frac{0.6477599 - 0.494778z^{-1} + 0.6477599z^{-2}}{1 - 1.0983341z^{-1} + 0.899076z^{-2}} \right].$$

As indicated above, the first section in the cascade is $H_1(z)$, followed by $H_2(z)$, and the last section is $H_3(z)$. This pairing is identical to that displayed in Figure 9.29 but the ordering is reversed. This can be easily verified by using the function [z,p,k] = sos2zp(sos), which computes the zeros and poles of each individual second-order section defined by the matrix sos.

Other MATLAB functions that are available for the analysis of second-order sections are

```
[num, den] = sos2tf(sos),  [A,B,C,D] = sos2ss(sos).
```

9.6.8   Signal-to-Noise Ratio in Low-Order IIR Filters

The output round-off noise variances of unscaled digital filters do not provide a realistic picture of the performances of these structures in practice since introduction of scaling multipliers can increase the number of error sources and the gain for the scaling transfer functions. It is therefore important to scale the digital filter structure first before its round-off noise performance is analyzed. In most applications, the round-off noise variance by itself is not sufficient, and a more meaningful result is obtained by computing instead the expression for the signal-to-round-off noise ratio (SNR) for performance evaluation. We illustrate such computation for the first-order and the second-order IIR structures considered earlier in Examples 9.7 and 9.8, respectively [Vai87c]. Most conclusions derived from the detailed analysis of these simple structures outlined here are also valid in the case of more complex structures. Moreover, the methods followed here can be easily extended to the general case.

First-order section   Consider first the causal unscaled first-order IIR filter of Figure 9.1. Its round-off noise variance was computed in Example 9.7 and is given by Eq. (9.98). To determine its signal-to-noise ratio, assume the input $x[n]$ to be a wide-sense-stationary (WSS) random signal with a uniform probability density function and a variance σ_x^2. The variance σ_y^2 of the output signal $y[n]$ generated by this input is then given by

$$\sigma_y^2 = \sigma_x^2 \left(\sum_{n=0}^{\infty} h^2[n] \right) = \frac{\sigma_x^2}{1 - \alpha^2}. \tag{9.144}$$

Taking the ratio of Eqs. (9.144) and (9.98), we arrive at the expression for the signal-to-noise ratio of the unscaled section as

$$\text{SNR} = \frac{\sigma_y^2}{\sigma_\gamma^2} = \frac{\sigma_x^2}{\sigma_0^2}, \tag{9.145}$$

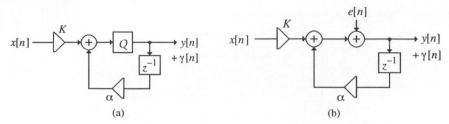

Figure 9.30   Scaled first-order section with quantizer, and its round-off noise model.

implying that the SNR is independent of α. However, this is not a valid result since the adder is likely to overflow in an unscaled structure. It is therefore important first to scale the structure and then to compute the SNR to arrive at a more meaningful result.

The scaled structure is as indicated in Figure 9.30 along with its round-off error analysis model assuming quantization after addition of all products. With the scaling multiplier present, the output signal power now becomes

$$\sigma_y^2 = \frac{K^2\sigma_x^2}{1-\alpha^2},$$
(9.146)

modifying the signal-to-noise ratio to

$$\text{SNR} = \frac{K^2\sigma_x^2}{\sigma_0^2}.$$
(9.147)

Since the scaling multiplier coefficient K depends on the pole location and the type of scaling rule being followed, the SNR will thus reflect this dependence. The scaling parameter K can be chosen according to the rules outlined in Section 9.6. To this end we need first to determine the scaling transfer function $F(z)$ of the unscaled structure. It follows from Figure 9.30(a) that

$$F(z) = H(z) = \frac{1}{1-\alpha z^{-1}},$$
(9.148)

with a corresponding impulse response

$$f[n] = \mathcal{Z}^{-1}\{F(z)\} = \alpha^n\mu[n].$$
(9.149)

Now, as indicated in Section 9.6, there are different scaling rules. To guarantee no overflow, we can use the scaling rule of Eq. (9.110) and arrive at

$$K = \frac{1}{\sum_{n=0}^{\infty}|f[n]|} = 1 - |\alpha|.$$
(9.150)

To evaluate the SNR, we need to know the type of input $x[n]$ being applied. If $x[n]$ is uniformly distributed with $|x[n]| \le 1$, its variance is given by

$$\sigma_x^2 = \tfrac{1}{3}.$$
(9.151)

Substituting Eqs. (9.150) and (9.151) in Eq. (9.147), we arrive at

$$\text{SNR} = \frac{(1-|\alpha|)^2}{3\sigma_0^2}.$$
(9.152)

Table 9.5   Signal-to-noise ratio of first-order IIR digital filters for different inputs.   Adapted from [Vai87c].

Scaling rule	Input type	SNR	Typical SNR, dB $(b = 12, |\alpha| = 0.95)$
No overflow	WSS, white uniform density	$\dfrac{(1 - |\alpha|)^2}{3\sigma_0^2}$	52.24
No overflow	WSS, white Gaussian density ($\sigma_x^2 = 1/9$)	$\dfrac{(1 - |\alpha|)^2}{9\sigma_0^2}$	47.97
No overflow	Sinusoid, known frequency	$\dfrac{1 - \alpha^2}{2\sigma_0^2}$	69.91

For a $(b + 1)$-bit signed fraction with round-off or two's-complementation truncation, $\sigma_0^2 = 2^{-2b}/12$. Substituting this figure in Eq. (9.152), we obtain the signal-to-noise ratio in dB as

$$\text{SNR}_{\text{dB}} = 20 \log_{10}(1 - |\alpha|) + 6.021 + 6.02b. \tag{9.153}$$

As a result, for a given transfer function, the SNR increases by 6 dB for each additional bit added to the register storing the product.

The above analysis can be carried out for other types of input and different scaling rules. We summarize in Table 9.5 the results for three different types of input for the scaling ensuring no overflow. Several conclusions can be made by examining this table. Independent of the type of input being applied, the SNR decreases rapidly as the pole moves closer to the unit circle. For a given transfer function, knowing the type of input being applied, the internal wordlength can be computed to achieve a desired SNR.

Second-order section   Consider next the causal unscaled second-order IIR filter of Figure 9.23(a). Its scaled version is indicated in Figure 9.31 along with the round-off noise analysis model, assuming again quantization after addition of all products. Now for a WSS input with an uniform probability density function and a variance σ_x^2, the signal power at the output is given by

$$\sigma_y^2 = \sigma_x^2 \left(K^2 \sum_{n=0}^{\infty} h^2[n] \right), \tag{9.154}$$

while the round-off noise power at the output is given by

$$\sigma_\gamma^2 = \sigma_0^2 \left(\sum_{n=0}^{\infty} h^2[n] \right). \tag{9.155}$$

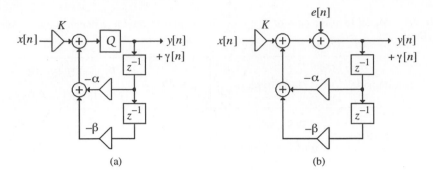

Figure 9.31   Scaled second-order section with quantizer, and its round-off noise model.

Therefore the signal-to-noise ratio of the scaled structure is of the form

$$\text{SNR} = \frac{\sigma_y^2}{\sigma_\gamma^2} = \frac{K^2 \sigma_x^2}{\sigma_0^2}. \tag{9.156}$$

To determine the appropriate value of the scaling multiplier K, we need to compute the scaling transfer function $F(z)$. It can be seen from Figure 9.31(a) that $F(z)$ is identical to the filter transfer function $H(z)$. If the poles of the transfer function are at $z = re^{-j\theta}$, then both transfer functions take the form

$$F(z) = H(z) = \frac{1}{1 + \alpha_1 z^{-1} + \alpha_2 z^{-2}} = \frac{1}{1 - 2r\cos\theta z^{-1} + r^2 z^{-2}}. \tag{9.157}$$

The corresponding impulse response is obtained by taking the inverse z-transform of the above, resulting in (Problem 3.68)

$$f[n] = h[n] = \frac{r^n \sin(n+1)\theta}{\sin\theta} \cdot \mu[n]. \tag{9.158}$$

To eliminate completely the overflow at the output in Figure 9.31, the scaling multiplier K must be chosen as

$$K = \frac{1}{\sum_{n=0}^{\infty} |h[n]|}. \tag{9.159}$$

The summation in the denominator of Eq. (9.159) with $h[n]$ given by Eq. (9.158) is difficult to compute analytically. However, it is possible to establish some bounds on the summation to provide a reasonable estimate of the value of the scaling multiplier coefficient K [Opp75]. To this end, note that the amplitude of the response of the unscaled second-order section of Figure 9.23(a) to a sinusoidal input at the resonant frequency $\omega = \theta$, i.e., $x[n] = \cos(\theta n)$, is given by

$$\left| H(e^{j\theta}) \right| = \left[\frac{1}{(1-r)^2(1 - 2r\cos\theta + r^2)} \right]^{1/2} \tag{9.160}$$

However, $\left| H(e^{j\theta}) \right|$ cannot be greater than $\sum_{n=0}^{\infty} |h[n]|$ since the latter is the largest possible value of output $y[n]$ for an input $x[n]$ with $|x[n]| \leq 1$. Moreover,

$$\sum_{n=0}^{\infty} |h[n]| = \frac{1}{\sin\theta} \sum_{n=0}^{\infty} r^n |\sin(n+1)\theta|$$

$$\leq \frac{1}{\sin\theta} \sum_{n=0}^{\infty} r^n = \frac{1}{(1-r)\sin\theta}. \tag{9.161}$$

A tighter upper bound on $\sum_{n=0}^{\infty} |h[n]|$ has been derived in [Unv75] and is given by

$$\sum_{n=0}^{\infty} |h[n]| \leq \frac{4}{\pi(1-r^2)\sin\theta}. \tag{9.162}$$

From Eqs. (9.159), (9.160), and (9.162), we therefore arrive at

$$(1-r)^2(1-2r\cos\theta+r^2) \geq K^2 \geq \frac{\pi^2(1-r^2)^2\sin^2\theta}{16}. \tag{9.163}$$

Following the technique carried out for the first-order section, we can derive the bounds on the SNR for the all-pole second-order section from Eqs. (9.156) and (9.163) for various types of inputs. As in the case of the first-order section, as the pole moves closer to the unit circle ($r \to 1$), the gain of the filter increases, causing the input signal to be scaled down significantly to avoid the overflow while at the same time boosting the output round-off noise. This type of interplay between round-off noise and dynamic range is a characteristic of all fixed-point digital filters [Jac70a].

In certain cases, it is possible to develop digital filter structures that have inherently the least quantization effects. In the following two sections we consider these structures.

9.7 Low-Sensitivity Digital Filters

In Section 9.2, we considered the effect of multiplier coefficient quantization on the performance of a digital filter. One major consequence is that the frequency response of the digital filter with quantized multiplier coefficients is different from that of the desired digital filter with unquantized coefficients, and in many cases, this difference may be significant enough to make the practical digital filter unsuitable in most applications. It is thus of interest to develop digital filter structures that are inherently less sensitive to coefficient quantization. To this end, the first approach advanced is based on the conversion of an inherently low sensitivity analog network composed of inductors, capacitors, and resistors to a digital filter structure by replacing each analog network component and their interconnections to a corresponding digital filter equivalent such that the overall structure "simulates" the analog prototype [Fet71], [Fet86]. The resulting digital filter structure is called a *wave digital filter*, which also shares some additional properties of its analog prototype. An alternative approach is to determine directly the conditions for low coefficient sensitivity to be satisfied by the digital filter structure and to develop realization methods that ensure that the final structure indeed satisfies these conditions [Vai84]. In this section we examine the latter approach.

Figure 9.32   A typical magnitude response of a bounded-real transfer function.

9.7.1   Requirements for Low Coefficient Sensitivity

Let the prescribed transfer function $H(z)$ be a *bounded real* (BR) function as defined in Section 4.3.8. That is, $H(z)$ is a causal stable real coefficient function characterized by a magnitude response $\left|H(e^{j\omega})\right|$ bounded above by unity, i.e.,

$$\left|H(e^{j\omega})\right| \leq 1. \tag{9.164}$$

Assume that $H(z)$ is such that at a set of frequencies ω_k the magnitude is exactly equal to 1:

$$\left|H(e^{j\omega_k})\right| = 1. \tag{9.165}$$

Since the magnitude function is bounded above by unity, the frequencies ω_k must be in the passband of the filter. A typical frequency response satisfying the above conditions is shown in Figure 9.32. Note that any causal stable transfer function can be scaled to satisfy conditions of Eqs. (9.164) and (9.165).

Let the digital filter structure $\mathcal{N}$ realizing $H(z)$ be characterized by a set of R multipliers with coefficients m_i. Moreover, let the nominal values of these multiplier coefficients, assuming infinite precision realization, be m_{i0}. Assume that the structure $\mathcal{N}$ is such that, regardless of actual values of the multiplier coefficients m_i in the immediate neighborhood of their design values m_{i0}, its transfer function remains bounded real, satisfying the condition of Eq. (9.164). Now, consider $\left|H(e^{j\omega_k})\right|$, the value of the magnitude response at $\omega = \omega_k$, which is equal to 1 when the multiplier values have infinite precision. Because of the assumption on $\mathcal{N}$ if a multiplier coefficient m_i is quantized, then $\left|H(e^{j\omega_k})\right|$, can only become less than 1 because of the BR condition of Eq. (9.164). Thus, a plot of $\left|H(e^{j\omega_k})\right|$, as a function of m_i will appear as indicated in Figure 9.33 with a zero-valued slope at $m_i = m_{i0}$:

$$\left.\frac{\partial \left|H(e^{j\omega_k})\right|}{\partial m_i}\right|_{m_i=m_{i0}} = 0, \tag{9.166}$$

implying that the first-order sensitivity of the magnitude function $|H(e^{j\omega})|$ with respect to each multiplier coefficient m_i is zero at all frequencies ω_k where $|H(e^{j\omega})|$ assumes its maximum value of unity. Since all frequencies ω_k, where the magnitude function is exactly equal to unity, are in the passband of the filter, and if these frequencies are closely spaced, we expect the sensitivity of the magnitude function to be very low at other frequencies in the passband.

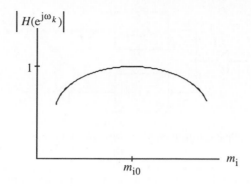

Figure 9.33 Illustration of the zero-sensitivity property.

A digital filter structure satisfying the above conditions for low passband sensitivity is called a *structurally bounded* system. Since the output energy of such a structure is less than the input energy for all finite energy inputs [see Eq. (4.122)], it is also called a *structurally passive* system. If Eq. (9.164) holds with an equality sign, the transfer function $H(z)$ is called a *lossless bounded real* (LBR) function, i.e., a stable allpass function. An allpass realization satisfying the LBR condition is therefore a *structurally lossless* or *LBR* system implying that the structure remains allpass under coefficient quantization.

We now outline methods for the realization of low passband sensitivity digital filters.

9.7.2 Low Passband Sensitivity IIR Digital Filter

In Section 6.10 we outlined a method for the realization of a large class of stable IIR transfer functions $G(z)$ in the form of a parallel allpass structure:

$$G(z) = \tfrac{1}{2}\left\{A_0(z) + A_1(z)\right\}, \tag{9.167}$$

where $A_0(z)$ and $A_1(z)$ are stable allpass transfer functions. Such a realization is possible if $G(z)$ is a BR function with a symmetric numerator and has a power-complementary BR transfer function $H(z)$ with an antisymmetric numerator [Vai86a]. In this case, $H(z)$ can be expressed as

$$H(z) = \frac{1}{2}\left\{A_0(z) - A_1(z)\right\}. \tag{9.168}$$

The allpass decompositions of Eqs. (9.167) and (9.168) permit the realization of the power-complementary pair $\{G(z), H(z)\}$ as indicated in Figure 9.34.

Now, on the unit circle, Eq. (9.167) becomes

$$G(e^{j\omega}) = \frac{1}{2}\left\{e^{j\theta_0(\omega)} + e^{j\theta_1(\omega)}\right\}. \tag{9.169}$$

since $A_0(z)$ and $A_1(z)$ are allpass functions with unity magnitude responses. Thus, if the allpass transfer functions are realized in structurally lossless forms, $|G(e^{j\omega})|$ will remain bounded above by unity. Moreover, at frequencies ω_k, where $|G(e^{j\omega_k})| = 1$, $\theta_0(\omega_k) = \langle \theta_1(\omega_k)\rangle_{2\pi}$, and $|e^{j\theta_0(\omega_k)} + e^{j\theta_1(\omega_k)}| = 2$. Or in other words, the realization

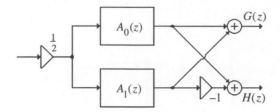

Figure 9.34   Parallel allpass realization of a power-complementary pair of transfer functions.

of Figure 9.34 is structurally passive implying low passband sensitivity with respect to multiplier coefficient quantization.

Section 6.10 describes a method for determining the two allpass transfer functions $A_0(z)$ and $A_1(z)$ for a given BR transfer function $H(z)$ satisfying the conditions given above. Once these allpass transfer functions have been determined, they can then be realized in structurally lossless forms using one of the two methods discussed in Section 6.7.

The following example illustrates the low passband sensitivity characteristics of a parallel allpass structure.

Example 9.12 Consider the realization of the fifth-order elliptic IIR lowpass filter of Example 7.19. The filter specifications are passband ripple of 0.5 dB, minimum stopband attenuation of 40 dB, and a passband edge at 0.4. Using a modified version of Program 7.2, we arrive at the desired transfer function given by

$$G(z) = \frac{\begin{array}{c} 0.0528168 + 0.0797082z^{-1} + 0.1294975z^{-2} + 0.1294975z^{-3} \\ + 0.0797082z^{-4} + 0.0528168z^{-5} \end{array}}{(1 - 0.4908777z^{-1})(1 - 0.7624319z^{-1} + 0.5390008z^{-2}) \atop \times (1 - 0.5573823z^{-1} + 0.8828417z^{-2})}$$

Its parallel allpass decomposition is given by

$$G(z) = \frac{1}{2}\left[\frac{(0.4908777 - z^{-1})(0.8828417 - 0.5573823z^{-1} + z^{-2})}{(1 - 0.4908777z^{-1})(1 - 0.5573823z^{-1} + 0.8828417z^{-2})} \right.$$
$$\left. + \frac{0.5390008 - 0.7624319z^{-1} + z^{-2}}{1 - 0.7624319z^{-1} + 0.5390008z^{-2}} \right]$$

A 5-multiplier realization of the above decomposition using a signed 7-bit fractional sign-magnitude representation of each multiplier coefficients is shown in Figure 9.35. The gain responses of the filter with infinite precision multiplier coefficients and that with the quantized coefficients are shown in Figure 9.36. The low passband sensitivity properties of the parallel allpass structure is evident from this figure. Figure 9.37 depicts the passband details of the gain response of a direct form realization of the above fifth-order lowpass filter with infinite-precision multiplier coefficients and that with the quantized coefficients. The poor passband sensitivity of the direct form realization is clearly seen in this plot.

It should be noted that the power-complementary transfer function $H(z)$ realized in the form of Eq. (9.168) also remains BR if the allpass filters are structurally LBR. Hence, it exhibits low sensitivity in its passband [which is the stopband of $G(z)$]. However, low

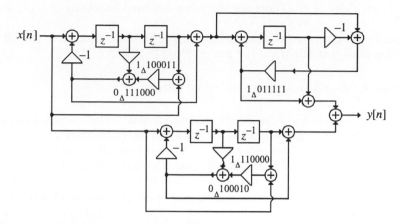

Figure 9.35   Parallel allpass realization of the lowpass filter using structurally lossless allpass sections with quantized multiplier coefficients.

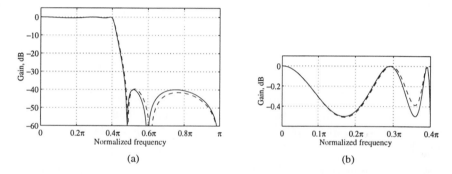

(a)   (b)

Figure 9.36   (a) Gain responses of the parallel allpass realization with infinite precision coefficients (shown with solid line), and with quantized coefficients (dashed line), and (b) passband details.

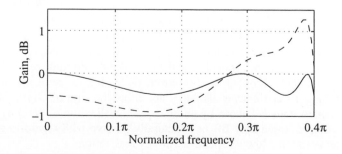

Figure 9.37   Passband responses of the direct form realization with infinite precision coefficients (shown with solid line), and with quantized coefficients (dashed line).

passband sensitivity of $H(z)$ does not imply low stopband sensitivity of $G(z)$ (Problem 9.30).

A number of other methods have been advanced for the realization of BR IIR digital transfer functions [Dep80], [Hen83], [Rao84], [Vai84], [Vai85a].

9.7.3   Low Passband Sensitivity FIR Digital Filter

In many applications linear-phase FIR filters are preferred. As indicated in Section 4.3.6, there are four types of linear-phase FIR filters of which the Type 1 filter is the most general and can realize any type of frequency response. We therefore restrict our attention here to a BR Type 1 FIR transfer function and outline a simple approach to realize it in a structurally passive form, thus ensuring low passband sensitivity to multiplier coefficients [Vai85b].

Now, from Eq. (4.98), the frequency response of a Type 1 FIR filter of degree N can be expressed as

$$H(e^{j\omega}) = e^{-j\omega N/2}\breve{H}(\omega) \tag{9.170}$$

where $\breve{H}(\omega)$, a real function of ω, is its amplitude response. Since $H(z)$ is a BR function, $\breve{H}(\omega) \leq 1$. Its delay-complementary filter $G(z)$ defined by (Section 4.6.1)

$$G(z) = z^{-N/2} - H(z), \tag{9.171}$$

has a frequency response given by

$$G(e^{j\omega}) = e^{-j\omega N/2}\left[1 - \breve{H}(\omega)\right] = e^{-j\omega N/2}\breve{G}(\omega), \tag{9.172}$$

where $\breve{G}(\omega) = 1 - \breve{H}(\omega)$ is its amplitude response. Amplitude responses of a typical delay-complementary FIR filter pair are depicted in Figure 9.38. It follows from this figure that at $\omega = \omega_k$, where $\left|H(e^{j\omega_k})\right| = 1$, $\breve{G}(\omega)$ has double zeros. Thus, $G(z)$ can be expressed as

$$G(z) = G_a(z)\prod_{k=1}^{L}\left(1 - 2\cos\omega_k z^{-1} + z^{-2}\right)^2 = G_a(z)G_b(z). \tag{9.173}$$

A delay-complementary implementation of $H(z)$ based on Eq. (9.171) is sketched in Figure 9.39, where the FIR filter $G(z)$ is realized according to Eq. (9.173) as a cascade of $G_a(z)$ and L fourth-order FIR sections, with the kth section having a transfer function $(1 - 2\cos\omega_k z^{-1} + z^{-2})^2$. Now if the multiplier coefficient $2\cos\omega_k$ of the kth section is quantized, its zeros are still double and remain on the unit circle. As a result, quantization of the coefficients in $G_b(z)$ does not change the sign of the amplitude response $\breve{G}(\omega)$, and in the passband of $H(z)$, $\breve{G}(\omega) \geq 0$. Moreover, $G_a(z)$ has no zeros on the unit circle, and quantization of its coefficients also does not affect the sign of $\breve{G}(\omega)$. Hence, $\breve{H}(\omega)$ continues to remain bounded above by unity, i.e., the realization of $H(z)$ as indicated in Figure 9.39 remains structurally BR or structurally passive with regard to all multiplier coefficients, resulting in a low-passband sensitivity realization.

Example 9.13 We consider here the low passband sensitivity realization of a lowpass FIR filter using the above method. The filter specifications are length 13 with a normalized passband edge

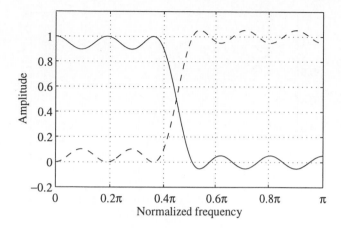

Figure 9.38   Amplitude responses of typical delay-complementary Type 1 FIR filters.

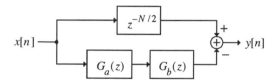

Figure 9.39   Low passband sensitivity realization of a Type 1 FIR filter $H(z)$.

at 0.5 and a normalized stopband edge at 0.6 with equal weights to passband and stopband ripples. Using Program 7_8 we first design the lowpass filter $H(z)$ and then form its delay-complementary filter as $G(z) = z^{-6} - H(z)$. Using the function roots in MATLAB, we next determine the roots of $G(z)$ which has six zeros on the unit circle: two at $z = 1$, a pair of complex conjugate zeros at $z = -0.26463064626566 \pm j0.96434984370664$, and a pair of complex conjugate zeros at $z = -0.27683551142484 \pm j0.96091732194510$. These unit circle zeros constitute the transfer function

$$G_b(z) = (1 - z^{-1})^2(1 - 0.52926129253132z^{-1} + z^{-2})$$
$$\times(1 - 0.55367102284968z^{-1} + z^{-2}).$$

By factoring out $G_b(z)$ from $G(z)$ using the function deconv in MATLAB, we arrive at the remaining term $G_a(z)$:

$$G_a(z) = 0.04107997195619 + 0.05197154435126z^{-1}$$
$$-0.12094731168744z^{-2} - 0.30704562224137z^{-3}$$
$$-0.12094731168744z^{-4} + 0.05197154435126z^{-5}$$
$$+0.04107997195619z^{-6}.$$

Next we quantize the coefficients of $G_a(z)$ and $G_b(z)$ by rounding the fractional part to two decimal digits, and from the quantized $G(z)$ determine again the delay complementary filter as indicated in Figure 9.39. Plots of the frequency responses of the original transfer function $H(z)$ and its

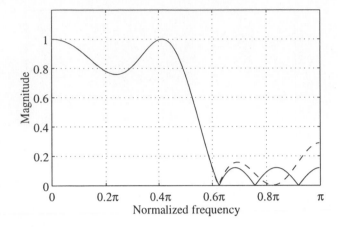

Figure 9.40 Magnitude responses of the original FIR lowpass filter (solid line) and its low sensitivity realization (dashed line).

realization as in Figure 9.39 with quantized coefficients are shown in Figure 9.40, verifying the low passband sensitivity behavior of the proposed realization.

Another novel approach to low passband sensitivity FIR filter is described in [Vai86b].

9.8   Reduction of Product Round-Off Errors

We have indicated earlier in Section 9.5 that in digital filters implemented using fixed-point arithmetic, the quantization of multiplication operations can be treated as a round-off noise at the output of the filter structure and can be analyzed using a statistical model of the quantization process. In many applications, this noise may decrease the output signal-to-noise ratio to an unacceptable level. It is thus of interest to investigate techniques that can reduce the output roundoff noise. In this section, we describe two possible solutions that require additional hardware. In critical applications, the cost of the additional hardware may be justified.

9.8.1   Error-Feedback Structures

The basic idea behind the error-feedback approach is to make use of the difference between the unquantized and quantized signal in reducing the round-off noise. This difference, called the error, is fed back to the digital filter structure in such a way as to not change the transfer function originally implemented by the structure while effectively decreasing the noise power [Cha81], [Hig84], [Mun81], [Tho77]. We illustrate the approach for a first-order and a second-order filter structure. The error-feedback concept is often used in designing high-precision oversampling A/D converters.

First-order section   Consider again the scaled first-order section of Figure 9.30(a). We also assume that all multiplier coefficients are signed $(b+1)$-bit fractions. The quantization

Figure 9.41   A first-order digital filter structure with error feedback.

error signal is then given by

$$e[n] = y[n] - v[n]. \tag{9.174}$$

We now modify the structure of Figure 9.30(a) as shown in Figure 9.41, where now the error signal is being fed back to the system through a delay and a multiplier with a coefficient β. In practice, the coefficient β is chosen to be simple integers or fractions, such as ± 1, ± 2, or ± 0.5, so that the multiplication can be simply performed using a shift operation and will not introduce an additional quantization error.

Analyzing Figure 9.41, we arrive at the expression for the transfer function of the digital filter structure with $y[n]$ as the output as

$$H(z) = \left.\frac{Y(z)}{X(z)}\right|_{E(z)=0} = \frac{K}{1 - \alpha z^{-1}}. \tag{9.175}$$

The noise transfer function $G(z)$ with the error feedback, with $y[n]$ as the output, is given by

$$G(z) = \left.\frac{Y(z)}{E(z)}\right|_{X(z)=0} = \frac{1 + \beta z^{-1}}{1 - \alpha z^{-1}}, \tag{9.176}$$

and without the error feedback ($\beta = 0$) is given by

$$G(z) = \frac{1}{1 - \alpha z^{-1}}. \tag{9.177}$$

Following steps similar to that outlined in Example 9.7, we arrive at the output noise variance of the error feedback structure as

$$\sigma_\gamma^2 = \frac{1 + 2\alpha\beta + \beta^2}{1 - \alpha^2}\sigma_0^2. \tag{9.178}$$

where σ_0^2 is the variance of the noise source $e[n]$. Note from the above that the output noise variance is a minimum when $\beta = -\alpha$. However, in practice $|\alpha| < 1$, and hence this choice for β will introduce an additional quantization noise source, making the analysis resulting in Eq. (9.178) invalid. Thus, from a practical point of view, it is more attractive to reduce the output round-off noise variance by choosing β as an integer with a value close to that of $-\alpha$, in which case the only noise source is due to the quantization of $\alpha y[n-1]$.

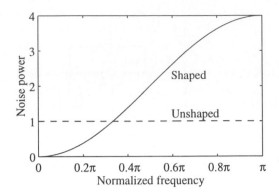

Figure 9.42   The normalized error power spectrum of the first-order section with and without error feedback.

For $|\alpha| < 0.5$, $\beta = 0$, implying no error feedback. However, in this case, the pole of $H(z)$ is far from the unit circle, and as a result, the noise variance is not high. For $|\alpha| \geq 0.5$, we choose $\beta = (-1)\mathrm{sgn}(\alpha)$.[3] Substituting this value of β in Eq. (9.178), we arrive at

$$\sigma_\gamma^2 = \frac{2}{1 + |\alpha|}\sigma_0^2. \tag{9.179}$$

Comparing Eqs. (9.179) and (9.178) with $\beta = 0$, we note that the introduction of error feedback has increased the signal-to-noise ratio (SNR) by a factor of $10\log_{10}[2(1 - |\alpha|)]$. The above increase in SNR is quite significant if the pole is closer to the unit circle. For example, if $|\alpha| = 0.99$, the improvement is about 17 dB, which is equivalent to about 3 bits of increased accuracy compared to the case without error feedback. The additional hardware requirements for the structure of Figure 9.41 are two new adders and an additional storage register.

Comparing the two expressions of Eqs. (9.176) and (9.177), we conclude that the noise transfer function with error feedback is given by that without the error feedback multiplied by the expression $(1 + \beta z^{-1})$. Equivalently, the error feedback circuit is *shaping* the error spectrum by modifying the input quantization noise $E(z)$ to $E_s(z) = (1 + \beta z^{-1})E(z)$. The output noise is generated by passing $E_s(z)$ through the usual noise transfer function of Eq. (9.177).

To examine the effect of noise spectrum shaping, consider the case of a narrowband lowpass first-order filter with $\alpha \to 1$. In this case, we choose $\beta = -1$, and as a result, $E_s(z)$ has a zero at $z = 1$ (i.e., $\omega = 0$). The power spectral density of the unshaped quantization noise $E(z)$ is σ_0^2, a constant. The power spectral density of the shaped noise source $E_s(z)$ is given by $4\sin^2(\omega/2)\sigma_0^2$ and is plotted in Figure 9.42 along with that of the unshaped case. The noise shaping redistributes the noise so as to move it mostly into the stopband of the lowpass filter, thus reducing the noise variance. Because of the noise redistribution caused by the error feedback circuit, this method of round-off noise reduction has also been called *error-spectrum shaping* approach in the literature [Hig84].

---

[3]$\mathrm{sgn}(\alpha) = +1$ for $\alpha \geq 0$ and $\mathrm{sgn}(\alpha) = -1$ for $\alpha < 0$.

Second-order section   The error-feedback approach for round-off noise reduction has also been applied to second-order IIR digital filter structures [Hig84], [Mun81]. One proposed structure obtained by modifying Figure 9.31(a) is indicated in Figure 9.43. The transfer function $H(z)$ of this structure is given by Eq. (9.157). It should be noted that the inclusion of the error-feedback circuit to the structure of Figure 9.31(a) does not affect either $H(z)$ or the scaling transfer function $F(z)$. Analyzing Figure 9.43, we arrive at the expression for the noise transfer function

$$G(z) = \frac{1 + \beta_1 z^{-1} + \beta_2 z^{-2}}{1 + \alpha_1 z^{-1} + \alpha_2 z^{-2}}. \tag{9.180}$$

The output round-off noise variance for a $\mathcal{L}_2$-scaling is given by

$$\sigma_\gamma^2 = (\|G\|_2)^2 \sigma_0^2. \tag{9.181}$$

Note that a choice of $\beta_1 = \alpha_1$ and $\beta_2 = \alpha_2$ makes $\|G\|_2 = 1$, yielding $\sigma_\gamma^2 = \sigma_0^2$, an apparent optimal solution. However, this choice for the multiplier coefficients in the error-feedback path introduces additional quantization noise sources that were not taken into account in the above analysis. As in the case of the first-order section with error-feedback, a more attractive solution is to make β_1 and β_2 integers with values close to α_1 and α_2, respectively. For example, for a narrowband lowpass transfer function, the poles are close to the unit circle and to the real axis, i.e., $r \approx 1$ and $\theta \approx 0$. In this case, α_1 is close to -2 and α_2 is close to 1. We therefore choose here $\beta_1 = -2$ and $\beta_2 = 1$, resulting in a noise transfer function

$$G(z) = \frac{1 - 2z^{-1} + z^{-2}}{1 + \alpha_1 z^{-1} + \alpha_2 z^{-2}}. \tag{9.182}$$

It has been shown by Vaidyanathan [Vai87c] that for a very narrowband lowpass filter with $r = 0.995$, $\theta = 0.07\pi$, and $b = 16$, the second-order error-feedback structure has a SNR that is approximately 25 dB higher than that without the error-feedback. A detailed comparison of the second-order section with and without the error-feedback for various types of inputs is left as an exercise (Problem 9.31).

It should be noted that here also the error-feedback circuit provides a noise shaping, as in the case of the first-order case. In fact, for the parameters indicated above, the noise transfer function with error-feedback is given by that without error-feedback multiplied by the expression $(1 - z^{-1})^2$. Or in other words, the error-feedback circuit is *shaping* the error spectrum by modifying the input quantization noise $E(z)$ to $E_s(z) = (1 - z^{-1})^2 E(z)$. The output noise is generated by passing $E_s(z)$ through the usual noise transfer function of Eq. (9.180) with $\beta_1 = \beta_2 = 0$. The power spectral density of the shaped noise source $E_s(z)$ is given by $16 \sin^4(\omega/2)\sigma_0^2$, whereas that of the unshaped case is simply σ_0^2. These power spectral densities have been plotted in Figure 9.44. Observe how the error-feedback lowers the noise in the passband by pushing it into the stopband of the filter.

9.8.2   State-Space Structures

As discussed earlier in this chapter, different realizations of a given transfer function exhibit different behavior with respect to wordlength quantizations, particularly with regard to multiplier coefficient sensitivity and product round-off error propagation. We have

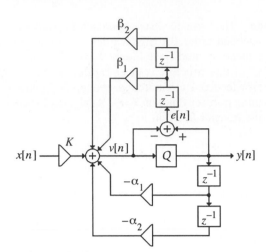

Figure 9.43   A second-order digital filter with error feedback.

outlined procedures to realize both IIR and FIR digital filters with low passband coefficient sensitivity. We also have shown how simple first-order and second-order IIR sections can be modified to reduce the product round-off errors. We consider here a more general approach of round-off noise minimization that is based on the state-space realizations discussed in Section 6.6.

The state-space description of a digital transfer function is given by Eqs. (2.94) and (2.95) repeated below for convenience:

$$\mathbf{s}[n+1] = \mathbf{A}\mathbf{s}[n] + \mathbf{B}x[n], \tag{9.183a}$$

$$y[n] = \mathbf{C}\mathbf{s}[n] + Dx[n], \tag{9.183b}$$

whose block diagram representation is shown in Figure 9.45. A direct realization of a digital transfer function based on the state-space representation of Eq. (9.183a) and Eq. (9.183b) has been referred to as the state-space structure in Section 6.6. In a state-space structure, therefore, each element of the state-space matrices ($\mathbf{A}$, $\mathbf{B}$, $\mathbf{C}$, D) is a multiplier coefficient. For example, Figure 6.31 shows the state-space realization of a second-order IIR filter described by the state equations of Eqs. (6.46) and (6.47).

The transfer function of the digital filter in terms of the state-space parameters is given in Eq. (4.56) and is repeated below for convenience:

$$H(z) = \mathbf{C}(z\mathbf{I} - \mathbf{A})^{-1}\mathbf{B} + D. \tag{9.184}$$

Without any loss of generality, we assume that $H(z)$ has been scaled to ensure that the output $y[n]$ does not overflow. In order to eliminate overflows at all internal adders, we need to ensure that after scaling, each element of the state variable sequence vector $\mathbf{s}[n]$ remains bounded by unity for all values of n.

If we denote the impulse response from the input to the output of the kth adder as $f_k[n]$, then from Eq. (9.183a) we obtain after some simple algebra

$$f_k[n] = [\mathbf{A}^{n-1}\mathbf{B}]_k \mu[n], \tag{9.185}$$

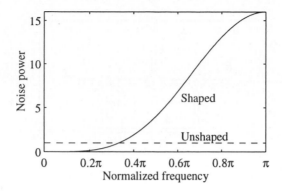

Figure 9.44   Normalized error power spectrum of the second-order section with and without error feedback.

Figure 9.45   Block diagram representation of a general state-space structure.

where $[\mathbf{A}^{n-1}\mathbf{B}]_k$ represents the kth element of the column vector $\mathbf{A}^{n-1}\mathbf{B}$. The z-transform $F_k(z)$ of $f_k[n]$ is the kth scaling transfer function.

After the state-space structure has been scaled, our next objective is to minimize the output noise variance due to the round-off of all products. An examination of Figure 9.45 reveals that the round-off noise at the output resulting from the quantization of products due to multipliers in $\mathbf{C}$ and D are of constant variances and, as a result, cannot be further minimized. Therefore, our objective here is to minimize the product round-off noise generated by the multipliers in $\mathbf{A}$ and $\mathbf{B}$. A noise model for analysis purpose based on this observation is sketched in Figure 9.46.

Next, from Eq. (9.183a) and Eq. (9.183b), the output $y[n]$ under zero-input condition is given by

$$y[n] = \mathbf{C}\mathbf{A}^n\mathbf{s}[0], \tag{9.186}$$

where $\mathbf{s}[0]$ is the state vector at $n = 0$. If we denote the impulse response from the kth state variable node to the filter output as $g_k[n]$, then it follows from Eq. (9.186) that

$$g_k[n] = [\mathbf{C}\mathbf{A}^n]_k \mu[n]. \tag{9.187}$$

The z-transform of $g_k[n]$ is the kth noise transfer function $G_k(z)$.

Figure 9.46   Noise model of a general state-space structure.

It is convenient to introduce two new matrices $\mathbf{K} = [K_{i\ell}]$ and $\mathbf{W} = [W_{i\ell}]$ given by

$$\mathbf{K} = \sum_{n=0}^{\infty} \mathbf{A}^n \mathbf{B}(\mathbf{A}^n \mathbf{B})^T, \qquad \mathbf{W} = \sum_{n=0}^{\infty} (\mathbf{CA}^n)^T \mathbf{CA}^n. \tag{9.188}$$

These matrices can be alternatively expressed in terms of the scaling transfer function vector $F(z)$ and the noise transfer function vector $G(z)$ as

$$\mathbf{K} = \frac{1}{2\pi j} \oint_C \mathbf{F}(z)\mathbf{F}^T(z^{-1})z^{-1}\,dz, \tag{9.189a}$$

$$\mathbf{W} = \frac{1}{2\pi j} \oint_C \mathbf{G}(z)\mathbf{G}^T(z^{-1})z^{-1}\,dz. \tag{9.189b}$$

It follows from the above equations that

$$K_{\ell\ell} = \sum_{n=0}^{\infty} f_\ell^2[n] = (\|F_\ell\|_2)^2, \tag{9.190a}$$

$$W_{\ell\ell} = \sum_{n=0}^{\infty} g_\ell^2[n] = (\|G_\ell\|_2)^2. \tag{9.190b}$$

The unscaled state-space structure has a total output noise variance due to product round-off given by

$$\sigma_\ell^2 = \sigma_0^2 \sum_{\ell=1}^{N} (\|G_\ell\|_2)^2 = \sigma_0^2 \sum_{\ell=1}^{N} W_{\ell\ell}. \tag{9.191}$$

We have shown earlier in Section 6.6.3 that equivalent state-space structures with identical transfer function $H(z)$ can be easily generated from a given state-space realization as in Eq. (9.186) by applying a similarity transformation to the state-variable vector:

$$\bar{\mathbf{s}}[n] = \mathbf{Q}^{-1}\mathbf{s}[n] \tag{9.192}$$

where $\mathbf{Q}$ is a nonsingular matrix. The new state-space description is given by Eqs. (6.52) and (6.53), repeated below for convenience:

$$\bar{\mathbf{s}}[n+1] = \bar{\mathbf{A}}\bar{\mathbf{s}}[n] + \bar{\mathbf{B}}x[n], \qquad (9.193\text{a})$$

$$y[n] = \bar{\mathbf{C}}\bar{\mathbf{s}}[n] + Dx[n], \qquad (9.193\text{b})$$

where

$$\bar{\mathbf{A}} = \mathbf{Q}^{-1}\mathbf{A}\mathbf{Q}, \qquad \bar{\mathbf{B}} = \mathbf{Q}^{-1}\mathbf{B}, \qquad \bar{\mathbf{C}} = \mathbf{C}\mathbf{Q}. \qquad (9.194)$$

After the transformation, the scaling transfer function vector and the noise transfer function vector become

$$\bar{\mathbf{F}}(z) = \mathbf{Q}\mathbf{F}(z), \qquad \bar{\mathbf{G}}(z) = \left(\mathbf{Q}^{-1}\right)^T \mathbf{G}(z). \qquad (9.195)$$

The state-space structure is scaled in the $\mathcal{L}_2$-sense if $\|\bar{F}_i\|_2 = 1$, $1 \leq i \leq N$. From Eq. (9.190a) this condition is equivalent to requiring

$$\bar{\mathbf{K}}_{ii} = \left(\|\bar{F}_i\|_2\right)^2 = 1, \qquad 1 \leq i \leq N. \qquad (9.196)$$

To this end, we choose

$$\mathbf{Q} = \mathrm{diag}\left[\sqrt{K_{11}} \quad \sqrt{K_{22}} \quad \cdots \quad \sqrt{K_{NN}}\right]. \qquad (9.197)$$

For the scaled structure, we then have $\bar{\mathbf{W}} = \mathbf{Q}^T\mathbf{W}\mathbf{Q}$ whose diagonal elements are given by

$$\bar{W}_{\ell\ell} = K_{\ell\ell}W_{\ell\ell}. \qquad (9.198)$$

The total output noise variance due to product round-off of the scaled state-space structure is now

$$\bar{\sigma}_\gamma^2 = \sigma_0^2 \sum_{\ell=1}^{N} K_{\ell\ell}W_{\ell\ell}. \qquad (9.199)$$

It has been shown that the necessary and sufficient conditions to ensure minimum output product round-off noise variance under an $\mathcal{L}_2$-norm scaling scheme are given by [Hwa76], [Mul76]

$$\bar{\mathbf{W}} = \mathbf{P}\bar{\mathbf{K}}\mathbf{P}, \qquad (9.200)$$

and

$$\bar{K}_{ii}\bar{W}_{ii} = \bar{K}_{\ell\ell}\bar{W}_{\ell\ell}, \qquad 1 \leq i, \ell \leq N, \qquad (9.201)$$

where $\mathbf{P}$ is a diagonal matrix of positive elements. It follows from the above conditions that for a state-space structure that is already scaled in the $\mathcal{L}_2$-sense, $\bar{K}_{\ell\ell} = 1$ for all ℓ, and hence, the structure exhibits a minimum output roundoff noise if and only if

$$\bar{W}_{ii} = \bar{W}_{\ell\ell}, \qquad 1 \leq i, \ell \leq N. \qquad (9.202)$$

From Eqs. (9.190b) and (9.202), we arrive at

$$\left(\|G_i\|_2\right)^2 = \left(\|G_\ell\|_2\right)^2, \qquad 1 \leq i, \ell \leq N, \qquad (9.203)$$

or in other words, each noise source makes equal contributions to the output noise.

Some comments with regard to the optimal state-space structures are in order here. First, the method described above leads to only a minimum round-off noise scaled state-space realization of a given transfer function. However, it is quite likely that there exists one or more other equivalent scaled structures (non-state-space) that has a round-off noise that is lower than the optimal state-space structure. Second, the state-space structure, in general, is computationally very intensive since it requires a total of $(N+1)^2$ multiplications for the realization of an Nth-order IIR transfer function in contrast to $(2N + 1)$ multiplications in a direct form implementation. The computational complexity of a state-space-based realization can be reduced by realizing the overall transfer function either as a cascade of optimum second-order sections or as a parallel connection of such sections. In the former case, the total number of multiplications reduces to $9(N/2)$, whereas in the latter case it reduces to $(4N + 1)$. However, in general, the latter realizations will lead to suboptimal structures compared to the optimal state-space structure.

It is difficult to develop an analytical solution for the optimal state-space structure for the general case. However, such a solution is possible for a second-order transfer function and has been advanced by a number of authors [Jac79], [Mil81], [Bar84], [Bom85]. We provide below the solution developed by Jackson, Lindgren, and Kim [Jac79]. They have shown that the second-order state-space structure with a minimum round-off output noise is characterized by the state-space parameters

$$\mathbf{A} = \begin{bmatrix} a_{11} & a_{12} \\ a_{21} & a_{11} \end{bmatrix}, \qquad \mathbf{B} = \begin{bmatrix} b_1 \\ b_2 \end{bmatrix}, \qquad \mathbf{C} = \begin{bmatrix} \gamma b_2 \\ \gamma b_1 \end{bmatrix}. \qquad (9.204)$$

A state-space realization satisfying the above conditions can be scaled to satisfy the $\mathcal{L}_2$-norm by applying the transformation

$$\mathbf{P} = \begin{bmatrix} \|F_1\|_2 & 0 \\ 0 & \|F_2\|_2 \end{bmatrix}, \qquad (9.205)$$

resulting in an optimal structure.

For a second-order transfer function

$$H(z) = p_0 + \frac{q_1 z + q_2}{z^2 + d_1 z + d_2} \qquad (9.206)$$

with complex conjugate poles, the parameters of the optimal state-space structure are given by [Jac79]

$$a_{11} = a_{22} = -\frac{d_1}{2}, \qquad (9.207a)$$

$$a_{12} = \frac{1}{q_1^2}(1 + q_2)(J_1 \pm J_2), \qquad (9.207b)$$

$$a_{21} = \frac{1}{1 + q_2}(J_1 \mp J_2), \qquad (9.207c)$$

$$b_1 = \frac{1 + q_2}{2}, \qquad b_2 = \frac{q_1}{2}, \qquad (9.207d)$$

$$c_1 = \frac{q_1}{1 + q_2}, \qquad b_2 = 1, \qquad (9.207e)$$

where

$$J_1 = q_2 - \frac{d_1 q_1}{2}, \qquad J_2 = \sqrt{q_2^2 - d_1 q_2 q_1 + d_2 q_1^2}. \qquad (9.208)$$

9.9 Limit Cycles in IIR Filters

So far we have treated the analysis of finite wordlength effects using a linear model of the system. However, a practical digital filter is a nonlinear system caused by the quantization of the arithmetic operations. Such nonlinearities may cause an IIR filter, which is stable under infinite precision, to exhibit an unstable behavior under finite precision arithmetic for specific input signals, such as zero or constant inputs. This type of instability usually results in an oscillatory periodic output called a *limit cycle*, and the system remains in this condition until an input of sufficiently large amplitude is applied to move the system into a more conventional operation. It should be noted here that limit cycles occur in IIR filters due to the presence of a feedback path. Such oscillations are absent in FIR structures which do not have any feedback path.

There are basically two types of limit cycles: (1) *granular* and (2) *overflow*. The former type of limit cycles is usually of low amplitude, whereas overflow oscillations have large amplitudes. We examine both types of limit cycles in this section.

9.9.1 Granular Limit Cycles

Two types of granular limit cycles have been observed in IIR digital filters: *inaccessible* and *accessible limit cycles* [Cla73]. The former type can appear only if the initial conditions of the digital filter at the time of starting pertains to that limit cycle, whereas in the second case, the limit cycle condition can be reached by starting the digital filter with initial conditions not pertaining to that limit cycle. We illustrate the generation of the limit cycle by analyzing the nonlinear behavior of a first-order and a second-order IIR digital filter.

Example 9.14 Consider the causal first-order digital filter of Figure 9.1 characterized by the difference equation of Eq. (9.1), where for stability $|\alpha| < 1$. We assume that the quantization operation being performed is rounding and it is carried out on the result of the multiplication, as indicated in Figure 9.47. In this case, Eq. (9.1) becomes the nonlinear difference equation

$$\hat{y}[n] = \mathcal{Q}\left(\alpha \hat{y}[n-1]\right) + x[n], \qquad (9.209)$$

where $\hat{y}[n]$ denotes the actual output of the filter. Without any loss of generality, we assume the digital filter is being implemented using a signed 5-bit fractional arithmetic with a quantization step of $\delta = 2^{-4}$. Table 9.6 shows the first seven output samples for two different pole positions for an impulse input with $x[0] = 0_\triangle 1101$ and $x[n] = 0$ for $n > 0$, and an initial condition $\hat{y}[-1] = 0$. Observe that the steady-state output in the first case is a nonzero constant, i.e., periodic with a period of 1, whereas in the second case it is periodic with a period of 2. On the other hand, with infinite precision, the ideal output goes to zero exponentially as $n \to \infty$.

The above types of oscillations at the output are called zero-input limit cycles, and the amplitude ranges of the oscillations are often called *dead bands* [Bla65]. A digital filter exhibiting limit cycles at its output can be modeled by a linear system with its poles on the unit circle [Jac69]. Using this representation, it is possible to determine the range of the dead band in low-order filters. For example, in the case of the first-order IIR filter considered above, we assume that the system

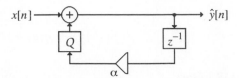

Figure 9.47   A first-order IIR filter with a quantizer.

Table 9.6   Limit cycle behavior of the first-order IIR digital filter.

n	$\alpha = 0_\Delta 1011,\ \hat{y}[-1] = 0$		$\alpha = 1_\Delta 1011,\ \hat{y}[-1] = 0$	
	$\alpha\hat{y}[n-1]$	$\hat{y}[n]$	$\alpha\hat{y}[n-1]$	$\hat{y}[n]$
0	0	$0_\Delta 1101$	0	$0_\Delta 1101$
1	$0_\Delta 10001111$	$0_\Delta 1001$	$1_\Delta 10001111$	$1_\Delta 1001$
2	$0_\Delta 01100011$	$0_\Delta 0110$	$0_\Delta 01100011$	$0_\Delta 0110$
3	$0_\Delta 01000010$	$0_\Delta 0100$	$1_\Delta 01000010$	$1_\Delta 0100$
4	$0_\Delta 00101100$	$0_\Delta 0011$	$0_\Delta 00101100$	$0_\Delta 0011$
5	$0_\Delta 00100001$	$0_\Delta 0010$	$1_\Delta 00100001$	$1_\Delta 0010$
6	$0_\Delta 00010110$	$0_\Delta 0001$	$0_\Delta 00010110$	$0_\Delta 0001$
7	$0_\Delta 00001011$	$0_\Delta 0001$	$1_\Delta 00001011$	$1_\Delta 0001$
8	$0_\Delta 00001011$	$0_\Delta 0001$	$0_\Delta 00001011$	$0_\Delta 0001$

under limit cycle condition has an effective pole at $z = 1$ when $\alpha > 0$ and at $z = -1$ when $\alpha < 0$. This implies that effectively

$$\mathcal{Q}\left(\alpha\hat{y}[n-1]\right) = \begin{cases} \hat{y}[n-1], & \alpha > 0, \\ -\hat{y}[n-1], & \alpha < 0. \end{cases} \qquad (9.210)$$

In addition, the quantization error due to rounding is bounded by $\pm\delta/2$, where δ is the quantization step, i.e.,

$$\left|\mathcal{Q}\left(\alpha\hat{y}[n-1]\right) - \alpha\hat{y}[n-1]\right| \le \frac{\delta}{2}. \qquad (9.211)$$

From the above two equations, we arrive at the dead band range of the first-order IIR filter as

$$\left|\hat{y}[n-1]\right| \le \frac{\delta}{2\left(1 - |\alpha|\right)}. \qquad (9.212)$$

As a result, if for any value of n, the output $\hat{y}[n-1]$ of the delay unit is in the above range with the input set to zero, then the system gets trapped into a limit cycle mode. For our numerical example cited above, the dead band range is $|\hat{y}[n-1]| \le 0.1$, which definitely is satisfied by the entries in Table 9.6.

The limit cycle generation can be easily illustrated on a computer. The MATLAB Program 9_4 given below can be used to study the limit cycle process.

```
% Program 9_4
% Limit Cycles in First-Order IIR Filter
```

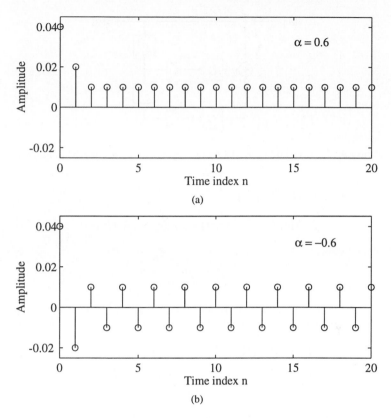

Figure 9.48 Illustration of limit cycles in a first-order IIR digital filter: (a) $\alpha = 0.6$, and (b) $\alpha = -0.6$.

```
alpha = 0.6;yi = 0; x = 0.04;
for n = 1:21
    y(n) = (round(100*alpha*yi)/100) + x;
    yi = y(n); x = 0;
end
k = 0:20;
stem(k,y)
ylabel('Amplitude');
xlabel('Time index n')
```

Figure 9.48 shows the plots of the first 21 samples of the output response of the first-order IIR digital filter of Eq. (9.1) implemented using a decimal fractional arithmetic with products rounded to two digits after the decimal point. The input is $x[n] = 0.04\delta[n]$ and the filter coefficients are $\alpha = \pm 0.6$. The initial condition $y[-1]$ is set to 0.

Example 9.15 In the case of a second-order IIR filter, a similar situation exists, except there are various types of limit cycles here. Consider the all-pole second-order IIR digital filter of Figure 9.23(a).

Figure 9.49   A second-order IIR section with quantizers after the multipliers.

It is described by the difference equation

$$y[n] = -\alpha_1 y[n-1] - \alpha_2 y[n-2] + x[n]. \tag{9.213}$$

The poles of this system are at

$$z = -\frac{\alpha_1}{2} - \frac{\sqrt{\alpha_1^2 - 4\alpha_2}}{2}, \tag{9.214}$$

which are complex for $\alpha_1^2 < 4\alpha_2$ and on the unit circle if $\alpha_2 = 1$. If we assume the products are first quantized by rounding, as indicated in Figure 9.49, the above difference equation becomes

$$\hat{y}[n] = \mathcal{Q}\left(\alpha_1 \hat{y}[n-1]\right) - \mathcal{Q}\left(\alpha_2 \hat{y}[n-2]\right) + x[n], \tag{9.215}$$

with $\hat{y}[n]$ denoting the actual output. One way the system of Figure 9.49 could be in the limit cycle mode with zero input is if its effective pole is on the unit circle. In this case,

$$\mathcal{Q}\left(\alpha_2 \hat{y}[n-2]\right) = \hat{y}[n-2]. \tag{9.216}$$

Moreover, because of rounding, we have

$$\left|\mathcal{Q}\left(\alpha_2 \hat{y}[n-2]\right) - \alpha_2 \hat{y}[n-2]\right| \le \frac{\delta}{2}. \tag{9.217}$$

From Eqs. (9.216) and (9.217), we arrive at the dead band region governing the limit cycle mode:

$$\left|\hat{y}[n-2]\right| \le \frac{\delta}{2\left|1+\alpha_2\right|}, \tag{9.218}$$

with α_1 determining the frequency of oscillation.

Limit cycles can also occur in second-order IIR filters if the system with quantizers has effective poles at $z = \pm 1$. In this mode the dead band is bounded by $2\delta/(1-|\alpha_1|+\alpha_2)$ [Jac69].

It should be noted that the limit cycles occurring with the amplitude bound of Eq. (9.218) are basically inaccessible limit cycles, and it is highly unlikely, in practice, that the digital filter will start with initial conditions pertaining to these limit cycles [Cla73]. On the other hand, for the second-order IIR filter structures of Figure 9.49, with arbitrary initial conditions, accessible limit cycles are highly likely to occur and their amplitude can only be bounded from below:

$$\left|\hat{y}[n-2]\right| \ge \frac{\delta}{2\left|1+\alpha_2\right|}. \tag{9.219}$$

Figure 9.50   A second-order IIR section with a quantizer after the accumulator.

Even though we have considered here the zero-input limit cycle generation in very simple IIR digital filter structures, limit cycles also occur in higher order structures. However, their analysis is almost impossible, except for the determination of the bounds on the amplitudes of the limit cycles [Lon73]. In some structures, periodic limit cycles have been observed with nonzero constant amplitudes and also for constant amplitude sinusoidal inputs.

9.9.2   Overflow Limit Cycles

As indicated earlier, limit-cycle-like oscillations can also result from overflow in digital filters implemented with finite-precision arithmetic. The amplitude of the overflow oscillations can cover the whole dynamic range of the register experiencing the overflow and are much more serious in nature than the granular type. We illustrate the generation of overflow oscillation by means of an example.

Example 9.16   We consider the causal all-pole second-order IIR digital filter of Figure 9.23(a). We assume its implementation using two's-complement arithmetic with a rounding of the sum of products by a single quantizer, as indicated in Figure 9.50. In this case, the linear difference equation of Eq. (9.213) describing the ideal filter reduces to the nonlinear difference equation given by

$$\hat{y}[n] = \mathcal{Q}\left(-\alpha_1 \hat{y}[n-1] - \alpha_2 \hat{y}[n-2] + x[n]\right), \tag{9.220}$$

where $\mathcal{Q}(\cdot)$ represents the rounding operation and $\hat{y}[n]$ denotes the actual output of the filter. All numbers are assumed to be signed 4-bit fractions. In performing the rounding, we round up a number that is exactly halfway.

Let the filter coefficients be given by $\alpha_1 = 1_\triangle 001 = -0.875_{10}$ and $\alpha_2 = 0_\triangle 111 = 0.875_{10}$. The initial conditions are assumed to be $\hat{y}[-1] = 0_\triangle 110 = 0.75_{10}$ and $\hat{y}[-2] = 1_\triangle 010 = -0.75_{10}$. We consider the zero-input case, i.e., $x[n] = 0$ for $n \geq 0$. We compute below the first few samples of the output $\hat{y}[n]$.

At $n = 0$ we get

$$\hat{y}[0] = \mathcal{Q}(-1_\triangle 001 \times 0_\triangle 110 - 0_\triangle 111 \times 1_\triangle 010)$$
$$= \mathcal{Q}(1_\triangle 010100) = 1_\triangle 011 = -0.625_{10}.$$

Next, at $n = 1$, we obtain

$$\hat{y}[1] = \mathcal{Q}(-1_\triangle 001 \times 1_\triangle 011 - 0_\triangle 111 \times 1_\triangle 110)$$
$$= \mathcal{Q}(10_\triangle 110011) = 0_\triangle 011 = 0.675_{10}.$$

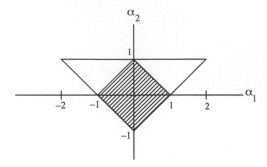

Figure 9.51   Coefficient ranges of a second-order direct form IIR filter to guarantee no overflow oscillations (shown by shaded region).

In the above, the sum of the two products has resulted in a carry bit to the left of the sign bit that is automatically lost, resulting in a positive number as indicated. Following the above procedure, we arrive at

$$\hat{y}[2] = 1_\triangle 010 = -0.75_{10},$$
$$\hat{y}[3] = 0_\triangle 110 = 0.75_{10},$$
$$\hat{y}[4] = 1_\triangle 011 = -0.625_{10}.$$

As is evident from these output samples, the digital filter exhibits a large amplitude overflow limit cycle of period 4 at its output.

For stability we have shown in Section 4.8.1 that the filter coefficients of the second-order direct form IIR structure of Figure 9.50 must remain inside the stability triangle of Figure 9.51. However, the structure can still get into a zero-input overflow oscillation mode for a large range of values of the filter constants, satisfying this stability constraint when implemented using two's-complement arithmetic with rounding. It has been shown that overflow limit cycles under zero-input cannot occur if the filter coefficients lie in the shaded region inside the stability triangle, as indicated in Figure 9.51 [Ebe69]. This region is defined by the relation

$$|\alpha_1| + |\alpha_2| < 1. \tag{9.221}$$

As the above condition is quite restrictive, we now examine other IIR structures that do not sustain limit cycles with less severe constraint on the pole locations.

9.9.3   Limit Cycle Free Structures

A number of authors have advanced structures that are limit cycle free when implemented using specific arithmetic schemes. The most general approach to the development of such structures is based on the state-space realization [Mil78]. Consider the state-space structure of Figure 9.45 characterized by the state-equations of Eqs. (9.183a) and (9.183b). Even though in an actual implementation the right-hand sides of both equations are quantized, the quantization errors caused by the quantization of the right-hand side of Eq. (9.183a) go through the feedback loop and are responsible for the generation of the limit cycles. We

Figure 9.52   The feedback loop with a quantizer in the state-space structure.

therefore redraw the feedback loop part of the state-space structure showing the quantizer block explicitly, as in Figure 9.52.

The equations governing the operation of the feedback loop are

$$\mathbf{v}[n + 1] = \mathbf{A}\mathbf{s}[n], \tag{9.222a}$$

$$\mathbf{s}[n + 1] = \mathcal{Q}(\mathbf{v}[n + 1]). \tag{9.222b}$$

We assume that Eq. (9.222a) is implemented with infinite precision. Moreover, the limit cycle can arise only because of the quantization operation carried out in Eq. (9.222b). It should be noted that the operation indicated in Eq. (9.222b) is carried out component-wise as a scalar operation:

$$s_\ell[n + 1] = \mathcal{Q}(v_\ell[n + 1]), \qquad \ell = 1, 2, \ldots, N. \tag{9.223}$$

We define a quantizer to be *passive* if

$$|\mathcal{Q}(x)| \leq |x|, \qquad \text{for all } x. \tag{9.224}$$

If x is inside the given dynamic range of the system, then for magnitude truncation (Section 9.1.1) it is evident that Eq. (9.224) holds, i.e., the quantizer is passive. If x is outside the dynamic range caused, for example, by overflow, it must be brought back to the range by following either the saturation arithmetic scheme or the two's-complement overflow scheme discussed in Section 8.6. As a result, magnitude truncation followed by either of the above two overflow handling schemes is again a passive quantizer.

The passivity of a scalar quantizer can be generalized to handle the matrix quantization operation indicated in Eq. (9.222b). It has been shown [Mil78] that in state-space structures both zero-input granular and overflow limit cycles can be totally eliminated using passive scalar quantizers if the following condition holds:

$$(\mathbf{A}\mathbf{w})^T \mathbf{A}\mathbf{w} < \mathbf{w}^T \mathbf{w}, \tag{9.225}$$

where $\mathbf{w}$ is any arbitrary nonzero vector. The above equation implies that the energy in the vector $\mathbf{A}\mathbf{w}$ is strictly less than the energy of the vector $\mathbf{w}$. A state-transition matrix $\mathbf{A}$ satisfying Eq. (9.225) is defined to be *strictly passive*.

Applying condition of Eq. (9.225) to Eq. (9.222a), we observe that for a strictly passive $\mathbf{A}$, the energy of $\mathbf{v}[n + 1]$ is strictly less than the energy of $\mathbf{s}[n]$. Moreover, if the scalar quantizers being used to generate $\mathbf{s}[n + 1]$ are assumed to be passive, then the energy of $\mathbf{s}[n + 1]$ must be less than or equal to the energy of $\mathbf{v}[n + 1]$. As a result, the energy of $\mathbf{s}[k]$ is a monotonically decreasing function of the time index k. In addition, if δ denotes the quantization level, the smallest amount of decrease in the energy levels of $\mathbf{s}[k]$ is δ^2.

Consequently, the energy of $\mathbf{s}[k]$ goes to zero after a finite number of iterations, ensuring that the state-space structure cannot sustain any limit cycles.

Let λ_{max} denote the eigenvalue of $\mathbf{A}$ with the largest magnitude.[4] It can be shown that the *induced $\mathcal{L}_2$-norm* $\|\mathbf{A}\|_2$ satisfy the the inequality [Gol89]

$$(\|\mathbf{A}\|_2)^2 = \max_{\mathbf{w}\neq 0} \frac{(\mathbf{Aw})^T \mathbf{Aw}}{\mathbf{w}^T \mathbf{w}} \leq |\lambda_{max}|^2 . \tag{9.226}$$

If the state-transition matrix $\mathbf{A}$ of a state-space realization satisfies the above condition with an equality sign, then it also satisfies Eq. (9.225). The corresponding realization is called a *minimum-norm structure*. Such structures with passive quantizers do not support zero-input limit cycles of either type. A special type of minimum-norm structure is the so-called *normal form structure* for which the following holds:

$$\mathbf{A}^T \mathbf{A} = \mathbf{A}\mathbf{A}^T . \tag{9.227}$$

The matrix $\mathbf{A}$ satisfying above condition and $\|\mathbf{A}\|_2 < 1$ is called a *normal matrix*.

Example 9.17   Consider the digital filter structure of Figure 9.53 [Yan82]. Analysis yields

$$s_1[n + 1] = cs_1[n] - cds_2[n] + cx[n], \tag{9.228a}$$
$$s_2[n + 1] = cds_1[n] + cs_2[n]. \tag{9.228b}$$

From the above, we easily identify the state transition matrix as

$$\mathbf{A} = \begin{bmatrix} c & -cd \\ cd & c \end{bmatrix}. \tag{9.229}$$

Comparing the determinant of $(z\mathbf{I} - \mathbf{A})$ with the denominator of a second-order IIR transfer function with poles at $z = re^{\pm j\theta}$ (with $r < 1$ for stability), we obtain

$$c = r \cos\theta, \qquad d = \tan\theta. \tag{9.230}$$

Substituting the above values in Eq. (9.229), we arrive at

$$\mathbf{A} = \begin{bmatrix} r\cos\theta & -r\sin\theta \\ r\sin\theta & r\cos\theta \end{bmatrix}. \tag{9.231}$$

Note that $\mathbf{A}^T\mathbf{A} = \mathbf{A}\mathbf{A}^T = r^2\mathbf{I}$. Since $r < 1$, we have $\|\mathbf{A}\|_2 = r < 1$. Therefore, the filter of Figure 9.53 is a normal form structure.

9.9.4   Random Rounding

A conceptually simple technique to suppress zero-input granular limit cycles in second-order IIR sections is called random rounding [But77], [Law78]. To explain the operation of this method, let the register storing the signal $v[n]$ prior to the quantization be as indicated in Figure 9.54. In the random rounding method, an uncorrelated binary random sequence $x[n]$ is generated externally by a random number generator, and the $(b + 1)$th bit a_{-b-1} of $v[n]$ is replaced with $x[n]$. The modified signal $v[n]$ is then passed through a passive quantizer. In addition to the increase in hardware complexity, this approach also results in a slightly higher quantization noise that is, however, distributed over the whole frequency range.

---

[4]Note that $|\lambda_{max}| < 1$ for a stable system.

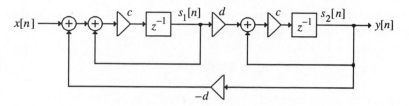

Figure 9.53   A second-order IIR section.

Figure 9.54   Register storing the signal to be quantized.

9.10   Round-Off Errors in FFT Algorithms

Since FFT is often employed in a number of digital signal processing applications, it is of interest to analyze the effects of finite wordlengths in FFT computations. The most critical error in the computation is that due to the arithmetic round-off errors. As in the earlier sections, we assume that the DFT computations are being carried out using fixed-point arithmetic, and we thus restrict our analysis to the effect of product round-off errors in the DFT computation via the FFT algorithms and to compare them with the errors generated in the direct DFT computation [Opp89], [Pro92], [Wel69]. The model for the round-off error analysis to be employed here is the same as in the case of LTI digital filters and is shown in Figure 9.18(b).

Now the multiplier coefficients W_N^{kn} and the signals in the DFT computation are, in general, complex numbers resulting in complex multiplications. Since each complex multiplication usually requires 4 real multiplications, there are 4 quantization errors per multiplication. If the DFT computation requires K complex multiplications, there are $4K$ sources of quantization errors in the computational structure. We make the usual assumptions about the statistical properties of the noise sources:

(a) All $4K$ errors are uncorrelated with each other and uncorrelated with the input sequence.

(b) The quantization errors are random variables uniformly distributed with a variance $\sigma_0^2 = 2^{-2b}/12$, assuming a signed b-bit fractional fixed-point arithmetic.

9.10.1   Direct DFT Computation

Recall that the N-point DFT $X[k]$ of a length-N complex sequence $x[n]$ is given by

$$X[k] = \sum_{n=0}^{N-1} x[n]W_N^{nk}, \qquad 0 \le k \le N - 1. \qquad (9.232)$$

Thus, the computation of a single DFT sample requires N complex multiplications, and hence, the total number of real multiplications for the computation of a single DFT sample is $4N$. As a result, there are $4N$ quantization error sources. The variance of the error in the computation of one DFT sample is therefore[5]

$$\sigma_\gamma^2 = 4N\sigma_0^2 = \frac{2^{-2b}N}{3}, \tag{9.233}$$

indicating that the output round-off error is proportional to the DFT length.

Now, the input sequence $x[n]$ must be scaled to avoid overflow in the computation of $X[k]$. From Eq. (9.232) it follows that

$$|X[k]| \leq \sum_{n=0}^{N-1} |x[n]| < N, \tag{9.234}$$

assuming that the input samples satisfy the dynamic range constraint $|x[n]| \leq 1$. To prevent overflow we need to ensure that

$$|X[k]| < 1, \tag{9.235}$$

which can be guaranteed by dividing each sample $x[n]$ in the input sequence by N.

To analyze the effect of the above scaling, assume the input to be a white noise sequence with each sample uniformly distributed in the range $(-1/N, 1/N)$ [Pro92]. The input signal power is then given by

$$\sigma_x^2 = \frac{(2/N)^2}{12} = \frac{1}{3N^2}. \tag{9.236}$$

The corresponding output signal power is

$$\sigma_X^2 = N\sigma_x^2 = \frac{1}{3N}. \tag{9.237}$$

As a result, the signal-to-noise ratio is

$$SNR = \frac{\sigma_X^2}{\sigma_\gamma^2} = \frac{2^{2b}}{N^2}. \tag{9.238}$$

The above expression indicates that the SNR has been reduced by a factor of N^2 due to scaling and the round-off error. The wordlength size needed to compute a DFT of given length with a desired SNR can be determined using Eq. (9.238) (Problem 9.36).

9.10.2   DFT Computation via FFT Algorithm

We now consider the round-off error analysis of the DFT computation based on an FFT algorithm. Without any loss of generality, we analyze the decimation-in-time radix-2 FFT algorithm. However, the results derived can be easily extended to other types of fast DFT algorithms.

---

[5] Strictly speaking, the output noise variance will be less than the value here since multiplications with W_N^0 and $W_N^{N/2}$ do not develop any error.

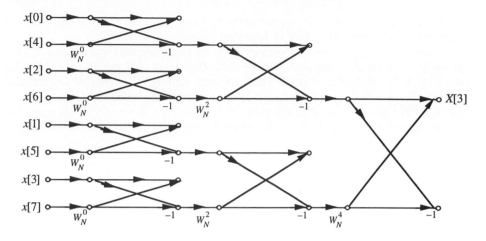

Figure 9.55   Reduced flow-graph for the computation of $X[3]$.

From the flow-graph of the DIT FFT algorithm given in Figure 8.25, it can be seen that the DFT samples are computed by a series of butterfly computations with a single complex multiplication per butterfly module. Some of the butterfly computations require multiplications by ± 1 or $\pm j$ that we do not treat separately here to simplify the analysis.

Consider now the computation of a single DFT sample as indicated in Figure 9.55. It follows from this figure that the computation of a single DFT sample involves $\nu = \log_2 N$ stages. The number of butterflies in a particular stage depend on its location in the computational chain with $N/2^r = 2^{\nu-r}$ butterflies in the rth stage, where $r = 1, 2, \ldots, \nu$. The total number of butterflies involved per DFT sample is therefore

$$1 + 2 + 2^2 + \cdots + 2^{\nu-2} + 2^{\nu-1} = 2^\nu - 1 = N - 1. \tag{9.239}$$

It also follows from Figure 9.55 that the quantization errors introduced at the rth stage appear at the output after propagating through $(r - 1)$ stages while getting multiplied by the twiddle factors at each subsequent stages. Since the magnitude of the twiddle factors is always unity, the variances of the quantization errors do not change while propagating to the output. The total number of error sources contributing to the output round-off error is $4(N - 1)$. Assuming that the quantization errors introduced in each butterfly are uncorrelated with those generated at other butterflies, the variance of the output round-off error is then

$$\sigma_\gamma^2 = 4(N - 1)\frac{2^{-2b}}{12} \cong \frac{2^{-2b}N}{3}, \tag{9.240}$$

It should be noted that the above expression for σ_γ^2 is identical to that derived for the direct DFT computation case given by Eq. (9.233). This is to be expected since the FFT algorithm does not alter the total number of complex multiplications to compute a single DFT sample, rather it organizes the computations more efficiently so that the number of multiplications to compute all N DFT samples is reduced.

Now, to prevent overflow at the output if we scale the input samples to satisfy the condition $|x[n]| < 1/N$, the SNR obtained in the FFT algorithm remains the same as given in Eq. (9.238). However, in the case of the FFT algorithm, the SNR can be improved by following a different scaling rule as described below.

Instead of scaling the input samples by $1/N$, we can scale the input signals at each stage by $1/2$. This scaling rule also guarantees that the output DFT samples are scaled by a factor of $(1/2)^{\nu} = 1/N$ as desired. However, each scaling by a factor of $1/2$ reduces the round-off noise variance by a factor of $1/4$. As a result, the round-off noise variances of the $4(2^{\nu-r})$ noise sources at the rth stage are reduced by a factor of $1/4^{r-1}$, while the noise propagates to the output. It can be shown that the total round-off noise variance at the output is now given by

$$\sigma_{\gamma}^2 = \frac{2}{3} \cdot 2^{-2b}(1 - \frac{1}{N}) \cong \frac{2}{3} \cdot 2^{-2b}, \tag{9.241}$$

assuming N to be large enough so that $1/N \ll 1$ (Problem 9.37). The SNR in this case reduces to

$$\frac{\sigma_X^2}{\sigma_{\gamma}^2} = \frac{2^{2b}}{2N}, \tag{9.242}$$

Hence, distribution of the scaling into each stage has increased the SNR by a factor of N. The formula given in Eq. (9.242) can be used to determine the wordlength required to achieve a specified SNR for computing DFT of a given length N (Problem 9.38).

9.11   Summary

This chapter is concerned with the effects of the finite wordlengths caused by the actual implementation of a digital signal processing algorithm that causes the results of the algorithm to be different, in general, from the desired ones obtained in the ideal case with infinite-precision wordlengths. To develop the appropriate models for the analysis of these effects prior to actual implementation, we first review the quantization process used to fit the infinite-precision data into a finite wordlength register and the resulting errors caused by it. The quantization of both fixed-point and floating-point numbers are considered. However, the discussion in the rest of the chapter is restricted to fixed-point implementations.

The effect of the quantization of the multiplier coefficients in the implementation of a digital filter is considered next as a coefficient sensitivity problem. Simple formulas are derived for such sensitivity analysis for the infinite-impulse response (IIR) filter and the finite-impulse response (FIR) filter.

In the digital processing of a continuous-time signal, the latter is sampled periodically by a sample-and-hold device and then converted into a digital form by an analog-to-digital (A/D) converter. A statistical model is developed for the analysis of the input quantization error caused by the A/D conversion and is used to derive the expression for the signal-to-quantization ratio as a function of the A/D converter wordlength. The A/D quantization error propagates to the output of the digital filter processing the digitized continuous-time signal, appearing as a noise added to the desired output and methods for the statistical analysis of the output error are provided.

The effect of product round-off in the fixed-point implementation of a digital filter is then analyzed using a statistical model. A statistical analysis of the output noise caused by the propagation of these internally generated errors also to the output of the digital filter is provided. An overflow may occur at certain internal nodes in a digital filter implemented in fixed-point arithmetic, which can result in a large amplitude oscillation at the filter output. Methods of scaling the internal signal variables with the aid of suitably placed scaling multipliers to minimize the probability of overflow are discussed. The performance of the cascade form of digital filters is particularly examined in depth to illustrate the effect of pole-zero pairing and the ordering of the low-order sections. A detailed analysis of the output signal-to-noise ratios of scaled first-order and second-order IIR digital filter sections is then provided.

Conditions for the low passband sensitivity realizations of IIR and FIR digital filters are next derived, and a method for such low sensitivity realization for each case is outlined. Two approaches to the reduction of round-off errors in IIR digital filter structures are then described.

The discretization process can also cause the occurrence of periodic oscillation, called limit cycles, at the output of an IIR digital filter under certain conditions. These limit cycles are difficult to analyze in the general case. Their existence is demonstrated here only for the first-order and second-order IIR digital filters. Conditions for the limit-cycle-free operation of a state-space structure is derived.

The chapter concludes with an analysis of round-off errors in the implementation of DFT and FFT algorithms.

9.12 Problems

9.1 Derive the ranges of the relative quantization errors of Table 9.2.

9.2 The digital filter structure of Figure 9.53 is more commonly known as a *modified coupled-form structure* [Yan82]. Another modified coupled-form structure is shown in Figure P9.1. Determine the transfer function of both structures and then compute their respective pole sensitivities. Compare these sensitivities with those of the structure of Figure 9.9 as given in Eq. (9.46).

Figure P9.1

9.3 Determine the transfer function of the second-order digital filter structure shown in Figure P9.2 and compute its pole sensitivities [Aga75]. Compare these sensitivities with those of the structures of Figure 9.9 and Figure P9.1.

9.4 A third-order elliptic lowpass transfer function

$$H(z) = \frac{0.05634(z+1)(z^2 + 1.0166z + 1)}{(z - 0.683)(z^2 - 1.4461z + 0.7857)},$$

is realized in (1) direct form, and (2) cascade form. Compute the pole-sensitivities of each structure.

9.5 Consider the digital filter structure of Figure P9.3 characterized by a transfer function $H(z) = Y_1/X_1$. Show that the sensitivity of $H(z)$ with respect to the multiplier coefficient α defined by $\partial H(z)/\partial \alpha$ is given by

$$\frac{\partial H(z)}{\partial \alpha} = F_\alpha(z) \cdot G_\alpha(z),$$

| Figure P9.2 | Figure P9.3 |

where $F_\alpha(z)$ is the scaling transfer function from the input node X_1 to the input node Y_2 of the multiplier α and $G_\alpha(z)$ is the noise transfer function from the output node X_2 of the multiplier α to the filter output node Y_1.

9.6 Show that the function $W_M(\omega)$ defined in Eq. (9.58) satisfies the following properties:

(a) $0 < W_M(\omega) \leq 1$,

(b) $W_M(0) = W_M(\pi) = 1$ for all M,

(c) $\lim_{M\to\infty} W_M(\omega) = \frac{1}{\sqrt{2}}$, $0 < \omega < \pi$.

9.7 Verify the SNR values given in Table 9.3.

9.8 Determine the output noise variance due to the propgation of the input quantization noise for each of the following causal IIR digital filters:

(a) $H_1(z) = \dfrac{(z+2)(z+1)}{(z+\frac{1}{2})(z+\frac{1}{3})}$,

(b) $H_2(z) = \dfrac{6(5z+2)(0.3z^2+0.6z+1)}{(2z+1)(3z+1)(z^2+0.6z+0.3)}$,

(c) $H_3(z) = \dfrac{(z+1)^2}{(z^2+0.5z+0.8)}$.

9.9 Determine the expression for the normalized output noise variance due to input quantization of the following digital filter realized in a parallel form

$$H(z) = C + \frac{A}{1-\alpha z^{-1}} + \frac{B}{1-\beta z^{-1}}.$$

Each section in the parallel structure is realized in direct form II. What is the value of the variance for $\alpha = 0.9$, $\beta = -0.8$, $A = 2$, $B = 3$, $C = 2$.

9.10 Realize the following transfer function

$$H(z) = \frac{(z+3)(z+5)}{(z+\frac{1}{3})(z+\frac{1}{2})},$$

in four different cascade forms with each first-order stage implemented in direct form II.

(a) Show the noise model for each unscaled structure for the computation of the product round-off noise at the output assuming quantization of products before addition assuming fixed-point implementation with either rounding or two's-complement truncation. Compute the normalized output round-off noise variance for each realization. Which cascade realization has the lowest round-off noise?

(b) Repeat part (a) assuming quantization after addition of product.

9.11 Realize the transfer function of Problem 9.10 in two different parallel forms with each first-order stage implemented in direct form II.

(a) Show the noise model for each unscaled structure for the computation of the product round-off noise at the output assuming quantization of products before addition and assuming fixed-point implementation with either rounding or two's-complement truncation. Compute the normalized output round-off noise variance for each realization. Which parallel realization has the lowest round-off noise?

(b) Repeat part (a) assuming quantization after addition of product.

9.12 Realize the following second-order transfer function

$$H(z) = \frac{z^{-1} + 0.2z^{-2}}{1 - 0.2z^{-1} - 0.24z^{-2}},$$

in (1) direct form, (2) cascade form, and (3) parallel form. Each section in the cascade and parallel structures is realized in direct form II. Show the noise model for each unscaled structure for the computation of the product round-off noise at the output assuming quantization of products before addition and assuming fixed-point implementation with either rounding or two's-complement truncation. Compute the product round-off noise variance for each realization. Which realization has the lowest round-off noise? *Note*: There are two cascade and two parallel realizations.

9.13 Realize the transfer function of Problem 9.12 in Gray-Markel form. (a) Show the noise model for the unscaled structure for the computation of the product round-off noise variance at the output assuming quantization of products before addition and compute its product round-off noise variance. (b) Repeat part (a) assuming quantization after addition of product.

9.14 One possible two-multiplier realization of a second-order Type 2 allpass transfer function

$$A_2(z) = \frac{d_1 d_2 + d_1 z^{-1} + z^{-2}}{1 + d_1 z^{-1} + d_1 d_2 z^{-2}}$$

is shown in Figure P9.4. Derive the expression for the normalized steady-state output noise variance due to product round-off assuming fixed-point implementation with either rounding or two's-complement truncation.

Figure P9.4

9.15 The allpass section of Figure P9.4 is employed to alter the phase response of the structure realizing the transfer function $H(z)$ of Problem 9.10 as indicated in Figure P9.5. The allpass equalizer of Figure P9.5 has a transfer function

$$A_2(z) = \frac{0.9 - 0.4z^{-1} + z^{-2}}{1 - 0.4z^{-1} + 0.9z^{-2}}.$$

Compute the normalized steady-state output noise variance due to product round-off of the phase-equalized structure of Figure P9.5 if $H(z)$ is realized in a cascade form with the lowest product round-off noise.

Figure P9.5

9.16 Consider the digital filter structure of Figure P9.6 that is assumed to be implemented using 9-bit signed two's-complement fixed-point arithmetic with all products quantized before additions. Draw the linear noise model of the unscaled system and compute its total output noise power.

Figure P9.6

9.17 Scale the first-order digital filter structure of Figure P9.7 using the $\mathcal{L}_2$-norm scaling rule.

Figure P9.7

9.18 Scale the second-order digital filter structure of Figure P9.8 using the $\mathcal{L}_2$-norm scaling rule.

Figure P9.8

9.19 Scale the structure realized in Problem 9.9 using the $\mathcal{L}_2$-norm scaling rule and then compute its output noise variance due to product round-off assuming quantization of products before addition.

9.20 Scale the structures realized in Problem 9.10 using the $\mathcal{L}_2$-norm scaling rule and then compute the output noise variances due to product round-off assuming quantization of products before addition. What would be the output noise variances if quantization is carried out after addition?

9.21 Scale the structures realized in Problem 9.11 using the $\mathcal{L}_2$-norm scaling rule and then compute the output noise variances due to product round-off assuming quantization of products before addition. What would be the output noise variances if quantization is carried out after addition?

9.22 Scale the structure realized in Problem 9.12 using the $\mathcal{L}_2$-norm scaling rule and then compute the output noise variance due to product round-off assuming quantization of products before addition. What would be the output noise variance if quantization is carried out after addition?

9.23 (a) Calculate the output noise variance of the digital filter structure of Figure P9.9(a) due to product round-off before addition. Assume all numbers are fractions and represented in a two's-complement fixed-point representation with a wordlength of $b + 1$ bits. Note that the multiplier '-1' does not generate noise. (b) Now consider the configuration of Figure P9.9(b), where the filters $A_1(z)$, $A_2(z)$, and $A_3(z)$ are implemented as in Figure P9.9(a) with the multiplier coefficients d_i replaced with d_1, d_2, and d_3, respectively. Calculate the output noise variance of this new digital filter structure due to product round-off before addition.

Figure P9.9

9.24 Consider the digital filter structure of Figure P9.10. Let σ_0^2 represent the total output noise variance due to product round-off in $G(z)$ and $A(z)$. If each delay in the realization of $G(z)$ is replaced with two delays, i.e., z^{-1} replaced with z^{-2}, but $A(z)$ is left unchanged, calculate the output noise variance due to product round-off in terms of σ_0.

Figure P9.10

9.25 Show that there are $(R!)^2$ different possible realizations of cascade of R second-order sections.

9.26 (a) What is the optimum pole-zero pairing and ordering of each of the following transfer functions for obtaining the smallest peak output noise due to product round-off under an $\mathcal{L}_2$-scaling rule? (b) Repeat part (a) if the objective is to minimize the output noise power due to product round-off under an $\mathcal{L}_\infty$-scaling rule.

(i) $H_1(z) = \dfrac{(z^2 + 0.8z + 0.2)(z^2 + 0.2z + 0.9)(z^2 + 0.3z + 0.5)}{(z^2 + 0.1z + 0.8)(z^2 + 0.2z + 0.4)(z^2 + 0.6z + 0.3)}$,

(ii) $H_2(z) = \dfrac{(z^2 + 0.1z + 0.7)(z^2 + 1.4z + 0.33)(z^2 + 0.5z + 0.6)}{(z^2 + 0.4z + 0.7)(z^2 - 0.2z + 0.9)(z^2 + 1.1z + 0.18)}$.

9.27 Derive the expressions for the SNR given in Table 9.5.

9.28 In Section 6.8.1, we showed that the first-order lowpass transfer function $H_{LP}(z)$ of Eq. (4.64) and the first-order highpass transfer function $H_{HP}(z)$ of Eq. (4.67) can be realized in the form of a parallel allpass structure of Figure 6.45, where $A_1(z)$ is a first-order allpass transfer function given by Eq. (6.89). The first-order allpass section can be implemented using a single multiplier α using any one of the four single-multiplier allpass structures of Figure 6.36. Show that the realization of $H_{LP}(z)$ in the form of Figure 6.45 exhibits low sensitivity in the passband at $\omega = 0$ with respect to the multiplier coefficient α, i.e.,

$$\frac{\partial \left| H_{LP}(e^{j\omega}) \right|}{\partial \alpha} \Bigg|_{\omega=0} = 0.$$

Likewise, show that the realization of $H_{HP}(z)$ in the form of Figure 6.45 exhibits low sensitivity in the passband at $\omega = \pi$ with respect to the multiplier coefficient α, i.e.,

$$\frac{\partial \left| H_{HP}(e^{j\omega}) \right|}{\partial \alpha} \Bigg|_{\omega=\pi} = 0.$$

9.29 In Section 6.8.2, we showed that the second-order bandpass transfer function $H_{BP}(z)$ of Eq. (4.68) and the second-order bandstop transfer function $H_{BS}(z)$ of Eq. (4.73) can be realized in the form of a parallel allpass structure of Figure 6.48, where $A_2(z)$ is a second-order allpass transfer function given by Eq. (6.92). The second-order allpass section can be implemented using two multipliers α and β by employing a cascaded lattice structure described in Section 6.72. Show that the realization of $H_{BP}(z)$ in the form of Figure 6.48 exhibits low sensitivity in the passband at the center frequency ω_0 with respect to the multiplier coefficients α and β, i.e.

$$\frac{\partial \left| H_{BP}(e^{j\omega}) \right|}{\partial \alpha} \Bigg|_{\omega=\omega_0} = 0, \qquad \frac{\partial \left| H_{BP}(e^{j\omega}) \right|}{\partial \beta} \Bigg|_{\omega=\omega_0} = 0.$$

Likewise, show that the realization of $H_{BS}(z)$ in the form of Figure 6.48 exhibits low sensitivity in the passband at $\omega = 0$ and $\omega = \pi$ with respect to the multiplier coefficients α and β, i.e.,

$$\frac{\partial \left| H_{BS}(e^{j\omega}) \right|}{\partial \alpha} \Bigg|_{\omega=0} = 0, \qquad \frac{\partial \left| H_{BS}(e^{j\omega}) \right|}{\partial \beta} \Bigg|_{\omega=0} = 0,$$

$$\frac{\partial \left| H_{BS}(e^{j\omega}) \right|}{\partial \alpha} \Bigg|_{\omega=\pi} = 0, \qquad \frac{\partial \left| H_{BS}(e^{j\omega}) \right|}{\partial \beta} \Bigg|_{\omega=\pi} = 0.$$

9.30 In the parallel allpass realization of a bounded real (BR) transfer function $G(z)$, passband of $G(z)$ is the stopband of its power-complementary transfer function $H(z)$. Show why low passband sensitivity of $G(z)$ does not imply low stopband sensitivity of $H(z)$.

9.31 Develop the expressions for the SNR of the second-order IIR filter structure of Figure 9.43 with and without error feedback for the following types of inputs: (a) WSS, white uniform density, (b) WSS, white Gaussian density ($\sigma_x = 1/3$), and (c) sinusoid with known frequency. The poles are at $z = (1 - \varepsilon)e^{\pm j\theta}$ with $\varepsilon \to 0$, $\theta \to 0$, and $\theta \gg \varepsilon$. Assume a $(b + 1)$-bit signed representations for the number representation.

9.32 Show that the coupled-form structure of Figure 9.9 will not support zero-input limit cycles under magnitude quantization.

9.33 Show that the modified coupled-form structure of Figure P9.1 will not support zero-input limit cycles under magnitude quantization.

9.34 Show that the second-order structure of Figure P9.11 will not support zero-input limit cycles under magnitude quantization [Mee76].

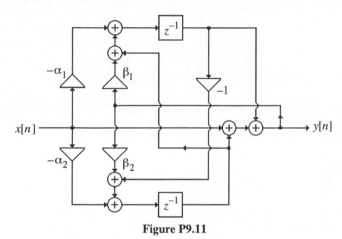

Figure P9.11

9.35 Determine the expression for the output round-off noise variance in the computation of a single sample of a length-N DFT using the Goertzel's algorithm implemented in a fractional signed $(b + 1)$-bit fixed-point arithmetic.

9.36 Determine the number of bits b to compute a single sample of a 512-point DFT of an input sequence of length 512 by direct computation with a SNR of 25 dB.

9.37 Derive Eq. (9.241).

9.38 Determine the number of bits needed to compute a single sample of a 512-point DFT of an input sequence of length 512 using a radix-2 decimation-in-time FFT algorithm with a SNR of 25 dB. Assume a distributed scaling to prevent overflow at the output.

9.13 MATLAB Exercises

M 9.1 Using MATLAB design an elliptic lowpass IIR filter with the following specifications: passband edge at 0.3π, stopband edge at 0.35π, passband ripple of 0.01 dB, and minimum stopband attenuation of 50 dB. Modify the transfer function by increasing its coefficients by +5%. Plot the magnitude responses of both filters in one figure. Use a solid line for the original filter and a dashed line for the modified structure. Next, using MATLAB develop the pole-zero plots of the two transfer functions in one composite figure. Use different symbols to identify the poles and zeros.

M 9.2 Determine the factored form of a fifth order elliptic lowpass transfer function with the following specifications: passband ripple of 1.0 dB, minimum stopband attenuation of 50 dB, and a passband edge at 0.4. From its pole locations, determine the optimum pole-zero pairing and their ordering to minimize the output noise power under a $\mathcal{L}_\infty$-scaling rule. Verify your result using MATLAB.

M 9.3 Realize the transfer function of Problem M9.2 as a parallel connection of two allpass filters and determine its power-complementary highpass transfer function. Plot on the same figure the gain responses of the two filters. Verify the low passband sensitivity of the parallel allpass realization by making a 1% change in the filter coefficients.

M 9.4 Write a MATLAB program to simulate a fourth-order transfer function of the form

$$H(z) = \frac{(1 + b_1 z^{-1} + b_2 z^{-2})(1 + b_3 z^{-1} + b_4 z^{-2})}{(1 + a_1 z^{-1} + a_2 z^{-2})(1 + a_3 z^{-1} + a_4 z^{-2})} \tag{9.243}$$

in a cascade form with each second-order section realized in direct form II. The input data to your program are the numerator coefficients $\{b_i\}$ and the denominator coefficients $\{a_i\}$ of the second-order sections. Using this program simulate two different cascade realizations of the following transfer function

$$H(z) = \frac{(1 - z^{-1})^2 (1 - 0.707 z^{-1} + z^{-2})}{(1 + 0.777 z^{-1} + 0.3434 z^{-2})(1 + 0.01877 z^{-1} + 0.801 z^{-2})}. \tag{9.244}$$

Using this program determine the impulse response of each pertinent scaling transfer functions and their approximate $\mathcal{L}_2$-norms. Based on this information, scale each realization using an $\mathcal{L}_2$-scaling rule and compute the total product roundoff noise variance for each scaled structure assuming rounding before addition.

M 9.5 Write a MATLAB program to simulate a fourth-order transfer function of the form given in Eq. (9.243) in parallel forms I and II. The input data to your program are the numerator coefficients $\{b_i\}$ and the denominator coefficients $\{a_i\}$. Using this program simulate the two different parallel realizations of the transfer function of Eq. (9.244). Determine the impulse response of each pertinent scaling transfer functions and their approximate $\mathcal{L}_2$-norms. Based on this information, scale each realization using an $\mathcal{L}_2$-scaling rule and compute the total product roundoff noise variance for each scaled structure assuming rounding before addition. Compare the output round-off noises of the two parallel structures with those of the cascade realizations of Problem M9.4.

M 9.6 Write a MATLAB program to simulate a fourth-order transfer function of the form given in Eq. (9.243) using the Gray-Markel method. The input data to your program are the numerator coefficients $\{b_i\}$ and the denominator coefficients $\{a_i\}$. Using this program simulate the Gray-Markel realization of the transfer function of Eq. (9.244). Determine the impulse response of each pertinent scaling transfer functions and their approximate $\mathcal{L}_2$-norms. Based on this information, scale the realization using an $\mathcal{L}_2$-scaling rule and compute the total product round-off noise variance for the scaled structure assuming rounding before addition. Compare the output roundoff noise of the Gray-Markel realization with those of the cascade realizations of Problems M9.4 and the two parallel structures of Problem M9.5.

CHAPTER 10

Multirate Digital Signal Processing

The digital signal processing structures discussed so far in this text belong to the class of single-rate systems since the sampling rates at the input and at the output and all internal nodes are the same. There are many applications where the signal of a given sampling rate needs to be converted into an equivalent signal with a different sampling rate. For example, in digital audio, three different sampling rates are presently employed: 32 kHz in broadcasting, 44.1 kHz in digital compact disk, and 48 kHz in digital audio tape (DAT) and other applications [Lag82]. Conversion of sampling rates of audio signals between these three different rates is often necessary in many situations. Another example is the pitch control of audio recordings usually performed by varying the tape recorder speed. However, such an approach in digital audio changes the sampling frequency of the digital signal and, as a result, conversion to the original sampling rate is needed [Lag82]. In the video applications, the sampling rates of NTSC and PAL composite video signals are, respectively, 14.3181818 MHz and 17.734475 MHz, whereas the sampling rates of the digital component video signal are 13.5 MHz and 6.75 MHz for the luminance and the color-difference signals, respectively [Lut91].[1] There are other applications where it is convenient (and often judicious) to have unequal rates of sampling at the filter input and output and at internal nodes. Examples of such sampling rate alterations are the oversampling A/D and D/A converters discussed in Sections 5.6.5 and 5.7.3, respectively, and analyzed in details in Sections 11.12 and 11.13, respectively. Additional applications of sampling rate alterations are also given in Sections 11.7 through 11.11.

To achieve different sampling rates at different stages, multirate digital signal processing systems employ the *down-sampler* and the *up-sampler*, the two basic sampling rate alteration devices in addition to the conventional elements such as the adder, the multiplier, and the delay. Discrete-time systems with unequal sampling rates at various parts of the system are called *multi-rate systems* and are the subject of discussion of this chapter.

We first examine the input-output relations of the up-sampler and the down-sampler both in the time-domain and transform-domain. As in many applications, cascade connections of the basic sampling rate alteration devices and digital filters are employed,

[1]CCIR Recommendation No. 601.

some basic cascade equivalences are then reviewed. For sampling rate alterations, the basic sampling rate alteration devices are invariably employed together with lowpass digital filters. The frequency response specifications of these filters are developed next. A computationally more efficient approach to sampling rate alteration is based on a multistage implementation that is then illustrated by means of a specific design problem. The polyphase decomposition of a sequence is reexamined next in the framework of multirate theory, and its application in developing computationally efficient sampling rate alteration systems is illustrated.

Many multirate systems employ a bank of filters with either a common input or a summed output. These filter banks are introduced next, and the design and computationally efficient implementation of a class of such filter banks are discussed. It is followed by an introduction of certain special types of transfer functions that are particularly attractive in the design of computationally efficient multirate systems. The rest of the chapter is devoted to a discussion of quadrature-mirror filter banks that find applications in signal compression and other areas.

10.1   The Basic Sample Rate Alteration Devices

The time-domain operations of the two basic sampling rate alteration devices are reexamined here again. However, it is instructive to analyze their operations in the frequency domain. This will point out why these devices must be used with additional filters. In addition, the frequency domain analysis provides the basic foundation for analyzing more complex multirate systems introduced in latter parts of the chapter.

10.1.1   Time-Domain Characterization

In up-sampling by an integer factor $L > 1$, $L - 1$ equidistant zero-valued samples are inserted by the up-sampler between two consecutive samples of the input sequence $x[n]$ to develop an output sequence $x_u[n]$ according to the relation

$$x_u[n] = \begin{cases} x[n/L], & n = 0, \pm L, \pm 2L, \ldots, \\ 0, & \text{otherwise.} \end{cases} \tag{10.1}$$

Note that the sampling rate of $x_u[n]$ is L times larger than that of the original sequence $x[n]$. The up-sampling operation is illustrated in the following example using MATLAB.

Example 10.1 Program 10_1 given below can be used to study the up-sampling of a sinusoidal input sequence. Its input data are the length of the input sequence, the up-sampling factor, and the frequency of the sinusoid in Hz. It then plots the input sequence and its up-sampled version.

```
% Program 10_1
% Illustration of Up-Sampling by an Integer Factor
%
clf;
N = input('Input length = ');
L = input('Up-sampling Factor = ');
fo = input('Input signal frequency = ');
```

Figure 10.1   Illustration of the up-sampling process.

$$x[n] \longrightarrow \boxed{\uparrow L} \longrightarrow x_u[n]$$

Figure 10.2   Representation of an up-sampler.

```
%  Generate the input sinusoidal sequence
n = 0:N-1;
x = sin(2*pi*fo*n);
% Generate the up-sampled sequence
y = zeros(1, L*length(x));
y([1: L: length(y)]) = x;
% Plot the input and the output sequences
subplot(2,1,1)
stem(n,x);
title('Input Sequence');
xlabel('Time index n');ylabel('Amplitude');
subplot(2,1,2)
stem(n,y(1:length(x)));
title(['Output sequence upsampled by', num2str(L)]);
xlabel('Time index n');ylabel('Amplitude');
```

Figure 10.1 shows the result obtained for a length-50 sinsusoidal sequence with a frequency of 0.12 Hz and with an up-sampling factor of 3.

The block-diagram representation of the up-sampler, also called a *sampling rate expander*, is shown in Figure 10.2.

Conversely, the down-sampling operation by an integer factor $M > 1$ on a sequence $x[n]$ consists of keeping every Mth sample of $x[n]$ and removing $M - 1$ in-between samples, generating an output sequence $y[n]$ according to the relation

$$y[n] = x[nM]. \tag{10.2}$$

This results in a sequence $y[n]$ whose sampling rate is $(1/M)$th of that of $x[n]$. Basically, all input samples with indices equal to an integer multiple of M are retained at the output and all others are discarded, as illustrated in the following example.

Example 10.2 We investigate the down-sampling of a sinusoidal input sequence using MATLAB. To this end we utilize Program 10_2 given below. Its input data are the length of the input sequence, the down-sampling factor, and the frequency of the sinusoid in Hz. It then plots the input sequence and its down-sampled version.

```
% Program 10_2
% Illustration of Down-Sampling by an Integer Factor
%
clf;
N = input('Output length = ');
M = input('Down-sampling Factor = ');
fo = input('Input signal frequency = ');
% Generate the input sinusoidal sequence
n = 0 : N-1;
m = 0 : N*M-1;
x = sin(2*pi*fo*m);
% Generate the down-sampled sequence
y = x([1 : M : length(x)]);
% Plot the input and the output sequences
subplot(2,1,1)
stem(n, x(1:N));
title('Input Sequence');
xlabel('Time index n');ylabel('Amplitude');
subplot(2,1,2)
stem(n, y);
title(['Output sequence downsampled by ',num2str(M)]);
ylabel('Amplitude');xlabel('Time index n');
```

The result obtained for a length-50 sinsusoidal sequence with a frequency 0.042 Hz and with an down-sampling factor of 3 are shown in Figure 10.3.

The schematic representation of the *down-sampler* or *sampling rate compressor* is shown in Figure 10.4.

The sampling periods involved have not been explicitly shown in Figures 10.2 and 10.4. This is in the interest of simplicity and in view of the fact that the mathematical theory of multirate systems can be understood without bringing T or the sampling frequency F_T into the picture. It is instructive in the beginning to explicitly see the time dimensions at various stages in the sampling rate alteration process, as indicated in Figure 10.5. In the remainder of this book, however, the explicit appearance of T or the sampling frequency F_T is not shown unless their actual values are relevant.

The up-sampler and the down-sampler are linear but time-varying discrete-time systems. The time-varying property of these devices is easy to show. Consider for example, the factor-of-M down-sampler defined by Eq. (10.2). Its output $y_1[n]$ for an input $x_1[n] = x[n - n_0]$ is then given by

$$y_1[n] = x_1[Mn] = x[Mn - n_0].$$

But from Eq. (10.2),

$$y[n - n_0] = x[M(n - n_0)] = x[Mn - Mn_0] \neq y_1[n].$$

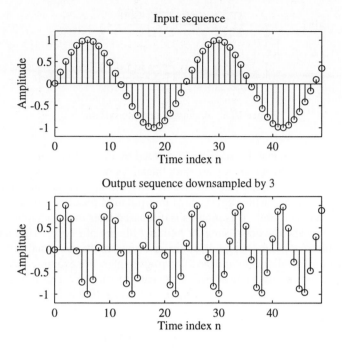

Figure 10.3   Illustration of the down-sampling process.

(b)

Figure 10.4   Representation of a down-sampler.

$$x[n] = x_a(nT) \longrightarrow \boxed{\downarrow M} \longrightarrow y[n] = x_a(nMT)$$

Input sampling   Output sampling

$$\text{frequency} = F_T = \frac{1}{T} \qquad \text{frequency} = F_T' = \frac{F_T}{M} = \frac{1}{T'}$$

$$x[n] = x_a(nT) \longrightarrow \boxed{\uparrow L} \longrightarrow x_u[n] = \begin{cases} x_a(nT/L), & n = 0, \pm L, \pm 2L, \dots \\ 0, & \text{otherwise} \end{cases}$$

Input sampling   Output sampling

$$\text{frequency} = F_T = \frac{1}{T} \qquad \text{frequency} = F_T' = LF_T = \frac{1}{T'}$$

Figure 10.5   The sampling rate alteration building blocks with sampling rates explicitly shown.

Figure 10.6   A simple multirate system.

Likewise, it can be shown that the up-sampler defined by Eq. (10.1) is also a time-varying device (Problem 10.1). However, they are both linear systems (Problem 10.2).

In practice, the zero-valued samples inserted by the up-sampler are replaced with appropriate nonzero values using some type of interpolation process in order that the new higher rate sequence be useful. This process is discussed later in this chapter.

The up-sampler and the down-sampler building blocks of Figures 10.2 and 10.4 are often used together in a number of applications involving multirate signal processing and are discussed in more details in this and the following chapter. For example, one application of using both types of sampling rate alteration devices is to achieve a sampling rate change by a rational number rather than an integer value. Example 10.3 below illustrates another application.

Example 10.3 Consider the multirate system of Figure 10.6. Its operation can be analyzed by writing down the relations between various signal variables and the input as depicted below:

n :	0	1	2	3	4	5	6	7	8
$x[n]$:	$x[0]$	$x[1]$	$x[2]$	$x[3]$	$x[4]$	$x[5]$	$x[6]$	$x[7]$	$x[8]$
$v[n]$:	$x[0]$	$x[2]$	$x[4]$	$x[6]$	$x[8]$	$x[10]$	$x[12]$	$x[14]$	$x[16]$
$w[n]$:	$x[-1]$	$x[1]$	$x[3]$	$x[5]$	$x[7]$	$x[9]$	$x[11]$	$x[13]$	$x[15]$
$v_u[n]$:	$x[0]$	0	$x[2]$	0	$x[4]$	0	$x[6]$	0	$x[8]$
$w_u[n]$:	$x[-1]$	0	$x[1]$	0	$x[3]$	0	$x[5]$	0	$x[7]$
$v_u[n-1]$:	0	$x[0]$	0	$x[2]$	0	$x[4]$	0	$x[6]$	0
$y[n]$:	$x[-1]$	$x[0]$	$x[1]$	$x[2]$	$x[3]$	$x[4]$	$x[5]$	$x[6]$	$x[7]$

It follows from the above that the output $y[n]$ of the multirate system of Figure 10.6 is given by $y[n] = v_u[n-1] + w_u[n]$, which is simply $x[n-1]$.

10.1.2   Frequency-Domain Characterization

We first derive the relations between the spectrums of the input and the output of a factor-of-2 up-sampler. From the input-output relation of the factor-of-L up-sampler given by Eq. (10.1), we arrive at the corresponding relation for the factor-of-2 up-sampler:

$$x_u[n] = \begin{cases} x[n/2], & n = 0, \pm 2, \pm 4, \ldots, \\ 0, & \text{otherwise.} \end{cases} \tag{10.3}$$

In terms of the z-transform, the input-output relation is then given by

$$X_u(z) = \sum_{n=-\infty}^{\infty} x_u[n] z^{-n} = \sum_{\substack{n=-\infty \\ n \text{ even}}}^{\infty} x[n/2] z^{-n}$$

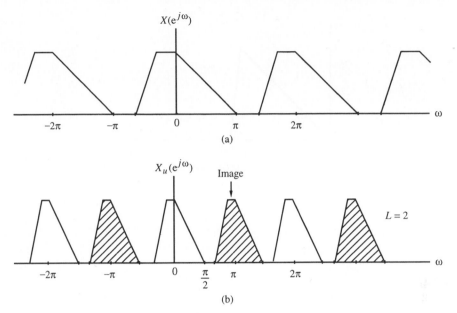

Figure 10.7 Effects of an up-sampler in the frequency domain. (a) Input spectrum, and (b) output spectrum for $L = 2$.

$$= \sum_{m=-\infty}^{\infty} x[m]z^{-2m} = X(z^2). \tag{10.4}$$

In a similar manner, we can show that for the factor-of-L up-sampler,

$$X_u(z) = X(z^L). \tag{10.5}$$

Let us examine the implication of the above relation on the unit circle. For $z = e^{j\omega}$ the above equation becomes $X_u(e^{j\omega}) = X(e^{j\omega L})$. Figure 10.7(a) shows the DTFT $X(e^{j\omega})$ that has been assumed to be a real function for convenience. Moreover, the DTFT $X(e^{j\omega})$ shown is not an even function of ω, implying that $x[n]$ is a complex sequence. The asymmetric response has been purposely chosen to illustrate more clearly the effect of up-sampling.

As shown in Figure 10.7(b), a factor-of-2 sampling rate expansion thus leads to a 2-fold repetition of $X(e^{j\omega})$, indicating that the Fourier transform is compressed by a factor of 2. This process is called *imaging* because we get an additional "image" of the input spectrum. In the case of a factor-of-L sampling rate expansion there will be $L-1$ additional images of the input spectrum in the baseband. Thus, a spectrum $X(e^{j\omega})$ bandlimited to the low-frequency region does not look like a low-frequency spectrum after up-sampling, because of the insertion of zero-valued samples between the nonzero samples of $x_u[n]$. Lowpass filtering of $x_u[n]$ removes the $L-1$ images and in effect "fills in" the zero-valued samples in $x_u[n]$ with interpolated sample values.

We next illustrate the frequency domain properties of the up-sampler using MATLAB. The input is a causal finite-length sequence with a bandlimited frequency response generated using the M-function fir2. The input to fir2 are as follows: length of the sequence

(a)

(b)

Figure 10.8   (a) Desired magnitude response and (b) corresponding time sequence.

is 100, the desired magnitude response vector mag = [0 1 0 0], and the vector of frequency points freq = [0 0.45 0.5 1]. The desired magnitude response is thus as indicated in Figure 10.8(a). A plot of the middle 61 samples of the signal generated is shown in Figure 10.8(b). To investigate the effect of up-sampling, we use Program 10_3 given below.

The input data called by the program is the up-sampling factor L. The program determines the output of the up-sampler and then plots the input and output spectrums, as indicated in Figure 10.9 for L = 5. It can be seen from this figure that, as expected, the output spectrum consists of a factor-of-5 compressed version of the input spectrum followed by L-1 = 4 images.

```
% Program 10_3
% Effect of Up-sampling in the Frequency Domain
% Use fir2 to create a bandlimited input sequence
freq = [0 0.45 0.5 1];
mag = [0 1 0 0];
x = fir2(99, freq, mag);
% Evaluate and plot the input spectrum
[Xz, w] = freqz(x, 1, 512);
plot(w/pi, abs(Xz)); grid
```

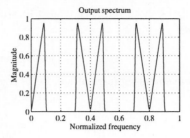

Figure 10.9   MATLAB-generated input and output spectrum of a factor-of-5 up-sampler.

```
xlabel('Normalized frequency'); ylabel('Magnitude');
title('Input spectrum');
pause
% Generate the up-sampled sequence
L = input('Type in the up-sampling factor = ');
y = zeros(1, L*length(x));
y([1: L: length(y)]) = x;
% Evaluate and plot the output spectrum
[Yz, w] = freqz(y, 1, 512);
plot(w/pi, abs(Yz)); grid
xlabel('Normalized frequency'); ylabel('Magnitude');
title('Output spectrum');
```

We now derive the relations between the spectrums of the input and the output of a down-sampler. Applying z-transform to the input-output relation given in Eq. (10.2), we arrive at

$$Y(z) = \sum_{n=-\infty}^{\infty} x[Mn]z^{-n}. \tag{10.6}$$

The expression on the right-hand side of the above equation cannot be directly expressed in terms of $X(z)$. To get around this problem, we define an intermediate sequence $x_{int}[n]$ whose sample values are same as that of $x[n]$ at values of n that are multiples of M and are zeros at other values of n:

$$x_{int}[n] = \begin{cases} x[n], & n = 0, \pm M, \pm 2M, \ldots, \\ 0, & \text{otherwise.} \end{cases} \tag{10.7}$$

Then,

$$Y(z) = \sum_{n=-\infty}^{\infty} x[Mn]z^{-n} = \sum_{n=-\infty}^{\infty} x_{int}[Mn]z^{-n}$$

$$= \sum_{k=-\infty}^{\infty} x_{int}[k]z^{-k/M} = X_{int}(z^{1/M}). \tag{10.8}$$

Now $x_{int}[n]$ can be formally related to $x[n]$ through $x_{int}[n] = c[n]x[n]$, where $c[n]$ is defined by

$$c[n] = \begin{cases} 1, & n = 0, \pm M, \pm 2M, \ldots, \\ 0, & \text{otherwise.} \end{cases} \tag{10.9}$$

A convenient representation of $c[n]$ is given by (Problem 10.4)

$$c[n] = \frac{1}{M} \sum_{k=0}^{M-1} W_M^{kn}, \tag{10.10}$$

where $W_M = e^{-j2\pi/M}$ is the quantity defined in Eq. (3.14). Substituting $x_{int}[n] = c[n]x[n]$ and making use of Eq. (10.10) in the z-transform of $x_{int}[n]$, we obtain

$$X_{int}(z) = \sum_{n=-\infty}^{\infty} c[n]x[n]z^{-n} = \frac{1}{M} \sum_{n=-\infty}^{\infty} \left(\sum_{k=0}^{M-1} W_M^{kn} \right) x[n]z^{-n}$$

$$= \frac{1}{M} \sum_{k=0}^{M-1} \left(\sum_{n=-\infty}^{\infty} x[n] W_M^{kn} z^{-n} \right) = \frac{1}{M} \sum_{k=0}^{M-1} X\left(z W_M^{-k}\right). \tag{10.11}$$

The desired input-output relation in the transform-domain for a factor-of-M down-sampler is then obtained by substituting Eq. (10.11) in Eq. (10.8), resulting in

$$Y(z) = \frac{1}{M} \sum_{k=0}^{M-1} X(z^{1/M} W_M^{-k}). \tag{10.12}$$

To understand the implication of the above relation, consider a factor-of-2 down-sampler with an input $x[n]$ whose spectrum is as shown in Figure 10.10(a). As before, for convenience we assume again $X(e^{j\omega})$ to be a real function with an asymmetric frequency response. From Eq. (10.12) we get

$$Y(e^{j\omega}) = \frac{1}{2} \left\{ X(e^{j\omega/2}) + X(-e^{j\omega/2}) \right\}. \tag{10.13}$$

The plot of $\frac{1}{2}X(e^{j\omega/2})$ is shown by the solid line in Figure 10.10(b). To determine the relation of the second term in Eq. (10.13) with respect to the first, we observe next that

$$X(-e^{j\omega/2}) = X(e^{j\frac{(\omega-2\pi)}{2}}), \tag{10.14}$$

indicating that the second term $X(e^{-j\omega/2})$ in Eq. (10.13) is obtained simply by shifting the first term $X(e^{j\omega/2})$ to the right by an amount 2π, as shown by the dotted lines in Figure 10.10(b). The plots of the two terms in Eq. (10.13) have an overlap, and hence, in general, the original "shape" of $X(e^{j\omega})$ is lost when $x[n]$ is down-sampled. This overlap causes the *aliasing* that takes place due to undersampling (i.e., down-sampling). There is no overlap, i.e., no aliasing, only if $X(e^{j\omega})$ is zero for $|\omega| \geq \pi/2$. Note that $Y(e^{j\omega})$ in Eq. (10.13) is indeed periodic with a period 2π, even though the stretched version $X(e^{j\omega})$ is periodic with a period 4π. For the general case, the situation is essentially the same, and

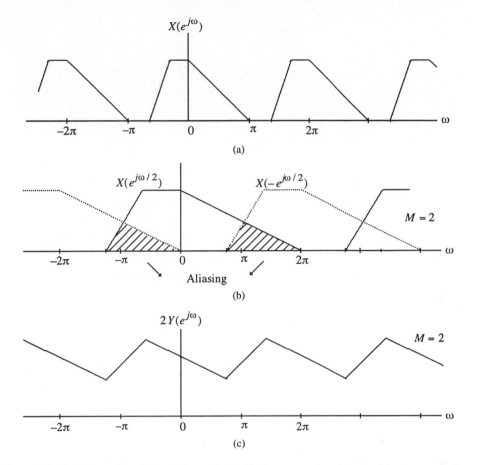

Figure 10.10 Illustration of the aliasing effect in the frequency domain caused by downsampling.

the relations between the Fourier transform of the output and the input of the factor-of-M down-sampler is given by

$$Y(e^{j\omega}) = \frac{1}{M} \sum_{k=0}^{M-1} X(e^{j\frac{(\omega-2\pi k)}{M}}). \tag{10.15}$$

The above relation implies that $Y(e^{j\omega})$ is a sum of M uniformly shifted and stretched versions of $X(e^{j\omega})$ then scaled by a factor $1/M$. Aliasing due to a factor-of-M down-sampling is absent if and only if the signal $x[n]$ is bandlimited to $\pm\pi/M$, as shown in Figure 10.11 for $M = 2$.

We next illustrate the aliasing effect caused by down-sampling using MATLAB. To investigate the effect of down-sampling, we use Program 10.4 given below. The input signal is again the signal generated using fir2 with a triangular magnitude response, as in Figure 10.8(a). However, here the frequency vector has been selected to be freq = [0 0.42 0.48 1] to ensure that there are no appreciable signal components above the

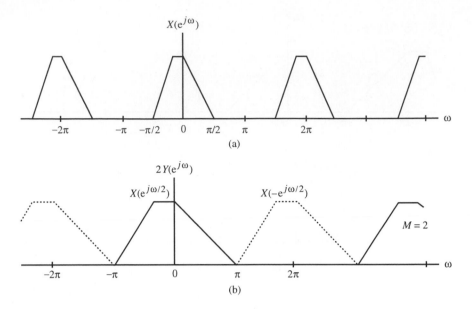

Figure 10.11   Effect of decimation in the frequency domain illustrating absence of aliasing.

normalized frequency of 0.5. The input data called by the program is the down-sampling factor M. The program generates the spectrums of the original input and the down-sampled output.

```
% Program 10_4
% Effect of Down-sampling in the Frequency Domain
% Use fir2 to create a bandlimited input sequence
freq = [0 0.42 0.48 1];
mag = [0 1 0 0];
x = fir2(101, freq, mag);
% Evaluate and plot the input spectrum
[Xz, w] = freqz(x, 1, 512);
plot(w/pi, abs(Xz)); grid
xlabel('Normalized frequency'); ylabel('Magnitude');
title('Input spectrum');
pause
% Generate the down-sampled sequence
M = input('Type in the down-sampling factor = ');
y = x([1: M: length(x)]);
% Evaluate and plot the output spectrum
[Yz, w] = freqz(y, 1, 512);
plot(w/pi, abs(Yz)); grid
xlabel('Normalized frequency'); ylabel('Magnitude');
title('Output spectrum');
```

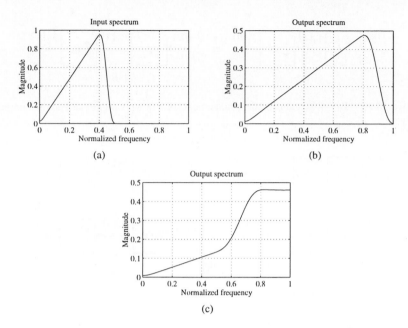

Figure 10.12   (a) Input spectrum, (b) output spectrum for a down-sampling factor of $M = 2$, and (c) output spectrum for a down-sampling factor of $M = 3$.

The plots generated by the above program are shown in Figure 10.12. The input spectrum is shown in Figure 10.12(a). Since the input signal is bandlimited to $\pi/2$, the output spectrum for a down-sampling factor of M $=$ 2 shown in Figure 10.12(b) is nearly of the same shape as the input spectrum, except it has been stretched by a factor of 2 in frequency and its magnitude is reduced by one-half as predicted by the factor $1/2$ in Eq. (10.13). On the other hand, the output spectrum for a down-sampling factor of M $=$ 3 shown in Figure 10.12(c) shows a severe distortion caused by the aliasing.

Any linear discrete-time multirate system can be analyzed in the transform-domain by using the input-output relations of the up-sampler and the down-sampler given by Eqs. (10.5) and (10.12), respectively. We illustrate the applications of these relations in latter parts of this chapter.

10.1.3   Cascade Equivalences

As we shall observe later, a complex multirate system is formed by an interconnection of the basic sampling rate alteration devices and the components of an LTI digital filter. In many applications, these devices appear in a cascade form. An interchange of the positions of the branches in a cascade often can lead to a computationally efficient realization. We investigate certain specific cascade connections and their equivalences, which leaves input-output relations invariant.

The basic sampling rate alteration devices can be used to change the sampling rate of a signal by an integer factor only. Therefore, to implement a fractional change in the sampling rate it follows that a cascade of a down-sampler and an up-sampler should be

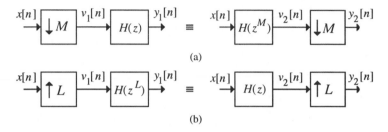

Figure 10.13   Two different cascade arrangements of a down-sampler and an up-sampler.

Figure 10.14   Cascade equivalences: (a) Equivalence #1, and (b) Equivalence #2.

used. It is of interest to determine the condition under which a cascade of a factor-of-M down-sampler with a factor-of-L up-sampler (Figure 10.13) is *interchangeable*, with no change in the input-output relation. It can be shown that this interchange is possible if and only if M and L are *relatively prime*, i.e., M and L do not have a common factor that is an integer $k > 1$ (Problem 10.6) [Vai90]. We discuss fractional changes in the sampling rate in Section 10.2.2.

Two other simple cascade equivalence relations are depicted in Figure 10.14. The validity of these equivalences can be readily established using Eqs. (10.5) and (10.12) and are left as exercises (Problem 10.7). These rules enable us to move the basic sampling rate alteration devices in multirate networks to more advantageous positions. Such rules are extremely useful in the design and analysis of more complicated systems, as we shall demonstrate later.

10.2   Filters in Sampling Rate Alteration Systems

From the sampling theorem introduced in Section 5.2, it is known that the sampling rate of a critically sampled discrete-time signal with a spectrum occupying the full Nyquist range cannot be reduced any further since such a reduction will introduce aliasing. Hence, the bandwidth of a critically sampled signal must first be reduced by lowpass filtering before its sampling rate is reduced by a down-sampler. Likewise, the zero-valued samples introduced by an up-sampler must be interpolated to more appropriate values for an effective sampling rate increase. As we shall show next, this interpolation can be simply achieved by digital lowpass filtering. We consider in this section some of the issues concerning these lowpass filters. We first develop the frequency response specifications of these lowpass filters. Next, we illustrate the decimation and interpolation of sequences using MATLAB. Finally, we investigate the computational complexity issues of the lowpass digital filters.

Figure 10.15   Filters in sampling rate alteration systems: (a) interpolator and (b) decimator.

10.2.1   Filter Specifications

Since up-sampling causes periodic repetition of the basic spectrum (Figure 10.7), the unwanted images in the spectra of the up-sampled signal $x_u[n]$ must be removed by using a lowpass filter $H(z)$, called the *interpolation filter*, as indicated in Figure 10.15(a). On the other hand, as indicated in Figure 10.15(b), prior to down-sampling the signal $v[n]$ should be band limited to $|\omega| < \pi/M$ by means of a lowpass filter $H(z)$, called the *decimation filter*, to avoid aliasing caused by down-sampling. The system of Figure 10.15(a) is often called an *interpolator* while the system of Figure 10.15(b) is called a *decimator*.[2]

The specifications for the lowpass filter in Figure 10.15 can now be derived. We first develop the specifications for the interpolation filter. Assume $x[n]$ has been obtained by sampling a bandlimited continuous-time signal $x_a(t)$ at the Nyquist rate. If $X_a(j\Omega)$ and $X(e^{j\omega})$ denote the Fourier transforms of $x_a(t)$ and $x[n]$, respectively, from Eq. (5.14a) it follows that these Fourier transforms are related through

$$X(e^{j\omega}) = \frac{1}{T_0} \sum_{k=-\infty}^{\infty} X_a\left(\frac{j\omega - j2\pi k}{T_0}\right), \tag{10.16}$$

where T_0 is the sampling period. Since the sampling is being done at the Nyquist rate, there is no overlap between the shifted spectras of $X_a(j\omega/T_0)$. If we instead sample $x_a(t)$ at a much higher rate $T = T_0/L$ yielding $y[n]$, its Fourier transform $Y(e^{j\omega})$ is related to $X_a(j\Omega)$ through

$$Y(e^{j\omega}) = \frac{1}{T} \sum_{k=-\infty}^{\infty} X_a\left(\frac{j\omega - j2\pi k}{T}\right) = \frac{L}{T_0} \sum_{k=-\infty}^{\infty} X_a\left(\frac{j\omega - j2\pi k}{(T_0/L)}\right). \tag{10.17}$$

On the other hand, if we pass $x[n]$ through a factor-of-L up-sampler generating $x_u[n]$, from Eq. (10.5) the relation between the Fourier transform $X_u(e^{j\omega})$ and $X(e^{j\omega})$ is given by

$$X_u(e^{j\omega}) = X(e^{j\omega L}). \tag{10.18}$$

It follows from Eqs. (10.16) to (10.18) that if $x_u[n]$ is passed through an ideal lowpass filter with a cutoff at π/L and a gain of L, the output of the filter will be precisely $y[n]$.

In practice, a transition band is provided to ensure the realizability and stability of the lowpass interpolation filter $H(z)$. Hence, the desired lowpass filter should have a stopband edge at $\omega_s = \pi/L$ and a passband edge ω_p close to ω_s to reduce the distortion

---

[2]It has been tacitly assumed here that the discrete-time signal to be interpolated or decimated has a lowpass frequency response and as a result, the desired interpolation or the decimation filter is a lowpass filter. However, if the discrete-time signal to be interpolated or decimated has a highpass (bandpass) frequency response, then the desired interpolation or the decimation filter is a highpass (bandpass) filter.

of the spectrum of the signal $x[n]$.[3] If ω_c denotes the highest frequency that need to be preserved in the signal to be interpolated, the passband edge ω_p of the lowpass filter should be at $\omega_p = \omega_c/L$. Summarizing, the specifications for lowpass interpolation filter are thus given by

$$\left| H(e^{j\omega}) \right| = \begin{cases} L, & |\omega| \le \omega_c/L, \\ 0, & \pi/L \le |\omega| \le \pi. \end{cases} \tag{10.19}$$

It should be noted that in many interpolation applications, one requirement is to ensure that the input samples are not changed at the output. This requirement can be satisfied by a Nyquist (Lth band) interpolation filter, which is discussed in Section 10.6.

In a similar manner, we can develop the specifications for the lowpass decimation filter that are given by

$$\left| H(e^{j\omega}) \right| = \begin{cases} 1, & |\omega| \le \omega_c/M, \\ 0, & \pi/M \le |\omega| \le \pi, \end{cases} \tag{10.20}$$

where ω_c denotes the highest frequency that need be preserved in the decimated signal.

The effects of decimation and interpolation in the frequency-domain are illustrated in Figures 10.16 and 10.17, respectively, for $M = 2$ and $L = 2$. Figure 10.16(a) shows the spectrum $X(e^{j\omega})$ of the input signal $x[n]$ and the spectrum $H(e^{j\omega})$ of the decimation filter. The passband edge and the stopband edge of the decimation filters are, respectively, $\omega_c/2$ and $\pi/2$, where ω_c is the highest frequency in the input signal that is to be preserved at the output of the decimator. From Eq. (10.13), the frequency response $Y(e^{j\omega})$ of the decimator output $y[n]$ is now given by

$$Y(e^{j\omega}) = \tfrac{1}{2} \left\{ V(e^{j\omega/2}) + V(-e^{j\omega/2}) \right\},$$

where $V(e^{j\omega})$ is the frequency response of the filter output $v[n]$. Due to prior filtering there is no overlap in these two spectra, resulting in no aliasing at the decimator output with the output spectrum $Y(e^{j\omega})$, as sketched in Figure 10.16(b). However, it is not possible to recover the original input signal $x[n]$ from the decimated version $y[n]$. The filter $H(z)$ is used to preserve $X(e^{j\omega})$ in the range $-\omega_c/2 < \omega < \omega_c/2$, and one can reconstruct this portion exactly from $y[n]$. On the other hand, if there is no filtering of the input $x[n]$ prior to down-sampling, the two components $\tfrac{1}{2}X(e^{j\omega/2})$ and $\tfrac{1}{2}X(-e^{j\omega/2})$ will overlap, resulting in a severe aliasing at the decimator output, as indicated in Figure 10.16(c). It should be noted that $x[n]$ corresponding to the frequency response $X(e^{j\omega})$ of Figure 10.16(a) can be recovered from its decimated version if the decimation factor here is 4/3. We discuss fractional rate sampling rate alteration in Section 10.2.2.

Likewise, in the case of the factor-of-2 interpolation, the spectrum $V(e^{j\omega})$ of the up-sampler output is given by $X(e^{j2\omega})$. Figure 10.17(b) shows $V(e^{j\omega})$ for an input spectrum indicated in Figure 10.17(a). The spectrum $Y(e^{j\omega})$ of the interpolator output obtained by filtering $v[n]$ is thus as sketched in Figure 10.17(c).

The design of $H(z)$ is a standard IIR or FIR lowpass filter design problem. Any of the techniques outlined in Chapter 7 can be applied here for the design of these lowpass filters.

---

[3]The meanings of ω_p, ω_s, δ_p, and δ_s in this chapter are precisely as in Section 7.1.1.

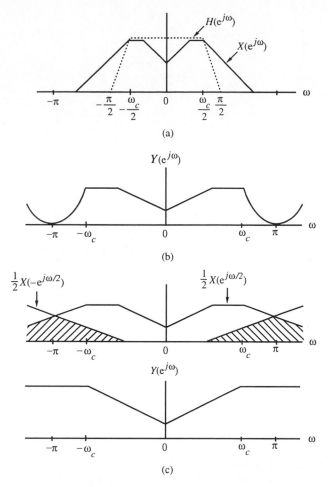

Figure 10.16 Spectrum of (a) the input $x[n]$, (b) the output $y[n]$ of the factor-of-2 decimator with $x[n]$ filtered, and (c) the output $y[n]$ of the factor-of-2 decimator with no filtering of $x[n]$ showing the effect of aliasing. The spectrum of the decimation filter $H(z)$ is shown in (a) with dotted lines.

10.2.2 Filters for Fractional Sampling Rate Alteration

A fractional change in the sampling rate can be achieved by cascading a factor-of-M decimator with a factor-of-L interpolator, where M and L are positive integers. Such a cascade is equivalent to a decimator with a decimation factor of M/L or, alternatively, to an interpolator with an interpolation factor of L/M. There are two possible such cascade connections, as sketched in Figure 10.18. Of these two, the scheme shown in Figure 10.18(b) is more efficient since only one of the filters, $H_u(z)$ or $H_d(z)$, is adequate to serve as the interpolation filter and the decimation filter, depending on which one of the two stopband frequencies, π/L or π/M, is a minimum. Hence, the desired configuration for the fractional sampling rate alteration is as indicated in Figure 10.18(c), where the

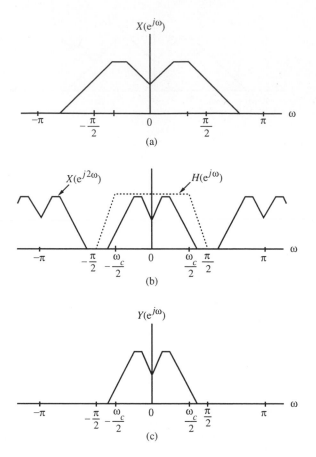

Figure 10.17   Spectrum of (a) the input $x[n]$, (b) the output $x[n]$ of the up-sampler, and (c) the output $y[n]$ of the factor-of-2 interpolator. The spectrum of the interpolation filter $H(z)$ is shown in (b) with dotted lines.

lowpass filter $H(z)$ has a normalized stopband cutoff frequency at

$$\omega_s = \min\left(\frac{\pi}{L}, \frac{\pi}{M}\right) \tag{10.21}$$

which suppresses the imaging caused by the interpolator while at the same time ensuring the absence of aliasing that would otherwise be caused by the decimator.

10.2.3   Computational Requirements

As indicated earlier, the lowpass decimation or the interpolation filter can be designed either as an FIR or an IIR digital filter. In the case of single-rate digital signal processing, IIR filters are, in general, computationally more efficient than the FIR digital filters, and are therefore preferred in applications where computational cost needs to be minimized. This

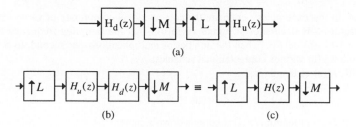

(b)   (c)

Figure 10.18   General schemes for increasing the sampling rate by L/M.

issue is not quite the same in the case of multirate digital signal processing, as elaborated next.

Consider the factor-of-M decimator structure of Figure 10.15(b). If the decimation filter $H(z)$ is an FIR filter of length N implemented in a direct form, then

$$v[n] = \sum_{m=0}^{N-1} h[m]x[n-m]. \qquad (10.22)$$

Now, the down-sampler keeps only every Mth sample of $v[n]$ at its output. As a result it is sufficient to compute $v[n]$ using Eq. (10.22) only for values of n that are multiples of M and skip the computations of the in-between $M-1$ samples. This leads to a factor-of-M savings in the computational complexity. If on the other hand, $H(z)$ in Figure 10.15(b) is an IIR filter of order K with a transfer function

$$H(z) = \frac{V(z)}{X(z)} = \frac{P(z)}{D(z)}, \qquad (10.23)$$

where

$$P(z) = \sum_{n=0}^{K} p_n z^{-n}, \qquad D(z) = 1 + \sum_{n=1}^{K} d_n z^{-n}, \qquad (10.24)$$

its noncanonic direct form implementation is given by

$$w[n] = -d_1 w[n-1] - d_2 w[n-2] -$$
$$\cdots - d_K w[n-K] + x[n], \qquad (10.25a)$$
$$v[n] = p_0 w[n] + p_1 w[n-1] + \cdots + p_K w[n-K]. \qquad (10.25b)$$

Since $v[n]$ is being down-sampled, it is sufficient to compute $v[n]$ using Eq. (10.25b) only for values of n that are integer multiples of M. However, the intermediate signal $w[n]$ in Eq. (10.25a) must be computed for all values of n. For example, in the computation of

$$v[M] = p_0 w[M] + p_1 w[M-1] + \cdots + p_K w[M-K],$$

$K+1$ successive values of $w[n]$ are *still* required. As a result, the savings in the computations in the case of an IIR filter is going to be less than a factor-of-M.

The following example provides a more detailed comparison.

Example 10.4 In this example we compare the computational complexity of various implementations of the factor-of-M decimator of Figure 10.15(b).[4] Let the sampling frequency for the input $x[n]$ in Figure 10.15(b) be F_T. Then the number of multiplications per second, to be denoted as $\mathcal{R}_M$, are as follows for various computational schemes:

For an FIR $H(z)$ of length N,

$$\mathcal{R}_{M,\text{FIR}} = N \times F_T.$$

For an FIR $H(z)$ of length N followed by a down-sampler,

$$\mathcal{R}_{M,\text{FIR–DEC}} = N \times F_T/M.$$

For an IIR $H(z)$ of order K,

$$\mathcal{R}_{M,\text{IIR}} = (2K+1) \times F_T.$$

For an IIR $H(z)$ of order K followed by a down-sampler,

$$\mathcal{R}_{M,\text{IIR–DEC}} = K \times F_T + (K+1) \times F_T/M.$$

Thus, in the FIR case, we save computations by a factor of M. In the IIR case we save by a factor of $M(2K+1)/[(M+1)K+1]$, which is not significant for large K. For $M=10$ and $K=9$, the savings is only by a factor of 1.9. We shall point out later that in certain cases the IIR filter can be computationally more efficient.

For the case of the interpolation filter in Figure 10.15(a), very similar arguments hold. If $H(z)$ is a FIR filter, then the computational savings is by a factor of L (since $v[n]$ has $L-1$ zeros between its two consecutive nonzero samples). On the other hand, computational savings is significantly less with IIR filters.

10.2.4   Sampling Rate Alteration Using MATLAB

The Signal Processing Toolbox of MATLAB includes three specific functions for sampling rate alteration. These are decimate, interp, and resample.

The function decimate can be employed to reduce the sampling rate of an input signal vector x by an integer factor M generating the output signal vector y and is available with four options :

```
y = decimate(x,M),        y = decimate(x,M,n),
y = decimate(x,M,'fir'),  y = decimate(x,M,N,'fir')
```

In the first option, the function employs an eighth-order Type 1 Chebyshev IIR lowpass filter by default and filters the input sequence in both directions to ensure zero-phase filtering. In the second option, it utilizes an order n Type 1 Chebyshev IIR lowpass filter where n should be less than 13 to avoid numerical instability. In the third option, it employs a 30-tap FIR lowpass filter which filters the input only in the forward direction. The FIR

---

[4]In this chapter the "computational complexity of an implementation" is taken to be equal to the number of multiplications required per second. Also, the symmetry of FIR impulse responses, which lead to about 50% computational savings, is ignored.

filter is designed with a stopband edge at π/M using the function `fir1` of MATLAB. Finally, in the last option, it designs and employs a length-N FIR lowpass filter.

We illustrate the application of the function `decimate` in the following example.

Example 10.5 We consider here the decimation of a sum of two sinusoidal sequences of normalized frequencies f_1 and f_2 by an arbitrary down-sampling factor M using Program 10_5 given below. The input data to the program are the length N of the input sequence $x[n]$, the down-sampling factor M, and the two normalized frequencies in Hz. The program uses a 30-tap FIR lowpass decimation filter designed to have a stopband edge at π/M.[5] It then plots the input and the output sequences as indicated in Figure 10.19 for $N = 100$, $M = 2$, $f_1 = 0.043$, and $f_2 = 0.031$.

```
% Program 10_5
% Illustration of Decimation Process
%
clf;
N = input('Length of input signal = ');
M = input('Down-sampling factor = ');
f1 = input('Frequency of first sinusoid = ');
f2 = input('Frequency of second sinusoid = ');
n = 0:N-1;
% Generate the input sequence
x = sin(2*pi*f1*n) + sin(2*pi*f2*n);
% Generate the decimated output sequence
y = decimate(x,M,'fir');
% Plot the input and the output sequences
subplot(2,1,1)
stem(n,x(1:N));
title('Input sequence');
xlabel('Time index n');ylabel('Amplitude');
subplot(2,1,2)
m=0:N/M-1;
stem(m,y(1:N/M));
title('Output sequence');
xlabel('Time index n'); ylabel('Amplitude');
```

The function `interp`[6] can be employed to increase the sampling rate of an input signal x by an integer factor L generating the output signal vector y. It is available with three options:

$$y = interp(x,L), \qquad y = interp(x,L,N,alpha),$$
$$[y,h] = interp(x,L,N,alpha)$$

[5]The group delay of the default decimation filter is 14.5 samples. For an integer valued group delay, a decimation filter of odd length should be used.

[6]`interp` is based on the interpolator design of Oetken et al [Oet75].

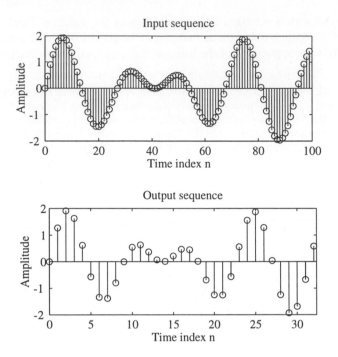

Figure 10.19   The input and output sequences of the factor-of-2 decimator of Example 10.5.

The lowpass filter designed by the program to fill in the zero-valued samples inserted by the up-sampling operation is a symmetric FIR filter that allows the original input samples to appear as is at the output and finds the missing samples by minimizing the mean-square errors between these samples and the ideal values. The length of the FIR filter is 2NL+1, where $N \leq 10$. The input signal is assumed to be bandlimited to the frequency range $0 \leq \omega \leq$ alpha, where alpha ≤ 0.5. In the first option, the default values of alpha and N are, respectively, 0.5, and 4. In the remaining two options, the bandedge alpha of the input signal and the length N of the interpolation filter are specified. In the last option, the output data contains in addition to the interpolated output vector y, the filter coefficients h.

The application of the function interp is considered in the following example.

Example 10.6 To illustrate the interpolation process, we take the input x to be a sum of two sinusoidal sequences. Program 10_6 is then used to find its interpolated output y for an up-sampling factor of L. The input data to the program are the length N of the input vector x, the up-sampling factor L, and the two normalized frequencies in Hz. The frequencies of the two sinusoids considered here are $f_1 = 0.043$ and $f_1 = 0.031$, while the interpolation factor is 2. The input length is 50. The output plots generated by this program are shown in Figure 10.20.

```
% Program 10_6
% Illustration of Interpolation Process
%
clf;
```

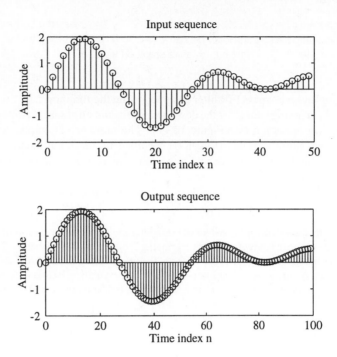

Figure 10.20   The input and output sequences of the factor-of-2 interpolator of Example 10.6.

```
N = input('Length of input signal = ');
L = input('Up-sampling factor = ');
f1 = input('Frequency of first sinusoid = ');
f2 = input('Frequency of second sinusoid = ');
% Generate the input sequence
n = 0:N-1;
x = sin(2*pi*f1*n) + sin(2*pi*f2*n);
% Generate the interpolated output sequence
y = interp(x,L);
% Plot the input and the output sequences
subplot(2,1,1)
stem(n,x(1:N));
title('Input sequence');
xlabel('Time index n'); ylabel('Amplitude');
subplot(2,1,2)
m=0:N*L-1;
stem(m,y(1:N*L));
title('Output sequence');
xlabel('Time index n'); ylabel('Amplitude');
```

Finally, the function `resample` in MATLAB can be utilized to increase the sampling rate of an input vector `x` by a ratio of two positive integers, `L/M`, generating an output vector `y`. There are five options available with this function:

```
y = resample(x,L,M),              y = resample(x,L,M,R),
y = resample(x,L,M,R,beta),       y = resample(x,L,M,h),
               [y,h] = resample(x,L,M)
```

The program employs a lowpass FIR filter designed using the function `fir1` with a Kaiser window to prevent aliasing caused by the down-sampling and eliminate the imaging caused by the up-sampling operation (see Figure 10.17). The number of samples used on both sides of the present input sample $x[n]$ can be specified by the input data R. The default value of R is 10. The design parameter β for selecting the Kaiser window can be specified through the input data `beta`.[7]

Example 10.7 demonstrates the application of the function `resample`.

Example 10.7 In this example, we consider the sampling rate increase by a ratio of two positive integers of a signal composed of two sinusoids employing Program 10_7 given below. The input data to the program are the length N of the input vector x, the up-sampling factor L, the down-sampling factor M, and the two frequencies. The up-sampling and down-sampling factors used here are 5 and 3, respectively. The frequencies of the two sinusoids considered here are $f_1 = 0.43$ and $f_1 = 0.31$. Figure 10.21 shows the output plots generated by this program.

```
% Program 10_7
% Illustration of Sampling Rate Alteration by
% a Ratio of Two Integers
%
clf;
N = input('Length of input signal = ');
L = input('Up-sampling factor = ');
M = input('Down-sampling factor = ');
f1 = input('Frequency of first sinusoid = ');
f2 = input('Frequency of second sinusoid = ');
%  Generate the input sequence
n = 0:N-1;
x = sin(2*pi*f1*n) + sin(2*pi*f2*n);
%  Generate the resampled output sequence
y = resample(x,L,M);
%  Plot the input and the output sequences
subplot(2,1,1)
stem(n,x(1:N));
title('Input sequence');
xlabel('Time index n'); ylabel('Amplitude');
subplot(2,1,2)
m=0:N*L/M-1;
stem(m,y(1:N*L/M));
title('Output sequence');
xlabel('Time index n'); ylabel('Amplitude');
```

---

[7] See Section 7.7.5.

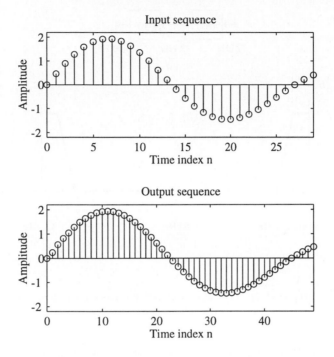

Figure 10.21   Illustration of sampling rate increase by a rational number 5/3.

(a)   (b)

Figure 10.22   Two-stage implementation of sampling rate alteration systems: (a) interpolator and (b) decimator.

10.3   Multistage Design of Decimator and Interpolator

The decimator and the interpolator of Figure 10.15 are single-stage structures since here the basic scheme for the implementation involves a single lowpass filter and a single sampling rate alteration device. If the interpolation factor L can be expressed as a product of two integers L_1 and L_2, the factor-of-L interpolator of Figure 10.15(a) can also be realized in two stages, as indicated in Figure 10.22(a). Likewise, the factor-of-M decimator of Figure 10.15(b) can be implemented in two stages, as shown in Figure 10.22(b), if the decimation factor M is a product of two integers M_1 and M_2. Of course, the design can involve more than two stages, depending on the number of factors used to express L and M, respectively. It turns out that, in general, the computational efficiency is improved significantly by designing the sampling rate alteration system as a cascade of several stages. We demonstrate this feature next by means of an example which also illustrates how to choose the specifications of the individual filters in a multistage design.

Figure 10.23   Illustration of the decimation filter design (Frequency response plots shown not to scale.

Example 10.8 Consider the design of a decimator for the reduction of the sampling rate of a signal from 12 kHz to 400 Hz [Figure 10.23(a)]. The specifications for the decimation filter $H(z)$ are assumed to be as follows: passband edge $F_p = 180$ Hz, stopband edge $F_s = 200$ Hz, passband ripple $\delta_p = 0.002$, and stopband ripple $\delta_s = 0.001$, and are shown in Figure 10.23(b). If $H(z)$ is required to be an equiripple linear-phase FIR filter, then its length N (i.e., order $N - 1$) needs to be estimated first. Recall from Section 7.1.4 that the length N can be determined from Eq. (7.13), which is repeated below for convenience:

$$N = \frac{-20\log_{10}\sqrt{\delta_p\delta_s} - 13}{14.6\Delta f} \qquad (10.26)$$

where $\Delta f = (F_s - F_p)/F_T$ is the normalized transition bandwidth.

Note that the above filter specifications are precisely the same as given in Example 7.2 for which the order was estimated using Program 7_1 written for MATLAB. The filter length obtained there is given by

$$N = 1808. \qquad (10.27)$$

Thus, the number of multiplications per second in the single stage implementation of the factor-of-30 decimator of Figure 10.23(a) is

$$\mathcal{R}_{M,H} = 1808 \times \frac{12000}{30} = 723,200. \qquad (10.28)$$

Let us now consider the implementation of $H(z)$ as a cascade realization in the form of $G(z^{15})F(z)$.[8] The parent filter $G(z)$ should now have specifications as shown in Figure 10.24(a). This corresponds to stretching the specifications in Figure 10.23(b) by 15. Figure 10.24(b) shows the magnitude response of $G(z^{15})$. The desired response of the $F(z)$ is also shown in the same figure.

---

[8]The implementation in the form of $G(z^{15})F(z)$ is known as the interpolated FIR (IFIR) realization [Neu84b].

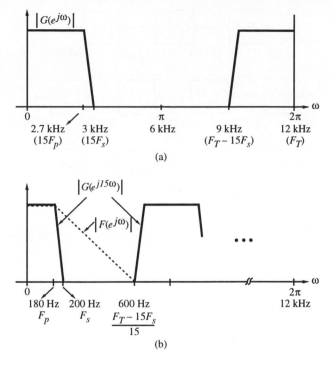

Figure 10.24 Decimation filter design based on the IFIR approach (frequency response plots not shown to scale).

The implementation is shown in Figure 10.25(a). Because of the cascade connection, the overall ripple of the cascade in dB is given by the sum of the passband ripples of $F(z)$ and $G(z^{15})$ in dB (see Chapter 7). This can be compensated for by designing $F(z)$ and $G(z)$ to have a passband ripple of $\delta_p = 0.001$ (rather than 0.002) each. On the other hand, the cascade of $F(z)$ and $G(z^{15})$ has a stopband at least as good as $F(z)$ or $G(z^{15})$ individually, so the specification for δ_s for both $F(z)$ and $G(z)$ can be retained as $\delta_s = 0.001$. Therefore, the specifications for the IFIR approach are as follows:

$$G(z): \quad \delta_p = 0.001, \quad \delta_s = 0.001, \quad \Delta f = \frac{300}{12000}, \tag{10.29a}$$

$$F(z): \quad \delta_p = 0.001, \delta_s = 0.001, \Delta f = \frac{420}{12000}. \tag{10.29b}$$

The filter lengths of $G(z)$ and $F(z)$ obtained using Program 7_1 are as follows:

$$\text{Length of } G(z) = 129, \tag{10.30a}$$

$$\text{Length of } F(z) = 92. \tag{10.30b}$$

It should be noted that Figure 10.25(a) has been obtained by replacing $H(z)$ in Figure 10.23(a) with $F(z)G(z^{15})$. The length of $H(z)$ for a direct implementation is 1808 from Eq. (10.27), whereas the length of $F(z)G(z^{15})$ is $93 + 15 \times 128 + 2 = 2015$. In other words, the length of the overall filter has been increased. However, the computational complexity (i.e., number of multiplications per second) of the new realization can be dramatically reduced by making use of the cascade equivalence #1 in Figure 10.14 as demonstrated next.

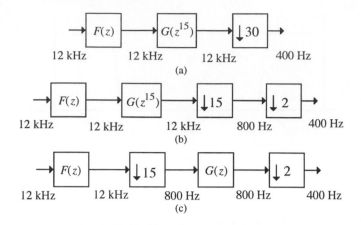

Figure 10.25   The steps in the two-stage realization of the decimator structure.

To this end we redraw Figure 10.25(a) as in Figure 10.25(b) by factoring the down-sampling factor M as $M_1 M_2$ with $M_1 = 15$ and $M_2 = 2$. Next, by invoking the equivalence of Figure 10.14(a), we replace it with Figure 10.25(c). From this figure, it can be seen that the implementation of $G(z)$ at the sampling rate 800 Hz followed by a down-sampling by a factor of 2 requires

$$\mathcal{R}_{M,G} = 129 \times \frac{800}{2} = 51,600 \qquad (10.31)$$

multiplications per second. The implementation of $F(z)$ followed by the down-sampler of factor 15 requires about

$$\mathcal{R}_{M,F} = 92 \times \frac{12000}{15} = 73,600 \qquad (10.32)$$

multiplications per second. Thus, $F(z)$ requires more multiplications per second than the model filter $G(z)$. The total complexity of the IFIR based implementation of Figure 10.25(c) is therefore

$$51,600 + 73,600 = 125,200 \qquad (10.33)$$

multiplications per second, which is about 5.78 times smaller than the complexity (Eq. (10.28)) of a direct implementation as in Figure 10.23(a).

Let us now implement the factor-of-15 decimator in the first stage of Figure 10.25(c) also in two stages. To this end we design the decimation filter $F(z)$ as an IFIR structure of the form $R(z)S(z^5)$. Following the same procedure as described above for the design of $H(z)$ in the form of $F(z)G(z^{15})$, we arrive at the specifications of the filters $S(z)$ and $R(z)$ as given below:

$$S(z): \quad \delta_p = 0.0005, \quad \delta_s = 0.001, \quad \Delta f = \frac{420}{12000} \times 5, \qquad (10.34a)$$

$$R(z): \quad \delta_p = 0.0005, \quad \delta_s = 0.001, \quad \Delta f = \frac{1620}{12000}. \qquad (10.34b)$$

The filter lengths of $S(z)$ and $R(z)$ obtained using Program 7_1 are as follows:

$$\text{Length of } S(z) = 20, \qquad (10.35a)$$

$$\text{Length of } R(z) = 26. \qquad (10.35b)$$

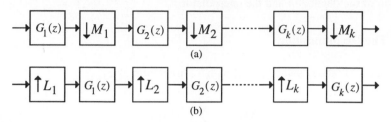

Figure 10.26   A three-stage realization of the decimator.

(a)

(b)

Figure 10.27   (a) A general multistage structure for decimation, (b) a general multistage structure for interpolation.

The final realization of the decimator is as indicated in Figure 10.26. From this figure, it can be seen that the implementation of $S(z)$ at the sampling rate 2.4 kHz followed by a down-sampling by a factor of 3 requires

$$\mathcal{R}_{M,S} = 20 \times \frac{2400}{3} = 16,000$$

multiplications per second, whereas the implementation of $R(z)$ at the sampling rate of 12 kHz followed by a down-sampling by a factor of 5 requires

$$\mathcal{R}_{M,R} = 26 \times \frac{12000}{5} = 62,400$$

multiplications per second. As a result, the total number of multiplications required in the three-stage realization is given by

$$\mathcal{R}_{M,G} + \mathcal{R}_{M,S} + \mathcal{R}_{M,R} = 137,200,$$

which is slightly higher than that required in the two-stage realization. As this example points out, a realization higher than two stages may not always lead to a computationally efficient realization.

In the general case of a k-stage implementation, as indicated in Figure 10.27(a), the decimation factor is $M = M_1 M_2 \dots M_k$. For a given M_k there are $k - 1$ quantities, $M_1, M_2, \dots, M_{k-1}$, which can be chosen in various ways with one combination leading to an optimum multistage realization of a decimator with the least computational complexity. The determination of an optimum realization depends on the selection of k, and the "best" combination and ordering of $M_1, M_2, \dots, M_k$ that minimizes the required number of multiplications per second [Cro83]. The corresponding multistage interpolator design is a very analogous problem [Figure 10.27(b)] [Cro83].

10.4   The Polyphase Decomposition

We have shown in Section 10.2.3 that a single-stage decimator or interpolator employing FIR lowpass filters can be computationally efficient since the necessary multiplications required to compute the output sample can be carried out only when needed. We demonstrated in the previous section that the computational requirements can be further decreased using a multistage design. Additional reduction in the computational complexity is possible by realizing the FIR filters using the polyphase decomposition described in Section 6.4.3. In certain cases, it is also possible to realize IIR decimation and interpolation filters in polyphase forms, resulting in reduced computational complexity realizations. We review the polyphase decomposition again here and illustrate its application in the efficient realization of the decimator and the interpolator.

10.4.1   The Decomposition

Consider an arbitrary sequence $\{x[n]\}$ with a z-transform $X(z)$:

$$X(z) = \sum_{n=-\infty}^{\infty} x[n]z^{-n} \tag{10.36}$$

We can rewrite $X(z)$ given above as

$$X(z) = \sum_{k=0}^{M-1} z^{-k} X_k(z^M) \tag{10.37}$$

where

$$X_k(z) = \sum_{n=-\infty}^{\infty} x_k[n]z^{-n} = \sum_{n=-\infty}^{\infty} x[Mn+k]z^{-n}, \qquad k = 0, 1, \ldots, M-1. \tag{10.38}$$

The subsequences $\{x_k[n]\}$ are called the polyphase components of the parent sequence $x[n]$, and the functions $X_k(z)$, given by the z-transform of $\{x_k[n]\}$, are called the *polyphase components* of $X(z)$ [Bel76]. The relation between the subsequences $\{x_k[n]\}$ and the original sequence $\{x[n]\}$ is given by

$$x_k[n] = x[Mn+k], \qquad k = 0, 1, 2, \ldots, M-1. \tag{10.39}$$

Equation (10.37) can be written in matrix form as

$$X(z) = \begin{bmatrix} 1 & z^{-1} & \cdots & z^{-(M-1)} \end{bmatrix} \begin{bmatrix} X_0(z^M) \\ X_1(z^M) \\ \vdots \\ X_{M-1}(z^M) \end{bmatrix}. \tag{10.40}$$

A multirate structural interpretation of the polyphase decomposition is given in Figure 10.28.

The polyphase decomposition of an FIR transfer function can be carried out by inspection as illustrated in Section 6.4.3, where we developed 2-branch and 3-branch polyphase decompositions of a length-9 FIR transfer function. Figure 6.18 illustrates the parallel realizations of an FIR transfer function based on the polyphase decomposition. In the following section, we consider several variations of the parallel realization of an FIR transfer function based on a polyphase decomposition.

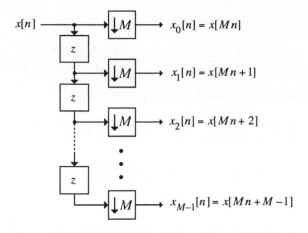

Figure 10.28   A structural interpretation of the M-band polyphase decomposition of a sequence $x[n]$.

The polyphase decomposition of an IIR transfer function $H(z) = P(z)/D(z)$, on the other hand, is not that straightforward. One way to arrive at an M-branch polyphase decomposition of $H(z)$ is to express it in the form $P'(z)/D'(z^M)$ by multiplying the denominator $D(z)$ and the numerator $P(z)$ with an appropriately chosen polynomial and then apply an M-branch polyphase decomposition to $P'(z)$. This approach is illustrated in the following example.

Example 10.9   Consider an IIR transfer function

$$H(z) = \frac{1 - 2z^{-1}}{1 + 3z^{-1}}.$$

To obtain a two-band decomposition of the form of Eq. (10.38) with $M = 2$, we rewrite $H(z)$ as

$$H(z) = \frac{\left(1 - 2z^{-1}\right)\left(1 - 3z^{-1}\right)}{\left(1 + 3z^{-1}\right)\left(1 - 3z^{-1}\right)} = \frac{1 - 5z^{-1} + 6z^{-2}}{1 - 9z^{-2}} = \frac{1 + 6z^{-2}}{1 - 9z^{-2}} + z^{-1}\frac{-5}{1 - 9z^{-2}}.$$

Therefore, $H(z) = E_0(z^2) + z^{-1}E_1(z^2)$, where

$$E_0(z) = \frac{1 + 6z^{-1}}{1 - 9z^{-1}}, \quad E_1(z) = \frac{-5}{1 - 9z^{-1}}.$$

Note that the above approach increases the overall order and the complexity of $H(z)$. However, when used in certain multirate structures, the approach may result in a more efficient structure. An alternative, more attractive approach is considered in the following example.

Example 10.10   Consider the transfer function of a fifth-order Butterworth lowpass filter with a 3-dB cutoff frequency at 0.5π:

$$H(z) = \frac{0.0527864045(1 + z^{-1})^5}{1 + 0.633436854z^{-2} + 0.05572809z^{-4}}.$$

It is easy to show that $H(z)$ can be expressed as

$$H(z) = \frac{1}{2}\left[\left(\frac{0.10557281 + z^{-2}}{1 + 0.10557281z^{-2}}\right) + z^{-1}\left(\frac{0.527864045 + z^{-2}}{1 + 0.527864045z^{-2}}\right)\right].$$

Therefore, we can express $H(z) = E_0(z^2) + z^{-1}E_1(z^2)$, where

$$E_0(z) = \frac{1}{2}\left(\frac{0.10557281 + z^{-1}}{1 + 0.10557281z^{-1}}\right), \qquad E_1(z) = \frac{1}{2}\left(\frac{0.527864045 + z^{-1}}{1 + 0.527864045z^{-1}}\right).$$

Note that in the above polyphase-like decomposition, the branch transfer functions are stable allpass functions. Moreover, the decomposition has not increased the order of the overall transfer function $H(z)$.

10.4.2   FIR Filter Structures Based on the Polyphase Decomposition

We have illustrated in Section 6.4.3 that a parallel realization of an FIR filter transfer function $H(z)$ can be obtained using a polyphase decomposition. As we shall point out later in this chapter, such a realization often results in computationally efficient structures in certain multirate applications. We revisit the polyphase decomposition based FIR filter realization again here using the more commonly used notation for the polyphase components and develop several other alternate realizations.

Consider first an M-branch polyphase decomposition of $H(z)$ given by

$$H(z) = \sum_{k=0}^{M-1} z^{-k} E_k(z^M). \tag{10.41}$$

A direct realization of the above is shown in Figure 10.29(a). The transpose of this realization is indicated in Figure 10.29(b). An alternative representation of the transpose structure of Figure 10.29(b) is obtained by using the notation

$$R_\ell(z^M) = E_{M-1-\ell}(z^M), \qquad \ell = 0, 1, \ldots, M-1, \tag{10.42}$$

resulting in the structure of Figure 10.30. The corresponding polyphase decomposition is thus given by

$$H(z) = \sum_{\ell=0}^{M-1} z^{-(M-1-\ell)} R_\ell(z^M). \tag{10.43}$$

To differentiate between the two decompositions of Eqs. (10.41) and (10.43), the former is usually called the *Type I polyphase decomposition*, while the latter is called *Type II polyphase decomposition*.

10.4.3   Computationally Efficient Interpolator and Decimator Structures

Computationally efficient decimator and interpolator structures employing linear-phase lowpass filters can be derived by applying a polyphase decomposition to the lowpass filters. We demonstrate this property next.

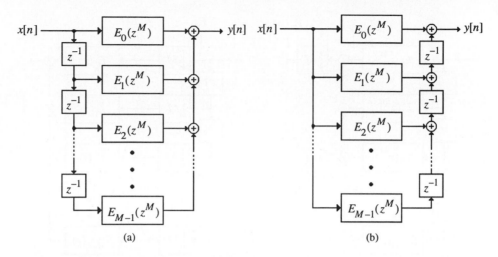

Figure 10.29   Realization of an FIR filter based on a Type I polyphase decomposition.

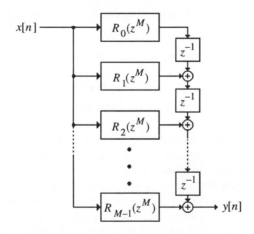

Figure 10.30   Realization of an FIR filter based on a Type II polyphase decomposition.

Consider first the use of the polyphase decomposition in the realization of the decimation filter of Figure 10.15(b). If the lowpass filter $H(z)$ is realized as in Figure 10.29(a), the overall decimator structure takes the form of Figure 10.31(a). By invoking the cascade equivalance #1 of Figure 10.14(a), this structure reduces to that indicated in Figure 10.31(b), which is computationally more efficient than the structure of Figure 10.15(b). To illustrate this point, assume that the decimation filter $H(z)$ of Figure 10.15(b) is a length-N FIR structure and the input sampling period $T = 1$. Since the decimator output $y[n]$ is obtained by down-sampling the filter output $v_1[n]$ by a factor-of-M, it is necessary only to compute $v[n]$ at $n = \ldots, -2M, -M, 0, M, 2M, \ldots$. The computational requirements are therefore N multiplications and $(N-1)$ additions per output sample being computed. However, as n increases, the stored signals in the delay registers change. As a result, all computations need to be completed in one sampling period, and for the following

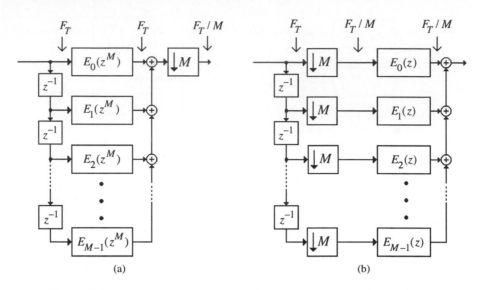

Figure 10.31   Decimator implementation based on a polyphase decomposition.

$(M - 1)$ sampling periods the arithmetic units remain idle. Now consider the structure of Figure 10.31(b). If the lengths of the subfilter $E_k(z)$ is N_k, then $N = \sum_{k=0}^{M-1} N_k$. The computational requirements of the kth subfilter are N_k multiplications and $N_k - 1$ additions per output sample, and that for the overall structure is therefore $\sum_{k=0}^{M-1} N_k = N$ multiplications and $\sum_{k=0}^{M-1}(N_k - 1) + (M - 1) = N - 1$ additions per decimator output sample. However, in the latter structure, the arithmetic units are operative at all instants of the output sampling period, which is M times that of the input sampling period.

Similar savings are also obtained in the case of interpolator structure employing polyphase decomposition in the realization of computationally efficient interpolators. Figure 10.32(a) shows the interpolator structure derived from Figure 10.15(a) by making use of the L-band Type I polyphase decomposition of the interpolation filter $H(z)$ and the cascade equivalence #2 of Figure 10.14(b). An alternative realization obtained using the Type II polyphase decomposition of the interpolation filter $H(z)$ and the cascade equivalence #2 is shown in Figure 10.32(b).

More efficient interpolator and decimator structures can be realized by exploiting the symmetry of the filter coefficients of $H(z)$ in the case of linear-phase filters. Consider for example the realization of a factor-of-3 ($M = 3$) decimator using a length-12 linear-phase FIR lowpass filter with a symmetric impulse response:

$$H(z) = h[0] + h[1]z^{-1} + h[2]z^{-2} + h[3]z^{-3} + h[4]z^{-4} + h[5]z^{-5} + h[5]z^{-6}$$
$$+ h[4]z^{-7} + h[3]z^{-8} + h[2]z^{-9} + h[1]z^{-10} + h[0]z^{-11}. \qquad (10.44)$$

A conventional polyphase decomposition of the above $H(z)$ yields the following subfilters:

$$E_0(z) = h[0] + h[3]z^{-1} + h[5]z^{-2} + h[2]z^{-3},$$
$$E_1(z) = h[1] + h[4]z^{-1} + h[4]z^{-2} + h[1]z^{-3}, \qquad (10.45)$$

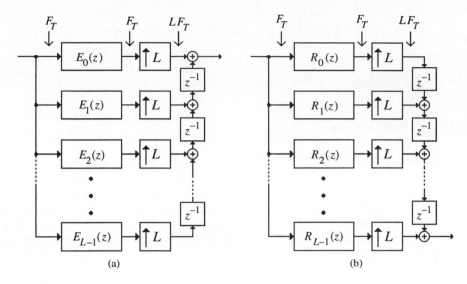

Figure 10.32 Computationally efficient interpolator structures: (a) Type I polyphase decomposition, and (b) Type II polyphase decomposition.

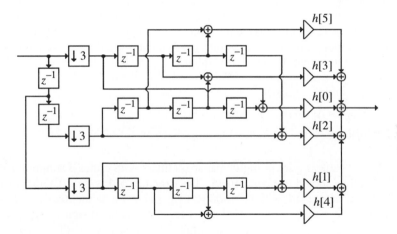

Figure 10.33 A computationally efficient realization of a factor-of-3 decimator exploiting the linear-phase symmetry of the length-12 decimation filter.

$$E_2(z) = h[2] + h[5]z^{-1} + h[3]z^{-2} + h[0]z^{-3}.$$

Note that the subfilter $E_1(z)$ still has a symmetric impulse response, whereas the impulse response of $E_2(z)$ is the mirror image of that of $E_0(z)$. These relations can be made use of in developing a computationally efficient realization using only 6 multipliers and 11 two-input adders, as depicted in Figure 10.33.

Figure 10.34   (a) A cascade multirate structure and (b) its equivalent form.

10.4.4   A Useful Identity

The cascade multirate digital filter structure of Figure 10.34(a) appears in a number of applications. If we express the transfer function $H(z)$ in its L-term Type I polyphase form $\sum_{k=0}^{L-1} z^{-k} E_k(z^L)$, it can be easily shown that the structure of Figure 10.34(a) is equivalent to the time-invariant digital filter of Figure 10.34(b), where $E_0(z)$ is the zeroth polyphase term (Problem 10.18) [Vai93]. This equivalence can be exploited in simplifying complex multirate networks containing cascade structures of the form of Figure 10.34(a) for analysis purposes.

10.5   Digital Filter Banks

So far, we have mostly concentrated on the design, realization, and applications of single-input, single-output digital filters. There are applications, as in the case of a spectrum analyzer, where it is desirable to separate a signal into a set of subband signals occupying, usually nonoverlapping, portions of the original frequency band. In other applications, it may be necessary to combine many such subband signals into a single composite signal occupying the whole Nyquist range. To this end, digital filter banks play an important role, and are the subject of discussion in this section.

10.5.1   Definitions

The digital filter bank is a set of digital bandpass filters with either a common input or a summed output, as shown in Figure 10.35. The structure of Figure 10.35(a) is called an M-band *analysis filter bank* with the subfilters $H_k(z)$ known as the *analysis filters*. It is used to decompose the input signal $x[n]$ into a set of M subband signals $v_k[n]$ with each subband signal occupying a portion of the original frequency band. (The signal is being "analyzed" by being separated into a set of narrow spectral bands.)

The *dual* of the above operation, whereby a set of subband signals $\hat{v}_k[n]$ (typically belonging to contiguous frequency bands) is combined into one signal $y[n]$ is called a *synthesis filter bank*. Figure 10.35(b) shows an L-band synthesis bank where each filter $F_k(z)$ is called a *synthesis filter*.

10.5.2   Uniform DFT Filter Banks

We now outline a simple technique for the design of a class of filter banks with equal passband widths. Let $H_0(z)$ represent a causal lowpass digital filter with an impulse response $h_0[n]$:

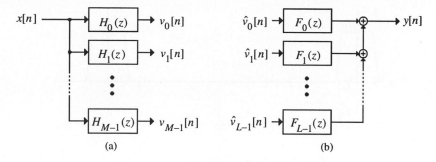

Figure 10.35   (a) Analysis filter bank, and (b) synthesis filter bank.

$$H_0(z) = \sum_{n=0}^{\infty} h_0[n]z^{-n}, \tag{10.46}$$

which we assume to be an IIR filter without any loss of generality. Let us now assume that $H_0(z)$ has its passband edge ω_p and stopband edge ω_s around π/M, where M is some arbitrary integer, as indicated in Figure 10.36(a). Now, consider the transfer function $H_k(z)$ whose impulse response $h_k[n]$ is defined to be

$$h_k[n] = h_0[n]W_M^{-kn}, \qquad k = 0, 1, \ldots, M-1, \tag{10.47}$$

where $W_M = e^{-j2\pi/M}$, as defined in Eq. (3.14). Thus,

$$H_k(z) = \sum_{n=0}^{\infty} h_k[n]z^{-n} = \sum_{n=0}^{\infty} h_0[n]\left(zW_M^k\right)^{-n}, \qquad k = 0, 1, \ldots, M-1, \tag{10.48}$$

i.e.,

$$H_k(z) = H_0\left(zW_M^k\right), \qquad k = 0, 1, \ldots, M-1, \tag{10.49}$$

with a corresponding frequency response

$$H_k(e^{j\omega}) = H_0(e^{j\left(\omega-\frac{2\pi k}{M}\right)}), \qquad k = 0, 1, \ldots, M-1. \tag{10.50}$$

In other words, the frequency response of $H_k(z)$ is obtained by shifting the response of $H_0(z)$ to the right, by an amount $2\pi k/M$. The responses of $H_1(z), H_2(z), \ldots, H_{M-1}(z)$ are shown in Figure 10.36(b). Note that the corresponding impulse responses $h_k[n]$ are, in general, complex and hence $|H_k(e^{j\omega})|$ does not necessarily exhibit symmetry with respect to zero frequency. Figure 10.36(b) therefore represents the responses of $M-1$ filters $H_1(z), H_2(z), \ldots, H_{M-1}(z)$, which are uniformly shifted versions of the response of the basic *prototype* filter $H_0(z)$ of Figure 10.36(a).

The M filters $H_k(z)$ defined by Eq. (10.49) could be used as the analysis filters in the analysis filter bank of Figure 10.35(a) or as the synthesis filters $F_k(z)$ in the synthesis filter bank of Figure 10.35(b).

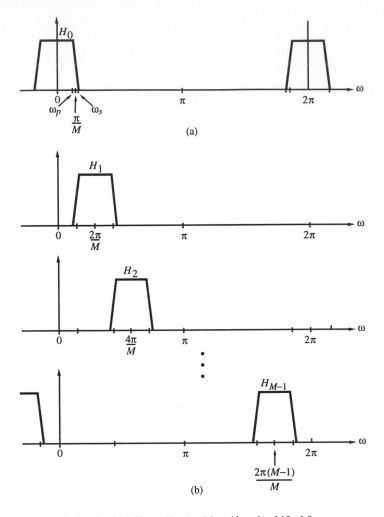

Figure 10.36   The bank of M filters $H_k(z)$ with uniformly shifted frequency responses.

Since the set of magnitude responses $|H_k(e^{j\omega})|$, $k = 0, 1, \ldots, M-1$, are uniformly shifted versions of a basic prototype $|H_0(e^{j\omega})|$, i.e.,

$$\left| H_k(e^{j\omega}) \right| = \left| H_0 \left(e^{j[\omega - (2\pi k/M)]} \right) \right|, \tag{10.51}$$

the filter bank obtained is called a *uniform filter bank*.

10.5.3   Polyphase Implementations of Uniform Filter Banks

Let Figure 10.35(a) represent a uniform filter bank with the M analysis filters $H_k(z)$ related through Eq. (10.49). The impulse response sequences $h_k[n]$ of the analysis filters are accordingly related as in Eq. (10.47). Instead of realizing each analysis filter as a separate filter, it is possible to develop a computationally more efficient realization of the above uniform filter bank, which is described next.

Let the lowpass prototype transfer function $H_0(z)$ be represented in its M-band polyphase form:

$$H_0(z) = \sum_{\ell=0}^{M-1} z^{-\ell} E_\ell(z^M), \tag{10.52}$$

where $E_\ell(z)$ is the ℓth polyphase component of $H_0(z)$:

$$E_\ell(z) = \sum_{n=0}^{\infty} e_\ell[n] z^{-n} = \sum_{n=0}^{\infty} h_0[\ell + nM] z^{-n}, \qquad 0 \le \ell \le M - 1. \tag{10.53}$$

Substituting z with $z W_M^k$ in Eq. (10.52), we arrive at the M-band polyphase decomposition of $H_k(z)$:

$$H_k(z) = \sum_{\ell=0}^{M-1} z^{-\ell} W_M^{-k\ell} E_\ell(z^M W_M^{kM}) = \sum_{\ell=0}^{M-1} z^{-\ell} W_M^{-k\ell} E_\ell(z^M),$$

$$k = 0, 1, \ldots, M - 1, \tag{10.54}$$

where we have used the identity $W_M^{kM} = 1$.

Note that Eq. (10.54) can be written in matrix form as

$$H_k(z) = \begin{bmatrix} 1 & W_M^{-k} & W_M^{-2k} & \cdots & W_M^{-(M-1)k} \end{bmatrix} \begin{bmatrix} E_0(z^M) \\ z^{-1} E_1(z^M) \\ z^{-2} E_2(z^M) \\ \vdots \\ z^{-(M-1)} E_{M-1}(z^M) \end{bmatrix}, \tag{10.55}$$

for $k = 0, 1, \ldots, M - 1$. All these M equations can be combined into a matrix equation as

$$\begin{bmatrix} H_0(z) \\ H_1(z) \\ H_2(z) \\ \vdots \\ H_{M-1}(z) \end{bmatrix}$$

$$= \begin{bmatrix} 1 & 1 & 1 & \cdots & 1 \\ 1 & W_M^{-1} & W_M^{-2} & \cdots & W_M^{-(M-1)} \\ 1 & W_M^{-2} & W_M^{-4} & \cdots & W_M^{-2(M-1)} \\ \vdots & \vdots & \vdots & \ddots & \vdots \\ 1 & W_M^{-(M-1)} & W_M^{-2(M-1)} & \cdots & W_M^{-(M-1)^2} \end{bmatrix} \begin{bmatrix} E_0(z^M) \\ z^{-1} E_1(z^M) \\ z^{-2} E_2(z^M) \\ \vdots \\ z^{-(M-1)} E_{M-1}(z^M) \end{bmatrix}, \tag{10.56}$$

which is equivalent to

$$\begin{bmatrix} H_0(z) \\ H_1(z) \\ H_2(z) \\ \vdots \\ H_{M-1}(z) \end{bmatrix} = M \mathbf{D}^{-1} \begin{bmatrix} E_0(z^M) \\ z^{-1} E_1(z^M) \\ z^{-2} E_2(z^M) \\ \vdots \\ z^{-(M-1)} E_{M-1}(z^M) \end{bmatrix}, \tag{10.57}$$

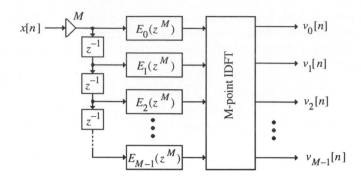

Figure 10.37   Polyphase implementation of a uniform DFT analysis filter bank where $H_k(z) = V_k(z)/X(z)$.

where **D** denotes the DFT matrix:

$$
\mathbf{D} = \begin{bmatrix}
1 & 1 & 1 & \cdots & 1 \\
1 & W_M^1 & W_M^2 & \cdots & W_M^{(M-1)} \\
1 & W_M^2 & W_M^4 & \cdots & W_M^{2(M-1)} \\
\vdots & \vdots & \vdots & \ddots & \vdots \\
1 & W_M^{(M-1)} & W_M^{2(M-1)} & \cdots & W_M^{(M-1)^2}
\end{bmatrix}.
\tag{10.58}
$$

An efficient implementation of the M-band analysis filter bank based on Eq. (10.57) is thus as shown in Figure 10.37, where the prototype lowpass filter $H_0(z)$ has been implemented in a polyphase form. The structure of Figure 10.37 is more commonly known as the *uniform DFT analysis filter bank*.

The computational complexity of Figure 10.37 is much smaller than that of a direct implementation, as in Figure 10.35(a). For example, an M-band uniform-DFT analysis filter bank based on an N-tap prototype lowpass filter requires a total of $(M/2)\log_2 M + N$ multipliers, whereas a direct implementation requires NM multiplications (Problem 10.22).

Following a development similar to that outlined above, we can derive the structure for a *uniform DFT synthesis filter bank*. Efficient realizations based on the Types I and II polyphase decompositions of the prototype lowpass filter $H_0(z)$ are indicated in Figure 10.38.

10.6   Nyquist Filters

In this section, we introduce a special type of lowpass filter with a transfer function that by design has certain zero-valued coefficients. Due to the presence of these zero-valued coefficients, these filters are by nature computationally more efficient than other lowpass filters of the same order. In addition, when used as interpolator filters, they preserve the nonzero samples of the up-sampler output at the interpolator output. These filters, called Lth band filters or Nyquist filters and discussed in this section, are often used both in single-rate and multirate signal processing. For example, they are usually preferred in decimator and interpolator design. Another application is in the design of a quadrature-mirror filter

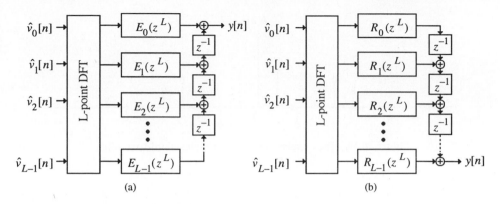

Figure 10.38   Uniform DFT synthesis filter bank.

bank, discussed in Section 10.7. A third application, described in Section 11.6, is in the design of a Hilbert transformer, which is employed for the generation of analytic signals.

10.6.1 Lth Band Filters

Consider the factor-of-L interpolator of Figure 10.15(a). The relation between the output and the input of the interpolator is given by

$$Y(z) = H(z)X(z^L). \tag{10.59}$$

If the interpolation filter $H(z)$ is realized in the L band polyphase form, then we have

$$H(z) = E_0(z^L) + z^{-1}E_1(z^L) + z^{-2}E_2(z^L) + \cdots + z^{-(L-1)}E_{L-1}(z^L).$$

Assume that the kth polyphase component of $H(z)$ is a constant, i.e., $E_k(z) = \alpha$:

$$\begin{aligned} H(z) = &\, E_0(z^L) + z^{-1}E_1(z^L) + \cdots + z^{-(k-1)}E_{k-1}(z^L) + \alpha z^{-k} \\ &+ z^{-(k+1)}E_{k+1}(z^L) + \cdots + z - (L-1)E_{L-1}(zL). \end{aligned} \tag{10.60}$$

Then we can express $Y(z)$ as

$$Y(z) = \alpha z^{-k}X(z^L) + \sum_{\substack{\ell=0 \\ \ell \neq k}}^{L-1} z^{-\ell}E_\ell(z^L)X(z^L). \tag{10.61}$$

As a result, $y[Ln] = \alpha x[n-k]$, i.e., the input samples appear at the output without any distortion at $n = k, k \pm L, k \pm 2L, \ldots$, whereas the in-between $(L-1)$ samples are determined by interpolation.

A filter with the above property is called a *Nyquist filter* or an *Lth band filter* and its impulse response has many zero-valued samples, making it computationally very attractive. For example, the impulse response of the Lth band filter obtained for $k = 0$ satisfies the following condition:

$$h[Ln] = \begin{cases} \alpha, & n = 0, \\ 0, & \text{otherwise.} \end{cases} \tag{10.62}$$

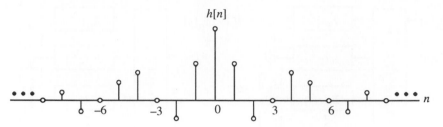

Figure 10.39   The impulse response of a typical third-band filter.

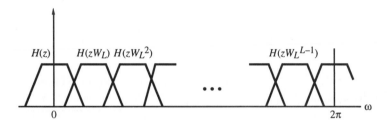

Figure 10.40   Frequency responses of $H(zW_L^k)$ for $k = 0, 1, \ldots, L-1$.

Figure 10.39 shows a typical impulse response of a third-band filter ($L = 3$). If $H(z)$ satisfies Eq. (10.60) with $k = 0$, i.e., $E_0(z) = \alpha$, then it can be shown that

$$\sum_{k=0}^{L-1} H(zW_L^k) = L\alpha = 1 \qquad \text{(assuming } \alpha = 1/L\text{).} \qquad (10.63)$$

Since the frequency response of $H(zW_L^k)$ is the shifted version $H(e^{j[\omega - (2\pi k/L)]})$ of $H(e^{j\omega})$, the sum of all of these L uniformly shifted versions of $H(e^{j\omega})$ add up to a constant (see Figure 10.40). Lth band filters can be either FIR or IIR filters.

10.6.2   Half-Band Filters

An Lth band filter for $L = 2$ is called a *half-band filter*. From Eq. (10.60) the transfer function of a half-band filter is thus given by

$$H(z) = \alpha + z^{-1}E_1(z^2), \qquad (10.64)$$

with its impulse response satisfying Eq. (10.62) with $L = 2$. The condition on frequency response given by Eq. (10.64) reduces to

$$H(z) + H(-z) = 1 \qquad \text{(assuming } \alpha = \tfrac{1}{2}\text{)} \qquad (10.65)$$

If $H(z)$ has real coefficients, then $H(-e^{j\omega}) = H(e^{j(\pi-\omega)})$, and Eq. (10.65) leads to

$$H(e^{j\omega}) + H(e^{j(\pi-\omega)}) = 1 \qquad (10.66)$$

Figure 10.41   Frequency response of a zero-phase half-band filter.

The above equality implies that $H(e^{j(\pi/2-\theta)})$ and $H(e^{j(\pi/2+\theta)})$ add up to unity for all θ. In other words, $H(e^{j\omega})$ exhibits a symmetry with respect to the half-band frequency $\pi/2$, thus justifying the name "half-band filter". Figure 10.41 illustrates this symmetry for a half-band lowpass filter for which the passband and stopband ripples are equal, i.e., $\delta_p = \delta_s$, and the passband and stopband bandedges are symmetric with respect to $\pi/2$, i.e., $\omega_p + \omega_s = \pi$.

An important attractive property of the half-band filter is that about 50% of the co-efficients of $h[n]$ are zero. This reduces the number of multiplications required in its implementation, making the filter computationally quite efficient. For example, if $N = 101$, an arbitrary Type 1 FIR transfer function requires about 50 multipliers, whereas a Type 1 half-band filter requires only about 25 multipliers.

An FIR half-band filter can be designed with linear phase. However, there is a constraint on its length. Consider a zero-phase half-band FIR filter for which $h[n] = ah^*[-n]$, with $|a| = 1$. Let the highest nonzero coefficient be $h[R]$. Then R is odd as a result of the condition of Eq. (10.62). Therefore, $R = 2K + 1$ for some integer K. Thus the length of the impulse response $h[n]$ is restricted to be of the form $2R + 1 = 4K + 3$ [unless $H(z)$ is a constant].

10.6.3   Design of Linear-Phase Lth Band FIR Filters

A lowpass linear-phase Nyquist Lth band FIR filter with a cutoff at $\omega_c = \pi/L$ and a good frequency response can be readily designed via the windowed Fourier series approach described in Section 7.7. In this approach the impulse response coefficients of the lowpass filter are chosen as

$$h[n] = h_{LP}[n] \cdot w[n], \tag{10.67}$$

where $h_{\text{LP}}[n]$ is the impulse response of an ideal lowpass filter with a cutoff at π/L, and $w[n]$ is a suitable window function. If

$$h_{\text{LP}}[n] = 0 \qquad \text{for } n = \pm L, \pm 2L, \ldots, \tag{10.68}$$

then Eq. (10.62) is indeed satisfied.

Now the impulse response $h_{LP}[n]$ of an ideal Lth band filter is obtained from Eq. (7.87) by substituting $\omega_c = \pi/L$ and is given by

$$h_{LP}[n] = \frac{\sin(\pi n/L)}{\pi n}, \quad -\infty \le n \le \infty. \tag{10.69}$$

It can be seen from the above that the impulse-response coefficients do indeed satisfy the condition of Eq. (10.68). Hence, an Lth band filter can be designed by applying a suitable window function to Eq. (10.69).

Likewise, an Lth band filter with a frequency response as in Figure 7.26(a) can also be designed from Eq. (7.114) by replacing ω_c with π/L, resulting in an impulse response

$$h_{LP}[n] = \begin{cases} \frac{1}{L}, & n = 0, \\ \frac{2\sin(\Delta\omega n/2)}{\Delta\omega n} \cdot \frac{\sin(\pi n/L)}{\pi n}, & |n| > 0, \end{cases} \tag{10.70}$$

which again is seen to satisfy the condition of Eq. (10.68). Other candidates for Lth band filter design are the lowpass filters of Eq. (7.115) and the raised cosine filter of Eq. (7.165) in Problem 7.22.

We illustrate the half-band lowpass filter design in the following example.

Example 10.11 Design a length-23 half-band ($L = 2$) linear-phase lowpass filter using the windowed Fourier series approach with a Hamming window. To this end we use the MATLAB Program 10_8 given below.

```
% Program 10_8
% Design of L-th Band FIR Filter Using the
% Windowed Fourier Series Approach
%
N = input('Type in the filter length = ');
L = input('Type in the value of L = ');
K = (N-1)/2;
n = -K:K;
% Generate the truncated impulse response of the ideal
% lowpass filter
b = sinc(n/L)/L;
% Generate the window sequence
win = hamming(N);
% Generate the coefficients of the windowed filter
c = b.*win';
% Plot the gain response of the windowed filter
[h,w] = freqz(fil,1,256);
g = 20*log10(abs(h));
plot(w/pi,g);grid
xlabel('Normalized frequency');ylabel('Gain, dB');
```

The input data requested by the program are the desired filter length N and the value of L. The program determines the impulse response coefficients of the Lth band FIR filter using the expression of Eq. (10.69), and computes and plots the gain response of the designed filter as shown in Figure 10.42. The filter coefficients are elements of the vector c and are given by:

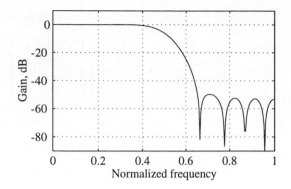

Figure 10.42   Gain response of a length-23 linear-phase half-band FIR filter.

```
h[-11] = h[11] = -0.00231498033043;   h[-10] = h[10] = 0;
 h[-9] = h[9]  =  0.00541209478301;    h[-8] = h[8]  = 0;
 h[-7] = h[7]  = -0.01586588771510;    h[-6] = h[6]  = 0;
 h[-5] = h[5]  =  0.03854508794537;    h[-4] = h[4]  = 0;
 h[-3] = h[3]  = -0.08925790518530;    h[-2] = h[2]  = 0;
 h[-1] = h[1]  =  0.31237874418292;     h[0] = 0.5.
```

As expected, the impulse response coefficients are equal to zero for $n = \pm 2, \pm 4, \ldots, \pm 10$.

An elegant method for the design of half-band linear-phase FIR filters is considered in Section 11.6.3. Also described there is a method for the design of half-band IIR filters. Several other design approaches have been advanced in the literature [Ren87], [Vai87a].

10.6.4   Relation Between Lth Band Filters and Power-Complementary Filters

Recall from Section 4.6.3, a set of filters is said to be power-complementary if their square magnitude responses add up to a constant. Consider an Lth band transfer function $H(z)$ represented in the L-band polyphase decomposition form given by

$$H(z) = \sum_{\ell=0}^{L-1} z^{-\ell} E_\ell(z^L) \tag{10.71}$$

Define a new transfer function $G(z) = \tilde{H}(z)$.[9] Then the set of transfer functions $\{E_0(z), E_1(z), \ldots, E_{L-1}(z)\}$ is power-complementary if and only if $G(z)$ is an Lth band filter [Vai90].

To prove the above property, define

$$H_r(z) = H(zW_L^{-r}) = \sum_{\ell=0}^{L-1} z^{-\ell} W_L^{\ell r} E_\ell(z^L) \tag{10.72}$$

---

[9] $\tilde{H}(z) = H(1/z^*)$.

for $0 \leq r \leq L - 1$. We can write the above set of equations in matrix form as

$$
\underbrace{\begin{bmatrix} H_0(z) \\ H_1(z) \\ \vdots \\ H_{L-1}(z) \end{bmatrix}}_{\mathbf{H}(z)} = \mathbf{D} \underbrace{\begin{bmatrix} 1 & 0 & \cdots & 0 \\ 0 & z^{-1} & \cdots & 0 \\ \vdots & \vdots & \ddots & \vdots \\ 0 & 0 & \cdots & z^{-(L-1)} \end{bmatrix}}_{\boldsymbol{\Lambda}(z)} \underbrace{\begin{bmatrix} E_0(z^L) \\ E_1(z^L) \\ \vdots \\ E_{L-1}(z^L) \end{bmatrix}}_{\mathbf{E}(z)} \tag{10.73}
$$

where $\mathbf{D}$ is the $L \times L$ DFT matrix satisfying $\mathbf{D}^{\oplus}\mathbf{D} = L\mathbf{I}$.[10] If the set of transfer functions $\{E_0(z), E_1(z), \dots, E_{L-1}(z)\}$ is power-complementary, then $\tilde{\mathbf{E}}(z)\mathbf{E}(z) = \gamma$ so that[11]

$$
\tilde{\mathbf{H}}(z)\mathbf{H}(z) = \tilde{\mathbf{E}}(z)\tilde{\boldsymbol{\Lambda}}(z)\mathbf{D}^{\oplus}\mathbf{D}\boldsymbol{\Lambda}(z)\mathbf{E}(z) = L\gamma. \tag{10.74}
$$

This in turn implies that the $\{H_0(z), H_1(z), \dots, H_{L-1}(z)\}$ is power-complementary. In other words, $G(z) = \tilde{H}(z)H(z)$ satisfies

$$
\sum_{k=0}^{L-1} G(zW_L^{-k}) = L\gamma \tag{10.75}
$$

so that it is an Lth band filter.

Conversely, assuming that $G(z)$ is an Lth band filter, we can prove that the set $\{E_0(z), E_1(z), \dots, E_{L-1}(z)\}$ is power-complementary simply by inverting the above matrix equation and carrying out an argument similar to that outlined above.

10.7   Two-Channel Quadrature-Mirror Filter Bank

In many applications, a discrete-time signal $x[n]$ is first split into a number of subband signals $\{v_k[n]\}$ by means of an analysis filter bank, the subband signals are then processed and finally combined by a synthesis filter bank resulting in an output signal $y[n]$. If the subband signals are bandlimited to frequency ranges much smaller than that of the original input signal, they can be down-sampled before processing. Because of the lower sampling rate, the processing of the down-sampled signals can be carried out more efficiently. After processing, these signals are up-sampled before being combined by the synthesis bank into a higher rate signal. The combined structure employed is called a *quadrature-mirror filter (QMF) bank*. If the down-sampling and the up-sampling factors are equal to the number of bands of the filter bank, then the output $y[n]$ can be made to retain some or all of the characteristics of the input $x[n]$ by properly choosing the filters in the structure. In this case, the filter bank is said to be a *critically sampled filter bank*. The most common application of this scheme is in the efficient coding of a signal $x[n]$ (see Section 11.8). Another possible application is in the design of an analog voice privacy system to provide secure telephone conversation, which is described in Section 11.7. In this section, we study a two-channel QMF bank.

---

[10]$\mathbf{D}^{\oplus}$ denotes the conjugate-transpose of $\mathbf{D}$.

[11]$\tilde{\mathbf{H}}(z)$ is the conjugate-transpose of $\mathbf{H}(1/z^*)$, i.e., $\tilde{\mathbf{H}}(z) = \mathbf{H}^{\oplus}(1/z^*)$.

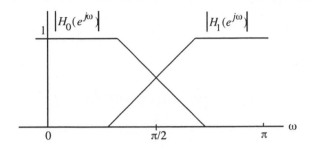

Figure 10.43   The two-channel quadrature-mirror filter (QMF) bank.

Figure 10.44   Typical frequency responses of the analysis filters.

10.7.1   The Filter Bank Structure

Figure 10.43 illustrates the basic two-channel quadrature-mirror filter (QMF) bank structure [Cro76a], [Est77]. Here, the input signal $x[n]$ is first passed through a two-band analysis filter bank containing the filters $H_0(z)$ and $H_1(z)$, which typically have lowpass and highpass frequency responses, respectively, with a cutoff frequency at $\pi/2$, as indicated in Figure 10.44. The subband signals $\{v_k[n]\}$ are then down-sampled by a factor of 2. In subband coding applications, coders are inserted after the down-sampler in each subband channel and each down-sampled subband signal is encoded by exploiting the special spectral properties of the signal, such as energy levels, perceptual importance, etc. (see Section 11.8). At the receiving end, decoders are used to produce approximations of the original down-sampled signals. The decoded signals are then up-sampled by a factor of 2 and passed through a two-band synthesis filter bank composed of the filters $G_0(z)$ and $G_1(z)$ whose outputs are then added yielding $y[n]$. The analysis and the synthesis filters in the QMF bank are chosen so as to ensure that the reconstructed output $y[n]$ is a reasonable replica of the input $x[n]$. In practice, various errors are generated in this scheme. In addition to the coding error, the QMF bank itself introduces several errors due to the sampling rate alterations and imperfect filters. We ignore the coding errors and investigate only the errors generated by the down-samplers and up-samplers in the filter bank and their effects on the performance of the system.

10.7.2   Analysis of the Two-Channel QMF Bank

It is convenient to analyze the filter bank in the z-domain. To this end, we make use of the input-output relations of the up-sampler and the down-sampler derived earlier in Section

10.1 and given by Eqs. (10.5) and (10.12). The expressions for the z-transforms of various intermediate signals in Figure 10.43 are then given by

$$V_k(z) = H_k(z)X(z), \tag{10.76a}$$

$$U_k(z) = \tfrac{1}{2}\left\{ V_k(z^{1/2}) + V_k(-z^{1/2}) \right\}, \tag{10.76b}$$

$$\hat{V}_k(z) = U_k(z^2). \tag{10.76c}$$

for $k = 0, 1$. From Eqs. (10.76a)–(10.76c), we obtain after some algebra

$$\hat{V}_k(z) = \tfrac{1}{2}\{V_k(z) + V_k(-z)\} = \tfrac{1}{2}\{H_k(z)X(z) + H_k(-z)X(-z)\}. \tag{10.77}$$

The reconstructed output of the filter bank is given by

$$Y(z) = G_0(z)\hat{V}_0(z) + G_1(z)\hat{V}_1(z). \tag{10.78}$$

Substituting Eq. (10.77) in the above, we obtain after some rearrangement the expression for the output of the filter bank as

$$
\begin{aligned}
Y(z) = {} & \tfrac{1}{2}\{H_0(z)G_0(z) + H_1(z)G_1(z)\}\, X(z) \\
& + \tfrac{1}{2}\{H_0(-z)G_0(z) + H_1(-z)G_1(z)\}\, X(-z).
\end{aligned}
\tag{10.79}
$$

The second term in the above equation is precisely due to the aliasing caused by sampling rate alteration. The above equation can be compactly expressed as

$$Y(z) = T(z)X(z) + A(z)X(-z), \tag{10.80}$$

where

$$T(z) = \tfrac{1}{2}\{H_0(z)G_0(z) + H_1(z)G_1(z)\}, \tag{10.81}$$

is called the *distortion transfer function*, and

$$A(z) = \tfrac{1}{2}\{H_0(-z)G_0(z) + H_1(-z)G_1(z)\}. \tag{10.82}$$

A matrix representation of Eq. (10.79) is given by:

$$Y(z) = \tfrac{1}{2}\,[\, X(z) \quad X(-z)\,]\begin{bmatrix} H_0(z) & H_1(z) \\ H_0(-z) & H_1(-z) \end{bmatrix}\begin{bmatrix} G_0(z) \\ G_1(z) \end{bmatrix}. \tag{10.83}$$

The 2×2 matrix in the above equation,

$$\mathbf{H}(z) = \begin{bmatrix} H_0(z) & H_1(z) \\ H_0(-z) & H_1(-z) \end{bmatrix}, \tag{10.84}$$

is called the *aliasing component (AC) matrix* [Smi85].

10.7.3   Errors in the QMF Bank

As noted in Section 10.1, the up-sampler and the down-sampler are linear time-varying components, and as a result, in general, the QMF structure of Figure 10.43 is a linear time-varying system. It can be shown also that it has a period of 2 (Problem 10.28). However, it is possible to choose the analysis and synthesis filters such that the aliasing effect is canceled, resulting in a linear, time-invariant (LTI) operation. To this end, we need to ensure that

$$2A(z) = H_0(-z)G_0(z) + H_1(-z)G_1(z) = 0. \tag{10.85}$$

There are various possible solutions of the above equation. For example, one solution is given by

$$G_0(z) = H_1(-z), \qquad G_1(z) = -H_0(-z). \tag{10.86}$$

If above relations hold, then Eq. (10.79) reduces to

$$Y(z) = T(z)X(z), \tag{10.87}$$

with

$$T(z) = \tfrac{1}{2}\left\{H_0(z)H_1(-z) - H_1(z)H_0(-z)\right\}. \tag{10.88}$$

On the unit circle, we have

$$Y(e^{j\omega}) = T(e^{j\omega})X(e^{j\omega}) = \left|T(e^{j\omega})\right| e^{j\phi(\omega)} X(e^{j\omega}) \tag{10.89}$$

If $T(z)$ is an allpass function, i.e., $\left|T(e^{j\omega})\right| = d \neq 0$, then

$$\left|Y(e^{j\omega})\right| = d \left|X(e^{j\omega})\right| \tag{10.90}$$

indicating that the output of the QMF bank has the same magnitude as that of the input, and the filter bank is said to be *magnitude preserving*. If $T(z)$ is linear phase, i.e.,

$$\phi(\omega) = \alpha\omega + \beta, \tag{10.91}$$

then

$$\arg\left\{Y(e^{j\omega})\right\} = \arg\left\{X(e^{j\omega})\right\} + \alpha\omega + \beta, \tag{10.92}$$

and the filter bank is said to be *phase preserving*. If an alias-free QMF bank has no amplitude and phase distortion, then it is called a *perfect reconstruction (PR)* QMF bank. In such a case, $T(z) = dz^{-n_0}$, resulting in

$$Y(z) = dz^{-n_0}X(z), \tag{10.93}$$

which in the time-domain is equivalent to

$$y[n] = dx[n - n_0] \tag{10.94}$$

for all possible inputs, indicating that the reconstructed output $y[n]$ is a scaled, delayed replica of the input.

10.7.4   An Alias-Free Realization

A very simple alias-free two-band QMF bank is obtained when

$$H_1(z) = H_0(-z). \tag{10.95}$$

The above condition, in the case of a real coefficient filter, implies

$$\left|H_1(e^{j\omega})\right| = \left|H_0(e^{j(\pi-\omega)})\right| \tag{10.96}$$

indicating that if $H_0(z)$ is a lowpass filter, then $H_1(z)$ is a highpass filter, and vice-versa. In fact, Eq. (10.96) indicates that $\left|H_1(e^{j\omega})\right|$ is a mirror image of $\left|H_0(e^{j\omega})\right|$ with respect to the $\pi/2$, the quadrature frequency. This has given rise to the name quadrature-mirror filter bank.

Substituting Eq. (10.95) in Eq. (10.86), we arrive at

$$G_0(z) = H_0(z), \quad G_1(z) = -H_1(z) = -H_0(-z). \tag{10.97}$$

Equations (10.95) and (10.97) imply that the two analysis filters and the two synthesis filters in the QMF bank are essentially determined from one transfer function $H_0(z)$. Moreover, Eq. (10.97) indicates that if $H_0(z)$ is a lowpass filter then $G_0(z)$ is also a lowpass filter, and $G_1(z)$ is a highpass filter. The distortion transfer function $T(z)$ of Eq. (10.88) in this case reduces to

$$T(z) = \tfrac{1}{2}\left\{ H_0^2(z) - H_1^2(z) \right\} = \tfrac{1}{2}\left\{ H_0^2(z) - H_0^2(-z) \right\} \tag{10.98}$$

A computationally efficient realization of the above alias-free two-channel QMF bank is obtained by realizing the analysis and the synthesis filters in polyphase form. Let the two-band Type I polyphase representation of $H_0(z)$ be given by

$$H_0(z) = E_0(z^2) + z^{-1}E_1(z^2). \tag{10.99a}$$

From Eq. (10.95) it follows then that

$$H_1(z) = E_0(z^2) - z^{-1}E_1(z^2). \tag{10.99b}$$

In matrix form Eqs. (10.99a) and (10.99b) can be expressed as

$$\begin{bmatrix} H_0(z) \\ H_1(z) \end{bmatrix} = \begin{bmatrix} 1 & 1 \\ 1 & -1 \end{bmatrix} \begin{bmatrix} E_0(z^2) \\ z^{-1}E_1(z^2) \end{bmatrix}. \tag{10.100}$$

Likewise the synthesis filters, in matrix form, can be expressed as

$$[\, G_0(z) \quad G_1(z) \,] = [\, z^{-1}E_1(z^2) \quad E_0(z^2) \,]\begin{bmatrix} 1 & 1 \\ 1 & -1 \end{bmatrix}. \tag{10.101}$$

Using Eqs. (10.100) and (10.101), we can redraw the two-channel QMF bank as shown in Figure 10.45(a), which can be further simplified using the cascade equivalences of Figure 10.14, resulting in the computationally efficient realization of Figure 10.45(b).

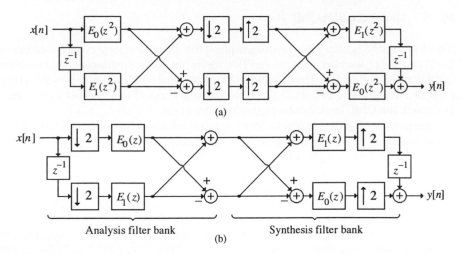

(a)

Analysis filter bank   (b)   Synthesis filter bank

Figure 10.45 Polyphase realization of the two-channel QMF bank. (a) Direct realization and (b) computationally efficient realization.

The expression for the distortion transfer function in this case, obtained by substituting Eqs. (10.99a) and (10.99b) in Eq. (10.88), is given by

$$T(z) = 2z^{-1}E_0(z^2)E_1(z^2).\qquad(10.102)$$

The following example illustrates the development of a very simple perfect reconstruction QMF bank.

Example 10.12 Consider a two-channel QMF bank with the analysis filter given by

$$H_0(z) = 1 + z^{-1}.$$

Its polyphase components are thus

$$E_0(z^2) = 1, \qquad E_1(z^2) = 1.$$

From Eq. (10.99b), the second analysis filter is therefore given by

$$H_1(z) = 1 - z^{-1},$$

and the corresponding synthesis filters for an alias-free realization from Eq. (10.101) are given by

$$G_0(z) = z^{-1}E_1(z^2) + E_0(z^2) = 1 + z^{-1},$$
$$G_1(z) = z^{-1}E_1(z^2) - E_0(z^2) = -1 + z^{-1}.$$

The distortion transfer function for this realization is given by

$$T(z) = 2z^{-1},$$

resulting in the realization of a perfect reconstruction QMF bank.

10.7.5   Alias-Free FIR QMF Bank

If the basic analysis filter $H_0(z)$ is a linear-phase FIR filter, then its polyphase components, $E_0(z)$ and $E_1(z)$, are also linear-phase FIR transfer functions. In this case, the two-channel QMF bank of Figure 10.43 has no phase distortion since here $T(z)$ given by Eq. (10.102) exhibits a linear-phase characteristic. However, in general, $|T(e^{j\omega})|$ is not a constant and, as a result, the QMF bank exhibits amplitude distortion.

We now develop the expression for the residual amplitude distortion and outline methods to minimize it. Let the prototype transfer function $H_0(z)$ be a length-N linear-phase real coefficient function given by

$$H_0(z) = \sum_{n=0}^{N-1} h_0[n]z^{-n}. \tag{10.103}$$

Note that $H_0(z)$ can be either a Type 1 or Type 2 linear-phase function since it has to be a lowpass filter. As a result, its impulse-response coefficients must satisfy the condition $h_0[n] = h_0[N - 1 - n]$, in which case we can write

$$H_0(e^{j\omega}) = e^{-j\omega(N-1)/2}\breve{H}_0(\omega) \tag{10.104}$$

where $\breve{H}_0(\omega)$ is the amplitude function, a real function of ω. By making use of above in Eq. (10.98) along with the property that $|H_0(e^{j\omega})|$ is an even function of ω, we can express the frequency response of the distortion transfer function as

$$T(e^{j\omega}) = \frac{e^{-j(N-1)\omega}}{2}\left\{\left|H_0(e^{j\omega})\right|^2 - (-1)^{N-1}\left|H_0(e^{j(\pi-\omega)})\right|^2\right\}. \tag{10.105}$$

From the above it can be seen that if N is odd, then $T(e^{j\omega}) = 0$ at $\omega = \pi/2$, implying severe amplitude distortion at the output of the filter bank. As a result, N must be chosen to be even, in which case Eq. (10.105) reduces to

$$\begin{aligned}
T(e^{j\omega}) &= \frac{e^{-j(N-1)\omega}}{2}\left\{\left|H_0(e^{j\omega})\right|^2 + \left|H_0(e^{j(\pi-\omega)})\right|^2\right\} \\
&= \frac{e^{-j(N-1)\omega}}{2}\left\{\left|H_0(e^{j\omega})\right|^2 + \left|H_1(e^{j\omega})\right|^2\right\}.
\end{aligned} \tag{10.106}$$

As can be seen from the above, the QMF bank will exhibit amplitude distortion unless $|T(e^{j\omega})|$ is almost constant for all values of ω. If $H_0(z)$ is a very good lowpass filter with $|H_0(e^{j\omega})| \cong 1$ in the passband and $|H_0(e^{j\omega})| \cong 0$ in the stopband, then $H_1(z)$ is a very good highpass filter with its passband coinciding with the stopband of $H_0(z)$, and vice-versa. As a result, $|T(e^{j\omega})| \cong 1/2$ in the passbands of $H_0(z)$ and $H_1(z)$. The amplitude distortion thus occurs primarily in the transition band of these filters, with the degree of distortion being determined by the amount of overlap between their squared-magnitude responses. This distortion can be minimized by controlling the overlap, which in turn can be controlled by appropriately choosing the passband edge of $H_0(z)$.

One way to minimize the amplitude distortion is to employ a computer-aided optimization method to iteratively adjust the filter coefficients $h_0[n]$ of $H_0(z)$ such that the constraint

$$\left|H_0(e^{j\omega})\right|^2 + \left|H_1(e^{j\omega})\right|^2 \cong 1 \tag{10.107}$$

is satisfied for all values of ω [Joh80]. To this end, the objective function ϕ to be minimized can be chosen as a linear combination of two functions: (1) stopband attenuation of $H_0(z)$ and (2) the sum of the square-magnitude responses of $H_0(z)$ and $H_1(z)$ as indicated in Eq. (10.107). One such objective function is given by

$$\phi = \alpha\phi_1 + (1 - \alpha)\phi_2, \tag{10.108}$$

where

$$\phi_1 = \int_{\omega_s}^{\pi} \left| H_0(e^{j\omega}) \right|^2 d\omega, \quad \phi_2 = \int_0^{\pi} \left(1 - \left| H_0(e^{j\omega}) \right|^2 - \left| H_1(e^{j\omega}) \right|^2 \right) d\omega, \tag{10.109}$$

and $0 < \alpha < 1$ and $\omega_s = (\pi/2) + \varepsilon$ for some small $\varepsilon > 0$. Note that since $|T(e^{j\omega})|$ is symmetric with respect to $\pi/2$, the second integral in Eq. (10.109) can be replaced with

$$2 \int_0^{\pi/2} \left(1 - \left| H_0(e^{j\omega}) \right|^2 - \left| H_1(e^{j\omega}) \right|^2 \right) d\omega.$$

After ϕ has been made very small by the minimization procedure, both ϕ_1 and ϕ_2 will also be very small. This in turn will make $H_0(z)$ have a magnitude response satisfying $|H_0(e^{j\omega})| \cong 1$ in its passband and $|H_0(e^{j\omega})| \cong 0$ in its stopband, as desired. Moreover, since the power-complementary condition of Eq. (10.107) will be satisfied approximately, the magnitude response of the power-complementary highpass filter $H_1(z)$ to the lowpass filter will satisfy $|H_1(e^{j\omega})| \cong 0$ in the passband of $H_0(z)$ and $|H_1(e^{j\omega})| \cong 1$ in the stopband of $H_0(z)$.

Using the above approach, Johnston has designed a large class of linear-phase FIR low-pass filters $H_0(z)$ meeting a variety of specifications and has tabulated their impulse response coefficients [Joh80], [Cro83], [Ans93]. The following example illustrates the performance of one such filter.

Example 10.13 The impulse-response coefficients of the length-12 linear-phase lowpass filter 12B of Johnston [Joh80] are given by

$h_0[0] = -0.006443977 = h_0[11],$		$h_0[1] = 0.02745539 = h_0[10],$	
$h_0[2] = -0.00758164 = h_0[9],$		$h_0[3] = -0.0913825 = h_0[8],$	
$h_0[4] = 0.09808522 = h_0[7],$		$h_0[5] = 0.4807962 = h_0[6],$	

We use MATLAB Program 10_9 to verify the performances of the above filter $H_0(z)$. The input data requested by the program are the first half of the filter coefficients. It determines the remaining half by using the function `fliplr`. The program computes the gain response of the above filter and that of its complementary highpass filter $H_1(z) = H_0(-z)$, as indicated in Figure 10.46(a). It then computes the amplitude distortion function $|H_0(e^{j\omega})|^2 + |H_1(e^{j\omega})|^2$ in dB as shown in Figure 10.46(b). From Figure 10.46(a), we note that the stopband edge frequency ω_s of filter 12B is about 0.71π, which corresponds to a transition bandwidth $(\omega_s - 0.5\pi)/2\pi = 0.105$. The minimum stopband attenuation is approximately 34 dB. We also observe from Figure 10.46(b) that the amplitude distortion function is indeed very close to 0 dB in both the passbands and the stopbands of the two filters, with a peak value of ± 0.02 dB.

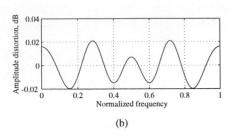

(a)   (b)

Figure 10.46   Johnston's 12B filter: (a) gain responses, and (b) reconstruction error in dB.

```
% Program 10_9
% Frequency Responses of Johnston's QMF Filters
% Type in prototype lowpass filter coefficients
B1 = input('Filter coefficients = ');
B1 = [B1 fliplr(B1)];
% Generate the complementary highpass filter
L = length(B1);
for k = 1:L
    B2(k) = ((-1)^k)*B1(k);
end
%  Compute the gain responses of the two filters
[H1z, w] = freqz(B1, 1, 256);
h1 = abs(H1z); g1 = 20*log10(h1);
[H2z, w] = freqz(B2, 1, 256);
h2 = abs(H2z); g2 = 20*log10(h2);
%  Plot the gain responses of the two filters
plot(w/pi, g1,'-',w/pi, g2,'--');grid
xlabel('Normalized frequency')
ylabel('Gain, dB')
pause
% Compute the sum of the square magnitude responses
for i = 1:256,
    sum(i) = (h1(i)*h1(i)) + (h2(i)*h2(i));
end
d =10*log10(sum);
% Plot the amplitude distortion
plot(w/pi,d);grid;
xlabel('Normalized frequency')
ylabel('Amplitude distortion, dB')
```

10.7.6   Alias-Free IIR QMF Bank

We now consider the design of an alias-free QMF bank employing IIR analysis and synthesis filters. Under the alias-free conditions of Eqs. (10.86) and (10.95), the distortion transfer function $T(z)$ of the two-channel QMF bank is given by $2z^{-1}E_0(z^2)E_1(z^2)$, as

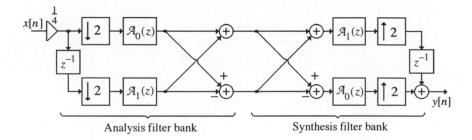

Figure 10.47   Magnitude-preserving two-channel QMF bank.

indicated in Eq. (10.102). If $T(z)$ is an allpass function, then its magnitude response is a constant and, as result, the corresponding QMF bank has no magnitude distortion [Vai87f]. Let the polyphase components $E_0(z)$ and $E_1(z)$ of $H_0(z)$ be expressed as

$$E_0(z) = \tfrac{1}{2}\mathcal{A}_0(z), \qquad E_1(z) = \tfrac{1}{2}\mathcal{A}_1(z), \qquad (10.110)$$

with $\mathcal{A}_0(z)$ and $\mathcal{A}_1(z)$ being stable allpass functions. Thus,

$$H_0(z) = \tfrac{1}{2}[\mathcal{A}_0(z^2) + z^{-1}\mathcal{A}_1(z^2)], \qquad (10.111a)$$

$$H_1(z) = \tfrac{1}{2}[\mathcal{A}_0(z^2) - z^{-1}\mathcal{A}_1(z^2)] \qquad (10.111b)$$

Substituting Eq. (10.110) in Eq. (10.100), we obtain the expressions for the analysis filters as

$$\begin{bmatrix} H_0(z) \\ H_1(z) \end{bmatrix} = \frac{1}{2}\begin{bmatrix} 1 & 1 \\ 1 & -1 \end{bmatrix}\begin{bmatrix} \mathcal{A}_0(z^2) \\ z^{-1}\mathcal{A}_1(z^2) \end{bmatrix}. \qquad (10.112)$$

The corresponding synthesis filters are obtained from Eq. (10.101) and are given by

$$[G_0(z) \quad G_1(z)] = \tfrac{1}{2}\begin{bmatrix} z^{-1}\mathcal{A}_1(z^2) & \mathcal{A}_0(z^2) \end{bmatrix}\begin{bmatrix} 1 & 1 \\ 1 & -1 \end{bmatrix}, \qquad (10.113)$$

which yields

$$G_0(z) = \tfrac{1}{2}[\mathcal{A}_0(z^2) + z^{-1}\mathcal{A}_1(z^2)] = H_0(z), \qquad (10.114a)$$

$$G_0(z) = \tfrac{1}{2}[-\mathcal{A}_0(z^2) + z^{-1}\mathcal{A}_1(z^2)] = -H_1(z). \qquad (10.114b)$$

The realization of the *magnitude-preserving* two-channel QMF bank shown in Figure 10.47 is obtained by making use of Eq. (10.110) in Figure 10.45(b).

From Eq. (10.111a), we observe that the lowpass transfer function $H_0(z)$ has a polyphase-like decomposition, except here the polyphase components are stable allpass transfer functions. The existence of this type of decomposition has been illustrated before in Example 10.10. It has been shown that a bounded real (BR) lowpass transfer function $H_0(z) = P_0(z)/D(z)$ of odd order, with no common factors between its numerator and the denominator, can be expressed as in Eq. (10.111a) if it satisfies the power-symmetry condition given by

$$\tilde{H}_0(z)H_0(z) + \tilde{H}_0(-z)H_0(-z) = 1, \qquad (10.115)$$

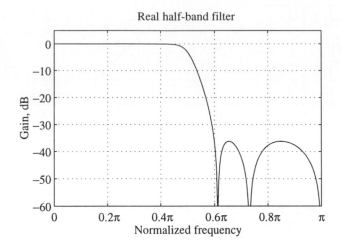

Figure 10.48   Gain response of the half-band filter of Example 10.14.

and the numerator $P_0(z)$ of $H_0(z)$ is a symmetric polynomial [Vai87f]. It can be easily verified that the transfer function $H(z)$ of Example 10.10 satisfies these conditions. It has also been shown that any odd-order elliptic lowpass half-band filter $H_0(z)$ with a frequency response specifications given by

$$1 - \delta_p \le \left| H_0(e^{j\omega}) \right| \le 1, \quad \text{for } 0 \le \omega \le \omega_p, \tag{10.116a}$$

$$\left| H_0(e^{j\omega}) \right| \le \delta_s, \quad \text{for } \omega_s \le \omega \le \pi, \tag{10.116b}$$

and satisfying the conditions $\omega_p + \omega_s = \pi$ and $\delta_s^2 = 4\delta_p(1 - \delta_p)$ can always be expressed as in Eq. (10.111a) [Vai87f]. The poles of the elliptic filter satisfying these two conditions lie on the imaginary axis. Using the pole-alternation property outlined in Section 6.10, we can readily identify the expressions for the two allpass transfer functions $\mathcal{A}_0(z)$ and $\mathcal{A}_1(z)$.

We illustrate the design of an IIR half-band filter using the above approach.

Example 10.14 The frequency response specifications of a real half-band elliptic filter $G(z)$ of odd order are given by $\omega_p = 0.4\pi$, $\omega_s = 0.6\pi$, and $\delta_s = 0.0155$. The corresponding passband ripple δ_p is found to be 0.00012013. In dB, the passband and stopband ripples are, respectively, Rp = 0.0010435178 and Rs = 36.193366. Using the function ellipord in MATLAB, we determine the minimum order of the elliptic lowpass filter to be 5. Next, we use the function ellip in MATLAB to determine the transfer function of the fifth-order half-band elliptic filter whose gain response is shown in Figure 10.48. We use the function tf2zp to determine its zero-pole locations and plot them using the function zplane, as indicated in Figure 10.49. The poles are at $z = 0$, $z = \pm j0.4866252630055$ and $z = \pm j0.84551995705794$.[12]

---

[12]Because of numerical accuracy problems, the real part of the poles are not exactly zero but are quite small in values and have been ignored.

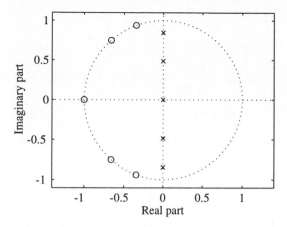

Figure 10.49   Pole-zero plot of the fifth-order elliptic IIR half-band lowpass filter of Example 10.14.

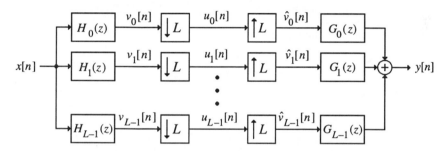

Figure 10.50   The basic *L*-channel QMF filter bank structure.

Using the pole-alternation property of the parallel allpass filter structure, we arrive at the transfer functions of the two allpass sections of the half-band filter $G(z)$:

$$\mathcal{A}_0(z^2) = \frac{z^{-2} + 0.2368041466}{1 + 0.2368041466z^{-2}}, \quad \mathcal{A}_1(z^2) = \frac{z^{-2} + 0.7149039978}{1 + 0.7149039978z^{-2}}. \qquad (10.117)$$

10.8   *L*-Channel QMF Bank

We now generalize the discussion of the previous section to the case of a QMF bank with more than two-channels. The basic structure of the *L*-channel QMF bank is shown in Figure 10.50.

10.8.1   Analysis of the *L*-Channel Filter Bank

We analyze the operation of the *L*-channel QMF bank of Figure 10.50 in the *z*-domain. The expressions for the *z*-transforms of various intermediate signals in Figure 10.50 are given by

$$V_k(z) = H_k(z)X(z), \tag{10.118a}$$

$$U_k(z) = \frac{1}{L} \sum_{\ell=0}^{L-1} H_k(z^{1/L}W_L^\ell)X(z^{1/L}W_L^\ell), \tag{10.118b}$$

$$\hat{V}_k(z) = U_k(z^L), \tag{10.118c}$$

where $k = 0, 1, \ldots, L-1$. The output of the QMF bank is given by

$$Y(z) = \sum_{k=0}^{L-1} G_k(z)\hat{V}_k(z). \tag{10.119}$$

From Eqs. (10.118a), (10.118b) and (10.119), we finally arrive at

$$Y(z) = \frac{1}{L} \sum_{\ell=0}^{L-1} X(zW_L^\ell) \sum_{k=0}^{L-1} H_k(zW_L^\ell)G_k(z), \tag{10.120}$$

which can be rewritten in compact form as

$$Y(z) = \sum_{\ell=0}^{L-1} a_\ell(z)X(zW_L^\ell), \tag{10.121}$$

where

$$a_\ell(z) = \frac{1}{L} \sum_{k=0}^{L-1} H_k(zW_L^\ell)G_k(z), \qquad 0 \le \ell \le L-1. \tag{10.122}$$

10.8.2   Error Caused by Sampling-Rate Alteration

On the unit circle the term $X(zW_L^\ell)$ becomes

$$X(e^{j\omega}W_L^\ell) = X(e^{j(\omega - \frac{2\pi\ell}{L})}). \tag{10.123}$$

Thus, from Eq. (10.121), the output spectrum $Y(e^{j\omega})$ is a weighted sum of $X(e^{j\omega})$ and its uniformly shifted versions $X(e^{j(\omega - 2\pi\ell/L)})$ for $\ell = 1, 2, \ldots, L-1$, which are caused by the sampling-rate alteration operations. The term $X(zW_L^\ell)$ is called the ℓth *aliasing term*, with $a_\ell(z)$ representing its *gain* at the output. In general, the QMF bank of Figure 10.50 is a linear, time-varying system with a period of L.

It follows from Eq. (10.121) that the aliasing effect can be completely eliminated at the output if and only if

$$a_\ell(z) = 0, \qquad 1 \le \ell \le L-1, \tag{10.124}$$

for all possible inputs $x[n]$. If Eq. (10.124) holds, then the L-channel QMF bank of Figure 10.50 becomes a linear time-invariant system with an input-output relation given by

$$Y(z) = T(z)X(z), \tag{10.125}$$

where $T(z)$ is the distortion transfer function given by

$$T(z) = a_0(z) = \frac{1}{L} \sum_{k=0}^{L-1} H_k(z) G_k(z). \qquad (10.126)$$

If $T(z)$ has a constant magnitude, then the system of Figure 10.50 is a *magnitude-preserving QMF bank*. If $T(z)$ has a linear phase, then the QMF bank has no phase distortion. Finally, if $T(z)$ is a pure delay, it is a *perfect reconstruction QMF bank*.

10.8.3   Matrix Representation

It is convenient to examine the operation of the *L*-channel QMF bank using a matrix formalism. Define

$$\mathbf{A}(z) = [a_0(z) \quad a_1(z) \quad \cdots \quad a_{L-1}(z)]^T. \qquad (10.127a)$$

$$\mathbf{g}(z) = [G_0(z) \quad G_1(z) \quad \cdots \quad G_{L-1}(z)]^T, \qquad (10.127b)$$

$$\mathbf{H}(z) = \begin{bmatrix} H_0(z) & H_1(z) & \cdots & H_{L-1}(z) \\ H_0(zW_L^1) & H_1(zW_L^1) & \cdots & H_{L-1}(zW_L^1) \\ \vdots & \vdots & \ddots & \vdots \\ H_0(zW_L^{L-1}) & H_1(zW_L^{L-1}) & \cdots & H_{L-1}(zW_L^{L-1}) \end{bmatrix}, \qquad (10.127c)$$

where the $L \times L$ matrix $\mathbf{H}(z)$ is called the *aliasing component (AC) matrix* [Smi85]. Using the above notations, Eq. (10.122) reduces to

$$L\mathbf{A}(z) = \mathbf{H}(z)\mathbf{g}(z). \qquad (10.128)$$

The aliasing cancellation condition can now be rewritten as

$$\mathbf{H}(z)\mathbf{g}(z) = \mathbf{t}(z), \qquad (10.129)$$

where

$$\mathbf{t}(z) = [La_0(z) \quad 0 \quad \cdots \quad 0]^T = [LT(z) \quad 0 \quad \cdots \quad 0]^T. \qquad (10.130)$$

Using Eq. (10.127a) and the notation

$$\mathbf{x}(z) = \begin{bmatrix} X(z) \\ X(zW_L^1) \\ \vdots \\ X(zW_L^{L-1}) \end{bmatrix}, \qquad (10.131)$$

we can rewrite Eq. (10.121) as

$$Y(z) = \mathbf{A}^T(z)\mathbf{x}(z), \qquad (10.132a)$$

or equivalently, as

$$Y(z) = \tfrac{1}{L}\mathbf{g}^T(z)\mathbf{H}^T(z)\mathbf{x}(z), \qquad (10.132b)$$

derived using Eq. (10.128).

From Eq. (10.129), it follows that by knowing the set of analysis filters $\{H_k(z)\}$, we can determine the desired set of synthesis filters $\{G_k(z)\}$ as

$$\mathbf{g}(z) = \mathbf{H}^{-1}(z)\mathbf{t}(z), \tag{10.133}$$

provided of course $[\det \mathbf{H}(z)] \neq 0$. Moreover, a perfect reconstruction QMF bank results if we set $T(z) = z^{-n_0}$ in the expression for $\mathbf{t}(z)$ in Eq. (10.130). In practice, the above approach is difficult to carry out for a number of reasons. A more practical solution to the design of perfect reconstruction QMF bank is obtained via the polyphase representation outlined next [Vai87d].

10.8.4   Polyphase Representation

Consider the Type I polyphase representation of the kth analysis filter $H_k(z)$:

$$H_k(z) = \sum_{\ell=0}^{L-1} z^{-\ell} E_{k\ell}(z^L), \qquad k = 0, 1, \ldots, L-1. \tag{10.134}$$

A matrix representation of the above set of equations is given by

$$\mathbf{h}(z) = \mathbf{E}(z^L)\mathbf{e}(z), \tag{10.135}$$

where

$$\mathbf{h}(z) = \begin{bmatrix} H_0(z) & H_1(z) & \cdots & H_{L-1}(z) \end{bmatrix}^T, \tag{10.136a}$$

$$\mathbf{e}(z) = \begin{bmatrix} 1 & z^{-1} & \cdots & z^{-(L-1)} \end{bmatrix}^T, \tag{10.136b}$$

and

$$\mathbf{E}(z) = \begin{bmatrix} E_{00}(z) & E_{01}(z) & \cdots & E_{0,L-1}(z) \\ E_{10}(z) & E_{11}(z) & \cdots & E_{1,L-1}(z) \\ \vdots & \vdots & \ddots & \vdots \\ E_{L-1,0}(z) & E_{L-1,1}(z) & \cdots & E_{L-1,L-1}(z) \end{bmatrix}. \tag{10.136c}$$

The matrix $\mathbf{E}(z)$ defined above is called the *Type I polyphase component matrix*. Figure 10.51(a) shows the Type I polyphase representation of the analysis filter bank.

Likewise, we can represent the L synthesis filters in a Type II polyphase form:

$$G_k(z) = \sum_{\ell=0}^{L-1} z^{-(L-1-\ell)} R_{\ell k}(z^L), \qquad k = 0, 1, \ldots, L-1. \tag{10.137}$$

In matrix form, the above set of L equations can be rewritten as

$$\mathbf{g}^T(z) = z^{-(L-1)}\tilde{\mathbf{e}}(z)\mathbf{R}(z^L), \tag{10.138}$$

where

$$\mathbf{g}(z) = \begin{bmatrix} G_0(z) & G_1(z) & \cdots & G_{L-1}(z) \end{bmatrix}^T, \tag{10.139a}$$

$$\tilde{\mathbf{e}}(z) = \begin{bmatrix} 1 & z & \cdots & z^{L-1} \end{bmatrix} = \mathbf{e}^T(z^{-1}), \tag{10.139b}$$

and

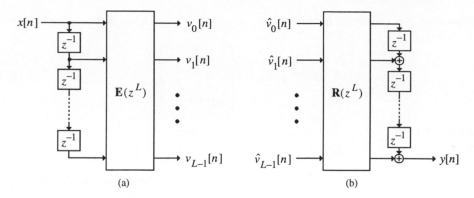

Figure 10.51 (a) Type I polyphase representation of the analysis filter bank and (b) Type II polyphase representation of the synthesis filter bank.

Figure 10.52 *L*-channel QMF bank structure based on the polyphase representations of the analysis and synthesis filter banks.

$$\mathbf{R}(z) = \begin{bmatrix} R_{00}(z) & R_{01}(z) & \cdots & R_{0,L-1}(z) \\ R_{10}(z) & R_{11}(z) & \cdots & R_{1,L-1}(z) \\ \vdots & \vdots & \ddots & \vdots \\ R_{L-1,0}(z) & R_{L-1,1}(z) & \cdots & R_{L-1,L-1}(z) \end{bmatrix}. \tag{10.139c}$$

The matrix $\mathbf{R}(z)$ defined above is called the *Type II polyphase component matrix*. Figure 10.51(b) shows the Type II polyphase representation of the synthesis filter bank.

Making use of the polyphase representations of Figure 10.51 in Figure 10.50 and the cascade equivalences of Figure 10.14, we arrive at an equivalent realization of the *L*-channel QMF bank shown in Figure 10.52.

The relation between the alias component matrix $\mathbf{H}(z)$ of Eq. (10.127c) and the Type I polyphase component matrix $\mathbf{E}(z)$ of Eq. (10.136c) can be easily established. From Eqs. (10.127c), (10.136a), and (10.135), we observe that

$$\mathbf{H}^T(z) = \begin{bmatrix} \mathbf{h}(z) & \mathbf{h}(zW_L^1) & \cdots & \mathbf{h}(zW_L^{L-1}) \end{bmatrix}$$

$$= \mathbf{E}(z^L) \begin{bmatrix} \mathbf{e}(z) & \mathbf{e}(zW_L^1) & \cdots & \mathbf{e}(zW_L^{L-1}) \end{bmatrix} \tag{10.140}$$

Now, from Eq. (10.136b), it follows that

$$
\mathbf{e}(zW_L^k) = \mathbf{\Delta}(z)
\begin{bmatrix}
1 \\
W_L^{-k} \\
\vdots \\
W_L^{-k(L-1)}
\end{bmatrix},
\tag{10.141}
$$

where

$$
\mathbf{\Delta}(z) = \operatorname{diag}\begin{bmatrix} 1 & z^{-1} & \cdots & z^{-(L-1)} \end{bmatrix}.
\tag{10.142}
$$

Making use of Eq. (10.142) in Eq. (10.140), we arrive at the desired result after some algebra:

$$
\mathbf{H}(z) = \mathbf{D}^{\circledast}\mathbf{\Delta}(z)\mathbf{E}^T(z^L),
\tag{10.143}
$$

where $\mathbf{D}$ is the $L \times L$ DFT matrix.

10.8.5   Condition for Perfect Reconstruction

If the polyphase component matrices of Figure 10.52 satisfy the relation

$$
\mathbf{R}(z)\mathbf{E}(z) = \mathbf{I},
\tag{10.144}
$$

where $\mathbf{I}$ is an $L \times L$ identity matrix, the structure of Figure 10.52 reduces to the one shown in Figure 10.53. Comparing Figure 10.53 with Figure 10.50, we note that the former can be considered as a special case of an L-channel QMF bank if we set

$$
H_k(z) = z^{-k}, \qquad G_k(z) = z^{-(L-1-k)}, \qquad 0 \le k \le L - 1.
\tag{10.145}
$$

Substituting the above in Eq. (10.122), we arrive at

$$
a_\ell(z) = \frac{1}{L}\sum_{k=0}^{L-1} z^{-k}W_L^{-\ell k}z^{-(L-1-k)} = z^{-(L-1)}\left(\frac{1}{L}\sum_{k=0}^{L-1} W_L^{-\ell k}\right).
\tag{10.146}
$$

From Eqs. (10.9) and (10.10), it follows that $a_0(z) = 1$ and $a_\ell(z) = 0$ for $\ell \neq 0$. Hence, from Eq. (10.126), we note that $T(z) = z^{-(L-1)}$, or in other words, the structure of Figure 10.53 is a perfect reconstruction L-channel QMF bank if condition of Eq. (10.144) is satisfied.

The analysis and synthesis filters of a perfect reconstruction filter bank of the form of Figure 10.52 can be easily determined as illustrated in the following example.

Example 10.15 The three-channel analysis/synthesis filter bank of Figure 10.54 is by construction a perfect reconstruction filter bank. The output of the filter bank is simply $y[n] = dx[n-2]$. Moreover, here, $\mathbf{E}(z^3) = \mathbf{P}$ and $\mathbf{R}(z^3) = d\mathbf{P}^{-1}$. Consider

$$
\mathbf{P} = \begin{bmatrix}
1 & 1 & 1 \\
1 & -1 & 1 \\
1 & 0 & -1
\end{bmatrix}.
$$

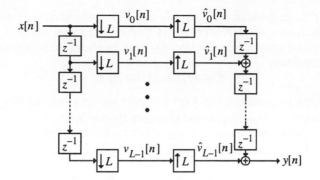

Figure 10.53   A simple perfect reconstruction multirate system.

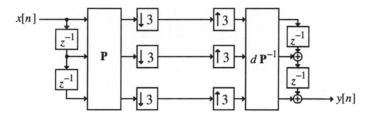

Figure 10.54   A 3-channel analysis/synthesis filter bank.

From Eqs. (10.135) and (10.136a)–(10.136c), we thus get

$$\begin{bmatrix} H_0(z) \\ H_1(z) \\ H_2(z) \end{bmatrix} = \begin{bmatrix} 1 & 1 & 1 \\ 1 & -1 & 1 \\ 1 & 0 & -1 \end{bmatrix} \begin{bmatrix} 1 \\ z^{-1} \\ z^{-2} \end{bmatrix},$$

resulting in the analysis filters

$$H_0(z) = 1 + z^{-1} + z^{-2}, \qquad H_1(z) = 1 - z^{-1} + z^{-2}, \qquad H_2(z) = 1 - z^{-2}.$$

Let $d = 4$. Then,

$$d\mathbf{P}^{-1} = \begin{bmatrix} 1 & 1 & 2 \\ 2 & -2 & 0 \\ 1 & 1 & -2 \end{bmatrix}.$$

Using Eqs. (10.138) and (10.139a–c), we arrive at

$$\begin{bmatrix} G_0(z) \\ G_1(z) \\ G_2(z) \end{bmatrix} = \begin{bmatrix} z^{-2} & z^{-1} & 1 \end{bmatrix} \begin{bmatrix} 1 & 1 & 2 \\ 2 & -2 & 0 \\ 1 & 1 & -2 \end{bmatrix},$$

leading to the synthesis filters $G_0(z) = 1 + 2z^{-1} + z^{-2}$, $G_1(z) = 1 - 2z^{-1} + z^{-2}$, $G_2(z) = -2 + 2z^{-2}$.

Now, for a given L-channel analysis filter bank, the polyphase matrix $\mathbf{E}(z)$ is known. From Eq. (10.144) it therefore follows that a perfect reconstruction L-channel QMF bank can be simply designed by constructing a synthesis filter bank with a polyphase matrix

$\mathbf{R}(z) = [\mathbf{E}(z)]^{-1}$. In general, it is not easy to compute the inverse of a rational $L \times L$ matrix. An alternative elegant approach is to design the analysis filter bank with an invertible polyphase matrix. For example, $\mathbf{E}(z)$ can be chosen to be a paraunitary matrix satisfying the condition

$$\tilde{\mathbf{E}}(z)\mathbf{E}(z) = c\mathbf{I}, \text{ for all } z, \qquad (10.147)$$

where $\tilde{\mathbf{E}}(z)$ is the paraconjugate of $\mathbf{E}(z)$ given by the transpose of $\mathbf{E}(z^{-1})$, with each coefficient replaced by its conjugate and choosing $\mathbf{R}(z) = \tilde{\mathbf{E}}(z)$.

For the design of a perfect reconstruction FIR L-channel QMF bank, the matrix $\mathbf{E}(z)$ can be expressed in a product form [Vai89]

$$\mathbf{E}(z) = \mathbf{E}_R(z)\mathbf{E}_{R-1}(z) \cdots \mathbf{E}_1(z)\mathbf{E}_0, \qquad (10.148)$$

where $\mathbf{E}_0$ is a constant unitary matrix, and

$$\mathbf{E}_\ell(z) = \mathbf{I} - \mathbf{v}_\ell[\mathbf{v}_\ell^*]^T + z^{-1}\mathbf{v}_\ell[\mathbf{v}_\ell^*]^T, \qquad (10.149)$$

in which $\mathbf{v}_\ell$ is a column vector of order L with unit norm, i.e., $[\mathbf{v}_\ell^*]^T \mathbf{v}_\ell = 1$. With this constraint on $\mathbf{E}(z)$, one can set up an appropriate objective function that can then be minimized to arrive at a set of L analysis filters meeting the desired passband and stopband specifications. To this end, a suitable objective function is given by

$$\phi = \sum_{k=0}^{L-1} \int_{k\text{-th stopband}} \left| H_k(e^{j\omega}) \right|^2 d\omega. \qquad (10.150)$$

The optimization parameters are the elements of $\mathbf{v}_\ell$ and $\mathbf{E}_0$.

Example 10.16 We consider the design of a 3-channel FIR perfect reconstruction QMF bank with a passband of width $\pi/3$ [Vai93]. The passband of the lowpass analysis filter $H_0(z)$ is from 0 to $\pi/3$, the passband of the bandpass analysis filter $H_1(z)$ is from $\pi/3$ to $2\pi/3$, and the passband of the highpass analysis filter $H_2(z)$ is from $2\pi/3$ to π. The objective function to be minimized here is thus of the form

$$\phi = \int_{\frac{\pi}{3}+\varepsilon}^{\pi} \left| H_0(e^{j\omega}) \right|^2 d\omega + \int_0^{\frac{\pi}{3}-\varepsilon} \left| H_1(e^{j\omega}) \right|^2 d\omega$$

$$+ \int_{\frac{2\pi}{3}+\varepsilon}^{\pi} \left| H_1(e^{j\omega}) \right|^2 d\omega + \int_0^{\frac{2\pi}{3}-\varepsilon} \left| H_2(e^{j\omega}) \right|^2 d\omega.$$

The impulse-response coefficients of the analysis filters $\{h_k[n]\}$, $k = 0, 1, 2$, of length 15 obtained by minimizing ϕ are given in [Vai93]. The gain responses of these filters are plotted in Figure 10.55. It should be noted that the coefficients of the corresponding synthesis filters $\{g_k[n]\}$, $k = 0, 1, 2$, are given by $g_k[n] = h_k[14 - n]$.

10.8.6   A Perfect Reconstruction QMF Bank Using IIR Filters

We have shown in Section 10.7.6 that a two-channel alias-free IIR QMF bank can be designed by choosing a polyphase decomposition of the half-band filters in which the polyphase components are stable allpass transfer functions, as given in Eq. (10.110). This

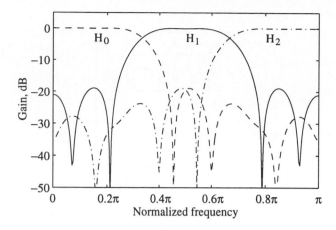

Figure 10.55 3-channel FIR QMF design example.

structure can be easily generalized to develop an alias-free L-channel IIR QMF bank. It has been shown that an Lth band stable IIR lowpass transfer function $H_0(z)$ can be expressed in a polyphase form given by

$$H_0(z) = \sum_{k=0}^{L-1} z^{-k} \mathcal{A}_k(z^L), \tag{10.151}$$

where $\mathcal{A}_k(z)$ are stable allpass transfer functions [Ren87]. The above filter can be used as a lowpass prototype to generate a uniform DFT analysis filter bank using the structure indicated in Figure 10.56. A transpose of this structure yields a uniform DFT synthesis filter bank, and a combination of the analysis filter bank of Figure 10.56 and its transpose with factor-of-L down-samplers and up-samplers results in an alias-free magnitude-preserving QMF bank.[13]

However, it is possible to achieve perfect reconstruction by using a different type of synthesis bank. The overall structure for perfect reconstruction is as indicated in Figure 10.57 [Hus90]. To verify its perfect reconstruction property, we can redraw this structure using the cascade equivalences of Figure 10.14, resulting in the equivalent structure of Figure 10.58. Now, the product of an $L \times L$ IDFT matrix with an $L \times L$ DFT matrix is an identity matrix. Then we are left with a cascade of an allpass with its inverse in each subband channel. Since $\mathcal{A}_k(z)\mathcal{A}_k(z^{-1}) = 1$, the structure of Figure 10.58 reduces to that shown in Figure 10.53, which is indeed a perfect reconstruction system.

Conditions for perfect reconstruction   Even though the L-channel analysis/synthesis filter bank of Figure 10.58 is computationally efficient, from a practical point of view a direct implementation of this structure is not feasible since the allpass filters in the synthesis bank are unstable. Moreover, the initial values of the state-variables in both the analysis and the synthesis allpass filters will also propagate to the output, making perfect

---

[13]The filter $H_0(z)$ of Eq. (10.151) has peaks in the stopband at $\omega_k = (2k + 1)\pi/L$ and particular care must be taken in its use as an uniform DFT filter bank and in a perfect reconstruction QMF bank.

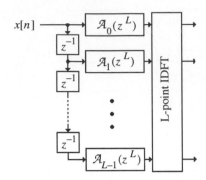

Figure 10.56   A uniform DFT analysis bank based on an allpass polyphase decompositin of an Lth band lowpass filter.

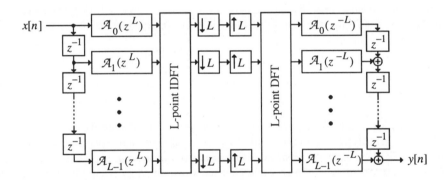

Figure 10.57   A perfect reconstruction L-channel analysis/syntehsis filter bank using allpass filters.

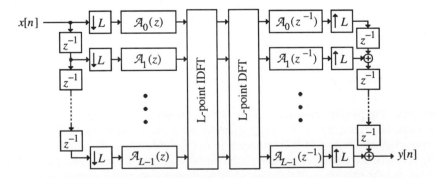

Figure 10.58   A computationally efficient realization of Figure 10.57.

Figure 10.59   Equivalent representation of the *k*th branch: (a) original structure, and (b) proposed implementation scheme.

reconstruction impossible. The solutions to these two problems are outlined next. To this end, we first assume the input sequence to be of finite extent and later outline a method to handle infinite-length sequences.

An equivalent representation of the *k*th channel in Figure 10.58 is as shown in Figure 10.59(a). If the input signal $x_k[n]$ is of length M, we can avoid the instability problem in implementing the synthesis branch allpass section by implementing it instead in a non-causal fashion, as indicated in Figure 10.59(b) [Bab91].[14] Here, only the first M samples of the output $v_k[n]$ of the analysis branch allpass are first computed. These are then time-reversed and fed into the identical allpass section $\mathcal{A}_k(z)$ of the synthesis bank, and again only the first M samples of its output $y_k[n]$ are computed. Finally, these outputs are time-reversed resulting in the desired output $\hat{x}_k[n]$.

We now investigate the conditions that need to be satisfied to completely eliminate the effect of the initial conditions at the output of the allpass in the synthesis branch. Let the allpass transfer function in both the analysis and synthesis branches be of Nth order. A state-space representation of the above allpass transfer function in the analysis branch is of the form

$$\mathbf{s}_a[n+1] = \mathbf{A}\mathbf{s}_a[n] + \mathbf{B}x[n], \qquad (10.152a)$$

$$v[n] = \mathbf{C}\mathbf{s}_a[n] + Dx[n], \qquad (10.152b)$$

where $\mathbf{s}_a[n]$ is the length-N state-variable vector of the analysis stage. The state-space representation of the allpass transfer function in the synthesis branch is identical in form to that of Eqs. (10.152a) and (10.152b) and is given by

$$\mathbf{s}_s[n+1] = \mathbf{A}\mathbf{s}_s[n] + \mathbf{B}u[n], \qquad (10.153a)$$

$$y[n] = \mathbf{C}\mathbf{s}_s[n] + Du[n], \qquad (10.153b)$$

where $\mathbf{s}_s[n]$ is the length-N state-variable vector of the synthesis stage.

It can be shown that by choosing the initial values of the state variables $\mathbf{s}_s[0]$ of the synthesis branch allpass section as a function of the final values of the state variables $\mathbf{s}_a[M]$ of the analysis branch allpass section according to [Bab91],

---

[14]See Section 4.3.5.

$$\mathbf{s}_s[0] = \begin{bmatrix} 0 & \cdots & 0 & 1 \\ 0 & \cdots & 1 & 0 \\ \cdots & \cdots & \cdot & \cdots \\ 1 & \cdots & 0 & 0 \end{bmatrix} \mathbf{s}_a[M], \qquad (10.154)$$

where M is the length of the input $x[n]$, the output $y[n]$ is precisely the time-reversed version of the input $x[n]$, i.e.,

$$y[n] = x[M - n - 1], \qquad \text{for } n = 0, 1, \ldots, M - 1. \qquad (10.155)$$

We verify the above condition for perfect reconstruction by means of computer simulation.

Example 10.17 Consider a second-order allpass section given by

$$A(z) = \frac{0.8 + 1.5z^{-1} + z^{-2}}{1 + 1.5z^{-1} + 0.8z^{-2}}.$$

The input sequence is given by $\{0 \quad 1 \quad 2 \quad 3 \quad 4 \quad 5\}$ and the initial conditions are $\{10.2 \quad 21.7\}$. To verify the perfect reconstruction condition of Eq. (10.154) we use the following MATLAB program.

```
% Program 10_10
% Verification of Perfect Reconstruction Condition
%
a = input('Type in denominator coefficients = ');
N = length(a) - 1;
k = [0: N];
b(1+k) = a(1+N-k);
x1 = input('Type in input sequence = ');
M = length(x1);
zi = input('Type in initial condition vector = ');
N = length (zi);
[yf1,zf1] = filter(b,a,x1,zi);
zf2 = fliplr(zf1);
x2 = fliplr(yf1);
yf2 = filter(b,a,x2,zf2);
yf3 = fliplr(yf2);
disp('input = ');disp(x1);
disp('output of first allpass = '); disp(yf1);
disp('output of second allpass = '); disp(yf2);
disp('time-reversed output of cascade = '); disp(yf3);
```

Note that in MATLAB the time-reversal operation is simply carried out using the function `fliplr`. When the program is run, it calls for the input samples, initial values of the state-variables, and the denominator coefficients. All input data must be entered in vector form inside square brackets. After completion of all executions, it displays the input sequence, the output of the analysis allpass section, the output of the synthesis allpass section, and finally, the time-reversed output of the synthesis allpass section as indicated below.

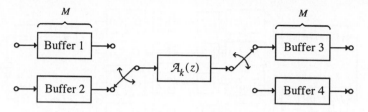

Figure 10.60 $A_k(z^{-1})$ implementation scheme.

```
input =
     0     1     2     3     4     5

output of analysis allpass =
   10.2000    7.2000   -15.8600    24.4300   -14.2570
   14.8415

output of synthesis allpass =
    5.0000    4.0000    3.0000    2.0000    1.0000
    0.0000

time-reversed output of cascade =
    0.0000    1.0000    2.0000    3.0000    4.0000
    5.0000
```

As can be seen from the above, the time-reversed output of the synthesis allpass is precisely the original input sequence.

It should be noted that the perfect reconstruction condition of Eq. (10.154) holds as long as the structures realizing the allpass sections in the analysis and synthesis branches are identical. This condition can be easily extended to the case when different structures are employed in realizing the two allpass sections in the kth branch [Bab91].

The basic perfect reconstruction IIR filter-based QMF structure outlined above for the processing of finite-length sequences can be easily modified for the processing of infinite-length sequences, such as speech, audio signals, using a double buffering scheme [Mit92b]. To illustrate the operation of this modification, we consider the kth synthesis branch of the modified QMF structure as shown in Figure 10.60, in which a double buffer of size M has been inserted both at the input and the output of the allpass filter $A_k(z)$. While the data is being read into one buffer, it is being read out of the other buffer in the reverse order in which it was read in earlier. Every time the buffer is switched at the input, the initial conditions in the allpass section of the synthesis branch is set according to Eq. (10.154) before the data is fed into the allpass in the reverse order. As a result, the modification described permits the decomposition of a continuous stream of input data by the analysis bank into blocks of length M, then coding or processing, and recombination by the synthesis bank into another continuous stream of data.

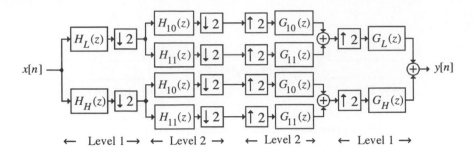

Figure 10.61   A two-level four-channel maximally-decimated QMF structure.

10.9   Multilevel Filter Banks

In Section 10.8, we analyzed the general L-channel QMF bank and developed the conditions to be satisfied by the analysis and synthesis filters for perfect reconstruction. It is also possible to develop a multiband analysis/synthesis filter bank by iterating a two-channel QMF bank. Moreover, if the two-band QMF bank is of the perfect reconstruction type, the generated multiband structure also exhibits perfect reconstruction property (Problem 10.44). In this section we consider this approach.

10.9.1   Filter Banks with Equal Passband Widths

By inserting a two-channel maximally decimated QMF bank in each channel of another two-channel maximally decimated QMF bank between the down-sampler and the up-sampler, we can generate a four-channel maximally decimated QMF bank, as shown in Figure 10.61. Since the analysis and the synthesis filter banks are formed like a tree, the overall system is often called a *tree-structured filter bank*. It should be noted that in the four-channel tree-structured filter bank of Figure 10.61, the two 2-channel QMF banks in the second level do not have to be identical. However, if they are different QMF banks with different analysis and synthesis filters, to compensate for the unequal gains and unequal delays of the two 2-channel systems, additional delays of appropriate values need to be inserted at the middle to ensure perfect reconstruction of the overall four-channel system.

An equivalent representation of the four-channel QMF system of Figure 10.61 is shown in Figure 10.62. The analysis and synthesis filters in the equivalent representation are related to those of the parent two-level tree-structured filter bank as follows:

$$H_0(z) = H_L(z)H_{10}(z^2), \qquad H_1(z) = H_L(z)H_{11}(z^2), \qquad (10.156a)$$
$$H_2(z) = H_H(z)H_{10}(z^2), \qquad H_3(z) = H_H(z)H_{11}(z^2), \qquad (10.156b)$$
$$G_0(z) = G_L(z)G_{10}(z^2), \qquad G_1(z) = G_L(z)G_{11}(z^2). \qquad (10.156c)$$
$$G_2(z) = G_H(z)G_{10}(z^2), \qquad G_3(z) = G_H(z)G_{11}(z^2). \qquad (10.156d)$$

Example 10.18 We illustrate the design of a four-channel QMF bank by iterating the two-channel QMF bank based on the filter 12B of Johnston [Joh80].

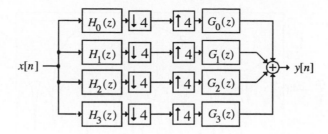

Figure 10.62 An equivalent representation of the four-channel QMF structure of Figure 10.61.

From the filter's impulse response given in Example 10.13, we compute the impulse response of each of the four analysis filters using Eqs. (10.156a) and (10.156b) and then determine the gain responses of each using Program 10_11 given below.

```
% Program 10_11
% Frequency Responses of Tree-Structured QMF
% Filters
%
clf;
% Type in prototype lowpass filter coefficients
B1 = input('Filter coefficients = ');
B1 = [B1 fliplr(B1)];
% Generate the complementary highpass filter
L = length(B1);
for k = 1:L
B2(k) = ((-1)^k)*B1(k);
end
% Determine the coefficients of the four filters
B10 = zeros(1, 2*length(B1));
B10([1: 2: length(B10)]) = B1;
B11 = zeros(1, 2*length(B2));
B11([1: 2: length(B11)]) = B2;
C0 = conv(B1, B10);C1 = conv(B1, B11);
C2 = conv(B2, B10);C3 = conv(B2, B11);
% Determine the frequency responses
[H0z, w] = freqz(C0, 1, 256);
h0 = abs(H0z);
M0 = 20*log10(h0);
[H1z, w] = freqz(C1, 1, 256);
h1 = abs(H1z);
M1 = 20*log10(h1);
[H2z, w] = freqz(C2, 1, 256);
h2 = abs(H2z);
M2 = 20*log10(h2);
[H3z, w] = freqz(C3, 1, 256);
h3 = abs(H3z);
M3 = 20*log10(h3);
plot(w/pi, M0,'-',w/pi, M1,'--',w/pi, M2,'--',w/pi,
```

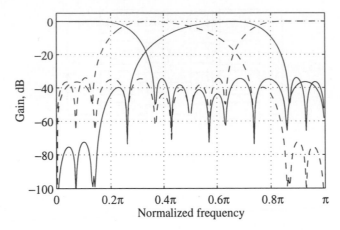

Figure 10.63   Gain responses of the four analysis filters of Example 10.18.

```
M3,'-');grid
xlabel('Normalized frequency')
ylabel('Gain, dB')
```

Figure 10.63 shows the gain responses of the four analysis filters of the final four-channel QMF bank.

From Eq. (10.156a)–(10.156d) it can be seen that each analysis filter $H_k(z)$ is a cascade of two filters, one with a single passband and a single stopband and the other with two passbands and two stopbands. The passband of the cascade is the frequency range where the passbands of the two filters overlap. On the other hand, the stopband of the cascade is formed from three different frequency ranges. In two of the frequency ranges, the passband of one coincides with the stopband of the other, while in the third range, the two stopbands overlap. As a result, the gain responses of the cascade in the three regions of the stopband are not equal, resulting in uneven stopband attenuation characteristic. This type of behavior of the gain response can also be seen in Figure 10.63 and should be taken into account in the design of the tree-structured filter bank.

By continuing the process described above, QMF banks with more than four channels can be easily constructed. It should be noted that the number of channels resulting from this approach is restricted to a power-of-2, i.e., $L = 2^\nu$. In addition, as illustrated by Figure 10.63, the filters in the analysis (synthesis) branch have passbands of equal width given by π/L. However, by a simple modification to the approach we can design QMF banks with analysis (synthesis) filters having passbands of unequal width as described next.

10.9.2   Filter Banks with Unequal Passband Widths

Consider the two-channel maximally decimated QMF bank of Figure 10.64(a). By inserting another two-channel maximally decimated QMF bank in the top subband channel between the down-sampler and the up-sampler at the position marked by a ∗, we arrive at a

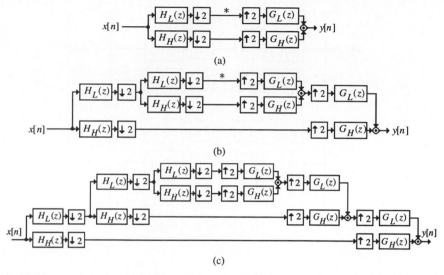

Figure 10.64 (a) A 2-channel QMF bank, (b) a 3-channel QMF bank derived from the 2-channel QMF bank, and(c) a 4-channel QMF bank derived from the 3-channel QMF bank.

three-channel maximally decimated QMF bank, as shown in Figure 10.64(b). The equivalent representation of the generated three-channel filter bank is indicated in Figure 10.65(a), where the analysis and synthesis filters are given by

$$H_0(z) = H_L(z)H_L(z^2), \quad H_1(z) = H_L(z)H_H(z^2), \quad H_2(z) = H_H(z),$$
$$G_0(z) = G_L(z)G_L(z^2), \quad G_1(z) = G_L(z)G_H(z^2), \quad G_2(z) = G_H(z). \quad (10.157)$$

Typical magnitude responses of the analysis filters of the two-channel QMF bank of Figure 10.64(a) and that of the derived three-channel filter of Figure 10.64(b) are sketched in Figure 10.66(a) and (b), respectively.

We can continue this process and generate a four-channel QMF bank from the three-channel QMF bank of Figure 10.64(b) by inserting a two-channel QMF bank in the top subband channel at the position marked by a $*$, resulting in the structure of Figure 10.64(c). Its equivalent representation is indicated in Figure 10.65(b), where

$$H_0(z) = H_L(z)H_L(z^2)H_L(z^4), \quad H_1(z) = H_L(z)H_L(z^2)H_H(z^4),$$
$$H_2(z) = H_L(z)H_H(z^2), \quad H_3(z) = H_H(z),$$
$$G_0(z) = G_L(z)G_L(z^2)G_L(z^4), \quad G_1(z) = G_L(z)G_L(z^2)G_H(z^4),$$
$$G_2(z) = G_L(z)G_H(z^2), \quad G_3(z) = G_H(z). \quad (10.158)$$

Figure 10.66(c) shows typical magnitude responses of the analysis (synthesis) filters of the four-channel QMF bank of Figure 10.64(c) derived from a parent two-channel QMF bank with magnitude responses as indicated in Figure 10.66(a).

Because of the unequal passband widths of the analysis and synthesis filters, these structures belong to the class of nonuniform QMF banks. The tree-structured filter banks of Figure 10.64 are also referred to as octave band QMF banks [Fli94]. Various other types

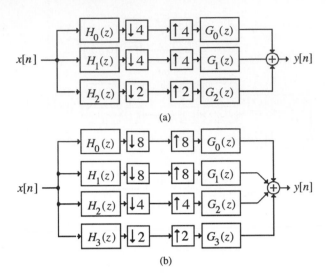

Figure 10.65   Maximally-decimated QMF banks with unequal passband width analysis (synthesis) filters.

of nonuniform filter banks can be generated by iterating branches of a parent uniform two-channel QMF in different forms. Nonuniform filter banks are often used in speech and image coding applications. The former application is discussed in Section 11.8.

10.10   Summary

The basic theory of multirate digital signal processing is introduced in this chapter along with the design of some useful multirate systems. The two basic sampling rate alteration devices are the up-sampler and the down-sampler. We first discuss the input-output relations of these devices, both in the time-domain and in the frequency-domain. We then describe several cascade equivalences that are used to develop computationally efficient implementation of a multirate system by permitting the movement of the up-sampler and down-sampler from one part of the system to another part.

The sampling rate converter is implemented using either the up-sampler or the down-sampler or both, and a lowpass digital filter. The design issues for the sampling rate conversion are next discussed. We demonstrate that a computationally efficient sampling rate converter can often be designed as a cascade of such converters. We also outline another approach to the implementation of a computationally efficient sampling rate converter that is based on the use of the polyphase decomposition of the lowpass digital filter.

The concept of analysis and synthesis filter banks is then introduced and a method for designing such filter banks from a prototype lowpass transfer function is described. A very special type of digital lowpass filter, called a Nyquist filter, which is particularly attractive for computationally efficient sampling rate converter implementation, is considered next.

The remaining part of the chapter is devoted to the analysis and design of the so-called quadrature-mirror filter (QMF) bank that is formed by a combination of an analysis filter bank with down-sampled outputs followed by a synthesis filter bank with up-sampled inputs. Conditions satisfied by the analysis and synthesis filters for an alias-free operation

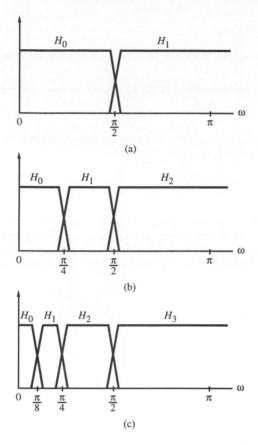

Figure 10.66 Magnitude responses of the analysis filters of a (a) 2-channel QMF bank, (b) 3-channel QMF bank derived from a 2-channel QMF bank, and (c) 4-channel QMF bank derived from a 3-channel QMF bank.

of the QMF bank are derived. Three types of QMF bank are defined and their design equations are developed.

Several other applications of multirate discrete-time systems are outlined in Chapter 11. These include a voice privacy system for telephone conversation (Section 11.7), subband coding of speech and audio signals (Section 11.8), transmultiplexers for signal conversion between frequency-division multiplex (FDM) to time-division multiplex (TDM) communication systems (Section 11.9), digital audio sampling rate conversion (Section 11.11), oversampling analog-to-digital conversion (Section 11.12), and oversampling digital-to-analog conversion (Section 11.13).

For additional details on multirate digital signal processing, we refer the reader to the texts by Akansu and Haddad [Aka92], Crochiere and Rabiner [Cro83], Fliege [Fli94], Vaidyanathan [Vai93], and Vetterli and Kovacevic [Vet95].

10.11   Problems

10.1 Show that the up-sampler defined by Eq. (10.1) is a time-varying system.

10.2 Show that the up-sampler defined by Eq. (10.1) and the down-sampler defined by Eq. (10.2) are linear systems.

10.3 Express the output $y[n]$ of Figure 10.6 as a function of the input $x[n]$. By simplifying the expression derived show that $y[n] = x[n-1]$.

10.4 Prove the identity of Eq. (10.10).

10.5 Consider the multirate structure of Figure P10.1(a) where $H_0(z)$, $H_1(z)$, and $H_2(z)$ are, respectively, ideal zero-phase real coefficient lowpass, bandpass, and highpass filters with frequency responses as indicated in Figure P10.1(b). If the input is a real sequence with a discrete-time Fourier transform as shown in Figure P10.1(c), sketch the discrete-time Fourier transforms of the outputs $y_0[n]$, $y_1[n]$, and $y_2[n]$.

Figure P10.1

10.6 Show that the two possible cascade configurations of factor-of-L up-sampler and a factor-of-M downsampler shown in Figure 10.13 are equivalent if and only if L and M are mutually prime.

10.7 Verify the cascade equivalences of Figure 10.14.

10.8 Develop an expression for the output $y[n]$ as a function of the input $x[n]$ for the multirate structure of Figure P10.2. [Hint: Replace the factor-of-12 down-sampler with a cascade of two down-samplers, and then make use of the identity of Figure 10.13.

$$x[n] \longrightarrow \boxed{\uparrow 4} \longrightarrow \boxed{\downarrow 12} \longrightarrow \boxed{\uparrow 3} \longrightarrow y[n]$$

Figure P10.2

10.9 Show that the transpose of a factor-of-M decimator is a factor-of-M interpolator if the transpose of a factor-of-M down-sampler is a factor-of-M up-sampler.

10.10 Show that the transpose of an M-channel analysis filter bank is an M-channel synthesis bank.

10.11 Develop an alternate two-stage design of the decimator of the example of Section 10.3 by designing the decimation filter in the form $H(z) = G(z^{10})I(z)$. Compare its computational requirements with the design in Section 10.3.

10.12 Determine the computational complexity of a single-stage decimator designed to reduce the sampling rate from 32 kHz to 800 Hz. The decimation filter is to be designed as an equiripple FIR filter with a passband edge at 350 Hz, a passband ripple of 0.003 and a stopband ripple of 0.004. Use the total multiplications per second as a measure of the computational complexity.

10.13 The decimator of Problem 10.12 is to be designed as a two-stage structure. Develop an optimum design with the smallest computational complexity.

10.14 Develop a two-band polyphase decomposition of each of the following IIR transfer functions:

(a) $H_1(z) = \dfrac{p_0 + p_1 z^{-1}}{1 + d_1 z^{-1}}, |d_1| < 1,$

(b) $H_2(z) = \dfrac{2 + 3.1z^{-1} + 1.5z^{-2}}{1 + 0.9z^{-1} + 0.8z^2},$

(c) $H_3(z) = \dfrac{2 + 3.1z^{-1} + 1.5z^{-2} + 4z^{-3}}{(1 - 0.5z^{-1})(1 + 0.9z^{-1} + 0.8z^2)}.$

10.15 Develop a computationally efficient realization of a factor-of-4 decimator employing a length-16 linear phase FIR filter.

10.16 Develop a computationally efficient realization of a factor-of-3 interpolator employing a length-15 linear phase FIR filter.

10.17 Let an ideal lowpass filter $H(z)$ with a cutoff at π/M be expressed as

$$H(z) \sum_{k=0}^{M-1} z^{-k} H_k(z^M).$$

Show that each polyphase subfilter $H_k(z)$ is an allpass.

10.18 Prove the identity shown in Figure 10.34.

10.19 Show that the set of M filters $H_k(z)$, $k = 0, 1, \ldots, M - 1$, of Eq. (10.54) satisfies the power-complementary property, i.e.,

$$\sum_{k=0}^{M-1} \left| H_k(e^{j\omega} W_M^k) \right|^2 = 1,$$

if the prototype lowpass filter $H_0(z)\tilde{H}(z)$ is an Mth band filter.

10.20 A generalization of the polyphase decomposition is considered in this problem. Let $H(z)$ be a causal FIR transfer function of degree $N - 1$ with N even:

$$H(z) = \sum_{n=0}^{N-1} h[n]z^{-n}.$$

(a) Show that $H(z)$ can be expressed in the form

$$H(z) = (1 + z^{-1})H_0(z^2) + (1 - z^{-1})H_1(z^2). \qquad (10.159)$$

(b) Express $H_0(z)$ and $H_1(z)$ in terms of the coefficients of the polyphase components $E_0(z)$ and $E_1(z)$. The decomposition of Eq. (10.159) is an example of a generalized polyphase decomposition, called the *structural subband decomposition* [Mit93].

(c) Show that the decomposition of Eq. (10.159) can be expressed in the form

$$H(z) = [1 \;\; z^{-1}] \begin{bmatrix} 1 & 1 \\ 1 & -1 \end{bmatrix} \begin{bmatrix} H_0(z^2) \\ H_1(z^2) \end{bmatrix}. \qquad (10.160)$$

The 2×2 matrix in Eq. (10.159) is called an Hadamard matrix of order 2 and denoted by $\mathbf{R}_2$.

(d) Show that if $N = 2^L$, then $H(z)$ can be expressed in the form

$$H(z) = [1 \;\; z^{-1} \ldots z^{-(L-1)}]\mathbf{R}_L \begin{bmatrix} H_0(z^L) \\ H_1(z^L) \\ \vdots \\ H_{L-1}(z^L) \end{bmatrix} \qquad (10.161)$$

where $\mathbf{R}_L$ is an $L \times L$ Hadamard matrix. Express the structural subband components $\{H_i(z)\}$ in terms of the polyphase components $\{E_i(z)\}$ of $H(z)$.

10.21 Develop a computationally efficient realization of a factor-of-4 decimator employing a length-16 linear phase FIR filter $H(z)$ and using a four-band structural subband decomposition of $H(z)$ in the form of Eq. (10.161).

10.22 Determine the total number of complex multiplications needed to implement an M-channel uniform DFT analysis filter bank designed from a N-tap prototype FIR filter in the form of Figure 10.37. What would be the total number of complex multiplications if each analysis filter in the filter bank is implemented individually in a direct form?

10.23 Consider the two-channel analysis filter bank structure of Figure P10.3 where $H_0(z)$ is a length-7 FIR filter with a transfer function given by

$$H_0(z) = h[0] + h[1]z^{-1} + h[2]z^{-2} + h[3]z^{-3} + h[4]z^{-4} + h[5]z^{-5} + h[6]^{-6}. \qquad (10.162)$$

If $H_1(z) = H_0(-z)$, develop a realization of this filter bank with 6 delays and 7 multipliers.

Figure P10.3

10.24 Consider the two-channel analysis filter bank structure of Figure P10.3 where $H_0(z)$ is an FIR filter of the form of Eq. (10.162). If $H_1(z)$ has a transfer function which is the mirror image of $H_0(z)$, i.e., $H_1(z) = z^{-6} H_0(z^{-1})$, develop a realization of this filter bank with only 7 multipliers.

10.25 The four-channel analysis filter bank of Figure P10.4(a), where **D** is a 4×4 DFT matrix, is characterized by the set of four transfer functions: $H_i(z) = Y_i(z)/X(z)$, $i = 0, 1, 2, 3$. Let the transfer functions of the four sub-filters be given by

$$E_0(z) = 1 - 0.2z^{-1} + 0.9z^{-2}, \quad E_1(z) = 2 + 1.3z^{-1} + 3z^{-2},$$
$$E_2(z) = 4 + 2.9z^{-1} + 5.1z^{-2}, \quad E_3(z) = 1 + 4.1z^{-1} - 1.5z^{-2}.$$

(a) Determine the expressions for the four transfer functions, $H_0(z)$, $H_1(z)$, $H_2(z)$, and $H_3(z)$.

(b) Assume that the analysis filter $H_2(z)$ has a magnitude response as indicated in Figure P10.4(b). Sketch the magnitude responses of the other three analysis filters.

(a)

(b)

Figure P10.4

10.26 Consider the analysis-synthesis filter bank of Figure P10.5. Develop the input-output relation of this structure in the z-domain. Let $H_0(z) = (1 + z^{-1})/2$ and $H_1(z) = (1 - z^{-1})/2$. Determine the synthesis filters $G_0(z)$ and $G_1(z)$ so that the structure of Figure P10.5 is a perfect reconstruction filter bank.

Figure P10.5

10.27 Let the analysis filters $H_0(z)$ and $H_1(z)$ of the structure of Figure P10.5 be power complementary FIR filters of order N each.

(a) Show that this structure becomes a perfect reconstruction filter bank if the synthesis filters $G_0(z)$ and $G_1(z)$ are chosen as

$$G_0(z) = z^{-N} H_0(z^{-1}), \qquad G_1(z) = z^{-N} H_1(z^{-1}), \qquad (10.163)$$

(b) Show that the synthesis filters are causal FIR filters if the analysis filters are causal.

(c) Show that the analysis and synthesis filters satisfying the perfect reconstruction condition cannot all be of linear phase.

10.28 Show that the two-channel QMF bank, in general, is a linear, time-varying system with a period of 2.

10.29 An approach to the implementation of a perfect reconstruction QMF bank using FIR filters is outlined in this problem. Let the analysis filters $H_0(z)$ and $H_1(z)$ of the 2-channel QMF bank of Figure 10.43 be power-complementary FIR filters of order $N - 1$ each. Choose the synthesis filters $G_0(z)$ and $G_1(z)$ according to Eq. (10.163). Determine the input-output relation of the corresponding QMF bank in the z-domain. Under what condition will this system be a perfect reconstruction filter bank?

10.30   (a) Decompose the third-order transfer function

$$G(z) = \frac{(1 + z^{-1})^3}{6 + 2z^{-2}},$$

in the form

$$G(z) = \tfrac{1}{2}\{A_0(z) + A_1(z)\},$$

where $A_0(z)$ and $A_1(z)$ are stable allpass transfer functions.

(b) Realize $G(z)$ as a parallel connection of allpass filters with $A_0(z)$ and $A_1(z)$ realized with the fewest number of multipliers.

(c) Determine the transfer function $H(z)$ which is power-complementary to $G(z)$.

(d) Sketch the magnitude responses of $G(z)$ and $H(z)$.

10.31 Let $G_a(s)$ denote the Nth order analog lowpass Butterworth transfer function with a 3-dB cutoff frequency at 1 rad/sec with N odd. Show that the corresponding digital Butterworth transfer function $H_0(z)$ obtained by a bilinear transformation is a half-band lowpass filter expressible in the form:

$$H_0(z) = \tfrac{1}{2}\{a_0(z^2) + z^{-1}a_1(z^2)\}$$

where $a_0(z)$ and $a_1(z)$ are stable allpass transfer functions.

10.32 Using the method of Problem 10.31, develop the transfer function of a fifth order lowpass half-band filter $H_0(z)$ and then determine its power-complementary transfer function $H_1(z)$. Develop a realization of a magnitude preserving two-channel QMF bank whose analysis filters are $H_0(z)$ and $H_1(z)$, and using no more than two multipliers

10.33 Let the order of the half-band lowpass IIR filter $H_0(z)$ of the QMF bank of Figure 10.43 be N where N is odd. (a) What is the total number of multipliers needed to implement the QMF bank of Figure 10.43? How many multiplications per second are needed to implement this structure? (b) If the QMF bank is magnitude preserving type, the analysis and synthesis filters can be realized as a sum of IIR allpass filters resulting in the structure of Figure 10.47. What is the total number of multipliers needed in the implementation of Figure 10.47. How many multiplications per second are needed to implement this structure?

10.34 Design a 4-channel perfect reconstruction QMF bank whose analysis filters are given by

$$\begin{bmatrix} H_0(z) \\ H_1(z) \\ H_2(z) \\ H_3(z) \end{bmatrix} = \begin{bmatrix} 1 & 3 & 3 & 1 \\ 1 & 1 & -1 & -1 \\ 1 & -1 & -1 & 1 \\ 1 & -3 & 3 & -1 \end{bmatrix} \begin{bmatrix} 1 \\ z^{-1} \\ z^{-2} \\ z^{-3} \end{bmatrix}.$$

Develop a computationally efficient realization of the filter bank.

10.35 Design a 3-channel perfect reconstruction QMF bank whose synthesis filters are given by

$$\begin{bmatrix} G_0(z) \\ G_1(z) \\ G_2(z) \end{bmatrix} = \begin{bmatrix} 1 & 2 & 3 \\ 3 & 1 & 4 \\ 2 & 1 & 1 \end{bmatrix} \begin{bmatrix} 1 \\ z^{-1} \\ z^{-2} \end{bmatrix}$$

Develop a computationally efficient realization of the filter bank.

10.36 Develop the state-space representation of the direct form II realization of the Nth order allpass transfer function

$$\mathcal{A}_k(z) = \frac{a_N + a_{N-1}z^{-1} + \cdots + a_1z^{-(N-1)} + z^{-N}}{1 + a_1z^{-1} + \cdots + a_{N-1}z^{-(N-1)} + a_Nz^{-N}},$$

in the form of Eqs. (10.152a) and (10.152b). Prove the following equalities relating the state-space parameter matrices $\mathbf{A}$, $\mathbf{B}$, $\mathbf{C}$, and D [Bab91]:

$$(a)\ \mathbf{APA} + \mathbf{BC} = \mathbf{P}, \quad (b)\ \mathbf{APB} + \mathbf{B}D = 0, \qquad\qquad (10.164a)$$
$$(c)\ \mathbf{CPA} + D\mathbf{C} = 0, \quad (d)\ \mathbf{CPB} + D^2 = 1, \qquad\qquad (10.164b)$$

where $\mathbf{P}$ is the $N \times N$ exchange matrix defined by

$$\mathbf{P} = \begin{bmatrix} 0 & \cdots & 0 & 1 \\ 0 & \cdots & 1 & 0 \\ \vdots & \ddots & \vdots & \vdots \\ 1 & \cdots & 0 & 0 \end{bmatrix}.$$

10.37 By making use of the equalities of Eqs. (10.164a) and (10.164b) show that if Eq. (10.154) holds, then the output $y[n]$ is the time-reversed version of a length-M input $x[n]$ as indicated in Eq. (10.155) [Bab91].

10.38 Show that, in general, the L-channel QMF bank is a linear, time-varying system with a period of L.

10.39 Consider the multirate system of Figure 10.51 where $\mathbf{P}(z) = \mathbf{R}(z)\mathbf{E}(z)$. Show that this structure is time-invariant with no aliasing if and only if $\mathbf{P}(z)$ is a pseudo-circulant matrix of the form [Vai88]:

$$\mathbf{P} = \begin{bmatrix} P_0(z) & z^{-1}P_1(z) & \cdots & z^{-1}P_{L-2}(z) & z^{-1}P_{L-1}(z) \\ z^{-1}P_{L-1}(z) & z^{-1}P_0(z) & \cdots & z^{-1}P_{L-3}(z) & z^{-1}P_{L-2}(z) \\ \vdots & \vdots & \ddots & \vdots & \vdots \\ z^{-1}P_2(z) & z^{-1}P_3(z) & \cdots & z^{-1}P_0(z) & z^{-1}P_1(z) \\ z^{-1}P_1(z) & z^{-1}P_2(z) & \cdots & z^{-1}P_{L-1}(z) & z^{-1}P_0(z) \end{bmatrix}. \qquad (10.165)$$

10.40 The structure of Figure P10.6 has been proposed for the computationally efficient implementation of FIR digital filters [Vet88].

(a) Show that the structure is alias-free and determine the overall transfer function $T(z) = Y(z)/X(z)$ in terms of $H_0(z)$ and $H_1(z)$.

(b) Determine the expression for $T(z)$ if

$$H_0(z^2) = \tfrac{1}{2}\{H(z) + H(-z)\}, \qquad H_1(z^2) = \tfrac{1}{2}\{H(z) - H(-z)\}.$$

(c) If $H(z)$ is a length $2K$ FIR filter, what are the lengths of the filters $H_0(z)$ and $H_1(z)$?

(d) Determine the computational efficiency of this structure.

Figure P10.6

10.41 (a) Show that the structure of Figure P10.7 is time-invariant with no aliasing if and only if the following matrix is pseudo-circulant as defined by Eq. (10.165) [Lin96]:

$$\mathbf{Q}(z) = \begin{bmatrix} 0 & 1 & 0 \\ z^{-1} & 0 & 1 \end{bmatrix} \mathbf{P}(z) \begin{bmatrix} 1 & 0 \\ 0 & 1 \\ z^{-1} & 0 \end{bmatrix}.$$

(b) Develop an equivalent realization of Figure P10.7 based on a critical down-sampling and critical up-sampling.

Figure P10.7

10.42 Analyze the structure of Figure P10.8 and determine its input-output relations. Comment on your results.

Figure P10.8

10.43 Let $R(z)$ be a half-band FIR transfer function of order $2M$ with a nonnegative amplitude response $\breve{R}(\omega)$ for all ω.[15] One way to construct $R(z)$ is to choose any half-band FIR transfer function $S(z)$ of order $2M$ with equal passband and stopband ripples, i.e., $\delta_p = \delta_s = \delta$ and set

$$R(z) = S(z) + \delta z^{-M}.$$

Show that $R(z)$ can be expressed in the form

$$R(z) = z^{-M} H_0(z) H_0(z^{-1}) = H_0(z)\hat{H}_0(z), \qquad (10.166)$$

where the impulse response $\hat{h}_0[n]$ of $\hat{H}_0(z)$ is the mirror image of the impulse response $h_0[n]$ of $H_0(z)$, i.e., $\hat{h}_0[n] = h_0[M - n]$. The transfer function $H_0(z)$ in Eq. (10.166) is called a *spectral factor* of $R(z)$.

10.44 Show that the 4-channel QMF bank of Figure 10.61 is a perfect reconstruction type if the two parent 2-channel QMF banks are of the perfect reconstruction type.

10.45 Show that the 3-channel and the 4-channel QMF banks of Figure 10.64(b) and 10.64(c), respectively, are of perfect reconstruction type if the two parent 2-channel QMF banks of Figure 10.64(a) are of the perfect reconstruction type.

10.12 MATLAB Exercises

M 10.1   (a) Modify Program 10_1 to study the operation of a factor-of-3 up-sampler on the following input sequences: (i) sum of two sinusoidal sequences of normalized frequencies 0.05 and 0.01 Hz, (ii) ramp sequence, and (iii) square wave sequence with various duty cycles. Choose the input length to be 75. Plot the input and the output sequences.

---

[15] See Eq. (4.94) for the definition of the amplitude response.

(b) Repeat part (a) for a factor-of-4 up-sampler.

M 10.2   (a) Modify Program 10_2 to study the operation of a factor-of-3 down-sampler on the following input sequences: (i) sum of two sinusoidal sequences of normalized frequencies 0.05 and 0.01 Hz, (ii) ramp sequence, and (iii) square wave sequence with various duty cycles. Choose the input length to be 50. Plot the input and the output sequences.

(b) Repeat part (a) for a factor-of-4 down-sampler.

M 10.3   (a) Use a vector of frequency points freq = [0 0.9 0.95 1] in Program 10_3 and run it with an up-sampling factor of $L = 7$. Comment on your results.

(b) Repeat part (a) for $L = 6$. Comment on your results.

M 10.4   (a) Use a vector of frequency points freq = [0 0.95 0.98 1] and a magnitude response vector mag = [1 0 0 0] in Program 10_3, and run it with an up-sampling factor of $L = 5$. Comment on your results.

(b) Repeat part (a) for $L = 8$. Comment on your results.

M 10.5   (a) Use a vector of frequency points freq = [0 0.22 0.25 1] and a magnitude response vector mag = [1 0 0 0] in Program 10_4, and run it with a down-sampling factor of $M = 4$. Comment on your results.

(b) Repeat part (a) for $M = 5$. Comment on your results.

M 10.6 Run Program 10_5 for the following input data: (a) $N = 120$, $M = 3$, $f_1 = 0.045$, $f_2 = 0.029$; (b) $N = 120$, $M = 4$, $f_1 = 0.045$, $f_2 = 0.029$. Comment on your results.

M 10.7 Run Program 10_6 for the following input data: (a) $N = 40$, $L = 3$, $f_1 = 0.045$, $f_2 = 0.029$; (b) $N = 30$, $L = 4$, $f_1 = 0.045$, $f_2 = 0.029$. Comment on your results.

M 10.8 Run Program 10_7 for the following input data: (a) $N = 30$, $L = 3$, $M = 2$, $f_1 = 0.045$, $f_2 = 0.029$; (b) $N = 40$, $L = 3$, $M = 5$, $f_1 = 0.045$, $f_2 = 0.029$. Comment on your results.

M 10.9 Design a length 60 linear-phase FIR lowpass filter with a cutoff frequency at $\pi/4$ using the windowed Fourier series approach. Express the transfer function in a 3-band polyphase decomposition. Using MATLAB compute and plot the frequency responses of each polyphase component. Show that all polyphase components have constant magnitude responses.

M 10.10 Design a seventh order IIR half-band elliptic lowpass filter satisfying the following specifications: $\omega_s = 0.55\pi$ and $\delta_s = 0.01$. Note that the half-band filter constraint is satisfied if $\omega_p + \omega_s = \pi$ and $(1 - \delta_p)^2 + \delta_s^2 = 1$. Realize the half-band filter using only 3 multipliers.

M 10.11 Design and realize a 4-channel uniform DFT analysis filter bank using a prototype linear-phase FIR filter of length 21. Design the prototype filter using the function remez of MATLAB. Assume a transition band of width 0.1π. Plot the magnitude responses of each filter on the same figure.

M 10.12 Design a 3-channel QMF bank in the form of Figure 10.64(b) by iterating the 2-channel QMF bank based on the Filter 16A of Johnston [Ans93], [Cro83], [Joh80]. Plot the gain responses of the three analysis filters, $H_0(z)$, $H_1(z)$ and $H_2(z)$ on the same figure. Comment on your results.

CHAPTER 11

Applications of Digital Signal Processing

As mentioned in Chapter 1, digital signal processing techniques are increasingly replacing conventional analog signal processing methods in many fields such as speech analysis and processing, radar and sonar signal processing, biomedical signal analysis and processing, telecommunications, geophysical signal processing, etc. Some typical applications of DSP chips are summarized in Table 11.1. A complete overview of these and other applications is beyond the scope of this book. Moreover, an understanding of many applications require a knowledge of the field where they are being used. We provide in this chapter, a few simple applications to provide a glimpse of the potential of DSP.

We first describe several applications of the discrete Fourier transform (DFT) introduced in Section 3.2. The first application considered is the detection of the frequencies of a pair of sinusoidal signals, called tones, employed in dial-pulse telephone signaling. Next we discuss the use of DFT in the determination of the spectral contents of a finite-length sequence. The effect of the DFT length and the windowing of the sequence are examined in detail here. In the following section, we introduce the concept of short-time Fourier transform (STFT) and discuss its application in the spectral analysis of nonstationary signals. Application of digital filtering methods to musical sound processing is considered next, and a variety of practical digital filter structures useful for the generation of certain audio effects, such as artificial reverberation, flanging, phasing, filtering and equalization, are introduced. The digital stereo generation for FM stereo transmission is treated in the following section. Generation of discrete-time analytic signals by means of a discrete-time Hilbert transformer is then considered, and several methods of designing these circuits are outlined along with an application. The design of a voice privacy system for the protection of privacy in telephone conversation is the subject of the following section. The basic scheme of the subband coding of speech and audio signals is reviewed next. The theory and design of transmultiplexers are discussed in the following section. One method of digital data transmission employing digital signal processing methods is introduced next.

Methods for the fractional sampling rate conversion of a discrete-time sequence are then outlined. The basic concepts behind the design of the oversampling A/D and D/A converters are reviewed in the final two sections.

Table 11.1 Typical applications of DSP chips.

General-purpose DSP	Graphics/imaging	Instrumentation
Digital filtering	3-D rotation	Spectrum analysis
Convolution	Robot vision	Function generation
Correlation	Image transmission/	Pattern matching
Hilbert transforms	compression	Seismic processing
Fast Fourier transforms	Pattern recognition	Transient analysis
Adaptive filtering	Image enhancement	Digital filtering
Windowing	Homomorphic processing	Phase-locked loops
Waveform generation	Workstations	
	Animation/digital map	

Voice/speech	Control	Military
Voice mail	Disk control	Secure communications
Speech vocoding	Servo control	Radar processing
Speech recognition	Robot control	Sonar processing
Speaker verification	Laser printer control	Image processing
Speech enhancement	Engine control	Navigation
Speech synthesis	Motor control	Missile guidance
Text to speech		Radio frequency modems

Telecommunications		Automotive
Echo cancellation	FAX	Engine control
ADPCM transcoders	Cellular telephone	Vibration analysis
Digital PBXs	Speaker phones	Antiskid brakes
Line repeaters	Digital speech	Adaptive ride control
Channel multiplexing	Interpolation (DSI)	Global positioning
1200 to 19200-bps Modems	X.25 packet switching	Navigation
Adaptive equalizers	Video conferencing	Voice commands
DTMF encoding/decoding	Spread spectrum	Digital radio
Data encryption	Communications	Cellular telephones

Consumer	Industrial	Medical
Radar detectors	Robotics	Hearing aids
Power tools	Numeric control	Patient monitoring
Digital audio/TV	Security access	Ultrasound equipment
Music synthesizer	Power line monitors	Diagnostic tools
Educational toys		Prosthetics
		Fetal monitors

Reprinted from K. S. Lin, ed., *Digital Signal Processing Applications with the TMS320 Family*, vol. 1, Prentice-Hall and Texas Instruments, 1987. Reprinted by Permission of Texas Instruments.

11.1 Dual-Tone Multifrequency Signal Detection

Dual-tone multifrequency (DTMF) signaling, increasingly being employed worldwide with push-button telephone sets, offers a high dialing speed over the dial-pulse signaling used in conventional rotary telephone sets. In recent years, DTMF signaling has also found applications requiring interactive control such as in voice mail, electronic mail (e-mail), telephone banking, and ATM machines.

A DTMF signal consists of a sum of two tones with frequencies taken from two mutually exclusive groups of preassigned frequencies. Each pair of such tones represents a unique number or a symbol. Decoding of a DTMF signal thus involves identifying the two tones in that signal and determining their corresponding number or symbol. The frequencies allocated to the various digits and symbols of a push-button keypad are internationally accepted standards and are shown in Figure 1.38 [ITU84]. The four keys in the last column of the keypad, as shown in this figure, are not yet available on standard handsets and are reserved for future use. Since the signaling frequencies are all located in the frequency band used for speech transmission, this is an *in-band system*. Interfacing with the analog input and output devices is provided by *codec* (coder/decoder) chips, or A/D and D/A converters (Sections 5.6 and 5.7).

Although a number of chips with analog circuitry are available for the generation and decoding of DTMF signals in a single channel, these functions can also be implemented digitally on DSP chips. Such a digital implementation surpasses analog equivalents in performance, since it provides better precision, stability, versatility, reprogrammability to meet other tone standards, and the scope for multichannel operation by time-sharing leading to a lower chip count.

The digital implementation of a DTMF signal involves adding two finite-length digital sinusoidal sequences with the latter simply generated by using look-up tables or by computing a polynomial expansion. The digital tone detection can be easily performed by computing the DFT of the DTMF signal and then measuring the energy present at the eight DTMF frequencies. The minimum duration of a DTMF signal is 40 ms. Thus, with a sampling rate of 8 kHz, there are at most $0.04 \times 8000 = 320$ samples available for decoding each DTMF digit. The actual number of samples used for the DFT computation is less than this number and is chosen so as to minimize the difference between the actual location of the sinusoid and the nearest integer value DFT index k.

The DTMF decoder computes the DFT samples closest in frequency to the eight DTMF fundamental tones and their respective second harmonics. In addition, a practical DTMF decoder also computes the DFT samples closest in frequency to the second harmonics corresponding to each of the fundamental tone frequencies. This latter computation is employed to distinguish between human voices and the pure sinusoids generated by the DTMF signal. In general, the spectrum of a human voice contains components at all frequencies including the above second harmonic frequencies. On the other hand, the DTMF signal generated by the handset has negligible second harmonics. The DFT computation scheme employed is a slightly modified version of Goertzel's algorithm, as described in Section 8.3.1 for the computation of the squared magnitudes of the DFT samples that are needed for the energy computation.

The DFT length N determines the frequency spacing between the locations of the DFT samples and the time it takes to compute the DFT sample. A large N makes the spacing smaller, providing higher resolution in the frequency domain but increases the

Table 11.2   DFT index values for DTMF tones and their second harmonics [Mar92].

Basic tone in Hz	Exact k value	Nearest integer k value	Absolute error in k
697	17.861	18	0.139
770	19.731	20	0.269
852	21.833	22	0.167
941	24.113	24	0.113
1209	30.981	31	0.019
1336	34.235	34	0.235
1477	37.848	38	0.152
1633	41.846	42	0.154

Second harmonic in Hz	Exact k value	Nearest integer k value	Absolute error in k
1394	35.024	35	0.024
1540	38.692	39	0.308
1704	42.813	43	0.187
1882	47.285	47	0.285
2418	60.752	61	0.248
2672	67.134	67	0.134
2954	74.219	74	0.219
3266	82.058	82	0.058

computation time. The frequency f_k in Hz corresponding to the DFT index (bin number) k is given by

$$f_k = \frac{kF_T}{N}, \qquad k = 0, 1, \ldots, N-1, \qquad (11.1)$$

where F_T is the sampling frequency. If the input signal contains a sinusoid of frequency f_{in} different from that given above, its DFT will contain not only large valued samples at values of k closest to Nf_{in}/F_T but also nonzero values at other values of k due to a phenomenon called leakage (See Example 8.10). To minimize the leakage it is desirable to choose N appropriately so that the tone frequencies fall as close as possible to a DFT bin, thus providing a very strong DFT sample at this index value relative to all other values. For an 8-kHz sampling frequency, the best value of the DFT length N to detect the eight fundamental DTMF tones has been found to be 205 and that for detecting the eight second harmonics is 201 [Mar92]. Table 11.2 shows the DFT index values closest to each of the tone frequencies and their second harmonics for these two values of N, respectively. Figure 11.1 shows 16 selected DFT samples computed using a 205-point DFT of a length-205 sinusoidal sequence for each of the fundamental tone frequencies.

The following MATLAB program demonstrates the DFT-based DTMF detection algorithm. It employs the function gfft(x,N,k) of Section 8.3.1 to calculate a single DFT sample using Goertzel's method. The input data is the telephone handset key symbol.

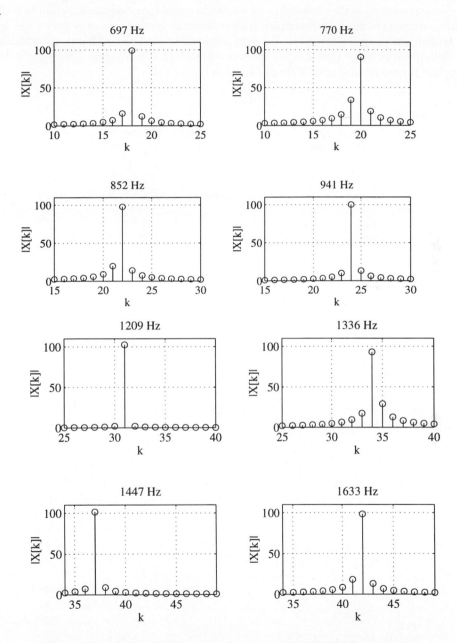

Figure 11.1   Selected DFT samples for each one of the DTMF tone signals.

The program generates a length-205 sequence consisting of two sinusoids of frequencies according to the convention shown in Figure 1.36. It then computes the eight DFT samples corresponding to the bin numbers of the fundamental tone frequencies given in Table 11.2 and displays these DFT samples and the decoded symbol. As only pure tones are generated, the program does not employ the test involving the second harmonic detection to distinguish between human voice and touch-tone digit. The outputs generated by this program for the input symbol # are displayed in Figure 11.2.

```
% Program 11_1
% Dual-Tone Multi-Frequency Tone Detection
% Using the DFT
%
clf
d = input('Type in the telephone digit = ', 's');
symbol = abs(d);
if symbol == 1
f1 = 697;f2 = 1209;
end
if symbol == 2
f1 = 697;f2 = 1336;
end
if symbol == 3
f1 = 697;f2 = 1477;
end
if symbol == 65
f1 = 697;f2 = 1633;
end
if symbol == 4
f1 = 770;f2 = 1209;
end
if symbol == 5
f1 = 770;f2 = 1336;
end
if symbol == 6
f1 = 770;f2 = 1477;
end
if symbol == 66
f1 = 770;f2 = 1633;
end
if symbol == 7
f1 = 852;f2 = 1209;
end
if symbol == 8
f1 = 852;f2 = 1336
end
if symbol == 9
f1 = 852;f2 = 1477;
```

```
end
if symbol == 67
f1 = 852;f2 = 1633;
end
if symbol == 42
f1 = 941;f2 = 1209;
end
if symbol == 0
f1 = 941;f2 = 1336;
end
if symbol == 35
f1 = 941;f2 = 1477;
end
if symbol == 68
f1 = 941;f2 = 1633;
end
n = 0:204;
x = sin(2*pi*n*f1/8000) + sin(2*pi*n*f2/8000);
k = [18 20 22 24 31 34 38 42];
for m = 1:8;
Fx(m) = gfft(x,205,k(m));
val = abs(Fx);
end
stem(k,val);grid
limit = 8100;
if val(1)*val(5) > limit
disp('Touch-Tone Symbol = 1');
end
if val(1)*val(6) > limit
disp('Touch-Tone Symbol = 2');
end
if val(1)*val(7) > limit
disp('Touch-Tone Symbol = 3');
end
if val(1)*val(8) > limit
disp('Touch-Tone Symbol = A');
end
if val(2)*val(5) > limit
disp('Touch-Tone Symbol = 4');
end
if val(2)*val(6) > limit
disp('Touch-Tone Symbol = 5');
end
if val(2)*val(7) > limit
disp('Touch-Tone Symbol = 6');
end
if val(2)*val(8) > limit
```

Touch-Tone Symbol =#

Figure 11.2   A typical output of Program 11_1.

```
disp('Touch-Tone Symbol = B');
end
if val(3)*val(5) > limit
disp('Touch-Tone Symbol = 7');
end
if val(3)*val(6) > limit
disp('Touch-Tone Symbol = 8');
end
if val(3)*val(7) > limit
disp('Touch-Tone Symbol = 9');
end
if val(3)*val(8) > limit
disp('Touch-Tone Symbol = C');
end
if val(4)*val(5) > limit
disp('Touch-Tone Symbol = *');
end
if val(4)*val(6) > limit
disp('Touch-Tone Symbol = 0');
end
if val(4)*val(7) > limit
disp('Touch-Tone Symbol = #');
end
if val(4)*val(8) > limit
disp('Touch-Tone Symbol = D');
end
```

11.2 Spectral Analysis Using DFT

An important application of digital signal processing methods is in determining in the discrete-time domain the frequency contents of a continuous-time signal, more commonly known as *spectral analysis*. Applications of digital spectral analysis can be found in many fields and are widespread. A number of different approaches have been advanced for spectral analysis, and most of these methods fall into one of two classes: (1) Fourier analysis based methods and (2) parametric model based methods. For a concise history of spectral analysis, we refer the reader to Kumaresan [Kum93], who also provides an overview of the whole field.

In this section, we consider only the first approach, which is conceptually simple and in many cases quite adequate. It is based on the following observation. If the continuous-time signal $g_a(t)$ is reasonably bandlimited, the discrete-time Fourier transform (DTFT) $G(e^{j\omega})$ of its discrete-time equivalent $g[n]$ should provide a good estimate of the spectral properties of $g_a(t)$. However, in most cases, $g_a(t)$ is defined for $-\infty < t < \infty$, and as result, $g[n]$ is of infinite extent and defined for $-\infty < n < \infty$. Moreover, the DTFT is a continuous function of ω and its exact evaluation is, in general, not feasible. A more practical approach, as implemented in most spectrum analyzers, is as follows. First, the continuous-time signal $g_a(t)$ is passed through an analog anti-aliasing filter before it is sampled to eliminate the effect of aliasing. The output of the filter is then converted into a discrete-time sequence equivalent $g[n]$. The infinite-length sequence $g[n]$ is windowed to make it into a finite-length sequence $\gamma[n]$ of length N, and its DTFT is next evaluated at a set of $R(R \geq N)$ discrete angular frequencies equally spaced in the range $0 \leq \omega \leq 2\pi$ by computing the R-point discrete Fourier transform (DFT) $\Gamma[k]$ of $\gamma[n]$. The DFT is usually computed using a FFT algorithm.

We examine the above approach in more detail to understand its limitations so that we can properly make use of the results obtained. In particular, we analyze here the effects of windowing and the evaluation of the frequency samples of the DTFT via the DFT. We neglect the effect of aliasing by assuming that the anti-aliasing filter has been designed appropriately. Moreover, we also assume that the A/D converter wordlength is large enough so that the A/D conversion noise can be neglected.

Before we can interpret the spectral content of $\Gamma(e^{j\omega})$, i.e., $G(e^{j\omega})$, from $\Gamma[k]$, we need to reexamine the relations between these transforms and their corresponding frequencies. Now, the relation between the R-point DFT $\Gamma[k]$ of $\gamma[n]$ and its DTFT $\Gamma(e^{j\omega})$ is given by

$$\Gamma[k] = \Gamma(e^{j\omega})\Big|_{\omega=2\pi k/R}, \qquad k = 0, 1, \ldots, R-1. \tag{11.2}$$

The normalized discrete-time angular frequency ω_k corresponding to the DFT bin number k (DFT frequency) is given by

$$\omega_k = \frac{2\pi k}{R}. \tag{11.3}$$

Likewise, the continuous-time frequency Ω_k corresponding to the DFT bin number k (DFT frequency) is given by

$$\Omega_k = \frac{2\pi k}{RT}. \tag{11.4}$$

To interpret the results of the DFT-based spectral analysis correctly, we first consider the frequency domain analysis of a sinusoidal sequence. Now an infinite-length sinusoidal sequence $g[n]$ of normalized angular frequency ω_0 is given by

$$g[n] = cos(\omega_0 n + \phi). \tag{11.5}$$

By expressing the above sequence as

$$g[n] = \tfrac{1}{2} \left(e^{j(\omega_0 n + \phi)} + e^{-j(\omega_0 n + \phi)} \right), \tag{11.6}$$

and making use of Table 3.1, we arrive at the expression for its DTFT as

$$G(e^{j\omega}) = \pi \sum_{\ell=-\infty}^{\infty} \left(e^{j\phi} \delta(\omega - \omega_0 + 2\pi\ell) + e^{-j\phi} \delta(\omega + \omega_0 + 2\pi\ell) \right). \tag{11.7}$$

Thus, the DTFT is a periodic function of ω with a period 2π containing two impulses in each period. In the frequency range, $-\pi \leq \omega \leq \pi$, there is an impulse at $\omega = \omega_0$ of complex amplitude $\pi e^{j\phi}$ and an impulse at $\omega = -\omega_0$ of complex amplitude $\pi e^{-j\phi}$.

To analyze $g[n]$ in the spectral domain using the DFT, we employ a finite-length version of the sequence given by

$$\gamma[n] = \cos(\omega_0 n + \phi), \qquad 0 \leq n \leq N - 1. \tag{11.8}$$

The computation of the DFT of a finite-length sinusoid has been considered in Example 8.10. In this example, using the MATLAB Program 8_9, we computed first the DFT of a length-32 sinusoid of frequency 10 Hz sampled at 64 Hz, as shown in Figure 8.31. As can be seen from this figure, there are only two nonzero DFT samples, one at bin $k = 5$ and the other at bin $k = 27$. From Eq. (11.4), bin $k = 5$ corresponds to frequency 10 Hz, while bin $k = 27$ corresponds to frequency 54 Hz, or equivalently, -10 Hz. Thus, the DFT has correctly identified the frequency of the sinusoid.

Next, using the same program, we computed the 32-point DFT of a length-32 sinusoid of frequency 11 Hz sampled at 64 Hz, as shown in Figure 8.32. This figure shows two strong peaks at bin locations $k = 5$ and $k = 6$ with nonzero DFT samples at other bin locations in the positive half of the frequency range. Note that the bin locations 5 and 6 correspond to frequencies 10 Hz and 12 Hz according to Eq. (11.4). Thus the frequency of the sinusoid being analyzed is exactly halfway between these two bin locations.

The phenomenon of the spread of energy from a single frequency to many DFT frequency locations as demonstrated by this figure is called *leakage*. To understand the cause of this effect, we recall that the DFT $\Gamma[k]$ of a length-N sequence $\gamma[n]$ is given by the samples of its discrete-time Fourier transform (DTFT) $\Gamma(e^{j\omega})$ evaluated at $\omega = 2\pi k/N$, $k = 0, 1, \ldots, N - 1$. Figure 11.3 shows the DTFT of the length-32 sinusoidal sequence of frequency 11 Hz sampled at 64 Hz. It can be seen that the DFT samples shown in Figure 8.32 are indeed obtained by the frequency samples of the plot of Figure 11.3.

To understand the shape of the DTFT shown in Figure 11.3 we observe that the sequence of Eq. (11.8) is a windowed version of the infinite-length sequence $g[n]$ of Eq. (11.6) obtained using a rectangular window $w[n]$:

$$w[n] = \begin{cases} 1, & 0 \leq n \leq N - 1, \\ 0, & \text{otherwise.} \end{cases} \tag{11.9}$$

Figure 11.3 DTFT of a sinusoidal sequence windowed by a rectangle window.

Hence, the DTFT $\Gamma(e^{j\omega})$ of $\gamma[n]$ is given by the frequency-domain convolution of the DTFT $G(e^{j\omega})$ of $g[n]$ with the DTFT $\Psi_R(e^{j\omega})$ of the rectangular window $w[n]$:

$$\Gamma(e^{j\omega}) = \frac{1}{2\pi}\int_{-\pi}^{\pi} G(e^{j\varphi})\Psi_R(e^{j(\omega-\varphi)})\,d\varphi, \tag{11.10}$$

where

$$\Psi_R(e^{j\omega}) = e^{-j\omega(N-1)/2}\frac{\sin(\omega N/2)}{\sin(\omega/2)}. \tag{11.11}$$

Substituting $G(e^{j\omega})$ from Eq. (11.7) into Eq. (11.10), we arrive at

$$\Gamma(e^{j\omega}) = \tfrac{1}{2}e^{j\phi}\Psi_R(e^{j(\omega-\omega_0)}) + \frac{1}{2}e^{-j\phi}\Psi_R(e^{j(\omega+\omega_0)}). \tag{11.12}$$

As indicated by the above equation, the DTFT $\Gamma(e^{j\omega})$ of the windowed sequence $\gamma[n]$ is a sum of the frequency shifted and amplitude scaled DTFT $\Psi_R(e^{j\omega})$ of the window $w[n]$ with the amount of frequency shifts given by $\pm\omega_0$. Now, for the length-32 sinusoid of frequency 11 Hz sampled at 64 Hz, the normalized frequency of the sinusoid is $11/64 = 0.172$. Hence, its DTFT is obtained by frequency shifting the DTFT $\Psi_R(e^{j\omega})$ of a length-32 rectangular window to the right and to the left by the amount $0.172 \times 2\pi = 0.344\pi$, adding both shifted versions, and then amplitude scaling by a factor 1/2. In the normalized angular frequency range 0 to 2π, which is one period of the DTFT, there are two peaks, one at 0.344π and the other at $2\pi(1 - 0.172) = 1.656\pi$, as verified by Figure 11.3. A 32-point DFT of this DTFT is precisely the DFT shown in Figure 8.32. The two peaks of the DFT at bin locations $k = 5$ and $k = 6$ are frequency samples of the main lobe located at the normalized frequency 0.172 on both sides of the peak. Likewise, the two peaks of the DFT at bin locations $k = 26$ and $k = 27$ are frequency samples of the main lobe located at the normalized frequency 0.828 on both sides of the peak. All other DFT samples are given by the samples of the sidelobes of the DTFT of the window causing the leakage of the frequency components at $\pm\omega_0$ to other bin locations with the amount of leakage determined by the relative amplitude of the mainlobe and the sidelobes. Since the relative sidelobe level $A_{s\ell}$, defined by the ratio in dB of the amplitude of the mainlobe to that of the largest sidelobe, of the rectangular window is very high, there is a considerable amount of leakage to the bin locations adjacent to the bins showing the peaks in Figure 8.32.

The above problem gets more complicated if the signal being analyzed has more than one sinusoid, as is typically the case. We illustrate the DFT-based spectral analysis approach by means of several examples. Through these examples we examine the effects of the length R of the DFT, the type of window being used, and its length N on the results of spectral analysis.

Example 11.1 In this example we examine the effect of the DFT length in spectral analysis. The signal to be analyzed in the spectral domain is given by

$$x[n] = \tfrac{1}{2}\sin(2\pi f_1 n) + \sin(2\pi f_2 n), \qquad 0 \le n \le N - 1. \tag{11.13}$$

Let the normalized frequencies of the two length-16 sinusoidal sequences be $f_1 = 0.22$ and $f_2 = 0.34$. We compute the DFT of their sum $x[n]$ for various values of the DFT length R. To this end, we use Program 11_2 given below whose input data are the length N of the signal, length R of the DFT, and the two frequencies f_1 and f_2. The program generates the two sinusoidal sequences, forms their sum, then computes the DFT of the sum and plots the DFT samples. In this example, we fix $N = 16$ and vary the DFT length R from 16 to 128. Note that when R > N, the M-file function fft(x,R) automatically zero-pads the sequence x with R-N zero-valued samples.

```
% Program 11_2
% Spectral Analysis of a Sum of Two Sinusoids
% Using the DFT
%
clf;
N = input('Signal length = ');
R = input('DFT length = ');
fr = input('Type in the sinusoid frequencies = ');
n = 0:N-1;
x = 0.5*sin(2*pi*n*fr(1)) + sin(2*pi*n*fr(2));
Fx = fft(x,R);
k = 0:R-1;
stem(k,abs(Fx));grid
xlabel('k'); ylabel('Magnitude');
title(['N = ',num2str(N),', R = ',num2str(R)]);
```

Figure 11.4(a) shows the magnitude $|X[k]|$ of the DFT samples of the signal $x[n]$ of Eq. (11.13) for $R = 16$. From the plot of the magnitude $|X(e^{j\omega})|$ of the DTFT given in Figure 11.4(b), it is evident that the DFT samples given in Figure 11.4(a) are indeed the frequency samples of the frequency response, as expected. As is customary, the horizontal axis in Figure 11.4(a) has been labeled in terms of the DFT frequency sample (bin) number k, where k is related to the normalized angular frequency ω through Eq. (11.3). Thus, $\omega = 2\pi \times 8/16 = \pi$ corresponds to $k = 8$ and $\omega = 2\pi \times 15/16 = 1.875\pi$ corresponds to $k = 15$.

From the plot of Figure 11.4(a) it is difficult to determine whether there is one or more sinusoids in the signal being examined, and the exact locations of the sinusoids. To increase the accuracy of the locations of the sinusoids, we increase the size of the DFT to 32 and recompute the DFT as indicated in Figure 11.4(c). In this plot there appears to be some concentrations around $k = 7$ and around $k = 11$ in the normalized frequency range from 0 to 0.5. Figure 11.4(d) shows the DFT

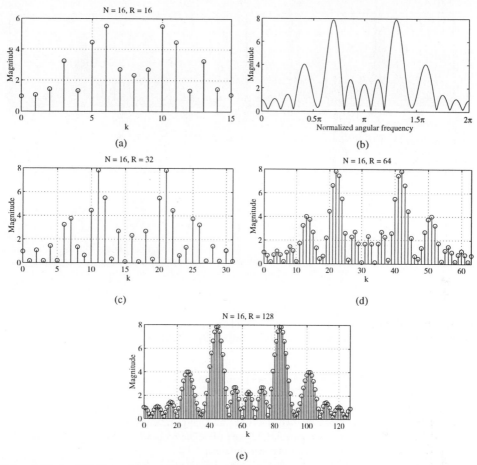

Figure 11.4 DFT based spectral analysis of a sum of two finite-length sinusoidal sequences of normalized frequencies 0.22 and 0.34, respectively, of length 16 each for various values of DFT lengths.

plot obtained for $R = 64$. In this plot there are two clear peaks occurring at $k = 13$ and $k = 22$ that correspond to the normalized frequencies of 0.2031 and 0.3438, respectively. To improve further the accuracy of the peak location, we compute next a 128-point DFT as indicated in Figure 11.4(e) in which the peak occurs around $k = 27$ and $k = 45$ corresponding to the normalized frequencies of 0.2109 and 0.3516, respectively. However, this plot also shows a number of minor peaks and it is not clear by examining this DFT plot whether additional sinusoids of lesser strengths are present in the original signal or not.

As this example points out, in general, an increase in the DFT length increases the resolution of the spectral analysis by reducing the spectral separation of adjacent DFT samples.

Example 11.2 In this example, we compute the DFT of a sum of two finite-length sinusoidal sequences as given by Eq. (11.13) with one of the sinusoids at a fixed frequency, while the other

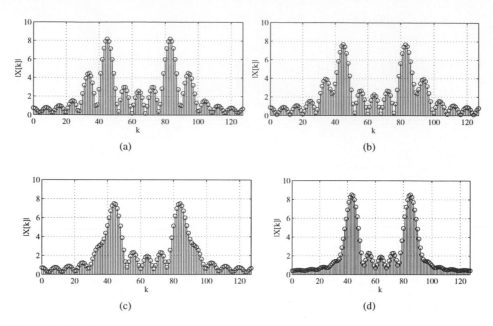

Figure 11.5   Illustration of the frequency resolution property. (a) $f_1 = 0.28$, $f_2 = 0.34$, (b) $f_1 = 0.29$, $f_2 = 0.34$, (c) $f_1 = 0.3$, $f_2 = 0.34$, and (d) $f_1 = 0.31$, $f_2 = 0.34$.

sinusoid is of varied frequency. Specifically, we keep $f_2 = 0.34$ and vary f_1 from 0.28 to 0.31. The length of the signal being analyzed is 16, while the DFT length is 128.

Figure 11.5 shows the plots of the DFTs computed, along with the frequencies of the sinusoids. As can be seen from these plots, the two sinusoids are clearly resolved in Figure 11.5(a) and (b), while they cannot be resolved in Figure 11.5(c) and (d). The reduced resolution occurs when the difference between the two frequencies becomes less than 0.04.

As indicated by Eq. (11.12), the DTFT of a length-N sinusoid of normalized angular frequency ω_1 is obtained by frequency translating the DTFT $\Psi_R(e^{j\omega})$ of a length-N rectangular window to the frequencies $\pm\omega_1$ and scaling their amplitudes appropriately. In the case of a sum of two length-N sinusoids of normalized angular frequencies ω_1 and ω_2, the DTFT is obtained by summing the DTFTs of the individual sinusoids. As the difference between the two frequencies become smaller, the mainlobes of the DTFTs of the individual sinusoids get closer and eventually overlap. If there is a significant overlap, it will be difficult to resolve the peaks. It follows therefore that the frequency resolution is essentially determined by the width Δ_{ML} of the mainlobe of the DTFT of the window.

Now from Table 7.3, the main lobe width Δ_{ML} of a length-N rectangular window is given by $4\pi/N$. In terms of normalized frequency, the main lobe width of a length-16 rectangular window is 0.125. Hence, two closely spaced sinusoids windowed with a rectangular window of length 16 can be clearly resolved if the difference in their frequencies is about half of the main lobe width, i.e., 0.0625.

Even though the rectangular window has the smallest main lobe width, it has the largest relative sidelobe amplitude and, as a consequence, causes considerable leakage. As seen from Examples 11.1 and 11.2, the large amount of leakage results in minor peaks that may be falsely identified as sinusoids. We now study the effect of windowing the signal with a Hamming window.[1]

Example 11.3 We compute the DFT of a sum of two sinusoids windowed by a Hamming window. The signal being analyzed is given by $x[n] \cdot w[n]$, where $x[n]$ is given by

$$x[n] = 0.85 \sin(2\pi f_1 n) + \sin(2\pi f_2 n),$$

and $w[n]$ is a Hamming window of length N. The two normalized frequencies are $f_1 = 0.22$ and $f_2 = 0.26$.

Figure 11.6(a) shows the 16-point DFT of the windowed signal with a window length of 16. As can be seen from this plot, the leakage has been reduced considerably, but it is difficult to resolve the two sinusoids. We next increase the DFT length to 64 while keeping the window length fixed at 16. The resulting plot is shown in Figure 11.6(b), indicating a substantial reduction in the leakage but with no change in the resolution. Now from Table 7.3, the main lobe width Δ_{ML} of a length-N Hamming window is $8\pi/N$. Thus, for $N = 16$, the normalized main lobe width is 0.25. Hence, with such a window, we can resolve two frequencies if their difference is of the order of half the main lobe width, i.e., 0.125 or better. In our example, the difference is 0.04 which is considerably smaller than this value.

In order to increase the resolution, we increase the window length to 32, which will reduce the main lobe width by half. Figure 11.6(c) shows its 32-point DFT. There now appears to be two peaks. Increasing the DFT size to 64 clearly separates the two peaks, as indicated in Figure 11.6(d). This separation becomes more visible with an increase in the DFT size to 256, as shown in Figure 11.6(e). Finally, Figure 11.6(f) shows the result obtained with a window length of 64 and a DFT length of 256.

It is clear from the above examples that performance of the DFT-based spectral analysis depends on several factors, the type of window being used and its length, and the size of the DFT. To improve the frequency resolution, one must use a window with a very small main lobe width, and to reduce the leakage, the window must have a very small relative sidelobe level. The mainlobe width can be reduced by increasing the length of the window. Furthermore, an increase in the accuracy of locating the peaks is achieved by increasing the size of the DFT. To this end, it is preferable to use a DFT length that is a power of 2 so that very efficient FFT algorithms can be employed to compute the DFT. Of course, an increase in the DFT size also increases the computational complexity of the spectral analysis procedure.

Often in practice, one or more sinusoidal sequences are corrupted with noise, and it is necessary to determine the frequencies of these sinusoids. In some cases, this problem can be solved by examining the DFT of the noise corrupted signal, as illustrated in the following example.

Example 11.4 Consider a sinusoidal signal $s[n]$ corrupted by a noise $\eta[n]$:

$$y[n] = s[n] + \eta[n] = \cos(0.1\pi n) + 6\nu[n], \tag{11.14}$$

where $\nu[n]$ is a uniformly distributed random noise with amplitude bounded by ± 1. The measure of noise corruption is usually given by a *signal-to-noise ratio* (SNR) defined by

$$\text{SNR} = 10 \log_{10} \frac{\sum_{n=0}^{N-1} s^2[n]}{\sum_{n=0}^{N-1} \eta^2[n]} \text{ dB}, \tag{11.15}$$

[1]For a review of some commonly used windows, see Sections 7.7.4 and 7.7.5

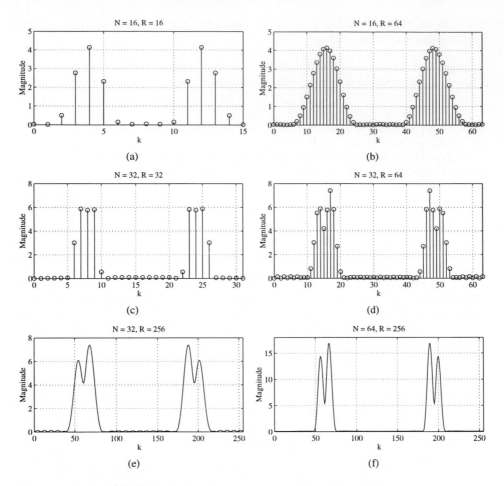

Figure 11.6   Spectral analysis using a Hamming window.

where N is the length of both the signal and the noise.

The MATLAB program used to analyze the signal of Eq. (11.14) is given below.

```
% Program 11_3
% DFT Analysis of a Noise Corrupted Signal
%
clf;
N = 256; R = 256;
n = 0:N-1;
x = sin(2*pi*n*0.05);
spwr = sum(x*x');
d = rand(1,N);
ave = sum(d)/256;
dw = d - ave;
```

```
y = x + 6*dw;
npwr = 36*sum(dw*dw');
Fy = fft(y,R);
k = 0:R-1;
plot(n,x);grid
xlabel('Time index n'); ylabel('Amplitude')
title('Uncorrupted sinusoid')
pause
plot(k,y);grid
xlabel('Time index n'); ylabel('Amplitude')
title('Noise corrupted sinusoid')
pause
clg
Fy = fft(y);
plot(k,abs(Fy));grid
xlabel('k'); ylabel('Magnitude')
SNR = 10*log10(spwr/npwr)
```

Figure 11.7(a) and (b) show, respectively, the uncorrupted sinusoid and the noise-corrupted sinusoid $y[n]$. As can be seen from Figure 11.7(b), it is difficult to tell by simply examining this plot that it is a noise-corrupted sinusoid. A 256-point DFT of $y[n]$ is shown in Figure 11.7(c), which clearly shows two peaks at bin locations $k = 13$ and $k = 243$ resulting from the DFT of the sinusoid. The SNR computed by the above program is -7.5511 dB. The normalized frequency of the sinusoid is therefore $13/256 = 0.0508$, which is seen to be very close to the actual value of 0.05.

11.3   Short-Term Discrete Fourier Transform

The discrete Fourier transform can be employed for the spectral analysis of a finite-length signal composed of sinusoidal components as long as the frequency, amplitude, and phase of each sinusoidal component are time-invariant and independent of the signal length. There are practical situations where the signal to be analyzed is instead nonstationary, for which these signal parameters are time-varying. An example of such a time-varying signal is the chirp signal given by

$$x[n] = A \cos(\omega_0 n^2) \tag{11.16}$$

and shown in Figure 11.8 for $\omega_0 = 10\pi \times 10^{-5}$. Note from Eq. (11.16) that the instantaneous frequency of $x[n]$ is given by $\omega_0 n$, which is not a constant but increases linearly with time. Speech, radar, and sonar signals are other examples of such nonstationary signals. To describe such signals in the frequency domain using a simple DFT of the complete signal will provide misleading results. To get around the time-varying nature of the signal parameters, an alternative approach would be to segment the sequence into a set of subsequences of short length, with each subsequence centered at uniform intervals of time and its DFT computed separately. If the subsequence length is reasonably small, it can be safely assumed to be stationary for practical purposes. As a result, the frequency-domain description of a long sequence is given by a set of short-length DFTs, i.e., a time-dependent DFT.

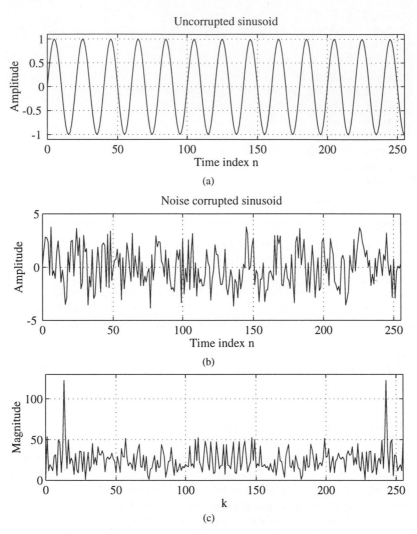

Figure 11.7   Spectral analysis of a noise-corrupted sinusoid.

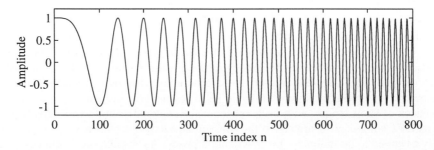

Figure 11.8   First 800 samples of a casual chirp signal $\cos(\omega_0 n^2)$ with $\omega_0 = 10\pi \times 10^{-5}$.

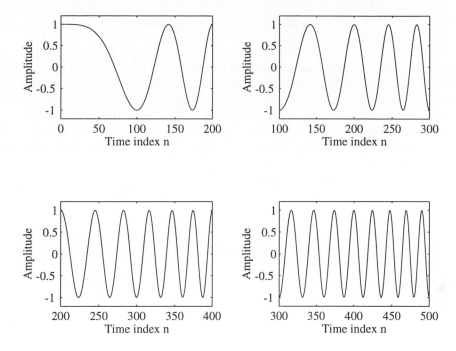

Figure 11.9 Examples of subsequences of the chirp signal of Figure 11.8 generated by a length 200 rectangular window.

To represent a nonstationary signal $x[n]$ in terms of a set of short-length subsequences, we can multiply it with a window $w[n]$ that is stationary with respect to time and move the signal through the window. For example, Figure 11.9 shows four segments of the chirp signal of Figure 11.8 as seen through a stationary rectangular window of length 200. As illustrated in this figure, the segments could be overlapping in time. A discrete-time Fourier transform of the short sequence obtained by windowing is called the short-term Fourier transform which is thus a function of the location of the window relative to the original long sequence and the frequency. In this section we review the basic concepts associated with this type of transform, study some of its properties, and point out one of its important application. A detailed exposition to this subject can be found in [All77], [Naw88], [Opp89], and [Rab78].

11.3.1 Definition

The *short-time Fourier transform* (STFT), also known as the *time-dependent Fourier transform*, of a sequence $x[n]$ is defined by

$$X_{\text{STFT}}(e^{j\omega}, n) = \sum_{m=-\infty}^{\infty} x[n-m]w[m]e^{-j\omega m}, \qquad (11.17)$$

where $w[n]$ is a suitably chosen window sequence. It should be noted that the function of the window is to extract a finite-length portion of the sequence $x[n]$ such that the spectral characteristics of the section extracted are approximately stationary over the duration of the window for practical purposes.

Note that if $w[n] = 1$, the definition of STFT given in Eq. (11.17) reduces to the conventional discrete-time Fourier transform (DTFT) of $x[n]$. However, even though the DTFT of $x[n]$ exists under certain well-defined conditions, the windowed sequence in Eq. (11.17) being finite in length ensures the existence of the STFT for any sequence $x[n]$. It should be noted also that, unlike the conventional DTFT, the STFT is a function of two variables: the integer variable time index n and the continuous frequency variable ω. It also follows from the definition of Eq. (11.17), that $X_{STFT}(e^{j\omega}, n)$ is a periodic function of ω with a period 2π.

To compute the inverse STFT we observe that $X_{STFT}(e^{j\omega}, n)$ is the DTFT of $x[n - m]w[m]$, and hence, we can determine $x[n - m]w[m]$ by performing an inverse DTFT operation according to

$$x[n - m]w[m] = \frac{1}{2\pi} \int_0^{2\pi} X_{STFT}(e^{j\omega}, n)e^{j\omega m} \, d\omega, \qquad -\infty < m < \infty \qquad (11.18)$$

From the above, we arrive at

$$x[n] = \frac{1}{2\pi \, w[0]} \int_0^{2\pi} X_{STFT}(e^{j\omega}, n) \, d\omega, \qquad (11.19)$$

and as a result, $x[n]$ can be determined if $w[0] \neq 0$. In fact, all the samples of the sequence $x[n]$ that are multiplied by nonzero samples of the window can be recovered in a similar manner using Eq. (11.19).

In most applications, the magnitude of the STFT is of interest. The display of the magnitude of the STFT is usually referred to as the *spectrogram*. However, since the STFT is a function of two variables, the display of its magnitude would normally require three dimensions. Often, it is plotted in two dimensions, with the magnitude represented by the darkness of the plot. Here, the white areas represent zero-valued magnitude while the gray areas represent non zero magnitudes, with the largest magnitudes being shown in black. In the STFT magnitude display, the vertical axis represents the frequency variable (ω) and the horizontal axis represents the time index (n). The STFT can also be visualized as a mesh plot in a three-dimensional coordinate frame in which the STFT magnitude is a point in the z-direction above the x-y plane.

Figure 11.10 shows the STFT of the chirp sequence of Eq. (11.16), with $\omega_0 = 10\pi \times 10^{-5}$ for a length of 20,000 samples computed using a Hamming window of length 200 in the above two forms. In Figure 11.10, the STFT for a given value of the time index n is essentially the DFT of a segment of a sinusoidal sequence. Recall from our discussion in Section 11.2 that the shape of the DFT of a such a segment is similar to that shown in Figure 11.3, with large nonzero-valued DFT samples around the frequency of the sinusoid and smaller nonzero-valued DFT samples at other frequency points. In the spectrogram plot, the large-valued DFT samples show up as narrow nearly black very short vertical lines while the other DFT samples show up as gray points. As the instantaneous frequency of the chirp signal increases linearly, the short black line moves up in the vertical direction and eventually, because of aliasing, the black line starts moving down in the vertical direction. As a result, the spectrogram of the chirp signal essentially appears as a thick line in the form of a triangular shape.

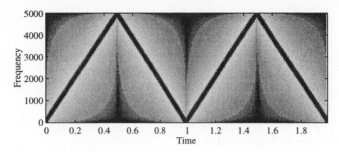

Figure 11.10 Spectrogram of a chirp signal.

11.3.2   Sampling in the Time and Frequency Dimensions

In practice, the STFT is computed at a finite set of discrete values of ω. Moreover, due to the finite length of the windowed sequence, the STFT is accurately represented by its frequency samples as long as the number of frequency samples is greater than the window length. Moreover, the portion of the sequence $x[n]$ inside the window can be fully recovered from the frequency samples of the STFT.

To be more precise, let the window be of length R defined in the range $0 \leq m \leq R-1$. We sample $X_{\text{STFT}}(e^{j\omega}, n)$ at N equally spaced frequencies $\omega_k = 2\pi k/N$, with $N \geq R$ as indicated below:

$$X_{\text{STFT}}[k, n] = X_{\text{STFT}}(e^{j\omega}, n)\Big|_{\omega=2\pi k/N} = X_{\text{STFT}}(e^{j2\pi k/N}, n)$$

$$= \sum_{m=0}^{R-1} x[n-m]w[m]e^{-j2\pi km/N}, \quad 0 \leq k \leq N-1. \quad (11.20)$$

It follows from the above, assuming $w[m] \neq 0$, that $X_{\text{STFT}}[k, n]$ is simply the DFT of $x[n-m]w[m]$. Note that $X_{\text{STFT}}[k, n]$ is a two-dimensional sequence and is periodic in k with a period N. Applying the IDFT, we thus arrive at

$$x[n-m]w[m] = \frac{1}{N} \sum_{k=0}^{N-1} X_{\text{STFT}}[k, n]e^{j2\pi km/N}, \quad 0 \leq m \leq R-1, \quad (11.21)$$

or in other words,

$$x[n-m] = \frac{1}{Nw[m]} \sum_{k=0}^{N-1} X_{\text{STFT}}[k, n]e^{j2\pi km/N}, \quad 0 \leq m \leq R-1, \quad (11.22)$$

verifying that the sequence values inside the window can be fully recovered from $X_{\text{STFT}}[k, n]$ as long as the number N of frequency samples is greater than or equal to the window length R. It should be evident by now that $x[n]$ for $-\infty < n < \infty$ can be fully recovered if $X_{\text{STFT}}(e^{j\omega}, n)$ or $X_{\text{STFT}}[k, n]$ is sampled also in the time dimension. More precisely, if we set $n = n_o$ in Eq. (11.22), we recover the signal in the interval $n_0 \leq n \leq n_0 + R - 1$ from $X_{\text{STFT}}[k, n_0]$. Likewise, by setting $n = n_0 + R$ in Eq. (11.22), we recover the signal in the interval $n_0 + R \leq n \leq n_0 + 2R - 1$ from $X_{\text{STFT}}[k, n_0 + R]$, and so on.

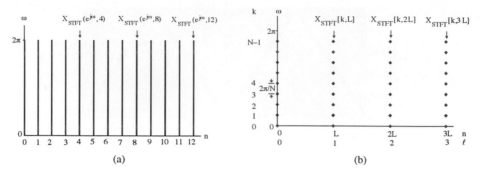

Figure 11.11   Sampling grid in the (ω, n)-plane for the sampled STFT $X_{\mathrm{STFT}}[k, \ell L]$ for $N = 9$ and $L = 4$.

The sampled STFT for a window defined in the region $0 \le m \le R - 1$ is given by

$$X_{\mathrm{STFT}}[k, \ell L] = X_{\mathrm{STFT}}(e^{j2\pi k/N}, \ell L) = \sum_{m=0}^{R-1} x[\ell L - m]w[m]e^{-j2\pi km/N}, \quad (11.23)$$

where ℓ and k are integers such that $-\infty < \ell < \infty$ and $0 \le k \le N - 1$. Figure 11.11 shows lines in the (ω, n)-plane corresponding to $X_{\mathrm{STFT}}(e^{j\omega}, n)$ and the grid of sampling points in the (ω, n)-plane for the case $N = 9$ and $L = 4$. As we have shown, it is possible to uniquely reconstruct the original signal from such a 2-D discrete representation provided $N \ge R \ge L$.

11.3.3   Window Selection

As in the case of the DFT-based spectral analysis of deterministic signals discussed in Section 11.2, in the STFT analysis of nonstationary signals, the window also plays an important role. Both the length and shape of the window are critical issues that need to be examined carefully.

The function of the window $w[n]$ is to extract a portion of the signal for analysis and ensure that the extracted section of $x[n]$ is approximately stationary. To this end, the window length R should be small, in particular for signals with widely varying spectral parameters. A decrease in the window length increases the time-resolution property of the STFT, whereas the frequency-resolution property of the STFT increases with an increase in the window length. A shorter window thus provides a *wideband spectrogram* while a longer window results in a *narrowband spectrogram*.

The two frequency-domain parameters characterizing the DTFT of a window are its main lobe width Δ_{ML} and the relative sidelobe amplitude $A_{s\ell}$. The former parameter determines the ability of the window to resolve two signal components in the vicinity of each other, while the latter controls the degree of leakage of one component into a nearby signal component. It thus follows that in order to obtain a reasonably good estimate of the frequency spectrum of a time-varying signal, the window should be chosen to have a very small relative sidelobe amplitude with a length chosen based on the acceptable accuracy of the frequency and time resolutions.

11.3.4 STFT Computation Using MATLAB

The Signal Processing Toolbox of MATLAB includes the function `specgram` for the computation of the STFT of a signal. There are a number of versions of this function, as given below:

```
      B = specgram(x),
      B = specgram(x,nfft),
  [B,f] = specgram(x,nfft,Fs),
  [B,f] = specgram(x,nfft,Fs),
      B = specgram(x,nfft,Fs,window),
      B = specgram(x,nfft,Fs,window,noverlap)
```

`B = specgram(x)` computes the STFT of the signal x specified as a vector using a DFT of length `nfft = min(256,length(x))` and a Hann window of length `nfft`. The sampling frequency used is `Fs = 2`, and the number of samples by which the consecutive segments overlap is given by `noverlap = length(window/2)`. For a real signal x, the DFT is computed at positive frequencies only, whereas for a complex x, the DFT is computed at both positive and negative frequencies. The column indices of B refer to the time position of the window with time increasing across the columns, and the row indices of B correspond to the frequency with the first row corresponding to $\omega = 0$, as indicated in Figure 11.11(b).

The parameters `nfft`, `Fs`, and `noverlap` can be specified along with the type of window, depending on the version of specgram being used. If a scalar integer is used for `window`, the function uses a Hann window of length given by this integer. The specified window length must be less than or equal to `nfft`.

The vector of frequencies f at which the DFT is computed is returned when `[B,f] = specgram(x,nfft,Fs)` is employed, whereas, both frequency and time vectors, f and t, respectively, are returned in the version `[B,f,t] = specgram(x,nfft,Fs)`. In the latter case, t is a column vector of scaled times and of length equal to the number of columns of B. The first element of t is always 0.

We illustrate the application of `specgram` in the following section.

11.3.5 Analysis of Speech Signals Using STFT

The short-term Fourier transform is often used in the analysis of speech, which is made up of nonstationary signals. As indicated in Section 1.3, the speech signal, generated by the excitation of the vocal tract, is composed of two types of basic waveforms: voiced and unvoiced sounds. A typical speech signal is shown in Figure 1.17. As can be seen from this figure, a segment of a speech over a small time interval can be considered as a stationary signal, and as a result, the DFT of the speech segment can provide a reasonable representation of the frequency-domain characteristic of the speech in this time interval. However, in the STFT analysis, the size of the window is critical since a shorter window developing a wideband spectrogram provides a better time resolution, whereas a longer window developing a narrowband spectrogram results in an improved frequency resolution. In order to provide a reasonably good estimate of the changes in the vocal tract and the excitation, a wideband spectrogram is preferable. To this end, the window size is selected to be approximately close to one pitch period, which is adequate for resolving the formants though not adequate to resolve the harmonics of the pitch frequencies. On the other hand, to resolve the harmonics of the pitch frequencies, a narrowband spectrogram with a window of size of several pitch periods is desirable.

The following example illustrates the STFT analysis of a speech signal.

Example 11.5 The mtlb.mat file in the *Signal Processing Toolbox* of MATLAB contains a speech signal of duration 4001 samples sampled at 7418 Hz. We compute its STFT using a Hamming window of length 256 with an overlap of 50 samples between consecutive windowed signals using Program 11_4 given below.

```
% Program 11_4
% Spectrogram of a Speech Signal
%
load mtlb
n = 1:4001;
plot(n-1,mtlb);
xlabel('Time index n');ylabel('Amplitude');
pause
nfft = input('Type in the window length = ');
ovlap = input('Type in the desired overlap = ');
specgram(mtlb,nfft,7418,hamming(nfft),ovlap)
```

The program first loads the mtlb.mat file and plots the signal, as indicated in Figure 11.12(a). It then requests the desired sizes of the window and the overlap between consecutive segments, computes the STFT using the function specgram, and displays a plot of the spectrogram. Figure 11.12(b) and (c) show, respectively, a narrowband spectrogram and a wideband spectrogram of the speech signal of Figure 11.12(a). The frequency and time resolution tradeoff between the two spectrograms of Figure 11.12 should be evident.

11.4   Musical Sound Processing

Recall from our discussion in Section 1.4.1 that almost all musical programs are produced in basically two stages. First, sound from each individual instrument is recorded in an acoustically inert studio on a single track of a multitrack tape recorder. Then, the signals from each track are manipulated by the sound engineer to add special audio effects and are combined in a mix-down system to finally generate the stereo recording on a two-track tape recorder [Ble78], [Ear86]. The audio effects are artificially generated using various signal processing circuits and devices, and they are increasingly being performed using digital signal processing techniques [Ble78], [Orf96].

Some of the special audio effects that can be implemented digitally are reviewed in this section.

11.4.1   Time-Domain Operations

As indicated in Section 1.4.1, the sound reaching the listener in a closed space, such as a concert hall, consists of several components: direct sound, early reflections, and reverberation. The early reflections are composed of several closely spaced echoes that

Figure 11.12   (a) A speech signal, and its (b) narrowband spectrogram, and (c) wideband spectrogram.

are basically delayed and attenuated copies of the direct sound, whereas the reverberation is composed of densely packed echoes. The sound recorded in an inert studio is different from that recorded inside a closed space, and as a result, the former does not sound "natural" to a listener. However, digital filtering can be employed to convert the sound recorded in an inert studio into a natural sounding one by artificially creating the echoes and adding it to the original signal.

Echoes are simply generated by delay units. For example, the direct sound and a single echo appearing R sampling period later can be simply generated by the FIR filter of Figure 11.13(a), which is characterized by the difference equation

$$y[n] = x[n] + \alpha x[n - R], \qquad |\alpha| < 1, \tag{11.24}$$

or equivalently, by a transfer function

$$H(z) = 1 + \alpha z^{-R}. \tag{11.25}$$

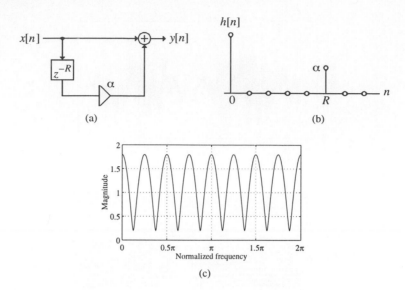

Figure 11.13   Single echo filter. (a) Filter structure, (b) typical impulse response, and (c) magnitude response for $R = 8$ and $\alpha = 0.8$.

Its impulse response is sketched in Figure 11.13(b). The magnitude response of a single echo FIR filter for $\alpha = 0.8$ and $R = 8$ is shown in Figure 11.13(c). The magnitude response exhibits R peaks and R dips in the range $0 \leq \omega < 2\pi$, with the peaks occurring at $\omega = 2\pi k/R$ and the dips occurring at $\omega = (2k+1)\pi/R, k = 0, 1, \ldots, R-1$. Because of the "comb"-like shape of the magnitude response, such a filter is also known as a *comb filter*. The maximum and minimum values of the magnitude response are given by $1 + \alpha = 1.8$ and $1 - \alpha = 0.2$, respectively.

To generate multiple echoes spaced R sampling periods apart with exponentially decaying amplitudes, one can use an FIR filter with a transfer function of the form

$$H(z) = 1 + \alpha z^{-R} + \alpha^2 z^{-2R} + \ldots + \alpha^{N-1} z^{-(N-1)R} = \frac{1 - \alpha^N z^{-NR}}{1 - \alpha z^{-R}}. \qquad (11.26)$$

An IIR realization of this filter is sketched in Figure 11.14(a). The impulse response of a multiple echo filter with $\alpha = 0.8$ for $N = 6$ and $R = 4$ is shown in Figure 11.14(b).

An infinite number of echoes spaced R sampling periods apart with exponentially decaying amplitudes can be created by an IIR filter with a transfer function of the form

$$H(z) = \frac{z^{-R}}{1 - \alpha z^{-R}}, \qquad |\alpha| < 1. \qquad (11.27)$$

Figure 11.15(a) shows one possible realization of the above IIR filter whose first 61 impulse response samples for $R = 4$ are indicated in Figure 11.15(b). The magnitude response of this IIR filter for $R = 7$ is sketched in Figure 11.15(c). The magnitude response exhibits R peaks and R dips in the range $0 \leq \omega < 2\pi$, with the peaks occurring at $\omega = 2\pi k/R$ and the dips occurring at $\omega = (2k+1)\pi/R, k = 0, 1, \ldots, R-1$. Because of the "comb"-like shape of the magnitude response, such a filter is also known as a comb filter. The

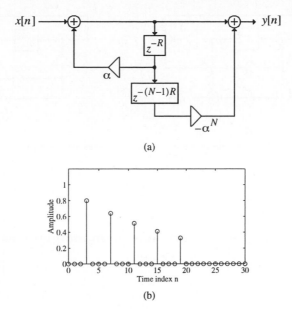

(a)

(b)

Figure 11.14 Multiple echo filter generating $N-1$ echoes. (a) The filter structure, and (b) impulse response with $\alpha = 0.8$ for $N = 6$ and $R = 4$.

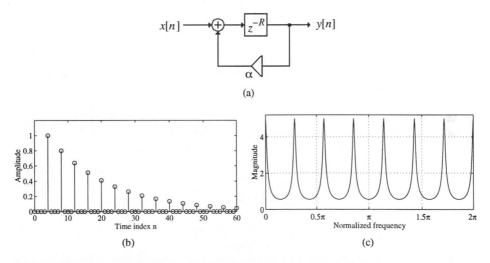

(a)

(b)   (c)

Figure 11.15 IIR filter generating infinite number of echoes. (a) The filter structure, (b) impulse response with $\alpha = 0.8$ for $R = 4$, and (c) magnitude response with $\alpha = 0.8$ for $R = 7$.

maximum and minimum values of the magnitude response are given by $1/(1 - \alpha) = 5$, and $1/(1 + \alpha) = 0.5556$, respectively.

The IIR comb filter of Figure 11.15(a) by itself does not provide natural sounding reverberations for two reasons [Sch62]. First, as can be seen from Figure 11.15(c), its magnitude response is not constant for all frequencies, resulting in a "coloration" of many

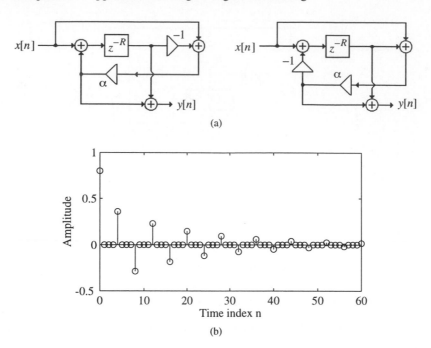

Figure 11.16   Allpass reverberator. (a) Block diagram representation, and (b) impulse response for with $\alpha = 0.8$ for $R = 4$.

musical sounds that are often unpleasant for listening purposes. Second, the output echo density, given by the number of echoes per second, generated by an unit impulse at the input, is much lower than that observed in a real room, thus causing a "fluttering" of the composite sound. It has been observed that approximately 1000 echoes per second are necessary to create a reverberation that sounds free of flutter [Sch62]. To develop a more realistic reverberation, a reverberator with an allpass structure, as indicated in Figure 11.16(a), has been proposed [Sch62].[2] Its transfer function is given by

$$H(z) = \frac{\alpha + z^{-R}}{1 + \alpha z^{-R}}, \qquad |\alpha| < 1. \tag{11.28}$$

In the steady-state, the spectral balance of the sound signal remains unchanged due to the unity magnitude response of the allpass reverberator.

The IIR comb filter of Figure 11.15(a) and the allpass reverberator of Figure 11.16(a) are basic reverberator units that are suitably interconnected to develop a natural sounding reverberation. Figure 11.17 shows one such interconnection composed of a parallel connection of four IIR echo generators in cascade with two allpass reverberators [Sch62]. By choosing different values for the delays in each section (obtained by adjusting R_i) and the multiplier constants α_i it is possible to arrive at a pleasant sounding reverberation, duplicating that occurring in a specific closed space, such as a concert hall.

---

[2]The structures shown here are the canonic single multiplier realization of a first-order allpass transfer function [Mit74a]. See also Section 6.7.1.

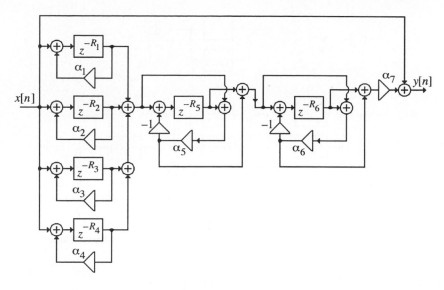

Figure 11.17 A proposed natural sounding reverberator scheme.

An interesting modification of the basic IIR comb filter of Figure 11.15(a) is obtained by replacing the multiplier α with a lowpass FIR or IIR filter $G(z)$, as indicated in Figure 11.18(a). It has a transfer function given by

$$H(z) = \frac{z^{-R}}{1 - z^{-R}G(z)}, \tag{11.29}$$

obtained by replacing α in Eq. (11.27) with $G(z)$. This structure has been referred to as the *teeth filter* and has been introduced to provide a natural tonal character to the artificial reverberation generated by it [Egg84]. This type of reverberator should be carefully designed to avoid the stability problem. To provide a reverberation with a higher echo density, the teeth filter has been used as a basic unit in a more complex structure such as that indicated in Figure 11.18(b).

Additional details concerning these composite and other such reverberator structures can be found in [Sch62], [Moo79].

There are a number of special sound effects that are often used in the mix-down process. One such effect is called *flanging*. Originally, it was created by feeding the same musical piece to two tape recorders and then combining their delayed outputs while varying the difference Δt between their delay times. One way to vary Δt is to slow down one of the tape recorders by placing the operator's thumb on the flange of the feed reel, which led to name flanging [Ear86]. The FIR comb filter of Figure 11.13(a) can be modified to create the flanging effect. In this case, the unit generating the delay of R samples, or equivalently, a delay of RT seconds, where T is the sampling period, is made a time-varying delay $\beta(n)$, as indicated in Figure 11.19. The corresponding input-output relation is then given by

$$y[n] = x[n] + \alpha x[n - \beta(n)]. \tag{11.30}$$

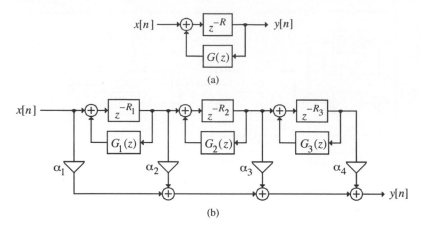

(a)

(b)

Figure 11.18   (a) Lowpass reverberator, and (b) a multitap reverberator structure.

Figure 11.19   Generation of a flanging effect.

Periodically varying the delay $\beta(n)$ between 0 and R with a low frequency ω_0 such as

$$\beta(n) = \frac{R}{2}\left(1 - \cos(\omega_0 n)\right),\qquad(11.31)$$

generates a flanging effect on the sound. It should be noted that as the value of $\beta(n)$ at an instant n in general is of noninteger value, in an actual implementation, the output sample value $y[n]$ should be computed using some type of interpolation method such as that outlined in Sections 11.11.2 and 11.11.3.

The *chorus* effect is achieved when several musicians are playing the same musical piece at the same time but with small changes in the amplitudes and small timing differences between their sounds. Such an effect can also be created synthetically by a *chorus generator* from the music of a single musician. A simple modification of the digital filter of Figure 11.19 leads to a structure that can be employed to simulate this sound effect. For example, the structure of Figure 11.20 can effectively create a chorus of four musicians from the music of a single musician. To achieve this effect, the delays $\beta_i(n)$ are randomly varied with very slow variations.

The *phasing* effect is produced by processing the signal through a narrowband notch filter with variable notch characteristics and adding a scaled portion of the notch filter output to the original signal as indicated in Figure 11.21 [Orf96]. The phase of the signal at the notch filter output can dramatically alter the phase of the combined signal, particularly around the notch frequency when it is varied slowly. The tunable notch filter can be implemented using the technique described in Section 6.8.2. The notch filter in Figure

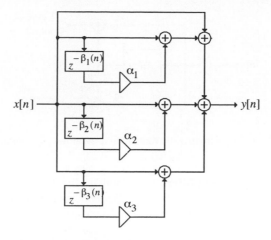

Figure 11.20 Generation of a chorus effect.

Figure 11.21 Generation of the phasing effect.

11.21 can be replaced with a cascade of tunable notch filters to provide an effect similar to flanging. However, in flanging, the swept notch frequencies are always equally spaced, whereas in phasing, the locations of the notch frequencies and their corresponding 3-dB bandwidth are varied independently.

11.4.2 Frequency-Domain Operations

The frequency responses of individually recorded instruments or musical sounds of performers are frequently modified by the sound engineer during the mix-down process. These effects are achieved by passing the original signals through an equalizer, briefly reviewed in Section 1.4.1. The purpose of the equalizer is to provide "presence" by peaking the mid-frequency components in the range of 1.5 to 3 kHz, and to modify the bass-treble relationships by providing "boost" or "cut" to components outside this range. It is usually formed from a cascade of first-order and second-order filters with adjustable frequency responses. The simple digital filters described in Section 4.3.2 can be employed for implementing these functions. We review these filters here to point out their specific properties that are suitable for musical sound processing. In addition, we describe some new structures with more flexible frequency responses.

Figure 11.22   A parametrically tunable first-order lowpass/highpass filter.

First-order filters and shelving filters   The transfer functions of a first-order lowpass filter and a highpass filter with tunable cutoff frequencies are given by Eqs. (4.70) and (4.73), respectively, and repeated below for convenience:

$$H_{LP}(z) = \frac{1-\alpha}{2} \cdot \frac{1+z^{-1}}{1-\alpha z^{-1}}, \tag{11.32a}$$

$$H_{HP}(z) = \frac{1+\alpha}{2} \cdot \frac{1-z^{-1}}{1-\alpha z^{-1}}. \tag{11.32b}$$

The 3-dB cutoff frequency ω_c for both transfer functions is related to the constant α through

$$\omega_c = \cos^{-1}\left(\frac{2\alpha}{1+\alpha^2}\right). \tag{11.33}$$

The two transfer functions of Eqs. (11.32a) and (11.32b) can be alternatively expressed as

$$H_{LP}(z) = \frac{1}{2}\left\{1 - A_1(z)\right\}, \tag{11.34a}$$

$$H_{HP}(z) = \frac{1}{2}\left\{1 + A_1(z)\right\}, \tag{11.34b}$$

where $A_1(z)$ is a first-order allpass transfer function given by

$$A_1(z) = \frac{\alpha - z^{-1}}{1 - \alpha z^{-1}}. \tag{11.35}$$

A composite realization of the above two transfer functions is sketched in Figure 11.22, where the first-order allpass transfer function $A_1(z)$ can be realized using any one of the single multiplier structures of Figure 6.36. Note that in this structure the 3-dB cutoff frequency ω_c of both filters is independently controlled by the multiplier constant α of the allpass section.

Combining the outputs of the structure of Figure 11.22 as indicated in Figure 11.23, we arrive at a low-frequency shelving filter that is characterized by a transfer function $G_1(z)$ given by

$$G_1(z) = \frac{K}{2}\left\{1 - A_1(z)\right\} + \frac{1}{2}\left\{1 + A_1(z)\right\}, \tag{11.36}$$

where K is a positive constant [Reg87b]. Figures 11.24 and 11.25 show the gain responses of $G_1(z)$ obtained by varying the multiplier constants K and α. Note that the parameter

Figure 11.23   Low-frequency shelving filter.

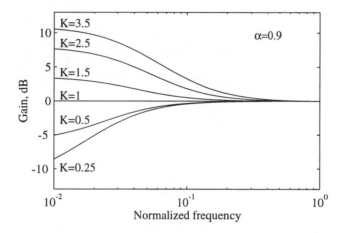

Figure 11.24   Gain responses of the low-frequency shelving filter of Figure 11.23 for various values of the parameter K with $\alpha = 0.9$.

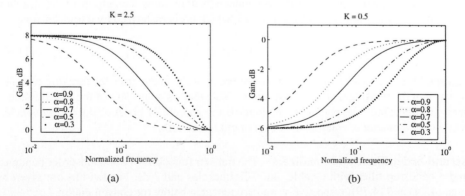

Figure 11.25   Gain responses of the low-frequency shelving filter of Figure 11.23 for various values of the α with two fixed values of the parameter K.

Figure 11.26   High-frequency shelving filter.

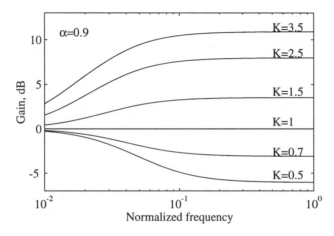

Figure 11.27   Gain responses of the high-frequency shelving filter of Figure 11.26 for various values of the parameter K with $\alpha = 0.9$.

K controls the amount of boost or cut at low frequencies, while the parameter α controls the boost or cut bandwidth.

A high-frequency shelving filter is obtained by combining the outputs of the structure of Figure 11.22 as indicated in Figure 11.26, which is characterized by a transfer function

$$G_1(z) = \frac{1}{2}\{1 - A_1(z)\} + \frac{K}{2}\{1 + A_1(z)\}. \tag{11.37}$$

Figures 11.27 and 11.28 show the gain responses of the above transfer function obtained by varying the multiplier constants K and α. Note that, as in the case of the low-frequency shelving filter, here the parameter K controls the amount of boost or cut at frequencies, while the parameter α controls the boost or cut bandwidth.

Second-order filters and equalizers   The transfer functions of a second-order bandpass and a bandstop filter with tunable cutoff frequencies and 3-dB bandwidths are given by Eqs. (4.74) and (4.79), respectively, and are repeated below for convenience:

$$H_{\mathrm{BP}}(z) = \frac{1-\alpha}{2} \cdot \frac{1 - z^{-2}}{1 - \beta(1+\alpha)z^{-1} + \alpha z^{-2}}, \tag{11.38a}$$

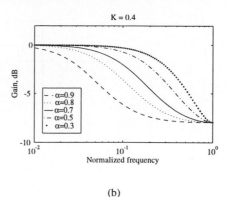

(a)   (b)

Figure 11.28   Gain responses of the high-frequency shelving filter of Figure 11.26 for various values of the parameter α with two fixed values of the parameter K.

$$H_{BS}(z) = \frac{1+\alpha}{2} \cdot \frac{1 - 2\beta z^{-1} + z^{-2}}{1 - \beta(1+\alpha)z^{-1} + \alpha z^{-2}}. \qquad (11.38b)$$

The center frequency ω_0 of the bandpass filter and the notch frequency ω_0 of the bandstop filter is related to the constant β through

$$\omega_0 = \cos^{-1}(\beta), \qquad (11.39)$$

while the 3-dB bandwidth $\Delta\omega_{3\,dB}$ of both transfer functions is related to the constant α through

$$\Delta\omega_{3\,dB} = \cos^{-1}\left(\frac{2\alpha}{1+\alpha^2}\right). \qquad (11.40)$$

A composite realization of both filters is as indicated in Figure 11.29(a) based on a sum of allpass decompositions of Eqs. (11.38a) and (11.38b):

$$H_{BP}(z) = \frac{1}{2}\left[1 - A_2(z)\right], \qquad (11.41a)$$

$$H_{BS}(z) = \frac{1}{2}\left[1 + A_2(z)\right], \qquad (11.41b)$$

where $A_2(z)$ is a second-order allpass transfer function given by

$$A_2(z) = \frac{\alpha - \beta(1+\alpha)z^{-1} + z^{-2}}{1 - \beta(1+\alpha)z^{-1} + \alpha z^{-2}}, \qquad (11.42)$$

and is realized using the cascaded lattice structure of Figure 11.29(b) for independent tuning of the center (notch) frequency ω_0 and the 3-dB bandwidth $\Delta\omega_{3\,dB}$.

As in the first-order case, a weighted combination of the outputs of the structure of Figure 11.29(a) results in a second-order equalizer (Figure 11.30) with a transfer function $G_2(z)$ given by

$$G_2(z) = \frac{K}{2}\{1 - A_2(z)\} + \frac{1}{2}\{1 + A_2(z)\}, \qquad (11.43)$$

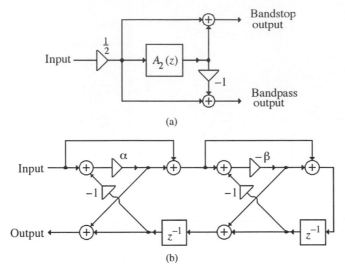

(a)

(b)

Figure 11.29   A parametrically tunable second-order bandpass/bandstop filter. (a) Overall structure, and (b) allpass section.

Figure 11.30   A parametrically tunable second-order equalizer.

where K is a positive constant [Reg87b]. It follows from the above discussion that the peak or the dip of the magnitude response occurs at the frequency ω_0, which is controlled independently by the parameter β according to Eq. (11.39), and the 3-dB bandwidth $\Delta\omega_{3\,dB}$ of the magnitude response is determined solely by the parameter α as indicated by Eq. (11.40). Moreover, the height of the peak or the dip of the magnitude response is given by $K = G_2(e^{j\omega_0})$. Figures 11.31 to 11.33 show the gain responses of $G_2(z)$ obtained by varying the parameters K, α, and β.

Higher-order equalizers   A graphic equalizer with tunable gain response can be built using a cascade of first-order and second-order equalizers with external control of the maximum gain values of each section in the cascade. Figure 11.34(a) shows the block diagram of a cascade of one first-order and three second-order equalizers with nominal frequency response parameters as indicated. Figure 11.34(b) shows its gain response for some typical values of the parameter K (maximum gain values) of the individual sections.

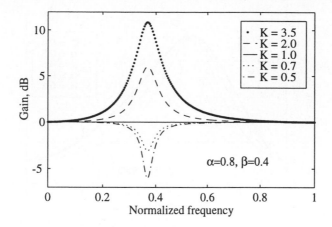

Figure 11.31 Gain responses of the second-order equalizer of Figure 11.30 for various values of the parameter K with $\alpha = 0.8$ and $\beta = 0.4$.

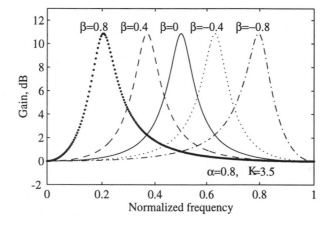

Figure 11.32 Gain responses of the second-order equalizer of Figure 11.30 for various values of the parameter β with $\alpha = 0.8$ and $K = 3.5$.

11.5 Digital FM Stereo Generation

Frequency-division multiplexing of stereo used in FM (frequency modulation) radio for the transmission of left and right channel audio signals has been described earlier in Section 1.4.3. We now consider the digital stereo generation for FM transmission. A simplified block diagram representation of a digital FM stereo generator is shown in Figure 11.35. As indicated here, the analog outputs of the left and right microphones, $s_L(t)$ and $s_R(t)$, are first converted into digital signals, $s_L[n]$ and $s_R[n]$, by means of individual analog-to-digital converters (ADCs). In practice, high-frequency components of the modulating signal have much smaller amplitudes, while low-frequency components have much larger amplitudes. However, smaller amplitude components produce a correspondingly smaller

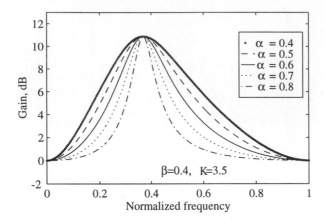

Figure 11.33   Gain responses of the second-order equalizer of Figure 11.30 for various values of the parameter α with $\beta = 0.4$ and $K = 3.5$.

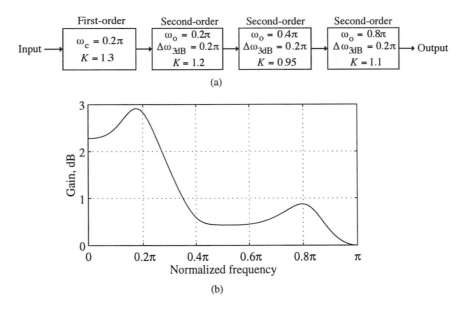

Figure 11.34   (a) Block diagram of a typical graphic equalizer, and (b) its gain response for the section parameter values shown.

frequency deviation. Consequently, the resulting FM signal does not fully utilize the bandwidth allotted for its transmission, lowering the signal-to-noise ratio considerably at the high-frequency end. The output SNR at the FM receiver is increased by emphasizing the higher frequencies of $s_L[n]$ and $s_R[n]$ by means of digital preemphasis filters, as indicated in Figure 11.35.

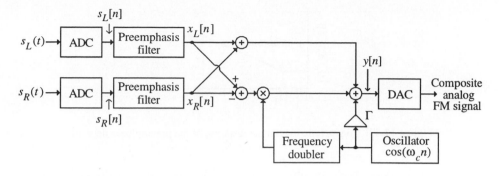

Figure 11.35 Block diagram representation of the FM stereo transmitter.

Figure 11.36 Spectrum of a composite stereo discrete-time broadcast signal.

The sum of the preemphasized left input discrete-time signal $x_L[n]$ and the preemphasized right input discrete-time signal $x_R[n]$ is transmitted in its baseband form for monophonic reception. The difference signal $x_L[n] - x_R[n]$ is transmitted by DSB-SC (double sideband, suppressed carrier) modulation[3] using a 38-kHz subcarrier. The transmitted multiplexed signal $y[n]$ includes the sum signal, the DSB-SC modulated difference signal, and a 19-kHz pilot carrier:

$$y[n] = (x_L[n] + x_R[n]) + (x_L[n] - x_R[n])\cos(2\omega_c n) + \Gamma\cos(\omega_c n), \qquad (11.44)$$

where $\omega_c = 2\pi F_c / F_T$ is the normalized angular frequency of the pilot carrier $F_c = 19$ kHz and F_T is the sampling frequency in Hz, which is typically 32 kHz. The 19-kHz pilot carrier is included to provide a reference for coherent demodulation at the stereo receiver. Figure 11.36 shows the power spectrum of a typical composite baseband signal $y[n]$. The composite signal $y[n]$ frequency-modulates the main carrier to generate the transmitted signal. The value of the gain constant Γ for the pilot signal is chosen such that the pilot is allotted about 10% of the peak frequency deviation. The original signal-power distribution is restored at the receiver output through a deemphasis network.

In the analog FM stereo generation, a first-order analog preemphasis network with a transfer function

$$G_a(s) = \frac{s + \Omega_1}{s + \Omega_1 + \Omega_2}, \qquad (11.45a)$$

---

[3] See Section 1.2.4.

Figure 11.37   Asymptotic magnitude response of the preemphasis network.

is employed [Pan65]. Typically, $\Omega_1 \ll \Omega_2$, in which case, the asymptotic magnitude response of the preemphasis network is of the form shown in Figure 11.37. In the frequency range of interest, the transfer function of the analog preemphasis network can be approximated as

$$G_a(s) \cong 1 + \frac{s}{\Omega_1},$$
(11.45b)

where $1/\Omega_1$ is typically 75 msec. Equation (11.45a) can be rewritten as

$$G_a(s) = \frac{s + K\lambda}{s + \lambda},$$
(11.46)

where K is the dc gain

$$K = \frac{\Omega_1}{\Omega_1 + \Omega_2},$$
(11.47)

and

$$\lambda = \Omega_1 + \Omega_2.$$
(11.48)

Now, $G_a(s)$ can be decomposed as [Reg87b]

$$H_a(s) = \frac{1}{2} \left[1 + A_a(s)\right] + \frac{K}{2} \left[1 - A_a(s)\right],$$
(11.49)

where $A_a(s)$ is a stable analog allpass transfer function:

$$A_a(s) = \frac{s - \lambda}{s + \lambda}.$$
(11.50)

The transfer function of the corresponding digital preemphasis network is obtained by applying the bilinear transformation of Eq. (7.43), repeated below for convenience:

$$s = \frac{2}{T} \left(\frac{1 - z^{-1}}{1 + z^{-1}}\right),$$
(11.51)

where $T = 1/F_T$. Applying the above transformation to Eq. (11.49), we arrive at

$$G(z) = \frac{1}{2} \left[1 + A_1(z)\right] + \frac{K}{2} \left[1 - A_1(z)\right],$$
(11.52)

where $A_1(z)$ is a stable first-order digital allpass transfer function

$$A_1(z) = \frac{\alpha - z^{-1}}{1 - \alpha z^{-1}}, \tag{11.53}$$

with the parameter α given by

$$\alpha = \frac{(2/T) - \lambda}{(2/T) + \lambda}. \tag{11.54}$$

The expression for $G(z)$ given by Eq. (11.52) is precisely the same as that of the first-order equalizer discussed in the previous section Eq. (11.36) with a realization of $G(z)$ as indicated in Figure 11.23. Now, from Eq. (11.51), the pole of the analog transfer function at $s = \lambda$ is mapped to the corresponding digital angular frequency ω_c according to

$$\lambda = \frac{2}{T} \tan(\omega_c/2). \tag{11.55}$$

The 19-kHz tone signal in Figure 11.35 can be generated using a table look-up method, using the trigonometric function approximation method of Section 8.8.1, or using a sine generator described in Section 6.11.

11.6 Discrete-Time Analytic Signal Generation

As discussed in Section 1.2.3, an analytic continuous-time signal has a zero-valued spectrum for all negative frequencies. Such a signal finds applications in single sideband analog communication systems and analog frequency-division multiplex systems. A discrete-time signal with a similar property finds applications in digital communication systems and is the subject of this section. We illustrate here the generation of an analytic signal $y[n]$ from a discrete-time real signal $x[n]$ and describe some of its applications.

Now, the Fourier transform $X(e^{j\omega})$ of a real signal $x[n]$, if it exists, is nonzero for both positive and negative frequencies. On the other hand, a signal $y[n]$ with a single-sided spectrum $Y(e^{j\omega})$ that is zero for negative frequencies must be a complex signal. Consider the complex analytic signal

$$y[n] = x[n] + j\hat{x}[n], \tag{11.56}$$

where $x[n]$ and $\hat{x}[n]$ are real. Its DTFT $Y(e^{j\omega})$ is given by

$$Y(e^{j\omega}) = X(e^{j\omega}) + j\hat{X}(e^{j\omega}), \tag{11.57}$$

where $\hat{X}(e^{j\omega})$ is the DTFT of $\hat{x}[n]$. Now, $x[n]$ and $\hat{x}[n]$ being real, their corresponding DTFTs are conjugate symmetric, i.e., $X(e^{j\omega}) = X^*(e^{-j\omega})$ and $\hat{X}(e^{j\omega}) = \hat{X}^*(e^{-j\omega})$ Hence, from Eq. (11.57) we obtain

$$X(e^{j\omega}) = \tfrac{1}{2}\left[Y(e^{j\omega}) + Y^*(e^{-j\omega})\right], \tag{11.58a}$$

$$j\hat{X}(e^{j\omega}) = \tfrac{1}{2}\left[Y(e^{j\omega}) - Y^*(e^{-j\omega})\right]. \tag{11.58b}$$

Since by assumption, $Y(e^{j\omega}) = 0$ for $-\pi \le \omega < 0$, we obtain from Eq. (11.58a)

$$Y(e^{j\omega}) = \begin{cases} 2X(e^{j\omega}), & 0 \le \omega < \pi, \\ 0, & -\pi \le \omega < 0. \end{cases} \tag{11.59}$$

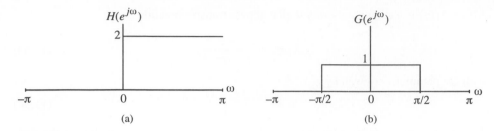

Figure 11.38   (a) Frequency response of the discrete-time filter generating an analytic signal, and (b) half-band lowpass filter.

Thus, the analytic signal $y[n]$ can be generated by passing $x[n]$ through a linear discrete-time system with a frequency response $H(e^{j\omega})$ given by

$$H(e^{j\omega}) = \begin{cases} 2, & 0 \le \omega < \pi, \\ 0, & -\pi \le \omega < 0, \end{cases} \tag{11.60}$$

as indicated in Figure 11.38(a).

11.6.1   The Discrete-Time Hilbert Transformer

We now relate the imaginary part $\hat{x}[n]$ of the analytic signal $y[n]$ to its real part $x[n]$. From Eq. (11.58b),

$$\hat{X}(e^{j\omega}) = \frac{1}{2j} \left[Y(e^{j\omega}) - Y^*(e^{-j\omega}) \right]. \tag{11.61}$$

For $0 \le \omega < \pi$, $Y(e^{-j\omega}) = 0$, and for $-\pi \le \omega < 0$, $Y(e^{j\omega}) = 0$. Using this property and Eq. (11.59) in the above, it can be easily shown that

$$\hat{X}(e^{j\omega}) = \begin{cases} -jX(e^{j\omega}), & 0 \le \omega < \pi, \\ jX(e^{j\omega}), & -\pi \le \omega < 0. \end{cases} \tag{11.62}$$

The above indicates that the imaginary part $\hat{x}[n]$ of the analytic signal $y[n]$ can be generated by passing its real part $x[n]$ through a linear discrete-time system with a frequency response $H_{\text{HT}}(e^{j\omega})$ given by

$$H_{\text{HT}}(e^{j\omega}) = \begin{cases} -j, & 0 \le \omega < \pi, \\ j, & -\pi \le \omega < 0. \end{cases} \tag{11.63}$$

The linear system defined by Eq. (11.63) is usually referred to as the ideal *Hilbert transformer*. Its output $\hat{x}[n]$ is called the *Hilbert transform* of its input $x[n]$. The basic scheme for the generation of an analytic signal $y[n] = y_{\text{re}}[n] + jy_{\text{im}}[n]$ from a real signal $x[n]$ is thus as indicated in Figure 11.39. Observe that $\left| H_{\text{HT}}(e^{j\omega}) \right| = 1$ for all frequencies and has a $-90°$ phase-shift for $0 \le \omega < \pi$ and a $+90°$ phase-shift for $-\pi \le \omega < 0$. As a result, an ideal Hilbert transformer is also called a 90°-*degree phase-shifter*.

The impulse response $h_{\text{HT}}[n]$ of the ideal Hilbert transformer is obtained by taking the inverse DTFT of $H_{\text{HT}}(e^{j\omega})$ and can be shown to be (Problem 7.14)

$$h_{\text{HT}}[n] = \begin{cases} 0, & \text{for } n \text{ even,} \\ \frac{2}{\pi n}, & \text{for } n \text{ odd.} \end{cases} \tag{11.64}$$

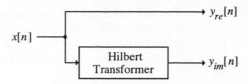

Figure 11.39   Generation of an analytic signal using a Hilbert transformer.

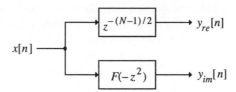

Figure 11.40   FIR realization of a complex half-band filter.

Since the ideal Hilbert transformer has a two-sided infinite-length impulse response defined for $-\pi < n < \pi$, it is an unrealizable system. Moreover, its transfer function $H_{\text{HT}}(z)$ exists only on the unit circle. We describe later two approaches for developing a realizable approximation.

11.6.2   Relation with Half-Band Filters

Consider the filter with a frequency response $G(e^{j\omega})$ obtained by shifting the frequency response $H(e^{j\omega})$ of Eq. (11.60) by $\pi/2$ radians and scaling by a factor $\frac{1}{2}$ (see Figure 11.38):

$$G(e^{j\omega}) = \tfrac{1}{2}H(e^{j(\omega+\pi/2)}) = \begin{cases} 1, & 0 < |\omega| < \frac{\pi}{2}, \\ 0, & \frac{\pi}{2} < |\omega| < \pi. \end{cases} \tag{11.65}$$

From our discussion in Section 10.6.2, we observe that $G(e^{j\omega})$ is a half-band lowpass filter. Because of the relation between $H(e^{j\omega})$ of Eq. (11.60) and the real coefficient half-band lowpass filter $G(e^{j\omega})$ of Eq. (11.65), the filter $H(e^{j\omega})$ has been referred to as a *complex half-band filter* [Reg93].

11.6.3   Design of the Hilbert Transformer

It also follows from the above relation that a complex half-band filter can be designed simply by shifting the frequency response of a half-band lowpass filter by $\pi/2$ radians and then scaling by a factor 2. Equivalently, the relation between the transfer functions of a complex half-band filter $H(z)$ and a real half-band lowpass filter $G(z)$ is given by

$$H(z) = j2G(-jz). \tag{11.66}$$

Design of the real half-band filter was briefly touched upon in Section 10.6.2. We outline below two other novel approaches.

FIR complex half-band filter   The real half-band filter design problem can be transformed into the design of a single passband FIR filter with no stopband, which can be easily designed using the popular Parks-McClellan algorithm[4] [Vai87b]. An inverse transformation on the realization of this wideband filter then yields the implementation of the desired real half-band filter.

Let the specifications of the real half-band lowpass filter $G(z)$ of length N be as follows: passband edge at ω_p, stopband edge at ω_s, passband ripple of δ_p, and stopband ripple of δ_s. As discussed in Section 10.6.2, the passband and stopband ripples of a linear-phase lowpass half-band FIR transfer function $G(z)$ are equal, i.e., $\delta_p = \delta_s = \delta$, and the length N is odd. Moreover, the passband and stopband edge frequencies are related through $\omega_p + \omega_s = \pi$. It has also been shown that $(N-1)/2$ must be odd.

Now, consider a wideband linear-phase filter $F(z)$ of degree $(N-1)/2$ with a passband from 0 to $2\omega_p$, a transition band from $2\omega_p$ to π, and a passband ripple of 2δ. Since $(N-1)/2$ is odd, $F(z)$ has a zero at $z = -1$. The wideband filter $F(z)$ can be designed using the Parks-McClellan method. Define

$$G(z) = \tfrac{1}{2}\left[z^{-(N-1)/2} + F(z^2) \right].   \tag{11.67}$$

It follows from Eq. (11.67) that $G(z)$ is indeed the desired half-band lowpass filter and has an impulse response

$$g[n] = \begin{cases} \tfrac{1}{2} f\left[\tfrac{n}{2}\right], & n \text{ even}, \\ 0, & n \text{ odd}, n \neq \tfrac{N-1}{2} \\ \tfrac{1}{2}, & n = \tfrac{N-1}{2}, \end{cases}   \tag{11.68}$$

where $f[n]$ is the impulse response of $F(z)$.

Substituting Eq. (11.67) in Eq. (11.66), we obtain

$$H(z) = j\left[(-jz)^{-(N-1)/2} + F(-z^2) \right] = z^{-(N-1)/2} + jF(-z^2).   \tag{11.69}$$

An FIR implementation of the complex half-band filter based on the above decomposition is thus as indicated in Figure 11.40. The linear-phase FIR filter $F(-z^2)$ is thus an approximation to a Hilbert transformer.

We illustrate the above approach by means of an example.

Example 11.6   Using MATLAB we design a wideband FIR filter $F(z)$ of degree 13 with a passband from 0 to 0.85π and an extremely small stopband from 0.9π to π. We use the function remez with a magnitude vector m = [1 1 0 0]. The weight vector used to weigh the passband and the stopband is wt = [2 0.05]. The magnitude responses of the wideband filter $F(z)$ and the Hilbert transformer $F(-z^2)$ are shown in Figure 11.41(a) and (b).

The FIR Hilbert transformer can be designed directly using MATLAB. To this end, the version of the function remez employed is either b = remez(N, f, m, 'Hilbert') or b = remez(N, f, m, wt, 'Hilbert'). The following example illustrates this approach.

---

[4]See Section 7.9.1.

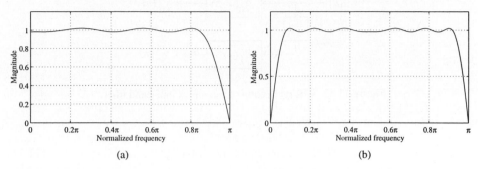

Figure 11.41   Magnitude responses of (a) the wideband FIR filter $F(z)$ and (b) the approximate Hilbert transformer $F(-z^2)$.

Figure 11.42   The magnitude response of the Hilbert transformer designed directly using MATLAB.

Example 11.7 We design a 26th order FIR Hilbert transformer with a passband from 0.1π to 0.9π. It should be noted that for the design of a Hilbert transformer, the first frequency point in the vector f containing the specified bandedges cannot be a 0. The magnitude response of the designed Hilbert transformer obtained using the statement b = remez(26, [0.1 0.9], [1 1], 'Hilbert') is indicated in Figure 11.42. It should be noted that due to numerical round-off problems, unlike the design of Example 11.6, the odd impulse-response coefficients of the Hilbert transformer here are not exactly zero.

IIR complex half-band filter   We indicated in Section 10.7.6 that a large class of stable IIR real coefficient half-band filters of odd order can be expressed as [Vai87f]

$$G(z) = \tfrac{1}{2}[\mathcal{A}_0(z^2) + z^{-1}\mathcal{A}_1(z^2)], \tag{11.70}$$

where $\mathcal{A}_0(z)$ and $\mathcal{A}_1(z)$ are stable allpass transfer functions. Substituting Eq. (11.70) in Eq. (11.66), we therefore arrive at

Figure 11.43   IIR realization of a complex half-band filter.

$$H(z) = \mathcal{A}_0(-z^2) + jz^{-1}\mathcal{A}_1(-z^2). \tag{11.71}$$

A realization of the complex half-band filter based on the above decomposition is thus as shown in Figure 11.43.

We illustrate the above approach to Hilbert transformer design in the following example.

Example 11.8 In Example 10.14, we designed a real half-band elliptic filter with the following frequency response specifications: $\omega_p = 0.4\pi$, $\omega_s = 0.6\pi$, and $\delta_s = 0.0155$. The transfer function of the real half-band filter $G(z)$ can be expressed as in Eq. (11.70), where the transfer functions of the two allpass sections are given in Eq. (10.117) and are repeated below for convenience:

$$\mathcal{A}_0(z^2) = \frac{z^{-2} + 0.2368041466}{1 + 0.2368041466z^{-2}}, \quad \mathcal{A}_1(z^2) = \frac{z^{-2} + 0.7149039978}{1 + 0.7149039978z^{-2}}. \tag{11.72}$$

The gain response of the complex half-band filter $H(z)$ obtained using Eq. (11.71) is sketched in Figure 11.44. Figure 11.45 shows the phase difference between the two allpass functions $\mathcal{A}_0(-z^2)$ and $z^{-1}\mathcal{A}_1(-z^2)$ of the complex half-band filter. Note that, as expected, the phase difference is 90° for most of the positive frequency range and 270° for most of the negative frequency range. In plotting the gain response of the complex half-band filter and the phase difference between its constituent two allpass sections, the M-file function `freqz(num,den,n,'whole')` has been used to compute the pertinent frequency response values over the whole normalized frequency range from 0 to 2π.

11.6.4   Single Sideband Modulation

For efficient transmission over long distances, a real low-frequency bandlimited signal $x[n]$, such as speech or music, is modulated by a very high-frequency sinusoidal carrier signal $\cos\omega_c n$, with the carrier frequency ω_c being less than half of the sampling frequency. The spectrum $V(e^{j\omega})$ of the resulting signal $v[n] = x[n]\cos\omega_c n$ is given by

$$V(e^{j\omega}) = \tfrac{1}{2}\left[X(e^{j(\omega-\omega_c)}) + X(e^{j(\omega+\omega_c)})\right]. \tag{11.73}$$

As indicated in Figure 11.46, if $X(e^{j\omega})$ is bandlimited to ω_M, the spectrum $V(e^{j\omega})$ of the modulated signal $v[n]$ has a bandwidth of $2\omega_M$ centered at $\pm\omega_c$. By choosing widely separated carrier frequencies, one can modulate a number of low-frequency signals to high-frequency signals, combine them by frequency-division multiplexing, and transmit over a common channel. The carrier frequencies are chosen appropriately to ensure that

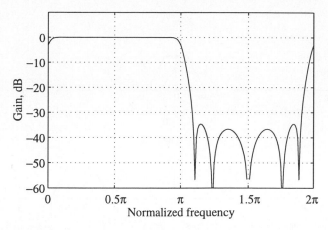

Figure 11.44 Gain responses of the complex half-band filter (normalized to 0 dB maximum gain).

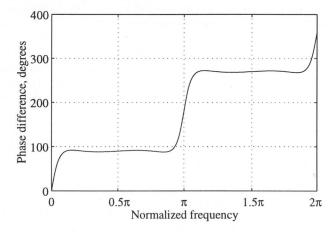

Figure 11.45 Phase difference between the two allpass sections of the complex half-band filter.

there is no overlap in the spectra of the modulated signals when combined by frequency-division multiplexing. At the receiving end, each of the modulated signals is then separated by a bank of bandpass filters of center frequencies corresponding to the different carrier frequencies.

It is evident from Figure 11.46 that, for a real low-frequency signal $x[n]$, the spectrum of its modulated version $v[n]$ is symmetric with respect to the carrier frequency ω_c. Thus, the portion of the spectrum in the frequency range from ω_c to $(\omega_c + \omega_M)$, called the *upper sideband*, has the same information content as the portion in the frequency range from $(\omega_c - \omega_M)$ to ω_c, called the *lower sideband*. Hence, for a more efficient utilization of the channel bandwidth, it is sufficient to transmit either the upper or the lower sideband signal. A conceptually simple way to eliminate one of the sidebands is to pass the modulated signal $v[n]$ through a sideband filter whose passband covers the frequency range of one of the sidebands.

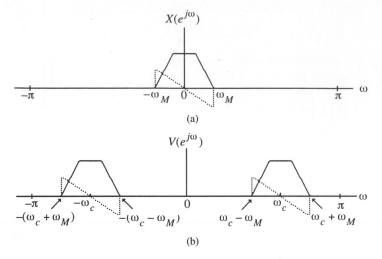

Figure 11.46   Spectra of a real signal and its modulated version. (Solid lines represent the real parts while dashed lines represent the imaginary parts.)

An alternative, often-preferred, approach for single sideband signal generation is by modulating the analytic signal whose real and imaginary parts are, respectively, the real signal and its Hilbert transform. To illustrate this approach, let $y[n] = x[n] + j\hat{x}[n]$ where $\hat{x}[n]$ is the Hilbert transform of $x[n]$. Consider

$$
\begin{aligned}
s[n] = y[n]e^{j\omega_c n} &= (y_{\text{re}}[n] + j y_{\text{im}}[n])\,(\cos \omega_c n + j \sin \omega_c n) \\
&= \left(x[n] \cos \omega_c n - \hat{x}[n] \sin \omega_c n\right) \\
&\quad + j \left(x[n] \sin \omega_c n + \hat{x}[n] \cos \omega_c n\right).
\end{aligned}
\tag{11.74}
$$

From Eq. (11.74), the real and imaginary parts of $s[n]$ are thus given by

$$
s_{\text{re}}[n] = x[n] \cos \omega_c n - \hat{x}[n] \sin \omega_c n, \tag{11.75a}
$$
$$
s_{\text{im}}[n] = x[n] \sin \omega_c n + \hat{x}[n] \cos \omega_c n. \tag{11.75b}
$$

Figure 11.47 shows the spectra of $x[n]$, $\hat{x}[n]$, $y[n]$, $s[n]$, $s_{\text{re}}[n]$, and $s_{\text{im}}[n]$. It therefore follows from these plots that a single sideband signal can be generated using either one of the modulation schemes described by Eq. (11.75a). A block diagram representation of the scheme of Eq. (11.75a) is sketched in Figure 11.48.

11.7   Voice Privacy System

Often there is a need to provide privacy to casual telephone conversation against eavesdropping. A number of methods have been proposed to design analog scramblers that provide speech privacy. We describe below one such method that employs digital signal processing methods [Cox83]. The proposed method is based on a simultaneous sequential time and frequency segment permutation (TFSP) scheme [Jay83].

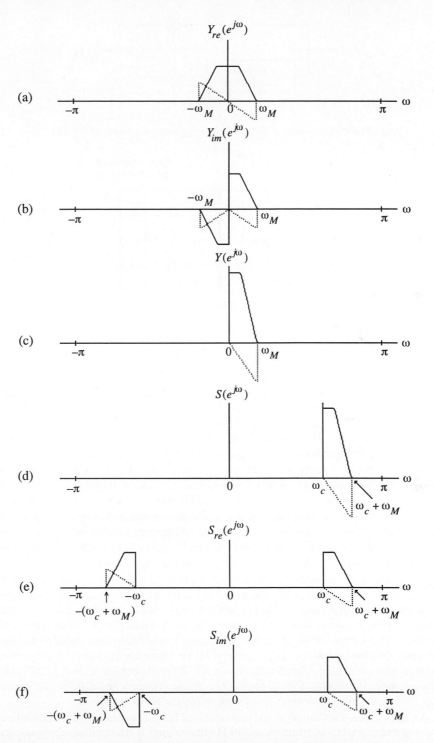

Figure 11.47 Illustration of the generation of single sidegand signals via Hilbert transform. (Solid lines represent the real parts while dashed lines represent the imaginary parts.)

Figure 11.48   Schematic of single sideband generation scheme of Eq. (11.75a).

Time-frequency
speech segment

F	$2F$	$3F$	–	–			BF
–	–						
2							
1	$F+1$						z

Frequency →

Time →

Figure 11.49   Sequential permutations of a time-frequency speech segment matrix.

The basic idea behind the TFSP scheme is to divide a long speech time segment into B contiguous blocks and then split each time segment block into F contiguous frequency subbands, resulting in a total of FB time-frequency segments as depicted in matrix form in Figure 11.49. These time-frequency segments are stored in the scrambler memory consisting of FB cells. The segment in the rth cell from the memory, with the location r chosen by a random number generator, is transmitted. Contents of all sth cells ($s > r$) are moved by one position to the left and the FBth cell is filled by the $(FB + 1)$th segment. The above random scrambling method is repeated until all the initial FB segments have been transmitted and the scrambler memory is filled with a new set of FB segments.

Figure 11.50 shows the block diagrams of the transmitter and the receiver parts of the proposed full-duplex analog voice privacy system. At the transmitting end, the analog voice signal $s_a(t)$ is first passed through an anti-aliasing lowpass filter and then converted into a digital signal $s[n]$ at the 8-kHz sampling rate. The digital signal $s[n]$ is next passed through a three-channel analysis filter bank and separated into three frequency bands, occupying, respectively, 0-500-Hz, 500-1000-Hz, and 2000-2500-Hz bands. Each subband signal is then down-sampled by a factor of 8, blocked into segments of size 5 ms each, resulting in the sequences $s_1[n]$, $s_2[n]$, and $s_3[n]$, respectively, and fed into a time-and-frequency block scrambler. These three subsequence blocks are then scrambled using a pseudo-random key generating the subsequences $x_1[n]$, $x_2[n$, and $x_3[n]$, which are up-sampled and combined by a three-channel synthesis bank into either a 500-2000-Hz band or a 2000-3500-Hz band, with each user getting one of these bands. The pseudo-random key used by the scrambler is also transmitted to the receiver for descrambling the blocks received.

An impulse is transmitted at start-up time and later, if necessary, for channel equalization at the receiver. In addition, a 250-Hz tone signal is transmitted continuously for frequency shift compensation. To accommodate this tone, the combined signal $x[n]$ is

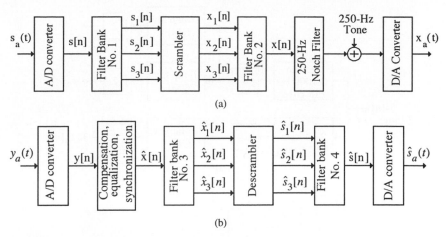

Figure 11.50 Block diagram of the full-duplex analog voice privacy system: (a) transmitter and (b) receiver.

first passed through a 250-Hz notch filter prior to the addition of the tone signal. The final signal is converted into an analog signal by means of a D/A converter operating at 8 kHz and transmitted over the channel as $x_a(t)$.

The analog signal $y_a(t)$ arriving at the receiver is, in general, different from the transmitted signal $x_a(t)$ due to imperfection in the channel. After its conversion to a digital form $y[n]$ by means of an 8 kHz A/D converter, it is processed by the compensation, equalization, and synchronization block to generate an output $\hat{x}[n]$ that is as close as possible to $x[n]$. The signal $\hat{x}[n]$ is then split into three subband signals, by a three-channel analysis filter bank (filter bank No. 3) that is designed to form a perfect reconstruction QMF bank structure with filter bank No. 2 (a synthesis filter bank) in the transmitter. Because of the QMF property, the filter bank outputs, $\hat{x}_1[n]$, $\hat{x}_2[n]$ and $\hat{x}_3[n]$ are almost identical to the scrambler output signals, $x_1[n]$, $x_2[n]$, and $x_3[n]$, in the transmitter. These signals are then fed into the descrambler, which generates the signals, $\hat{s}_1[n]$, $\hat{s}_2[n]$, and $\hat{s}_3[n]$ that are almost identical to their parent signals, $s_1[n]$, $s_2[n]$, and $s_3[n]$, at the input of the scrambler in the transmitter. The descrambler output signals are then combined by a synthesis filter bank (filter bank No. 4) whose output is converted into an analog form by a D/A converter operating at 8 kHz. Filter bank No. 1 and filter bank No. 4 form a perfect reconstruction QMF structure.

The transmitter and the receiver in the half-duplex system are essentially similar in structure to those in the full-duplex system depicted in Figure 11.51, except here the filter bank pairs, Nos. 1 and 4, and Nos. 2 and 3, are a 5-band QMF bank each. In both QMF banks, there are two additional subbands occupying the 1000-1500-Hz band and the 1500-2000-Hz band. Moreover, in the half-duplex system, the five bands are added into a single 500-3000-Hz band.

The FIR filters employed in the QMF filter banks of the above voice privacy system are from the filter family designed by Johnston and discussed in Section 10.8.5 [Joh80]. Figure 11.51 shows the structures of filter banks Nos. 1 and 4, while Figure 11.52 shows the structures of filter banks Nos. 2 and 3. The filter labeled 64-tap QMF in Figure 11.52 is a half-band lowpass or highpass filter depending on the band it is being used.

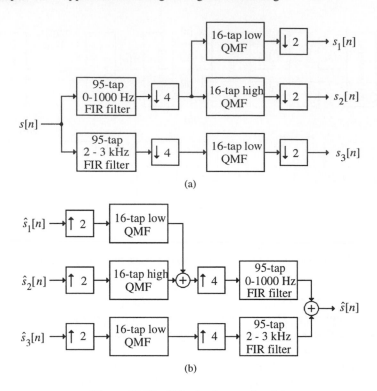

Figure 11.51   Filter banks nos. 1 and 4.

The structures of the analysis and the synthesis filter banks employed in the half-duplex system are depicted in Figure 11.53. Here, in the transmitter, the 2.5-3-kHz band is eliminated, whereas in the receiver, the 0-500-Hz band is eliminated. In both parts, a five-band QMF bank is used.

Further details on the design and implementation of the above voice privacy system can be found in Cox and Tribolet [Cox83]. In particular, it discusses the design of the scrambler and the descrambler, synchronization and equalization issues, DSP chip implementation considerations, and also includes simulation results. A comparison of four different methods for analog speech privacy is discussed in Jayant et al. [Jay81].

11.8   Subband Coding of Speech and Audio Signals

In many applications, the digital signal obtained by the analog-to-digital conversion of an analog signal has to be transmitted over a channel with a limited bandwidth or stored for future use in a storage medium with limited capacity. Signal compression methods are being increasingly employed to increase the efficiency of transmission and/or storage. For example, the speech signal in a telephone communication system has a bandwidth of about 3.4 kHz. For digital transmission, it is sampled at a rate of 8 kHz and then converted by means of an 8-bit A/D converter, resulting in a 64,000 bits per second or, equivalently, 64 kb/sec digital signal. Up to 24 such 64 kb/sec digital signals can be transmitted over the

Figure 11.52   Filter banks nos. 2 and 3.

T-1 carrier channel in the United States by time-division multiplexing. By compressing the speech signal to a rate of 32 kb/sec, the number of telephone signals sharing the channel can be doubled to a total of 48. Likewise, the storage capacity of a compact disk can be increased by compressing the digitized music.

One of the most popular schemes for signal compression is the subband coding that employs the quadrature-mirror filter bank discussed in Sections 10.8 and 10.9, and efficiently encodes the signal by exploiting the nonuniform distribution of signal energy in the frequency band. In this section, we review the principles of subband coding of speech and audio signals and describe one specific coding system.

A simplified representation of the subband coding scheme is shown in Figure 11.54. As indicated here, in this scheme, the input signal $x[n]$ at the transmitting end is decomposed into a set of narrowband signals occupying contiguous frequency bands by means of an analysis filter bank. These narrowband signals are then down-sampled, producing the subband signals. Each subband signal is next compressed by an encoder, and all compressed subband signals are multiplexed and sent over the channel to the receiver. At the receiving end, the composite signal is first demultiplexed into a set of subband signals. Each subband signal is then decompressed by a decoder, up-sampled, and passed through a synthesis filter bank. The outputs of the synthesis filter bank are finally combined to produce a signal $y[n]$ that is an acceptable replica of the original input signal $x[n]$. The subband coding offers better compression ratio than the direct compression of the original signal $x[n]$ since it allocates different number of bits to each subband signal by taking advantage of its spectral characteristics resulting in a lower average bit rate per sample.

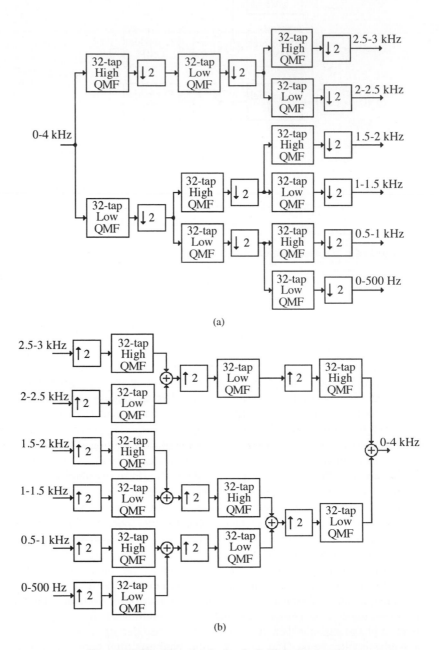

(a)

(b)

Figure 11.53   Filter banks for the half-duplex system.

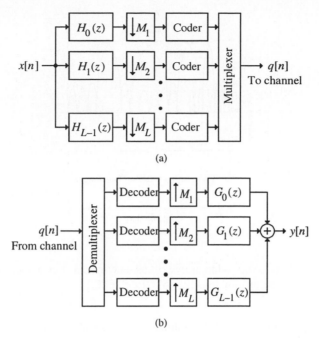

Figure 11.54 The basic subband coding and decoding scheme.

Subband coding of 7-kHz wide-band audio at 56 kb/sec based on a 5-band QMF bank has been investigated by Richardson and Jayant [Ric86]. The five bands occupy the frequency ranges 0-875 Hz, 875-1750 Hz, 1750-3500 Hz, 3500-5250 Hz, and 5250-7000 Hz. The gain responses of the five filters used are shown in Figure 11.55 along with the sum of the squares of their magnitude responses.

The five-band partition has been obtained using a three-stage cascade of two-channel QMF banks as indicated in Figure 11.56, which also shows the sampling rate of the input and the subband signals along with the order of the filters used. The encoding delay caused by the subband filtering operations is around 10 msec. The subband signals have been encoded using an adaptive differential PCM (ADPCM) scheme [Jay74]. The bit allocations used are 5 bits/sample for each of the two low-frequency channels, 3 bits/sample for the highest frequency channel, and 4 bits/sample for the two intermediate-frequency channels. The total bit rate of the subband ADPCM coder is thus 56 kb/sec, which is equivalent to an average bit rate of 4 bits/sample of the input signal at the 14-kHz sampling rate.

Further details of the above subband coder along with a comparison of its performance with that of two other coding schemes can be found in [Ric86].

11.9   Transmultiplexers

In the United States and most other countries, the telephone service employs two types of multiplexing schemes to transmit multiple low-frequency voice signals over a wideband channel. In the *frequency-division multiplex* (FDM) telephone system, multiple analog voice signals are first modulated by single sideband (SSB) modulators onto several sub-

Figure 11.55   (a) Gain responses of the analysis filters and (b) sum of magnitude square responses in dB. (Reproduced with permission from [Ric86] ©1986 IEEE.)

carriers, combined, and transmitted simultaneously over a common wideband channel. To avoid *cross-talk*, the subcarriers are chosen to ensure that the spectra of the modulated signals do not overlap. At the receiving end, the modulated subcarrier signals are separated by analog bandpass filters, demodulated to reconstruct the individual voice signals. On the other hand, in the *time-division multiplex* (TDM) telephone system, the voice signals are first converted into digital signals by sampling and A/D conversion. The samples of the digital signals are time-interleaved by a digital multiplexer, and the combined signal is transmitted. At the receiving end, the digital voice signals are separated by a digital demultiplexer and then passed through a D/A converter and an analog reconstruction filter to recover the original analog voice signal.

The TDM system is usually employed for short-haul communication, while the FDM scheme is preferred for long-haul transmission. Until the telephone service becomes all digital, it is necessary to translate signals between the two formats. This is achieved by the transmultiplexer system discussed next.

The transmultiplexer is a multiinput, multioutput, multirate structure, as shown in Figure 11.57. It is essentially exactly opposite that of the QMF filter bank of Figure 10.50 and consists of a synthesis filter bank at the input end followed by an analysis filter bank at the output end. To determine the input-output relation of the transmultiplexer, consider one typical path from the kth input to the ℓth output as indicated in Figure 11.58(a) [Vai93]. Invoking the identity of Section 10.4.4, we note that the structure of Figure 11.58(a) is equivalent to that shown in Figure 11.58(b), consisting of an LTI branch with a transfer function $F_{k\ell}(z)$ that is the zeroth polyphase component of $H_k(z)G_\ell(z)$. The input-output relation of the transmultiplexer is therefore given by

$$Y_k(z) = \sum_{\ell=0}^{L-1} F_{k\ell}(z)X_\ell(z), \quad 0 \le k \le L-1. \tag{11.76}$$

Figure 11.56   (a) Three-stage realization of the five-band analysis bank and (b) its equivalent representation.

Denoting

$$\mathbf{Y}(z) = \begin{bmatrix} Y_0(z) & Y_1(z) & \cdots & Y_{L-1}(z) \end{bmatrix}^T, \qquad (11.77\text{a})$$

$$\mathbf{X}(z) = \begin{bmatrix} X_0(z) & X_1(z) & \cdots & X_{L-1}(z) \end{bmatrix}^T, \qquad (11.77\text{b})$$

we can rewrite Eq. (11.76) as

$$\mathbf{Y}(z) = \mathbf{F}(z)\mathbf{X}(z), \qquad (11.78)$$

where $\mathbf{F}(z)$ is an $L \times L$ matrix whose (k, ℓ)th element is given by $F_{k\ell}(z)$. The objective of the transmultiplexer design is to ensure that $y_k[n]$ is a reasonable replica of $x_k[n]$. If $y_k[n]$ contains contributions from $x_r[n]$ with $r \neq n$, then there is cross-talk between these two channels. It follows from Eq. (11.78) that cross-talk is totally absent if $F(z)$ is a diagonal matrix, in which case Eq. (11.78) reduce to

$$Y_k(z) = F_{kk}(z)X_k(z), \quad k = 0, 1, \ldots, L - 1. \qquad (11.79)$$

As in the case of the QMF bank, we can define three types of transmultiplexer systems. It is a phase-preserving system, if $F_{kk}(z)$ is a linear-phase transfer function for all values of

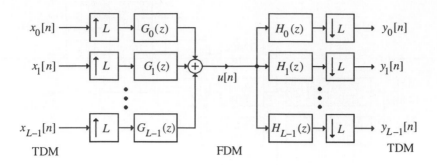

Figure 11.57   The basic L-channel transmultiplexer structure.

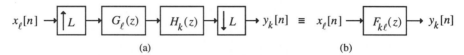

Figure 11.58   The k, ℓ-path of the L-channel transmultiplexer structure.

k. Likewise, it is a magnitude-preserving system, if $F_{kk}(z)$ is an allpass function. Finally, for a perfect reconstruction transmultiplexer,

$$F_{kk}(z) = \alpha_k z^{-n_k}, \quad k = 0, 1, \ldots, L - 1, \tag{11.80}$$

where n_k is an integer and α_k is a nonzero constant. For a perfect reconstruction system, $y_k[n] = \alpha_k x_k[n - n_k]$.

The perfect reconstruction condition can also be derived in terms of the polyphase components of the synthesis and analysis filter banks of the transmultiplexer of Figure 11.57, as shown in Figure 11.59(a) [Kol91]. Using the cascade equivalences of Figure 10.14, we arrive at the equivalent representation indicated in Figure 11.59(b). Note that the structure in the center part of this figure is a special case of the system of Figure 11.57, where $G_\ell(z) = z^{-(L-1-\ell)}$ and $H_k(z) = z^{-k}$, with $\ell, k = 0, 1, \ldots, L - 1$. Here the zeroth polyphase component of $H_{\ell+1}(z)G_\ell(z)$ is z^{-1} for $\ell = 0, 1, \ldots, L - 2$, the zeroth polyphase component of $H_0(z)G_{L-1}(z)$ is 1, and the zeroth polyphase component of $H_k(z)G_\ell(z)$ is 0 for all other cases. As a result, a simplified equivalent representation of Figure 11.59(b) is as shown in Figure 11.60.

The transfer matrix characterizing the transmultiplexer is thus given by

$$\mathbf{M}(z) = \mathbf{E}(z) \begin{bmatrix} \mathbf{0} & 1 \\ z^{-1}\mathbf{I}_{L-1} & \mathbf{0} \end{bmatrix} \mathbf{R}(z), \tag{11.81}$$

where $\mathbf{I}_{L-1}$ is an $(L - 1) \times (L - 1)$ identity matrix. Now, for a perfect reconstruction system it is sufficient to ensure that

$$\mathbf{M}(z) = d z^{-n_0} \mathbf{I}_L, \tag{11.82}$$

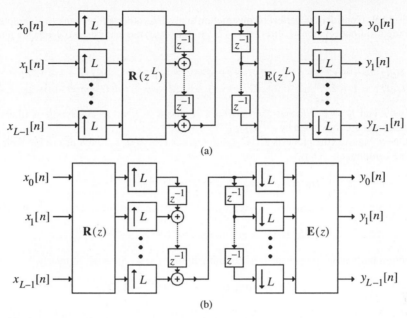

(a)

(b)

Figure 11.59   (a) Polyphase representation of the L-channel transmultiplexer, and (b) its computationally efficient realization.

Figure 11.60   Simplified equivalent circuit of Figure 11.59.

where n_0 is a positive integer. From Eqs. (11.81) and (11.82) we arrive at the condition for perfect reconstruction in terms of the polyphase components as

$$\mathbf{R}(z)\mathbf{E}(z) = dz^{-m_0} \begin{bmatrix} \mathbf{0} & \mathbf{I}_{L-1} \\ z^{-1} & \mathbf{0} \end{bmatrix},\tag{11.83}$$

where m_0 is a suitable positive integer.

It is possible to develop a perfect reconstruction transmultiplexer from a perfect reconstruction QMF bank with analysis filters $H_\ell(z)$ and synthesis filters $G_\ell(z)$, with a distortion transfer function given by $T(z) = dz^{-K}$, where d is a nonzero constant and K is a positive integer. It can be shown that a perfect reconstruction transmultiplexer can then be designed using the analysis filters $H_\ell(z)$ and synthesis filters $z^{-R}G_\ell(z)$, where R is a positive integer less than L such that $R + K$ is a multiple of L [Kol91]. We illustrate this approach in the following example.

Example 11.9 Consider the perfect reconstruction analysis/synthesis filter bank of Example 10.11 with an input-output relation $y[n] = 2x[n - 2]$. In this case, the analysis and synthesis filters are given by

$$H_0(z) = 1 + z^{-1} + z^{-2}, \quad H_1(z) = 1 - z^{-1} + z^{-2}, \quad H_2(z) = 1 - z^{-2},$$
$$G_0(z) = 1 + 2z^{-1} + z^{-2}, \quad G_1(z) = 1 - 2z^{-1} + z^{-2}, \quad G_2(z) = -2 + 2z^{-2}.$$

Here, $d = 2$ and $K = 2$. We thus choose $R = 1$ so that $R + K = 3$. The synthesis filters of the transmultiplexer are thus given by $z^{-1}G_\ell(z)$.

We now examine the products $z^{-1}G_\ell(z)H_k(z)$, for $\ell, k = 0, 1, 2$, and determine their zeroth polyphase components. Thus,

$$z^{-1}G_0(z)H_0(z) = z^{-1} + 3z^{-2} + 4z^{-3} + 3z^{-4} + z^{-5},$$

whose zeroth polyphase component is given by $4z^{-3}$, and hence, $y_0[n] = 4x_0[n - 3]$. Likewise,

$$z^{-1}G_1(z)H_1(z) = z^{-1} - 3z^{-2} + 4z^{-3} - 3z^{-4} + z^{-5},$$

with a zeroth polyphase component $4z^{-3}$ resulting in $y_1[n] = 4x_1[n - 3]$. Similarly,

$$z^{-1}G_2(z)H_2(z) = -2z^{-1} + 4z^{-3} - 2z^{-5},$$

whose zeroth polyphase component is again $4z^{-3}$, implying $y_2[n] = 4x_2[n - 3]$. It can be shown that the zeroth polyphase components for all other products $z^{-1}G_\ell(z)H_k(z)$, with $\ell \neq k$, is 0, indicating a total absence of crosstalk between channels.

In a typical TDM to FDM format translation, 12 digitized speech signals are interpolated by a factor of 12, modulated by single-sideband modulation, digitally summed, and then converted into an FDM analog signal by D/A conversion. At the receiving end, the analog signal is converted into a digital signal by A/D conversion and passed through a bank of 12 single-sideband demodulators whose outputs are then decimated, resulting in the low-frequency speech signals. The speech signals have a bandwidth of 4 kHz and are sampled at an 8-kHz rate. The FDM analog signal occupies the band 60 kHz to 108 kHz, as illustrated in Figure 11.61. The interpolation and the single-sideband modulation can be performed by up-sampling and appropriate filtering. Likewise, the single-sideband demodulation and the decimation can be implemented by appropriate filtering and down-sampling.

11.10   Discrete Multitone Transmission of Digital Data

Binary data are normally transmitted serially as a pulse train, as indicated in Figure 11.62 (a). However, in order to faithfully extract the information transmitted, the receiver requires complex equalization procedures to compensate for channel imperfection and to make full use of the channel bandwidth. For example, the pulse train of Figure 11.62(a) arriving at the receiver may appear as indicated in Figure 11.62(b). To alleviate the problems encountered with the transmission of data as a pulse train, frequency-division multiplexing with overlapping subchannels has been proposed. In such a system, each binary digit a_r, $r = 0, 1, 2, \ldots, N - 1$, modulates a subcarrier sinusoidal signal $\cos(2\pi rt/T)$, as indicated in Figure 11.62(c) for the transmission of the data of Figure 11.62(a), and then the modulated subcarriers are summed and transmitted as one composite analog signal. At the receiver, the analog signal is passed through a bank of coherent demodulators whose

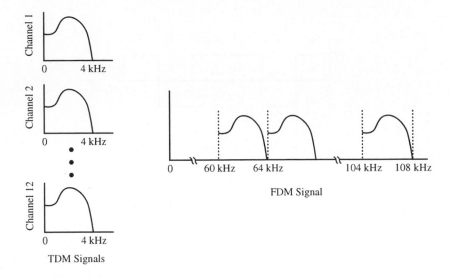

Figure 11.61   Spectrums of TDM signals and the FDM signal.

outputs are tested to determine the digits transmitted. This is the basic idea behind the multicarrier modulation/demodulation scheme for digital data transmission.

A widely used form of the multicarrier modulation is the discrete multitone transmission (DMT) scheme in which the modulation and demodulation processes are implemented via the discrete Fourier transform (DFT) efficiently realized using fast Fourier transform (FFT) methods. This approach leads to an all digital system, eliminating the arrays of sinusoidal generators and the coherent demodulators [Cio91], [Pel80].

We outline here the basic idea behind the DMT scheme. Let $\{a_k[n]\}$ and $\{b_k[n]\}$, $k = 1, 2, \ldots, M - 1$, be two $M - 1$ real-valued data sequences operating at a sampling rate of F_T that are to be transmitted. Define a new set of complex sequence $\{\alpha_k[n]\}$ of length $N = 2M$ according to

$$\alpha_k[n] = \begin{cases} 0, & k = 0, \\ a_k[n] + jb_k[n], & 1 \le k \le \frac{N}{2} - 1, \\ 0, & k = \frac{N}{2}, \\ a_{N-k}[n] - jb_{N-k}[n], & \frac{N}{2} + 1 \le k \le N - 1. \end{cases} \qquad (11.84)$$

We apply an inverse DFT, and the above set of N sequences is transformed into another new set of N signals $\{u_\ell[n]\}$ given by

$$u_\ell[n] = \frac{1}{N} \sum_{k=0}^{N-1} \alpha_k[n] W_N^{-\ell k}, \qquad \ell = 0, 1, \ldots, N - 1, \qquad (11.85)$$

where $W_N = e^{-j2\pi/N}$. Note that the method of generation of the complex sequence $\{\alpha_k[n]\}$ ensures that its IDFT $\{u_\ell[n]\}$ will be a real sequence. Each of these N signals is then up-sampled by factor of N and time-interleaved, generating a composite signal

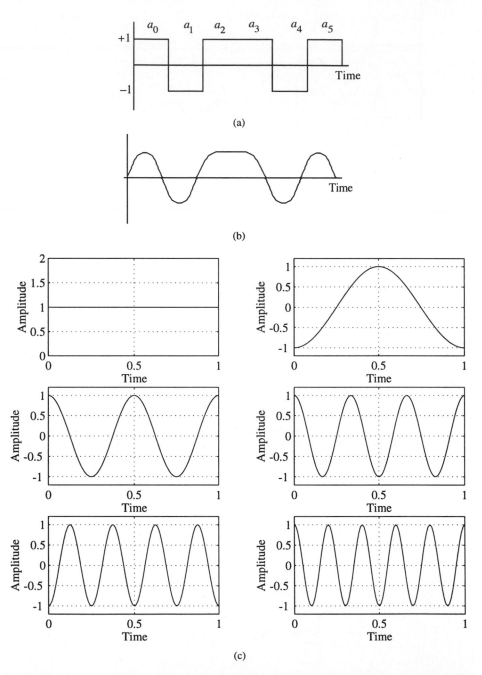

Figure 11.62   (a) Serial binary data stream, (b) baseband serially transmitted signal at the receiver, and (c) signals generated by modulating a set of subcarriers by the digits of the pulse train in (a).

$\{x[n]\}$ operating at a rate NF_T that is assumed to be equal to $2F_c$. The composite signal is converted into an analog signal $x_a(t)$ by passing it through a D/A converter followed by an analog reconstruction filter. The analog signal $x_a(t)$ is then transmitted over the channel.

At the receiver, the received analog signal $y_a(t)$ is passed through an analog anti-aliasing filter and then converted into a digital signal $\{y[n]\}$ by an S/H circuit followed by an A/D converter operating at a rate $NF_T = 2F_c$. The received digital signal is then deinterleaved by a delay chain containing $N-1$ unit delays whose outputs are next down-sampled by a factor of N, generating the set of signals $\{v_\ell[n]\}$. Applying DFT to these N signals, we finally arrive at N signals $\{\beta_k[n]\}$

$$\beta_k[n] = \sum_{\ell=0}^{N-1} v_\ell[n] W_N^{\ell k}, \qquad k = 0, 1, \ldots, N-1. \tag{11.86}$$

Figure 11.63 shows schematically the overall DMT scheme. If we assume the frequency response of the channel to have a flat passband, and assume the analog reconstruction and anti-aliasing filters to be ideal lowpass filters, then neglecting the nonideal effects of the D/A and the A/D converters, we can assume $y[n] = x[n]$. Hence, the interleaving circuit of the DMT structure at the transmitting end connected to the deinterleaving circuit at the receiving end is identical to the same circuit in the transmultiplexer stricture of Figure 11.59(b) (with $L = N$). From the equivalent representation given in Figure 11.60, it follows then that

$$\begin{aligned} v_k[n] &= u_{k-1}[n-1], \qquad k = 1, 2, \ldots, N-2, \\ v_0[n] &= u_{N-1}[n], \end{aligned} \tag{11.87}$$

or in other words,

$$\begin{aligned} \beta_k[n] &= \alpha_{k-1}[n-1], \qquad k = 1, 2, \ldots, N-2, \\ \beta_0[n] &= \alpha_{N-1}[n]. \end{aligned} \tag{11.88}$$

Transmission channels, in general, have a bandpass frequency response $H_{\text{ch}}(f)$ with a magnitude response dropping to zero at some frequency F_c. In some cases, in the passband of the channel, the magnitude response, instead of being flat, drops very rapidly outside its passband, as indicated in Figure 11.64. For reliable digital data transmission over such a channel and its recovery at the receiving end, the channel's frequency response needs to be compensated by essentially a highpass equalizer at the receiver. However, such an equalization also amplifies high-frequency noise that is invariably added to the data signal as it passes through the channel.

For a large value of the DFT length N, the channel can be assumed to be composed of a series of contiguous narrow bandwidth bandpass subchannels. If the bandwidth is reasonably narrow, the corresponding bandpass subchannel can be considered to have an approximately flat magnitude response, as indicated by the dotted lines in Figure 11.64, and the channel can be approximately characterized by a single complex number given by the value of its frequency response at $\omega = 2\pi k/N$. The values can be determined by first transmitting a known training signal of unmodulated carriers and generating the respective channel frequency response samples. The real data samples are then divided by these complex numbers at the receiver to compensate for channel distortion.

Further details on the performance of the above DMT scheme under nonideal conditions can be found in [Bin90], [Cio91], [She95].

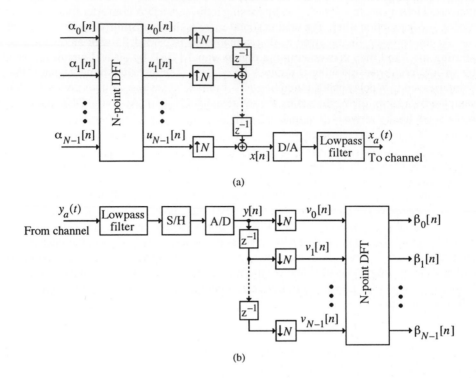

(a)

(b)

Figure 11.63   The DMT scheme. (a) Transmitter, and (b) receiver.

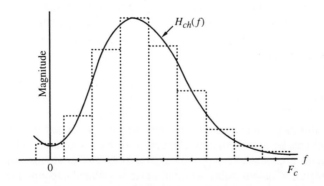

Figure 11.64   Frequency response of a typical bandlimited channel.

11.11   Digital Audio Sampling Rate Conversion

As indicated in Chapter 10, fractional sampling rate conversion of digital signals is often required in professional audio and video applications. A classical approach to such conversion is to convert the input digital signal into an analog signal by means of a digital-to-analog (D/A) converter followed by an analog lowpass filter, resample the analog signal at the desired sampling rate, and convert the resampled signal into a digital form by means of an analog-to-digital (A/D) converter. To reduce the effect of aliasing to a minimum, the analog reconstruction filter must have a very sharp cutoff with sufficiently large stopband attenuation and very small passband ripple, in addition to exhibiting nearly linear phase response in the passband. It is difficult to build an economical analog filter with such stringent requirements. Moreover, to achieve the desired accuracy in the output digital signal, both the D/A converter and the A/D converter must have large enough resolutions. An alternative approach is to use an all-digital conversion technique, the basics of which have been discussed in Section 10.2.2.

The complexity of the design of the fractional sampling rate converter depends on the ratio of the sampling rates between the input and the output digital signals. For example, in digital audio applications, the three different sampling frequencies employed are 44.1 kHz, 32 kHz, and 48 kHz. As a consequence, there are three different values for the sampling rate conversion factor - 2:3, 147:160, and 320:441. Likewise, in digital video applications, the sampling rates of composite video signals are 14.3181818 MHz and 17.734475 MHz, whereas the sampling rates of the digital component video signal are 13.5 MHz and 6.75 MHz for the luminance and the color-difference signals, respectively, for the NTSC and PAL systems. Here, the sampling rates for the component and the NTSC composite video signals are related by a ratio 35:33, whereas the sampling rates for the component and the PAL composite video signals are related by a ratio 709,379:540,000. There are applications, such as in the pitch control of audio signals, where the ratio is irrational, in which case, there is no periodic relation between the sampling instants of the input and the output digital signals.

The design of a fractional sampling rate converter for conversion between two digital signals related by a ratio of two low-valued integers can be carried out using the method discussed in Section 10.2.2. However, for a sampling rate conversion factor that is a ratio of two very large integers or a irrational number, the design is somewhat more complex.

In this section we discuss the design of fractional sampling rate converters for both of the above two cases as encountered in audio applications [Cuc91], [Lag81], [Ram84]. Similar procedure is followed in video applications [Lut91].

11.11.1   Conversion Between 32-kHz and 48-kHz Sampling Rates

The conversion of a digital audio signal of 48-kHz sampling rate to one of 32-kHz sampling rate requires the design of a fractional sampling rate decimator with a decimation factor of 2/3. Its basic form is as indicated in Figure 11.65(a), where the lowpass filter $H(z)$ has a stopband edge at $\pi/3$. Note that in this structure, the filter $H(z)$ operates at the 96-kHz rate. We now outline the development of a computationally efficient realization in which all filters operate at the 16-kHz rate [Hsi87], [Vai90]. Replacing the lowpass filter with its Type II polyphase realization and then making use of the cascade equivalence of Figure 10.14(b), we arrive at the equivalent realization indicated in Figure 11.65(b) in which now

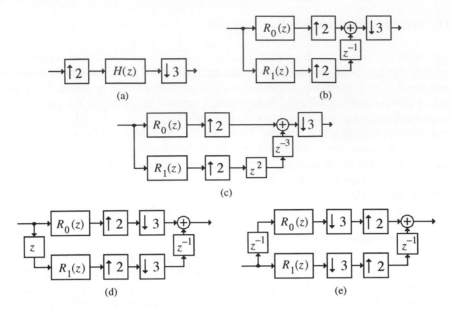

(a)

(b)

(c)

(d)

(e)

Figure 11.65   Various steps in the development of the 48–32-kHz sampling rate converter.

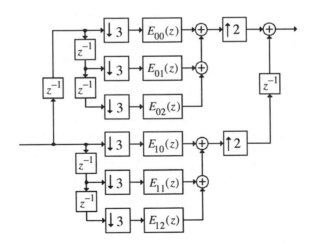

Figure 11.66   Computationally efficient realization of the 48–32-kHz sampling rate converter.

the filters $R_0(z)$ and $R_1(z)$ operate at the input sampling rate of 48 kHz. Simple block diagram manipulations and use of the cascade equivalences of Figure 10.13 lead finally to an equivalent realization depicted in Figure 11.65(e).

The computational efficiency of the sampling rate converter can be improved further by realizing the filters $R_0(z)$ and $R_1(z)$ in Type I polyphase forms and then applying the cascade equivalence of Figure 10.14(a). The final realization is as shown in Figure 11.66, in which all filters now operate at the 16 kHz rate.

Figure 11.67 Sampling rate alteration based on conversion of input digital signal to analog form followed by a resampling at the desired output rate.

The transpose of the structure of Figure 11.66 yields the realization of a fractional rate interpolator with an interpolation factor of 3/2 and can be used as the 32-48-kHz sampling rate converter (Problem 11.14).

11.11.2 Ideal Arbitrary-Rate Sampling Rate Converter

As indicated earlier, an all-digital design of a sampling rate converter with an arbitrary conversion factor is not that simple. In particular, the design is quite difficult and expensive when the conversion factor is a ratio of two very large integers or an irrational number. In principle, a sampling rate conversion by an arbitrary conversion factor can be implemented simply by passing the input digital signal through an ideal analog reconstruction lowpass filter whose output is resampled at the desired output rate, as indicated in Figure 11.67 [Ram84]. If the impulse response of the analog lowpass filter is denoted by $g_a(t)$, the output of the filter is then given by

$$\hat{x}_a(t) = \sum_{\ell=-\infty}^{\infty} x[\ell]g_a(t - \ell T). \tag{11.89}$$

If the analog filter is chosen to bandlimit its output to the frequency range $F_g < F_T'/2$, its output $\hat{x}_a(t)$ can then be resampled at the rate F_T' and hence, the output $y[n]$ of the resampler at new time instants $t = nT'$ is given by

$$y[n] = \hat{x}_a(nT') = \sum_{\ell=-\infty}^{\infty} x[\ell]g_a(nT' - \ell T). \tag{11.90}$$

Since the impulse response $g_a(t)$ of an ideal lowpass analog filter is of infinite duration and $g_a(nT' - \ell T)$ has to be computed at each output sampling instant, implementation of the ideal bandlimited interpolation algorithm of Eq. (11.90) in exact form is not practical. Thus, approximations to this ideal interpolation algorithm are usually employed in practice.

The basic interpolation problem based on a finite weighted sum of input signal samples can then be stated as follows: Given $N + 1$ input signal samples, $x[k]$, $k = 0, 1, \ldots, N$, obtained by sampling an analog signal $x_a(t)$ at $t = t_k = t_0 + kT_{in}$, determine the sample value $x_a(t_0 + \alpha T_{in}) = y[\alpha]$ at time $t' = t_0 + \alpha T_{in}$, where $0 \le \alpha \le N$. Figure 11.68 illustrates the interpolation process by an arbitrary factor.

We next describe a commonly employed interpolation algorithm based on a finite weighted sum of input samples.

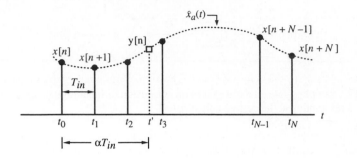

Figure 11.68   Interpolation by an arbitrary factor.

11.11.3   Lagrange Interpolation Algorithm

In this approach, a polynomial approximation $\hat{x}_a(t)$ to $x_a(t)$ is defined as

$$\hat{x}_a(t) = \sum_{k=0}^{N} P_k(t)x[n+k], \tag{11.91}$$

where $P_k(t)$ are the Lagrange polynomials given by

$$P_k(t) = \prod_{\substack{\ell=0, \\ \ell \neq k}}^{N} \left(\frac{t - t_\ell}{t_k - t_\ell} \right), \qquad 0 \leq k \leq N. \tag{11.92}$$

Since

$$P_k(t_r) = \begin{cases} 1, & k = r \\ 0, & k \neq r, \end{cases} \qquad 0 \leq r \leq N, \tag{11.93}$$

it follows from Eqs. (11.91) to (11.93) that

$$\hat{x}_a(t_k) = x_a(t_k), \qquad 0 \leq k \leq N. \tag{11.94}$$

From Eq. (11.91), the value of $x_a(t)$ at an arbitrary value $t' = t_0 + \alpha T_{in}$ is given by

$$\hat{x}_a(t') = \hat{x}_a(t_0 + \alpha T_{in}) = y[n] = \sum_{k=0}^{N} P_k(\alpha)x[n+k], \tag{11.95}$$

where

$$P_k(\alpha) = P_k(t_0 + \alpha T_{in}) = \prod_{\substack{\ell=0, \\ \ell \neq k}}^{N} \left(\frac{\alpha - \ell}{k - \ell} \right), \qquad 0 \leq k \leq N. \tag{11.96}$$

We illustrate in the following example the application of the Lagrange interpolation algorithm in designing a fractional-rate sampling-rate converter.

Figure 11.69 The input and output sample locations of an up-sampler with a conversion factor of 3/2.

Example 11.10 Consider the design of a fractional-rate interpolator with an interpolation factor of 3/2. For simplicity, we make use of a third-order polynomial approximation, $N = 3$, in which case Eq. (11.95) reduces to

$$y[n] = P_0(\alpha)x[n] + P_1(\alpha)x[n+1] + P_2(\alpha)x[n+2] + P_3(\alpha)x[n+3]. \qquad (11.97)$$

Here, the Lagrange polynomials $P_k(\alpha)$ are given by

$$P_0(\alpha) = -\frac{(\alpha - 1)(\alpha - 2)(\alpha - 3)}{6}, \qquad (11.98a)$$

$$P_1(\alpha) = \frac{\alpha(\alpha - 2)(\alpha - 3)}{2}, \qquad (11.98b)$$

$$P_2(\alpha) = -\frac{\alpha(\alpha - 1)(\alpha - 3)}{2}, \qquad (11.98c)$$

$$P_3(\alpha) = \frac{\alpha(\alpha - 1)(\alpha - 2)}{6}. \qquad (11.98d)$$

obtained from Eq. (11.92) with $N = 3$.

Figure 11.69 shows the locations of the samples of the input and the output sequences for an interpolator with a conversion factor of 3/2, where input sample locations are marked with a circle and output sample locations are marked with a square. The locations of the output samples $y[0]$, $y[1]$, and $y[2]$ in the input sample domain are marked with an arrow.

Consider the computations of $y[n]$, $y[n+1]$, and $y[n+2]$ using 4 input samples, $x[n]$ through $x[n+3]$. To this end we require a third order Lagrange polynomial, i.e., $N = 3$ in Eq. (11.96). From Figure 11.69 it follows that, for the computation of $y[n]$, the value of α in Eq. (11.97), to be labeled α_0, is 0. Substituting this value of α in Eqs. (11.98a)–(11.98d), we arrive at the desired values of the filter coefficients as

$$P_0(\alpha_0) = -\frac{(0-1)(0-2)(0-3)}{6} = 1, \qquad (11.99a)$$

$$P_1(\alpha_0) = \frac{0(0-2)(0-3)}{2} = 0, \qquad (11.99b)$$

$$P_2(\alpha_0) = -\frac{0(0-1)(0-3)}{2} = 0, \qquad (11.99c)$$

$$P_3(\alpha_0) = \frac{0(0-1)(0-2)}{6} = 0. \qquad (11.99d)$$

Next, we observe from Figure 11.69 that for the computation of $y[n+1]$ the value of α, to be labeled α_1, is equal to 2/3. Using this value of α in Eqs. (11.98a)–(11.98d), we obtain the new set of values for the interpolator filter coefficients as

$$P_0(\alpha_1) = \tfrac{14}{81} = 0.1728, \qquad P_1(\alpha_1) = \tfrac{28}{27} = 1.037, \qquad (11.100a)$$

$$P_2(\alpha_1) = -\tfrac{7}{27} = -0.2593, \qquad P_3(\alpha_1) = \tfrac{4}{81} = 0.0494. \qquad (11.100b)$$

Finally, for the computation of $y[n+2]$, the value of α, to be labeled α_2, is 4/3, which leads to the filter coefficients.

$$P_0(\alpha_2) = -\tfrac{5}{81} = -0.0617, \qquad P_1(\alpha_2) = \tfrac{20}{27} = 0.7407, \qquad (11.101a)$$

$$P_2(\alpha_2) = \tfrac{10}{27} = 0.3704, \qquad P_3(\alpha_2) = -\tfrac{4}{81} = -0.0494. \qquad (11.101b)$$

Combining Eqs. (11.97) to (11.101b) into matrix form, we arrive at the input-output relation of the interpolation filter as

$$\begin{bmatrix} y[n] \\ y[n+1] \\ y[n+2] \end{bmatrix} = \begin{bmatrix} P_0(\alpha_0) & P_1(\alpha_0) & P_2(\alpha_0) & P_3(\alpha_0) \\ P_0(\alpha_1) & P_1(\alpha_1) & P_2(\alpha_1) & P_3(\alpha_1) \\ P_0(\alpha_2) & P_1(\alpha_2) & P_2(\alpha_2) & P_3(\alpha_2) \end{bmatrix} \begin{bmatrix} x[n] \\ x[n+1] \\ x[n+2] \\ x[n+3] \end{bmatrix}$$

$$= \mathbf{H} \begin{bmatrix} x[n] \\ x[n+1] \\ x[n+2] \\ x[n+3] \end{bmatrix}, \qquad (11.102)$$

where $\mathbf{H}$ is the block coefficient matrix, which for the above factor of 3/2 interpolator design is given by

$$\mathbf{H} = \begin{bmatrix} 1 & 0 & 0 & 0 \\ 0.1728 & 1.0370 & -0.2593 & 0.0494 \\ -0.0617 & 0.7407 & 0.3704 & -0.0494 \end{bmatrix}. \qquad (11.103)$$

It should be evident from an examination of Figure 11.69 that the filter coefficients to compute $y[n+3]$, $y[n+4]$, and $y[n+5]$ are again given by the coefficients given in Eqs. (11.99a)–(11.99d), (11.100a) and (11.100b), and (11.101a) and (11.101b), respectively. Or in other words, the desired interpolation filter is a time-varying filter with a period of three samples. A realization of the above factor of 3/2 interpolator based on an implementation of Eq. (11.102) is indicated in Figure 11.70(a). Note that, in practice, the overall system delay will be three sample periods, and as a result, the output sample $y[n]$ actually appears at the time index $n+3$.

An alternative realization of the above fractional rate interpolator in the form of a time-varying FIR filter is indicated in Figure 11.70(b). The filter coefficients of the fifth order time-varying FIR filter have a period of 3 and are assigned the values as indicated in Figure 11.70(c).

Figure 11.71 shows the plots of the input and the output of the above fractional-rate interpolator for a sinusoidal input of frequency 0.05 Hz sampled at 1 Hz rate along with that of the error sequence defined by the sample-wise difference between the output sequence and the actual sine values.

11.11.4   Practical Considerations

A direct design of a fractional sampling rate converter in most applications is impractical since the length of the time-varying filter needed is very long and the corresponding filter coefficient calculations in real time are thus nearly impossible. As a result, a fractional sampling rate converter is almost always realized in a hybrid form consisting of a digital sampling rate converter with an integer-valued conversion factor followed by an "analog" fractional rate converter as indicated in Figure 11.72 [Ram84]. The digital sampling rate converter, of course, if need be, can be implemented in a multi-stage form.[5]

---

[5]See Section 10.3.

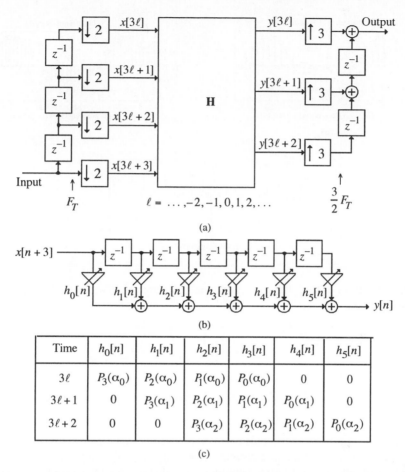

Figure 11.70 Implementation of a fractional rate interpolator with a conversion factor of 3/2. (a) Block digital filter implementation, (b) implementation using a time-varying FIR interpolation filter, (c) coefficients of the time-varying filter as a function of sample index.

11.12   Oversampling A/D Converter

For the digital processing of an analog continuous-time signal, the signal is first passed through a sample-and-hold circuit whose output is then converted into a digital form by means of an A/D converter. However, according to the sampling theorem, discussed in Section 5.2.1, a bandlimited continuous-time signal with a lowpass spectrum can be fully recovered from its uniformly sampled version if it is sampled at a sampling frequency that is at least twice the highest frequency contained in the analog signal. If this condition is not satisfied, the original continuous-time signal cannot be recovered from its sampled version because of aliasing. To prevent aliasing, the analog signal is thus passed through an analog anti-aliasing lowpass filter prior to sampling, which enforces the condition of the sampling theorem. The passband cutoff frequency of the lowpass filter is chosen equal to the frequency of the highest signal frequency component that needs to be preserved at

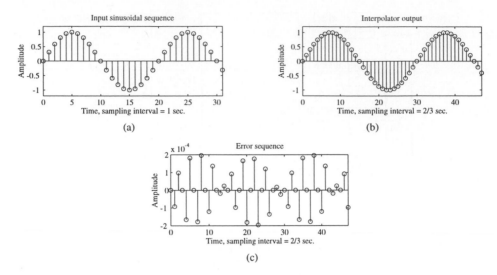

Figure 11.71   Plots of the interpolator output and the error sequence for a sinusoidal input of frequency 0.05 Hz.

Figure 11.72   Hybrid form of a fractional rate sampling converter.

the output. The anti-aliasing filter also cuts off all out-of-band signal components and any high-frequency noise that may be present in the original analog signal, which otherwise would alias into the baseband after sampling. The filtered signal is then sampled at a rate that is at least twice that of the cutoff frequency.

Let the signal band of interest be the frequency range $0 \leq f \leq F_m$. Then, the Nyquist rate is given by $F_N = 2F_m$. Now, if the sampling rate F_T is the same as the Nyquist rate, we need to design an anti-aliasing lowpass filter with a very sharp cutoff in its frequency response, satisfying the requirements as given by Eq. (5.57).[6] This requires the design of a very high-order anti-aliasing filter structure built with high-precision analog components, and it is usually difficult to implement such a filter in VLSI technology. Moreover, such a filter also introduces undesirable phase distortion in its output. An alternative approach mentioned in Section 5.4 is to sample the analog signal at a rate much higher than the Nyquist rate, use a fast low-resolution A/D converter, and then decimate the digital output of the converter to the Nyquist rate. This approach relaxes the sharp cutoff requirements of the analog anti-aliasing filter, resulting in a simpler filter structure that can be built using low-precision analog components while requiring fast, more complex digital signal processing hardware at later stages. The overall structure is not only amenable to VLSI fabrication but also can be designed to provide linear phase response in the signal band of interest.

---

[6]Recall that $F_T = 1/T$, where T is the sampling period.

The oversampling approach is an elegant application of multirate digital signal processing and is increasingly being employed in the design of high-resolution A/D converters for many practical systems [Can92], [Fre94]. In this section, we analyze the quantization noise performance of the conventional A/D converter and show analytically how the oversampling approach decreases the quantization noise power in the signal band of interest [Fre94]. We then show that further improvement in the noise performance of an oversampling A/D converter can be obtained by employing a sigma-delta ($\Sigma\Delta$) quantization scheme.[7] For simplicity we restrict our discussion to the case of a basic first-order sigma-delta quantizer.

To illustrate the above property, consider a b-bit A/D converter operating at F_T Hz. Now, for a full-scale peak-to-peak input analog voltage of R_{FS}, the smallest voltage step represented by b bits is

$$\Delta V = \frac{R_{FS}}{2^b - 1} \cong \frac{R_{FS}}{2^b}.$$ (11.104)

From Eq. (9.73), the RMS quantization noise power σ_e^2 of the error voltage, assuming a uniform distribution of the error between $-\Delta V/2$ and $\Delta V/2$, is given by

$$\sigma_e^2 = \frac{(\Delta V)^2}{12}.$$ (11.105)

The RMS noise voltage, given by σ_e, therefore has a flat spectrum in the frequency range from 0 to $F_T/2$.

The noise power per unit bandwidth, called the *noise density*, is then given by

$$P_{e,n} = \frac{(\Delta V)^2/12}{F_T/2} = \frac{(\Delta V)^2}{6 F_T}.$$ (11.106)

A plot of the noise densities for two different sampling rates are shown in Figure 11.73, where the shaded portion indicates the signal band of interest. As can be seen from this figure, the total amount of noise in the signal band of interest for the high sampling rate case is smaller than that for the low sampling rate case. The total noise in the signal band of interest, called the *in-band noise power*, is given by

$$P_{\text{total}} = \frac{(R_{FS}/2^b)^2}{12} \cdot \frac{F_m}{F_T/2}.$$ (11.107)

It is interesting to compute the needed wordlength β of the A/D converter operating at the Nyquist rate in order that its total noise in the signal band of interest be equal to that of a b-bit A/D converter operating at a higher rate. Substituting $F_T = 2F_m$ and replacing b with β in Eq. (11.107), we arrive at

$$P_{total} = \frac{(R_{FS}/2^\beta)^2}{12} = \frac{(R_{FS}/2^b)^2}{12} \cdot \frac{F_m}{F_T/2},$$ (11.108)

which leads to the desired relation

$$\beta = b + \frac{1}{2}\log_2 M,$$ (11.109)

---

[7]Also known as delta-sigma A/D converter.

Figure 11.73   A/D converter noise density.

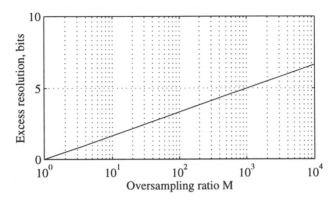

Figure 11.74   Increase in resolution as a function of the oversampling ratio M.

where $M = F_T/2F_m$ denotes the *oversampling ratio* (OSR). Thus, $\beta - b$ denotes the increase in the resolution of a b-bit converter whose oversampled output is filtered by an ideal brick-wall lowpass filter. A plot of the increase in resolution as a function of the oversampling ratio is shown in Figure 11.74. For example, for an OSR of $M = 1000$, an 8-bit oversampling A/D converter has an effective resolution equal to that of a 13-bit A/D converter operating at the Nyquist rate. Note that Eq. (11.109) implies that the increase in the resolution is $1/2$-bit per doubling of the OSR.

We now illustrate the improvement in the noise performance obtained by employing a sigma-delta ($\Sigma\Delta$) quantization scheme. The sigma-delta A/D converter briefly introduced in Section 5.6.5 and is shown in block diagram form in Figure 11.75 for convenience. This figure also indicates the sampling rates at various stages of the structure. It should be noted here that the 1-bit output samples of the quantizer after decimation become b-bit samples at the output of the sigma-delta A/D converter due to the filtering operations involving b-bit multiplier coefficients of the Mth band digital lowpass filter.

Since the oversampling ratio M is typically very large in practice, the sigma-delta A/D converter is most useful in low-frequency applications such as digital telephony, digital audio, and digital spectrum analyzers. For example, Figure 11.76 shows the block diagram of a typical compact disk encoding system used to convert the input analog audio signal

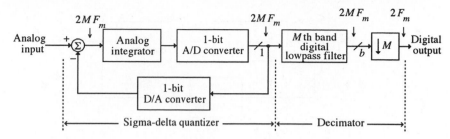

Figure 11.75   Oversampling sigma-delta A/D converter structure.

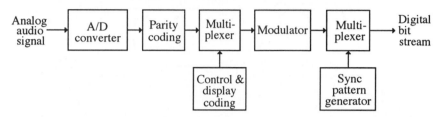

Figure 11.76   Compact disc encoding system for one channel of a stereo audio.

into a digital bit stream that is then applied to generate the master disk [Hee82]. Here the oversampling sigma-delta A/D converter employed has a typical input sampling rate of 3175.2 kHz and an output sampling rate of 44.1 kHz [Kam86].

To understand the operation of the sigma-delta A/D converter of Figure 11.75, we need to study the operation of the sigma-delta quantizer shown in Figure 11.77(a). To this end it is convenient to use the discrete-time equivalent circuit of Figure 11.77(b), where the integrator has been replaced with an accumulator.[8] Here, the input $x[n]$ is a discrete-time sequence of analog samples developing an output sequence of binary-valued samples $y[n]$. From this diagram, we observe that, at each discrete instant of time, the circuit forms the difference (Δ) between the input and the delayed output, which is accumulated by a summer (Σ) whose output is then quantized by a one-bit A/D converter, i.e., a comparator.

Even though the input-output relation of the sigma-delta quantizer is basically nonlinear, the low-frequency content of the input $x_c(t)$ can be recovered from the output $y[n]$ by passing it through a digital lowpass filter. This property can be easily shown for a constant input analog signal $x_a(t)$ with a magnitude less than $+1$. In this case, the output $w[n]$ of the accumulator is a bounded sequence with sample values equal to either -1 or $+1$. This can happen only if the input to the accumulator has an average value of zero. Or in other words, the average value of $w[n]$ must equal to the average value of the input $x[n]$ [Sch91]. The following two examples illustrate the operation of a sigma-delta quantizer.

Example 11.11   We first consider the operation for the case of a constant input signal using MATLAB. To this end we can use Program 11_5 given below.

```
% Program 11_5
% Sigma-Delta Quantizer Operation
```

---

[8]In practice, the integrator is implemented as a discrete-time switched-capacitor integrator.

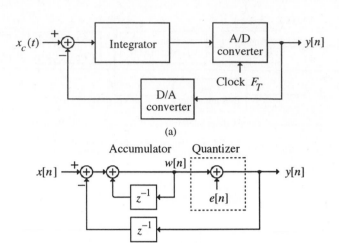

Figure 11.77   Sigma-delta quantization scheme.

```
%
N = input('Type in the length of input sequence = ');
n = 1:1:N;
m = n-1;
A = input('Type in the input amplitude = ');
x = A*ones(1,N);
plot(m,x);
axis([0 N-1 -1.2 1.2]);
xlabel('Time'); ylabel('Amplitude');
title('Input analog signal');
pause
y = zeros(1,N+1);
v0 = 0;
    for k = 2:1:N+1;
        v1 = x(k-1) - y(k-1) + v0;
        y(k) = sign(v1);
        v0 = v1;
    end
yn = y(2:N+1);
axis([0 N-1 -1.2 1.2]);
stem(m, yn);
xlabel('Time'); ylabel('Amplitude');
title('Output of sigma-delta modulator');
ave = sum(yn)/N;
disp('Average value of output is = ');disp(ave)'
```

The plots generated by this program are the input and output waveforms of the sigma-delta quantizer of Figure 11.77(a) and are shown in Figure 11.78. The program also prints the average value of the output as indicated below:

Figure 11.78 Input and output waveforms of the sigma-delta quantizer of Figure 11.77(a) for a constant input.

```
Average value of output is =
    0.8095
```

which is very close to the amplitude 0.8 of the constant input. It can be easily verified that the average value of the output gets closer to the amplitude of the constant input as the length of the input increases.

Example 11.12 We now verify the operation of the sigma-delta A/D converter for a sinusoidal input of frequency 0.01 Hz using MATLAB. To this end we make use of the Program 11_6 given below. Because of the short length of the input sequence, the filtering operation is performed here in the DFT domain [Sch91].

```
% Program 11_6
% Sigma-Delta A/D Converter Operation
%
wo = 2*pi*0.01;
N = input('Type in the length of input sequence = ');
n = 1:1:N;
m = n-1;
A = input('Type in the amplitude of the input = ');
x = A*cos(wo*m);
axis([0 N-1 -1.2 1.2]);
plot(m,x);
xlabel('Time'); ylabel('Amplitude');
title('Input analog signal');
pause
y = zeros(1,N+1);
v0 = 0;
for k = 2:1:N+1;
v1 = x(k-1) - y(k-1) + v0;
    if v1 >= 0;
      y(k) = 1;
    else
      y(k) = -1;
    end
v0 = v1;
end
```

Figure 11.79   Input and output waveforms of the sigma-delta quantizer of Figure 11.77(a) with a sine wave input.

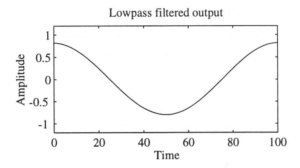

Figure 11.80   The lowpass filtered version of the waveform of Figure 11.79(b).

```
yn = y(2:N+1);
axis([0 N-1 -1.2 1.2]);
stairs(m, yn);
xlabel('Time'); ylabel('Amplitude');
title('Output of sigma-delta quantizer');
Y = fft(yn);
pause
H = [1 1 0.5 zeros(1,N-5) 0.5 1];
YF = Y.*H;
out = ifft(YF);
axis([0 N-1 -1.2 1.2]);
plot(m,out);
xlabel('Time'); ylabel('Amplitude');
title('Lowpass filtered output');
```

Figure 11.79 shows the input and output waveforms of the sigma-delta quantizer of Figure 11.77(a) for a sinusoidal input. Figure 11.80 depicts the lowpass filtered version of the output signal shown in Figure 11.79(b). As can be seen from these figures, the filtered output is nearly an exact replica of the input.

It follows from Fig. 11.77(b) that the output $y[n]$ of the quantizer is given by

$$y[n] = w[n] + e[n], \tag{11.110}$$

where

$$w[n] = x[n] - y[n-1] + w[n-1]. \tag{11.111}$$

From the above equations, we obtain after some algebra

$$y[n] = x[n] + (e[n] - e[n-1]), \tag{11.112}$$

where the quantity inside the parentheses represents the noise due to sigma-delta modulation. The noise transfer function is simply $G(z) = (1 - z^{-1})$. The power spectral density of the modulation noise is therefore given by

$$P_y(f) = \left| G(e^{j2\pi fT}) \right|^2 P_e(f) = 4\sin^2\left(\frac{2\pi fT}{2}\right) P_e(f), \tag{11.113}$$

where we have assumed the power spectral density $P_e(\omega)$ of the quantization noise to be the one-sided power spectral density defined for positive frequencies only. For a random signal input $x[n]$, $P_e(f)$ is constant for all frequencies and is given by

$$P_e(f) = \frac{(\Delta V)^2/12}{F_T/2}. \tag{11.114}$$

Substituting the above in Eq. (11.113), we arrive at the power spectral density of the output noise, given by

$$P_y(f) = \frac{2}{3}\frac{(\Delta V)^2}{F_T}\sin^2(\pi fT). \tag{11.115}$$

The noise-shaping provided by the sigma-delta quantizer is similar to that encountered in the first-order error-feedback structures of Section 9.8.1 and shown in Figure 9.41. For a very large OSR, as is usually the case, the frequencies in the signal band of interest are much smaller than F_T, the sampling frequency. Thus, we can approximate $P_y(f)$ of Eq. (11.115) as

$$P_y(f) \cong \frac{2}{3}\frac{(\Delta V)^2}{F_T}(\pi fT)^2 = \frac{2}{3}\pi^2(\Delta V)^2 T^3 f^2, \qquad f \ll F_T. \tag{11.116}$$

From the above, the in-band noise power of the sigma-delta A/D converter is thus given by

$$P_{\text{total,sd}} = \int_0^{F_m} P_y(f)\, df = \frac{2}{3}\pi^2(\Delta V)^2 T^3 \int_0^{F_m} f^2\, df = \frac{2}{9}\pi^2(\Delta V)^2 T^3 (F_m)^3. \tag{11.117}$$

It is instructive to compare the noise performance of the sigma-delta A/D converter with that of a direct oversampling A/D converter operating at a sampling rate of F_T with a signal band of interest from dc to F_m. From Eq. (11.108), the in-band noise power of the latter is given by

$$P_{\text{total,os}} = \frac{1}{6}(\Delta V)^2 T F_m. \tag{11.118}$$

The improvement in the noise performance is therefore given by

$$10\log_{10}\left(\frac{P_{\text{total,os}}}{P_{\text{total,sd}}}\right) = 10\log_{10}\left(\frac{3M^2}{\pi^2}\right) = -5.1718 + 20\log_{10}(M) \text{ dB}, \tag{11.119}$$

Figure 11.81   A very simple factor-of-N decimator structure.

where we have used $M = F_T/2F_m$ to denote the OSR. For example, for an OSR of $M = 1000$, the improvement in the noise performance using the sigma-delta modulation scheme is about 55 dB. In this case, the increase in the resolution is about 1.5 bits per doubling of the OSR.

The improved noise performance of the sigma-delta A/D converter results from the shape of $|G(e^{j2\pi fT})|$, which decreases the noise power spectral density in-band ($0 \le f \le F_m$) while increasing it outside the signal band of interest ($f > F_m$). Since this type of converter also employs oversampling, it requires a less stringent analog anti-aliasing filter.

The A/D converter of Figure 11.75 employs a single-loop feedback and is often referred to as a first-order sigma-delta converter. Multiple feedback loop modulation schemes have been advanced to reduce the in-band noise further. However, the use of more than two feedback loops may result in unstable operation of the system, and care must be taken in the design to ensure stable operation [Can92].

As indicated in Figure 11.75, the quantizer output is passed through an Mth band lowpass digital filter whose output is then down-sampled by a factor of M to reduce the sampling rate to the desired Nyquist rate. The function of the digital lowpass filter is to eliminate the out-of-band quantization noise and the out-of-band signals that would be aliased into the passband by the down-sampling operation. As a result, the filter must exhibit a very sharp cutoff frequency response with a passband edge at F_m. This necessitates the use of a very high-order digital filter. In practice, it is preferable to use a filter with a transfer function having simple, preferably integer-valued, coefficients to reduce the cost of hardware implementation and to permit all multiplication operations to be carried out at the down-sampled rate. In addition, most applications require the use of linear-phase digital filters, which can be easily implemented using FIR filters.

The simplest lowpass FIR filter is the moving average filter of Eq. (2.49), repeated below for convenience:[9]

$$H(z) = 1 + z^{-1} + z^{-2} + \cdots + z^{-(N-1)}. \tag{11.120}$$

A more convenient form of the above transfer function for realization purposes is given by

$$H(z) = \frac{1 - z^{-N}}{1 - z^{-1}}, \tag{11.121}$$

also known as a *recursive running-sum filter* or a *boxcar filter*. A realization of a factor-of-N decimator based on the decimation filter of Eq. (11.121) is sketched in Figure 11.81.[10]

---

[9]For simplicity, we have ignored the scale factor of $1/N$ which is needed to provide a dc gain of 0 dB.

[10]The integrator overload caused by the input adder overflow can be easily handled with binary arithmetic.

Figure 11.82 A two-stage CIC decimator structure.

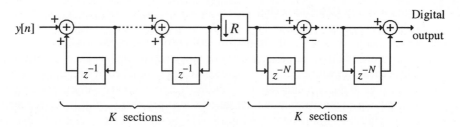

Figure 11.83 A CIC decimator structure with cascaded sections.

Since the decimator based on a running-sum filter does not provide sufficient out-of-band attenuation, often a multistage decimator formed by a cascade of the running-sum decimators, more commonly known as *cascaded integrator comb* (CIC) filters, is used in practice [Hog81]. The structure of a two-stage CIC decimator is shown in Figure 11.82. It can be easily shown that the structure is equivalent to a factor-of-R decimator with a length-RN running sum decimation filter. Further flexibility in the design of a CIC decimator is obtained by including K feedback paths before and K feedforward paths after the down-sampler, as indicated in Figure 11.83. The corresponding transfer function is then given by

$$H(z) = \left(\frac{1 - z^{-RN}}{1 - z^{-1}} \right)^K .\qquad(11.122)$$

The parameters N and K can be adjusted for a given down-sampling factor R to yield the desired out-of-band attenuation.

In some applications, it may be preferable to use a multistage decimation process in which all but the last stage employs the CIC decimators in various forms, followed by an FIR lowpass filter providing a much sharper cutoff before the final down-sampling. For example, in digital telephone applications, the following IIR transfer function provides a very good frequency response and can be used to design a factor-of-4 decimator [Can92]:

$$H(z) = (1 - \tfrac{3}{2}z^{-1} + z^{-2})(1 - \tfrac{11}{8}z^{-1} + \tfrac{5}{8}z^{-2})(1 - \tfrac{5}{4}z^{-1} + z^{-2})$$
$$\times (1 - \tfrac{101}{64}z^{-1} + \tfrac{7}{8}z^{-2}) \cdot \frac{(1 + z^{-1})(1 + z^{-2})(1 - z^{-5})}{(1 - z^{-1})} .\qquad(11.123)$$

Further details on first and higher order sigma-delta converters can be found in Candy and Temes [Can92].

Figure 11.84   Block diagram representation of an oversampling sigma-delta D/A converter.

11.13   Oversampling D/A Converter

As indicated earlier in Section 5.1, the digital-to-analog conversion process consists of two steps: the conversion of input digital samples into a staircase continuous-time waveform by means of a D/A converter with a zero-order hold at its output, followed by an analog lowpass reconstruction filter. If the sampling rate F_T of the input digital signal is same as the Nyquist rate, the analog lowpass reconstruction filter must have a very sharp cutoff in its frequency response, satisfying the requirements of Eq. (5.66). As in the case of the anti-aliasing filter, this involves the design of a very high-order analog reconstruction filter requiring high-precision analog circuit components. To get around the above problem, here also an oversampling approach is often used, in which case a wide transition band can be tolerated in the frequency response of the reconstruction filter allowing its implementation using low-precision analog circuit components while requiring more complex digital interpolation filter at the front end.

Further improvement in the performance of an oversampling D/A converter is obtained by employing a digital sigma-delta 1-bit quantizer at the output of the digital interpolator, as indicated in Figure 5.41 repeated in Figure 11.84 for convenience [Can86], [Lar93]. The quantizer extracts the MSB from its input and subtracts the remaining LSBs, the quantization noise, from its input. The MSB output is then fed into a 1-bit D/A converter and passed through an analog lowpass reconstruction filter to remove all frequency components beyond the signal band of interest. Since the signal band occupies a very small portion of the base band of the high-sample-rate signal, the reconstruction filter in this case can have a very wide transition band, permitting its realization with a low-order filter that, for example, can be implemented using a Bessel filter to provide an approximately linear phase in the signal band.[11]

The spectrum of the quantized 1-bit output of the digital sigma-delta quantizer is nearly the same as that of its input. Moreover, it also shapes the quantization noise spectrum by moving the noise power out of the signal band of interest. To verify this result analytically, consider the sigma-delta quantizer shown separately in Figure 11.85. It follows from this figure that the input-output relation of the quantizer is given by

$$y[n] - e[n] = x[n] - e[n-1],$$

or equivalently, by

$$y[n] = x[n] + e[n] - e[n-1], \tag{11.124}$$

where $y[n]$ is the MSB of the nth sample of the adder output, and $e[n]$ is the nth sample of the quantization noise composed of all bits except the MSB. From Eq. (11.124) it can be

---

[11] See Section 5.3.5.

Figure 11.85   The sigma-delta quantizer.

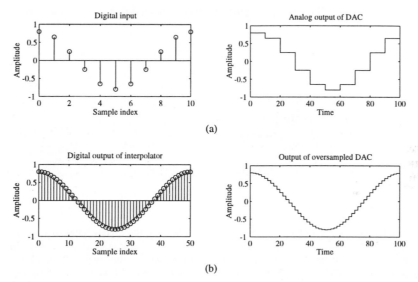

(a)

(b)

Figure 11.86   Input and output signals of (a) lower-rate D/A converter and (b) oversampling D/A converter.

seen that the transfer function of the quantizer with no quantization noise is simply unity and the noise transfer function is given by $G(z) = 1 - z^{-1}$, which is same as that for the first-order sigma-delta modulator employed in the oversampling A/D converter discussed in the previous section.

The following example illustrates by computer simulation the operation of a sigma-delta D/A converter for a discrete-time sinusoidal input sequence.

Example 11.13   Let the input to the D/A converter be a sinusoidal sequence of frequency 100 Hz operating at a sampling rate F_T of 1 kHz. Figure 11.86(a) shows the digital input sequence and the analog output generated by a D/A converter operating at F_T from this input. Figure 11.86(b) depicts the interpolated sinusoidal sequence operating at a higher sampling rate of 5 kHz obtained by passing the low-sampling-rate sinusoidal signal through a factor-of-5 digital interpolator and the corresponding analog output generated by a D/A converter operating at $5F_T$ rate. If we compare the two D/A converter outputs, we can see that the staircase waveform of the oversampling D/A converter output is much smoother with smaller jumps than that of the lower rate D/A converter output. Thus, the oversampling D/A converter output has considerably smaller high-frequency components in contrast to the lower rate D/A converter. This fact can be easily verified by examining their spectras.

Figure 11.87   Lowpass filtered output signals of (a) conventional D/A converter and (b) oversampling D/A converter.

Figure 11.88   Filtered output signals of the conventional D/A converter employing a sharp cutoff lowpass filter.

The high-frequency components in the baseband outside the signal band of interest can be removed by passing the D/A converter output through an analog lowpass filter, which also eliminates any left over replicas of the baseband not completely removed by the zero-order hold in the D/A converter. Since the signal band of interest occupies a small portion of the baseband, the replicas of the signal band immediately outside the baseband are widely separated from the signal band inside the baseband. Hence, the lowpass filter can be designed with a very wide transition band. Moreover, due to reduced high-frequency components in the D/A converter output caused by oversampling, the stopband attenuation also does not have to be very large. On the other hand, the replicas of the signal band in the spectrum of the output of the low-rate D/A converter are closely spaced, and the high-frequency components are relatively large in amplitudes. In this case, the lowpass filter must have a sharp cutoff with much larger stopband attenuation to effectively remove the undesired components in the D/A converter output.

Figures 11.87 shows the filtered outputs of the conventional lower rate and oversampled D/A converters when the same lowpass filter with a wide transition band is used in both cases. As can be seen from this figure, the analog output in the case of the low-rate D/A converter still contains some high-frequency components, while that in the case of the oversampled D/A converter is very close to a perfect sinusoidal signal. A much better output response is obtained in the case of a conventional D/A converter if a sharp cutoff lowpass filter is employed, as indicated in Figure 11.88.

Example 11.14 In this example, we verify using MATLAB the operation of the sigma-delta D/A converter for a sinusoidal input sequence of frequency 100 Hz operating at a sampling rate F_T of 5 kHz. The signal is clearly oversampled since the sampling rate is much higher than the Nyquist rate of 200 Hz. Program 11_7 given below first generates the input digital signal, then generates a two-valued digital signal by quantizing the output of the sigma-delta quantizer, and finally, develops the output of the D/A converter by lowpass filtering the quantized output. As in

the case of the sigma-delta converter of Example 11.13, the filtering operation here has also been performed in the DFT domain due to the short length of the input sequence [Sch91].

```
% Program 11_7
% Sigma-Delta D/A Converter Operation
%
clf;
% Generate the input sinusoidal sequence
N = input('Type in length of the input sequence = ');
A = input('Type in amplitude of the input = ');;
w0 = 2*pi*0.02;
n = 1:N;
m = n-1;
x = A*cos(w0*m);
axis([0 N -1 1]);
stem(m,x);
xlabel('Time index'); ylabel('Amplitude');
title ('Input digital signal');
pause
% Generation of quantized output
x = (x)/(A);
y = zeros(1,N+1);
a = zeros(1,N+1);
e = 0;
for k = 2:N+1
   a(k) = x(k-1) - e;
     if a(k) >= 0,
          y(k) = 1;
     else
          y(k) = -1;
     end
   e = y(k) - a(k);
end
yn = y(2:N+1);
axis([0 N -1.2 1.2]);
stem(m, yn); % Plot the quantized output
xlabel('Time'); ylabel('Amplitude');
title ('Digital output of sigma-delta quantizer');
pause
Y = fft(yn);
H = [1 1 0.5 zeros(1,N-5) 0.5 1];% Lowpass filter
YF = Y.*H; % Filtering in the DFT domain
out = ifft(YF);
plot(m,out);
xlabel('Time'); ylabel('Amplitude');
title ('Lowpass filtered analog output');
```

Figure 11.89 shows the digital input signal, the quantized digital output of the sigma-delta quantizer, and the filtered output of the D/A converter generated by this program. As can be seen from these plots, the lowpass filtered output is nearly a scaled replica of the desired sinusoidal analog signal.

(a)

(b)

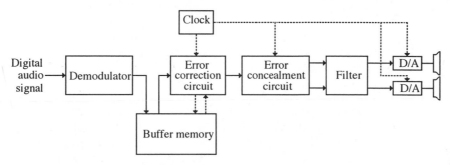

(c)

Figure 11.89   Input and output waveforms of the sigma-delta quantizer of Figure 11.84.

Figure 11.90   Signal processing part of a player.

One of the most common applications of the oversampling sigma-delta D/A converter is in the compact disc players. Figure 11.90 depicts the block diagram of the basic components in the signal processing part of a CD player, where typically a factor-of-4 oversampling D/A converter is employed for each audio channel [Goe82]. Here, the 44.1-kHz input digital audio signal is interpolated first by a factor of 4 to the 176.4-kHz rate and then converted into an analog audio signal.

11.14   Summary

The discrete Fourier transform (DFT) is a widely used digital signal processing algorithm. One of the reasons for its widespread use is the availability of fast Fourier transform (FFT) algorithms for its computation. Earlier in Section 3.5 we discussed the implementation of high-speed convolution using the DFT. It also plays a major role in the implementation of many functions in the Signal Processing Toolbox of MATLAB. Three other applications

are considered in this chapter. The first application discussed is in the efficient and robust detection of dual-tone, multifrequency (DTMF) tones employed in dial-pulse telephone signaling, ATM machines, voice mails, etc. The next application treated here is in the spectral analysis of stationary signals and is the basis of most commercial spectrum analyzers. For the spectral analysis of nonstationary signals, such as speech and radar signals, the DFTs of small windowed segments of these signals are computed, and a three-dimensional display of the resulting spectrums, called spectrograms, are employed. This type of signal analysis is more popularly called the short-term Fourier transform, or the time-dependent Fourier transform which is covered next in this chapter.

Digital signal processing methods are increasingly being employed in digital audio applications for spectral shaping and for the generation of special audio effects. A number of these applications are outlined. These are followed by a review of digital stereo generation for FM stereo transmission and a discussion on the use of digital filtering for preemphasis of audio signals.

The discrete-time analytic signal has a zero-valued spectrum for all negative frequencies and, as a result, is a complex signal. Such a signal can be generated from a real signal by passing the latter through a discrete-time Hilbert transformer. Several methods for designing the Hilbert transformer are described. One application of the analytic signal considered in this chapter is in the design of a digital single sideband communication system.

The next three applications treated involve multirate digital signal processing. These include the design of a voice privacy system for telephone communication, subband coding of speech and audio signals for signal compression, and the design of transmultiplexers for interconnecting frequency-division multiplex (FDM) and time-division multiplex (TDM) communication systems. Each of these applications require the use of the so-called quadrature mirror-filter (QMF) banks.

A method for the efficient digital data transmission based on DSP techniques is discussed next. This method makes use of the discrete Fourier transform (DFT) computation and, as a result, can be implemented using fast Fourier transform (FFT) algorithms.

An important problem is the conversion of a discrete-time signal with a prescribed sampling rate into another "equivalent" discrete-time signal with a different sampling rate. This type of conversion involving a fractional sampling rate change is needed in practical situations. A commonly used method for such conversion discussed here is based on the Lagrange interpolation technique.

The remaining two applications considered in this chapter are the oversampling sigma-delta analog-to-digital (A/D) and digital-to-analog (D/A) converter design. Such converters are being increasingly employed in many systems because of their improved signal-to-noise ratios.

This chapter has touched upon a variety of practical applications of digital signal processing. There are numerous other applications with many requiring a knowledge of other subjects, and as a result, they are beyond the scope of this book. Additional applications can be found in various other books; see, for example, [Bab95], [Fre94], [Lin87], [Mar92], [Opp78], [Pap90], [She95].

11.15   Problems

11.1 A speech signal is sampled at the 8-kHz rate. We wish to analyze, using the DFT, the spectrum of the sampled speech of length 256.

(a) If we take a 256-point DFT of this segment, what would be the resolution of the DFT samples?

(b) Describe a technique to achieve a 16-Hz resolution of the DFT samples.

(c) Describe a method to achieve a 128-Hz resolution using a minimum length DFT.

11.2 A bandlimited continuous-time real signal $g_a(t)$ is sampled at a rate of F_T Hz where $F_T = 2F_m$ with F_m denoting the highest frequency contained in $g_a(t)$. An R-point DFT $G[k]$ of the sequence $g[n]$ obtained by sampling $g_a(t)$ is next computed.

(a) If $F_m = 4$ kHz and $R = 1024$, determine the contiuous-time frequencies corresponding to the DFT sample indices $k = 200, 350$, and 824.

(b) If $F_m = 6$ kHz and $R = 1024$, determine the contiuous-time frequencies corresponding to the DFT sample indices $k = 200, 350$, and 824.

(c) If $F_m = 4$ kHz and $R = 512$, determine the contiuous-time frequencies corresponding to the DFT sample indices $k = 100, 200$, and 312.

11.3 A bandlimited continuous-time real signal $g_a(t)$ is sampled at a rate of F_T Hz, where $F_T = 2F_m$ with F_m denoting the highest frequency contained in $g_a(t)$. An R-point DFT $G[k]$ of the sequence $g[n]$ obtained by sampling $g_a(t)$ is next computed.

(a) If $F_m = 4$ kHz, what are the values of R and the sampling rate F_T to provide a DFT resolution of 3 Hz with no aliasing?

(b) If R has to be a power-of-2 to make use of efficient FFT algorithms for the computation of the DFT, what should then be the value of R to achieve the resolution of 3 Hz if $g_a(t)$ is sampled at a rate to produce no aliasing?

11.4 Let $X_{\text{STFT}}(e^{j\omega}, n)$ and $Y_{\text{STFT}}(e^{j\omega}, n)$ denote the short-term Fourier transforms of two sequences $x[n]$ and $y[n]$, obtained by applying the same window sequence $w[n]$. Prove the following properties of the STFT:

(a) Linearity property: If $g[n] = \alpha x[n] + \beta y[n]$, then $G_{\text{STFT}}(e^{j\omega}, n) = \alpha X_{\text{STFT}}(e^{j\omega}, n) + \beta Y_{\text{STFT}}(e^{j\omega}, n)$ where $G_{\text{STFT}}(e^{j\omega}, n)$ is the STFT of $g[n]$ obtained using the window $w[n]$.

(b) Shifting property: If $y[n] = x[n - n_0]$, then $Y_{\text{STFT}}(e^{j\omega}, n) = X_{\text{STFT}}(e^{j\omega}, n - n_0)$, where n_0 is any positive or negative integer.

(c) Modulation property: If $y[n] = e^{j\omega_0}x[n]$, then $Y_{\text{STFT}}(e^{j\omega}, n) = X_{\text{STFT}}(e^{j(\omega - \omega_0)}, n)$.

11.5 An alternate definition of the short-term Fourier transform is given by

$$\bar{X}_{\text{STFT}}(e^{j\omega}, n) = \sum_{m=-\infty}^{\infty} x[m]w[n - m]e^{-j\omega m}. \qquad (11.125)$$

Express $\bar{X}_{\text{STFT}}(e^{j\omega}, n)$ in terms of the short-term Fourier transform $X_{\text{STFT}}(e^{j\omega}, n)$ defined in Eq. (11.17). What are the main differences between the two definitions?

11.6 Show that the inverse STFT can also be computed using the following expression [Naw88]:

$$x[n] = \frac{1}{2\pi W(0)} \int_0^{2\pi} \sum_{\ell=-\infty}^{\infty} \bar{X}_{\text{STFT}}(e^{j\omega}, \ell)e^{j\omega n}\, d\omega, \qquad (11.126)$$

where $W(0) = \sum_{n=-\infty}^{\infty} w[n]$.

11.7 Show that the sampled short-term discrete Fourier transform $X_{\text{STFT}}[k, n]$ defined by Eq. (11.20) can be interpreted as a bank on N linear time-invariant filters as indicated in Figure P11.1 whose outputs $y_k[n]$ are precisely $X_{\text{STFT}}[k, n]$, $k = 0, 1, \ldots, N - 1$. Determine the expressions for the impulse responses $h_k[n]$ of these N filters.

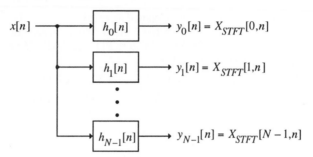

Figure P11.1

11.8 Let $X_{\text{STFT}}(e^{j\omega}, n)$ denote the STFT of a real sequence $x[n]$. Denote the inverse DTFT of $\left|X_{\text{STFT}}(e^{j\omega}, n)\right|^2$ as $r[k, n]$. What is the relation between $r[k, n]$ and $x[n]$?

11.9 The short-term autocorrelation function of a determinstic sequence $x[n]$ is defined by [Rab78]

$$\varphi_{\text{ST}}[k, n] = \sum_{m=-\infty}^{\infty} x[m]w[n - m]x[m + k]w[n - k - m], \qquad (11.127)$$

where $w[n]$ is an appropriately chosen window sequence.

(a) Show that $\varphi_{\text{ST}}[k, n]$ is an even function of k, i.e. $\varphi_{\text{ST}}[k, n] = \varphi_{\text{ST}}[-k, n]$

(b) Show that $\varphi_{\text{ST}}[k, n]$ can be computed using the digital filter structure of Figure P11.2, where $h_k[n]$ is the impulse response of an LTI discrete-time system. Determine the expression for $h_k[n]$.

Figure P11.2

11.10 The structure of Figure P11.3 has been proposed as an allpass reverberator [Sch62]. Develop an equivalent realization with the smallest number of multipliers.

Figure P11.3

11.11 The structure of Figure P11.4 has been proposed as an allpass reverberator with a variable ratio of direct-to-reverberated sound, where the box labeled "allpass reverberator" is typically a cascade of the allpass reverberators of the form of Figure P11.3 [Sch62]. Develop an equivalent realization with the smallest number of multipliers.

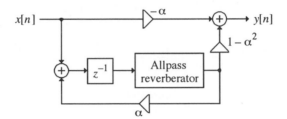

Figure P11.4

11.12 A generalization of the first-order equalizers of Eqs. (11.36) and (11.37) is given by

$$G_1(z) = \frac{K_1}{2}\{1 - A_1(z)\} + \frac{K_2}{2}\{1 + A_1(z)\}, \tag{11.128}$$

where $A_1(z)$ is the first-order allpass transfer function of Eq. (11.35). Develop the pertinent design equations for this equalizer and plot the gain responses for various values of the filter parameters K_1, K_2, and α.

11.13 A generalization of the second-order equalizer of Eq. (11.43) is given by

$$G_2(z) = \frac{K_1}{2}\{1 - A_2(z)\} + \frac{K_2}{2}\{1 + A_2(z)\}, \tag{11.129}$$

where $A_2(z)$ is the second-order allpass transfer function of Eq. (11.42). Develop the pertinent design equations for this equalizer and plot the gain responses for various values of the filter parameters K_1, K_2, α, and β.

11.14 Develop a computationally efficient structure for realizing a fractional-rate interpolator with an interpolation factor of 3/2 by taking the transpose of Figure 11.66.

11.16   MATLAB Exercises

M 11.1 Verify the dual-tone multifrequency (DTMF) tone detection Program 11_1 by running it for various telephone key symbols as inputs.

M 11.2 Using Program 11_2 analyze the spectral contents of a length-16 sequence $x[n]$ composed of a sum of two sinusoidal sequences of the form of Eq. (11.13) with frequencies $f_1 = 0.68$ and $f_2 = 0.8$ for DFT lengths $R = 16, 32, 64$, and 128, respectively. Comment on your results.

M 11.3 Using Program 11_2 analyze the spectral contents of a length-16 sequence $x[n]$ composed of a sum of two sinusoidal sequences of the form of Eq. (11.13) with frequencies $f_1 = 0.68$ and $f_2 = 0.8, 0.77, 0.74$, and 0.71, respectively. Use a DFT length $R = 128$. Comment on your results.

M 11.4 Modify Program 11_2 to investigate the effect of a tapered window on the spectral analysis of a sequence $x[n]$ composed of a sum of two sinusoidal sequences with closely spaced frequencies. Let $x[n]$ be of the form $x[n] = \sin(2\pi f_1 n) + 0.9 \sin(2\pi f_2 n)$, where $f_1 = 0.68$ and $f_2 = 0.71$. Use a length-N Hamming window, where $N = 16, 32, 64$, and 128, respectively. Comment on your results.

M 11.5 Repeat Problem M11.4 using a Hann window.

M 11.6 Repeat Problem M11.4 using a Blackman window.

M 11.7 Modify Program 11_3 to perform a DFT-based spectral analysis of a noise-corrupted sinusoidal sequence of frequency 0.11. Are you able to detect the frequency of the sinusoid? Justify your answer.

M 11.8 Repeat Problem M11.7 for a frequency of 0.16.

M 11.9 Design using the function remez a wideband FIR filter $F(z)$ of degree 17 with a pass-band from 0 to 0.9π and a stopband from 0.95π to π. Use an appropriate weight vector to weigh the passband and the stopband of the wideband filter. Plot the magnitude responses of $F(z)$ and its corresponding Hilbert transformer $F(-z^2)$. How many multipliers are needed to implement the Hilbert transformer?

M 11.10 Design using the function remez an Hilbert transformer of degree 34 with a passband from 0.05π to 0.95π. Compare its computational complexity with that designed in Problem M11.9.

M 11.11 Design a real coefficient half-band elliptic filter $G(z)$ with the following specifications: $\omega_p = 0.46\pi$, $\omega_s = 0.64\pi$, and $\delta_s = 0.014$. Plot its pole-zero locations using the function z-plane. Express $G(z)$ in the form of Eq. (11.70) and determine the transfer functions of the allpass sections $\mathcal{A}_0(z^2)$ and $\mathcal{A}_1(z^2)$. Next, determine the expression for the complex half-band filter $H(z)$ according to Eq. (11.71). Plot the gain responses of the real half-band filter $G(z)$ and the complex half-band filter $H(z)$, along with the phase difference between $\mathcal{A}_0(-z^2)$ and $z^{-1}\mathcal{A}_1(-z^2)$.

M 11.12 Design a fractional rate interpolator for a sampling rate alteration factor of 4/5 using the Lagrange interpolation approach. Verify its operation by interpolating a sinusoidal sequence.

M 11.13 Verify the operation of the sigma-delta quantizer on a constant amplitude discrete-time input sequence using Program 11_4. Choose different values for the length N of the sequence and its amplitude A.

M 11.14 Verify the operation of the sigma-delta A/D converter on a discrete-time sinusoidal input sequence using Program 11_5. Choose different values for the length N of the sequence, its amplitude A, and angular frequency ω_0.

M 11.15 Verify the operation of the sigma-delta D/A converter on a discrete-time sinusoidal input sequence using Program 11_6. Choose different values for the length N of the sequence, its amplitude A, and angular frequency ω_0.

Bibliography

[Abr72]   M. Abramowitz and I.A. Stegun, editors. *Handbook of Mathematical Functions*. Dover Publications, New York NY, 1972.

[Aca83]   A. Acampora. Wideband picture detail restoration in a ditigal NTSC comb-filter system. *RCA Engineer*, 28(5):44–47, September/October 1983.

[Aga75]   R.C. Agarwal and C.S. Burrus. New recursive digital filter structures having very low sensitivity and low roundoff noise. *IEEE Trans. on Circuits and Systems*, CAS-22:921–927, March 1975.

[Aga77]   R.C. Agarwal and J.W. Cooley. New algorithms for digital convolution. *IEEE Trans. on Acoustics, Speech, and Signal Processing*, ASSP-25:392–410, October 1977.

[Aka92]   A.N. Akansu and R.A. Haddad. *Multiresolution Signal Decomposition*. Academic Press, New York NY, 1992.

[All77]   J.B. Allen and L.R. Rabiner. A unified approach to short-term Fourier analysis and synthesis. *Proc. IEEE*, 65:1558–1564, November 1977.

[All80]   H.G. Alles. Music synthesis using real time digital techniques. *Proc. IEEE*, 68:436–449, April 1980.

[Ans93]   R. Ansari and B. Liu. Multirate signal processing. In S.K. Mitra and J.F. Kaiser, editors, *Handbook for Digital Signal Processing*, chapter 14, pages 981–1084. Wiley-Interscience, New York NY, 1993.

[Ant93]   A. Antoniou. *Digital Filters: Analysis, Design, and Applications*. McGraw-Hill, New York NY, second edition, 1993.

[Bab91]   H. Babic, S.K. Mitra, C. Creusere, and A. Das. Perfect reconstruction recursive QMF banks for subband image coding. In *Proc. 25th Asilomar Conference on Signals, Systems and Computers*, pages 746–750, Pacific Grove, California, November 1991.

[Bab95]   J. Babst, editor. *Digital Signal Processing Applications Using the ADSP-2100 Family*, Prentice-Hall, Englewood Cliffs NJ, 1995.

[Bar84]   C.W. Barnes. On the design of optimal state-space realizations of second-order digital filters. *IEEE Trans. on Circuits and Systems*, CAS-31:602–608, July 1984.

[Bel76]   M. Bellanger, G. Bonnerot, and M. Coudreuse. Digital filtering by polyphase network: Application to sample rate alteration and filter banks. *IEEE Trans. Acoustics, Speech & Signal Processing*, ASSP-24:109–114, April 1976.

[Bel84]   M. Bellanger. *Digital Processing of Signals*. Wiley, New York NY, 1984.

[Ben48]   W.R. Bennett. Spectra of quantized signals. *Bell System Technical Journal*, 27:446–472, 1948.

[Bin90]   J.A.C. Bingham. Multicarrier modulation for data transmission: An idea whose time has come. *IEEE Communications Magazine*, pages 5–14, May 1990.

[Bla65]   R.B. Blackman. *Linear Data Smoothing and Prediction in Theory and Practice*. Addison-Wesley, Reading MA, 1965.

[Ble78]   B. Blesser and J.M. Kates. Digital processing in audio signals. In A.V. Oppenheim, editor, *Applications of Digital Signal Processing*, chapter 2. Prentice-Hall, Englewood Cliffs NJ, 1978.

[Bol93]   B.A. Bolt. *Earthquakes*. W.H. Freeman, New York NY, 1993.

[Bom85]   B.W. Bomer. New second-order state-space structures for realizing low round-off noise digital filters. *IEEE Trans. on Acoustics, Speech and Signal Processing*, ASSP-33:106–110, February 1985.

[Boo51]   A.D. Booth. A signed binary multiplication technique. *Quart. J. Mech. Appl. Math.*, 4(Part 2):236–240, 1951.

[Box70]   G.E.P. Box and G.M. Jenkins. *Time Series Analysis: Forecasting and Control*. Holden-Day, San Francisco CA, 1970.

[Bra83]   R.N. Bracewell. The discrete Hartley transform. *Journal of the Optical Society of America*, 73:1832–1835, December 1983.

[Bur72]   C.S. Burrus. Block realization of digital filters. *IEEE Trans. on Audio & Electracoustics*, AU-20:230–235, October 1972.

[Bur73]   R.S. Burrington. *Handbook of Mathematical Tables and Formulas*. McGraw-Hill, New York NY, 5th edition, 1973.

[Bur77]   C.S. Burrus. Index mappings for multidimensional formulation of the DFT and convolution. *IEEE Trans. on Acoustics, Speech, and Signal Processing*, ASSP-25:239–242, June 1977.

[Bur81]   C.S. Burrus and P.W. Eschenbacher. An in-place, in-order prime factor FFT algorithm. *IEEE Trans. on Acoustics, Speech and Signal Processing*, ASSP-29:806–817, April 1981.

[Bur92]   C.S. Burrus, A.W. Soewito, and R.A. Gopinath. Least squared error FIR filter design with transition bands. *IEEE Trans. on Signal Processing*, 40:1327–1340, 1992.

[But77]   M. Buttner. Elimination of limit cycles in digital filters with very low increase in the quantization noise. *IEEE Trans. on Circuits & Systems*, CAS-24:300–304, 1977.

[Cad73]   J.A. Cadzow. *Discrete-Time Systems*. Prentice-Hall, Englewood Cliffs NJ, 1973.

[Cad87]   J.A. Cadzow. *Foundations of Digital Signal Processing and Data Analysis*. Macmillan, New York NY, 1987.

[Can86]   J.C. Candy and A-N. Huynh. Double interpolation for digital-to-analog conversion. *IEEE Trans. on Communications*, COM-34:77–81, January 1986.

[Can92]   J.C. Candy and G.C. Temes. Oversampling methods for A/D and D/A conversion. In J.C. Candy and G.C. Temes, editors, *Oversampling Delta-Sigma Data Converters*, pages 1–25. IEEE Press, New York NY, 1992.

[Cha73]   D.S.K. Chan and L.R. Rabiner. Analysis of quantization errors in the direct form for finite impulse response digital filters. *IEEE Trans. on Audio and Electroacoustics*, AU-21:354–366, August 1973.

[Cha81]   T.L. Chang and S.A. White. An error cancellation digital filter structure and its distributed arithmetic implementation. *IEEE Trans. on Circuits and Systems*, CAS-28:339–342, April 1981.

[Cha90]   R. Chassaing and D.W. Horning. *Digital Signal Processing with the TMS320C25*. Wiley, New York NY, 1990.

[Chr66]   E. Christian and E. Eisenmann. *Filter Design Tables and Graphs*. Wiley, New York NY, 1966.

[Cio91]   J.M. Cioffi. *A multicarrier primer*. ANSI T1E1.4 Committee Contribution, Boca Raton FL, November 1991.

[Cio93]   J.M. Cioffi and Y-S Byun. Adaptive filtering. In S.K. Mitra and J.F. Kaiser, editors, *Handbook for Digital Signal Processing*, chapter 15, pages 1085–1142. Wiley Interscience, New York NY, 1993.

[Cla73]   T.A.C.M. Classen, W.F.G. Mecklenbräuker, and J.B.H. Peek. Some remarks on the classifications of limit cycles in digital filters. *Philips Research Reports*, 28:297–305, August 1973.

[Coh86]   A. Cohen. *Biomedical Signal Processing*, volume II. CRC Press, Boca Raton FL, 1986.

[Con70]   A.C. Constantinides. Spectral transformations for digital filters. *Proc. IEE*, 117:1585–1590, August 1970.

[Coo65]   J.W. Cooley and J.W. Tukey. An algorithm for the machine calculation of complex Fourier series. *Math. Computation*, 19:297–301, 1965.

[Cou83]   L.W. Couch II. *Digital and Analog Communication Systems*. Macmillan, New York NY, 1983.

[Cox83]   R.V. Cox and J.M. Tribolet. Analog voice privacy systems using TFSP scrambling: Full duplex and half duplex. *Bell System Technical Journal*, 62:47–61, January 1983.

[Cro75]   R.E. Crochiere and A.V. Oppenheim. Analysis of linear digital networks. *Proc. IEEE*, 62:581–595, April 1975.

[Cro76a]   A. Croisier, D. Esteban, and C. Galand. Perfect channel splitting by use of interpolation/decimation/tree decomposition techniques. In *Proc. International Symposium on Information Science and Systems*, Patras, Greece, 1976.

[Cro76b]   R.E. Crochiere and L.R. Rabiner. On the properties of frequency transformations for variable cutoff linear phase filters. *IEEE Trans. on Circuits & Systems*, CAS-23:684–686, 1976.

[Cro83]   R.E. Crochiere and L.R. Rabiner. *Multirate Digital Signal Processing*. Prentice-Hall, Englewood Cliffs NJ, 1983.

[Cuc91]   S. Cucchi, F. Desinan, G. Parladori, and G. Sicuranza. DSP implementation of arbitrary sampling frequency conversion for high quality sound application. In *Proc. IEEE International Conference on Acoustics, Speech and Signal Processing*, pages 3609–3612, Toronto Canada, May 1991.

[Dan74]   R.W. Daniels. *Approximation Methods for Electronic Filter Design*. McGraw-Hill, New York NY, 1974.

[Dar76]   G. Daryanani. *Principles of Active Nework Synthesis and Design*. Wiley, New York NY, 1976.

[DeF88]   D.J. DeFatta, J.G. Lucas, and W.S. Hodgkiss. *Digital Signal Processing: A System Design Approach*. Wiley, New York NY, 1988.

[Dep80]   E. Deprettere and P. DeWilde. Orthogonal cascade realization of a real multiport digital filter. *International Journal on Circuit Theory Appl*, 8:245–277, 1980.

[Dre90]   J.O. Drewery. Digital filtering of television signals. In C.P. Sandbank, editor, *Digital Television*, chapter 5. Wiley, New York NY, 1990.

[Dug80]   J.P. Dugre, A.A.L. Beex, and L.L. Scharf. Generating covariance sequences and the calculation of quantization and roundoff error variances in digital filters. *IEEE Trans. on Acoustics, Speech and Signal Processing*, ASSP-28:102–104, 1980.

[Dug82]   J.P. Dugre and E.I. Jury. A note on the evaluation of complex integrals using filtering interpretations. *IEEE Trans. on Acoustics, Speech and Signal Processing*, ASSP-30:804–807, 1982.

[Duh86]   P. Duhamel. Implementation of 'split-radix' FFT algorithms for complex, real, and real-symmetric data. *IEEE Trans. on Acoustics, Speech, and Signal Processing*, ASSP-34:285–295, April 1986.

[Dut80]   D.L. Duttweiler. Bell's echo-killer chip. *IEEE Spectrum*, 17:34–37, October 1980.

[Ear76]   J. Eargle. *Sound Recording*. Van Nostrand Reinhold, New York NY, 1976.

[Ear86]   J.M. Eargle. *Handbook of Recording Engineering*. Van Nostrand Reinhold, New York NY, 1986.

[Ebe69]   P.M. Ebert, J.E. Mazo, and M.G. Taylor. Overflow oscillations in digital filters. *Bell System Technical Journal*, 48:2999–3020, November 1969.

[Egg84]   L.D.J. Eggermont and P.J. Berkhout. Digital audio circuits: Computer simulations and listening tests. *Philips Technical Review*, 41(3):99–103, 1983/84.

[Est77]   D. Esteban and C. Galand. Application of quadrature mirror filters to split-band voice coding schemes. In *Proc. IEEE International Conference on Acoustics, Speech and Signal Processing*, pages 191–195, May 1977.

[Fad93]   J. Fadavi-Ardekani and K. Mondal. Software considerations. In S.K. Mitra and J.F. Kaiser, editors, *Handbook for Digital Signal Processing*, chapter 12, pages 783–905. Wiley-Interscience, New York NY, 1993.

[Fet71]   A. Fettweis. Digital filter structures related to classical filter networks. *Archiv Elektrotechnik und Übertragungstechnik*, 25:79–81, 1971.

[Fet86]   A. Fettweis. Wave digital filters: Theory and practice. *Proc. IEEE*, 74:270–316, February 1986.

[Fla79]   J.L. Flanagan, M.R. Schroeder, B.S. Atal, R.E. Chochiere, N.S. Jayant, and J.M. Tribolet. Speech coding. *IEEE Trans. on Communications*, COM-27:710–737, April 1979.

[Fli94]   N.J. Fliege. *Multirate Digital Signal Processing*. Wiley, New York NY, 1994.

[Fre78]   S.L. Freeny, J.F. Kaiser, and H.S. McDonald. Some applications of digital signal processing in telecommunications. In A.V. Oppenheim, editor, *Applications of Digital Signal Processing*, chapter 1, pages 1–28. Prentice-Hall, Englewood Cliffs NJ, 1978.

[Fre94]   M.E. Frerking. *Digital Signal Processing in Communication Systems*. Van Nostrand Reinhold, New York NY, 1994.

[Gab87]   R.A. Gabel and R.A. Roberts. *Signals and Linear Systems*. Wiley, New York NY, 1987.

[Gar80]   P. Garde. Allpass crossover systems. *Journal of the Audio Engineering Society*, 28:575–584, September 1980.

[Gas85]   L. Gaszi. Explicit formulas for lattice wave digital filters. *IEEE Trans. on Circuits & Systems*, CAS-32:68–88, January 1985.

[Goe58]   G. Goertzel. An algorithm for evaluation of finite trigonometric series. *American Mathematical Monthly*, 65:34–35, January 1958.

[Goe82]   D. Goedhart, R.J. Van de Plassche, and E.F. Stikvoort. Digital-to-analog conversion in playing a compact disc. *Philips Technical Review*, 40(6):174–179, 1982.

[Gol69a]   B. Gold and K.L. Jordan. A direct search procedure for designing finite-duration impulse response filters. *IEEE Trans. Audio & Electroacoustics*, AU-17:33–36, March 1969.

[Gol69b]   B. Gold and C.M. Radar. *Digital Processing of Signals*. McGraw-Hill, New York, NY, 1969.

[Gol73]   R. Golden. Digital filters. In G.C. Temes and S.K. Mitra, editors, *Modern Filter Theory & Design*, chapter 12, pages 505–557. Wiley, New York NY, 1973.

[Gol89]   G.H. Golub and C.F. Van Loan. *Matrix Computation*. The Johns Hopkins University Press, Baltimore MD, 2nd edition, 1989.

[Gon87]   R.C. Gonzalez and P. Wintz. *Digital Image Processing*. Addison Wesley, Reading MA, 1987.

[Gra73]   A.H. Gray Jr. and J.D. Markel. Digital lattice and ladder filter synthesis. *IEEE Trans. Audio & Electroacoustics*, AU-21:491–500, December 1973.

[Had91]   R.A. Haddad and T.W. Parsons. *Digital Signal Processing: Theory, Applications, and Hardware*. Computer Science Press, New York NY, 1991.

[Ham89]   R.W. Hamming. *Digital Filters*. Prentice-Hall, Englewood Cliffs NJ, third edition, 1989.

[Hay70]   S.S. Haykin and R. Carnegie. New method of synthesising linear digital filters based on convolution integral. *IEE Proc.*, 117:1063–1072, June 1970.

[Hee82]   J.P.J. Heemskerk and K.A.S. Immink. Compact disc: System aspects and modulation. *Philips Technical Review*, 40(6):157–165, 1982.

[Hel68]   H.D. Helms. Nonrecursive digital filters: Design method for achieving specifications on frequency response. *IEEE Trans. Audio Electroacoustics*, AU-16:336–342, September 1968.

[Hen83]   D. Henrot and C.T. Mullis. A modular and orthogonal digital filter structure for parallel processing. In *Proc. IEEE International Conference Acoustics, Speech, and Signal Processing*, pages 623–626, 1983.

[Her70]   O. Hermann and H.W. Schüssler. Design of nonrecursive digital filters with minimum phase. *Electronics Letters*, 6:329–330, 1970.

[Her73]   O. Herrmann, L.R. Rabiner, and D.S.K. Chan. Practical design rules for optimum finite impulse response lowpass digital filters. *Bell System Tech. J.*, 52:769–799, 1973.

[Hig84]   W.E. Higgins and D.C. Munson Jr. Optimal and suboptimal error spectrum shaping for cascade form digital filters. *IEEE Trans. on Circuits & Systems*, CAS-31:429–437, May 1984.

[Hir73]   K. Hirano, T. Saito, S. Nishimura, and S.K. Mitra. Time-sharing realization of Butterworth digital filters. In *Monograph of the Circuits & Systems Group*. Institution of Electronic and Communication Engineers (Japan), No. CST 73–59, December 1973. (In Japanese).

[Hir74]   K. Hirano, S. Nishimura, and S.K. Mitra. Design of digital notch filters. *IEEE Trans. on Circuits & Systems*, CAS-21:540–546, July 1974.

[Hog81]   E.B. Hogenauer. An economical class of digital filters for decimation and interpolation. *IEEE Trans. Acoustics, Speech and Signal Processing*, ASSP-29:155–162, April 1981.

[Hsi87]   C-C. Hsiao. Polyphase filter matrix for rational sampling rate conversions. In *Proc. IEEE International Conference on Acoustics, Speech and Signal Processing*, pages 2173–2176, Dallas TX, April 1987.

[Hub89]   D.M. Huber and R.A. Runstein. *Modern Recording Techniques*. Howard W Sams, Indianapolis IN, 3rd edition, 1989.

[Hus90]   H. Husøy and T.A. Ramstad. Application of an efficient parallel IIR filter bank to image subband coding. *Signal Processing*, 20:279–292, August 1990.

[Hwa76]   S.Y. Hwang. Round-off noise in state-space digital filtering. *IEEE Trans. on Acoustics, Speech and Signal Processing*, ASSP-24:256–262, June 1976.

[Hwa79]   K. Hwang. *Computer Arithmetic: Principles, Architecture and Designs*. Wiley, New York NY, 1979.

[Ife93]   E.C. Ifeachor and B.W. Jervis. *Digital Signal Processing: A Practical Approach*. Addison-Wesley, Reading MA, 1993.

[IEEE85]   Institute of Electrical and Electronic Engineers. IEEE Standard for Binary Floating-Point Arithmetic, 1985.

[ITU84]   International Telecommunicaion Union. *CCITT Red Book*, volume VI. Fascicle VI.1, October 1984.

[Jac69]   L.B. Jackson. An analysis of limit cycles due to multiplicative rounding in recursive digital filters. In *Proc. 7th Allerton Conference on Circuit and System Theory*, pages 69–78, Monticello IL, 1969.

[Jac70a]   L.B. Jackson. On the interaction of roundoff noise and dynamic range in digital filters. *Bell System Technical Journal*, 49:159–184, February 1970.

[Jac70b]   L.B. Jackson. Roundoff-noise analysis for fixed-point digital filters realized in cascade or parallel form. *IEEE Trans. on Audio and Electroacoustics*, AU-18:107–122, June 1970.

[Jac79]   L.B. Jackson, A.G. Lindgren, and Y. Kim. Optimal synthesis of second-order state-space structures for digital filters. *IEEE Trans. on Circuits and Systems*, CAS-26:149–153, March 1979.

[Jac91]   L.B. Jackson. *Signals, Systems, and Transforms*. Addison-Wesley, Reading MA, 1991.

[Jac96]   L.B. Jackson. *Digital Filters and Signal Processing*. Kluwer, Boston MA, third edition, 1996.

[Jai89]   A.K. Jain. *Fundamentals of Digital Image Processing*. Prentice-Hall, Englewood Cliffs NJ, 1989.

[Jar88]   P. Jarske, Y. Neuvo, and S.K. Mitra. A simple approach to the design of FIR digital filters with variable characteristics. *Signal Processing*, 14:313–326, 1988.

[Jay74]   N.S. Jayant. Digital coding of speech waveforms. *Proc. IEEE*, 62:611–632, May 1974.

[Jay81]   N.S. Jayant, J. McDermott, S.W. Christiansen, and A.M. Quinn. A comparison of four methods for analog speech privacy. *IEEE Trans. on Communications*, COM-29:18–23, January 1981.

[Jay83]   N.S. Jayant, R.V. Cox, J. McCermott, and A.M. Quinn. Analog scramblers for speech-based sequential permutations in time and frequency. *Bell System Technical Journal*, 62:25–46, January 1983.

[Jen91]   Y-C. Jenq. Digital convolution algorithm for pipelining multiprocessor system. *IEEE Trans. on Computers*, C-30:966–973, December 1991.

[Joh80]   J.D. Johnston. A filter family designed for use in quadrature mirror filter banks. In *Proc. IEEE International Conference on Acoustics, Speech and Signal Processing*, pages 291–294, April 1980.

[Joh89]   J.R. Johnson. *Introduction to Digital Signal Processing*. Prentice-Hall, Englewood Cliffs NJ, 1989.

[Jur81]   R.K. Jurgen. Detroit bets on electronics to stymie Japan. *IEEE Spectrum*, 18:29–32, July 1981.

[Kai66]   J.F. Kaiser. Digital filters. In F. Kuo and J.F. Kaiser, editors, *System Analysis by Digital Computers*, chapter 7. Wiley, New York NY, 1966.

[Kai74]   J.F. Kaiser. Nonrecursive digital filter design using the I_0-sinh window function. In *Proc. 1974 IEEE International Symposium on Circuits & Systems*, pages 20–23, San Francisco CA, April 1974.

[Kai77]   J.F. Kaiser and R.W. Hamming. Sharpening the response of a symmetric nonrecursive filter by multiple use of the same filter. *IEEE Trans. on Acoustics, Speech, & Signal Processing*, ASSP-25:415–422, October 1977.

[Kai80]   J.F. Kaiser. On a simple algorithm to calculate the 'energy' of a signal. *Proc. IEEE International Conference on Acoustics, Speech and Signal Processing*, pages 381–384, Albuquerque NM, April 1980.

[Kam86]   J.J. Van der Kam. A digital 'decimating' filter for analog-to-digital conversion of hi-fi audio signals. *Philips Technical Review*, 42:230–238, 1986.

[Kan71]   E.P.F. Kan and J.K. Aggarwal. Error analysis in digital filters employing floating point arithmetic. *IEEE Trans. on Circuit Theory*, CT-18:678–686, November 1971.

[Kel62]   J.L. Kelly Jr. and C. Lochbaum. Speech synthesis. In *Proc. Stockholm Speech Communication Seminar*, Stockholm, Sweden, September 1962. Royal Institute of Technology.

[Knu69]   D.E. Knuth. *The Art of Computer Programming: Volume 2 – Seminumerical Algorithms*. Addison-Wesley, Reading MA, 2nd edition, 1969.

[Kol77]   D.P. Kolba and T.W. Parks. A prime factor FFT algorithm using high speed convolution. *IEEE Trans. on Acoustics, Speech, and Signal Processing*, ASSP-25:281–294, August 1977.

[Kol91]   R.D. Kolipillai, T.Q. Nguyen, and P.P. Vaidyanathan. Some results in the theory of cross-talk free transmultiplexers. *IEEE Trans. Signal Processing*, 39:2174–2183, October 1991.

[Kor93]   I. Koren. *Computer Arithmetic Algorithms*. Prentice-Hall, Englewood Cliffs NJ, 1993.

[Kra94]   T.P. Krauss, L. Shure, and J. Little. *Signal Processing Toolbox for use with* MATLAB. The Mathworks Inc, South Natick MA, 1994.

[Kum93]   R. Kumaresan. Spectral analysis. In S.K. Mitra and J.F. Kaiser, editors, *Handbook for Digital Signal Processing*, chapter 16, pages 1143–1242. Wiley-Interscience, New York NY, 1993.

[Lag81]   R. Lagadec and H.O. Kunz. A universal digital sampling frequency converter for digital audio. In *Proc. IEEE International Conference on Acoustics, Speech and Signal Processing*, pages 595–598, Atlanta GA, April 1981.

[Lag82]   R. Lagadec, D. Pelloni, and D. Weiss. A 2-channel, 16-bit digital sampling rate converter for professional digital audio. In *Proc. IEEE International Conference on Acoustics, Speech and Signal Processing*, pages 93–96, April 1982.

[Lar93]   L.E. Larson and G.C. Temes. Signal conditioning and interface circuits. In S.K. Mitra and J.F. Kaiser, editors, *Handbook for Digital Signal Processing*, chapter 10, pages 677–720. Wiley-Interscience, New York NY, 1993.

[Lat87]   B.P. Lathi. *Signals and Systems*. Berkley-Cambridge, Carmichael CA, 1987.

[Law78]   V.B. Lawrence and K.V. Mina. Control of limit cycle oscillations in second-order recursive digital filters by constrained random quantization. *IEEE Trans. on Acoustics, Speech, and Signal Processing*, ASSP-26:127–134, February 1978.

[Ler83]   E.L. Lerner. Electronically synthesized music. *IEEE Spectrum*, 17:46–51, June 1983.

[Lig94]   M.A. Lightstone, S.K. Mitra, I-S. Lin, S. Bagchi, P. Jarski, and Y. Neuvo. Efficient frequency-sampling design of one- and two-dimensional FIR filters using structural subband decomposition. *IEEE Trans. Circuits & Systems – Part II*, 41:189–201, March 1994.

[Lim90]   J.S. Lim. *Two-Dimensional Signal and Image Processing*. Prentice-Hall, Englewood Cliffs NJ, 1990.

[Lin87]   K.S. Lin, editor. *Digital Signal Processing Applications with the TMS320 Family*. Prentice-Hall, Englewood Cliffs NJ, 1987.

[Lin96]   I-S. Lin and S.K. Mitra. Overlapped block digital filtering. *IEEE Trans. on Circuits and Systems, II: Analog and Digital Signal Processing*, 43:586–596, August 1996.

[Liu69]   B. Liu and T. Kaneko. Error analysis of digital filters realized in floating-point arithmetic. *Proc. IEEE*, 57:1735–1747, October 1969.

[Lon73]   J.L. Long and T.N. Trick. An absolute bound on limit cycles due to roundoff errors in digital filters. *IEEE Trans. on Audio and Electroacoustics*, AU-21:27–30, February 1973.

[Lut91]   A. Luthra and G. Rajan. Sampling rate conversion of video signals. *SMPTE J.*, pages 869–879, November 1991.

[Mak75]   J. Makhoul. Linear prediction: A tutorial review. *Proc. IEEE*, 62:561–580, April 1975.

[Mar92]   A. Mar, editor. *Digital Signal Processing Applications Using the ADSP-2100 Family*. Prentice-Hall, Englewood Cliffs NJ, 1992.

[Mas60]   S.J. Mason and H.J. Zimmerman. *Electronic Circuits, Signals and Systems*, pages 122–123. Wiley, New York, NY, 1960.

[Mee76]   K. Meerkötter. Realization of limit cycle-free second-order digital filters. In *Proc. 1976 IEEE International Symposium on Circuits and Systems*, pages 295–298, 1976.

[Mes82]   D.G. Messerschmitt. Echo cancellation in speech and data transmission. *IEEE J. on Selected Areas in Communications*, SAC-2:283–297, March 1982.

[Mil78]   W.L. Mills, C.T. Mullis, and R.A. Roberts. Digital filter realizations without overflow oscillations. *IEEE Trans. on Acoustics, Speech and Signal Processing*, ASSP-26:334–338, August 1978.

[Mil81]   W.L. Mills, C.T. Mullis, and R.A. Roberts. Low roundoff noise and normal realizations of fixed-point IIR digital filters. *IEEE Trans. on Acoustics, Speech and Signal Processing*, ASSP-29:893–903, August 1981.

[Mit72]   S.K. Mitra and R.J. Sherwood. Canonic realizations of digital filters using the continued-fraction expansion. *IEEE Trans. Audio Electroacoustics*, AU-20:185–194, August 1972.

[Mit73a]   S.K. Mitra. On reciprocal digital two-pairs. *Proc. IEEE (Letters)*, 61:1647–1648, November 1973.

[Mit73b]   S.K. Mitra and R.J. Sherwood. Digital ladder networks. *IEEE Trans. Audio Electroacoustics*, AU-21:30–36, February 1973.

[Mit74a]   S.K. Mitra and K. Hirano. Digital allpass networks. *IEEE Trans. on Circuits & Systems*, CAS-21:688–700, 1974.

[Mit74b]   S.K. Mitra, K. Hirano, and H. Sakaguchi. A simple method of computing the input quantization and the multiplication round-off errors in digital filters. *IEEE Trans. on Acoustics, Speech, and Signal Processing*, ASSP-22:326–329, October 1974.

[Mit74c]   S.K. Mitra and R.J. Sherwood. Estimation of pole-zero displacements of a digital filter due to coefficient quantization. *IEEE Trans. on Circuits & Systems*. CAS-21:116–124, January 1974.

[Mit75]   S.K. Mitra, K. Hirano, and K. Furuno. Digital sine-cosine generator. In *Proc. Second Florence International Conference on Digital Signal Processing*, pages 142–149, Florence, Italy, September 1975.

[Mit77a]   S.K. Mitra and C.S. Burrus. A simple efficient method for the analysis of structures of digital and analog systems. *Archiv für Elektrotechnik und Übertrangungstechnik*, 31:33–36, 1977.

[Mit77b]   S.K. Mitra, P.S. Kamat, and D.C. Huey. Cascaded lattice realization of digital filters. *International Journal on Circuit Theory & Applications*, 5:3–11, 1977.

[Mit77c]   S.K. Mitra, K. Mondal, and J. Szczupak. An alternate parallel realization of digital transfer functions. *Proc. IEEE (Letters)*, 65:577–578, April 1977.

[Mit80]   S.K. Mitra. *An Introduction to Digital and Analog Integrated Circuits, and Applications*. Harper & Row, New York NY, 1980.

[Mit90a]   S.K. Mitra, K. Hirano, S. Nishimura, and K. Sugahara. Design of digital bandpass/bandstop digital filters with tunable characteristics. *Frequenz*, 44:117–121, March/April 1990.

[Mit90b]   S.K. Mitra, Y. Neuvo, and H. Roivainen. Design and implementation of recursive digital filters with variable characteristics. *International Journal on Circuit Theory and Applications*, 18:107–119, 1990.

[Mit92a]   S.K. Mitra, S. Chakrabarti, and E. Abreu. The nonuniform discrete Fourier transform and its signal processing applications. *Proc. European Signal Processing Conference – EUSIPCO'92*, pages 909–912, Brussels, Belgium, August 1992.

[Mit92b]   S.K. Mitra, C.D. Creseure, and H. Babic. A novel implementation of perfect reconstruction QMF banks using IIR filters for subband coding applications. In *Proc. 1992 International Symposium on Circuits and Systems*, pages 2312–2315, San Diego CA, May 1992.

[Mit93]   S.K. Mitra, A. Mahalanabis, and T. Saramäki. A generalized structural subband decomposition of FIR filters and its application in efficient FIR filter design and implementation. *IEEE Trans. on Circuits & Systems II: Analog & Digital Signal Processing*, 40:363–374, June 1993.

[Moo77]   J.A. Moorer. Signal processing aspects of computer music: A survey. *Proc. IEEE*, 65:1108–1137, August 1977.

[Moo79]   J.A. Moorer. About this reverberation business. *Computer Music Journal*, 3(2):13–28, 1979.

[Mul76]   C.T. Mullis and R.A. Roberts. Synthesis of miniumum round-off noise fixed point digital filters. *IEEE Trans. on Circuits and Systems*, CAS-23:551–562, September 1976.

[Mun81]   D.C. Munson, Jr., and B. Liu. Narrowband recursive filters with error spectrum shaping. *IEEE Trans. on Circuits and Systems*, CAS-28:160–163, February 1981.

[Naw88]   S.H. Nawab and T.F. Quatieri. Short-time Fourier transform. In J.S. Lim and A.V. Oppenheim, editors, *Advanced Topics in Signal Processing*, chapter 6. Prentice-Hall, Englewood Cliffs NJ, 1988.

[Neu84a]   Y. Neuvo and S.K. Mitra. Complementary IIR digital filters. In *Proc. IEEE International Symposium on Circuits and Systems*, pages 234–237, Montreal, Canada, May 1984.

[Neu84b]   Y. Neuvo, C-Y. Dong, and S.K. Mitra. Interpolated finite impulse response filters. *IEEE Trans. on Acoustics, Speech, and Signal Processing*, ASSP-32:563–570, June 1984.

[Oet75]   G. Oetken, T.W. Parks, and H.W. Schüssler. New results in the design of digital interpolators. *IEEE Trans. on Acoustics, Speech & Signal Processing*, ASSP-23:301–309, June 1975.

[Opp75]   A.V. Oppenheim and R.W. Schafer. *Digital Signal Processing*. Prentice-Hall, Englewood Cliffs NJ, 1975.

[Opp76]   A.V. Oppenheim, W.F.G. Mecklenbräuker, and R.M. Mersereau. Variable cutoff linear phase digital filters. *IEEE Trans. on Circuits & Systems*, CAS-23:199–203, 1976.

[Opp78]   A.V. Oppenheim, editor. *Applications of Digital Signal Processing*. Prentice-Hall, Englewood Cliffs NJ, 1978.

[Opp83]   A.V. Oppenheim and A.S. Willsky. *Signals and Systems*. Prentice-Hall, Englewood Cliffs NJ, 1983.

[Opp89]   A.V. Oppenheim and R.W. Schafer. *Discrete-Time Signal Processing*. Prentice-Hall, Englewood Cliffs NJ, 1989.

[Orf96]   S.J. Orfanidis. *Introduction to Signal Processing*. Prentice-Hall, Englewood Cliffs NJ, 1996.

[Pan65]   P.F. Panter. *Modulation, Noise, and Spectral Analysis*. McGraw-Hill, New York NY, 1965.

[Pap65]   A. Papoulis. *Probability, Random Variables, and Stochastic Processes*. McGraw-Hill, New York NY, 1965.

[Pap90]   P. Papamichalis. *Digital Signal Processing Applications with the TMS320 Family, Theory, Algorithms, and Implementations*, volume 3. Prentice-Hall, Englewood Cliffs NJ, 1990.

[Par60]   E. Parzen. *Modern Probability Theory and Its Applications*. Wiley, New York NY, 1960.

[Par72]   T.W. Parks and J.H. McClellan. Chebyshev approximation for nonrecursive digital filters with linear phase. *IEEE Trans. on Circuit Theory*, CT-19:189–194, 1972.

[Par87]   T.W. Parks and C.S. Burrus. *Digital Filter Design*. Wiley, New York NY, 1987.

[Pou87]   K. Poulton, J.J. Corcoran, and T. Hornak. A 1-GHz 6-bit ADC system. *IEEE J. on Solid-State Circuits*, SC-22:962–970, December 1987.

[Pee87]   P.Z. Peebles. *Probability, Random Variables, and Random Signal Principles*. McGraw-Hill, New York NY, second edition, 1987.

[Pel80]   A. Peled and A. Ruiz. Frequency domain data transmission using reduced computational complexity algorithm. In *Proc. IEEE International Conference on Acoustics, Speech and Signal Processing*, pages 964–967, Denver CO, April 1980.

[Pri80]   D.H. Pritchard. A CCD comb filter for color TV receiver picture enhancement. *RCA Review*, 41:3–28, 1980.

[Pro92]   J.G. Proakis and D.G. Manolakis. *Digital Signal Processing: Principles, Algorithms and Applications*. Prentice-Hall, Englewood Cliffs NJ, second edition, 1992.

[Rab69]   L.R. Rabiner, R.W. Schafer, and C.M. Rader. The chirp-z transform algorithm. *IEEE Trans. on Audio and Electroacoustics*, AU-17:86–92, June 1969.

[Rab70]   L.R. Rabiner, B. Gold, and C.A. McGonegal. An approach to the approxi-mation problem for nonrecursive digital filters. *IEEE Trans. Audio Electroa-coustics*, AU-18:83–106, June 1970.

[Rab73]   L.R. Rabiner. Approximate design relationships for lowpass FIR digital filters. *IEEE Trans. on Audio & Electroacoustics*, AU-21:456–460, 1973.

[Rab75]   L.R. Rabiner and B. Gold. *Theory and Application of Digital Signal Process-ing*. Prentice-Hall, Englewood Cliffs NJ, 1975.

[Rab78]   L.R. Rabiner and R.W. Schafer. *Digital Processing of Speech Signals*. Prentice-Hall, Englewood Cliffs NJ, 1978.

[Ram84]   T. Ramstad. Digital methods for conversion between arbitrary sampling fre-quencies. *IEEE Trans. Acoustics, Speech and Signal Processing*, ASSP-32:577–591, June 1984.

[Rao84]   S.K. Rao and T. Kailath. Orthogonal digital filters for VLSI implementation. *IEEE Trans. on Circuits and Systems*, CAS-31:933–945, 1984.

[Reg87a]   P.A. Regalia, S.K. Mitra, and J. Fadavi. Implementation of real digital fil-ters using complex arithmetic. *IEEE Trans. on Circuits and Systems*, CAS-34:345–353, April 1987.

[Reg87b]   P.A. Regalia and S.K. Mitra. Tunable digital frequency response equalization filters. *IEEE Trans. Acoustics, Speech, & Signal Processing*, ASSP-35:118–120, February 1987.

[Reg87c]   P.A. Regalia and S.K. Mitra. A class of magnitude complementary loud-speaker crossovers. *IEEE Trans. on Acoustics, Speach and Signal Processing*, ASSP-35:1509–1515, November 1987.

[Reg88]   P.A. Regalia, S.K. Mitra, and P.P. Vaidyanathan. The digital allpass network: A versatile signal processing building block. *Proc. IEEE*, 76:19–37, January 1988.

[Reg93]   P.A. Regalia. Special filter designs. In S.K. Mitra and J.F. Kaiser, editors, *Handbook for Digital Signal Processing*, chapter 13, pages 967–980. Wiley-Interscience, New York NY, 1993.

[Ren87]   M. Renfors and T. Saramäki. Recursive n-th band digital filters, Parts I and II. *IEEE Trans. on Circuits & Systems*, CAS-34:24–51, January 1987.

[Ric86]   E.B. Richardson and N.S. Jayant. Subband coding with adaptive prediction for 56 kbits/s audio. *IEEE Trans. Acoustics, Speech and Signal Processing*, ASSP-34:691–696, August 1986.

[Rob80]   E.A. Robinson and S. Treitel. *Geophysical Signal Analysis*. Prentice-Hall, Englewood Cliffs NJ, 1980.

[Ros75]   J.P. Rossi. Digital television image enhancement. *SMPTE J.*, 84:545–551, July 1975.

[San67]   I.W. Sandberg. Floating-point-roundoff accumulation in digital filter realiza-
           tion. *Bell System Technical Journal*, 46:1775–1791, October 1967.

[Sar93]   T. Saramäki. Finite impulse response filter design. In S.K. Mitra and J.F.
           Kaiser, editors, *Handbook for Digital Signal Processing*, chapter 4, pages
           155–278. Wiley-Interscience, New York NY, 1993.

[Sch62]   M.R. Schroeder. Natural sounding artificial reverberation. *Journal of the
           Audio Engineering Society*, 10:219–223, 1962.

[Sch72]   H.W. Schüssler. On structures for nonrecursive digital filters. *Archiv für
           Electrotechnik und Überstragunstechnik*, 26:255–258, June 1972.

[Sch75]   M. Schwartz and L. Shaw. *Signal Processing: Discrete Spectral Analysis,
           Detection, and Estimation*. McGraw-Hill, New York NY, 1975.

[Sch91]   R. Schreier. *Noise-Shaped Coding*. PhD thesis, University of Toronto, Toronto
           Canada, 1991.

[Sha81]   A.F. Shackil. Microprocessors and the MD. *IEEE Spectrum*, 18:45–49, April
           1981.

[She95]   K. Shenoi. *Digital Signal Processing in Telecommunication*. Prentice-Hall,
           Englewood Cliffs NJ, 1995.

[Sko62]   M.I. Skolnik. *Introduction to Radar Systems*. McGraw-Hill, New York NY,
           1962.

[Skw65]   J.K. Skwirzynski. *Design Theory and Data for Electrical Filters*. Van Nos-
           trand Reinhold, New York NY, 1965.

[Slu84]   R.J. Sluyter. Digitization of speech. *Philips Technical Review*, 41(7/8):201–
           223, 1983–84.

[Smi85]   M.J.T. Smith and T.P. Barnwell III. A unifying framework for analy-
           sis/synthesis systems based on maximally decimated filter banks. In *Proc.
           IEEE Conf. on Acoustics, Speech and Signal Processing*, pages 521–524,
           Tampa FL, March 1985.

[Sta94]   H. Stark and J.W. Woods. *Probability, Random Processes, and Estimation
           Theory for Engineers*. Prentice-Hall, Englewood Cliffs NJ, second edition,
           1994.

[Ste93]   K. Steiglitz. Mathematical foundations of signal processing. In S.K. Mitra
           and J.F. Kaiser, editors, *Handbook for Digital Signal Processing*, chapter 2,
           pages 57–99. Wiley-Interscience, New York NY, 1993.

[Sto66]   T.G. Stockham Jr. High speed convolution and correlation. *1966 Spring Joint
           Computer Conference, AFIPS Proc*, 28:229–233, 1966.

[Str94]   N. Strobel and S.K. Mitra. Quadratic filters for image contrast enhancement.
           In *Proc. 28th Annual Asilomar Conference on Circuits, Systems and Comput-
           ers*, pages 208–212, Pacific Grove CA, November 1994.

[Stu88]   R.D. Sturm and D.E. Kirk. *First Principles of Discrete Systems and Signal Processing*. Addison-Wesley, Reading MA, 1988.

[Szc75]   J. Szczupak and S.K. Mitra. Detection, location, and removal of delay-free loops in a digital filter configuration. *IEEE Trans. Acoustics, Speech & Signal Processing*, ASSP-23:558–562, December 1975.

[Szc88]   J. Szczupak, S.K. Mitra, and J. Fadavi-Ardekani. A computer-based synthesis method of structurally LBR digital allpass networks. *IEEE Trans. on Circuits & Systems*, CAS-35:755–760, 1988.

[Tem73]   G.C. Temes and S.K. Mitra, editors. *Modern Filter Theory & Design*. Wiley, New York NY, 1973.

[Tem77]   G.C. Temes and J.W. LaPatra. *Introduction to Circuit Synthesis and Design*. McGraw-Hill, New York NY, 1977.

[The92]   C.W. Therrien. *Discrete Random Signals and Statistical Signal Processing*. Prentice-Hall, Englewood Cliffs NJ, 1992.

[Tho77]   T. Thong and B. Liu. Error spectrum shaping in narrowband recursive digital filters. *IEEE Trans. on Acoustics, Speech, and Signal Processing*, ASSP-25:200–203, 1977.

[Tom81]   W.J. Tompkins and J.G. Webster, editors. *Design of Microcomputer-Based Medical Instrumentation*. Prentice-Hall, Englewood Cliffs NJ, 1981.

[Tri79]   J.M. Tribolet. *Seismic Applications of Homomorphic Signal Processing*. Prentice-Hall, Englewood Cliffs NJ, 1979.

[Urk58]   H. Urkowitz. An extension to the theory of airborne moving target indicators. *IRE Trans*, ANE-5:210–214, December 1958.

[Unv75]   Z. Unver and K. Abdullah. A tighter practical bound on quantization errors in second-order digital filters with complex conjugate poles. *IEEE Trans. on Circuits and Systems*, CAS-22:632–633, July 1975.

[Vai84]   P.P. Vaidyanathan and S.K. Mitra. Low passband sensitivity digital filters: A generalized viewpoint and synthesis procedures. *Proc. IEEE*, 72:404–423, 1984.

[Vai85a]   P.P. Vaidyanathan. The doubly terminated lossless digital two-pair in digital filtering. *IEEE Trans. on Circuits and Systems*, CAS-32:197–200, 1985.

[Vai85b]   P.P. Vaidyanathan and S.K. Mitra. Very low-sensitivity FIR filter implementation using 'structural passivity' concept. *IEEE Trans. on Circuits and Systems*, CAS-32:360–364, April 1985.

[Vai86a]   P.P. Vaidyanathan, S.K. Mitra, and Y. Neuvo. A new approach to the realization of low sensitivity IIR digital filters. *IEEE Trans. on Acoustics, Speech, and Signal Processing*, ASSP-34:350–361, April 1986.

[Vai86b]   P.P. Vaidyanathan. Passive cascaded lattice structures for low-sensitivity FIR filter design, with application to filter banks. *IEEE Trans. on Circuits and Systems*, CAS-33:1045–1064, November 1986.

[Vai87a]   P.P. Vaidyanathan and T.Q. Nguyen. Eigenfilters: A new approach to least-squares FIR filter design and applications including Nyquist filters. *IEEE Trans. on Circuits and Systems*, CAS-34:11–23, January 1987.

[Vai87b]   P.P. Vaidyanathan and T.Q. Nguyen. A 'TRICK' for the design of FIR half-band filters. *IEEE Trans. on Circuits and Systems*, CAS-34:11–23, January 1987.

[Vai87c]   P.P. Vaidyanathan. Low-noise and low-sensitivity digital filters. In D.F. Elliot, editor, *Handbook of Digital Signal Processing*, chapter 5. Academic Press, New York NY, 1987.

[Vai87d]   P.P. Vaidyanathan. Quadrature mirror filter banks, M-band extension and perfect reconstruction techniques. *IEEE ASSP Magazine*, 4:4–20, 1987.

[Vai87e]   P.P. Vaidyanathan and S.K. Mitra. A unified structural interpretation and tutorial review of stability test procedures for linear systems. *Proc. IEEE*, 75:478–497, April 1987.

[Vai87f]   P.P. Vaidyanathan, P.A. Regalia, and S.K. Mitra. Design of doubly-complementary IIR digital filters using a single complex allpass filter, with multirate applications. *IEEE Trans. on Circuits and Systems*, CAS-34:378–389, April 1987.

[Vai88]   P.P. Vaidyanathan and S.K. Mitra. Polyphase networks, block digital filtering, LPTV systems, and alias-free QMF banks: A unified approach based on pseudo-circulants. *IEEE Trans. on Acoustics, Speech and Signal Processing*, ASSP-36:381–391, March 1988.

[Vai89]   P.P. Vaidyanathan, T.Q. Nguyen, Z. Doganata, and Saramäki. Improved technique for design of perfect reconstruction FIR QMF filter banks with lossless polyphase matrices. *IEEE Trans. Acoustics, Speech, & Signal Processing*, ASSP-37:1042–1046, July 1989.

[Vai90]   P.P. Vaidyanathan. Multirate digital filters, filter banks, polyphase networks, and applications: A tutorial. *Proc. IEEE*, 78:56–93, January 1990.

[Vai93]   P.P. Vaidyanathan. *Multirate Systems and Filter Banks*. Prentice Hall, Englewood Cliffs NJ, 1993.

[Vet88]   M Vetterli. Running FIR and IIR filtering using multirate filter banks. *IEEE Trans. on Acoustics, Speech and Signal Processing*, ASSP-36:381–391, May 1988.

[Vet95]   M. Vetterli and J. Kovacevic. *Wavelets and Subband Coding*. Prentice-Hall, Englewood Cliffs NJ, 1995.

[Vla69]   J. Vlach. *Computerized Approximation and Synthesis of Linear Networks.* Wiley, New York NY, 1969.

[Wei69a]   C.J. Weinstein and A.V. Oppenheim. A comparison of roundoff noise in fixed point and floating point digital filter realizations. *Proc. IEEE*, 57:1181–1183, June 1969.

[Wei69b]   C.J. Weinstein. Roundoff noise in floating point fast Fourier transform computation. *IEEE Trans. on Audio and Electrocacoustics*, AU-17:209–215, September 1969.

[Wel69]   P.D. Welch. A fixed-point fast Fourier transform error analysis. *IEEE Trans. on Audio & Electroacoustics*, AU-17:153–157, June 1969.

[Whi58]   W.D. White. Synthesis of comb filters. *Proc. National Electronics Conference*, pages 279–285, 1958.

[Whi71]   S.A. White. New method of synthetizing linear digital filters based on convolution integral. *IEE Proc. (Corr.)*, 118:348, February 1971.

[Wid56]   B. Widrow. A study of rough amplitude quantization by means of Nyquist sampling theory. *IRE Trans. on Circuit Theory*, CT-3:266–276, December 1956.

[Wid61]   B. Widrow. Statistical analysis of amplitude-quantized sampled-data systems. *AIEE Trans. (Appl Industry)*, 81:555–568, January 1961.

[Wor89]   J.M. Worham. *Sound Recording Handbook.* Howard W Sams, Indianapolis IN, 1989.

[Yan82]   G-T. Yan and S.K. Mitra. Modified coupled form digital-filter structures. *Proc. IEEE (Letters)*, 70:762–763, July 1982.

[Yel96]   D. Yellin and E. Weinstein. Multichannel signal separation: Methods and analysis. *IEEE Trans. on Acoustics, Speech and Signal Processing*, ASSP-44:106–118, January 1996.

[Zie83]   R.E. Ziemar, W.H. Tranter, and D.R. Fannin. *Signals and Systems: Continuous and Discrete.* Macmillan, NY, 1983.

[Zve67]   A.I. Zverev. *Handbook of Filter Synthesis.* Wiley, New York NY, 1967.

Index